ROCKET PROPULSION ELEMENTS

Rocket Propulsion Elements

Seventh Edition

GEORGE P. SUTTON
Consultant
Formerly Laboratory Associate
Lawrence Livermore National Laboratory
and formerly
Executive Director, Engineering
Rocketdyne, now The Boeing Company

OSCAR BIBLARZ
Professor
Department of Aeronautics and Astronautics
Naval Postgraduate School

A Wiley-Interscience Publication
JOHN WILEY & SONS, INC.
New York / Chichester / Weinheim / Brisbane / Singapore / Toronto

Library of Congress Cataloging-in-Publication Data:
Sutton, George Paul.
    Rocket propulsion elements : an introduction to the engineering of rockets / by George P. Sutton, Oscar Biblarz.—7th ed.
        p.   cm.
    "A Wiley-Interscience publication."
    Includes bibliographical references and index.
    ISBN 0-471-32642-9 (cloth : alk. paper)
    1. Rocket engines. I. Biblarz, Oscar. II. Title
TL782.S8   2000
629.47′5—dc21   00-027334

Printed in the United States of America.
10 9

CONTENTS

PREFACE

This new edition concentrates on the subject of rocket propulsion, its basic technology, performance, and design rationale. The intent is the same as in previous editions, namely to provide an introduction to the subject, an understanding of basic principles, a description of their key physical mechanisms or designs, and an appreciation of the application of rocket propulsion to flying vehicles.

The first five chapters in the book cover background and fundamentals. They give a classification of the various propulsion systems with their key applications, definitions, basic thermodynamics and nozzle theory, flight performance, and the thermochemistry of chemical propellants. The next nine chapters are devoted to chemical propulsion, namely liquid rocket engines and solid rocket motors. We devote almost half of the book to these two, because almost all past, current, and planned future rocket-propelled vehicles use them. Hybrid rocket propulsion, another form of using chemical combustion energy, has a separate chapter. The new longer chapter on electric propulsion has been extensively revised, enlarged, and updated. Chapters 16–18 and 20 apply to all types of propulsion, namely thrust vector control, selection of a rocket propulsion system for specific applications, testing of propulsion systems, and behavior of chemical rocket exhaust plumes. Only a little space is devoted to advanced new concepts, such as nuclear propulsion or solar thermal propulsion, because they have not yet been fully developed, have not yet flown, and may not have wide application.

The book attempts to strike a balance between theory, analysis, and practical design or engineering tasks; between propulsion system and nonpropulsion system subjects, which are related (such as testing, flight performance, or exhaust plumes); and between rocket systems and their key components and materials. There is an emphasis on up-to-date information on current propulsion systems and the relation between the propulsion system, the flight vehicle, and the needs of the overall mission or flight objectives.

The new edition has more pages and extensive changes compared with the sixth edition. We have expanded the scope, reorganized the existing subject matter into a more useful form or logical sequence in some of the chapters, and updated various data. About one-third of the book is new or extensively revised text and figures. This new version has been heavily edited, upgraded, and improved. Altogether we count about 2500 changes, additions, new or rewritten sections or paragraphs, inserts, clarifications, new illustrations, more data, enlarged tables, new equations, more specific terminology, or new references. We have deleted the chapter on heat transfer that was in the sixth edition, because we learned that it was not being used often and is somewhat out of date. Instead we have added revised small specific sections on heat transfer to several chapters. A new chapter on liquid propellant thrust chambers was added, because this component is the heart of liquid propellant rocket engines.

Here are some of the topics that are new or completely revised. New sections or subsections include engine structures, two-step nozzles, multiple nozzles, gas properties of gas generator or preburner gases, classification of engine valves, a promising new monopropellant, gaseous rocket propellants, propellant additives, materials and fabrication of solid propellant motors, launch vehicles, elliptical orbits, new sample design calculations, vortex instability in solid rocket motors, design of turbopumps, design of liquid propellant engines, insensitive munitions requirements, aerospike rocket engines, solid rocket motor nozzles, and plume signatures. In addition there are new figures, for example, the payload variation with orbit altitude or inclination angle, some recently developed rocket propulsion systems, the design of shortened bell-shaped nozzle contours, and the expander engine cycle, and new tables, such as different flight maneuvers versus the type of rocket propulsion system, list of mission requirements, and the physical and chemical processes in rocket combustion. There are new paragraphs on rocket history, four additional nozzle loss factors, use of venturi in feed systems, extendible nozzles, and water hammer.

In the last couple of decades rocket propulsion has become a relatively mature field. The development of the more common propulsion systems is becoming routine and the cost of new ones is going down. For example, much R&D was done on many different chemical propellants, but just a few are used, each for specific applications. Although some investigations on new propellants or new propellant ingredients are still under way, a new propellant has not been introduced for a rocket production application in the last 25 years. Most of the new propulsion systems are uprated, improved, or modified versions of existing proven units in the chemical propulsion and electrical propulsion areas. There are only a few novel engines or motors, and some are mentioned in this book. We have therefore placed emphasis on describing several of the proven existing modern rocket propulsion systems and their commonly used propellants, because they are the heritage on which new ones will be based. It is not possible in any one book to mention all the varieties, types, and designs of propulsion systems, their propellants, or materials of construction, and we therefore selected some of the most commonly used ones. And we discuss the process of uprating or modifying them, because this is different from the design process for a truly new unit.

The number of countries that develop or produce rocket propulsion systems has gone from three in 1945 to at least 35 today, a testimony to proliferation and the rising interest in the subject. There are today more colleges that teach rocket propulsion than before. Prior editions of this book have been translated into three languages, Russian, Chinese, and more recently (1993) Japanese. People outside of the U.S. have made some excellent contributions to the rocket field and the authors regret that we can mention only a few in this book.

We have had an ongoing disparity about units. Today in U.S. propulsion companies, most of the engineering and design and almost all the manufacturing is still being done in English engineering (EE) units (foot or inch, pounds, seconds). Many of the technical papers presented by industry authors use EE units. Papers from university authors, government researchers, and from a few companies use the SI (International Standard—metric) units. If a customer demands SI units, some companies will make new drawings or specifications especially for this customer, but they retain copies with EE units for in-house use. The planned transition to use exclusively SI units is complex and proceeding very slowly in U.S. industry. Therefore both sets of units are being used in this revised edition with the aim of making the book comfortable for colleges and professionals in foreign countries (where SI units are standard) and to practicing engineers in the U.S. who are used in the EE system. Some tables have both units, some sections have one or the other.

The use of computers has changed the way we do business in many fields. We have developed computer programs for many an engineering analysis, computer-aided design, computer-aided manufacturing, business and engineering transactions, test data collection, data analysis or data presentation, project management, and many others. In fact computers are used extensively in some companies to design new propulsion devices. Therefore we identify in this book the places where computer programs will be helpful and we mention this often. However, we do not discuss specific programs, because they take up too much space, become obsolete in a short time without regular upgrading, some do not have a way to provide help to a user, and some of the better programs are company proprietary and thus not available.

The first edition of this book was issued in 1949. With this new revised seventh edition this is probably the longest active aerospace book (51 years) that has been upgraded regularly and is still being actively used in industry and universities. To the best of the authors' knowledge the book has been or is being used as a college text in 45 universities worldwide. It is a real satisfaction to the authors that a very large number of students and engineers were introduced to this subject through one of the editions of this book.

The book has three major markets: it has been used and is still used as a college text. It contains more material and more student problems than can be given in a one-semester course. This then allows the choosing of selected portions of the book to fit the student's interest. A one-term course might consist of a review of the first four or five chapters, followed by a careful study of Chapters 6, 10, 11, 14, and 19, a brief scanning of most of the other chapters, and the detailed study of whatever additional chapter(s) might have appeal. The book also has been used to indoctrinate engineers new to the propulsion business and to serve as a reference to experienced engineers, who want to look up some topic, data, or equation.

We have tried to make the book easier to use by providing (1) a much more detailed table of contents, so the reader can find the chapter or section of interest, (2) an expanded index, so specific key words can be located, and (3) five appendices, namely a summary of key equations, a table of the properties of the atmosphere, conversion factors and constants, and two derivations of specific equations.

All rocket propellants are hazardous materials. The authors and the publisher recommend that the reader do not work with them or handle them without an exhaustive study of the hazards, the behavior, and the properties of each propellant, and rigorous safety training, including becoming familiar with protective equipment. Safety training is given routinely to employees by organizations in this business. Neither the authors nor the publisher assume any responsibility for actions on rocket propulsion taken by readers, either directly or indirectly. The information presented in this book is insufficient and inadequate for conducting rocket propulsion experiments or operations.

Professor Oscar Biblarz of the Naval Postgraduate School joins George P. Sutton as a co-author in this edition. We both shared in the preparation of the manuscript and the proofreading. Terry Boardman of Thiokol Propulsion (a division of Cordant Technologies) join as a contributing author; he prepared Chapter 15 (hybrid rocket propulsion) and the major portion of the section on rocket motor nozzles in Chapter 14.

We gratefully acknowledge the help and contributions we have received in preparing this edition. Terrence H. Murphy and Mike Bradley of The Boeing Company, Rocketdyne Propulsion and Power, contributed new data and perspective drawings to the chapters on rocket propulsion with liquid propellants. Warren Frick of Orbital Sciences Corporation provided valuable data on satellite payloads for different orbits. David McGrath, Thomas Kirschner, and W. Lloyd McMillan of Thiokol Propulsion (a division of Cordant Technologies, Inc.) answered questions and furnished data on solid propellant rocket motors. Carl Stechman of Kaiser-Marquardt furnished design information on a small bipropellant thruster. Carl Pignoli and Pat Mills of Pratt & Whitney (a United Technologies Company) gave us engine data and permission to copy data on turbopumps and upper-stage space engines with extendible nozzle skirts.

Kathleen F. Hodge and Gary W. Joseph of the Space and Technology Division of TRW, Inc., gave data on a pressurized storable propellant rocket engine and a jet tab attitude control system. Oscar Biblarz acknowledges his colleagues David W. Netzer, Brij N. Agrawal, and Sherif Michael who, together with many students, have been an integal part of the research and educational environment at the Naval Postgraduate School. Craig W. Clauss of Atlantic Research Corporation (a unit of Sequa Corporation) helped with electric propulsion.

George P. Sutton
Los Angeles, California

Oscar Biblarz
Monterey, California

COVER ILLUSTRATIONS

The color illustrations on the cover show several rocket propulsion systems, each at a different scale. Below we briefly describe these illustrations and list the page numbers, where more detail can be found.

The front cover shows the rocket nozzles at the aft end of the winged Space Shuttle, shortly after takeoff. The two large strap-on solid rocket motors (see page 545) have brightly glowing white billowy exhaust plumes. The three Space Shuttle main engines (page 199) have essentially transparent plumes, but the hot regions, immediately downstream of strong shock waves, are faintly visible. The two darker-colored nozzles of the thrust chambers of the orbital maneuvering system and the small dark nozzle exit areas (pointing upward) of three of the thrusters of the reaction control system of the Space Shuttle (see page 208) are not firing during the ascent of the Shuttle.

The back cover shows (from top to bottom) small illustrations of (1) an image of a stress/strain analysis model (see page 461) of a solid propellant rocket motor grain and case, (2) a small storable bipropellant thruster of about 100 lbf thrust (page 307), (3) a three-quarter section of a solid propellant rocket motor (page 9), and (4) an experimental aerospike rocket engine (page 298) during a static firing test.

CHAPTER 1

CLASSIFICATION

Propulsion in a broad sense is the act of changing the motion of a body. Propulsion mechanisms provide a force that moves bodies that are initially at rest, changes a velocity, or overcomes retarding forces when a body is propelled through a medium. *Jet propulsion* is a means of locomotion whereby a reaction force is imparted to a device by the momentum of ejected matter.

Rocket propulsion is a class of jet propulsion that produces thrust by ejecting stored matter, called the *propellant*. *Duct propulsion* is a class of jet propulsion and includes turbojets and ramjets; these engines are also commonly called air-breathing engines. Duct propulsion devices utilize mostly the surrounding medium as the "working fluid", together with some stored fuel. Combinations of rockets and duct propulsion devices are attractive for some applications and are described in this chapter.

The *energy source* most useful to rocket propulsion is *chemical combustion*. Energy can also be supplied by *solar radiation* and, in the past, also by *nuclear reaction*. Accordingly, the various propulsion devices can be divided into *chemical propulsion, nuclear propulsion*, and *solar propulsion*. Table 1–1 lists many of the important propulsion concepts according to their energy source and type of propellant or working fluid. Radiation energy can originate from sources other than the sun, and theoretically can cover the transmission of energy by microwave and laser beams, electromagnetic waves, and electrons, protons, and other particle beams from a transmitter to a flying receiver. Nuclear energy is associated with the transformations of atomic particles within the nucleus of atoms and can be of several types, namely, fission, fusion, and decay of radioactive species. Other energy sources, both internal (in the vehicle) and external, can be considered. The energy form

TABLE 1–1. Energy Sources and Propellants for Various Propulsion Concepts

Propulsion Device	Energy Source[a]			Propellant or Working Fluid
	Chemical	Nuclear	Solar	
Turbojet	D/P	TFD		Fuel + air
Turbo–ramjet	TFD			Fuel + air
Ramjet (hydrocarbon fuel)	D/P	TFD		Fuel + air
Ramjet (H_2 cooled)	TFD			Hydrogen + air
Rocket (chemical)	D/P	TFD		Stored propellant
Ducted rocket	TFD			Stored solid fuel + surrounding air
Electric rocket	D/P	TFD	D/P	Stored propellant
Nuclear fission rocket		TFD		Stored H_2
Nuclear fusion rocket		TFND		Stored H_2
Solar heated rocket			TFD	Stored H_2
Photon rocket (big light bulb)		TFND		Photon ejection (no stored propellant)
Solar sail			TFD	Photon reflection (no stored propellant)

[a]D/P, developed and/or considered practical; TFD, technical feasibility has been demonstrated, but development is incomplete; TFND, technical feasibility has not yet been demonstrated.

found in the output of a rocket is largely the kinetic energy of the ejected matter; thus the rocket converts the input from the energy source into this form. The ejected mass can be in a solid, liquid, or gaseous state. Often a combination of two or more of these is ejected. At very high temperatures it can also be a plasma, which is an electrically activated gas.

1.1. DUCT JET PROPULSION

This class, also called air-breathing engines, comprises devices which have a duct to confine the flow of air. They use oxygen from the air to burn fuel stored in the flight vehicle. The class includes turbojets, turbofans, ramjets, and pulse-jets. This class of propulsion is mentioned primarily to provide a comparison with rocket propulsion and a background for combination rocket–duct engines, which are mentioned later. Several textbooks, such as Refs. 1–1 and 1–2, contain a discussion of duct jet propulsion fundamentals. Table 1–2 compares several performance characteristics of specific chemical rockets with those of typical turbojets and ramjets. A high specific impulse is directly related to a long flight range and thus indicates the superior range capability of air breather engines over chemical rockets at relatively low altitude. The uniqueness of the rocket, for example, high thrust to weight, high thrust to frontal

TABLE 1–2. Comparison of Several Characteristics of a Typical Chemical Rocket and Two Duct Propulsion Systems

Feature	Rocket Engine or Rocket Motor	Turbojet Engine	Ramjet Engine
Thrust-to-weight ratio, typical	75:1	5:1, turbojet and afterburner	7:1 at Mach 3 at 30,000 ft
Specific fuel consumption (pounds of propellant or fuel per hour per pound of thrust)[a]	8–14	0.5–1.5	2.3–3.5
Specific thrust (pounds of thrust per square foot frontal area)[b]	5000 to 25,000	2500 (Low Mach at sea level)	2700 (Mach 2 at sea level)
Thrust change with altitude	Slight increase	Decreases	Decreases
Thrust vs. flight speed	Nearly constant	Increases with speed	Increases with speed
Thrust vs. air temperature	Constant	Decreases with temperature	Decreases with temperature
Flight speed vs. exhaust velocity	Unrelated, flight speed can be greater	Flight speed always less than exhaust velocity	Flight speed always less than exhaust velocity
Altitude limitation	None; suited to space travel	14,000–17,000 m	20,000 m at Mach 3 30,000 m at Mach 5 45,000 m at Mach 12
Specific impulse typical[c] (thrust force per unit propellant or fuel weight flow per second)	270 sec	1600 sec	1400 sec

[a]Multiply by 0.102 to convert to kg/hr-N.
[b]Multiply by 47.9 to convert to N/m².
[c]Specific impulse is a performance parameter and is defined in Chapter 2.

area, and thrust independence of altitude, enables extremely long flight ranges to be obtained in rarefied air and in space.

The *turbojet engine* is the most common of ducted engines. Figure 1–1 shows the basic elements.

At supersonic flight speeds above Mach 2, the *ramjet engine* (a pure duct engine) becomes attractive for flight within the atmosphere. Thrust is produced by increasing the momentum of the air as it passes through the ramjet, basically as is accomplished in the turbojet and turbofan engines but without compressors or turbines, Figure 1–2 shows the basic components of one type of ramjet. Ramjets with subsonic combustion and hydrocarbon fuel have an upper speed limit of approximately Mach 5; hydrogen fuel, with hydrogen cooling, raises this to at least Mach 16. Ramjets depend on rocket boosters, or some other method (such as being launched from an aircraft) for being accelerated to near their design flight speed to become functional. The primary applications have been in shipboard and ground-launched antiaircraft missiles. Studies of a hydrogen-fueled ramjet for hypersonic aircraft look promising. The supersonic flight vehicle is a combination of a ramjet-driven high-speed airplane and a one- or two-stage rocket booster. It can travel at speeds up to a Mach number of 25 at altitudes of up to 50,000 m.

1.2. ROCKET PROPULSION

Rocket propulsion systems can be classified according to the type of energy source (chemical, nuclear, or solar), the basic function (booster stage, sustainer, attitude control, orbit station keeping, etc.), the type of vehicle (aircraft, missile, assisted take-off, space vehicle, etc.), size, type of propellant, type of construction, or number of rocket propulsion units used in a given vehicle. Each is treated in more detail in subsequent chapters.

Another way is to classify by the method of producing thrust. A thermodynamic expansion of a gas is used in the majority of practical rocket propulsion concepts. The internal energy of the gas is converted into the kinetic energy of the exhaust flow and the thrust is produced by the gas pressure on the surfaces exposed to the gas, as will be explained later. This same thermo-

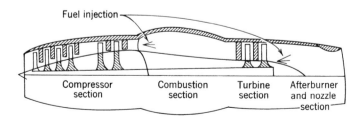

FIGURE 1–1. Simplified schematic diagram of a turbojet engine.

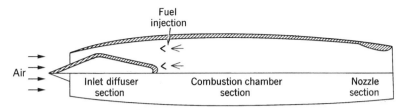

FIGURE 1–2. Simplified diagram of a ramjet with a supersonic inlet (converging and diverging flow passage).

dynamic theory and the same generic equipment (nozzle) is used for jet propulsion, rocket propulsion, nuclear propulsion, laser propulsion, solar-thermal propulsion, and some types of electrical propulsion. Totally different methods of producing thrust are used in other types of electric propulsion or by using a pendulum in a gravity gradient. As described below, these electric systems use magnetic and/or electric fields to accelerate electrically charged molecules or atoms at very low densities. It is also possible to obtain a very small acceleration by taking advantage of the difference in gravitational attraction as a function of altitude, but this method is not explained in this book.

The Chinese developed and used solid propellant in rocket missiles over 800 years ago and military bombardment rockets were used frequently in the eighteenth and nineteenth centuries. However, the significant developments of rocket propulsion took place in the twentieth century. Early pioneers included the Russian Konstantin E. Ziolkowsky, who is credited with the fundamental rocket flight equation and his 1903 proposals to build rocket vehicles. The German Hermann Oberth developed a more detailed mathematical theory; he proposed multistage vehicles for space flight and fuel-cooled thrust chambers. The American Robert H. Goddard is credited with the first flight using a liquid propellant rocket engine in 1926. An early book on the subject was written by the Viennese engineer Eugen Sänger. For rocket history see Refs. 1–3 to 1–7.

Chemical Rocket Propulsion

The energy from a high-pressure combustion reaction of propellant chemicals, usually a fuel and an oxidizing chemical, permits the heating of reaction product gases to very high temperatures (2500 to 4100°C or 4500 to 7400°F). These gases subsequently are expanded in a nozzle and accelerated to high velocities (1800 to 4300 m/sec or 5900 to 14,100 ft/sec). Since these gas temperatures are about twice the melting point of steel, it is necessary to cool or insulate all the surfaces that are exposed to the hot gases. According to the physical state of the propellant, there are several different classes of chemical rocket propulsion devices.

Liquid propellant rocket engines use liquid propellants that are fed under pressure from tanks into a thrust chamber.* A typical pressure-fed liquid propellant rocket engine system is schematically shown in Fig. 1–3. The liquid *bipropellant* consists of a liquid oxidizer (e.g., liquid oxygen) and a liquid fuel (e.g., kerosene). A *monopropellant* is a single liquid that contains both oxidizing and fuel species; it decomposes into hot gas when properly catalyzed. A large turbopump-fed liquid propellant rocket engine is shown in Fig. 1–4. Gas pressure feed systems are used mostly on low thrust, low total energy propulsion systems, such as those used for attitude control of flying vehicles, often with more than one thrust chamber per engine. Pump-fed liquid rocket systems are used typically in applications with larger amounts of propellants and higher thrusts, such as in space launch vehicles.

In the *thrust chamber* the propellants react to form hot gases, which in turn are accelerated and ejected at a high velocity through a supersonic nozzle, thereby imparting momentum to the vehicle. A nozzle has a converging section, a constriction or throat, and a conical or bell-shaped diverging section as further described in the next two chapters.

Some liquid rocket engines permit repetitive operation and can be started and shut off at will. If the thrust chamber is provided with adequate cooling capacity, it is possible to run liquid rockets for periods exceeding 1 hour, dependent only on the propellant supply. A liquid rocket propulsion system requires several precision valves and a complex feed mechanism which includes propellant pumps, turbines, or a propellant-pressurizing device, and a relatively intricate combustion or thrust chamber.

In *solid propellant rocket motors*[†] the propellant to be burned is contained within the combustion chamber or *case*. The solid propellant charge is called the *grain* and it contains all the chemical elements for complete burning. Once ignited, it usually burns smoothly at a predetermined rate on all the exposed internal surfaces of the grain. Initial burning takes place at the internal surfaces of the cylinder perforation and the four slots. The internal cavity grows as propellant is burned and consumed. The resulting hot gas flows through the supersonic nozzle to impart thrust. Once ignited, the motor combustion proceeds in an orderly manner until essentially all the propellant has been consumed. There are no feed systems or valves (see Fig. 1–5).

Liquid and solid propellants, and the propulsion systems that use them, are discussed in Chapters 6 to 10 and 11 to 14, respectively. Liquid and solid propellant rocket propulsion systems are compared in Chapter 17.

---

*The term *thrust chamber*, used for the assembly of the injector, nozzle, and chamber, is preferred by several official agencies and therefore has been used in this book. However, other terms, such as *thrust cylinder* and *combustor*, are still used in the literature. For small spacecraft control rockets the term *thruster* is commonly used and this term will be used in some sections of this book.

[†]Historically the word *engine* is used for a liquid propellant rocket propulsion system and the word *motor* is used for solid propellant rocket propulsion. They were developed originally by different groups.

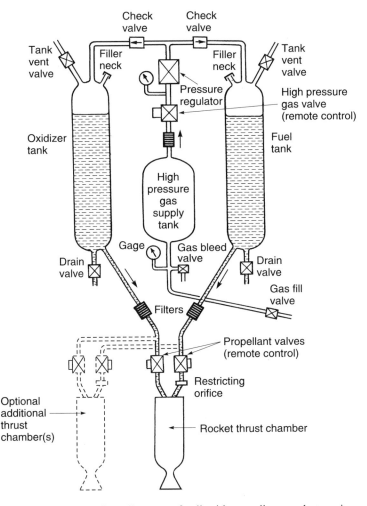

FIGURE 1–3. Schematic flow diagram of a liquid propellant rocket engine with a gas pressure feed system. The dashed lines show a second thrust chamber, but some engines have more than a dozen thrust chambers supplied by the same feed system. Also shown are components needed for start and stop, controlling tank pressure, filling propellants and pressurizing gas, draining or flushing out remaining propellants, tank pressure relief or venting, and several sensors.

Gaseous propellant rocket engines use a stored high-pressure gas, such as air, nitrogen, or helium, as their working fluid or propellant. The stored gas requires relatively heavy tanks. These cold gas engines have been used on many early space vehicles as attitude control systems and some are still used today. Heating the gas by electrical energy or by combustion of certain mono-propellants improves the performance and this has often been called *warm gas propellant rocket propulsion.*

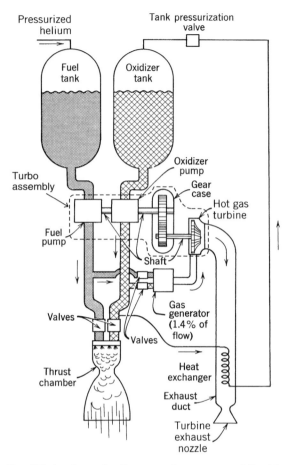

FIGURE 1–4. Simplified schematic diagram of one type of liquid propellant rocket engine with a turbopump feed system and a separate gas generator, which generates warm gas for driving the turbine. Not shown are components necessary for controlling the operation, filling, venting, draining, or flushing out propellants, filters or sensors. The turbopump assembly consists of two propellant pumps, a gear case, and a high speed turbine.

Hybrid propellant rocket propulsion systems use both a liquid and a solid propellant. For example, if a liquid oxidizing agent is injected into a combustion chamber filled with solid carbonaceous fuel grain, the chemical reaction produces hot combustion gases (see Fig. 1–6). They are described further in Chapter 15.

There are also chemical rocket propulsion *combination systems* that have both solid and liquid propellants. One example is a pressurized liquid propellant system that uses a solid propellant to generate hot gases for tank pressurization; flexible diaphragms are necessary to separate the hot gas and the reactive liquid propellant in the tank.

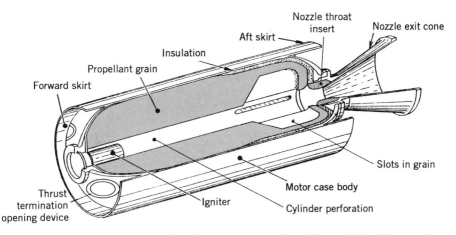

Nozzle throat insert

Nozzle exit cone

Aft skirt

Insulation

Propellant grain

Forward skirt

Thrust termination opening device

Igniter

Motor case body

Cylinder perforation

Slots in grain

FIGURE 1–5. Simplified perspective three-quarter section of a typical solid propellant rocket motor with the propellant grain bonded to the case and the insulation layer and with a conical exhaust nozzle. The cylindrical case with its forward and aft hemispherical domes form a pressure vessel to contain the combustion chamber pressure. Adapted with permission from Reference 11–1.

Combinations of Ducted Jet Engines and Rocket Engines

The Tomahawk surface-to-surface missile uses two stages of propulsion in sequence. The solid propellant rocket booster lifts the missile away from its launch platform and is discarded after its operation. A small turbojet engine sustains the low level flight at nearly constant speed toward the target.

A *ducted rocket*, sometimes called an *air-augmented rocket*, combines the principles of rocket and ramjet engines; it gives higher performance (specific impulse) than a chemical rocket engine, while operating within the earth's atmosphere. Usually the term *air-augmented rocket* denotes mixing of air with the rocket exhaust (fuel-rich for afterburning) in proportions that enable the propulsion device to retain the characteristics typifying a rocket engine, for example, high static thrust and high thrust-to-weight ratio. In contrast, the ducted rocket often is like a ramjet in that it must be boosted to operating speed and uses the rocket components more as a fuel-rich gas generator (liquid, solid, or hybrid), igniter, and air ejecter pump.

The principles of the rocket and ramjet can be combined so that the two propulsion systems operate in sequence and in tandem and yet utilize a common combustion chamber volume as shown in Fig. 1–7. The low-volume configuration, known as an *integral rocket–ramjet*, can be attractive in air-launched missiles using ramjet propulsion (see Ref. 1–8). The transition from the rocket to the ramjet requires enlarging the exhaust nozzle throat (usually by ejecting rocket nozzle parts), opening the ramjet air inlet–combustion chamber interface, and following these two events with the normal ramjet starting sequence.

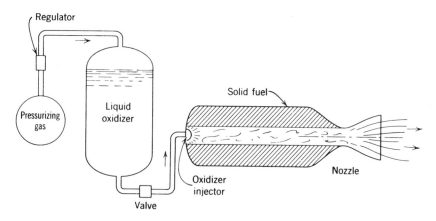

FIGURE 1–6. Simplified schematic diagram of a typical hybrid rocket engine. The relative positions of the oxidizer tank, high pressure gas tank, and the fuel chamber with its nozzle depend on the particular vehicle design.

A *solid fuel ramjet* uses a grain of solid fuel that gasifies or ablates and reacts with air. Good combustion efficiencies have been achieved with a patented boron-containing solid fuel fabricated into a grain similar to a solid propellant and burning in a manner similar to a hybrid rocket propulsion system.

Nuclear Rocket Engines

Three different types of nuclear energy sources have been investigated for delivering heat to a working fluid, usually liquid hydrogen, which subsequently can be expanded in a nozzle and thus accelerated to high ejection velocities (6000 to 10,000 m/sec). However, none can be considered fully developed today and none have flown. They are the *fission reactor*, the

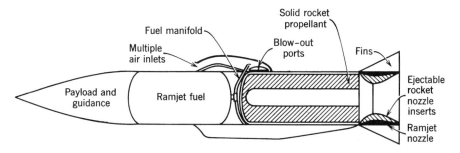

FIGURE 1–7. Elements of an air-launched missile with integral rocket–ramjet propulsion. After the solid propellant has been consumed in boosting the vehicle to flight speed, the rocket combustion chamber becomes the ramjet combustion chamber with air burning the ramjet liquid fuel.

radioactive isotope decay source, and the *fusion reactor*. All three types are basically extensions of liquid propellant rocket engines. The heating of the gas is accomplished by energy derived from transformations within the nuclei of atoms. In chemical rockets the energy is obtained from within the propellants, but in nuclear rockets the power source is usually separate from the propellant.

In the *nuclear fission reactor rocket*, heat can be generated by the fission of uranium in the solid reactor material and subsequently transferred to the working fluid (see Refs. 1–9 to 1–11). The *nuclear fission rocket* is primarily a high-thrust engine (above 40,000 N) with specific impulse values up to 900 sec. Fission rockets were designed and tested in the 1960s. Ground tests with hydrogen as a working fluid culminated in a thrust of 980,000 N (210,000 lb force) at a graphite core nuclear reactor level of 4100 MW with an equivalent altitude-specific impulse of 848 sec and a hydrogen temperature of about 2500 K. There were concerns with the endurance of the materials at the high temperature (above 2600 K) and intense radiations, power level control, cooling a reactor after operation, moderating the high-energy neutrons, and designing lightweight radiation shields for a manned space vehicle.

In recent years there have been renewed interest in nuclear fission rocket propulsion primarily for a potential manned planetary exploration mission. Studies have shown that the high specific impulse (estimated in some studies at 1100 sec) allows shorter interplanetary trip transfer times, smaller vehicles, and more flexibility in the launch time when planets are not in their optimum relative position.

In the *isotope decay engine* a radioactive material gives off radiation, which is readily converted into heat. Isotope decay sources have been used successfully for generating electrical power in space vehicles and some have been flown as a power supply for satellites and deep space probes. The released energy can be used to raise the temperature of a propulsive working fluid such as hydrogen or perhaps drive an electric propulsion system. It provides usually a lower thrust and lower temperature than the other types of nuclear rocket. As yet, isotope decay rocket engines have not been developed or flown.

Fusion is the third nuclear method of creating nuclear energy that can heat a working fluid. A number of different concepts have been studied. To date none have been tested and many concepts are not yet feasible or practical. Concerns about an accident with the inadvertent spreading of radioactive materials in the earth environment and the high cost of development programs have to date prevented a renewed experimental development of a large nuclear rocket engine. Unless there are some new findings and a change in world attitude, it is unlikely that a nuclear rocket engine will be developed or flown in the next few decades, therefore no further discussion of it is given in this book.

Electric Rocket Propulsion

In all electric propulsion the source of the electric power (nuclear, solar radiation receivers, or batteries) is physically separate from the mechanism that produces the thrust. This type of propulsion has been handicapped by heavy and inefficient power sources. The thrust usually is low, typically 0.005 to 1 N. In order to allow a significant increase in the vehicle velocity, it is necessary to apply the low thrust and thus a small acceleration for a long time (weeks or months) (see Chapter 19 and Refs. 1–12 and 1–13).

Of the three basic types, electrothermal rocket propulsion most resembles the previously mentioned chemical rocket units; propellant is heated electrically (by heated resistors or electric arcs) and the hot gas is then thermodynamically expanded and accelerated to supersonic velocity through an exhaust nozzle (see Fig. 1–8). These electrothermal units typically have thrust ranges of 0.01 to 0.5 N, with exhaust velocities of 1000 to 5000 m/sec, and ammonium, hydrogen, nitrogen, or hydrazine decomposition product gases have been used as propellants.

The two other types—the electrostatic or ion propulsion engine and the electromagnetic or magnetoplasma engine—accomplish propulsion by different principles and the thermodynamic expansion of gas in a nozzle, as such, does not apply. Both will work only in a vacuum. In an ion rocket (see Fig. 1–9) a working fluid (typically, xenon) is ionized (by stripping off electrons) and then the electrically charged heavy ions are accelerated to very high velocities (2000 to 60,000 m/sec) by means of electrostatic fields. The ions are subsequently electrically neutralized; they are combined with electrons to prevent the buildup of a space charge on the vehicle.

In the magnetoplasma rocket an electrical plasma (an energized hot gas containing ions, electrons, and neutral particles) is accelerated by the interaction between electric currents and magnetic fields and ejected at high velocity

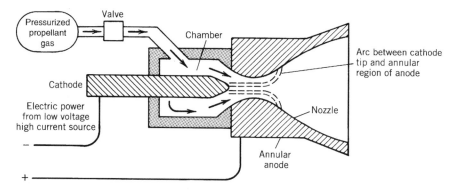

FIGURE 1–8. Simplified schematic diagram of arc-heating electric rocket propulsion system. The arc plasma temperature is very high (perhaps 15,000 K) and the anode, cathode, and chamber will get hot (1000 K) due to heat transfer.

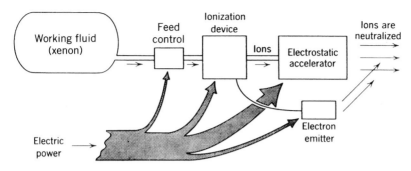

FIGURE 1–9. Simplified schematic diagram of a typical ion rocket, showing the approximate distribution of the electric power.

(1000 to 50,000 m/sec). There are many different types and geometries. A simple pulsed (not continuously operating) unit with a solid propellant is shown in Fig. 1–10. This type has had a good flight record as a spacecraft attitude control engine.

Other Rocket Propulsion Concepts

Several technologies exist for harnessing solar energy to provide the power for spacecraft and also to propel spacecraft using electrical propulsion. Solar cells generate electric power from the sun's radiation. They are well developed and have been successful for several decades. Most electric propulsion systems have used solar cells for their power supply.

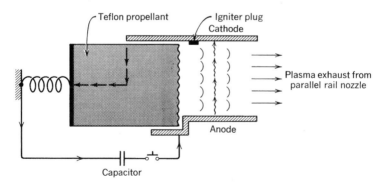

FIGURE 1–10. Simplified diagram of a rail accelerator for self-induced magnetic acceleration of a current-carrying plasma. When the capacitor is discharged, an arc is struck at the left side of the rails. The high current in the plasma arc induces a magnetic field. The action of the current and the magnetic field causes the plasma to be accelerated at right angles to both the magnetic field and the current, namely in the direction of the rails. Each time the arc is created a small amount of solid propellant (Teflon) is vaporized and converted to a small plasma cloud, which (when ejected) gives a small pulse of thrust. Actual units can operate with many pulses per second.

An attractive concept, the *solar thermal rocket*, has large diameter optics to concentrate the sun's radiation (e.g., by lightweight precise parabolic mirrors or Fresnel lenses) onto a receiver or optical cavity. Figure 1–11 shows one concept and some data is given in Table 2–1. The receiver is made of high temperature metal (such as tungsten or rhenium) and has a cooling jacket or heat exchanger. It heats a working fluid, usually liquid hydrogen, up to perhaps 2500°C and the hot gas is controlled by hot gas valves and exhausted through one or more nozzles. The large mirror has to be pointed toward the sun and this requires the mirror to be adjustable in its orientation. Performance can be two to three times higher than that of a chemical rocket and thrust levels in most studies are low (1 to 10 N). Since large lightweight optical elements cannot withstand drag forces without deformation, the optical systems are deployed outside the atmosphere. Contamination is negibible, but storage or refueling of liquid hydrogen is a challenge. Problems being investigated include rigid, lightweight mirror or lens structures, operational life, minimizing hydrogen evaporation, and heat losses to other spacecraft components. To date the solar thermal rocket has not yet provided the principal thrust of a flying spacecraft.

The *solar sail* is another concept. It is basically a big photon reflector surface. The power source for the solar sail is the sun and it is external to the vehicle (see Ref. 1–14). Approaches using nuclear explosions and pulsed nuclear fusion have been analyzed (Refs. 1–15 and 1–16), but are not yet feasible. Concepts for transmitting radiation energy (by lasers or microwaves) from earth stations to satellites have been proposed, but are not yet developed.

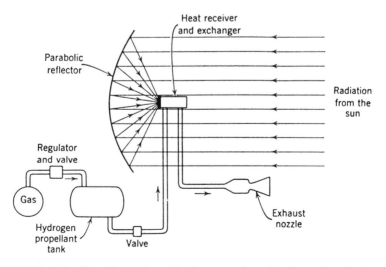

FIGURE 1-11. Simplified schematic diagram of a solar thermal rocket concept.

International Rocket Propulsion Effort

Active development or production of rocket propulsion systems is currently under way in more than 30 different countries. Some of them have made significant and original contributions to the state of the art of the technologies. There is mention in this book of a few foreign rocket units and their accomplishments and references to international rocket literature. Although most of the data in this book are taken from U.S. rocket experience, this is not intended to minimize foreign achievements.

At the time of this writing the major international program was the *International Space Station* (ISS), a multi-year cooperative effort with major contributions from the USA and Russia and active participation by several other nations. This manned orbital space station is used for conducting experiments and observations on a number of research projects.

1.3. APPLICATIONS OF ROCKET PROPULSION

Because the rocket can reach a performance unequaled by other prime movers, it has its own fields of application and does not usually compete with other propulsion devices. Examples of important applications are given below and discussed further in Chapter 4.

Space Launch Vehicles

Between the first space launch in 1957 and the end of 1998 approximately 4102 space launch attempts have taken place in the world and all but about 129 were successful (see Ref. 1–17). *Space launch vehicles* or *space boosters* can be classified broadly as expendable or recoverable/reusable. Other bases of classification are the type of propellant (storable or cryogenic liquid or solid propellants), number of stages (single-stage, two-stage, etc.), size/mass of payloads or vehicles, and manned or unmanned. Figure 1–12 shows the Titan III-C space launch vehicle, one member of the Titan family of storable propellant space launch vehicles, which is used extensively for boosting satellites into synchronous earth orbit or into escape trajectories for planetary travel. This heavy-duty launch vehicle consists of the basic 2-stage Titan III standard launch vehicle (liquid propellant rockets) supplemented by two solid propellant "strap-on motors." A fourth stage, known as the transtage, permits a wide variety of maneuvers, orbit changes, and trajectory transfers to be accomplished with the payload, which can be one or more satellites or spacecraft.

Each space launch vehicle has a specific space flight objective, such as an earth orbit or a moon landing. It uses between two and five stages, each with its own propulsion system, and each is usually fired sequentially after the lower stage is expended. The number of stages depends on the specific space trajectory, the number and types of maneuvers, the energy content of a unit mass of

FIGURE 1–12. Titan III launch vehicle shortly after lift-off, with bright radiant exhaust gas. Two solid propellant rocket motors, each providing about 2.4 million pounds of thrust, boost the first stage, which also gets a sustained thrust of 470,000 pounds from two liquid rocket engines. The second stage has 100,000 pounds of thrust from a single liquid rocket engine, and one version of the third stage has two liquid rocket engines, each at 16,000 pounds of thrust.

the propellant, and other factors. The initial stage, usually called the booster stage, is the largest and it is operated first; this stage is then separated from the ascending vehicle before the second-stage rocket propulsion system is ignited and operated. As will be explained in Chapter 4, adding an extra stage permits a significant increase in the payload (such as more scientific instruments or more communications gear).

Each stage of a multistage launch vehicle is essentially a complete vehicle in itself and carries its own propellant, its own rocket propulsion system or systems, and its own control system. Once the propellant of a given stage is expended, the dead mass of that stage (including empty tanks, cases, instruments, etc.) is no longer useful in providing additional kinetic energy to the succeeding stages. By dropping off this useless mass it is possible to accelerate the final stage with its useful payload to a higher terminal velocity than would be attained if multiple staging were not used. Both solid propellant and liquid propellant rocket propulsion systems have been used for low earth orbits.

A single stage to orbit vehicle, attractive because it avoids the costs and complexities of staging, is expected to have improved reliability (simple structures, fewer components), and some versions may be recoverable and reusable. However, its payload is relatively very small. A low earth orbit (say 100 miles altitude) can only be achieved with such a vehicle if the propellant performance is very high and the structure is efficient and low in mass. Liquid propellants such as liquid hydrogen with liquid oxygen are usually chosen.

The missions and payloads for space launch vehicles are many, such as military (reconnaissance satellites, command and control satellites), non-military government (weather observation satellites, GPS or geopositioning satellites), space exploration (space environment, planetary missions), or commercial (communication satellites). Forecasts indicate that a large number of future commercial communications satellites will be needed.

Table 1–3 lists several important U.S. launch vehicles and their capabilities and Table 1–4 gives data on the Space Shuttle, which is really a combination of launch vehicle, spacecraft, and a glider. It can be seen that the thrust levels are highest for booster or first stages and are relatively high for upper stages (thousands of pounds). Only for the attitude control system of the vehicle (also called reaction control in Table 1–4) are the thrust levels low (from a fraction of a pound for small spacecraft to as high as about 1000 pounds thrust in the space shuttle vehicle). Frequent propulsion starts and stops are usually required in these applications.

Spacecraft

Depending on their missions, *spacecraft* can be categorized as earth satellites, lunar, interplanetary, and trans-solar types, and as manned and unmanned spacecraft. Rocket propulsion is used for both primary propulsion (i.e.,

TABLE 1-3. Selected United States Space Launch Vehicles

Name	Stage	Number of Engines or Motors per Stage	Thrust kN	Thrust lbf	Propellants	Launch Mass (metric tons)	Two-stage Payload Weight 100 n-mi (185 km) Orbit kg	lbf	Three-stage Payload Weight Geosynchronous Orbit kg	lbf
Titan 34D	0	2	10,750	2,400,000 vac	Solid composite	1091	13,600	30,000	1820	4000
	1	2	2370	529,000 }	N₂O₄/N₂H₄ + UDMH					
	2	1	452	101,000 }						
	3	1	107	23,800	Solid composite					
Delta II 6925	0	6 + 3	443.5	Each 97,000 SL	Solid composite	132	2545	5600	1454	3200
	1	1	927	207,000 SL	LO₂/RP-1					
			1037	231,700 vac						
	2	1	43.2	9645	N₂O₄/N₂H₄ – UDMH					
	3	1	67.6	15,100 vac	Solid composite					
Atlas Centaur	½	2	Each 829 SL	Each 185,000 SL	LO₂/RP-1	141	2772	6100	1545	3400
	1	1	269	60,000	LO₂/RP-1					
	2	2	Each 74 vac	Each 16,500 vac	LO₂/LH₂					
Pegasus (air-launched)	1	1	726	163,000	Solid	23.1	490 (Three stages)	1078	NA	NA
	2	1	196	44,200	Solid					
	3	1	36	8060	Solid					

"SL" refers to sea level and "vac" refers to altitude or vacuum conditions.

TABLE 1–4. Propulsion Systems for the Space Shuttle

Vehicle Section	Propulsion System (No. of Units)	Number of Starts and Typical Burn Time	Propellant and Specific Impulse	Thrust	Mission
Shuttle orbiter	Space Shuttle main engine (3)	Start at launch 8.4 min duration Life: 55 starts and 7.5 hr	Liquid hydrogen–liquid oxygen 4464 N-sec/kg (455 sec)	1670 kN each (375,000 lb) at sea level 2100 kN each (470,000 lbf) at space vacuum Throttled 109 to 65% of rated power	Lift orbiter off ground and accelerate to orbit velocity. Individual engines can be shut down to reduce thrust level.
	Orbital maneuver systems (2)	3 to 10 starts/mission; designed for 1000 starts, 100 flights, 15 hours of cumulative time	See Note 1; $I_s = 313$ sec	27 kN each (6000 lbf) in vacuum	Insert orbiter vehicle into earth orbit, correct orbit, abort, and deorbit maneuver.
	Reaction control system, 38 primary thrusters, 6 vernier thrusters	Multiple operations; thousands of starts; duration from a few milliseconds to seconds	See Note 1; $I_s = 280$–304 sec, depending on nozzle area ratio	Primary thruster 3870 N each (870 lbf), vernier thruster 106.8 N each (25 lbf)	Small vehicle velocity adjustments and attitude control during orbit insertion, on orbit corrections, rendezvous, and reentry.
Solid rocket boosters (SRBs)	Attached to external tank; multisection, 2 units	Single start at launch 2 min	See Note 2	14,700 kN each, or 3.3×10^6 lbf each	Boost Shuttle vehicle to about 5500 km/hr
	Separation rocket motors; 16 units	4 each at forward frustum and aft skirt; 0.66 sec, nominal	Solid propellant; $I_s = 250$ sec	97,840 N each or 22,000 lbf	Move SRB away from vehicle after cut-off

Notes:
1. MMH, monomethylhydrazine and NTO, nitrogen tetroxide.
2. 70% Ammonium perchlorate; 16% aluminum; 12% polybutadiene acrylic acid binder; 2% epoxy curing agent.

along the flight path, such as for orbit insertion or orbit change maneuvers) and secondary propulsion functions in these vehicles. Some of the *secondary propulsion* functions are attitude control, spin control, momentum wheel and gyro unloading, stage separation, and the settling of liquids in tanks. A spacecraft usually has a series of different rocket propulsion systems, some often very small. For spacecraft attitude control about three perpendicular axes, each in two rotational directions, the system must allow the application of pure torque for six modes of angular freedom, thus requiring a minimum of 12 thrust chambers. Some missions require as few as four to six rocket units whereas the more complex manned spacecraft have 40 to 80 rocket units in all of its stages. Often the small *attitude control rockets* must give pulses or short bursts of thrust, necessitating thousands of restarts.

Table 1–5 presents a variety of spacecraft along with their weights, missions, and propulsion. Although only U.S. launch vehicles are listed in this table, there are also launch vehicles developed by France, the European Space Agency, Russia, Japan, China, India, and Israel that have successfully launched payloads into satellite orbits. They use rocket propulsion systems that were developed in their own countries.

The U.S. Space Shuttle program, using technology and experience from the X-15 rocket-powered research airplane, the Mercury and Gemini orbital flights, the Apollo lunar flight program, and Skylab, provided the first *reusable spacecraft* that lands on a runway. Figure 1–13 shows the basic configuration of the Space Shuttle, which consists of two stages, the booster and the orbiter. It shows all the 67 rocket propulsion systems of the shuttle. The orbiter is really a reusable combination vehicle, namely a spacecraft combined with a glider. The two solid propellant rocket motors are the largest in existence; they are equipped with parachutes for sea recovery of the burned-out motors. The large liquid oxygen/liquid hydrogen (LO_2/LH_2) external tank is jettisoned and expended just before orbit insertion (see Ref. 1–18). Details of several of these Space Shuttle rocket propulsion systems are given elsewhere in this book. The Space Shuttle accomplishes both civilian and military missions of placing satellites in orbit, undertaking scientific exploration, and repairing, servicing, and retrieving satellites.

A reusable single stage to orbit, experimental vehicle with a novel rocket engine is currently (1997) under development in the USA. It is a combination launch vehicle and spacecraft. The design takes advantage of advances in lightweight structures, a clever lifting aerodynamic body concept, and a tailored novel rocket engine that requires little space and fits well into the flight vehicle. This engine, known as a linear aerospike, has a novel configuration and is described further in Chapter 8.

The majority of spacecraft have used liquid propellant engines, with solid propellant boosters. Several spacecraft have operated successfully with electrical propulsion for attitude control. Electrical propulsion systems will probably also be used for some primary and secondary propulsion missions on long-duration space flights, as described in Chapter 19.

TABLE 1–5. Selected United States Spacecraft

| Name | Space Maneuver Propulsion | | | Remarks |
	Thrust (lbf)	Propellants[a]	Weight (lbf)	
Mariner 69	50 (primary) 1.0 (secondary)	Hydrazine monopropellant Hydrazine monopropellant	1100	Flyby of Venus/Mercury
Pioneer 10, 11	50 (primary)	Hydrazine monopropellant	570	Fly to Jupiter and beyond
Viking	600 (primary) 5.0 (secondary)	Hydrazine monopropellant Hydrazine monopropellant	7500	Mars orbiter with soft lander
Nimbus 5	0.5 (secondary)	Stored nitrogen	1700	Weather satellite
Apollo command and service module	20,500 (primary) 100 lbf 16 units 93 lbf 6 units (secondary)	N_2O_4/50:50 UDMH $-N_2H_4$ N_2O_4/MMH	64,500	Manned lunar landing
Space Shuttle orbiter	Two 6000-lbf units (primary) 38 units @ 900 lbf (secondary) Six 25-lbf units (secondary)	N_2O_4/MMH N_2O_4/MMH N_2O_4/MMH	150,000	Reusable spacecraft with runway landing
Fleet Communications Satellite	0.1 (secondary)	Hydrazine monopropellant	1854	UHF communications
Photo Recon	4.0 (secondary)	Hydrazine monopropellant	25,000	Radio/photo communications
Intelsat V communication satellite	0.10	Hydrazine	4180	Resistojet, electric propulsion for N-S station keeping
Deep Space I (DS1)	0.02 (primary)	Xenon	1070	Ion propulsion engine for asteroid fly-by

[a]N_2O_4, nitrogen tetroxide (oxidizer); MMH, monomethylhydrazine (fuel); 50:50 UDMH–N_2H_4 is a 50% mixture of unsymmetrical dimethylhydrazine and hydrazine.

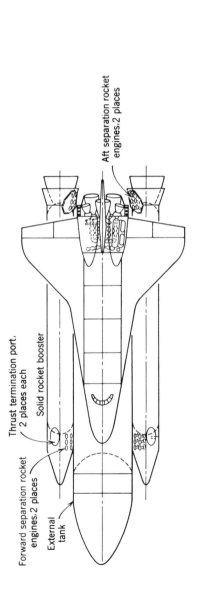

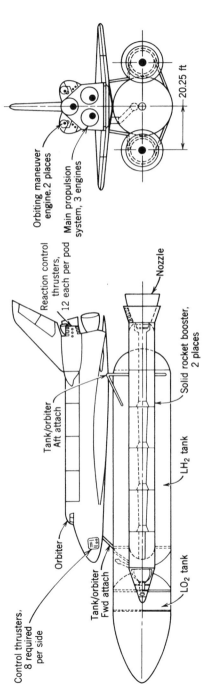

FIGURE 1–13. Simplified sketch of the Space Shuttle vehicle. The Shuttle Orbiter—the delta-winged vehicle about the size of a medium-range jet liner—is a reusable, cargo carrying, spacecraft–airplane combination that takes off vertically and lands horizontally like a glider. Each shuttle orbiter was designed for a minimum of 100 missions and can carry as much as 65,000 lb of payload to a low Earth orbit, and a crew of up to four members and 10 passengers. It can return up to 25,000 lb of payload back to Earth.

TABLE 1-6. Selected United States Missiles

Mission Category	Name	Diameter (ft)	Length (ft)	Propulsion	Launch Weight (lb)
Surface-to-surface (long range)	Minuteman III	6.2	59.8	3 stages, solid	78,000
	Poseidon	6.2	34	2 stages, solid	65,000
	Titan II	10	103	2 stages, liquid	330,000
Surface-to-air (or to missile)	Chaparral	0.42	9.5	1 stage, solid	185
	Improved Hawk	1.2	16.5	1 stage, solid	1398
	Standard Missile	1.13	15 or 27	2 stage, solid	1350/2996
	Redeye	0.24	4	1 stage, solid	18
	Patriot	1.34	1.74	1 stage, solid	1850
Air-to-surface	Maverick	1.00	8.2	1 stage, solid	475
	Shrike	0.67	10	1 stage, solid	400
	SRAM	1.46	14	2 staged grains	2230
Air-to-air	Falcon	0.6	6.5	1 stage, solid	152
	Phoenix	1.25	13	1 stage, solid	980
	Sidewinder	0.42	9.5	1 stage, solid	191
	Sparrow	0.67	12	1 stage, solid	515
Antisubmarine	Subroc	1.75	22	1 stage, solid	4000
Battlefield	Lance	1.8	20	2 stages, liquid	2424
Support (surface-to-surface, short range)	Hellfire (antitank)	0.58	5.67	1 stage, solid	95
	Pershing II	3.3	34.5	2 stages, solid	10,000
	Tow (antitank)	0.58	3.84	1 stage, solid	40
cruise missile (subsonic)	Tomahawk	1.74	21	solid booster + turbofan	3900

TABLE 1-7. Typical Propulsion Characteristics of Some Rocket Applications

Application	Type of Propellant	Thrust Profile	Typical Duration	Maximum Acceleration[a]
Large space launch vehicle booster	Solid or cryogenic liquid	Nearly constant thrust	2–8 min	2–6 g_0
Antiaircraft or antimissile-missile	Solid, some with liquid terminal divert stage	High thrust boost, decreasing thrust sustain phase	2–75 sec each	5 to 20 g_0, but can be up to 100 g_0
Spacecraft orbit maneuvers	Storable liquid or cryogenic liquid	Restartable	Up to 10 min cumulative duration	0.2–6 g_0
Air launched guided missile	Solid	High thrust boost phase with low thrust or decreasing thrust for sustain phase; sometimes 2 pulses	Boost: 2–5 sec Sustain: 10–30 sec	Up to 25 g_0
Battlefield support—surface launched	Solid	Same as above	Up to 2 min each stage	Up to 10 g_0
Rocket assisted projectile, gun launched	Solid	Increase and then decrease in thrust	A few sec	Up to 20,000 g_0
Spacecraft attitude control—large vehicles	Storable liquid (monopropellant or bipropellant); electric propulsion; xenon	Many restarts (up to 60,000); pulsing	Up to 1 hr cumulative duratiaon	Less than 0.1 g_0
Spacecraft attitude control—small vehicle	Cold or warm gas or storable liquid, electric propulsion	Same	Up to 40 min cumulative	Same
Reusable main engines for space shuttle	Cryogenic liquid (O_2/H_2)	Variable thrust, many flights with same engine	8 min, over 7 hr cumulative in several missions	Several g_0
Single stage to orbit (has not yet flown)	Cryogenic liquid (O_2/H_2)	Throttled to lower thrust	6–10 min	4–7 g_0
Lunar landing	Storable bipropellant	10:1 thrust variation	4 min	Several g_0
Weather sounding rocket	Solid	Single burn period—often decreasing thrust	5–50 sec	Up to 15 g_0
Antitank	Solid	Single burn period	0.2–3 sec	Up to 20 g_0

24

Missiles and Other Applications

Military missiles can be classified as shown in Table 1–6. Rocket propulsion for new U.S. missiles uses now almost exclusively solid propellant rocket motors. They can be *strategic missiles*, such as long-range ballistic missiles (800 to 9000 km range) which are aimed at military targets within an enemy country, or *tactical missiles*, which are intended to support or defend military ground forces, aircraft, or navy ships.

The term *surface launch* can mean a launch from the ground, the ocean surface (from a ship), or from underneath the sea (submarine launch). Some tactical missiles, such as the air-to-surface SRAM missile, have a two-pulse solid propellant motor, where two separate, insulated grains are in the same motor case; the time interval before starting the second pulse can be timed to control the flight path or speed profile. Most countries now have tactical missiles in their military inventories, and many of these countries have a capability to produce their own rocket propulsion systems that are used to propel them.

Other applications of rockets include primary engines for research airplanes, assist-take-off rockets for airplanes, ejection of crew escape capsules and stores, personnel "propulsion belts,"and propulsion for target drones, weather sounding rockets, signal rockets, decoy rockets, spin rockets, vernier rockets, underwater rockets for torpedoes and missiles, the throwing of lifelines to ships, and "Fourth of July" rockets.

Tables 1–6 and 1–7 show some parameters of rocket propulsion devices for different applications. The selection of the best rocket propulsion system type and design for any given application is a complex process involving many factors, including system performance, reliability, propulsion system size, and compatibility, as described in Chapter 17. Comparisons and evaluations of many of these criteria are discussed in this book. Many factors, such as development, production or operating costs, available technology, and service life, though beyond the scope of this book, enter strongly into such a selection.

REFERENCES

1–1. G. C. Oates, *Aerothermodynamics of Gas Turbines and Rocket Propulsion*, American Institute of Aeronautics and Astronautics, Washington, DC, Revised 1988, 452 pages.

1–2. H. Cohen, G. F. C. Rogers, and H. I. H. Saravanamuttoo, *Gas Turbine Theory*, 3rd ed., Longman Scientific and Technical, New York, 1987, 414 pages.

1–3. K. E. Ziolkowsky, *Space Investigations by Means of Propulsive Spaceships* (in Russian), Kaluga, Saint Petersburg, 1914.

1–4. E. C. Goddard and G. E. Pendray. (Eds.), *The Papers of Robert H. Goddard*, three volumes, McGraw Hill Book Company, 1970. It includes the treatise "A

Method of Reaching Extreme Altitudes," originally published as Smithsonian Miscellaneous Collections, Vol. 71, No. 2, 1919.

1–5. Hermann Oberth, *Die Rakete zu den Planetenräumen* (By Rocket into Planetary Space), R. Oldenburg, Munich, 1923.

1–6. E. Sänger, *Raketenflugtechnik* (Rocket Flight Technology), R. Oldenburg, Munich, 1933.

1–7. W. von Braun and F. Ordway, *History of Rocketry and Space Travel*, 3rd ed., Thomas Y. Crowell, New York, 1974.

1–8. F. F. Webster, "Integral Rocket/Ramjet Propulsion—Flight Data Correlation and Analysis Technique," *Journal of Spacecraft*, Vol. 19, No. 4, July–August 1982.

1–9. R. W. Bussard and R. D. DeLauer, *Nuclear Rocket Propulsion*, McGraw-Hill Book Company, New York, 1958.

1–10. "Nuclear Thermal Rockets; Next Step in Space" (collection of three articles), *Aerospace America*, June 1989, pp. 16–29.

1–11. D. Buden, "Nuclear Rocket Safety," *Acta Astronautica*, Vol. 18, 30 Years of Progress in Space, 1988, pp. 217–224.

1–12. R. C. Finke (Ed.), *Electric Propulsion and its Application to Space Missions*, Vol. 79, Progress in Aeronautics and Astronautics, American Institute of Aeronautics and Astronautics, New York, 1981.

1–13. R. G. Jahn, *Physics of Electric Propulsion*, McGraw-Hill Book Company, New York, 1968, 339 pages.

1–14. T. Svitek et al., "Solar Sails as Orbit Transfer Vehicle—Solar Sail Concept Study—Phase II Report," *AIAA Paper 83-1347*, 1983.

1–15. V. P. Ageev et al., "Some Characteristics of the Laser Multi-pulse Explosive Type Jet Thruster," *Acta Astronautica*, Vol. 8, No. 5–6, 1981, pp. 625–641.

1–16. R. A. Hyde, "A Laser Fusion Rocket for Interplanetary Propulsion," *Preprint UCRL 88857*, Lawrence Livermore National Laboratory, Livermore, CA, September 1983

1–17. T. D. Thompson (Ed.), *TRW Space Log*, Vol. 32 to 34., TRW Space and Electronics Group, TRW, Inc., Redondo Beach, CA., 1996 and 1997–1998.

1–18. National Aeronautics and Space Administration, *National Space Transportation System Reference*, Vol. 1, *Systems and Facilities*, U.S. Government Printng Office, Washington, DC, June 1988.

CHAPTER 2

DEFINITIONS AND FUNDAMENTALS

Rocket propulsion is an exact but not a fundamental subject, and there are no basic scientific laws of nature peculiar to propulsion. The basic principles are essentially those of mechanics, thermodynamics, and chemistry.

Propulsion is achieved by applying a force to a vehicle, that is, accelerating the vehicle or, alternatively, maintaining a given velocity against a resisting force. This propulsive force is obtained by ejecting propellant at high velocity. This chapter deals with the definitions and the basic relations of this propulsive force, the exhaust velocity, and the efficiencies of creating and converting the energy and other basic parameters. The symbols used in the equations are defined at the end of the chapter. Wherever possible the American Standard letter symbols for rocket propulsion (as given in Ref. 2–1) are used.

2.1. DEFINITIONS

The *total impulse* I_t is the thrust force F (which can vary with time) integrated over the burning time t.

$$I_t = \int_0^t F \, dt \tag{2–1}$$

For constant thrust and negligible start and stop transients this reduces to

$$I_t = Ft \tag{2–2}$$

I_t is proportional to the total energy released by all the propellant in a propulsion system.

The *specific impulse* I_s is the total impulse per unit weight of propellant. It is an important figure of merit of the performance of a rocket propulsion system, similar in concept to the miles per gallon parameter used with automobiles. A higher number means better performance. Values of I_s are given in many chapters of this book and the concept of an optimum specific impulse for a particular mission is introduced later. If the total mass flow rate of propellant is $\dot{m}$ and the standard acceleration of gravity at sealevel g_0 is 9.8066 m/sec^2 or 32.174 ft/sec^2, then

$$I_s = \frac{\int_0^t F \, dt}{g_0 \int \dot{m} \, dt} \qquad (2\text{--}3)$$

This equation will give a time-averaged specific impulse value for any rocket propulsion system, particularly where the thrust varies with time. During transient conditions (during start or the thrust buildup period, the shutdown period, or during a change of flow or thrust levels) values of I_s can be obtained by integration or by determining average values for F and $\dot{m}$ for short time intervals. For constant thrust and propellant flow this equation can be simplified; below, m_p is the total effective propellant mass.

$$I_s = I_t/(m_p g_0) \qquad (2\text{--}4)$$

In Chapter 3 there is further discussion of the specific impulse. For constant propellant mass flow $\dot{m}$, constant thrust F, and negligibly short start or stop transients:

$$\begin{aligned} I_s &= F/(\dot{m} g_0) = F/\dot{w} \\ &\quad I_t/(m_p g_0) = I_t/w \end{aligned} \qquad (2\text{--}5)$$

The product $m_p g_0$ is the total effective propellant weight w and the weight flow rate is $\dot{w}$. The concept of weight relates to the gravitational attraction at or near sea level, but in space or outer satellite orbits, "weight" signifies the mass multiplied by an arbitrary constant, namely g_0. In the *Système International* (SI) or metric system of units I_s can be expressed simply in "seconds," because of the use of the constant g_0. In the USA today we still use the English Engineering (EE) system of units (foot, pound, second) in many of the chemical propulsion engineering, manufacturing, and test operations. In many past and current US publications, data and contracts, the specific impulse has units of thrust (lbf) divided by weight flow rate of propellants (lbf/sec), simplified as seconds. The numerical value of I_s is the same in the EE and the SI system of units. However, the units of I_s do not represent a measure of elapsed time, but a thrust force per unit "weight"-flow-rate. In this book the symbol I_s is used for the specific impulse, as listed in Ref. 2–1. For solid propellant systems the symbol I_{sp} is sometimes used, as listed in Ref. 2–2.

In a rocket nozzle the actual exhaust velocity is not uniform over the entire exit cross-section and does not represent the entire thrust magnitude. The velocity profile is difficult to measure accurately. For convenience a uniform axial velocity c is assumed which allows a one-dimensional description of the problem. This *effective exhaust velocity* c is the average equivalent velocity at which propellant is ejected from the vehicle. It is defined as

$$c = I_s g_0 = F/\dot{m} \qquad (2\text{–}6)$$

It is given either in meters per second or feet per second. Since c and I_s differ only by an arbitrary constant, either one can be used as a measure of rocket performance. In the Russian literature c is generally used.

In solid propellant rockets it is difficult to measure the propellant flow rate accurately. Therefore, the specific impulse is often calculated from total impulse and the propellant weight (using the difference between initial and final motor weights and Eq. 2–5). In turn the total impulse is obtained from the integral of the measured thrust with time, using Eq. 2–1. In liquid propellant units it is possible to measure thrust and instantaneous propellant flow rate and thus to use Eq. 2–3 for calculation of specific impulse. Eq. 2–4 allows another definition for specific impulse, namely, the amount of impulse imparted to a vehicle per unit sea-level weight of propellant expended.

The term *specific propellant consumption* refers to the reciprocal of the specific impulse and is not commonly used in rocket propulsion. It is used in automotive and duct propulsion systems. Typical values are listed in Table 1–2.

The mass ratio **MR** of a vehicle or a particular vehicle stage is defined to be the final mass m_f (after rocket operation has consumed all usable propellant) divided by m_0 (before rocket operation). The various terms are depicted in Fig. 4–1.

$$\mathbf{MR} = m_f/m_0 \qquad (2\text{–}7)$$

This applies to a single or a multi-stage vehicle; for the latter, the overall mass ratio is the product of the individual vehicle stage mass ratios. The final mass m_f is the mass of the vehicle after the rocket has ceased to operate when all the useful propellant mass m_p has been consumed and ejected. The final vehicle mass m_f includes all those components that are not useful propellant and may include guidance devices, navigation gear, payload (e.g., scientific instruments or a military warhead), flight control systems, communication devices, power supplies, tank structure, residual or unusable propellant, and all the propulsion hardware. In some vehicles it can also include wings, fins, a crew, life support systems, reentry shields, landing gears, etc. Typical values of **MR** can range from 60% for some tactical missiles to less than 10% for some unmanned launch vehicle stages. This mass ratio is an important parameter in analyzing flight performance, as explained in Chapter 4. When **MR** is applied to a single stage, then its upper stages become the "payload."

The *propellant mass fraction* ζ indicates the fraction of propellant mass m_p in an initial mass m_0. It can be applied to a vehicle, a stage of a vehicle or to a rocket propulsion system.

$$\zeta = m_p/m_0 \tag{2-8}$$

$$\zeta = (m_0 - m_f)/m_0 = m_p/(m_p + m_f) \tag{2-9}$$

$$m_0 = m_f + m_p \tag{2-10}$$

When applied to a rocket propulsion system, the mass ratio **MR** and propellant fraction ζ are different from those that apply to a vehicle as described above. Here the initial or loaded mass m_0 consists of the inert propulsion mass (the hardware necessary to burn and store the propellant) and the effective propellant mass. It would exclude masses of nonpropulsive components, such as payload or guidance devices. For example, in a liquid propellant rocket engine the final or inert propulsion mass m_f would include the propellant feed tanks, the pressurization system (with turbopump and/or gas pressure system), one or more thrust chambers, various piping, fittings and valves, an engine mount or engine structure, filters and some sensors. The *residual or unusable remaining propellant* is usually considered to be part of the final inert mass m_f, as it will be in this book. However, some rocket propulsion manufacturers and some literature assign residuals to be part of the propellant mass m_p. When applied to a rocket propulsion system, the value of the propellant mass fraction ζ indicates the quality of the design; a value of, say, 0.91 means that only 9% of the mass is inert rocket hardware and this small fraction contains, feeds, and burns a substantially larger mass of propellant. A high value of ζ is desirable.

The *impulse-to-weight* ratio of a complete propulsion system is defined as the total impulse I_t divided by the initial or propellant-loaded vehicle weight w_0. A high value indicates an efficient design. Under our assumptions of constant thrust and negligible start and stop transients, it can be expressed as

$$\frac{I_t}{w_0} = \frac{I_t}{(m_f + m_p)g_0} \tag{2-11}$$

$$= \frac{I_s}{m_f/m_p + 1} \tag{2-12}$$

The *thrust to weight* ratio F/w_0 expresses the acceleration (in multiples of the earth's surface acceleration of gravity) that the engine is capable of giving to its own loaded propulsion system mass. For constant thrust the maximum value of the thrust to weight ratio, or maximum acceleration, occurs just before termination or burnout because the vehicle mass has been diminished by the mass of useful propellant. Values of F/w are given in Table 2–1. The *thrust to weight ratio* is useful to compare different types of rocket systems.

Example 2–1. A rocket projectile has the following characteristics:

Initial mass	200 kg
Mass after rocket operation	130 kg
Payload, nonpropulsive structure, etc.	110 kg
Rocket operating duration	3.0 sec
Average specific impulse of propellant	240 sec

Determine the vehicle's mass ratio, propellant mass fraction, propellant flow rate, thrust, thrust-to-weight ratio, acceleration of vehicle, effective exhaust velocity, total impulse, and the impulse-to-weight ratio.

SOLUTION. Mass ratio of vehicle (Eq. 2–8) $\mathbf{MR} = m_f/m_0 = 130/200 = 0.65$; mass ratio of rocket system $\mathbf{MR} = m_f/m_0 = (130 - 110)/(200 - 110) = 0.222$. Note that the empty and initial masses of the propulsion system are 20 and 90 kg, respectively.
   The propellant mass fraction (Eq. 2–9) is

$$\zeta = (m_0 - m_f)/m_0 = (90 - 20)/90 = 0.778$$

The propellant mass is $200 - 130 = 70$ kg. The propellant mass flow rate is $\dot{m} = 70/3 = 23.3$ kg/sec,
   The thrust (Eq. 2–5) is

$$F = I_s \dot{w} = 240 \times 23.3 \times 9.81 = 54{,}857 \text{ N}$$

The thrust-to-weight ratio of the vehicle is

$$\text{initial value } F/w_0 = 54{,}857/(200 \times 9.81) = 28$$
$$\text{final value } 54{,}857/(130 \times 9.81) = 43$$

The maximum acceleration of the vehicle is $43 \times 9.81 = 421$ m/sec^2. The effective exhaust velocity (Eq. 2–6) is

$$c = I_s g_0 = 240 \times 9.81 = 2354 \text{ m/sec}$$

The total impulse (Eqs. 2–2 and 2–5) is

$$I_t = I_s w = 240 \times 70 \times 9.81 = 164{,}808 \text{ N-sec}$$

This result can also be obtained by multiplying the thrust by the duration. The impulse-to-weight ratio of the propulsion system (Eq. 2–11) is

$$I_t/w_0 = 164{,}808/[(200 - 110)9.81] = 187$$

2.2. THRUST

The thrust is the force produced by a rocket propulsion system acting upon a vehicle. In a simplified way, it is the reaction experienced by its structure due to the ejection of matter at high velocity. It represents the same phenomenon that pushes a garden hose backwards or makes a gun recoil. In the latter case, the forward momentum of the bullet and the powder charge is equal to the recoil or rearward momentum of the gun barrel. Momentum is a vector quantity and is defined as the product of mass times velocity. All ship propellers and oars generate their forward push at the expense of the momentum of the water or air masses, which are accelerated towards the rear. Rocket propulsion differs from these devices primarily in the relative magnitude of the accelerated masses and velocities. In rocket propulsion relatively small masses are involved which are *carried within* the vehicle and ejected at high velocities.

The thrust, due to a change in momentum, is given below. A derivation can be found in earlier editions of this book. The thrust and the mass flow are constant and the gas exit velocity is uniform and axial.

$$F = \frac{dm}{dt} v_2 = \dot{m}v_2 = \frac{\dot{w}}{g_0} v_2 \qquad (2\text{–}13)$$

This force represents the total propulsion force when the nozzle exit pressure equals the ambient pressure.

The pressure of the surrounding fluid (i..e, the local atmosphere) gives rise to the second contribution that influences the thrust. Figure 2–1 shows schematically the external pressure acting uniformly on the outer surface of a rocket chamber and the gas pressures on the inside of a typical thermal rocket engine. The size of the arrows indicates the relative magnitude of the pressure forces. The axial thrust can be determined by integrating all the pressures acting on areas that can be projected on a plane normal to the nozzle axis. The forces acting radially outward are appreciable, but do not contribute to the axial thrust because a rocket is typically an axially symmetric chamber. The conditions prior to entering the nozzle are essentially stagnation conditions.

Because of a fixed nozzle geometry and changes in ambient pressure due to variations in altitude, there can be an imbalance of the external environment or atmospheric pressure p_3 and the local pressure p_2 of the hot gas jet at the exit plane of the nozzle. Thus, for a steadily operating rocket propulsion system moving through a homogeneous atmosphere, the total thrust is equal to

$$F = \dot{m}v_2 + (p_2 - p_3)A_2 \qquad (2\text{–}14)$$

The first term is the *momentum thrust* represented by the product of the propellant mass flow rate and its exhaust velocity relative to the vehicle. The second term represents the *pressure thrust* consisting of the product of the cross-sectional area at the nozzle exit A_2 (where the exhaust jet leaves the

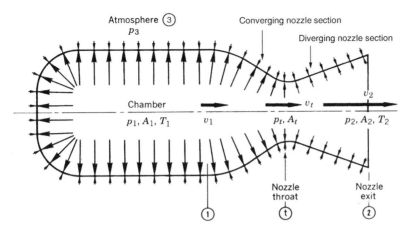

FIGURE 2–1. Pressure balance on chamber and nozzle interior walls is not uniform. The internal gas pressure (indicated by length of arrows) is highest in the chamber (p_1) and decreases steadily in the nozzle until it reaches the nozzle exit pressure p_2. The external or atmospheric pressure p_3 is uniform. At the throat the pressure is p_t. The four subscripts (shown inside circles) refer to the quantities A, v, T, and p at specific locations.

vehicle) and the difference between the exhaust gas pressure at the exit and the ambient fluid pressure. If the exhaust pressure is less than the surrounding fluid pressure, the pressure thrust is negative. Because this condition gives a low thrust and is undesirable, the rocket nozzle is usually so designed that the exhaust pressure is equal or slightly higher than the ambient fluid pressure.

When the ambient atmosphere pressure is equal to the exhaust pressure, the pressure term is zero and the thrust is the same as in Eq. 2–13. In the vacuum of space $p_3 = 0$ and the thrust becomes

$$F = \dot{m}v_2 + p_2 A_2 \qquad (2\text{–}15)$$

The pressure condition in which the exhaust pressure is exactly matched to the surrounding fluid pressure ($p_2 = p_3$) is referred to as the rocket nozzle with *optimum expansion ratio*. This is further elaborated upon in Chapter 3.

Equation 2–14 shows that the thrust of a rocket unit is independent of the flight velocity. Because changes in ambient pressure affect the pressure thrust, there is a variation of the rocket thrust with altitude. Because atmospheric pressure decreases with increasing altitude, the thrust and the specific impulse will increase as the vehicle is propelled to higher altitudes. This change in pressure thrust due to altitude changes can amount to between 10 and 30% of the overall thrust, as is shown for a typical rocket engine in Fig. 2–2. Table 8–1 shows the sea level and high altitude thrust for several rocket engines. Appendix 2 gives the properties of the Standard Atmosphere (ambient pressure).

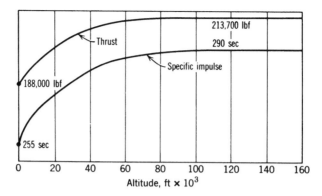

FIGURE 2–2. Altitude performance of RS 27 liquid propellant rocket engine used in early versions of the Delta launch vehicle.

2.3. EXHAUST VELOCITY

The *effective exhaust velocity* as defined by Eq. 2–6 applies to all rockets that thermodynamically expand hot gas in a nozzle and, indeed, to all mass expulsion systems. From Eq. 2–14 and for constant propellant mass flow this can be modified to

$$c = v_2 + (p_2 - p_3)A_2/\dot{m} \tag{2–16}$$

Equation 2–6 shows that c can be determined from thrust and propellant flow measurements. When $p_2 = p_3$, the effective exhaust velocity c is equal to the average actual exhaust velocity of the propellant gases v_2. When $p_2 \neq p_3$ then $c \neq v_2$. The second term of the right-hand side of Eq. 2–16 is usually small in relation to v_2; thus the effective exhaust velocity is usually close in value to the actual exhaust velocity. When $c = v_2$ the thrust (from Eq. 2–14) can be rewritten as

$$F = (\dot{w}/g_0)v_2 = \dot{m}c \tag{2–17}$$

The *characteristic velocity* has been used frequently in the rocket propulsion literature. Its symbol c^*, pronounced "cee-star," is defined as

$$c^* = p_1 A_t/\dot{m} \tag{2–18}$$

The characteristic velocity c^* is used in comparing the relative performance of different chemical rocket propulsion system designs and propellants; it is easily determined from measured data of $\dot{m}$, p_1, and A_t. It relates to the efficiency of the combustion and is essentially independent of nozzle characteristics.

However, the specific impulse I_s and the effective exhaust velocity c are functions of the nozzle geometry, such as the nozzle area ratio A_2/A_t, as shown in Chapter 3. Some values of I_s and c^* are given in Tables 5–4 and 5–5.

Example 2–2. The following measurements were made in a sea level test of a solid propellant rocket motor:

Burn duration	40 sec
Initial mass before test	1210 kg
Mass of rocket motor after test	215 kg
Average thrust	62,250 N
Chamber pressure	7.00 MPa
Nozzle exit pressure	0.070 MPa
Nozzle throat diameter	0.0855 m
Nozzle exit diameter	0.2703 m

Determine $\dot{m}$, v_2, c^*, c, and I_s at sea level, and c and I_s at 1000 and 25,000 m altitude. Assume an invariant thrust and mass flow rate and negligible short start and stop transients.

SOLUTION. The mass flow rate $\dot{m}$ is determined from the total propellant used (initial motor mass − final motor mass) and the burn time.

$$\dot{m} = (1210 - 215)/40 = 24.9 \text{ kg/sec}$$

The nozzle areas at the throat and exit are

$$A_t = \pi D^2/4 = \pi \times 0.0855^2/4 = 0.00574 \text{ m}^2$$
$$A_2 = \pi D^2/4 = \pi \times 0.2703^2/4 = 0.0574 \text{ m}^2$$

Equation 2–14 is to be solved for v_2, the actual average exhaust velocity.

$$v_2 = F/\dot{m} - (p_2 - p_3)A_2/\dot{m}$$
$$= 62,250/24.9 - (0.070 - 0.1013)10^6 \times 0.0574/24.9$$
$$= 2572 \text{ m/sec}$$

The characteristic velocity and effective exhaust velocity are found from Eqs. 2–6 and 2–18 for sea level conditions.

$$c^* = p_1 A_t/\dot{m} = 7.00 \times 10^6 \times 0.00574/24.9 = 1613 \text{ m/sec}$$
$$I_s = F/\dot{m}g_0 = 62,250/(24.9 \times 9.81) = 255 \text{ sec}$$
$$c = I_s g_0 = 255 \times 9.81 = 2500 \text{ m/sec}$$

For altitudes of 1000 and 25,000 m the ambient pressure (see Appendix 2) is 0.0898 and 0.00255 MPa. From Eq. 2–16 the altitude values of c can be obtained.

$$c = v_2 + (p_2 - p_3)A_2/\dot{m}$$

At 1000 m altitude,

$$c = 2572 + (0.070 - 0.0898) \times 10^6 \times 0.0574/24.9 = 2527 \text{ m/ sec}$$
$$I_s = 2527/9.81 = 258 \text{ sec}$$

At 25,000 m altitude,

$$c = 2572 + (0.070 - 0.00255) \times 10^6 \times 0.0574/24.9 = 2727 \text{ m/ sec}$$
$$I_s = 2727/9.80 = 278 \text{ sec}$$

2.4. ENERGY AND EFFICIENCIES

Although efficiencies are not commonly used directly in designing rocket units, they permit an understanding of the energy balance of a rocket system. Their definitions are arbitrary, depending on the losses considered, and any consistent set of efficiencies, such as the one presented in this section, is satisfactory in evaluating energy losses. As stated previously, two types of energy conversion processes occur in any propulsion system, namely, the generation of energy, which is really the conversion of stored energy into available energy and, subsequently, the conversion to the form in which a reaction thrust can be obtained. The kinetic energy of ejected matter is the form of energy useful for propulsion. The *power of the jet* P_{jet} is the time rate of expenditure of this energy, and for a constant gas ejection velocity v this is a function of I_s and F

$$P_{jet} = \tfrac{1}{2}\dot{m}v^2 = \tfrac{1}{2}\dot{w}g_0 I_s^2 = \tfrac{1}{2}Fg_0 I_s = \tfrac{1}{2}Fv_2 \tag{2–19}$$

The term *specific power* is sometimes used as a measure of the utilization of the mass of the propulsion system including its power source; it is the jet power divided by the loaded propulsion system mass, P_{jet}/m_0. For electrical propulsion systems which carry a heavy, relatively inefficient energy source, the specific power can be much lower than that of chemical rockets. The energy input from the energy source to the rocket propulsion system has different forms in different rocket types. For chemical rockets the energy is created by combustion. The maximum energy available per unit mass of chemical propellants is the heat of the combustion reaction Q_R; the *power input to a chemical engine* is

$$P_{chem} = \dot{m}Q_R J \tag{2–20}$$

where J is a conversion constant which depends on the units used. A large portion of the energy of the exhaust gases is unavailable for conversion into kinetic energy and leaves the nozzle as residual enthalpy. This is analogous to the energy lost in the high-temperature exhaust gases of internal combustion engines.

The *combustion efficiency* for chemical rockets is the ratio of the actual and the ideal heat of reaction per unit of propellant and is a measure of the source efficiency for creating energy. Its value is high (approximately 94 to 99%), and it is defined in Chapter 5. When the power input P_{chem} is multiplied by the combustion efficiency, it becomes the power available to the propulsive device, where it is converted into the kinetic power of the exhaust jet. In electric propulsion the analogous efficiency is the power conversion efficiency. For solar cells it has a low value; it is the efficiency for converting solar radiation energy into electric power (10 to 20%).

The *power transmitted to the vehicle* at any one time is defined in terms of the thrust of the propulsion system F and the vehicle velocity u:

$$P_{vehicle} = Fu \qquad (2\text{--}21)$$

The *internal efficiency* of a rocket propulsion system is an indication of the effectiveness of converting the system's energy input to the propulsion device into the kinetic energy of the ejected matter; for example, for a chemical unit it is the ratio of the kinetic power of the ejected gases expressed by Eq. 2–19 divided by the power input of the chemical reaction as given in Eq. 2–20. Internal efficiencies are used in Example 2–3. The energy balance diagram for a chemical rocket (Fig. 2–3) shows typical losses. The internal efficiency can be expressed as

$$\eta_{int} = \frac{\text{kinetic power in jet}}{\text{available chemical power}} = \frac{\frac{1}{2}\dot{m}v^2}{\eta_{comb}P_{chem}} \qquad (2\text{--}22)$$

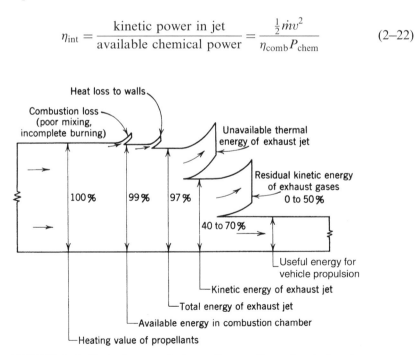

FIGURE 2–3. Typical energy balance diagram for a chemical rocket.

Typical values of η_{int} are listed later in Example 2–3.

The *propulsive efficiency* (Fig. 2–4) determines how much of the kinetic energy of the exhaust jet is useful for propelling a vehicle. It is also used often with duct jet engines and is defined as

$$\eta_P = \frac{\text{vehicle power}}{\text{vehicle power} + \text{residual kinetic jet power}}$$

(2–23)

$$= \frac{Fu}{Fu + \frac{1}{2}(\dot{w}/g_0)(c - u)^2} = \frac{2u/c}{1 + (u/c)^2}$$

where F is the thrust, u the absolute vehicle velocity, c the effective rocket exhaust velocity with respect to the vehicle, $\dot{w}$ the propellant weight flow rate, and η_p the propulsive efficiency. The propulsive efficiency is a maximum when the forward vehicle velocity is exactly equal to the exhaust velocity. Then the residual kinetic energy and the absolute velocity of the jet are zero and the exhaust gases stand still in space.

While it is desirable to use energy economically and thus have high efficiencies, there is also the problem of minimizing the expenditure of ejected mass, which in many cases is more important than minimizing the energy. In nuclear reactor energy and some solar energy sources, for example, there is an almost unlimited amount of heat energy available; yet the vehicle can only carry a limited amount of working fluid. Economy of mass expenditures of working fluid can be obtained if the exhaust velocity is high. Because the specific impulse is proportional to the exhaust velocity, it is a measure of this propellant mass economy.

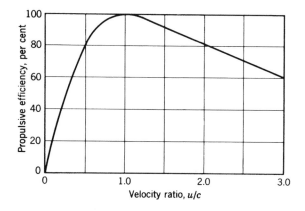

FIGURE 2–4. Propulsive efficiency at varying velocities.

2.5. TYPICAL PERFORMANCE VALUES

Typical values of representative performance parameters for different types of rocket propulsion are given in Table 2–1 and in Fig. 2–5.

Chemical rockets have relatively low values of specific impulse, relatively light machinery (i.e., low engine weight), a very high thrust capability, and therefore high acceleration and high specific power. At the other extreme, the ion propulsion devices have a very high specific impulse, but they must carry a heavy electrical power source with them to deliver the power necessary for high ejection velocities. The very low acceleration potential for the electrical propulsion units and those using solar radiation energy usually requires a long period for accelerating and thus these systems are best used for missions where the flight time is long. The low thrust values of electrical systems imply that they are not useful in fields of strong gravitational gradients (for takeoff or landing) but are best used in a true space flight mission.

The chemical systems (solid and liquid propellant rockets) are fully developed and widely used for many different vehicle applications. They are described in Chapters 5 to 15. Electrical propulsion has been in operation in many space flight applications (see Chapter 19). Some of the other types are still in their exploratory or development phase, but may become useful.

Example 2–3. As a comparison of different propulsion systems, compute the energy input and the propellant flow required for 100 N thrust with several types of propulsion systems.

SOLUTION. From Equations 2–13 and 2–19,

$$\dot{m} = F/(I_s g_0)$$
$$\text{power input} = P_{jet}/\eta_{int} = \tfrac{1}{2}\dot{m}v_2^2/\eta_{int}$$

From Table 2–1 typical values of I_s and from experience typical internal efficiencies were selected. Depending on the propellant and the design, these values may vary somewhat. The equations above were solved for $\dot{m}$ and the power input as indicated in the table below.

Engine Type	η_{int}	I_s	v_2 (m/sec)	$\dot{m}$ (kg/sec)	Power Input (kW)
Chemical rocket	0.50	300	2940	0.0340	294
Nuclear fission	0.50	800	7840	0.0128	787
Arc—electrothermal	0.50	600	5880	0.0170	588
Ion electrostatic	0.90	2000	19,600	0.0051	1959

More than half a megawatt of power is needed for the last three propulsion systems, but the propellant flows are small. The data for the last two types are illustrative, but hypothetical. To date the largest experimental units have been about 120 kW for arcjets and perhaps 10 kW with ion propulsion. Although thruster designs for megawatt-level units are feasible, it is unlikely that the needed flight-qualified electrical power generator would be available in the next decade.

TABLE 2–1. Ranges of Typical Performance Parameters for Various Rocket Propulsion Systems

Engine Type	Specific Impulse[a] (sec)	Maximum Temperature (°C)	Thrust-to-Weight Ratio[b]	Propulsion Duration	Specific Power[c] (kW/kg)	Typical Working Fluid	Status of Technology
Chemical—solid or liquid bipropellant	200–410	2500–4100	10^{-2}–100	Seconds to a few minutes	10^{-1}–10^3	Liquid or solid propellants	Flight proven
Liquid monopropellant	180–223	600–800	10^{-1}–10^{-2}	Seconds to minutes	0.02–200	N_2H_4	Flight proven
Nuclear fission	500–860	2700	10^{-2}–30	Seconds to minutes	10^{-1}–10^3	H_2	Development was stopped
Resistojet	150–300	2900	10^{-2}–10^{-4}	Days	10^{-3}–10^{-1}	H_2, N_2H_4	Flight proven
Arc heating—electrothermal	280–1200	20,000	10^{-4}–10^{-2}	Days	10^{-3}–1	N_2H_4,H_2,NH_3	Flight proven
Electromagnetic including Pulsed Plasma (PP)	700–2500	—	10^{-6}–10^{-4}	Weeks	10^{-3}–1	H_2 Solid for PP	Flight proven
Hall effect	1000–1700	—	10^{-4}	Weeks	10^{-1}–5×10^{-1}	Xe	Flight proven
Ion—electrostatic	1200–5000	—	10^{-6}–10^{-4}	Months	10^{-3}–1	Xe	Several have flown
Solar heating	400–700	1300	10^{-3}–10^{-2}	Days	10^{-2}–1	H_2	In development

[a] At p_1 = 1000 psia and optimum gas expansion at sea level ($Ip_2 = p_3 = 14.7$ psia).
[b] Ratio of thrust force to full propulsion system sea level weight (with propellants, but without payload).
[c] Kinetic power per unit exhaust mass flow.

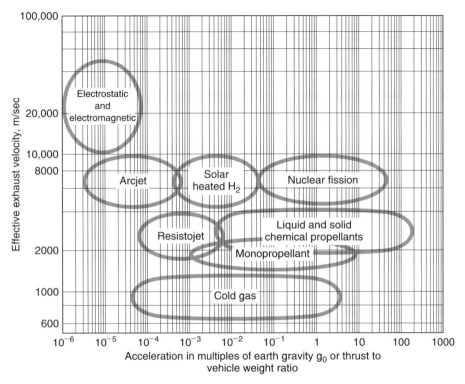

FIGURE 2–5. Exhaust velocities as a function of typical vehicle accelerations. Regions indicate approximate performance values for different types of propulsion systems. The mass of the vehicle includes the propulsion system, but the payload is assumed to be zero.

PROBLEMS

When solving problems, three appendixes (see end of book) may be helpful:

Appendix 1. Conversion Factors and Constants
Appendix 2. Properties of the Earth's Standard Atmosphere
Appendix 3. Summary of Key Equations

1. Prove that the value of the reaction thrust F equals twice the total dynamic pressure across the area A for an incompressible fluid as shown below.

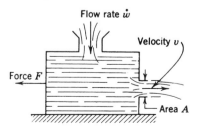

2. The following data are given for a certain rocket unit: thrust, 8896 N; propellant consumption, 3.867 kg/sec; velocity of vehicle, 400 m/sec; energy content of propellant, 6.911 MJ/kg. Assume 100% combustion efficiency.

   Determine (a) the effective velocity; (b) the kinetic jet energy rate per unit flow of propellant; (c) the internal efficiency; (d) the propulsive efficiency; (e) the overall efficiency; (f) the specific impulse; (g) the specific propellant consumption.

   Answers: (a) 2300 m/sec; (b) 2.645 MJ-sec/kg; (c) 38.3%; (d) 33.7%; (e) 13.3%; (f) 234.7 sec; (g) 0.00426 sec^{-1}.

3. A certain rocket has an effective exhaust velocity of 7000 ft/sec; it consumes 280 lbm/sec of propellant mass, each of which liberates 2400 Btu/lbm. The unit operates for 65 sec. Construct a set of curves plotting the propulsive, internal, and overall efficiencies versus the velocity ratio u/c $(0 < u/c < 1.0)$. The rated flight velocity equals 5000 ft/sec. Calculate (a) the specific impulse; (b) the total impulse; (c) the mass of propellants required; (d) the volume that the propellants occupy if their average specific gravity is 0.925.

   Answers: (a) 217.5 sec; (b) 3,960,000 lbf-sec; (c) 18,200 lbm; (d) 315 ft^3.

4. For the rocket in Problem 2, calculate the specific power, assuming a propulsion system dry mass of 80 kg and a duration of 3 min.

5. For the values given in Table 2–1 for the various propulsion systems, calculate the total impulse for a fixed propellant mass of 2000 kg.

6. A jet of fluid hits a stationary flat plate in the manner shown below.

   (a) If there is 50 kg of fluid flowing per minute at an abolute velocity of 200 m/sec, what will be the force on the plate?
   Answer: 167 N.

   (b) What will this force be when the plate moves in the direction of flow at $u = 50$ km/h?
   Answer: 144 N.

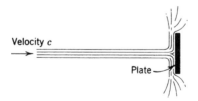

7. Plot the variation of the thrust and specific impulse against altitude, using the atmospheric pressure information given in Appendix 2, and the data for the Minuteman first-stage rocket thrust chamber in Table 11–3. Assume that $p_2 = 8.66$ psia.

8. Derive an equation relating the mass ratio **MR** and the propellant mass fraction.
   Answer: $\zeta = 1 - $ **MR**.

SYMBOLS (English engineering units are given in parentheses)

A	area, m^2 (ft^2)
A_t	nozzle throat area, m^2 (ft^2)
A_2	exist area of nozzle, m^2 (ft^2)
c	effective velocity, m/sec (ft/sec)
c^*	characteristic velocity, m/sec (ft/sec)
E	energy, J (ft-lbf)
F	thrust force, N (lbf)
g_0	standard sea level acceleration of gravity, 9.80665 m/sec^2 (32.174 ft/sec^2)
I_s	specific impulse, sec
I_t	impulse or total impulse, N-sec (lbf-sec)
J	conversion factor or mechanical equivalent of heat, 4.184 J/cal or 1055 J/Btu or 778 ft-lbf/Btu.
m	mass, kg (slugs) (1 slug = mass of 32.174 lb of weight at sea level)
$\dot{m}$	mass flow rate, kg/sec (lbm/sec)
m_f	final mass (after rocket propellant is ejected), kg (lbm or slugs)
m_p	propellant mass, kg (lbm or slugs)
m_0	initial mass (before rocket propellant is ejected), kg (lbm or slugs)
MR	mass ratio (m_f/m_0)
p	pressure, pascal [Pa] or N/m^2 (lbf/ft^2)
p_3	ambient or atmospheric pressure, Pa (lbf/ft^2)
p_2	rocket gas pressure at nozzle exit, Pa (lbf/ft^2)
p_1	chamber pressure, Pa (lbf/ft^2)
P	power, J/sec (ft-lbf/sec)
P_s	specific power, J/sec-kg (ft-lbf/sec-lbm)
Q_R	heat of reaction per unit propellant, J/kg (Btu/lbm)
t	time, sec
u	vehicle velocity, m/sec (ft/sec)
v_2	gas velocity leaving the rocket, m/sec (ft/sec)
w	weight, N or $kg\text{-}m/sec^2$ (lbf)
$\dot{w}$	weight flow rate, N/sec (lbf/sec)
w_0	initial weight, N or $kg\text{-}m/sec^2$ (lbf)

Greek Letters

ζ	propellant mass fraction
η	efficiency
η_{comb}	combustion efficiency
η_{int}	internal efficiency
η_p	propulsive efficiency

REFERENCES

2–1. "American National Standard Letter Symbols for Rocket Propulsion," *ASME Publication Y 10.14*, 1959.

2–2. "Solid Propulsion Nomenclature Guide," *CPIA Publication 80*, Chemical Propulsion Information Agency, Johns Hopkins University, Laurel, MD., May 1965, 18 pages.

CHAPTER 3

NOZZLE THEORY AND THERMODYNAMIC RELATIONS

Thermodynamic relations of the processes inside a rocket nozzle and chamber furnish the mathematical tools needed to calculate the performance and determine several of the key design parameters of rocket propulsion systems. They are useful as a means of evaluating and comparing the performance of various rocket systems; they permit the prediction of the operating performance of any rocket unit that uses the thermodynamic expansion of a gas, and the determination of several necessary design parameters, such as nozzle size and generic shape, for any given performance requirement. This theory applies to chemical rocket propulsion systems (both liquid and solid propellant types), nuclear rockets, solar heated and resistance or arc heated electrical rocket systems, and to any propulsion system that uses the expansion of a gas as the propulsive mechanism for ejecting matter at high velocity.

These thermodynamic relations, which are fundamental and important in analysis and design of rocket units, are introduced and explained in this chapter. The utilization of these equations should give the reader a basic understanding of the thermodynamic processes involved in rocket gas behavior and expansion. A knowledge of elementary thermodynamics and fluid mechanics on the part of the reader is assumed (see Refs. 1–1, 3–1, 3–2, and 3–3). This chapter also addresses different nozzle configurations, non-optimum performance, energy losses, nozzle alignment, variable thrust and four different ways for establishing nozzle performance parameters.

3.1. IDEAL ROCKET

The concept of ideal rocket propulsion systems is useful because the relevant basic thermodynamic principles can be expressed as simple mathematical relationships, which are given in subsequent sections of this chapter. These equations theoretically describe a quasi-one-dimensional nozzle flow, which corresponds to an idealization and simplification of the full two- or three-dimensional equations and the real aerothermochemical behavior. However, with the assumptions and simplifications stated below, they are very adequate for obtaining useful solutions to many rocket propulsion systems. For chemical rocket propulsion the measured actual performance is usually between 1 and 6% below the calculated ideal value. In designing new rockets, it has become accepted practice to use ideal rocket parameters which can then be modified by appropriate corrections, such as those discussed in Section 5 of this chapter. An *ideal rocket* unit is one for which the following assumptions are valid:

1. The working substance (or chemical reaction products) is *homogeneous*.
2. All the species of the working fluid are *gaseous*. Any condensed phases (liquid or solid) add a negligible amount to the total mass.
3. The working substance obeys the *perfect gas law*.
4. There is no *heat transfer* across the rocket walls; therefore, the flow is adiabatic.
5. There is no appreciable *friction* and all *boundary layer* effects are neglected.
6. There are no *shock waves* or *discontinuities* in the nozzle flow.
7. The *propellant flow* is *steady* and *constant*. The expansion of the working fluid is uniform and steady, without vibration. Transient effects (i.e., start up and shut down) are of very short duration and may be neglected.
8. All exhaust gases leaving the rocket have an *axially directed velocity*.
9. The gas velocity, pressure, temperature, and density are all uniform across any section normal to the nozzle axis.
10. *Chemical equilibrium* is established within the rocket chamber and the gas composition does not change in the nozzle (frozen flow).
11. Stored propellants are at room temperature. Cryogenic propellants are at their boiling points.

These assumptions permit the derivation of a simple, quasi-one-dimensional theory as developed in subsequent sections. Later in this book we present more sophisticated theories or introduce correction factors for several of the items on the list, and they allow a more accurate determination of the simplified analysis. The next paragraph explains why these assumptions cause only small errors.

For a liquid propellant rocket the idealized theory postulates an injection system in which the fuel and oxidizer are mixed perfectly so that a homogeneous working substance results. A good rocket injector can approach this condition closely. For a solid propellant rocket unit, the propellant must essentially be homogeneous and uniform and the burning rate must be steady. For nuclear, solar-heated or arc-heated rockets, it is assumed that the hot gases are uniform in temperature at any cross-section and steady in flow. Because chamber temperatures are typically high (2500 to 3600 K for common propellants), all gases are well above their respective saturation conditions and actually follow the perfect gas law very closely. Postulates 4, 5, and 6 above allow the use of the *isentropic expansion* relations in the rocket nozzle, thereby describing the maximum conversion of heat to kinetic energy of the jet. This also implies that the nozzle flow is thermodynamically reversible. Wall friction losses are difficult to determine accurately but they are usually small in nozzles. Except for very small chambers, the energy lost as heat to the walls of the rocket is usually less than 1% (occasionally up to 2%) of the total energy and can therefore be neglected. Short-term fluctuations of the steady propellant flow rate and pressure are usually less than 5% of the rated value, their effect on rocket performance is small and can be neglected. In well-designed supersonic nozzles, the conversion of thermal energy into directed kinetic energy of the exhaust gases proceeds smoothly and without normal shocks or discontinuities; thus the flow expansion losses are generally small.

Some companies and some authors do not include all or the same eleven items listed above in their definition of an ideal rocket. For example, instead of assumption 8 (all nozzle exit velocity is axially directed), some use a conical exit nozzle with a 15° half-angle as their base configuration in their ideal nozzle; this discounts the divergence losses, which are described later in this chapter.

3.2. SUMMARY OF THERMODYNAMIC RELATIONS

In this section we review briefly some of the basic relationships needed for the development of the nozzle flow equations. Rigorous derivations and discussions of these relations can be found in many thermodynamics or fluid dynamics texts, such as Refs. 3–1 and 3–2.

The principle of *conservation of energy* can be readily applied to the adiabatic, no shaft-work process inside the nozzle. Furthermore, without shocks or friction, the flow entropy change is zero. The concept of *enthalpy* is useful in flow systems; the enthalpy comprises the *internal thermal energy* plus the *flow work* (or work performed by the gas at a velocity v in crossing a boundary). For ideal gases the enthalpy can conveniently be expressed as the product of the specific heat c_p times the absolute temperature T (the specific heat at constant pressure is formally defined as the partial derivative of the enthalpy with respect to temperature at constant pressure). Under the above assumptions, the total or stagnation enthalpy per unit mass h_0 is constant, i.e.,

$$h_0 = h + v^2/2J = \text{constant} \tag{3-1}$$

In the above, J is the mechanical equivalent of heat which is inserted only when thermal units (i.e., the Btu and calorie) are mixed with mechanical units (i.e., the ft-lbf and the joule). In SI units (kg, m, sec) the value of J is one. In the English Engineering system of units another constant (see Appendix 1) has to be provided to account for the mass units (i.e., the lbm). The conservation of energy for isentropic flow between any two sections x and y shows that the decrease in enthalpy or thermal content of the flow appears as an increase of kinetic energy since and any changes in potential energy may be neglected.

$$h_x - h_y = \frac{1}{2}(v_y^2 - v_x^2)/J = c_p(T_x - T_y) \tag{3-2}$$

The principle of *conservatism of mass* in a steady flow with a single inlet and single outlet is expressed by equating the mass flow rate $\dot{m}$ at any section x to that at any other section y; this is known in mathematical form as the continuity equation. Written in terms of the cross-sectional area A, the velocity v, and the specific volume V,

$$\dot{m}_x = \dot{m}_y \equiv \dot{m} = Av/V \tag{3-3}$$

The *perfect gas law* is written as

$$p_x V_x = RT_x \tag{3-4}$$

where the gas constant R is found from the universal gas constant R' divided by the molecular mass $\mathfrak{M}$ of the flowing gas mixture. The molecular volume at standard conditions becomes 22.41 m³/kg-mol or ft³/lb-mol and it relates to a value of $R' = 8314.3$ J/kg-mole-K or 1544 ft-lbf/lb-mole-R. One often finds Eq. 3-3 written in terms of density ρ which is the reciprocal of the specific volume V. The specific heat at constant pressure c_p, the specific heat at constant volume c_v, and their ratio k are constant for perfect gases over a wide range of temperatures and are related.

$$k = c_p/c_v \tag{3-5a}$$
$$c_p - c_v = R/J \tag{3-5b}$$
$$c_p = kR/(k-1)J \tag{3-6}$$

For an *isentropic flow process* the following relations hold between any points x and y:

$$T_x/T_y = (p_x/p_y)^{(k-1)/k} = (V_y/V_x)^{k-1} \tag{3-7}$$

During an isentropic nozzle expansion the pressure drops substantially, the absolute temperature drops somewhat less, and the specific volume increases. When a flow is stopped isentropically the prevailing conditions are known as *stagnation conditions* and are designated by the subscript "0". Sometimes the word "total" is used instead of stagnation. As can be seen from Eq. 3–1 the stagnation enthalpy consists of the sum of the static or local enthalpy and the fluid kinetic energy. The stagnation temperature T_0 is found from the energy equation as

$$T_0 = T + v^2/(2c_p J) \tag{3–8}$$

where T is the absolute fluid static temperature. In adiabatic flows, the stagnation temperature remains constant. The relationship of the stagnation pressure to the local pressure in the flow can be found from the previous two equations:

$$p_0/p = [1 + v^2/(2c_p J T)]^{k/(k-1)} = (V/V_0)^k \tag{3–9}$$

When the local velocity comes close to zero, the local temperature and pressure will approach the stagnation pressure and stagnation temperature. In a combustion chamber, where the gas velocity is small, the local combustion pressure is essentially equal to the stagnation pressure. The *velocity of sound a* or the acoustic velocity in ideal gases is independent of pressure. It is defined as

$$a = \sqrt{kRT} \tag{3–10}$$

In the English Engineering (EE) system the value of R has to be corrected and the constant g_0 is added. Equation 3–10 becomes $\sqrt{g_0 kRT}$. This correction factor must be applied wherever R is used in EE units. The *Mach number M* is a dimensionless flow parameter and is used to define the ratio of the flow velocity v to the local acoustic velocity a.

$$M = v/a = v/\sqrt{kRT} \tag{3–11}$$

A Mach number less than one corresponds to subsonic flow and greater than one to supersonic flow. When the Mach number is equal to one then the flow is moving at precisely the velocity of sound. It is shown later that at the throat of all supersonic nozzles the Mach number must be equal to one. The relation between stagnation temperature and Mach number can now be written from Eqs. 3–2, 3–7, and 3–10 as

$$T_0 = T\left[1 + \tfrac{1}{2}(k-1)M^2\right] \tag{3–12}$$

or

$$M = \sqrt{\frac{2}{k-1}\left(\frac{T_0}{T}-1\right)}$$

T_0 and p_0 designate the stagnation values of the temperature and pressure. Unlike the temperature, the stagnation pressure during an adiabatic nozzle expansion remains constant only for isentropic flows. It can be computed from

$$p_0 = p\left[1 + \tfrac{1}{2}(k-1)M^2\right]^{k/(k-1)} \tag{3-13}$$

The *area ratio* for a nozzle with isentropic flow can be expressed in terms of Mach numbers for any points x and y within the nozzle. This relationship, along with those for the ratios T/T_0 and p/p_0, is plotted in Fig. 3-1 for $A_x = A_t$ and $M_x = 1.0$. Otherwise,

$$\frac{A_y}{A_x} = \frac{M_x}{M_y}\sqrt{\left\{\frac{1+[(k-1)/2]M_y^2}{1+[(k-1)/2]M_x^2}\right\}^{(k+1)/(k-1)}} \tag{3-14}$$

As can be seen from Fig. 3-1, for subsonic flow the chamber contraction ratio A_1/A_t can be small, with values of 3 to 6, and the passage is convergent. There is no noticeable effect from variations of k. In solid rocket motors the chamber area A_1 refers to the flow passage or port cavity in the virgin grain. With supersonic flow the nozzle section diverges and the area ratio becomes large very quickly; the area ratio is significantly influenced by the value of k. The area ratio A_2/A_t ranges between 15 and 30 at $M = 4$, depending on the value of k. On the other hand, pressure ratios depend little on k whereas temperature ratios show more variation.

The average *molecular mass* $\mathfrak{M}$ of a mixture of gases is the sum of all the molar fractions n_i multiplied by the molecular mass of each chemical species $(n_i\mathfrak{M}_i)$ and then divided by the sum of all molar mass fractions. This is further elaborated upon in Chapter 5. The symbol $\mathfrak{M}$ is used to avoid confusion with M for the Mach number. In many pieces of rocket literature $\mathfrak{M}$ is called molecular weight.

Example 3–1. An ideal rocket chamber is to operate at sea level using propellants whose combustion products have a specific heat ratio k of 1.30. Determine the required chamber pressure and nozzle area ratio between throat and exit if the nozzle exit Mach number is 2.40. The nozzle inlet Mach number may be considered to be negligibly small.

SOLUTION. For optimum expansion the nozzle exit pressure should be equal to the atmospheric pressure which has the value 0.1013 MPa. If the chamber velocity is small, the chamber pressure is equal to the total or stagnation pressure, which is, from Eq. 3-13,

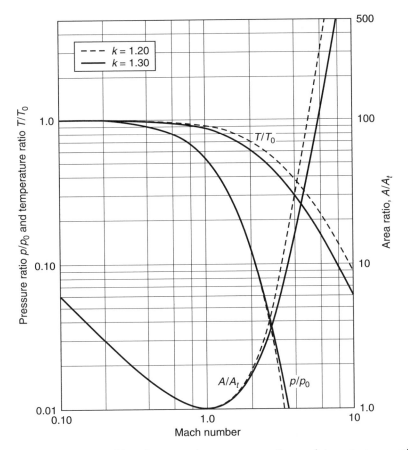

FIGURE 3–1. Relationship of area ratio, pressure ratio, and temperature ratio as functions of Mach number in a De Laval nozzle for the subsonic and supersonic nozzle regions.

$$p_0 = p\left[1 + \tfrac{1}{2}(k - 1)M^2\right]^{k/(k-1)}$$

$$= 0.1013\left[1 + \tfrac{1}{2} \times 0.30 \times 2.40^2\right]^{1.3/0.3} = 1.51 \text{ MPa}$$

The nozzle area is determined from Eq. 3–14 by setting $M_t = 1.0$ at the throat (see also Fig. 3–1):

$$\frac{A_2}{A_t} = \frac{1.0}{2.40}\sqrt{\left(\frac{1 + 0.15 \times 2.4^2}{1 + 0.15}\right)^{2.3/0.3}} = 2.64$$

3.3. ISENTROPIC FLOW THROUGH NOZZLES

In a converging–diverging nozzle a large fraction of the thermal energy of the gases in the chamber is converted into kinetic energy. As will be explained, the gas pressure and temperature drop dramatically and the gas velocity can reach values in excess of two miles per second. This is a reversible, essentially isentropic flow process and its analysis is described here. If a nozzle inner wall has a flow obstruction or a wall protrusion (a piece of weld splatter or slag), then the kinetic gas enery is locally converted back into thermal energy essentially equal to the stagnation temperature and stagnation pressure in the chamber. Since this would lead quickly to a local overheating and failure of the wall, nozzle inner walls have to be smooth without any protrusion. Stagnation conditions can also occur at the leading edge of a jet vane (described in Chapter 16) or at the tip of a gas sampling tube inserted into the flow.

Velocity

From Eq. 3–2 the nozzle exit velocity v_2 can be found:

$$v_2 = \sqrt{2J(h_1 - h_2) + v_1^2} \tag{3.15a}$$

This equation applies to ideal and non-ideal rockets. For constant k this expression can be rewritten with the aid of Eqs. 3–6 and 3–7. The subscripts 1 and 2 apply to the nozzle inlet and exit conditions respectively:

$$v_2 = \sqrt{\frac{2k}{k-1} RT_1 \left[1 - \left(\frac{p_2}{p_1}\right)^{(k-1)/k} \right] + v_1^2} \tag{3.15b}$$

This equation also holds for any two points within the nozzle. When the chamber section is large compared to the nozzle throat section, the chamber velocity or nozzle approach velocity is comparatively small and the term v_1^2 can be neglected. The chamber temperature T_1 is at the nozzle inlet and, under isentropic conditions, differs little from the stagnation temperature or (for a chemical rocket) from the combustion temperature. This leads to an important simplified   expression of the exhaust velocity v_2, which is often used in the analysis.

$$\begin{aligned}
v_2 &= \sqrt{\frac{2k}{k-1} RT_1 \left[1 - \left(\frac{p_2}{p_1}\right)^{(k-1)/k} \right]} \\
&= \sqrt{\frac{2k}{k-1} \frac{R'T_0}{\mathfrak{M}} \left[1 - \left(\frac{p_2}{p_1}\right)^{(k-1)/k} \right]}
\end{aligned} \tag{3–16}$$

It can be seen that the exhaust velocity of a nozzle is a function of the pressure ratio p_1/p_2, the ratio of specific heats k, and the absolute temperature at the nozzle inlet T_1, as well as the gas constant R. Because the gas constant for any particular gas is inversely proportional to the molecular mass $\mathfrak{M}$, the exhaust velocity or the specific impulse are a function of the ratio of the absolute nozzle entrance temperature divided by the molecular mass, as is shown in Fig. 3–2. This ratio plays an important role in optimizing the mixture ratio in chemical rockets.

Equations 2–14 and 2–15 give the relations between the velocity v_2, the thrust F, and the specific impulse I_s; it is plotted in Fig. 3–2 for two pressure ratios and three values of k. Equation 3–16 indicates that any increase in the gas temperature (usually caused by an increase in energy release) or any decrease of the molecular mass of the propellant (usually achieved by using light molecular mass gases rich in hydrogen content) will improve the performanace of the rocket; that is, they will increase the specific impulse I_s or the exhaust velocity v_2 or c and, thus, the performance of the vehicle. The influences of the pressure ratio across the nozzle p_1/p_2 and of the specific heat ratio k are less pronounced. As can be seen from Fig. 3–2, performance increases

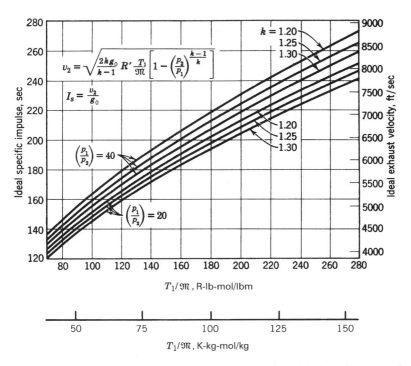

FIGURE 3–2. Specific impulse and exhaust velocity of an ideal rocket at optimum nozzle expansion as functions of the absolute chamber temperature T_1 and the molecular mass $\mathfrak{M}$ for several values of k and p_1/p_2.

with an increase of the pressure ratio; this ratio increases when the value of the chamber pressure p_1 increases or when the exit pressure p_2 decreases, corresponding to high altitude designs. The small influence of k-values is fortuitous because low molecular masses are found in diatomic or monatomic gases, which have the higher values of k.

For comparing specific impulse values from one rocket system to another or for evaluating the influence of various design parameters, the value of the pressure ratio must be standardized. A chamber pressure of 1000 psia (6.894 MPa) and an exit pressure of 1 atm (0.1013 MPa) are generally in use today.

For *optimum expansion* $p_2 = p_3$ and the effective exhaust velocity c (Eq. 2–16) and the ideal rocket exhaust velocity are related, namely

$$v_2 = (c_2)_{opt} \tag{3–17}$$

and c can be substituted for v_2 in Eqs. 3–15 and 3–16. For a fixed nozzle exit area ratio, and constant chamber pressure, this optimum condition occurs only at a particular altitude where the ambient pressure p_3 happens to be equal to the nozzle exhaust pressure p_2. At all other altitudes $c \neq v_2$.

The maximum theoretical value of the nozzle outlet velocity is reached with an infinite expansion (exhausting into a vacuum).

$$(v_2)_{max} = \sqrt{2kRT_0/(k - 1)} \tag{3–18}$$

This maximum theoretical exhaust velocity is finite, even though the pressure ratio is infinite, because it represents the finite thermal energy content of the fluid. Such an expansion does not happen, because, among other things, the temperature of many of the working medium species will fall below their liquefaction or the freezing points; thus they cease to be a gas and no longer contribute to the gas expansion.

Example 3–2. A rocket operates at sea level ($p = 0.1013$ MPa) with a chamber pressure of $p_1 = 2.068$ MPa or 300 psia, a chamber temperature of $T_1 = 2222$ K, and a propellant consumption of $\dot{m} = 1$ kg/sec. (Let $k = 1.30$, $R = 345.7$ J/kg-K). Show graphically the variation of A, v, V, and M, with respect to pressure along the nozzle. Calculate the ideal thrust and the ideal specific impulse.

SOLUTION. Select a series of pressure values and calculate for each pressure the corresponding values of v, V, and A. A sample calculation is given below. The initial specific volume V_1 is calculated from the equation of state of a perfect gas, Eq. 3–4:

$$V_1 = RT_1/p_1 = 345.7 \times 2222/(2.068 \times 10^6) = 0.3714 \text{ m}^3/\text{kg}$$

In an isentropic flow at a point of intermediate pressure, say at $p_x = 1.379$ MPa or 200 psi, the specific volume and the temperature are, from Eq. 3–7,

$$V_x = V_1(p_1/p_x)^{1/k} = 0.3714(2.068/1.379)^{1/1.3} = 0.5072 \text{ m}^3/\text{kg}$$
$$T_x = T_1(p_x/p_1)^{(k-1)/k} = 2222(1.379/2.068)^{0.38/1.3} = 2023 \text{ K}$$

The calculation of the velocity follows from Eq. 3–16:

$$v_x = \sqrt{\frac{2kRT_1}{k-1}\left[1 - \left(\frac{p_x}{p_1}\right)^{(k-1)/k}\right]}$$

$$= \sqrt{\frac{2 \times 1.30 \times 345.7 \times 2222}{1.30 - 1}\left[1 - \left(\frac{1.379}{2.068}\right)^{0.2307}\right]} = 771 \text{ m/sec}$$

The cross-sectional area is found from Eq. 3–3:

$$A_x = \dot{m}_x V_x/v_x = 1 \times 0.5072/771 = 658 \text{ cm}^2$$

The Mach number M is, using Eq. 3–11,

$$M_x = v_x/\sqrt{kRT_x} = 771/\sqrt{1.30 \times 345.7 \times 1932} = 0.8085$$

Figure 3–3 shows the variations of the velocity, specific volume, area, and Mach number with pressure in this nozzle. At optimum expansion the ideal exhaust velocity v_2 is equal to the effective exhaust velocity c and, from Eq. 3–16, it is calculated to be 1827 m/sec. Therefore, the thrust F and the specific impulse can be determined from Eqs. 2–6 and 2–14:

$$F = \dot{m}\, v_2 = 1 \times 1827 = 1827 \text{ N}$$
$$I_s = c/g_0 = 1827/9.80 = 186 \text{ sec}$$

A number of interesting deductions can be made from this example. Very *high gas velocities* (over 1 km/sec) can be obtained in rocket nozzles. The *temperature drop* of the combustion gases flowing through a rocket nozzle is appreciable. In the example given the temperature changed 1117°C in a relatively short distance. This should not be surprising, for the increase in the kinetic energy of the gases is derived from a decrease of the enthalpy, which in turn is proportional to the decrease in temperature. Because the exhaust gases are still very hot (1105 K) when leaving the nozzle, they contain considerable thermal energy not available for conversion into kinetic energy of the jet.

Nozzle Flow and Throat Condition

The required nozzle area decreases to a *minimum* (at 1.130 MPa or 164 psi pressure in the previous example) and then increases again. Nozzles of this type (often called De Laval nozzles after their inventor) consist of a convergent section followed by a divergent section. From the continuity equation, the

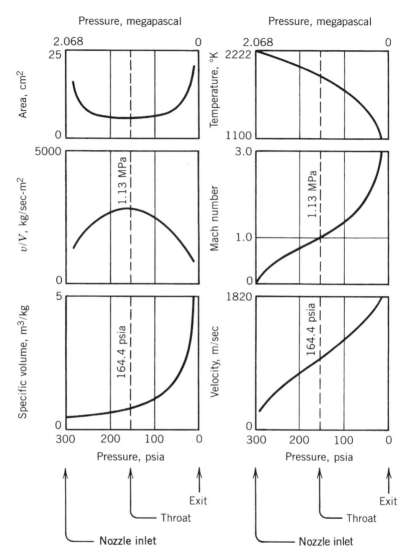

FIGURE 3–3. Typical variation of cross-sectional area, temperature, specific volume, and velocity with pressure in a rocket nozzle.

area is inversely propportional to the ratio v/V. This quantity has also been plotted in Fig. 3–3. There is a maximum in the curve of v/V because at first the velocity increases at a greater rate than the specific volume; however, in the divergent section, the specific volume increases at a greater rate.

The minimum nozzle area is called the *throat area*. The ratio of the nozzle exit area A_2 to the throat area A_t is called the nozzle area expansion ratio and is designated by the Greek letter ϵ. It is an important nozzle design parameter.

$$\epsilon = A_2/A_t \tag{3–19}$$

The maximum gas flow per unit area occurs at the throat where there is a unique gas pressure ratio which is only a function of the ratio of specific heats k. This pressure ratio is found by setting $M = 1$ in Eq. 3–13.

$$p_t/p_1 = [2/(k + 1)]^{k/(k-1)} \tag{3–20}$$

The throat pressure p_t for which the isentropic mass flow rate is a maximum is called the *critical pressure*. Typical values of this critical pressure ratio range between 0.53 and 0.57. The flow through a specified rocket nozzle with a given inlet condition is less than the maximum if the pressure ratio is larger than that given by Eq. 3–20. However, note that this ratio is not that across the entire nozzle and that the maximum flow or choking condition (explained below) is always established internally at the throat and not at the exit plane. The nozzle inlet pressure is very close to the chamber stagnation pressure, except in narrow combustion chambers where there is an appreciable drop in pressure from the injector region to the nozzle entrance region. This is discussed in Section 3.5. At the point of critical pressure, namely the throat, the Mach number is one and the values of the specific volume and temperature can be obtained from Eqs. 3–7 and 3–12.

$$V_t = V_1[(k + 1)/2]^{1/(k-1)} \tag{3–21}$$

$$T_t = 2T_1/(k + 1) \tag{3–22}$$

In Eq. 3–22 the nozzle inlet temperature T_1 is very close to the combustion temperature and hence close to the nozzle flow stagnation temperature T_0. At the critical point there is only a mild change of these properties. Take for example a gas with $k = 1.2$; the critical pressure ratio is about 0.56 (which means that p_t equals almost half of the chamber pressure p_1); the temperature drops only slightly ($T_t = 0.91T_1$), and the specific volume expands by over 60% ($V_t = 1.61V_1$). From Eqs. 3–15, 3–20, and 3–22, the critical or throat velocity v_t is obtained:

$$v_t = \sqrt{\frac{2k}{k+1}RT_1} = a_t = \sqrt{kRT} \tag{3–23}$$

The first version of this equation permits the throat velocity to be calculated directly from the nozzle inlet conditions without any of the throat conditions being known. At the nozzle throat the critical velocity is clearly also the sonic velocity. The divergent portion of the nozzle permits further decreases in pressure and increases in velocity under supersonic conditions. If the nozzle is cut off at the throat section, the exit gas velocity is sonic and the flow rate remains a maximum. The sonic and supersonic flow condition can be attained only if the critical pressure prevails at the throat, that is, if p_2/p_1 is equal to or less than the quantity defined by Eq. 3–20. There are, therefore, three different types of nozzles: subsonic, sonic, and supersonic, and these are described in Table 3–1.

The supersonic nozzle is the one used for rockets. It achieves a high degree of conversion of enthalpy to kinetic energy. The ratio between the inlet and exit pressures in all rockets is sufficiently large to induce supersonic flow. Only if the absolute chamber pressure drops below approximately 1.78 atm will there be subsonic flow in the divergent portion of the nozzle during sea-level operation. This condition occurs for a very short time during the start and stop transients.

The velocity of sound is equal to the propagation speed of an elastic pressure wave within the medium, sound being an infinitesimal pressure wave. If, therefore, sonic velocity is reached at any point within a steady flow system, it is impossible for a pressure disturbance to travel past the location of sonic or supersonic flow. Thus, any partial obstruction or disturbance of the flow downstream of the nozzle throat with sonic flow has no influence on the throat or upstream of it, provided that the disturbance does not raise the downstream pressure above its critical value. It is not possible to increase the throat velocity or the flow rate in the nozzle by further lowering the exit pressure or even evacuating the exhaust section. This important condition is often described as *choking* the flow. It is always established at the throat and not the nozzle exit plane. *Choked flow* through the critical section of a supersonic nozzle may be derived from Eqs. 3–3, 3–21, and 3–23. It is equal to the mass flow at any section within the nozzle.

TABLE 3–1. Nozzle Types

	Subsonic	Sonic	Supersonic
Throat velocity	$v_1 < a_t$	$v_t = a_t$	$v_t = a_t$
Exit velocity	$v_2 < a_2$	$v_2 = v_t$	$v_2 > v_t$
Mach number	$M_2 < 1$	$M_2 = M_t = 1.0$	$M_2 > 1$
Pressure ratio	$\dfrac{p_1}{p_2} < \left(\dfrac{k+1}{2}\right)^{k/(k-1)}$	$\dfrac{p_1}{p_2} = \dfrac{p_1}{p_t} = \left(\dfrac{k+1}{2}\right)^{k/(k-1)}$	$\dfrac{p_1}{p_2} > \left(\dfrac{k+1}{2}\right)^{k/(k-1)}$
Shape			

$$\dot{m} = \frac{A_t v_t}{V_t} = A_t p_1 k \frac{\sqrt{[2/(k+1)]^{(k+1)/(k-1)}}}{\sqrt{kRT_1}} \tag{3-24}$$

The mass flow through a rocket nozzle is therefore proportional to the throat area A_t and the chamber (stagnation) pressure p_1; it is also inversely proportional to the square root of $T/\mathfrak{M}$ and a function of the gas properties. For a supersonic nozzle the *ratio between the throat and any downstream area* at which a pressure p_x prevails can be expressed as a function of the pressure ratio and the ratio of specific heats, by using Eqs. 3–4, 3–16, 3–21, and 3–23, as follows:

$$\frac{A_t}{A_x} = \frac{V_t v_x}{V_x v_t} = \left(\frac{k+1}{2}\right)^{1/(k-1)} \left(\frac{p_x}{p_1}\right)^{1/k} \sqrt{\frac{k+1}{k-1}\left[1 - \left(\frac{p_x}{p_1}\right)^{(k-1)/k}\right]} \tag{3-25}$$

When $p_x = p_2$, then $A_x/A_t = A_2/A_t = \epsilon$ in Eq. 3–25. For low-altitude operation (sea level to about 10,000 m) the nozzle area ratios are typically between 3 and 25, depending on chamber pressure, propellant combinations, and vehicle envelope constraints. For high altitude (100 km or higher) area ratios are typically between 40 and 200, but there have been some as high as 400. Similarly, an expression for the ratio of the velocity at any point downstream of the throat with the pressure p_x, and the throat velocity may be written from Eqs. 3–15 and 3–23:

$$\frac{v_x}{v_t} = \sqrt{\frac{k+1}{k-1}\left[1 - \left(\frac{p_x}{p_1}\right)^{(k-1)/k}\right]} \tag{3-26}$$

These equations permit the direct determination of the velocity ratio or the area ratio for any given pressure ratio, and vice versa, in ideal rocket nozzles. They are plotted in Figs. 3–4 and 3–5, and these plots allow the determination of the pressure ratios given the area or velocity ratios. When $p_x = p_2$, Eq. 3–26 describes the velocity ratio between the nozzle exit area and the throat section. When the exit pressure coincides with the atmospheric pressure ($p_2 = p_3$, see Fig. 2–1), these equations apply for optimum nozzle expansion. For rockets that operate at high altitudes, not too much additional exhaust velocity can be gained by increasing the area ratio above 1000. In addition, design difficulties and a heavy inert nozzle mass make applications above area ratios of about 350 marginal.

Appendix 2 is a table of several properties of the Earth's atmosphere with agreed-upon standard values. It gives ambient pressure for different altitudes. These properties can vary somewhat from day to day (primarily because of solar activity) and between hemispheres. For example, the density of the atmosphere at altitudes between 200 and 3000 km can change by more than an order of magnitude, affecting satellite drag.

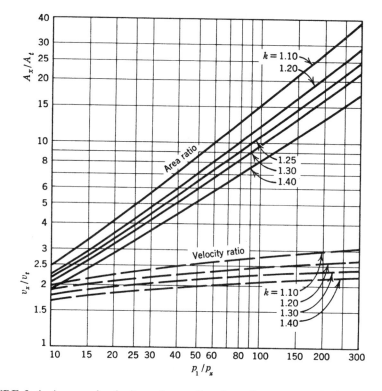

FIGURE 3–4. Area and velocity ratios as function of pressure ratio for the diverging section of a supersonic nozzle.

Example 3–3. Design a nozzle for an ideal rocket that has to operate at 25 km altitude and give 5000 N thrust at a chamber pressure of 2.068 MPa and a chamber temperature of 2800 K. Assuming that $k = 1.30$ and $R = 355.4$ J/kg-K, determine the throat area, exit area, throat velocity, and exit temperature.

SOLUTION. At 25 km the atmospheric pressure equals 0.002549 MPa (in Appendix 2 the ratio is 0.025158 which must be multiplied by the pressure at sea level or 0.1013 MPa). The pressure ratio is

$$p_2/p_1 = p_3/p_1 = 0.002549/2.068 = 0.001232 = 1/811.3$$

The critical pressure, from Eq. 3–20, is

$$p_t = 0.546 \times 2.068 = 1.129 \text{ MPa}$$

The throat velocity, from Eq. 3–23, is

$$v_t = \sqrt{\frac{2k}{k+1}RT_1} = \sqrt{\frac{2 \times 1.30}{1.3+1}355.4 \times 2800} = 1060 \text{ m/sec}$$

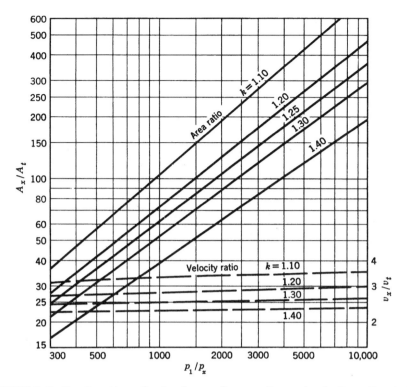

FIGURE 3–5. Continuation of prior figure of area ratios and velocity ratios, but for higher pressure ratios in a supersonic nozzle.

The ideal exit velocity is found from Eq. 3–16 or Fig. 3–5, using a pressure ratio of 811.3:

$$v_2 = \sqrt{\frac{2k}{k-1} RT_1 \left[1 - \left(\frac{p_2}{p_1} \right)^{(k-1)/k} \right]}$$

$$= \sqrt{\frac{2 \times 1.30}{1.30 - 1} 355.4 \times 2800 \times 0.7869} = 2605 \text{ m/sec}$$

An approximate value of this velocity can also be obtained from the throat velocity and Fig. 3–4. The ideal propellant consumption for optimum expansion conditions is

$$\dot{m} = F/v_2 = 5000/2605 = 1.919 \text{ kg/sec}$$

The specific volume at the entrance to the nozzle equals

$$V_1 = RT_1/p_1 = 355.4 \times 2800/(2.068 \times 10^6) = 0.481 \text{ m}^3/\text{kg}$$

At the throat and exit sections the specific volumes are obtained from Eqs. 3–21 and 3–7:

$$V_t = V_1 \left(\frac{k+1}{2}\right)^{1/(k-1)} = 0.481 \left(\frac{2.3}{2}\right)^{1/0.3} = 0.766 \text{ m}^3/\text{kg}$$

$$V_2 = V_1 \left(\frac{p_1}{p_2}\right)^{1/k} = 0.481(2.068/0.002549)^{0.7692} = 83.15 \text{ m}^3/\text{kg}$$

The areas at the throat and exit sections and the nozzle area ratio A_2/A_t are

$$A_t = \dot{m}V_t/v_t = 1.919 \times 0.766/1060 = 13.87 \text{ cm}^2$$
$$A_2 = \dot{m}V_2/v_2 = 1.919 \times 83.15/2605 = 612.5 \text{ cm}^2$$
$$\epsilon = A_2/A_t = 612.5/13.87 = 44.16$$

An approximate value of this area ratio can also be obtained directly from Fig. 3–5 for $k = 1.30$ and $p_1/p_2 = 811.2$. The exit temperature is given by

$$T_2 = T_1(p_2/p_1)^{(k-1)/k} = 2800(0.002549/2.068)^{0.2307} = 597 \text{ K}$$

Thrust and Thrust Coefficient

The efflux of the propellant gases or the momentum flux-out causes the thrust or reaction force on the rocket structure. Because the flow is supersonic, the pressure at the exit plane of the nozzle may be different from the ambient pressure and the pressure thrust component adds to the momentum thrust as given by Eq. 2–14:

$$F = \dot{m}v_2 + (p_2 - p_3)A_2 \tag{2–14}$$

The maximum thrust for any given nozzle operation is found in a vacuum where $p_3 = 0$. Between sea level and the vacuum of space, Eq. 2–14 gives the variation of thrust with altitude, using the properties of the atmosphere such as those listed in Appendix 2. Figure 2–2 shows a typical variation of thrust with altitude. To modify values calculated for optimum operating conditions ($p_2 = p_3$) for given values of p_1, k, and A_2/A_t, the following expressions may be used. For the thrust,

$$F = F_{\text{opt}} + p_1 A_t \left(\frac{p_2}{p_1} - \frac{p_3}{p_1}\right)\frac{A_2}{A_t} \tag{3–27}$$

For the specific impulse, using Eqs. 2–5, 2–18, and 2–14,

$$I_s = (I_s)_{\text{opt}} + \frac{c^*\epsilon}{g_0}\left(\frac{p_2}{p_1} - \frac{p_3}{p_1}\right) \tag{3–28}$$

If, for example, the specific impulse for a new exit pressure p_2 corresponding to a new area ratio A_2/A_t is to be calculated, the above relations may be used.

Equation 2–14 can be expanded by modifying it and substituting v_2, v_t and V_t from Eqs. 3–16, 3–21, and 3–23.

$$F = \frac{A_t v_t v_2}{V_t} + (p_2 - p_3)A_2$$

$$= A_t p_1 \sqrt{\frac{2k^2}{k-1}\left(\frac{2}{k+1}\right)^{(k+1)/(k-1)}\left[1 - \left(\frac{p_2}{p_1}\right)^{(k-1)/k}\right]} + (p_2 - p_3)A_2 \tag{3–29}$$

The first version of this equation is general and applies to all rockets, the second form applies to an ideal rocket with k being constant throughout the expansion process. This equation shows that the thrust is proportional to the throat area A_t and the chamber pressure (or the nozzle inlet pressure) p_1 and is a function of the pressure ratio across the nozzle p_1/p_2, the specific heat ratio k, and of the pressure thrust. It is called the ideal thrust equation. The thrust coefficient C_F is defined as the thrust divided by the chamber pressure p_1 and the throat area A_t. Equations 2–14, 3–21, and 3–16 then give

$$C_F = \frac{v_2^2 A_2}{p_1 A_t V_2} + \frac{p_2}{p_1}\frac{A_2}{A_t} - \frac{p_3}{p_1}\frac{A_2}{A_t}$$

$$= \sqrt{\frac{2k^2}{k-1}\left(\frac{2}{k+1}\right)^{(k+1)/(k-1)}\left[1 - \left(\frac{p_2}{p_1}\right)^{(k-1)/k}\right]} + \frac{p_2 - p_3}{p_1}\frac{A_2}{A_t} \tag{3–30}$$

The thrust coefficient C_F is a function of gas property k, the nozzle area ratio ϵ, and the pressure ratio across the nozzle p_1/p_2, but independent of chamber temperature. For any fixed pressure ratio p_1/p_3, the thrust coefficient C_F and the thrust F have a peak when $p_2 = p_3$. This peak value is known as the *optimum thrust coefficient* and is an important criterion in nozzle design considerations. The use of the thrust coefficient permits a simplification to Eq. 3–29:

$$F = C_F A_t p_1 \tag{3–31}$$

Equation 3–31 can be solved for C_F and provides the relation for determining the thrust coefficient experimentally from measured values of chamber pressure, throat diameter, and thrust. Even though the thrust coefficient is a function of chamber pressure, it is not simply proportional to p_1, as can be seen from Eq. 3–30. However, it is directly proportional to throat area. The thrust coefficient can be thought of as representing the amplification of thrust due to the gas expanding in the supersonic nozzle as compared to the thrust that would be exerted if the chamber pressure acted over the throat area only.

The thrust coefficient has values ranging from about 0.8 to 1.9. It is a convenient parameter for seeing the effects of chamber pressure or altitude variations in a given nozzle configuration, or to correct sea-level results for flight altitude conditions.

Figure 3–6 shows the variation of the *optimum expansion* ($p_2 = p_3$) thrust coefficient for different pressure ratios p_1/p_2, values of k, and area ratio ϵ. The complete thrust coefficient is plotted in Figs 3–7 and 3–8 as a function of pressure ratio p_1/p_3 and area ratio for $k = 1.20$ and 1.30. These two sets of curves are useful in solving various nozzle problems for they permit the evaluation of under- and over-expanded nozzle operation, as explained below. The values given in these figures are ideal and do not consider such losses as divergence, friction or internal expansion waves.

When p_1/p_3 becomes very large (e.g., expansion into near-vacuum), then the thrust coefficient approaches an asymptotic maximum as shown in Figs. 3–7 and 3–8. These figures also give values of C_F for any mismatched nozzle ($p_2 \neq p_3$), provided the nozzle is flowing full at all times, that is, the working fluid does not separate or break away from the walls. Flow separation is discussed later in this section.

Characteristic Velocity and Specific Impulse

The *characteristic velocity* c^* was defined by Eq. 2–18. From Eqs. 3–24 and 3–31 it can be shown that

$$c^* = \frac{p_1 A_t}{\dot{m}} = \frac{I_s g_0}{C_F} = \frac{c}{C_F} = \frac{\sqrt{kRT_1}}{k\sqrt{[2/(k+1)]^{(k+1)/(k-1)}}} \tag{3-32}$$

It is basically a function of the propellant characteristics and combustion chamber design; it is independent of nozzle characteristics. Thus, it can be used as a figure of merit in comparing propellant combinations and combustion chamber designs. The first version of this equation is general and allows the determination of c^* from experimental data of $\dot{m}$, p_1, and A_t. The last version gives the maximum value of c^* as a function of gas properties, namely k, the chamber temperature, and the molecular mass $\mathfrak{M}$, as determined from the theory in Chapter 5. Some values of c^* are shown in Tables 5–4 and 5–5. The term c^*-*efficiency* is sometimes used to express the degree of completion of the energy release and the creation of high temperature, high pressure gas in the chamber. It is the ratio of the actual value of c^*, as determined from measurements, and the theoretical value (last part of Eq. 3–32), and typically has a value between 92 and 99.5 percent.

Using Eqs. 3–31 and 3–32, the thrust itself may now be expressed as the mass flow rate times a function of the combustion chamber (c^*) times a function of the nozzle expansion C_F),

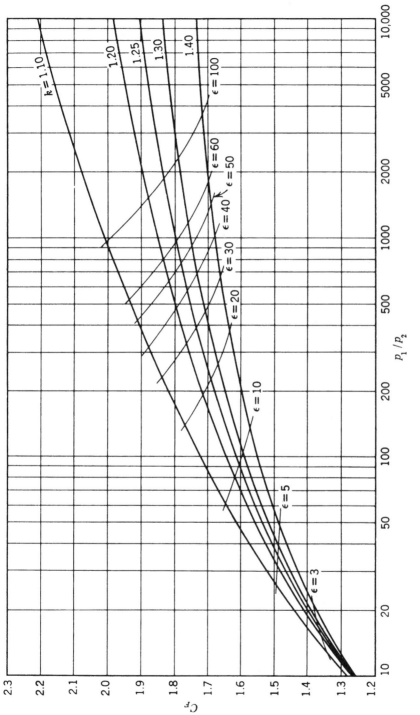

FIGURE 3-6. Thrust coefficient C_F as a function of pressure ratio, nozzle area ratio, and specific heat ratio for optimum expansion conditions ($p_2 = p_3$).

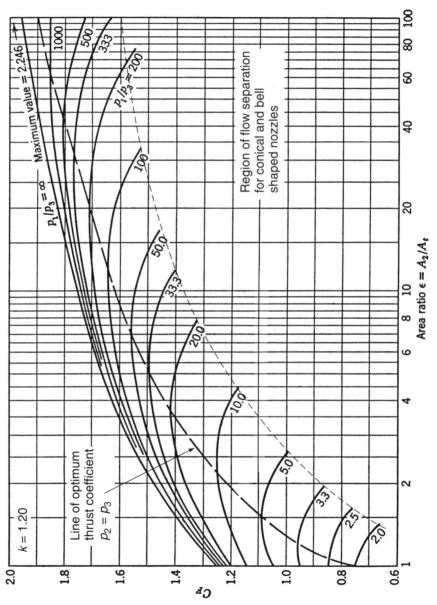

FIGURE 3–7. Thrust coefficient C_F versus nozzle area ratio for $k = 1.20$.

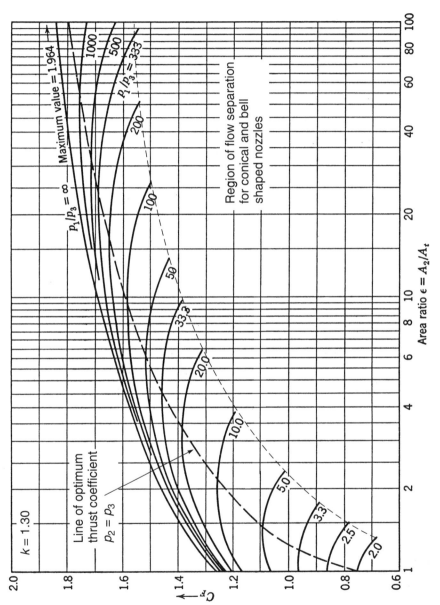

FIGURE 3–8. Thrust coefficient C_F versus nozzle area ratio for $k = 1.30$.

$$F = C_F \dot{m} c^* \qquad (3\text{–}33)$$

Some authors use a term called the *discharge coefficient* C_D which is merely the reciprocal of c^*. Both C_D and the characteristic exhaust velocity c^* are used primarily with chemical rocket propulsion systems.

The influence of *variations in the specific heat ratio* k on various parameters (such as $c, c^* A_2/A_t, v_2/v_t$, or I_s) is not as large as the changes in chamber temperature, pressure ratio, or molecular mass. Nevertheless, it is a noticeable factor, as can be seen by examining Figs. 3–2 and 3–4 to 3–8. The value of k is 1.67 for monatomic gases such as helium and argon, 1.4 for cold diatomic gases such as hydrogen, oxygen, and nitrogen, and for triatomic and beyond it varies between 1.1 and 1.3 (methane is 1.11 and ammonia and carbon dioxide 1.33). In general, the more complex the molecule the lower the value of k; this is also true for molecules at high temperatures when their vibrational modes have been activated. The average values of k and $\mathfrak{M}$ for typical rocket exhaust gases with several constituents depend strongly on the composition of the products of combustion (chemical constituents and concentrations), as explained in Chapter 5. Values of k and $\mathfrak{M}$ are given in Tables 5–4, 5–5, and 5–6.

Example 3–4. What is percentage variation in thrust between sea level and 25 km for a rocket having a chamber pressure of 20 atm and an expansion area ratio of 6? (Use $k = 1.30$.)

SOLUTION.   At sea level: $p_1/p_3 = 20/1.0 = 20$; at 25 km: $p_1/p_3 = 20/0.0251 = 754$ (see Appendix 2).

Use Eq. 3–30 or Fig. 3–8 to determine the thrust coefficient (hint: use a vertical line on Fig. 3–8 corresponding to $A_2/A_t = 6.0$). At sea level: $C_F = 1.33$. At 25 km: $C_F = 1.64$. The thrust increase $= (1.64 - 1.33)/1.33 = 23\%$.

Under- and Over-Expanded Nozzles

An *under-expanded nozzle* discharges the fluid at an exit pressure greater than the external pressure because the exit area is too small for an optimum area ratio. The expansion of the fluid is therefore incomplete within the nozzle, and must take place outside. The nozzle exit pressure is higher than the local atmospheric pressure.

In an *over-expanded nozzle* the fluid attains a lower exit pressure than the atmosphere as it has an exit area too large for optimum. The phenomenon of over-expansion for a supersonic nozzle is shown in Fig. 3–9, with typical pressure measurements of superheated steam along the nozzle axis and different back pressures or pressure ratios. Curve AB shows the variation of pressure with the optimum back pressure corresponding to the area ratio. Curves AC and AD show the variation of pressure along the axis for increasingly higher external pressures. The expansion within the nozzle proceeds normally for the

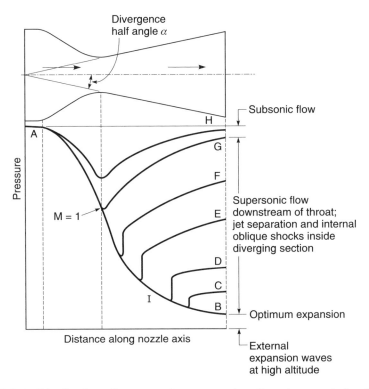

FIGURE 3–9. Distribution of pressures in a converging–diverging nozzle for different flow conditions. Inlet pressure is the same, but exit pressure changes. Based on experimental data from A. Stodala.

initial portion of the nozzle. At point I on curve AD, for example, the pressure is lower than the exit pressure and a sudden rise in pressure takes place which is accompanied by the *separation* of the flow from the walls (separation is described later).

The non-ideal behavior of nozzles is strongly influenced by the presence of compression waves or shock waves inside the diverging nozzle section, which are strong compression discontinuities and exist only in supersonic flow. The sudden pressure rise in the curve ID is such a compression wave. Expansion waves, also strictly supersonic phenomena, match the flow from a nozzle exit to lower ambient pressures. Compression and expansion waves are described in Chapter 18.

The different possible flow conditions in a supersonic nozzle are as follows:

1. When the external pressure p_3 is below the nozzle exit pressure p_2, the nozzle will flow full but will have external expansion waves at its exit (i.e., under-expansion). The expansion of the gas inside the nozzle is incomplete and the value of C_F and I_s, will be less than at optimum expansion.

2. For external pressures p_3 slightly higher than the nozzle exit pressure p_2, the nozzle will continue to flow full. This occurs until p_2 reaches a value between about 25 and 40% of p_3. The expansion is somewhat inefficient and C_F and I_s will have lower values than an optimum nozzle would have. Shock waves will exist outside the nozzle exit section.

3. For higher external pressures, separation of the flow will take place inside the divergent portion of the nozzle. The diameter of the supersonic jet will be smaller than the nozzle exit diameter. With steady flow, separation is typically axially symmetric. Figs. 3–10 and 3–11 show diagrams of separated flows. The axial location of the separation plane depends on the local pressure and the wall contour. The point of separation travels downstream with decreasing external pressure. At the nozzle exit the flow in the center portion remains supersonic, but is surrounded by an annular shaped section of subsonic flow. There is a discontinuity at the separation location and the thrust is reduced, compared to a nozzle that would have been cut off at the separation plane. Shock waves exist outside the nozzle in the external plume.

4. For nozzles in which the exit pressure is just below the value of the inlet pressure, the pressure ratio is below the critical pressure ratio (as defined by Eq. 3–20) and subsonic flow prevails throughout the entire nozzle. This condition occurs normally in rocket nozzles for a short time during the start and stop transients.

The method for estimating pressure at the location of the separation plane inside the diverging section of a supersonic nozzle has usually been empirical. Reference 3–4 shows separation regions based on collected data for several dozen actual conical and bell-shaped nozzles during separation. Reference 3–5 describes a variety of nozzles, their behavior, and methods used to estimate the location and the pressure at separation. Actual values of pressure for the over-expanded and under-expanded regimes described above are functions of the specific heat ratio and the area ratio (see Ref. 3–1).

The axial thrust direction is not usually altered by separation, because a steady flow usually separates uniformly over a cross-section in a divergent nozzle cone of conventional rocket design. During transients, such as start and stop, the separation may not be axially symmetric and may cause momentary but large side forces on the nozzle. During a normal sea-level transient of a large rocket nozzle (before the chamber pressure reaches its full value) some momentary flow oscillations and non-symmetric separation of the jet can occur during over-expanded flow operation. Reference 3–4 shows that the magnitude and direction of transient side forces can change rapidly and erratically. The resulting side forces can be large and have caused failures of nozzle exit cone structures and thrust vector control gimbal actuators. References 3–5 and 3–6 discuss techniques for estimating these side forces.

When the flow separates, as it does in a highly over-expanded nozzle, the thrust coefficient C_F can be estimated if the point of separation in the nozzle is known. Thus, C_F can be determined for an equivalent smaller nozzle with an exit area equal to that at the point of separation. The effect of separation is to increase the thrust and the thrust coefficient over the value that they would have if separation had not occurred. Thus, with separated gas flow, a nozzle designed for high altitude (large value of ϵ) would have a larger thrust at sea level than expected, but not as good as an optimum nozzle; in this case separation may actually be desirable. With separated flow a large and usually heavy portion of the nozzle is not utilized and the nozzle is bulkier and longer than necessary. The added engine weight and size decrease flight performance. Designers therefore select an area ratio that will not cause separation.

Because of uneven flow separation and potentially destructive side loads, sea-level static tests of an upper stage or a space propulsion system with a high area ratio over-expanded nozzle are usually avoided; instead, a sea-level test nozzle with a much smaller area ratio is substituted. However, actual and simulated altitude testing (in an altitude test facility similar to the one described in Chapter 20) would be done with a nozzle having the correct large area ratio. The ideal solution that avoids separation at low altitudes and has high values of C_F at high altitudes is a nozzle that changes area ratio in flight. This is discussed at the end of this section.

For most applications, the rocket system has to operate over a range of altitudes; for a fixed chamber pressure this implies a range of nozzle pressure ratios. The condition of optimum expansion ($p_2 = p_3$) occurs only at one altitude, and a nozzle with a fixed area ratio is therefore operating much of the time at either over-expanded or under-expanded conditions. The best nozzle for such an application is not necessarily one that gives optimum nozzle gas expansion, but one that gives the largest vehicle flight performance (say, total impulse, or specific impulse, or range, or payload); it can often be related to a time average over the powered flight trajectory.

Example 3–5. Use the data from Example 3–4 ($p_1 = 20$ atm, $\epsilon = 6.0$, $k = 1.30$) but instead use an area ratio of 15. Compare the altitude performance of the two nozzles with different ϵ by plotting their C_F against altitude. Assume no shocks inside the nozzle.

SOLUTION. For the $\epsilon = 15$ case, the optimum pressure ratio $p_1/p_3 = p_1/p_2$, and from Fig. 3–6 or 3–8 this value is about 180; $p_3 = 20/180 = 0.111$ atm, which occurs at about 1400 m altitude. Below this altitude the nozzle is over-expanded. At sea level, $p_1/p_3 = 20$ and $p_3 = 1$ atm. As shown in Fig. 3–10, separation would occur. From other similar nozzles it is estimated that separation will occur approximately at a cross-section where the total pressure is about 40% of p_3, or 0.4 atm. The nozzle would not flow full below an area ratio of about 6 or 7 and the gas jet would only be in the center of the exit area. Weak shock waves and jet contraction would then raise the exhaust jet's pressure to match the one atmosphere external pressure. If the jet had not separated, it would have reached an exit pressure of 0.11 atm, but this is an unstable condition that could not be maintained at sea level. As the vehicle gains altitude, the separation plane would

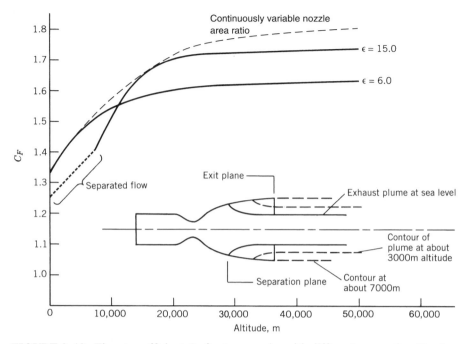

FIGURE 3–10. Thrust coefficient C_F for two nozzles with different area ratios. One has jet separation below about 7000 m altitude. The fully expanded exhaust plume is not shown in the sketch.

gradually move downstream until, at an altitude of about 7000 m, the exhaust gases would occupy the full nozzle area.

The values of C_F can be obtained by following a vertical line for $\epsilon = 15$ and $\epsilon = 6$ in Fig. 3–8 for different pressure ratios, which correspond to different altitudes. Alternatively, Eq. 3–30 can be used for better accuracy. Results are similar to those plotted in Fig. 3–10. The lower area ratio of 6 gives a higher C_F at low altitudes, but is inferior at the higher altitudes. The larger nozzle gives a higher C_F at higher altitudes.

Figure 3–11 shows a comparison of altitude and sea-level behavior of three nozzles and their plumes at different area ratios for a typical three-stage satellite launch vehicle. When fired at sea-level conditions, the nozzle of the third stage with the highest area ratio will experience flow separation and suffer a major performance loss; the second stage will flow full but the external plume will contrast; since $p_2 < p_3$ there is a loss in I_s and F. There is no effect on the first stage nozzle.

Example 3–6. A rocket engine test gives the following data: thrust $F = 53,000$ lbf, propellant flow $\dot{m} = 208$ lbm/sec, nozzle exit area ratio $A_2/A_t = 10.0$, atmospheric

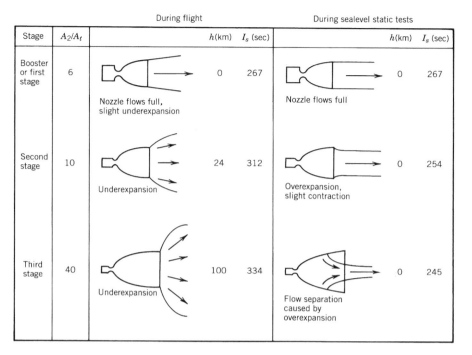

		During flight			During sealevel static tests	
Stage	A_2/A_t		h(km)	I_s (sec)		h(km)	I_s (sec)
Booster or first stage	6	Nozzle flows full, slight underexpansion	0	267	Nozzle flows full	0	267
Second stage	10	Underexpansion	24	312	Overexpansion, slight contraction	0	254
Third stage	40	Underexpansion	100	334	Flow separation caused by overexpansion	0	245

FIGURE 3-11. Simplified sketches of exhaust gas behavior of three typical rocket nozzles for a three-stage launch vehicle. The first vehicle stage has the biggest chamber and the highest thrust but the lowest nozzle area ratio, and the top or third stage usually has the lower thrust but the highest nozzle area ratio.

pressure at test station (the nozzle flows full) $p_3 = 13.8$ psia, and chamber pressure $p_1 = 620$ psia. The test engineer also knows that the theoretical specific impulse is 289 sec at the standard reference conditions of $p_1 = 1000$ psia and $p_3 = 14.7$ psia, and that $k = 1.20$. Correct the value of the thrust to sea-level expansion and the specific impulse corresponding. Assume the combustion temperature and k do not vary significantly with chamber pressure; this is realistic for certain propellants.

SOLUTION. The actual pressure ratio was $p_1/p_3 = 620/13.8 = 44.9$; the ideal pressure ratio at standard conditions would have been equal to $1000/14.7 = 68.0$ and the actual pressure ratio for expansion to sea level would have been $620/14.7 = 42.1$. The thrust coefficient for the test conditions is obtained from Fig. 3–7 or from Eq. 3–30 as $C_F = 1.52$ (for $p_1/p_3 = 44.9$, $\epsilon = 10$ and $k = 1.20$). The thrust coefficient for the corrected sea-level conditions is similarly found to be 1.60. The thrust at sea level would have been $F = 53,000 (1.60/1.52) = 55,790$ lbf. The specific impulse would have been

$$I_s = F/\dot{w} = 53,000/208(1.60/1.52) = 268 \text{ sec}$$

The specific impulse can be corrected in proportion to the thrust coefficient because k, T, and therefore c^* do not vary with p_1; I_s is proportional to c if $\dot{m}$ remains constant. The theoretical specific impulse is given for optimum expansion, i.e., for a nozzle area ratio other than 10.0. From Fig. 3–6 or 3–7 and for $p_1/p_2 = 68.0$ the thrust coefficient is 1.60 and its optimum area ratio approximately 9.0. The corrected specific impulse is accordingly 255 (1.60/1.51) = 270 sec. In comparison with the theoretical specific impulse of 289 sec, this rocket has achieved 270/289 or 93.5% of its maximum performance.

Figs. 3–10 and 3–11 suggest that an ideal design for an ascending (e.g., launch) rocket vehicle would have a "rubber-like" diverging section that could be lengthened so that the nozzle exit area could be made larger as the ambient pressure is reduced. The design would then allow the rocket vehicle to attain its maximum performance at all altitudes as it ascends. As yet we have not achieved a simple mechanical hardware design with this full altitude compensation similar to "stretching rubber." However, there are a number of practical nozzle configurations that can be used to alter the flow shape with altitude and obtain maximum performance. They are discussed in the next section.

Influence of Chamber Geometry

When the chamber has a cross section that is larger than about four times the throat area ($A_1/A_t > 4$), the chamber velocity v_1, can be neglected, as was mentioned in explaining Eqs. 3–15 and 3–16. However, vehicle space or weight constraints often require smaller thrust chamber areas for liquid propellant engines and grain design considerations lead to small void volumes or small perforations or port areas for solid propellant motors. Then v_1 can no longer be neglected as a contribution to the performance. The gases in the chamber expand as heat is being added. The energy necessary to accelerate these expanding gases within the chamber will also cause a pressure drop and an additional energy loss. This acceleration process in the chamber is adiabatic (no heat transfer) but not isentropic. This loss is a maximum when the chamber diameter is equal to the nozzle diameter, which means that there is no converging nozzle section. This has been called a *throatless rocket motor* and has been used in a few tactical missile booster applications, where there was a premium on minimum inert mass and length. The flight performance improvement due to inert mass savings supposedly outweighs the nozzle performance loss of a throatless motor. Table 3–2 lists some of the performance penalties for three chamber area ratios.

Because of this pressure drop within narrow chambers, the chamber pressure is lower at the nozzle entrance than it would be if A_1/A_t had been larger. This causes a small loss in thrust and specific impulse. The theory of this loss is given in Ref. 3–7.

TABLE 3–2. Estimated Losses for Small-Diameter Chambers

Chamber-to-Throat Area Ratio	Throat Pressure (%)	Thrust Reduction (%)	Specific Impulse Reduction (%)
∞	100	0	0
3.5	99	1.5	0.31
2.0	96	5.0	0.55
1.0	81	19.5	1.34

$k = 1.20$; $p_1/p_2 = 1000$.

3.4. NOZZLE CONFIGURATIONS

A number of different proven nozzle configurations are available today. This section describes their geometries and performance. Other chapters (6, 8, 11, 14, and 16) discuss their materials, heat transfer, or application, and mention their requirements, design, construction, and thrust vector control. Nozzles and chambers are usually of circular cross section and have a converging section, a throat at the narrowest location (minimum cross section), and a diverging section. Nozzles can be seen in Figs. 1–4, 1–5, 1–8, 2–1, 3–11 to 3–13, 3–15, 10–2 to 10–5, 10–16, 11–1 to 11–3, and 14–6 to 14–8. Refs. 3–5 and 3–8 describe many nozzle configurations.

The *converging nozzle section* between the chamber and the nozzle throat has never been critical in achieving high performance. The subsonic flow in this section can easily be turned at very low pressure drop and any radius, cone angle, wall contour curve, or nozzle inlet shape is satisfactory. A few small attitude control thrust chambers have had their nozzle at 90 degrees from the combustion chamber axis without any performance loss. The *throat contour* also is not very critical to performance, and any radius or other curve is usually acceptable. The pressure gradients are high in these two regions and the flow will adhere to the walls. The principal difference in the different nozzle configurations is found in the diverging supersonic-flow section, as described below. The wall surface throughout the nozzle should be smooth and shiny to minimize friction, radiation absorption, and convective heat transfer due to surface roughness. Gaps, holes, sharp edges, or protrusions must be avoided.

Six different nozzle configurations are shown in Fig. 3–12 and each will be discussed. The first three sketches show conical and bell-shaped nozzles. The other three have a center body inside the nozzle and have excellent altitude compensation. Although these last three have been ground tested, to date none of them has flown in a space launch vehicle. The lengths of several nozzle types are compared in Fig. 3–13. The objectives of a good nozzle configuration are to obtain the highest practical I_s, minimize inert nozzle mass, and conserve length

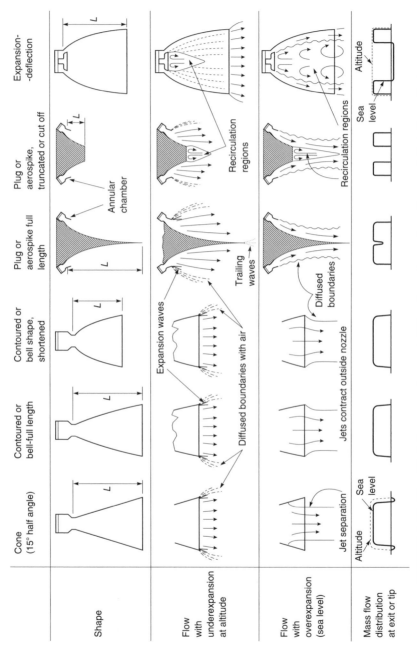

FIGURE 3–12. Simplified diagrams of several different nozzle configurations and their flow effects.

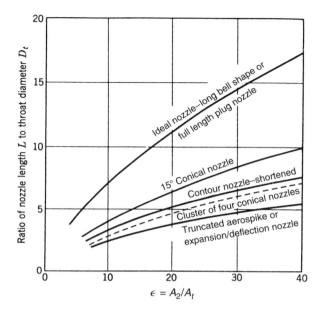

FIGURE 3–13. Length comparison of several types of nozzles. (*Taken in part from G. V. R. Rao, "Recent Developments in Rocket Nozzle Configurations,"* American Rocket Society Journal, *Vol. 31, No. 11, November 1961.*)

(shorter nozzles can reduce vehicle length, vehicle structure, and vehicle inert mass).

Cone- and Bell-Shaped Nozzles

The *conical nozzle* is the oldest and perhaps the simplest configuration. It is relatively easy to fabricate and is still used today in many small nozzles. A theoretical correction factor λ can be applied to the nozzle exit momentum of an ideal rocket with a conical nozzle exhaust. This factor is the ratio between the momentum of the gases in a nozzle with a finite nozzle angle 2α and the momentum of an ideal nozzle with all gases flowing in an axial direction:

$$\lambda = \frac{1}{2}(1 + \cos\alpha) \qquad (3\text{–}34)$$

The variation of λ with different values of α is shown in Table 3–3 for any nozzle that has uniform mass flow per unit exit area. For ideal rockets $\lambda = 1.0$. For a rocket nozzle with a divergence cone angle of $30°$ (half angle $\alpha = 15°$), the exit momentum and therefore the exhaust velocity will be 98.3% of the velocity calculated by Eq. 3–15b. Note that the correction factor λ only applies

TABLE 3–3. Nozzle Angle Correction Factor for Conical Nozzles

Nozzle Cone Divergence Half Angle, α (deg)	Correction Factor, λ
0	1.0000
2	0.9997
4	0.9988
6	0.9972
8	0.9951
10	0.9924
12	0.9890
14	0.9851
15	0.9830
16	0.9806
18	0.9755
20	0.9698
22	0.9636
24	0.9567

to the first term (the momentum thrust) in Eqs. 2–14, 3–29, and 3–30 and not to the second term (pressure thrust).

A small nozzle divergence angle causes most of the momentum to be axial and thus gives a high specific impulse, but the long nozzle has a penalty in rocket propulsion system mass, vehicle mass, and also design complexity. A large divergence angle gives short, lightweight designs, but the performance is low. There is an optimum conical nozzle shape and length (typically between 12 and 18 degrees half angle) and it is usually a compromise which depends on the specific application and flight path.

The *bell-shaped* or *contour nozzle* (see Figs. 3–12 and 3–13) is probably the most common nozzle shape today. It has a high angle expansion section (20 to 50°) right behind the nozzle throat; this is followed by a gradual reversal of nozzle contour slope so that at the nozzle exit the divergence angle is small, usually less than a 10° half angle. It is possible to go to large divergence angles immediately behind the throat (20 to 50°) because the high relative pressure, the large pressure gradient, and the rapid expansion of the working fluid do not allow separation in this region unless there are discontinuities in the nozzle contour. The expansion in the supersonic bell nozzle is more efficient than in a simple straight cone of similar area ratio and length, because the wall contour is designed to minimize losses, as explained later in this section. For the past several decades most of the nozzles have been bell shaped.

A change of flow direction of a supersonic gas in an expanding wall geometry can only be achieved through expansion waves. An expansion wave occurs at a thin surface, where the flow velocity increases and changes its flow direction slightly, and where the pressure and temperature drop. These wave surfaces are at an oblique angle to the flow. As the gas passes through the throat, it undergoes a series of these expansion waves with essentially no loss of energy. In the bell-shaped nozzle shown in Fig. 3–14 these expansions occur internally in the flow between the throat and the inflection location I; the area is steadily increasing like a flare on a trumpet. The contour angle θ_i is a maximum at the inflection location. Between the inflection point I and the nozzle exit E the flow area is still increasing, but at a diminishing rate, allowing further gas expansion and additional expansion waves. However, the contour of the nozzle wall is different and the change in cross-sectional area per unit length is decreasing. The purpose of this last segment of the contoured nozzle is to have a low divergence loss as the gas leaves the nozzle exit plane. The angle at the exit θ_e is small, usually less than 10°. The difference between θ_i and θ_e is called the turn-back angle. When the gas flow is turned in the opposite direction (between points I and E) oblique compression waves will occur. These compression waves are thin surfaces where the flow undergoes a mild shock, the flow is turned, and the velocity is actually reduced slightly. Each of these multiple compression waves causes a small energy loss. By carefully determining the wall contour (by an analysis that uses a mathematical tool called the method of characteristics), it is possible to balance the oblique expansion waves with the oblique compression waves and minimize the energy loss. The analysis leading to the nozzle contour is presented in Chapter 20.33 of Ref. 3–3 and also in Refs. 3–8 to 3–11; it is based on supersonic aerodynamic flow, the method of characteristics (Ref. 3–1), and the properties of the expanding gas. Most of the rocket organizations have computer codes for this analysis. The radius of curvature or the contour shape at the throat region have an influence on the contour of the diverging bell-shaped nozzle section.

The length of a bell nozzle is usually given a fraction of the length of a reference conical nozzle with a 15° half angle. An 80% bell nozzle has a length (distance between throat plane and exit plane) that is 20% shorter than a comparable 15° cone of the same area ratio. Ref. 3–9 shows the original presentation by Rao of the method of characteristics applied to shorter bell nozzles. He also determined that a parabola was a good approximation for the bell-shaped contour curve (Ref. 3–3, Section 20.33), and parabolas have actually been used in some nozzle designs. The top part of Fig. 3–14 shows that the parabola is tangent (θ_i) at point I and has an exit angle (θ_e) at point E and a length L that has to be corrected for the curve TI. These conditions allow the parabola to be determined by simple geometric analysis or geometric drawing. A throat approach radius of 1.5 r_t and a throat expansion radius of 0.4 r_t were used. If somewhat different radii had been used, the results would have been only slightly different. The middle set of curves gives the relation between length, area ratio, and the two angles of the bell contour. The bottom set of curves gives the correction factors, equivalent to the λ factor for conical nozzles, which are to be applied to the thrust coefficient or the exhaust velocity, provided the nozzles are at optimum expansions, that is, $p_2 = p_3$.

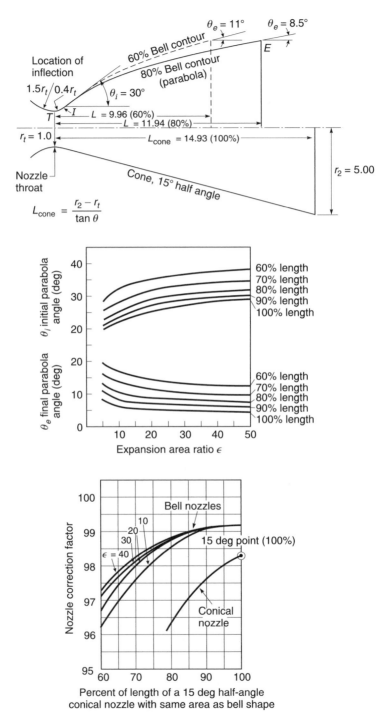

TABLE 3–4. Data on Several Bell-Shaped Nozzles

Area Ratio	10	25	50
Cone (15° *Half Angle*)			
Length (100%)[a]	8.07	14.93	22.66
Correction factor λ	0.9829	0.9829	0.9829
80% Bell Contour			
Length[a]	6.45	11.94	18.12
Correction factor λ	0.985	0.987	0.988
Approximate half angle at inflection point and exit (degrees)	25/10	30/8	32/7.5
60% Bell Contour			
Length[a]	4.84	9.96	13.59
Correction factor λ	0.961	0.968	0.974
Approximate half angle at inflection point and exit (degrees)	32.5/17	36/14	39/18

[a]The length is given in dimensionless form as a multiple of the throat radius, which is one.

Table 3–4 shows data for parabolas developed from this figure, which allow the reader to apply this method and check the results. The table shows two shortened bell nozzles and a conical nozzle, each for three area ratios. It can be seen that as the length has been decreased, the losses are higher for the shorter length and slightly higher for small nozzle area ratios. A 1% improvement in the correction factor gives about 1% more specific impulse (or thrust) and this difference can be significant in many applications. The reduced length is an important benefit, and it is usually reflected in an improvement of the vehicle mass ratio. The table and Fig. 3–14 show that bell nozzles (75 to 85% length) are just as efficient as or slightly more efficient than a longer 15° conical nozzle (100% length) at the same area ratio. For shorter nozzles (below 70% equivalent length) the energy losses due to internal oblique shock waves become substantial and such short nozzles are not commonly used today.

For solid propellant rocket motor exhausts with small solid particles in the gas (usually aluminum oxide), and for exhausts of certain gelled liquid propellants, there is an impingement of these solid particles against the nozzle wall in

---

FIGURE 3–14. Top sketch shows comparison sketches of nozzle inner wall surfaces for a 15° conical nozzle, an 80% length bell nozzle, a 60% length bell nozzle, all at an area ratio of 25. The lengths are expressed in multiples of the throat radius r_t, which is one here. The middle set of curves shows the initial angle θ_i and the exit angle θ_e as functions of the nozzle area ratio and percent length. The bottom curves show the nozzle losses in terms of a correction factor. Adapted and copied with permission of AIAA from Ref. 6–1.

the reversing curvature section between I and E in Fig. 3–14. While the gas can be turned by oblique waves to have less divergence, the particles (particularly the larger particles) have a tendency to move in straight lines and hit the walls at high velocity. The resulting abrasion and erosion of the nozzle wall can be severe, especially with the ablative and graphite materials that are commonly used. This abrasion by hot particles increases with turn-back angle. If the turn-back angle and thus also the inflection angle θ_i are reduced, the erosion can become acceptable. Typical solid rocket motors flying today have values of inflection angles between 20 and 26° and turn-back angles of 10 to 15°. In comparison, current liquid rocket engines without entrained particles have inflection angles between 27 and 50° and turn-back angles of between 15 and 30°. Therefore the performance enhancement caused by using a bell-shaped nozzle (high value of correction factor) is somewhat lower in solid rocket motors with solid particles in the exhaust.

The ideal bell-shaped nozzle (minimum loss) is long, equivalent to a conical nozzle of perhaps 10 to 12°, as seen in Fig. 3–12. It has about the same length as a full-length aerospike nozzle. This is usually too long for reasonable vehicle mass ratios.

Two-Step Nozzles. Several modifications of a bell-shaped nozzle have evolved that allow full or almost complete altitude compensation; that is, they achieve maximum performance at more than a single altitude. Figure 3–15 shows three concepts for a two-step nozzle, one that has an initial low area ratio A_2/A_t for operation at or near the earth's surface and a larger second area ratio that improves performance at high altitudes. See Ref. 3–5.

The *extendible nozzle* requires actuators, a power supply, mechanisms for moving the extension into position during flight, fastening and sealing devices. It has successfully flown in several solid rocket motor nozzles and in a few liquid engine applications, where it was deployed prior to ignition. Although only two steps are shown, there have been versions with three steps; one is shown in Fig. 11-3. As yet it has not made the change in area ratio during rocket firing. The principal concerns are a reliable rugged mechanism to move the extension into position, the hot gas seal between the nozzle sections, and the extra weight involved.

The *droppable insert concept* avoids the moving mechanism and gas seal but has a potential stagnation temperature problem at the joint. It requires a reliable release mechanism, and the ejected insert creates flying debris. To date it has little actual test experience. See Ref. 3–12.

The *dual bell nozzle concept* uses two shortened bell nozzles combined into one with a bump or inflection point between them, as shown in Fig. 3–15. During ascent it functions first at the lower area ratio, with separation occurring at the inflection point. As altitude increases and the gas expands further, the flow attaches itself downstream of this point, with the flow filling the full nozzle exit section and operating with the higher area ratio at higher performance. There is a small performance penalty for a compromised bell nozzle

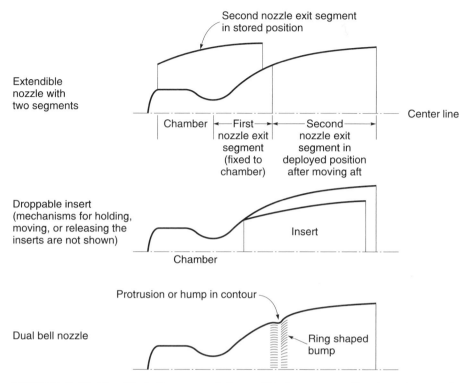

FIGURE 3–15. Simplified diagrams of three altitude-compensating two-step nozzle concepts.

contour with a circular bump. To date there has been little experience with this concept.

Nozzles with Aerodynamic Boundaries

The group of two-step nozzle concepts described above corresponds to the performance represented by upper portions of the two fixed area ratio nozzle curves shown in Fig. 3–10; the performance of a continuously varying nozzle with full altitude compensation is shown by the dashed curve. When integrated over the flight time, the extra performance is important for high velocity missions such as the single stage to orbit application. The three nozzles shown on the right side of Fig. 3–12 offer full altitude compensation and are discussed next. Refs. 3–5 and 3–8 give more information.

The *plug nozzle* or *aerospike nozzle* has an annular doughnut-shaped chamber with an annular nozzle slot. An alternate version has a number of individual small chambers (each with low area ratio short nozzles, a round throat, and a rectangular exit) arranged in a circle around a common plug or spike. The outside aerodynamic boundary of the gas flow in the divergent section of the nozzle is the interface between the hot gas and the ambient air; there is no outer wall as in a conical or bell-shaped nozzle. As the external or ambient pressure is reduced during the ascending flight, this gas boundary expands outward, causes a change in pressure distribution on the central spike, and allows an automatic and continuous altitude compensation. The aerospike contour with the minimum flow losses turns out to be very long, similar in length to an optimum bell nozzle as shown in Figs. 3–12 and 3–13. The mass flow per unit exit area is relatively uniform over the cross section and the divergence losses are minimal.

If the central plug is cut off or truncated and the wall contour is slightly altered, then the nozzle will be very short, as shown in Fig. 3–13; it will have some internal supersonic waves and will show a small but real loss in thrust compared to a nozzle with a full central spike. The pressure distribution and the heat transfer intensity vary on the inner contoured spike wall surface. Figure 8–14 shows a typical pressure distribution over the contoured spike surface at high and low altitudes.

The pressure in the recirculating trapped gas of the subsonic region below the bottom plate also exerts a thrust force. The losses caused by the cut-off spike can be largely offset by injecting a small amount of the gas flow (about 1% of total flow) through this base plate into the recirculating region, thus enhancing the back pressure on the base plate. The advantages of the truncated aerospike are short length (which helps to reduce the length and mass of the flight vehicle), full altitude compensation, no flow separation from the wall at lower altitudes, and ease of vehicle/engine integration for certain vehicle configurations.

The *linear aerospike nozzle* is a variation of the round axisymmetric aerospike nozzle. Basically, it is an unrolled version of the circular configuration. It is explained further in Chapter 8.2.

In the *expansion deflection nozzle* (Fig. 3–12) the flow from the chamber is directed radially outward away from the nozzle axis. The flow is turned on a curved contour outer diverging nozzle wall. The nozzle has been shortened and has some internal oblique shock wave losses. The hot gas flow leaving the chamber expands around a central plug. The aerodynamic interface between the ambient air and gas flow forms an inner boundary of the gas flow in the diverging nozzle section. As the ambient pressure is reduced, the hot gas flow fills more and more of the nozzle diverging section. Altitude compensation is achieved by this change in flow boundary and by changes in the pressure distribution on the outer walls.

Multiple Nozzles. If a single large nozzle is replaced by a cluster of smaller nozzles on a solid motor (all at the same cumulative thrust), then it is possible to reduce the nozzle length. Similarly, if a single large thrust chamber of a liquid engine is replaced by several smaller thrust chambers, the nozzle length will be shorter, reducing the vehicle length and thus the vehicle structure and inert mass. Russia has pioneered a set of four thrust chambers, each with 25%

of the total thrust, assembled next to each other and fed from the same liquid propellant feed system. This quadruple thrust chamber arrangement has been used effectively on many large Russian space launch vehicles and missiles. As seen in Fig. 3–13, this cluster is about 30% shorter than a single large thrust chamber. The vehicle diameter at the cluster nozzle exit is somewhat larger, the vehicle drag is somewhat higher, and there is additional engine complexity and engine mass.

3.5. REAL NOZZLES

In a real nozzle the flow is really two-dimensional, but axisymmetric. For simple single nozzle shapes the temperatures and velocities are not uniform over any one section and are usually higher in the central region and lower near the periphery. For example, the surface where the Mach number is one is a plane at the throat for an ideal nozzle; for two-dimensional flow it is typically a slightly curved surface somewhat downstream of the throat. If the velocity distribution is known, the average value of v_2 can be determined for an axisymmetric nozzle as a function of the radius r.

$$(v_2)_{\text{average}} = \frac{2\pi}{A_2} \int_0^{r_2} v_2 r \; dr \tag{3–35}$$

The 11 assumptions and simplifications listed in Section 1 of this chapter are only approximations that allow relatively simple algorithms and simple mathematical solutions to the analysis of real rocket nozzle phenomena. For most of these assumptions it is possible either (1) to use an empirical correction factor (based on experimental data) or (2) to develop or use a more accurate algorithm, which involves more detailed understanding and simulation of energy losses, the physical or chemical phenomena, and also often a more complex theoretical analysis and mathematical treatment. Some of these approaches are mentioned briefly in this section.

Compared to an ideal nozzle, the real nozzle has energy losses and energy that is unavailable for conversion into kinetic energy of the exhaust gas. The principal losses are listed below and several of these are discussed in more detail.

1. The *divergence of the flow* in the nozzle exit sections causes a loss, which varies as a function of the cosine of the divergence angle as shown by Eq. 3–34 and Table 3–3 for conical nozzles. The losses can be reduced for bell-shaped nozzle contours.

2. Small chamber or port area cross sections relative to the throat area or *low nozzle contraction ratios* A_1/A_t cause pressure losses in the chamber and reduce the thrust and exhaust velocity slightly. See Table 3–2.

3. Lower flow velocity in the *boundary layer* or wall friction can reduce the effective exhaust velocity by 0.5 to 1.5%.

4. *Solid particles* or *liquid roplets* in the gas can cause losses up to 5%, as described below.

5. *Unsteady combustion* and oscillating flow can account for a small loss.

6. *Chemical reactions in nozzle flow* change gas properties and gas temperatures, giving typically a 0.5% loss. See Chapter 5.

7. There is lower performance during *transient pressure operation*, for example during start, stop, or pulsing.

8. For uncooled nozzle materials, such as fiber reinforced plastics or carbon, the gradual *erosion* of the *throat region* increases the throat diameter by perhaps 1 to 6% during operation. In turn this will reduce the chamber pressure and thrust by about 1 to 6% near the end of the operation and cause a slight reduction in specific impulse of less than 0.7%.

9. *Non-uniform gas composition* can reduce performance (due to incomplete mixing, turbulence, or incomplete combustion regions).

10. Using *real gas properties* can at times change the gas composition, the value of k and $\mathfrak{M}$, and this can cause a small loss in performance, say 0.2 to 0.7%.

11. Operation at non-optimum nozzle expansion area ratio can reduce thrust and specific impulse. There is no loss if the vehicle always flies at the altitude for optimum nozzle expansion ($p_2 = p_3$). If it flies with a fixed nozzle area ratio at higher or lower altitudes, then there is a loss (during a portion of the flight) by up to 15% in thrust compared to a nozzle with altitude compensation, as can be seen in Figs. 3–7 and 3–8. It also reduces performance by 1 to 5%.

Boundary Layer

Real nozzles have a viscous *boundary layer* next to the nozzle walls, where the gas velocities are much lower than the free-stream velocities in the inviscid flow regions. An enlarged schematic view of a boundary layer is shown in Fig. 3–16. Immediately next to the wall the flow velocity is zero and then the boundary layer can be considered as being built up of successive annular-shaped thin layers of increasing velocity until the free-stream velocity is reached. The low-velocity flow close to the wall is laminar and subsonic, but in the higher-velocity regions of the boundary layer the flow is supersonic and can become turbulent. The local temperature in part of the boundary layer can be substantially higher than the free-stream temperature because of the conversion of kinetic energy into thermal energy as the local velocity is slowed down and as heat is created by viscous friction. The layer right next to the wall will be cooler because of heat transfer to the wall. The gaseous

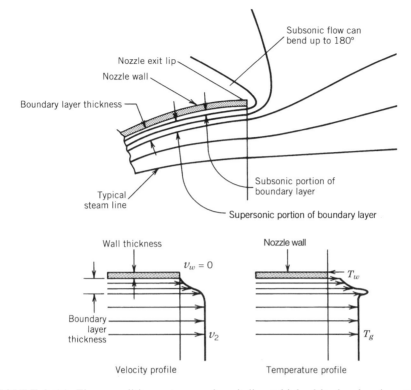

FIGURE 3–16. Flow conditions at a nozzle exit lip at high altitude, showing streamlines, boundary layer, velocity and temperature profiles.

boundary layer has a profound effect on the overall heat transfer to nozzle and chamber walls. It also has an effect on the rocket performance, particularly in applications with relatively long nozzles with high nozzle area ratios, where a relatively high proportion of the total mass flow (2 to 25%) can be in the lower-velocity region of the boundary layer. The high gradients in pressure, temperature, or density and the changes in local velocity (direction and magnitude) influence the boundary layer. Scaling laws for boundary layer phenomena have not been reliable.

Theoretical approaches to boundary layer performance effects can be found in Chapters 26 to 28 of Reference 3–1 and in Reference 1–1. A truly satisfactory theoretical analysis of boundary layers in rocket nozzles has not yet been developed. Fortunately, the overall effect of boundary layers on rocket performance has been small. For most rocket nozzles the loss seldom exceeds 1% of specific impulse.

Multiphase Flow

In some rockets the gaseous working fluid contains many small liquid droplets and/or solid particles that must be accelerated by the gas. They give up heat to the gas during the expansion in a nozzle. This, for example, occurs with solid propellants (see Chapter 12) or some gelled liquid propellants (Chapter 7), which contain aluminum powder that forms small oxide particles in the exhaust. It can also occur with ion oxide catalysts, or propellants containing beryllium, boron, or zirconium.

In general, if the particles are very small (typically with diameters of 0.005 mm or less), they will have almost the same velocity as the gas and will be in thermal equilibrium with the nozzle gas flow. Thus, as the gases give up kinetic energy to accelerate the particles, they gain thermal energy from the particles. As the particle diameters become larger, the mass (and thus the inertia) of the particle increases as the cube of its diameter; however, the drag force increases only as the square of the diameter. Larger particles therefore do not move as fast as the gas and do not give heat to the gas as readily as do smaller particles. The larger particles have a lower momentum than an equivalent mass of smaller particles and they reach the nozzle exit at a higher temperature than the smaller particles, thus giving up less thermal energy.

It is possible to derive a simple theoretical approach for correcting the performance (I_s, c, or c^*) as shown below and as given in Refs. 3–13 and 3–14. It is based on the assumption that specific heats of the gases and the particles are constant throughout the nozzle flow, that the particles are small enough to move at the same velocity as the gas and are in thermal equilibrium with the gas, and that particles do not exchange mass with the gas (no vaporization or condensation). Expansion and acceleration occur only in the gas and the volume occupied by the particles is negligibly small compared to the gas volume. If the amount of particles is small, the energy needed to accelerate the particles can be neglected. There are no chemical reactions.

The enthalpy h, the specific volume V, and the gas constant R can be expressed as functions of the particle fraction β, which is the mass of particles (liquid and/or solid) divided by the total mass. Using the subscripts g and s to refer to the gas or solid state, the following relationships then apply:

$$h = (1 - \beta)(c_p)_g T + \beta c_s T \tag{3–36}$$

$$V = V_g(1 - \beta) \tag{3–37}$$

$$p = R_g T / V_g \tag{3–38}$$

$$R = (1 - \beta)R_g \tag{3–39}$$

$$k = \frac{(1 - \beta)c_p + \beta c_s}{(1 - \beta)c_v + \beta c_s} \tag{3–40}$$

These relations are then used in the formulas for simple one-dimensional nozzle flow, such as Eq. 2–16, 3–15, or 3–32. The values of specific impulse or characteristic velocity will decrease as β, the percent of particles, is increased. For very small particles (less than 0.01 mm in diameter) and small values of β (less than 6%) the loss in specific impulse is often less than 2%. For larger particles (over 0.015 mm diameter) and larger values of β this theory is not helpful and the specific impulse can be 10 to 20% less than the I_s value without flow lag. The actual particle sizes and distribution depend on the specific propellant, the combustion, the particular particle material, and the specific rocket propulsion system, and usually have to be measured (see Chapters 12 and 18). Thus adding a metal, such as aluminum, to a solid propellant will increase the performance only if the additional heat release can increae the combustion temperature T_1 sufficiently so that it more than offsets the decrease caused by particles in the exhaust.

With very-high-area-ratio nozzles and a low nozzle exit pressure (high altitude or space vacuum) it is possible to condense some of the propellant ingredients that are normally gases. As the temperature drops sharply in the nozzle, it is possible to condense gaseous species such as H_2O, CO_2, or NH_3 and form liquid droplets. This causes a decrease in the gas flow per unit area and the transfer of the latent heat of vaporization to the remaining gas. The overall effect on performance is small if the droplet size is small and the percent of condensed gas mass is moderate. It is also possible to form a solid phase and precipitate fine particles of snow (H_2O) or frozen fog of other species.

Other Phenomena and Losses

The *combustion process* is really not steady. Low- and high-frequency oscillations in chamber pressure of up to perhaps 5% of rated value are usually considered as smooth-burning and relatively steady flow. Gas properties (k, $\mathfrak{M}$, c_p) and flow properties (v, V, T, p, etc.) will also oscillate with time and will not necessarily be uniform across the flow channel. These properties are therefore only "average" values, but it is not always clear what kind of an average they are. The energy loss due to nonuniform unsteady burning is difficult to assess theoretically. For smooth-burning rocket systems they are negligibly small, but they become significant for larger-amplitude oscillations.

The *composition of the gas* changes somewhat in the nozzle, chemical reactions occur in the flowing gas, and the assumption of a uniform or "frozen" equilibrium gas composition is not fully valid. A more sophisticated analysis for determining performance with changing composition and changing gas properties is described in Chapter 5. The thermal energy that is carried out of the nozzle ($\dot{m} c_p T_2$) is unavailable for conversion to useful propulsive (kinetic) energy, as is shown in Fig. 2–3. The only way to decrease this loss is to reduce the nozzle exit temperature T_2 (larger nozzle area ratio), but even then it is a large loss.

When the operating durations are short (as, for example, with antitank rockets or pulsed attitude control rockets which start and stop repeatedly), the start and stop *transients* are a significant portion of the total operating time. During the transient periods of start and stop the average thrust, chamber pressure, or specific impulse will be lower in value than those same parameters at steady full operating conditions. This can be analyzed in a step-by-step process. For example, during startup the amount of propellant reacting in the chamber has to equal the flow of gas through the nozzle plus the amount of gas needed to fill the chamber to a higher pressure; alternatively, an empirical curve of chamber pressure versus time can be used as the basis of such a calculation. The transition time is very short in small, low-thrust propulsion systems, perhaps a few milliseconds, but it can be longer (several seconds) for large propulsion systems.

Performance Correction Factors

In this section we discuss semiempirical correction factors that have been used to estimate the test performance data from theoretical, calculated performance values. An understanding of the theoretical basis also allows correlations between several of the correction factors and estimates of the influence of several parameters, such as pressure, temperature, or specific heat ratio.

The *energy conversion efficiency* is defined as the ratio of the kinetic energy per unit of flow of the actual jet leaving the nozzle to the kinetic energy per unit of flow of a hypothetical ideal exhaust jet that is supplied with the same working substance at the same initial state and velocity and expands to the same exit pressure as the real nozzle. This relationship is expressed as

$$e = \frac{(v_2)_a^2}{(v_2)_i^2} = \frac{(v_2)_a^2}{(v_1)_a^2 + c_p(T_1 - T_2)} \tag{3-41}$$

where e denotes the energy conversion efficiency, v_1 and v_2 the velocities at the nozzle inlet and exit, and $c_p T_1$ and $c_p T_2$ the respective enthalpies for an ideal isentropic expansion. The subscripts a and i refer to actual and ideal conditions, respectively. For many practical applications, $v_1 \rightarrow 0$ and the square of the expression given in Eq. 3–16 can be used for the denominator.

The *velocity correction factor* ζ_v is defined as the square root of the energy conversion efficiency $\sqrt{e}$. Its value ranges between 0.85 and 0.99, with an average near 0.92. This factor is also approximately the ratio of the actual specific impulse to the ideal or theoretical specific impulse.

The *discharge correction factor* ζ_d is defined as the ratio of the mass flow rate in a real rocket to that of an ideal rocket that expands an identical working fluid from the same initial conditions to the same exit pressure (Eq. 2–17).

$$\zeta_d = (\dot{m}_a/\dot{m}_i) = \dot{m}_a(c/F_i) \tag{3-42}$$

and, from Eq. 3–24,

$$\zeta_d = \frac{\dot{m}_a\sqrt{kRT_1}}{A_t p_1 k\sqrt{[2/(k+1)]^{(k+1)/(k-1)}}}$$

The value of this discharge correction factor is usually larger than 1 (1.0 to 1.15); the actual flow is larger than the theoretical flow for the following reasons:

1. The molecular weight of the gases usually increases slightly when flowing through a nozzle, thereby changing the gas density.
2. Some heat is transferred to the nozzle walls. This lowers the temperature in the nozzle, and increases the density and mass flow slightly.
3. The specific heat and other gas properties change in an actual nozzle in such a manner as to slightly increase the value of the discharge correction factor.
4. Incomplete combustion can increase the density of the exhaust gases.

The actual thrust is usually lower than the thrust calculated for an ideal rocket and can be found by an empirical *thrust correction* factor ζ_F:

$$F_a = \zeta_F F_i = \zeta_F C_F p_1 A_t = \zeta_F c_i \dot{m}_i \qquad (3\text{--}43)$$

where

$$\zeta_F = \zeta_v \zeta_d = F_a/F_i \qquad (3\text{--}44)$$

Values of ζ_F fall between 0.92 and 1.00 (see Eqs. 2–6 and 3–31). Because the thrust correction factor is equal to the product of the discharge correction factor and the velocity correction factor, any one can be determined if the other two are known.

Example 3–7. Design a rocket nozzle to conform to the following conditions:

Chamber pressure	20.4 atm = 2.068 MPa
Atmospheric pressure	1.0 atm
Chamber temperature	2861 K
Mean molecular mass of gases	21.87 kg/kg-mol
Ideal specific impulse	230 sec (at operating conditions)
Specific heat ratio	1.229
Desired thrust	1300 N

Determine the following: nozzle throat and exit areas, respective diameters, actual exhaust velocity, and actual specific impulse.

SOLUTION. The theoretical thrust coefficient is found from Eq. 3–30. For optimum conditions $p_2 = p_3$. By substituting $k = 1.229$ and $p_1/p_2 = 20.4$, the thrust coefficient is $C_F = 1.405$. This value can be checked by interpolation between the values of C_F

obtained from Figs. 3–7 and 3–8. The throat area is found using $\zeta_F = 0.96$, which is based on test data.

$$A_t = F/(\zeta_F C_F p_1) = 1300/(0.96 \times 1.405 \times 2.068 \times 10^6) = 4.66 \text{ cm}^2$$

The throat diameter is then 2.43 cm. The area expansion ratio can be determined from Fig. 3–5 or Eq. 3–25 as $\epsilon = 3.42$. The exit area is

$$A_2 = 4.66 \times 3.42 = 15.9 \text{ cm}^2$$

The exit diameter is therefore 4.50 cm. The theoretical exhaust velocity is

$$v_2 = I_s g_0 = 230 \times 9.81 = 2256 \text{ m/sec}$$

By selecting an empirical velocity correction factor ζ_v such as 0.92 (based on prior related experience), the actual exhaust velocity will be equal to

$$(v_2)_a = 2256 \times 0.92 = 2076 \text{ m/sec}$$

Because the specific impulse is proportional to the exhaust velocity, its actual value can be found by multiplying the theoretical value by the velocity correction factor ζ_v.

$$(I_s)_a = 230 \times 0.92 = 212 \text{ sec}$$

3.6. FOUR PERFORMANCE PARAMETERS

In using values of thrust, specific impulse, propellant flow, and other performance parameters, one must be careful to specify or qualify the conditions under which a specific number is presented. There are at least four sets of performance parameters and they are often quite different in concept and value, even when referring to the same rocket propulsion system. Each performance parameter, such as F, I_s, c, v_2 and/or $\dot{m}$, should be accompanied by a clear definition of the conditions under which it applies, namely:

a. Chamber pressure; also, for slender chambers, the location where this pressure prevails (e.g., at nozzle entrance).
b. Ambient pressure or altitude or space (vacuum).
c. Nozzle expansion area ratio and whether this is an optimum.
d. Nozzle shape and exit angle (see Table 3–3).
e. Propellants, their composition or mixture ratio.
f. Key assumptions and corrections made in the calculations of the theoretical performance: for example, was frozen or shifting equilibrium used in the analysis? (This is described in Chapter 5.)
g. Initial temperature of propellants.

1. *Theoretical performance values* are defined in Chapters 2, 3, and 5 and generally apply to ideal rockets, but usually with some corrections. Most organizations doing nozzle design have their own computer programs, often different programs for different nozzle designs, different thrust levels, or operating durations. Most are two dimensional and correct for the chemical reactions in the nozzle using real gas properties, and correct for divergence. Many also correct for one or more of the other losses mentioned above. For example, programs for solid propellant motor nozzles can include losses for throat erosion and multiphase flow; for liquid propellant engines it may include two or more concentric zones, each at different mixtures ratios and thus with different gas properties. Nozzle wall contour analysis with expansion and compression waves may use a finite element analysis and/or a method of characteristics approach. Some of the more sophisticated programs include viscous boundary layer effects and heat transfer to the walls. Typically these computer simulation programs are based on computer fluid dynamics finite element analyses and on the basic Navier–Stokes relationships. Most companies also have simpler, one-dimensional computer programs which may include one or more of the above corrections; they are used frequently for preliminary estimates or proposals.

2. *Delivered*, that is, *actually measured, performance values* are obtained from static tests or flight tests of full-scale propulsion systems. Again, the conditions should be explained (e.g., define p_1, A_2/A_t, T_1, etc.) and the measured values should be corrected for instrument deviations, errors, or calibration constants. Flight test data need to be corrected for aerodynamic effects, such as drag. Often empirical coefficients, such as the thrust correction factor, the velocity correction factor, and the mass discharge flow correction factors are used to convert the theoretical values of item 1 above to approximate actual values and this is often satisfactory for preliminary estimates. Sometimes subscale propulsion systems are used in the development of new rocket systems and then scale factors are used to correct the measured data to full-scale values.

3. *Performance values at standard conditions* are corrected values of items 1 and 2 above. These standard conditions are generally rigidly specified by the customer. Usually they refer to conditions that allow ready evaluation or comparison with reference values and often they refer to conditions that can be easily measured and/or corrected. For example, to allow a good comparison of specific impulse for several propellants or rocket propulsion systems, the values are often corrected to the following standard conditions (see Examples 3–4 and 3–5):

a. $p_1 = 1000$ psia or 6.894×10^6 Pa.

b. $p_2 = p_3 = 14.69$ psia (sea level) or 1.0132×10^5 Pa or 0.10132 MPa.

c. Area ratio is optimum, $p_2 = p_3$.

d. Nozzle divergence half angle $\alpha = 15°$ for conical nozzles, or some agreed-upon value.

e. Specific propellant, its design mixture ratio and/or propellant composition.

f. Propellant initial temperature: 21°C (sometimes 20 or 25°C) or boiling temperature, if cryogenic.

A rocket propulsion system is generally designed, built, tested, and delivered in accordance with some predetermined requirements or *specifications*, usually in formal documents often called the *rocket engine or rocket motor specifications*. They define the performance as shown above and they also define many other requirements. More discussion of these specifications is given as a part of the selection process for propulsion systems in Chapter 17.

4. Rocket manufacturers are often required by their customers to deliver rocket propulsion systems with a *guaranteed minimum performance*, such as minimum F or I_s or both. The determination of this value can be based on a nominal value (items 1 or 2 above) diminished by all likely losses, including changes in chamber pressure due to variation of pressure drops in injector or pipelines, a loss due to nozzle surface roughness, propellant initial ambient temperatures, manufacturing variations from rocket to rocket (e.g., in grain volume, nozzle dimensions, or pump impeller diameters, etc.). This minimum value can be determined by a probabilistic evaluation of these losses and is then usually validated by actual full-scale static and flights tests.

3.7. NOZZLE ALIGNMENT

When the thrust line or direction does not intersect the center of mass of a flying vehicle, a turning moment will tend to rotate a vehicle in flight. Turning moments are desirable and necessary for the controlled turning or attitude control of a vehicle as is routinely done by means of the deflection of the thrust vector, aerodynamic fins, or by separate attitude control rocket engines. However, this turning is undesirable when its magnitude or direction is not known; this happens when a fixed nozzle of a major propulsion system has its thrust axis misaligned. A large high-thrust booster rocket system, even if misaligned by a very small angle (less than $\frac{1}{2}$°), can cause major upsetting turning moments for the firing duration. If not corrected or compensated, such a small misalignment can cause the flight vehicle to tumble and/or deviate from the intended flight path. For this moment not to exceed the vehicle's compensating attitude control capability, it is necessary to align the nozzle axis of all propulsion systems with fixed (non-gimbal) nozzles very accurately. Normally, the geometric axis of the nozzle diverging exit surface geometry is taken to be the thrust axis. Special alignment fixtures are usually needed to orient the nozzle axis to be within less than ±0.25° of the intended line to the vehicle's center of gravity and to position the center of a large nozzle throat to be on the vehicle centerline, say within 1 or 2 mm. See Ref. 3–15.

There are other types of misalignments: (1) irregularities in the nozzle geometry (out of round, protuberances, or unsymmetrical roughness in the surface); (2) transient misalignments during start to stop; (3) uneven deflection of the propulsion system or vehicle structure under load; and (4) irregularities in the gas flow (faulty injector, uneven burning rate in solid propellants). For simple unguided rocket vehicles it has been customary to rotate or spin the vehicle to prevent the misalignment from being in one direction only or to even out the misalignment during powered flight.

In the cramped volume of spacecraft or upper stage launch vehicles, it is sometimes not possible to accommodate the full length of a large-area-ratio nozzle within the available vehicle envelope. In this case the nozzles are cut off at an angle at the vehicle surface, which allows a compact installation. Figure 3–17 shows a diagram of two (out of four) roll control thrusters whose nozzle exit conforms to the vehicle contour. The thrust direction of a *scarfed nozzle* is

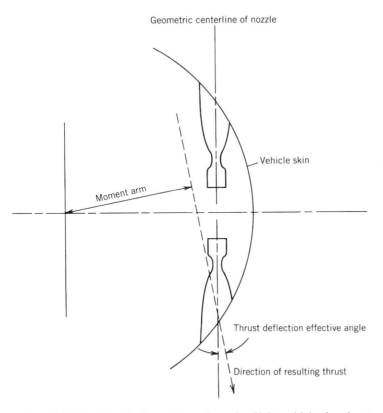

FIGURE 3–17. Simplified partial section of a flight vehicle showing two attitude control thrusters with scarfed nozzles to fit a cylindrical vehicle envelope.

no longer on the nozzle axis centerline, as it is with fully symmetrical nozzles, and the nozzle exit flow will not be axisymmetric. Reference 3-16 shows how to estimate the performance and thrust direction of scarfed nozzles

3.8. VARIABLE THRUST

Only a few applications require a change in thrust during flight. Equations 3–30, 3–24, and 3–31 show that the thrust is directly proportional to the throat area A_t, the chamber pressure p_1, or the mass flow rate $\dot{m}$, but it is a weak function of C_F, which in turn depends on k, the altitude, a pressure ratio, and A_2/A_t. These equations show how the thrust may be varied and imply how other performance parameters may be affected by such variation. For liquid propellant rockets the mass flow to the chamber can be decreased (by throttling valves in the propellant feed system) while the chamber geometry and the nozzle throat area are unchanged. The reduced mass flow will cause an almost linear decrease in p_1 and thus an almost linear decrease of F. The combustion temperature does change slightly but it does not enter into the above relations. The specific impulse would also decrease slightly. Thus, there is a small performance penalty for throttling the thrust. A two-to-one thrust decrease has been achieved with throttle valves in a liquid propellant rocket engine. Random throttling of liquid propellant engines and their design features are discussed in Chapter 8.5.

Another way of varying the thrust is to change the throat area simultaneously with throttling the flow (by inserting a moveable contoured pintle or tapered plug into the nozzle); in this case the chamber pressure p_1 can remain reasonably constant. This throttling method has been used on liquid propellant engines (e.g., a ten-to-one thrust change on a moon landing rocket) and in a few experimental solid propellant motors.

Random thrust control requires a control system and special hardware; one example is discussed in Chapter 10.5. Random throttling of production solid propellant motors has not been achieved as yet in flight. A repeatable, programmed variation of thrust for solid propellants is possible and is discussed in Chapter 11.3. For solid propellants, a predetermined variation of mass flow rate has been achieved by clever grain geometric design, which changes the burning area at different stages during the operation. This is useful in many air-launched military rockets. Liquid propellant rockets are the most appropriate choice for randomly variable thrust rockets, as has been amply demonstrated in missions such as the lunar landings.

PROBLEMS

1. Certain experimental results indicate that the propellant gases of a liquid oxygen–gasoline reaction have a mean molecular mass of 23.2 kg/kg-mol and a specific heat ratio of 1.22. Compute the specific heat at constant pressure and at constant volume, assuming a perfect gas.

2. The actual conditions for an optimum expansion nozzle operating at sea level are given below. Calculate v_2, T_2, and C_F. The mass flow $\dot{m} = 3.7$ kg/sec; $p_1 = 2.1$ MPa; $T_1 = 2585°$K; $\mathfrak{M} = 18.0$ kg/kg-mol; and $k = 1.30$.

3. A certain nozzle expands a gas under isentropic conditions. Its chamber or nozzle entry velocity equals 70 m/sec, its final velocity 1500 m/sec. What is the change in enthalpy of the gas? What percentage of error is introduced if the initial velocity is neglected?

4. Nitrogen at $500°$C ($k = 1.38$, molecular mass is 28.00) flows at a Mach number of 2.73. What are its actual and its acoustic velocity?

5. The following data are given for an optimum rocket:

Average molecular mass	24 kg/kg-mol
Chamber pressure	2.533 MPa
External pressure	0.090 MPa
Chamber temperature	2900 K
Throat area	0.00050 m^2
Specific heat ratio	1.30

Determine (a) throat velocity; (b) specific volume at throat; (c) propellant flow and specific impulse; (d) thrust; (e) Mach number at throat.

6. Determine the ideal thrust coefficient for Problem 5 by two methods.

7. A certain ideal rocket with a nozzle area ratio of 2.3 and a throat area of 5 in.^{2} delivers gases at $k = 1.30$ and $R = 66$ ft-lbf/lbm-°R at a design chamber pressure of 300 psia and a constant chamber temperature of 5300 R against a back pressure of 10 psia. By means of an appropriate valve arrangement, it is possible to throttle the propellant flow to the thrust chamber. Calculate and plot against pressure the following quantities for 300, 200, and 100 psia chamber pressure: (a) pressure ratio between chamber and atmosphere; (b) effective exhaust velocity for area ratio involved; (c) ideal exhaust velocity for optimum and actual area ratio; (d) propellant flow; (e) thrust; (f) specific impulse; (g) exit pressure; (h) exit temperature.

8. For an ideal rocket with a characteristic velocity $c^* = 1500$ m/sec, a nozzle throat diameter of 18 cm, a thrust coefficient of 1.38, and a mass flow rate of 40 kg/sec, compute the chamber pressure, the thrust, and the specific impulse.

9. For the rocket unit given in Example 3–2 compute the exhaust velocity if the nozzle is cut off and the exit area is arbitrarily decreased by 50%. Estimate the losses in kinetic energy and thrust and express them as a percentage of the original kinetic energy and the original thrust.

10. What is the maximum velocity if the nozzle in Example 3–2 was designed to expand into a vacuum? If the expansion area ratio was 2000?

11. Construction of a variable-area nozzle has often been considered to make the operation of a rocket thrust chamber take place at the optimum expansion ratio at any altitude. Because of the enormous design difficulties of such a device, it has never been successfully realized. Assuming that such a mechanism can eventually be constructed, what would have to be the variation of the area ratio with altitude (plot up to 50 km) if such a rocket had a chamber pressure of 20 atm? Assume that $k = 1.20$

12. Design a supersonic nozzle to operate at 10 km altitude with an area ratio of 8.0. For the hot gas take $T_0 = 3000$ K, $R = 378$ J/kg–K and $k = 1.3$. Determine the exit Mach number, exit velocity, and exit temperature, as well as the chamber pressure. If this chamber pressure is doubled, what happens to the thrust and the exit velocity? Assume no change in gas properties. How close to optimum nozzle expansion is this nozzle?

13. The German World War II A–4 propulsion system had a sea level thrust of 25,400 kg and a chamber pressure of 1.5 MPa. If the exit pressure is 0.084 MPa and the exit diameter 740 mm, what is the thrust at 25,000 m?

14. Derive Eq. 3–34. (*Hint*: Assume that all the mass flow originates at the apex of the cone.) Calculate the nozzle angle correction factor for a conical nozzle whose divergence half angle is 13°.

15. For Example 3–2, determine (*a*) the actual thrust; (*b*) the actual exhaust velocity; (*c*) the actual specific impulse; (*d*) the velocity correction factor. Assume that the thrust correction factor is 0.985 and the discharge correction factor is 1.050.

16. An ideal rocket has the following characteristics:

Chamber pressure	27.2 atm
Nozzle exit pressure	3 psia
Specific heat ratio	1.20
Average molecular mass	21.0 lbm/lb-mol
Chamber temperature	4200°F

Determine the critical pressure ratio, the gas velocity at the throat, the expansion area ratio, and the theoretical nozzle exit velocity.
Answers: 0.5645; 3470 ft/sec; 14; and 8570 ft/sec.

17. For an ideal rocket with a characteristic velocity c^* of 1220 m/sec, a mass flow rate of 73.0 kg/sec, a thrust coefficient of 1.50, and a nozzle throat area of 0.0248 m², compute the effective exhaust velocity, the thrust, the chamber pressure, and the specific impulse.
Answers: 1830 m/sec; 133,560 N; 3.590 × 10⁶ N/m²; 186.7 sec.

18. Derive equations 3–24 and 3–25.

19. A propulsion system with a thrust of 400,000 N is expected to have a maximum thrust misalignment α of ±0.50 degrees and a horizontal off-set d of the thrust vector of 0.125 in. as shown in this sketch. One of four small reaction control thrust chambers will be used to counteract the disturbing torque. What should be its maximum thrust level and best orientation? Distance of vernier gymbal to CG is 7 m.

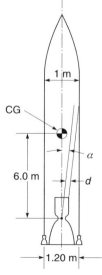

SYMBOLS

A	area, m^2 (ft^2)
c	effective exhaust velocity, m/sec (ft/sec)
c_p	specific heat at constant pressure, J/kg-K (Btu/lbm-R)
c_s	specific heat of solid, J/kg-K (Btu/lbm-R)
c_v	specific heat at constant volume, J/kg-K (Btu/lbm-R)
c^*	characteristic velocity, m/sec (ft/sec)
C_F	thrust coefficient
C_D	discharge coefficient $(1/c^*)$, sec/m (sec/ft)
d	total derivative
D	diameter, m (ft)
e	energy conversion efficiency
F	thrust, N (lbf)
g_0	standard sea level gravitational acceleration, 9.8066 m/sec^2 (32.174 ft/sec^2)
h	enthalpy per unit mass, J/kg (Btu/lbm)
I_s	specific impulse, sec or N-sec^3/kg-m (lbf-sec/lbm)
J	mechanical equivalent of heat; $J = 4.186$ J/cal in SI units or 1 Btu = 777.9 ft-lbf
k	specific heat ratio
L	length of nozzle, m (ft)
$\dot{m}$	mass flow rate, kg/sec (lbm/sec)
M	mach number
$\mathfrak{M}$	molecular mass, kg/kg-mol (or molecular weight, lbm/lb-mol)
n_i	molar fraction of species i
p	pressure, N/m^2 (lbf /ft^2 or lbf/in.^{2})
R	gas constant per unit weight, J/kg-K (ft-lbf/lbm-R) $(R = R'/\mathfrak{M})$
R'	universal gas constant, 8314.3 J/kg mol-K (1544 ft-lb/lb mol-R)
T	absolute temperature, K (R)
v	velocity, m/sec (ft/sec)
V	specific volume, m^3/kg (ft^3/lbm)
$\dot{w}$	propellant weight flow rate, N/sec (lbf/sec)

Greek Letters

α	half angle of divergent conical nozzle section
β	mass fraction of solid particles
ϵ	area ratio A_2/A_t
ζ_d	discharge correction factor
ζ_F	thrust correction factor
ζ_v	velocity correction factor
λ	divergence angle correction factor for conical nozzle exit

Subscripts

a	actual
g	gas
i	ideal, or a particular species in a mixture
max	maximum
opt	optimum nozzle expansion
s	solid
sep	point of separation
t	throat
x	any plane within rocket nozzle
y	any plane within rocket nozzle
0	stagnation or impact condition
1	nozzle inlet or chamber
2	nozzle exit
3	atmospheric or ambient

REFERENCES

3–1. A. H. Shapiro, *The Dynamics and Thermodynamics of Compressible Fluid Flow*, Vols. 1 and 2, The Ronald Press Company, New York, 1953 and M. J. Zucrow and J. D. Hoffman, *Gas Dynamics*, Vols. I and II, John Wiley & Sons, 1976 (has section on nozzle analysis by method of characteristics).

3–2. M. J. Moran and H. N. Shapiro, *Fundamentals of Engineering Thermodynamics*, Third edition, John Wiley & Sons, 1996; also additional text, 1997.

3–3. H. H. Koelle (Ed.), *Handbook of Astronautical Engineering*, McGraw-Hill Book Company, New York, 1961.

3–4. T. V. Nguyen and J. L. Pieper, "Nozzle Separation Prediction Techniques and Controlling Techniques," AIAA paper, 1996.

3–5. G. Hagemann, H. Immich, T. V. Nguyen, and D. E. Dumnov, "Advanced Rocket Nozzles," *Journal of Propulsion and Power*, Vol. 14, No. 5, pp. 620–634, AIAA, 1998.

3–6. M. Frey and G. Hagemann, "Flow Separation and Side-Loads in Rocket Nozzles," *AAIA Paper 99-2815*, June 1999.

3–7. G. P. Sutton, "Flow through a Combustion Zone," Section of Chapter 3, *Rocket Propulsion Elements*, John Wiley & Sons, Second, third, and fourth editions, 1956, 1963, and 1976.

3–8. J. A. Muss, T. V. Nguyen, E. J. Reske, and D. M. McDaniels, "Altitude Compensating Nozzle Concepts for RLV," *AIAA Paper 97-3222*, July 1997.

3–9. G. V. R. Rao, "Recent Developments in Rocket Nozzle Configurations," *ARS Journal*, Vol. 31, No. 11, November 1961, pp. 1488–1494; and G. V. R. Rao, "Exhaust Nozzle Contour for Optimum Thrust," *Jet Propulsion*, Vol. 28, June 1958, pp. 377–382.

3–10. J. M. Farley and C. E. Campbell, "Performance of Several Method-of-Characteristics Exhaust Nozzles," *NASA TN D-293*, October 1960.

3–11. J. D. Hoffman, "Design of Compressed Truncated Perfect Nozzles," *Journal of Propulsion and Power*, Vol. 3, No. 2, March–April 1987, pp. 150–156.

3–12. G. P. Sutton, *Stepped Nozzle*, U.S. Patent 5,779,151, 1998.

3–13. F. A. Williams, M. Barrère, and N. C. Huang, "Fundamental Aspects of Solid Propellant Rockets," *AGARDograph 116*, Advisory Group for Aerospace Research and Development, NATO, October 1969, 783 pages.

3–14. M. Barrère, A. Jaumotte, B. Fraeijs de Veubeke, and J. Vandenkerckhove, *Rocket Propulsion*, Elsevier Publishing Company, Amsterdam, 1960.

3–15. R. N. Knauber, "Thrust Misalignments of Fixed Nozzle Solid Rocket Motors," *AIAA Paper 92–2873*, 1992.

3–16. J. S. Lilley, "The Design and Optimization of Propulsion Systems Employing Scarfed Nozzles," *Journal of Spacecraft and Rockets*, Vol. 23, No. 6, November–December 1986, pp. 597–604; and J. S. Lilley, "Experimental Validation of a Performance Model for Scarfed Nozzles," *Journal of Spacecraft and Rockets*, Vol. 24, No. 5, September–October 1987, pp. 474–480.

CHAPTER 4

FLIGHT PERFORMANCE

This chapter deals with the performance of rocket-propelled vehicles such as missiles, spacecraft, space launch vehicles, or projectiles. It is intended to give the reader an introduction to the subject from a rocket propulsion point of view. Rocket propulsion systems provide forces to a flight vehicle and cause it to accelerate (or decelerate), overcome drag forces, or change flight direction. They are usually applied to several different flight regimes: (1) flight within the atmosphere (air-to-surface missiles or sounding rockets); (2) near-space environment (earth satellites); (3) lunar and planetary flights; and (4) sun escape; each is discussed further. References 4–1 to 4–4 give background on some of these regimes. The appendices give conversion factors, atmosphere properties, and a summary of key equations. The chapters begins with analysis of simplified idealized flight trajectories, then treats more complex flight path conditions, and discusses various flying vehicles.

4.1. GRAVITY-FREE, DRAG-FREE SPACE FLIGHT

This simple rocket flight analysis applies to an outer space environment, where there is no air (thus no drag) and essentially no significant gravitational attraction. The flight direction is the same as the thrust direction (along the axis of the nozzle), namely, a one-dimensional, straight-line acceleration path; the propellant mass flow $\dot{m}$, and thus the thrust F, remain constant for the propellant burning duration t_p. For a constant propellant flow the flow rate is m_p/t_p, where m_p is the total usable propellant mass. From Newton's second law and for an instantaneous vehicle mass m and a vehicle velocity u.

$$F = m \, du/dt \qquad (4\text{--}1)$$

For a rocket where the propellant flow rate is constant the instantaneous mass of the vehicle m can be expressed as a function of the initial mass of the full vehicle m_0, m_p, t_p, and the instantaneous time t.

$$m = m_0 - \frac{m_p}{t_p} t = m_0 \left(1 - \frac{m_p}{m_0} \frac{t}{t_p} \right) \qquad (4\text{--}2)$$

$$= m_0 \left(1 - \zeta \frac{t}{t_p} \right) = m_0 \left[1 - (1 - \mathbf{MR}) \frac{t}{t_p} \right] \qquad (4\text{--}3)$$

Equation 4–3 expresses the vehicle mass in a form useful for trajectory calculations. The vehicle mass ratio **MR** and the propellant mass fraction ζ have been defined by Eqs. 2–7 and 2–8. They are related by

$$\zeta = 1 - \mathbf{MR} \qquad (4\text{--}4)$$

A definition of the various masses is shown in Fig. 4–1. The initial mass at takeoff m_0 equals the sum of the useful propellant mass m_p plus the empty or final vehicle mass m_f; m_f in turn equals the sum of the inert masses of the engine system (such as nozzles, tanks, cases, or unused, residual propellant), plus the guidance, control, electronics, and related equipment, and the payload.

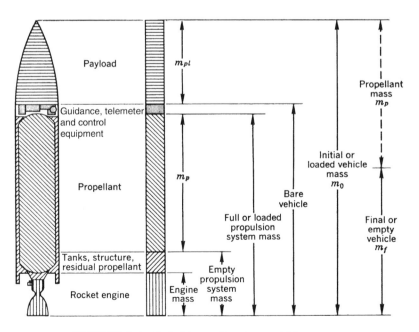

FIGURE 4–1. Definitions of various vehicle masses.

For constant propellant flow $\dot{m}$ and a finite propellant burning time the total propellant mass m_p is $\dot{m}t_p$ and the instantaneous vehicle mass $m = m_0 - \dot{m}t$. Equation 4–1 can be written as

$$du = (F/m)dt = (c\dot{m}/m)\ dt$$
$$= \frac{(c\dot{m})\ dt}{m_0 - m_p t/t_p} = \frac{c(m_p/t_p)\ dt}{m_0(1 - m_p t/m_0 t_p)} = \frac{c\zeta/t_p}{1 - \zeta t/t_p}\ dt$$

Integration leads to the maximum vehicle velocity at propellant burnout u_p that can be attained in a gravity-free vacuum. When $u_0 \neq 0$ it is often called the velocity increment Δu.

$$\Delta u = -c\ \ln(1 - \zeta) + u_0 = c\ln(m_0/m_f) + u_0 \tag{4–5}$$

If the initial velocity u_0 is assumed to be zero, then

$$\begin{aligned} u_p = \Delta u &= -c\ln(1 - \zeta) = -c\ln[m_0/(m_0 - m_p)] \\ &= -c\ln \mathbf{MR} = c\ln(1/\mathbf{MR}) \\ &= c\ln(m_0/m_f) \end{aligned} \tag{4–6}$$

This is the maximum velocity increment Δu that can be obtained in a gravity-free vacuum with constant propellant flow, starting from rest with $u_0 = 0$. The effect of variations in c, I_s, and ζ on the flight velocity increment are shown in Fig. 4–2. An alternate way to write Eq. 4–6 uses e, the base of the natural logarithm.

$$e^{\Delta u/c} = 1/\mathbf{MR} = m_0/m_f \tag{4–7}$$

The concept of the maximum attainable flight velocity increment Δu in a gravity-free vacuum is useful in understanding the influence of the basic parameters. It is used in comparing one propulsion system or vehicle with another, one flight mission with another, or one proposed upgrade with another possible design improvement.

From Eq. 4–6 it can be seen that *propellant mass fraction* has a logarithmic effect on the vehicle velocity. By increasing this ratio from 0.80 to 0.90, the interplanetary maximum vehicle velocity in gravitationless vacuum is increased by 43%. A mass fraction of 0.80 would indicate that only 20% of the total vehicle mass is available for structure, skin, payload, propulsion hardware, radios, guidance system, aerodynamic lifting surfaces, and so on; the remaining 80% is useful propellant. It requires careful design to exceed 0.85; mass fraction ratios approaching 0.95 appear to be the probable practical limit for single-stage vehicles and currently known materials. When the mass fraction is 0.90, then $\mathbf{MR} = 0.1$ and $1/\mathbf{MR} = 10.0$. This marked influence of mass fraction or mass ratio on the velocity at power cutoff, and therefore also the range,

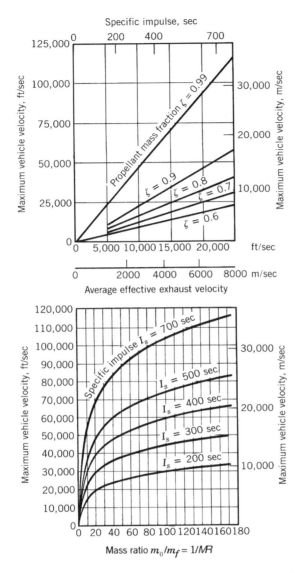

FIGURE 4–2. Maximum vehicle velocity in a gravitationless, drag-free space for different mass ratios and specific impulses (plot of Eq. 4–6). Single-state vehicles can have values of $1/\mathbf{MR}$ up to about 20 and multistage vehicles can exceed 200.

not only is true of interplanetary spaceships in a vacuum but applies to almost all types of rocket-powered vehicles. For this reason, importance is placed on saving inert mass on every vehicle component, including the propulsion system.

Equation 4–6 can be modified and solved for the effective propellant mass m_p required to achieve a desired velocity increment for a given initial takeoff mass or a final burnout mass of the vehicle. The final mass consists of the payload, the structural mass of the vehicle, the empty propulsion system mass (which includes residual propellant), plus a small additional mass for guidance, communications, and control devices. Here $m_p = m_0 - m_f$.

$$m_p = m_f(e^{\Delta u/c} - 1) = m_0(1 - e^{(-\Delta u/c)}) \tag{4–8}$$

The flight velocity increment u_p is proportional to the effective exhaust velocity c and, therefore, to the specific impulse. Thus any improvement in I_s (such as better propellants, more favorable nozzle area ratio, or higher chamber pressure) reflects itself in improved vehicle performance, provided that such an improvement does not also cause an excessive increase in rocket propulsion system inert mass, which causes a decrease in the effective propellant fraction.

4.2. FORCES ACTING ON A VEHICLE IN THE ATMOSPHERE

The external forces commonly acting on vehicles flying in the earth's atmosphere are thrust, aerodynamic forces, and gravitational attractions. Other forces, such as wind or solar radiation pressure, are small and generally can be neglected for many simple calculations.

The *thrust* is the force produced by the power plant, such as a propeller or a rocket. It usually acts in the direction of the axis of the power plant, that is, along the propeller shaft axis or the rocket nozzle axis. The thrust force of a rocket with constant mass flow has been expressed by Eq. 2–6 as a function of the effective exhaust velocity c and the propellant flow rate $\dot{m}$. In many rockets the mass rate of propellant consumption $\dot{m}$ is essentially constant, and the starting and stopping transients are usually very short and can be neglected. Therefore, the thrust is

$$F = c\dot{m} = cm_p/t_p \tag{4–9}$$

As explained in Chapter 3, for a given propellant the value of the effective exhaust velocity c or specific impulse I_s depends on the nozzle area ratio and the altitude. The value of c can increase by a relatively small factor of between 1.2 and 1.6 as altitude is increased.

The *drag* D is the *aerodynamic force* in a direction opposite to the flight path due to the resistance of the body to motion in a fluid. The *lift* L is the aerodynamic force acting in a direction normal to the flight path. They are expressed as functions of the flight speed u, the mass density of the fluid in which the vehicle moves ρ, and a typical surface area A.

$$L = C_L \tfrac{1}{2} \rho A u^2 \tag{4–10}$$

$$D = C_D \tfrac{1}{2} \rho A u^2 \tag{4–11}$$

C_L and C_D are lift and drag coefficients, respectively. For airplanes and winged missiles the area A is understood to mean the wing area. For wingless missiles or space launch vehicles it is the maximum cross-sectional area normal to the missile axis. The lift and drag coefficients are primarily functions of the vehicle configuration, flight Mach number, and angle of attack, which is the angle between the vehicle axis (or the wing plane) and the flight direction. For low flight speeds the effect of Mach number may be neglected, and the drag and lift coefficients are functions of the angle of attack. The variation of the drag and lift coefficients for a typical supersonic missile is shown in Fig. 4–3. The values of these coefficients reach a maximum near a Mach number of unity. For wingless vehicles the angle of attack α is usually very small ($0 < \alpha < 1°$). The density and other properties of the atmosphere are listed in Appendix 2. The density of the earth's atmosphere can vary by a factor up to two (for altitudes of 300 to 1200 km) depending on solar activity and night-to-day temperature variations. This introduces a major unknown in the drag. The aerodynamic forces are affected by the flow and pressure distribution of the rocket exhaust gases, as explained in Chapter 18.

For space launch vehicles and ballistic missiles the drag loss, when expressed in terms of Δu, is typically 5 to 10% of the final vehicle velocity increment. This relatively low value is due to the fact that the air density is low at high altitudes, when the velocity is high, and at low altitudes the air density is high but the flight velocity and thus the dynamic pressure are low.

Gravitational attraction is exerted upon a flying space vehicle by all planets, stars, the moon, and the sun. Gravity forces pull the vehicle in the direction of the center of mass of the attracting body. Within the immediate vicinity of the earth, the attraction of other planets and bodies is negligibly small compared to the earth's gravitational force. This force is the *weight*.

If the variation of gravity with the geographical features and the oblate shape of the earth are neglected, the acceleration of gravity varies inversely as the square of the distance from the earth's center. If R_0 is the radius of the earth's surface and g_0 the acceleration on the earth's surface at the earth's effective radius R_0, the gravitational attraction g is

$$
\begin{aligned}
g &= g_0(R_0/R)^2 \\
&= g_0[R_0/(R_0 + h)]^2
\end{aligned}
\tag{4–12}
$$

where h is the altitude. At the equator the earth's radius is 6378.388 km and the standard value of g_0 is 9.80665 m/sec². At a distance as far away as the moon, the earth's gravity acceleration is only about $3.3 \times 10^{-4}\, g_0$.

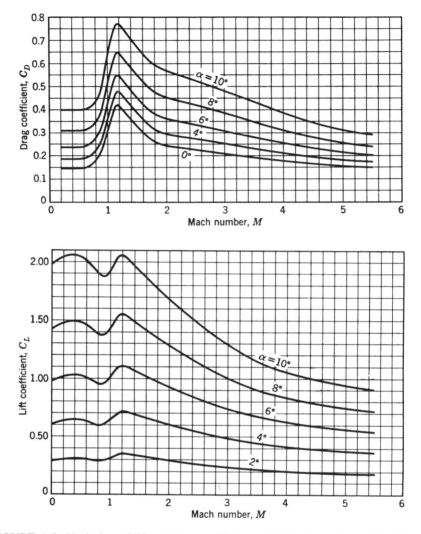

FIGURE 4–3. Variation of lift and drag coefficient with Mach number of the German V-2 missile based on body cross-sectional area with jet off and without exhaust plume effects at several angles of attack α.

4.3. BASIC RELATIONS OF MOTION

For a vehicle that flies within the proximity of the earth, the gravitational attraction of all other heavenly bodies may usually be neglected. Let it be assumed that the vehicle is moving in rectilinear equilibrium flight and that all control forces, lateral forces, and moments that tend to turn the vehicle are zero. The trajectory is two-dimensional and is contained in a fixed plane. The vehicle has wings that are inclined to the flight path at an angle of attack α and that give a lift in a direction normal to the flight path. The direction of flight does not coincide with the direction of thrust. Figure 4–4 shows these conditions schematically.

Let θ be the angle of the flight path with the horizontal and ψ the angle of the direction of thrust with the horizontal. In the direction of the flight path the product of the mass and the acceleration has to equal the sum of all forces, namely the propulsive, aerodynamic, and gravitational forces:

$$m(du/dt) = F\cos(\psi - \theta) - D - mg\sin\theta \qquad (4\text{–}13)$$

The acceleration perpendicular to the flight path is $u(d\theta/dt)$; for a constant value of u and the instantaneous radius R of the flight path it is u^2/R. The equation of motion in a direction normal to the flight velocity is

$$mu(d\theta/dt) = F\sin(\psi - \theta) + L - mg\cos\theta \qquad (4\text{–}14)$$

By substituting from Equations 4–10 and 4–11, these two basic equations can be solved for the accelerations as

$$\frac{du}{dt} = \frac{F}{m}\cos(\psi - \theta) - \frac{C_D}{2m}\rho u^2 A - g\sin\theta \qquad (4\text{–}15)$$

$$u\frac{d\theta}{dt} = \frac{F}{m}\sin(\psi - \theta) + \frac{C_L}{2m}\rho u^2 A - g\cos\theta \qquad (4\text{–}16)$$

No general solution can be given to these equations, since t_p, u, C_D, C_L, p, θ, or ψ can vary independently with time, mission profile, or altitude. Also, C_D and C_L are functions of velocity or Mach number. In a more sophisticated analysis other factors may be considered, such as the propellant used for nonpropulsive purposes (e.g., attitude control or flight stability). See Refs. 4–1 to 4–5 for a

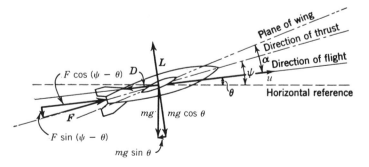

FIGURE 4–4. Two-dimensional free-body force diagram for flying vehicle with wings and fins.

background of flight performance in some of the flight regimes. Different flight performance parameters are maximized or optimized for different rocket flight missions or flight regimes, such as Δu, range, time-to-target, or altitude. Rocket propulsion systems are usually tailored to fit specific flight missions.

Equations 4–15 and 4–16 are general and can be further simplified for various special applications, as shown in subsequent sections. Results of such iterative calculations of velocity, altitude, or range using the above two basic equations often are adequate for rough design estimates. For actual trajectory analyses, navigation computation, space-flight path determination, or missile-firing tables, this two-dimensional simplified theory does not permit sufficiently accurate results. The perturbation effects, such as those listed in Section 4.6 of this chapter, must then be considered in addition to drag and gravity, and digital computers are necessary to handle the complex relations. An arbitrary division of the trajectory into small elements and a step-by-step or numerical integration to define a trajectory are usually indicated. The more generalized three-body theory includes the gravitational attraction among three masses (for example, the earth, the moon, and the space vehicle) and is considered necessary for many space-flight problems (see Refs. 4–2 and 4–3). When the propellant flow and the thrust are not constant, the form and the solution to the equations above become more complex.

A form of Eqs. 4–15 and 4–16 can also be used to determine the actual thrust or actual specific impulse during actual vehicle flights from accurately observed trajectory data, such as from optical or radar tracking data. The vehicle acceleration (du/dt) is essentially proportional to the net thrust and, by making an assumption or measurement on the propellant flow (which usually varies in a predetermined manner) and an analysis of aerodynamic forces, it is possible to determine the rocket propulsion system's actual thrust under flight conditions.

When integrating Eqs. 4–15 and 4–16 one can obtain actual histories of velocities and distances traveled and thus complete trajectories. The more general case requires six equations; three for translation along each of three perpendicular axes and three for rotation about these axes. The choice of coordinate systems and the reference points can simplify the mathematical solutions (see Refs. 4–2 and 4–4).

For a wingless rocket projectile, a space launch vehicle, or a missile with constant thrust and propellant flow, these equations can be simplified. In Fig. 4–5 the flight direction θ is the same as the thrust direction and lift forces for a symmetrical, wingless, stably flying vehicle can be assumed to be zero of zero angle of attack. For a two-dimensional trajectory in a single plane (no wind forces) and a stationary earth, the acceleration in the direction of flight is as follows:

$$\frac{du}{dt} = \frac{c\zeta/t_p}{1 - \zeta t/t_p} - g\sin\theta - \frac{C_D \frac{1}{2}\rho u^2 A/m_0}{1 - \zeta t/t_p} \tag{4–17}$$

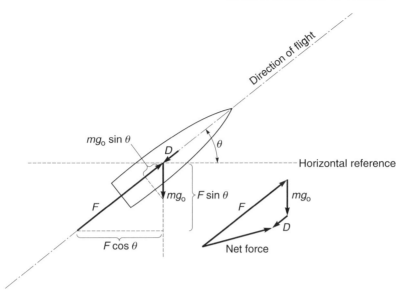

FIGURE 4–5. Simplified free-body force diagram for vehicle without wings or fins. The force vector diagram shows the net force on the vehicle.

A force vector diagram in Fig. 4–5 shows the net force (by adding thrust, drag and gravity vectors) to be at an angle to the flight path, which will be curved. These types of diagram form the basis for iterative trajectory numerical solutions.

The relationships in this Section 4.3 are for a two-dimensional flight path, one that lies in a single plane. If maneuvers out of that plane are also made (e.g., due to solar attraction, thrust misalignment, or wind) then the flight paths become three-dimensional and another set of equations will be needed to describe these flights. Reference 4–1 describes equations for the motion of rocket projectiles in the atmosphere in three dimensions. It requires energy and forces to push a vehicle out of its flight plane. Trajectories have to be calculated accurately in order to reach the intended flight objective and today almost all are done with the aid of a computer. A good number of *computer programs* for analyzing flight trajectories exit and are maintained by aerospace companies or Government agencies. Some are two-dimensional, relatively simple, and are used for making preliminary estimates or comparisons of alternative flight paths, alternative vehicle designs, or alternative propulsion schemes. Several use a stationary flat earth, while others use a rotating curved earth. Three-dimensional programs also exit, are used for more accurate flight path analyses, include some or all perturbations, orbit plane changes, or flying at angles of attack. As explained in Ref. 4–3, they are more complex.

If the flight trajectory is vertical (as for a sounding rocket), Eq. 4–17 is the same, except that $\sin \theta = 1.0$, namely

$$\frac{du}{dt} = \frac{c\zeta/t_p}{1 - \zeta t/t_p} - g - \frac{C_D \frac{1}{2}\rho u^2 A/m_0}{1 - \zeta t/t_p} \tag{4-18}$$

The velocity at the end of burning can be found by integrating between the limits of $t = 0$ and $t = t_p$ when $u = u_0$ and $u = u_p$. The first two terms can readily be integrated. The last term is of significance only if the vehicle spends a considerable portion of its time within the atmosphere. It can be integrated graphically or by numerical methods, and its value can be designated as $BC_D A/m_0$ such that

$$B = \int_0^{t_p} \frac{\frac{1}{2}\rho u^2}{1 - \zeta t/t_p} dt$$

The cutoff velocity or velocity at the end of propellant burning u_p is then

$$u_p = -\bar{c}\ln(1 - \zeta) - \bar{g}t_p - \frac{BC_D A}{m_0} + u_0 \tag{4-19}$$

where u_0 is the initial velocity, such as may be given by a booster, $\bar{g}$ is an average gravitational attraction evaluated with respect to time and altitude from Eq. 4–12, and $\bar{c}$ is a time average of the effective exhaust velocity, which is a function of altitude.

There are always a number of trade-offs in selecting the best trajectory for a rocket projectile. For example, there is a trade-off between burning time, drag, payload, maximum velocity, and maximum altitude (or range). Reference 4–6 describes the trade-offs between payload, maximum altitude, and flight stability for a sounding rocket.

If aerodynamic forces outside the earth's atmosphere are neglected (operate in a vacuum) and no booster or means for attaining an initial velocity ($u_0 = 0$) is assumed, the velocity at the end of the burning reached in a vertically ascending trajectory will be

$$\begin{aligned} u_p &= -\bar{c}\ln(1 - \zeta) - \bar{g}t_p \\ &= -\bar{c}\ln \mathbf{MR} - \bar{g}t_p \\ &= \bar{c}\ln(1/\mathbf{MR}) - \bar{g}t_p \end{aligned} \tag{4-20}$$

The first term is usually the largest and is identical to Eq. 4–6. It is directly proportional to the effective rocket exhaust velocity and is very sensitive to changes in the mass ratio. The second term is always negative during ascent, but its magnitude is small if the burning time t_p is short or if the flight takes place in high orbits or in space where $\bar{g}$ is comparatively small.

For a flight that is not following a vertical path, the gravity loss is a function of the angle between the flight direction and the local horizontal; more specifically, the gravity loss is the integral of $g \sin\theta\, dt$, as shown by Eq. 4–15.

For the simplified two-dimensional case the net acceleration a for vertical takeoff at sea level is

$$a = (F_0 g_0 / w_0) - g_0 \qquad (4-21)$$
$$a/g_0 = (F_0/w_0) - 1 \qquad (4-22)$$

where a/g_0 is the *initial takeoff acceleration* in multiples of the sea level gravitational acceleration g_0, and F_0/w_0 is the thrust-to-weight ratio at takeoff. For large surface-launched vehicles, this initial-thrust-to-initial-weight ratio has values between 1.2 and 2.2; for small missiles (air-to-air, air-to-surface, and surface-to-air types) this ratio is usually larger, sometimes even as high as 50 or 100. *The final* or *terminal acceleration* a_f of a vehicle in vertical ascent usually occurs just before the rocket engine is shut off and before the propellant is completely consumed.

$$a_f/g_0 = (F_f/w_f) - 1 \qquad (4-23)$$

In a gravity-free environment this equation becomes $a_f/g_0 = F_f/w_f$. In rockets with constant propellant flow the final acceleration is usually also the maximum acceleration, because the vehicle mass to be accelerated has its minimum value just before propellant exhaustion, and for ascending rockets the thrust usually increases with altitude. If this terminal acceleration is too large (and causes overstressing of the structure, thus necessitating an increase in structure mass), then the thrust can be designed to a lower value for the last portion of the burning period.

Example 4–1. A simple single-stage rocket for a rescue flare has the following characteristics and its flight path nomenclature is shown in the sketch.

Launch weight	4.0 lbf
Useful propellant mass	0.4 lbm
Effective specific impulse	120 sec
Launch angle (relative to horizontal)	80°
Burn time (with constant thrust)	1.0 sec

Drag is to be neglected, since the flight velocities are low. Assume no wind. Assume the local acceleration of gravity to be equal to the sea level g_0 and invariant throughout the flight.

Solve for the initial and final acceleration of powered flight, the maximum trajectory height, the time to reach maximum height, the range or distance to impact, and the angle at propulsion cutoff and at impact.

SOLUTION. Divide the flight path into three portions: the powered flight for 1 sec, the unpowered ascent after cutoff, and the free-fall descent. The thrust is obtained from Eq. 2–5:

$$F = I_s w/t_p = 120 \times 0.4/1 = 48 \text{ lbf}$$

The initial accelerations along the x and y directions are, from Eq. 4.22,

$$(a_0)_y = g_0[(F \sin\theta/w) - 1] = 32.2[(48/4)\sin 80° - 1] = 348 \text{ ft/sec}^2$$
$$(a_0)_x = g_0(F/w)\cos\theta = 32.2(48/4)\cos 80° = 67.1 \text{ ft/sec}^2$$

The initial acceleration in the flight direction is

$$a_0 = \sqrt{(a_0)_x^2 + (a_0)_y^2} = 354.4 \text{ ft/sec}^2$$

The direction of thrust and the flight path are the same. The vertical and horizontal components of the velocity u_p at the end of powered flight is obtained from Eq. 4–20. The vehicle mass has been diminished by the propellant that has been consumed.

$$(u_p)_y = c\ln(w_0/w_f)\sin\theta - g_0 t_p = 32.2 \times 120\ln(4/3.6)0.984 - 32.2 = 375 \text{ ft/sec}$$
$$(u_p)_x = c\ln(w_0/w_f)\cos\theta = 32.2 \times 120\ln(4/3.6)0.1736 = 70.7 \text{ ft/sec}$$

The trajectory angle with the horizontal at rocket cutoff for a dragless flight is

$$\tan^{-1}(375/70.7) = 79.3°$$

Final acceleration is $a_f = Fg_0/w = 48 \times 32.2/3.6 = 429 \text{ ft/sec}^2$. For the short duration of the powered flight the coordinates at propulsion burnout y_p and x_p can be calculated approximately by using an average velocity (50% of maximum) for the powered flight.

$$y_p = \tfrac{1}{2}(u_p)_y t_p = \tfrac{1}{2} \times 375 \times 1.0 = 187.5 \text{ ft}$$
$$x_p = \tfrac{1}{2}(u_p)_x t_p = \tfrac{1}{2} \times 70.7 \times 1.0 = 35.3 \text{ ft}$$

The unpowered part of the trajectory has a zero vertical velocity at its zenith. The initial velocities, the x and y values for this parabolic trajectory segment, are those of propulsion termination ($F = 0, u = u_p, x = x_p, y = y_p$); at the zenith $(u_y)_z = 0$.

$$(u_y)_z = 0 = -g_0(t_z - t_p) + (u_p)_y \sin\theta$$

At this zenith $\sin \theta = 1.0$. Solving for t_z yields

$$t_z = t_p + (u_p)_y/g_0 = 1 + 375/32.2 = 12.6 \text{ sec}$$

The trajectory maximum height or zenith can be determined:

$$y_z = y_p + (u_p)_y(t_z - t_p) - \tfrac{1}{2}g_0(t_z - t_p)^2$$
$$= 187.5 + 375(11.6) - \tfrac{1}{2}32.2(11.6)^2 = 2370 \text{ ft}$$

The range during ascent to the zenith point is

$$x_z = (u_p)_x(t_z - t_p) + x_p$$
$$= 70.7 \times 11.6 + 35.3 = 855 \text{ ft}$$

The time of flight for the descent is, using $y_z = \tfrac{1}{2}g_0 t^2$,

$$t = \sqrt{2y_z/g_0} = \sqrt{2 \times 2370/32.2} = 12.1 \text{ sec}$$

The final range or x distance to the impact point is found by knowing that the initial horizontal velocity at the zenith $(u_z)_x$ is the same as the horizontal velocity at propulsion termination $(u_p)_x$:

$$x_f = (u_p)_x(t_{\text{descent}}) = 70.7 \times 12.1 = 855 \text{ ft}$$

The total range for ascent and descent is $855 + 855 = 1710$. The time to impact is $12.6 + 12.1 = -24.7$ sec. The vertical component of the impact or final velocity u_f is

$$u_f = g_0(t_f - t_z) = 32.2 \times 12.1 = 389.6 \text{ ft/sec}$$

The impact angle θ_f can be found:

$$\theta_f = \tan^{-1}(389.6/70.7) = 79.7°$$

If drag had been included, it would have required an iterative solution for finite elements of the flight path and all velocities and distances would be somewhat lower in value. A set of flight trajectories for a sounding rocket is given in Ref. 4–5.

4.4. EFFECT OF PROPULSION SYSTEM ON VEHICLE PERFORMANCE

This section gives several methods for improving flight vehicle performance. Most of these enhancements, listed below, are directly influenced by the selection or design of the propulsion system. A few of the flight vehicle performance improvements do not depend on the propulsion system. Most of those listed below apply to all missions, but some are peculiar to some missions only.

1. The *effective exhaust velocity* c or the *specific impulse* I_s usually have a direct effect on the vehicle's flight performance. For example the vehicle final velocity increment Δu can be inceased by a higher I_s. This can be done by using a more energetic propellant (see Chapter 7 and 12), by a higher chamber pressure and, for upper stages operating at high altitudes, also by a larger nozzle area ratio.

2. The mass ratio m_0/m_f has a logarithmic effect. It can be increased in several ways. One way is by reducing the final mass m_f, which consists of the inert hardware plus the nonusable, residual propellant mass. Reducing the inert mass implies lighter structures, smaller payloads, lighter guidance/control devices, or less unavailable residual propellant; this means going to stronger structural materials at higher stresses, more efficient power supplies, or smaller electronic packages. During design there is always great emphasis to reduce all hardware masses and the residual propellants to their practical minima. Another way is to increase the initial mass, namely by increasing the thrust and adding more propellant, but with a minimum increase in the structure or propulsion system masses. It is possible to improve the effective mass ratio greatly by using two or more stages, as will be explained in Section 4.7.

3. Reducing the burning time (i.e., increasing the thrust level) will reduce the gravitational loss. However, the higher acceleration usually requires more structural and propulsion system mass, which in turn causes the mass ratio to be less favorable.

4. The *drag*, which can be considered as a negative thrust, can be reduced in at least four ways. The drag has several components: (a) The form drag depends on the aerodynamic shape. A slender pointed nose or sharp, thin leading edges of fins or wings have less drag than a stubby, blunt shape. (b) A vehicle with a small cross-sectional area has less drag. A propulsion design that can be packaged in a long, thin shape will be preferred. (c) The drag is proportional to the cross-sectional or frontal vehicle area. A higher propellant density will decrease the propellant volume and therefore will allow a smaller cross section. (d) The skin drag is caused by the friction of the air flowing over all the vehicle's outer surfaces. A smooth contour and a polished surface are usually better. The skin drag is also influenced by the propellant density, because it gives a smaller volume and thus a lower surface area. (e) The base drag is the fourth component; it is a function of the local ambient air pressure acting over the surface of the vehicle's base or bottom plate. It is influenced by the nozzle exit design (exit pressure) and the geometry of the vehicle base design. It is discussed further in Chapter 18.

5. The length of the propulsion nozzle often is a significant part of the overall vehicle or stage length. As was described in Chapter 3, there is an optimum nozzle contour and length, which can be determined by

trade-off analysis. A shorter nozzle length allows a somewhat shorter vehicle; on many designs this implies a somewhat lighter vehicle structure and a slightly better vehicle mass ratio.

6. The final vehicle velocity at propulsion termination can be increased by increasing the initial velocity u_0. By launching a satellite in an eastward direction the rotational speed of the earth is added to the final satellite orbital velocity. This tangential velocity of the earth is about 464 m/sec or 1523 ft/sec at the equator and about 408 m/sec or 1340 ft/sec for an easterly launch at Kennedy Space Center (latitude of 28.5° north). Conversely, a westerly satellite launch has a negative initial velocity and thus requires a higher-velocity increment. Another way to increase u is to launch a spacecraft from a satellite or an aircraft, which increases the initial vehicle velocity and allows launching in the desired direction, or to launch an air-to-surface missile from an airplane.

7. For vehicles that fly in the atmosphere it is possible to increase the range when aerodynamic lift is used to counteract gravity and reduce gravity losses. Using a set of wings or flying at an angle of attack increases the lift, but is also increases the drag. This lift can also be used to increase the maneuverability and trajectory flexibility.

8. When the flight velocity u is close to the rocket's effective exhaust velocity c, the propulsive efficiency is the highest (Eq. 2–23) and more of the rocket exhaust gas energy is transformed into the vehicle's flight energy. Trajectories where u is close in value to c for a major portion of the flight therefore need less propellant.

Several of these influencing parameters can be optimized. Therefore, for every mission of flight application there is an optimum propulsion system design and the propulsion parameters that define the optimum condition are dependent on vehicle or flight parameters.

4.5. SPACE FLIGHT

Newton's law of gravitation defines the attraction of gravitational force F_g between two bodies in space as follows:

$$F_g = Gm_1m_2/R^2 = \mu m_2/R^2 \qquad (4\text{--}24)$$

Here G is the universal gravity constant ($G = 6.670 \times 10^{-11}$ m³/ kg-sec²), m_1 and m_2 are the masses of the two attracting bodies (such as the earth and the moon, the earth and a spacecraft, or the sun and a planet) and R is the distance between their centers of mass. The earth's gravitational constant μ is the product of Newton's universal constant G and the mass of the earth m_1 (5.974×10^{24} kg). It is $\mu = 3.98600 \times 10^{14}$ m³/sec².

The rocket offers a means for escaping the earth for lunar and interplanetary travel, for escaping our solar system, and for creating a stationary or moving station in space. The flight velocity required to escape from the earth can be found by equating the kinetic energy of a moving body to the work necessary to overcome gravity, neglecting the rotation of the earth and the attraction of other celestial bodies.

$$\tfrac{1}{2}mu^2 = m \int g \, dR$$

By substituting for g from Eq. 4–12 and by neglecting air friction the following relation for the escape velocity is obtained:

$$u_e = R_0\sqrt{\frac{2g_0}{R_0 + h}} = \sqrt{\frac{2\mu}{R}} \qquad (4\text{–}25)$$

Here R_0 is the effective earth radius (6374.2 km), h is the orbit altitude above sea level, and g is the acceleration of gravity at the earth surface (9.806 m/sec). The spacecraft radius R measured from the earth's center is $R = R_0 + h$. The velocity of escape at the earth's surface is 11,179 m/sec or 36,676 ft/sec and does not vary appreciably within the earth's atmosphere, as shown by Fig. 4–6. Escape velocities for surface launch are given in Table 4–1 for the sun, the

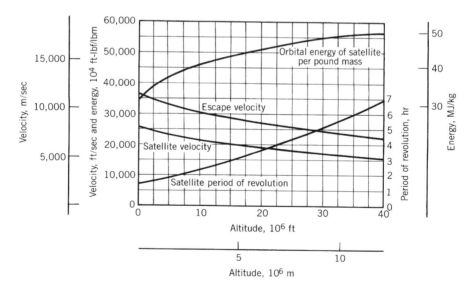

FIGURE 4–6. Orbital energy, orbital velocity, period of revolution, and earth escape velocity of a space vehicle as a function of altitude for circular satellite orbits. It is based on a spherical earth and neglects the earth's rotation and atmospheric drag.

TABLE 4-1. Characteristic Data for Several Heavenly Bodies

Name	Mean Radius of Orbit (million km)	Period of Revolution		Mean Diameter (km)	Relative Mass (Earth = 1.0)	Specific Gravity	Acceleration of Gravity at Surface (m/sec²)	Escape Velocity at Surface (m/sec)
Sun	—	—		1,393,000	332,950	1.41	273.4	616,000
Moon	0.383	27.3	days	3475	0.012	3.34	1.58	2380
Mercury	57.87	87.97	days	4670	0.06	5.5	3.67	4200
Venus	108.1	224.70	days	12,400	0.86	5.3	8.67	10,300
Earth	149.6	365.256	days	12,742	1.00[a]	5.52	9.806	11,179
Mars	227.7	686.98	days	6760	0.15	3.95	3.749	6400
Jupiter	777.8	11.86	yr	143,000	318.4	1.33	26.0	59,700
Saturn	1486	29.46	yr	121,000	95.2	0.69	11.4	35,400
Uranus	2869	84.0	yr	47,100	15.0	1.7	10.9	22,400
Neptune	4475	164.8	yr	50,700	17.2	1.8	11.9	31,000
Pluto	5899	284.8	yr	5950	0.90	4	7.62	10,000

Source: in part from Refs 4–2 and 4–3.
[a] Earth mass is 5.976×10^{24} kg.

planets, and the moon. Launching from the earth's surface at escape velocity is not practical. As a vehicle ascends through the earth's atmosphere, it is subject to severe aerodynamic heating and dynamic pressures. A practical launch vehicle has to traverse the atmosphere at relatively low velocity and accelerate to the high velocities beyond the dense atmosphere. For example, during a portion of the Space Shuttle's ascent, its main engines are actually throttled to a lower thrust to avoid excessive pressure and heating. Alternatively, an escape vehicle can be launched from an orbiting space station or from an orbiting Space Shuttle.

A rocket spaceship can become a *satellite* of the earth and revolve around the earth in a fashion similar to that of the moon. Satellite orbits are usually elliptical and some are circular. Low earth orbits, typically below 500 km altitude, are designated by the letters LEO. Satellites are useful as communications relay stations for television or radio, weather observation, or reconnaissance observation. The altitude of the orbit is usually above the earth's atmosphere, because this minimizes the expending of energy to overcome the drag which pulls the vehicle closer to the earth. The effects of the radiation in the Van Allen belt on human beings and sensitive equipment sometimes necessitate the selection of an earth orbit at low altitude.

For a circular trajectory the velocity of a satellite must be sufficiently high so that its centrifugal force balances the earth's gravitational attraction.

$$mu_s^2/R = mg$$

For a circular orbit, the satellite velocity u_s is found by using Eq. 4–12,

$$u_s = R_0\sqrt{g_0/(R_0 + h)} = \sqrt{\mu/R} \qquad (4\text{–}26)$$

which is smaller than the escape velocity by a factor of $\sqrt{2}$. The period τ in seconds of one revolution for a circular orbit relative to a stationary earth is

$$\tau = 2\pi(R_0 + h)/u_s = 2\pi(R_0 + h)^{3/2}/(R_0\sqrt{g_0}) \qquad (4\text{–}27)$$

The energy E necessary to bring a unit of mass into a circular satellite orbit neglecting drag, consists of kinetic and potential energy, namely,

$$E = \tfrac{1}{2}u_s^2 + \int_{R_0}^{R} g\, dR$$
$$= \tfrac{1}{2}R_0^2\frac{g_0}{R_0 + h} + \int_{R_0}^{R} g_0\frac{R_0^2}{R^2}\, dR = \tfrac{1}{2}R_0 g_0\frac{R_0 + 2h}{R_0 + h} \qquad (4\text{–}28)$$

The escape velocity, satellite velocity, satellite period, and satellite orbital energy are shown as functions of altitude in Fig. 4–6.

A satellite circulating around the earth at an altitude of 300 miles or 482.8 km has a velocity of about 7375 m/sec or 24,200 ft/sec, circles a stationary earth in 1.63 hr, and ideally requires an energy of 3.35×10^7 J to place 1 kg of spaceship mass into its orbit. An equatorial satellite in a circular orbit at an altitude of 6.611 earth radii (about 26,200 miles, 42,200 km, or 22,700 nautical miles) has a period of revolution of 24 hr. It will appear stationary to an observer on earth. This is known as a *synchronous* satellite in *geo-synchronous earth orbit*, usually abbreviated as GEO. It is used extensively for communications satellite applications. In Section 4.7 on launch vehicles we will describe how the payload of a given space vehicle diminishes as the orbit circular altitude is increased and as the inclination (angle between orbit plane and earth equatorial plane) is changed.

Elliptical Orbits

The circular orbit described above is a special case of the more general elliptic orbit shown in Fig. 4–7; here the earth (or any other heavenly body around which another body is moving) is located at one of the focal points of this ellipse. The equations of motion may be derived from Kepler's laws, and the elliptical orbit can be described as follows, when expressed in polar coordinates:

$$u = \left[\mu \left(\frac{2}{R} - \frac{1}{a} \right) \right]^{1/2} \tag{4-29}$$

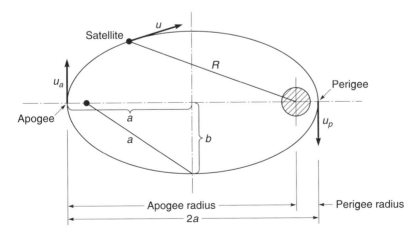

FIGURE 4–7. Elliptical orbit; the attracting body is at one of the focal points of the ellipse.

where u is the velocity of the body in the elliptical orbit, R is the instantaneous radius from the center of the attracting body (a vector quantity, which changes direction as well as magnitude), a is the major axis of the ellipse, and μ is the earth's gravitational constant with a value of 3.986×10^{14} m^3/sec^2. The symbols are defined in Fig. 4–7. From this equation it can be seen that the velocity u_p is a maximum when the moving body comes closest to its focal point at the orbit's *perigee* and that its velocity u_a is a minimum at its *apogee*. By substituting for R in Eq. 4–29, and by defining the ellipse's shape factor e as the *eccentricity of the ellipse*, $e = \sqrt{a^2 - b^2}/a$, then the apogee and perigee velocities can be expressed as

$$u_a = \sqrt{\frac{\mu(1 - e)}{a(1 + e)}} \tag{4–30}$$

$$u_b = \sqrt{\frac{\mu(1 + e)}{a(1 - e)}} \tag{4–31}$$

Another property of an elliptical orbit is that the product of velocity and instantaneous radius remains constant for any location a or b on the ellipse, namely, $u_a R_a = u_b R_b = uR$. The exact path that a satellite takes depends on the velocity (magnitude and vector orientation) with which it is started or injected into its orbit.

For interplanetary transfers the ideal mission can be achieved with minimum energy in a simple transfer ellipse, as suggested originally by Hohmann (see Ref. 4–6). Assuming the planetary orbits about the sun to be circular and coplanar, it can be demonstrated that the path of minimum energy is an ellipse tangent to the planetary orbits as shown in Fig. 4–8. This operation requires a velocity increment (relatively high thrust) at the initiation and another at ter-

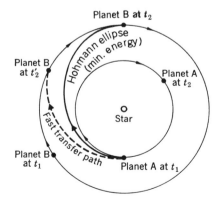

FIGURE 4–8. Schematic diagram of interplanetary transfer paths. These same transfer maneuvers apply when going from a low-altitude earth satellite orbit to a higher orbit.

mination; both increments are the velocity differences between the respective circular planetary velocities and the perigee and apogee velocity which define the transfer ellipse. The thrust levels at the beginning and end maneuvers of the Hohmann ellipse must be high enough to give a short operating time and the acceleration of at least $0.01 \, g_0$, but preferably more. With electrical propulsion these accelerations would be about $10^{-5} \, g_0$, the operating time would be weeks or months, and the best transfer trajectories would be very different from a Hohmann ellipse; they are described in Chapter 19.

The departure date or the *relative positions of the launch planet and the target planet* for a planetary transfer mission is critical, because the spacecraft has to meet with the target planet when it arrives at the target orbit. The Hohmann transfer time $(t_2 - t_1)$ starting on earth is about 116 hours to go to the moon and about 259 days to Mars. If a faster orbit (shorter transfer time) is desired (see dashed lines in Fig. 4–8), it requires more energy than a Hohmann transfer ellipse. This means a larger vehicle with a larger propulsion system that has more total impulse. There also is a *time window* for a launch of a spacecraft that will make a successful rendezvous. For a Mars mission an earth-launched spacecraft may have a launch time window of more than two months. A Hohmann transfer ellipse or a faster transfer path apply not only to planetary flight but also to earth satellites, when an earth satellite goes from one circular orbit to another (but within the same plane). Also, if one spacecraft goes to a rendezvous with another spacecraft in a different orbit, the two spacecraft have to be in the proper predetermined positions prior to the launch for simultaneously reaching their rendezvous. When the launch orbit (or launch planet) is not in the same plane as the target orbit, then additional energy will be needed by applying thrust in a direction normal to the launch orbit plane.

Example 4–2. A satellite is launched from a circular equatorial parking orbit at an altitude of 160 km into a coplanar circular synchronous orbit by using a Hohmann transfer ellipse. Assume a homogeneous spherical earth with a radius of 6374 km. Determine the velocity increments for entering the transfer ellipse and for achieving the synchronous orbit at 42,200 km altitude. See Fig. 4–8 for the terminology of the orbits.

SOLUTION.   The orbits are $R_A = 6.531 \times 10^6$ m; $R_B = 48.571 \times 10^6$ m. The major axis a of the transfer ellipse

$$a_{te} = \tfrac{1}{2}(R_A + R_B) = 27.551 \times 10^6 \text{ m/sec}$$

The orbit velocities of the two satellites are

$$u_A = \sqrt{\mu / R_A} = [3.986005 \times 10^{14}/6.571 \times 10^6]^{\frac{1}{2}} = 7788 \text{ m/sec}$$
$$u_B = \sqrt{\mu / R_B} = 2864.7 \text{ m/sec}$$

The velocities needed to enter and exit the transfer ellipse are

$$(u_{te})_A = \sqrt{\mu}[(2/R_A) - (1/a)]^{\frac{1}{2}} = 10,337 \text{ m/sec}$$
$$(u_{te})_B = \sqrt{\mu}[(2/R_B) - (1/a)]^{1/2} = 1394 \text{ m/sec}$$

The changes in velocity going from parking orbit to ellipse and from ellipse to final orbit are:

$$\Delta u_A = |(u_{te})_A - u_A| = 2549 \text{ m/sec}$$
$$\Delta u_B = |u_B - (u_{te})_B| = 1471 \text{ m/sec}$$

The total velocity change for the transfer maneuvers is:

$$\Delta u_{total} = \Delta u_A + \Delta u_B = 4020 \text{ m/sec}$$

Figure 4–9 shows the elliptical transfer trajectory of a ballistic missile or a satellite ascent vehicle. During the initial powered flight the trajectory angle is adjusted by the guidance system to an angle that will allows the vehicle to reach the apogee of its elliptical path exactly at the desired orbit altitude. For the ideal satellite *orbit injection* the simplified theory assumes an essentially instantaneous application of the total impulse as the ballistic trajectory reaches its apogee or zenith. In reality the rocket propulsion system operates over a finite time, during which gravity losses and changes in altitude occur.

Deep Space

Lunar and *interplanetary* missions include circumnavigation, landing, and return flights to the moon, Venus, Mars, and other planets. The energy necessary to escape from earth can be calculated as $\frac{1}{2}mv_e^2$ from Eq. 4–25. It is 6.26×10^7 J/kg, which is more than that required for a satellite. The gravitational attraction of various heavenly bodies and their respective escape velocities depends on their masses and diameters; approximate values are listed in Table 4–1. An idealized diagram of an interplanetary landing mission is shown in Fig. 4–10.

The *escape from the solar system* requires approximately 5.03×10^8 J/kg. This is eight times as much energy as is required for escape from the earth. There is technology to send small, unmanned probes away from the sun to outer space; as yet there needs to be an invention and demonstrated proof of a long duration, novel, rocket propulsion system before a mission to the nearest star can be achieved. The trajectory for a spacecraft to escape from the sun is either a parabola (minimum energy) or a hyperbola.

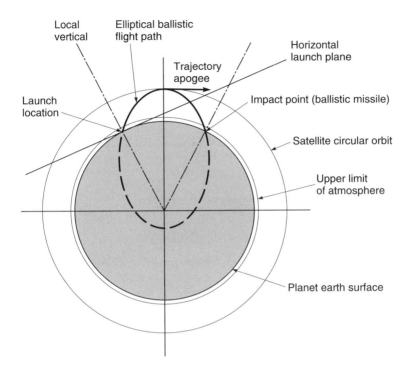

FIGURE 4–9. Long-range ballistic missiles follow an elliptical free-flight trajectory (in a drag-free flight) with the earth's center as one of the focal points. The surface launch is usually vertically up (not shown here), but the trajectory is quickly tilted during early powered flight to enter into the ellipse trajectory. The ballistic range is the arc distance on the earth's surface. For satellites, another powered flight period occurs (called orbit injection) just as the vehicle is at its elliptical apogee (as indicated by the velocity arrow), causing the vehicle to enter an orbit.

Perturbations

This section gives a brief discussion of the disturbing torques and forces which cause perturbations or deviations from any space flight path or satellite's flight trajectory. For a more detailed treatment of flight paths and their perturbations, see Refs. 4–2 and 4–3. A system is needed to measure the satellite's position and deviation from the intended flight path, to determine the needed periodic correction maneuver and then to counteract, control, and correct them. Typically, the corrections are performed by a set of small reaction control thrusters which provide predetermined total impulses into the desired directions. These corrections are needed throughout the life of the spacecraft (for 1 to 20 years) to overcome the effects of the disturbances and maintain the intended flight regime.

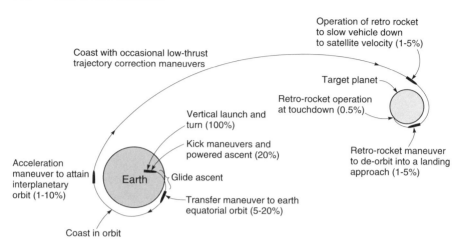

FIGURE 4–10. Schematic diagram of typical powered flight maneuvers during a hypothetical interplanetary mission with a landing. The numbers indicate typical thrust magnitudes of the maneuvers in percent of launch takeoff thrust. This is not drawn to scale. Heavy lines show powered flight segments.

Perturbations can be cateogirzed as short-term and long-term. The daily or orbital period oscillating forces are called *diurnal* and those with long periods are called *secular*.

High-altitude each satellites (36,000 km and higher) experience perturbing forces primarily as gravitational pull from the sun and the moon, with the forces acting in different directions as the satellite flies around the earth. This third-body effect can increase or decrease the velocity magnitude and change its direction. In extreme cases the satellite can come very close to the third body, such as the moon, and undergo what is called a hyperbolic maneuver that will radically change the trajectory. This encounter can be used to increase or decrease the energy of the satellite and intentionally change the velocity and the shape of the orbit.

Medium- and low-altitude satellites (500 to 35,000 km) experience perturbations because of the earth's oblateness. The earth bulges in the vicinity of the equator and a cross section through the poles is not entirely circular. Depending on the inclination of the orbital plane to the earth equator and the altitude of the satellite orbit, two perturbations result: (1) the regression of the nodes, and (2) shifting of the apsides line (major axis). Regression of the nodes is shown in Fig. 4–11 as a rotation of the plane of the orbit in space, and it can be as high as 9° per day at relatively low altitudes. Theoretically, regression does not occur in equatorial orbits.

Figure 4–12 shows an exaggerated shift of the apsidal line, with the center of the earth remaining as a focus point. This perturbation may be visualized as the movement of the prescribed elliptical orbit in a fixed plane. Obviously, both the apogee and perigee points change in position, the rate of change being a func-

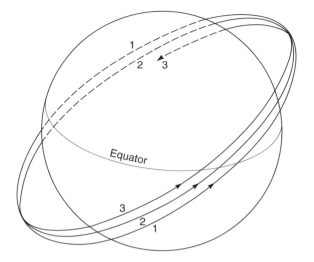

FIGURE 4–11. The regression of nodes is shown as a rotation of the plane of the orbit. The direction of the movement will be opposite to the east–west components of the earth's satellite motion.

tion of the satellite altitude and plane inclination angle. At an apogee altitude of 1000 nautical miles (n.m.) and a perigee of 100 n.m. in an equatorial orbit, the apsidal drift is approximately 10° per day.

Satellites of modern design, with irregular shapes due to protruding antennas, solar arrays, or other asymmetrical appendages, experience torques and forces that tend to perturb the satellite's position and orbit throughout its orbital life. The principal torques and forces result from the following factors:

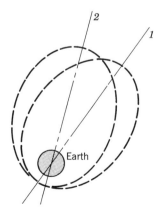

FIGURE 4–12. Shifting of the apsidal line of an elliptic orbit from position 1 to 2 because of the oblateness of the earth.

1. *Aerodynamic drag.* This factor is significant at orbital altitudes below 500 km and is usually assumed to cease at 800 km above the earth. Reference 4–7 gives a detailed discussion of aerodynamic drag which, in addition to affecting the attitude of unsymmetrical vehicles, causes a change in elliptical orbits known as apsidal drift, a decrease in the major axis, and a decrease in eccentricity of orbits about the earth.

2. *Solar radiation.* This factor dominates at high altitudes (above 800 km) and is due to impingement of solar photons upon satellite surfaces. The solar radiation pressure p (N/m^2) on a given surface of the satellite in the vicinity of the earth exposed to the sun can be determined as

$$p = 4.5 \times 10^{-6} \cos \theta [(1 - k_s) \cos \theta + 0.67 k_d] \tag{4–32}$$

where θ is the angle (degrees) between the incident radiation vector and the normal to the surface, and k_s and k_d are the specular and diffuse coefficients of reflectivity. Typical values are 0.9 and 0.5, respectively, for k_s and k_d on the body and antenna, and 0.25 and 0.01 respectively, for k_s and k_d with solar array surfaces. The radiation intensity varies as the square of the distance from the sun (see Ref. 4–8). The torque T on the vehicle is given by $T = pAl$, where A is the projected area and l is the offset distance between the spacecraft's center of gravity and the center of solar pressure.

3. *Gravity gradients.* Gravitational torque in spacecraft results from a variation in the gravitational force on the distributed mass of a spacecraft. Determination of this torque requires knowledge of the gravitational field and the distribution of spacecraft mass. This torque decreases as a function of the orbit radius and increases with the offset distances of masses within the spacecraft (including booms and appendages), it is most significant in large spacecraft or space stations operating in relatively low orbits (see Ref. 4–9).

4. *Magnetic field.* The earth's magnetic field and any magnetic moment within the satellite interact to produce torque. The earth's magnetic field precesses about the earth's axis but is very weak (0.63 and 0.31 gauss at poles and equator, respectively). This field is continually fluctuating in direction and intensity because of magnetic storms and other influences. Since the field strength decreases with $1/R^3$ with the orbital altitude, magnetic field forces are often neglected in the preliminary design of satellites (see Ref. 4–10).

5. *Internal accelerations.* Deployment of solar array panels, the shifting of propellant, movement of astronauts or other mass within the satellite, or the "unloading" of reaction wheels produce torques and forces.

We can categorize satellite propulsion needs according to function as listed in Table 4–2, which shows the total impulse "budget" applicable to a typical

TABLE 4.2. Propulsion Functions and Total Impulse Needs of a 2000-lbm Geosynchronous Satellite with a 7-Year Life

Function	Total Impulse (N-sec)
Acquisition of orbit	20,000
Attitude control (rotation)	4,000
Station keeping, E–W	13,000
Station keeping, N–S	270,000
Repositioning (Δu, 200 ft/sec)	53,000
Control apsidal drift (third body attraction)	445,000
Deorbit	12,700
Total	817,700

high altitude, elliptic orbit satellite. The control system designer often distinguishes two different kinds of stationary-keeping orbit corrections needed to keep the satellite in a synchronous position. The east–west correction refers to a correction that moves the point at which a satellite orbit intersects the earth's equatorial plane in an east or west direction; it usually corrects forces caused largely by the oblateness of the earth. The north–south correction counteracts forces usually connected with the third-body effects of the sun and the moon.

In many satellite missions the gradual changes in orbit caused by perturbation forces are not of concern. However, in certain missions it is necessary to compensate for these perturbing forces and maintain the satellite in a specific orbit and in a particular position in that orbit. For example, a synchronous communications satellite in a GEO needs to maintain its position and its orbit, so it will be able to (1) keep covering a specific area of the earth or communicate with the same stations on earth within its line of sight, and (2) not become a hazard to other satellites in this densely occupied synchronous equatorial orbit. Another example is a LEO communications satellite system with several coordinated satellites; here at least one satellite has to be in a position to receive and transmit RF signals to specific points on earth. Their orbits, and the positions of these several satellites with respect to each other, need to be controlled and maintained (see Refs. 4–11 to 4–13).

Orbit maintenance means applying small correcting forces and torques periodically; for GEO it is typically every few months. Typical velocity increments for the orbit maintenance of synchronous satellites require a Δu between 10 and 50 m/sec per year. For a satellite mass of about 2000 kg a 50 m/sec correction for a 10-year orbit life would need a total impulse of about 100,000 N-sec, which corresponds to a propellant mass of 400 to 500 kg (about a quarter of the satellite mass) if done by a small monopropellant or bipropellant thrust. It would require much less propellant if electrical propulsion were used, but in some spacecraft the inert mass of the power supply would increase.

Mission Velocity

A convenient way to describe the magnitude of the energy requirement of a space mission is to use the concept of the *mission velocity*. It is the sum of all the flight velocity increments needed to attain the mission objective. In the simplified sketch of a planetary landing mission of Fig. 4–10, it is the sum of all the Δu velocity increments shown by the heavy lines (rocket-powered flight segments) of the trajectories. Even though some of the velocity increments were achieved by retro-action (a negative propulsion force to decelerate the flight velocity), these maneuvers required energy and their absolute magnitude is counted in the mission velocity. The initial velocity from the earth's rotation (464 m/sec at the equator and 408 m/sec at a launch station at 28.5° latitude) does not have to be provided by the vehicle's propulsion systems. For example, the required mission velocity for launching at Cape Kennedy, bringing the space vehicle into an orbit at 110 km, staying in orbit for a while, and then entering a de-orbit maneuver has the Δu components shown in Table 4–3.

The required mission velocity is the sum of the absolute values of all translation velocity increments that have forces going through the center of gravity of the vehicle (including turning maneuvers) during the flight of the mission. It is the theoretical hypothetical velocity that can be attained by the vehicle in a gravity-free vacuum, if all the propulsive energy of the momentum-adding thrust chambers in all stages were to be applied in the same direction. It is useful for comparing one flight vehicle design with another and as an indicator of the mission energy.

The required mission velocity has to be equal to the "supplied" mission velocity, that is, the sum of all the velocity increments provided by the propulsion systems of each of the various vehicle stages. The total velocity increment to be "supplied" by the shuttle's propulsion systems for the shuttle mission described below (solid rocket motor strap-on boosters, main engines and, for orbit injection, also the increment from the orbital maneuvering system—all shown in Fig. 1–13) has to equal or exceed 9621 m/sec. With chemical propulsion systems and a single stage, we can achieve a space mission velocity of 4000

TABLE 4–3. Space Shuttle Incremental Flight Velocity Breakdown

Ideal satellite velocity	7790 m/sec
Δu to overcome gravity losses	1220 m/sec
Δu to turn the flight path from the vertical	360 m/sec
Δu to counteract aerodynamic drag	118 m/sec
Orbit injection	145 m/sec
Deorbit maneuver to re-enter atmosphere and aerodynamic braking	60 m/sec
Correction maneuvers and velocity adjustments	62 m/sec
Initial velocity provided by the earth's rotation at 28.5° latitude	−408 m/sec
Total required mission velocity	9347 m/sec

to 13,000 m/sec, depending on the payload, vehicle design, and propellant. With two stages it can be between perhaps 12,000 and 22,000 m/sec.

Rotational maneuvers, described later, do not change the flight velocity and are not usually added to the mission velocity requirements. Also, maintaining a satellite in orbit against long-term perturbing forces (see prior section) is often not counted as part of the mission velocity. However, the designers need to provide additional propulsion capability and propellants for these purposes. These are often separate propulsion systems, called reaction control systems.

Typical vehicle velocities required for various interplanetary missions have been estimated as shown in Table 4–4. By starting interplanetary journeys from a space satellite station, a considerable saving in this vehicle velocity can be achieved, namely, the velocity necessary to achieve the earth-circling satellite orbit. As the space-flight objective becomes more ambitious, the mission velocity is increased. For a given single or multistage vehicle it is possible to increase the vehicle's terminal velocity, but usually only at the expense of payload. Table 4–5 shows some typical ranges of payload values for a given multistage vehicle as a percentage of a payload for a relatively simple earth orbit. Thus a vehicle capable of putting a substantial payload into a near-earth orbit can only land a very small fraction of this payload on the moon, since it has to have additional upper stages, which displace payload mass. Therefore, much larger vehicles are required for space flights with high mission velocities if compared to a vehicle of less mission velocity but identical payload. The values listed in Tables 4–4 and 4–5 are only approximate because they depend on specific vehicle design features, the propellants used, exact knowledge of the

TABLE 4–4. Vehicle Mission Velocities for Typical Interplanetary Missions

Mission	Ideal Velocity (km/sec)	Approximate Actual Velocity (1000 m/sec)
Satellite orbit around earth (no return)	7.9–10	9.1–12.5
Escape from earth (no return)	11.2	12.9
Escape from moon	2.3	2.6
Earth to moon (soft landing on moon, no return)	13.1	15.2
Earth to Mars (soft landing)	17.5	20
Earth to Venus (soft landing)	22	25
Earth to moon (landing on moon and return to earth[a])	15.9	17.7
Earth to Mars (landing on Mars, and return to earth [a])	22.9	27

[a]Assumes air braking within atmospheres.

TABLE 4–5. Relative Payload–Mission Comparison Chart for High-Energy Chemical Multistage Rocket Vehicles

Mission	Relative Payload[a] (%)
Earth satellite	100
Earth escape	35–45
Earth 24-hr orbit	10–25
Moon landing (hard)	35–45
Moon landing (soft)	10–20
Moon circumnavigation (single fly-by)	30–42
Moon satellite	20–30
Moon landing and return	1–4
Moon satellte and return	8–15
Mars flyby	20–30
Mars satellite	10–18
Mars landing	0.5–3

[a]300 nautical miles (555.6 km) earth orbit is 100% reference.

trajectory–time relation, and other factors that are beyond the scope of this short treatment. Further information on space flight can be found in Refs. 4–2 to 4–4 and 4–11 to 4–13.

For example, for a co-planar earth–moon and return journey it is necessary to undertake the following steps in sequence and provide an appropriate velocity increment for each. This is similar in concept to the diagram for interplanetary flight of Fig. 4–10. For the ascent from the earth and the entry into an earth satellite orbit, the vehicle has to be accelerated ideally to approximately 7300 m/sec; to change to the transfer orbit requires roughly another 2900 m/sec; to slow down and put the spacecraft into an approach to the moon (retro-action) and enter into an orbit about the moon is about 1000 m/sec; and to land on the moon is about another 1600 m/sec. The ascent from the moon and the entry into an earth return orbit is about 2400 m/sec. Aerodynamic drag is used to slow down the earth reentry vehicle and this maneuver does not require the expenditure of propellant. Adding these together and allowing 300 m/sec for various orbit adjustments comes to a total of about 14,500 m/sec, which is the approximate cumulative total velocity needed for the mission. Tables 4–3 and 4–4 compare very rough values of mission velocities and payloads for several space missions.

4.6. FLIGHT MANEUVERS

In this section we describe different flight maneuvers and relate them to specific propulsion system types. The three categories of *maneuvers* are:

1. In *translation maneuvers* the rocket propulsion thrust vector goes through the center of gravity of the vehicle. The vehicle momentum is changed in the direction of the flight velocity. An example of several powered (translational maneuvers) and unpowered (coasting) segments of a complex space flight trajectory is shown in schematic, simplified form in Fig. 4–10. To date, most maneuvers have used chemical propulsion systems.

2. In *truly rotational maneuvers* there is no net thrust acting on the vehicle. These are true couples that apply only torque. It requires four thrusters to be able to rotate the vehicle in either direction about any one axis (two thrusters apart, firing simultaneously, but in opposite directions). These types of maneuver are usually provided by reaction control systems. Most have used multiple liquid propellant thrusters, but in recent years many space missions have used electrical propulsion.

3. A combination of categories 1 and 2, such as a large misaligned thrust vector that does not go exactly through the center of gravity of the vehicle. The misalignment can be corrected by changing the vector direction of the main propulsion system (thrust vector control) during powered flight or by applying a simultaneous compensating torque from a separate reaction control system.

The following types of *space flight maneuvers and vehicle accelerations* use rocket propulsion. All propulsion operations are controlled (started, monitored, and stopped) by the vehicle's guidance and control system.

   a. *First stage* and its *upper stage propulsion systems* add momentum during launch and ascent. They require rocket propulsion of high or medium thrusts and limited durations (typically 0.7 to 8 minutes). To date all have used chemical propulsion systems. They constitute the major mass of the space vehicle and are discussed further in the next section.

   b. *Orbit injection or transferring from one orbit to another* requires accurately predetermined total impulses. It can be performed by the main propulsion system of the top stage of the launch vehicle. More often it is done by a separate propulsion system at lower thrust levels than the upper stages in item (a) above. Orbit injection can be a single thrust operation after ascent from an earth launch station. If the flight path is a Hohmann transfer ellipse (minimum energy) or a faster transfer orbit, then two thrust application periods are necessary, one at the beginning and one at the end of the transfer path. For injection into earth orbit, the thrust levels are typically between 200 and 45,000 N or 50 and 11,000 lbf, depending on the payload size transfer time, and the specific orbit. If the new orbit is higher, then the thrusts are applied in the flight direction. If the new orbit is at a lower altitude, then the thrusts must be applied in a direction opposite to the flight velocity vector. The transfer orbits can also be achieved with a very low thrust level (0.001 to 1 N) using an electric propulsion system, but the flight paths will be very different (multi-loop spiral) and the transfer duration will be much longer. This is explained in Chapter 19. Similar maneuvers are also performed with lunar or interplanetary flight missions, as the planetary landing mission shown schematically in Fig. 4–10.

c. *Velocity vector adjustment and minor in-flight correction maneuvers* are usually performed with low thrust, short duration and intermittent (pulsing) operations, using a reaction control system with multiple small liquid propellant thrusters, both for translation and rotation. The vernier rockets on a ballistic missile are used to accurately calibrate the terminal velocity vector for improved target accuracy. The reaction control rocket systems in a space launch vehicle will allow accurate orbit injection adjustment maneuvers after it is placed into orbit by another, less accurate propulsion system. Mid-course guidance-directed *correction maneuvers* for the trajectories of deep space vehicles fall also into this category. Propulsion systems for *orbit maintenance maneuvers*, also called *station keeping maneuvers* (to overcome perturbing forces), keeping a spacecraft in its intended orbit and orbital position and are also considered to be part of this category.

d. *Reentry and landing maneuvers* can take several forms. If the landing occurs on a planet that has an atmosphere, then the drag of the atmosphere will slow down the reentering vehicle. For an elliptical orbit the drag will progressively reduce the perigee altitude and the perigee velocity on every orbit. Landing at a precise, preplanned location requires a particular velocity vector at a predetermined altitude and distance from the landing site. The vehicle has to be rotated into the right position and orientation, so as to use its heat shield correctly. The precise velocity magnitude and direction prior to entering the denser atmosphere are critical for minimizing the heat transfer (usually to the vehicle's heat shield) and to achieve touchdown at the intended landing site or, in the case of ballistic missiles, the intended target. This usually requires a relatively minor maneuver (low total impulse). If there is very little or no atmosphere (for instance, landing on the moon or Mercury), then a reverse thrust has to be applied during descent and touchdown. The rocket propulsion system usually has variable thrust to assure a soft landing and to compensate for the decrease in vehicle mass as propellant is consumed during descent. The lunar landing rocket engine, for example, had a 10 to 1 thrust variation.

e. *Rendezvous and docking* involve both rotational and translational maneuvers of small reaction control thrusters. Rendezvous and its time windows were discussed on page 123. Docking (sometimes called lock-on) is the linking up of two spacecraft and requires a gradual gentle approach (low thrust, pulsing node thrusters) so as not to damage the spacecraft.

f. *A change of plane of the flight trajectory* requires the application of a thrust force (through the vehicle center of gravity) in a direction normal to the original plane of the flight path. This is usually performed by a propulsion system that has been rotated (by the reaction control system) into the proper orientation. This maneuver is done to change the plane of a satellite orbit or when going to a planet, such as Mars, whose orbit is inclined to the plane of the earth's orbit.

g. *Simple rotational maneuvers* rotate the vehicle on command into a specific angular position so as to orient or point a telescope, instrument, solar panel, or antenna for purposes of observation, navigation, communication, or solar power reception. Such a maneuver is also used to keep the orientation of a satellite in a specific direction; for example, if an antenna needs to be continuously pointed at the center of the earth, then the satellite needs to be rotated around its own axis once every satellite revolution. Rotation is also used to point a nozzle of the primary propulsion system into its intended direction just prior to its start. It can also provide for achieving flight stability, or for correcting angular oscillations, that would otherwise increase drag or cause tumbling of the vehicle. Spinning or rolling a vehicle will improve flight stability, but will also average out the misalignment in a thrust vector. If the rotation needs to be performed quickly, then a chemical multi-thruster reaction control system is used. If the rotational changes can be done over a long period of time, then an electrical propulsion system with multiple thrusters is often preferred.

h. *De-orbiting and disposal of used or spent spacecraft* is required today to remove space debris. The spent spacecraft should not become a hazard to other spacecraft. A relatively small thrust will cause the vehicle to go to a low enough elliptical orbit so that atmospheric drag will cause further slowing. In the dense regions of the atmosphere the reentering, expended vehicle will typically break up or overheat (burn up).

i. *Emergency or alternative mission.* If there is a malfunction in a spacecraft and it is decided to abort the mission, such as a premature quick return to the earth without pursuing the originally intended mission, then some of the rocket engines can be used for an alternate mission. For example, the main rocket engine in the Apollo lunar mission service module is normally used for retroaction to attain a lunar orbit and for return from lunar orbit to the earth; it can be used for emergency separation of the payload from the launch vehicle and for unusual midcourse corrections during translunar coast, enabling an emergency earth return.

Table 4–6 lists the maneuvers that have just been described, together with some others, and shows the various types of rocket propulsion system (as mentioned in Chapter 1) that have been used for each of these maneuvers. The table omits several propulsion systems, such as solar thermal or nuclear

TABLE 4–6. Types of Rocket Propulsion System Commonly Used for Different Flight Maneuvers

Flight Maneuvers and Applications ↓	High thrust, liquid propellant rocket engine, with turbopump	Medium to low thrust, liquid propellant rocket engine	Pulsing liquid propellant, multiple small thrusters	Large solid propellant rocket motor, often segmented	Medium to small solid propellant motors	Arc jet, resisto jet	Ion propulsion, Electromagnetic propulsion	Pulsed plasma jet
Launch vehicle booster	××			××				
Strap-on motor/engine	××			××				
Upper stages of launch vehicle	××	××		×	××			
Satellite orbit injection and transfer orbits		××			××	×		
Flight velocity adjustments, Flight path corrections, Orbit raising		×	××			×	×	
Orbit/position maintenance, rotation of spacecraft			××			××	×	×
Docking of two spacecraft			××					
Reentry and landing, Emergency maneuvers		×	×					
Deorbit		×	×		×	×		
Deep space, Sun escape		×	×				×	
Tactical missiles					××			
Strategic missiles	×	×	×	××				
Missile defense			×	××	××			

Legend: × = in use: ×× = preferred for use.

rocket propulsion, because these have not yet flown in a real space mission. The electrical propulsion systems have very high specific impulse (see Table 2–1), which makes them very attractive for deep space missions, but they can be applied only to missions with sufficiently long thrust action time for reaching the desired vehicle velocity with very small acceleration. The items with a double mark "××" have been the preferred methods in recent years.

Reaction Control System

The functions of a reaction control system have been described in the previous section on flight maneuvers. They are used for the maneuvers identified by paragraphs c, e, and g. In some vehcle designs they are also used for tasks described in b, part of d, and f, if the thrust levels are low.

A *reaction control system* (RCS), often called an *auxiliary rocket propulsion system*, is needed to provide for trajectory corrections (small Δu additions), as well as correcting the rotational or attitude position of almost all spacecraft and all major launch vehicles. If only rotational maneuvers are made, it has been called an *attitude control system*. The nomenclature has not been consistent throughout the industry or the literature.

An RCS can be incorporated into the payload stage and each of the stages of a multiple stage vehicle. In some missions and designs the RCS is built into only the uppermost stage; it operates throughout the flight and provides the control torques and forces for all the stages. Liquid propellant rocket engines with multiple thrusters have been used for almost all launch vehicles and the majority of all spacecraft. Cold gas systems were used with early spacecraft design. In the last decade an increasing number of electrical propulsion systems have been used, primarily on spacecraft, as described in Chapter 19. The life of an RCS may be short (when used on an individual vehicle stage), or it may see use throughout the mission duration (perhaps 10 years) when part of an orbiting spacecraft.

The vehicle attitude has to be controlled about three mutually perpendicular axes, each with two degrees of freedom (clockwise and counterclockwise rotation), giving a total of six degrees of rotational freedom. *Pitch* control raises or lowers the nose of the vehicle, *yaw* torques induce a motion to the right or the left side, and *roll* torques will rotate the vehicle about its axis, either clockwise or counterclockwise. In order to apply a true torque it is necessary to use two thrust chambers of exactly equal thrust and equal start and stop times, placed an equal distance from the center of mass. Figure 4–13 shows a simple spherical spacecraft attitude control system; thrusters $x - x$ or $x' - x'$ apply torques that rotate about the X-axis. There is a minimum of 12 thrusters in this system, but some spacecraft with geometrical or other limitations on the placement of these nozzles or with provisions for redundancy may actually have more than 12. The same system can, by operating a different set of nozzles, also provide translation forces; for example, if one each of the thrust units x and x' were operated simultaneously, the resulting forces would propel the vehicle in the direction of the Y-axis. With clever design it is possible to use fewer thrusters.

An RCS usually contains the following major subsystems: (1) sensing devices for determining the attitude, velocity, and position of the vehicle with respect to a reference direction at any one time, such as provided by gyroscopes, star-trackers, or radio beacons; (2) a control-command system that compares the actual space and rotary position with the desired or programmed position and issues command signals to change the vehicle position within a desired time period; and (3) devices for changing the angular position, such as a set of high-speed gyroscopic wheels and a set of attitude control thrust-providing devices. See Refs. 4–12 and 4–14.

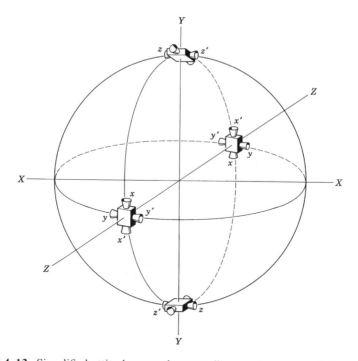

FIGURE 4–13. Simplified attitude control system diagram for spacecraft. It requires 12 thrusters (identified as x, y, z) to allow the application of pure torques about three perpendicular axes. The four unlabeled thrusters are needed for translation maneuvers along the z axis. They are shown here in four clusters.

A precise attitude angular correction can also be achieved by the use of an inertial or high-speed rotating reaction wheel, which applies torque when its rotational speed is increased or decreased. While these wheels are quite simple and effective, the total angular momentum change they can supply is generally small. By using a pair of supplementary attitude control thrust rocket units it is possible to unload or respin each wheel so it can continue to supply small angular position corrections as needed.

The torque T of a pair of thrust chambers of thrust F and a separation distance l is applied to give the vehicle with an angular or rotational moment of inertia M_a an angular acceleration of magnitude α:

$$T = Fl = M_a \alpha \tag{4–33}$$

For a cylinder of equally distributed mass $M_a = \frac{1}{2}mr^2$ and for a homogeneous sphere it is $M_a = \frac{2}{5}mr^2$. The largest possible practical value of moment arm l will minimize the thrust and propellant requirements. If the angular acceleration is constant over a time period t, the vehicle will move at an angular speed ω and through a displacement angle θ, namely

$$\omega = \alpha t \quad \text{and} \quad \theta = \frac{1}{2}\alpha t^2 \tag{4–34}$$

Commonly a control system senses a small angular disturbance and then commands an appropriate correction. For this detection of an angular position change by an accurate sensor it is actually necessary for the vehicle to undergo a slight angular displacement. Care must be taken to avoid overcorrection and hunting of the vehicle position or the control system. For this reason many spacecraft require extremely short multiple pulses (0.010 to 0.030 sec) and low thrust (0.01 to 100 N) (see Refs. 4–13 and 4–14).

Reaction control systems can be characterized by the magnitude of the total impulse, the number, thrust level, and direction of the thrusters, and by their duty cycles. The *duty cycle* refers to the number of thrust pulses, their operating times, the times between thrust applications, and the timing of these short operations during the mission operating period. For a particular thruster, a 30% duty cycle means an average active cumulative thrust period of 30% during the propulsion system's flight duration. These propulsion parameters can be determined from the mission, the guidance and control approach, the desired accuracy, flight stability, the likely thrust misalignments of the main propulsion systems, the three-dimensional flight path variations, the perturbations to the trajectory, and several other factors. Some of these parameters are often difficult to determine.

4.7. FLIGHT VEHICLES

As mentioned, the vast majority of rocket propelled vehicles are simple, single stage, and use solid propellant rocket motors. Most are used in military applications, as described in the next section. This section discusses more sophisticated multistage space launch vehicles and mentions others, such as large ballistic missiles (often called strategic missiles) and some sounding rockets. All have some intelligence in their guidance and navigation system. The total number of multistage rocket vehicles produced world wide in the last few years has been between 140 and 220 per year.

A single stage to orbit (LEO) is limited in the payload it can carry. Figure 4–2 shows that a high-performance single-stage vehicle with a propellant fraction of 0.95 and an average I_s of 400 sec can achieve an ideal terminal velocity of about 12,000 m/sec without payload. If the analysis includes drag and gravity forces, a somewhat higher value of I_s, maneuvers in the trajectory, and an attitude control system, it is likely that the payload would be between 0.2 and 1.4 percent of the gross take-off mass, depending on the design. For a larger percentage of payload, and for ambitious missions, we use vehicles with two or more stages as described here.

Multistage Vehicles

Multistep or *multistage rocket vehicles* permit higher vehicle velocities, more payload for space vehicles, and improved performance for long-range ballistic missiles. After the useful propellant is fully consumed in a particular stage, the remaining empty mass of that expended stage is dropped from the vehicle and the operation of the propulsion system of the next step or stage is started. The last or top stage, which is usually the smallest, carries the payload. The empty mass of the expended stage or step is separated from the remainder of the vehicle, because it avoids the expenditure of additional energy for further accelerating a useless mass. As the number of steps is increased, the initial takeoff mass can be decreased; but the gain in a smaller initial mass becomes less apparent when the total number of steps is large. Actually, the number of steps chosen should not be too large, because the physical mechanisms become more numerous, complex, and heavy. The most economical number of steps is usually between two and six, depending on the mission. Several different multistage launch vehicle configurations have been used successfully and four are shown in Fig. 4–14. Most are launched vertically, but a few have been launched from an airplane, such as the three-stage Pegasus space vehicle.

The payload of a multistage rocket is essentially proportional to the takeoff mass, even though the payload is only a very small portion of the initial mass. If a payload of 50 kg requires a 6000-kg multistage rocket, a 500-kg payload would require a 60,000-kg rocket unit with an identical number of stages, and a similar configuration with the same payload fraction. When the operation of the upper stage is started, immediately after thrust termination of the lower stage, then the total ideal velocity of a multistage vehicle of tandem or series-stage arrangement is simply the sum of the individual stage velocity increments. For n stages, the final velocity increment Δu_f is

$$\Delta u_f = \sum_1^n \Delta u = \Delta u_1 + \Delta u_2 + \Delta u_3 + \cdots \tag{4–35}$$

The individual velocity increments are given by Eq. 4–6. For the simplified case of a vacuum flight in a gravity-free field this can be expressed as

$$\Delta u_f = c_1 \ln(1/\mathbf{MR}_1) + c_2 \ln(1/\mathbf{MR}_2) + c_3 \ln(1/\mathbf{MR}_3) + \cdots \tag{4–36}$$

This equation defines the maximum velocity an ideal multistage vehicle can attain in a gravity-free vacuum environment. For more accurate actual trajectories the individual velocity increments can be determined by integrating Eqs. 4–15 and 4–16, which consider drag and gravity losses. Other losses or trajectory perturbations can also be included, as mentioned earlier in this chapter. Such an approach requires numerical solutions.

For two- or three-stage vehicles the overall vehicle mass ratio (initial mass at takeoff to final mass of last stage) can reach values of over 100 (corresponding to an equivalent single-stage propellant mass fraction ζ of 0.99). Figure 4–2 can be thus divided into regions for single- and multistage vehicles.

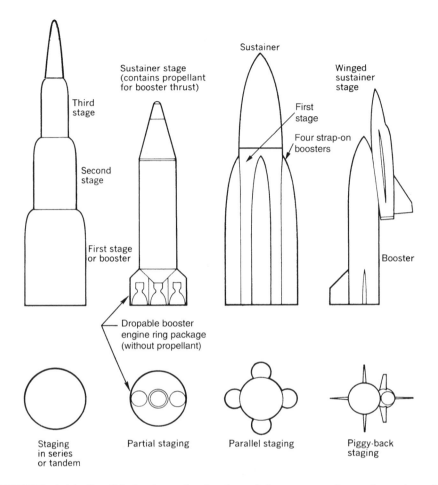

FIGURE 4–14. Simplified schematic sketches of four geometric configurations for assembling individual stages into a launch vehicle. The first is very common and the stages are stacked vertically on top of each other, as in the Minuteman long-range missile or the Delta launch vehicle. Partial staging was used on early versions of the Atlas; it allows all engines to be started at launching, thus avoiding a start during flight, and it permits the shut-off of engines on the launch stand if a failure is sensed prior to lift-off. The two booster engines, arranged in a doughnut-shaped assembly, are dropped off in flight. In the third sketch there are two or more separate "strap-on" booster stages attached to the bottom stage of a vertical configuration and this allows an increase in vehicle performance. The piggy-back configuration concept on the right is used in the Space Shuttle.

For multistage vehicles the stage mass ratios, thrust levels, propulsion durations, and the location or travel of the center of gravity of the stages are usually optimized, often using a complex trajectory computer program. The high specific impulse rocket engine (e.g., using hydrogen–oxygen propellants) is normally employed in upper stages of space launch vehicles, because a small increase in specific impulse is more effective there than in lower stages.

Example 4–3. A two-stage planetary exploration vehicle is launched from a high-orbit satellite into a gravity-free vacuum trajectory. The following notations are used and explained in the diagram.

m_0 = initial mass of vehicle (or stage) at launch

m_p = useful propellant mass of stage

m_i = initial mass of stage(s)

m_f = final mass of stage (after rocket operation); it includes the empty propulsion system with its residual propellant, the structures of the vehicle and the propulsion system, the control, guidance, and payload masses.

m_{pl} = payload mass; it includes the guidance, control and communications equipment, antennas, scientific instruments, research apparatus, power supply, solar panels, sensors, etc.

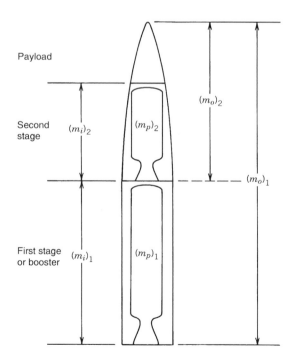

Subscripts 1 and 2 refer to first and second stages. The following are given:

Flight and velocity increment in gravity-free vacuum	6200 m/sec
Specific impulse, I_s	310 sec
Effective exhaust velocity, c (all stages)	3038 m/sec
Initial launch vehicle mass	4500 kg
Propellant mass fraction, ζ (each stage)	0.88
Structural mass fraction, $(1 - \zeta)$ (each stage)	0.12

Determine the payload for two cases: (1) when the two stage masses are equal, and (2) when the mass ratios of the two stages are equal.

SOLUTION. For launch the takeoff mass (m_0) equals the loaded first-stage mass $(m_i)_1$ plus the loaded second-stage mass $(m_i)_2$ plus the payload (m_{pl}). The propellant mass fraction ζ is 0.88. For case (1) the first and second stages are identical. Thus

$$m_i = (m_i)_1 = (m_i)_2$$

$$m_p = (m_p)_1 = (m_p)_2 = 0.88 m_i$$

$$(m_p)_1 = 0.88(m_i)_1$$

$$(m_0)_1 = 4500 \text{ kg} = 2m_i + m_{pl}$$

$$e^{\Delta u/c} = e^{6200/3038} = 7.6968 = \frac{(m_0)_1}{(m_0)_1 - (m_p)_1} \cdot \frac{(m_0)_2}{(m_0)_2 - (m_p)_2}$$

From these relationships it is possible to solve for the payload mass m_{pl}, which is 275 kg.

$$m_i = (4500 - 275)/2 = 2113 \text{ kg each stage}$$

$$m_p = 0.88 m_i = 1855 \text{ kg each stage}$$

For case (2) the mass ratios of the two stages are the same. The mass ratio $(1/\text{MR})$ was defined by

$$m_0/m_f = (m_0)_1/[(m_0)_1 - [(m_p)_1] = (m_0)_2/[(m_0)_2 - (m_p)_2]$$

$$(m_0)_1 = 4500 = (m_i)_1 + (m_i)_2 + m_{pl}$$

$$e^{\Delta u/c} = 7.6968 = \{4500/[4500 - (m_p)_1]\}^2$$

Solving for the first-stage propellant mass gives $(m_p)_1 = 2878$ kg.

$$(m_i)_1 = (m_p)_1/0.88 = 3270 \text{ kg}$$

$$(m_0)_2 = (m_i)_2 + m_{pl} = 4500 - 3270 = 1230 \text{ kg}$$

$$e^{\Delta u/c} = 7.6968 = \{1230/[1230 - (m_p)_2]\}^2; \quad (m_p)_2 = 786.6 \text{ kg}$$

$$(m_i)_2 = (m_p)_2/0.88 = 894 \text{ kg}$$

The payload m_{pl} is $1230 - 894 = 336$ kg. This is about 22% larger than the payload of 275 kg in the first case. When the mass ratios of the stages are equal, the payload is a maximum for gravity-free vacuum flight and the distribution of the masses between the stages is optimum. For a single-stage vehicle with the same take-off mass and same propellant fraction, the payload is substantially less. See Problem 4–13.

If a three-stage vehicle had been used in Example 4–3 instead of a two-stage version, the payload would have been even larger. However, the theoretical payload increase will only be about 8 or 10%. A fourth stage gives an even smaller theoretical improvement; it would add only 3 to 5% to the payload. The amount of potential performance improvement diminishes with each added stage. Each additional stage means extra complications in an actual vehicle (such as a reliable separation mechanism, an interstage structure, joints or couplings in a connecting pipes and cables, etc.), requires additional inert mass (increasing the mass ratio **MR**), and compromises the overall reliability. Therefore, the minimum number of stages that will meet the payload and the Δu requirements is usually selected.

The flight paths taken by the vehicles in the two simplified cases of Example 4–3 are different, since the time of flight and the acceleration histories are different. One conclusion from this example applies to all multistage rocket-propelled vehicles; for each mission there is an optimum number of stages, an optimum distribution of the mass between the stages, and there is usually also an optimum flight path for each design, where a key vehicle parameter such as payload, velocity increment, or range is a maximum.

Launch Vehicles

Usually the *first or lowest stage*, often called a *booster stage*, is the largest and it requires the largest thrust and largest total impulse. All stages need chemical propulsion to achieve the desired thrust-to-weight ratio. These thrusts usually become smaller with each subsequent stage, also known as *upper stage* or *sustainer stage*. The thrust magnitudes depend on the mass of the vehicle, which in turn depends on the mass of the payload and the mission. Typical actual configurations are shown by simple sketches in Fig. 4–14. There is an optimum size and thrust value for each stage in a multistage vehicle and the analysis to determine these optima can be quite complex.

Many heavy launch vehicles have two to six *strap-on solid propellant motor boosters*, which together form a supplementary first stage strapped on or mounted to the first stage of the launch vehicle (Space Shuttle, Titan, Delta, Atlas, Ariane). This is shown in the third sketch of Fig. 4–14. The Russians have used liquid propellant strap-on boosters on several vehicles, because they give better performance. Boosters operate simultaneously with the first stage and, after they burn out, they are usually separated and dropped off before completion of the first stage's propulsive operation. This has also been called a *half stage* or *zero stage*, as in Table 1–3.

There is a variety of existing launch vehicles. The smaller ones are for low payloads and low orbits; the larger ones usually have more stages, are heavier, more expensive, have larger payloads, or higher mission velocities. The vehicle cost increases with the number of stages and the initial vehicle launch mass. Once a particular launch vehicle has been proven to be reliable, it is usually modified and uprated to allow improvements in its capability or mission flexibility. Each of the stages of a space launch vehicle can have several rocket engines, each with specific missions or maneuvers. The Space Shuttle system has 67 different rockets which are shown schematically in Fig. 1–13. In most cases each rocket engine is used for a specific maneuver, but in many cases the same engine is used for more than one specific purpose; the small reaction control thrusters in the Shuttle serve, for example, to give attitude control (pitch, yaw, and roll) during orbit insertion and reentry, for counteracting internal shifting of masses (astronaut movement, extendible arm), small trajectory corrections, minor flight path adjustments, docking, and precise pointing of scientific instruments.

The *spacecraft* is that part of a launch vehicle that carries the payload. It is the only part of the vehicle that goes into orbit or deep space and some are designed to return to earth. The final major space maneuver, such as orbit injection or planetary landing, often requires a substantial velocity increment; the propulsion system, which provides the force for this maneuver, may be integrated with the spacecraft, or it may be part of a discardable stage, just below the spacecraft. Several of the maneuvers described in Section 4–6 can often be accomplished by propulsion systems located in two different stages of a multistage vehicle. The selection of the most desirable propulsion systems, and the decision on which of the several propulsion systems will perform specific maneuvers, will depend on optimizing performance, cost, reliability, schedule, and mission flexibility as described in Chapter 17.

When a space vehicle is launched from the earth's surface into an orbit, it flies through three distinct trajectory phases. (1) Most are usually launched vertically and then undergo a turning maneuver while under rocket power to point the flight velocity vector into the desired direction. (2) The vehicle then follows a free-flight (unpowered) ballistic trajectory (usually elliptical), up to its apex. Finally (3) a satellite needs an extra push from a chemical rocket system up to add enough total impulse or energy to accelerate it to orbital velocity. This last maneuver is also known as *orbit insertion*. During the initial powered flight the trajectory angle and the thrust cut-off velocity of the last stage are adjusted by the guidance system to a velocity vector in space that will allow the vehicle to reach the apogee of its elliptic path exactly at the desired orbit altitude. As shown in Fig. 4–9, a *multistage ballistic missile* follows the same two ascent flight phases mentioned above, but it then continues its elliptical ballistic trajectory all the way down to the target.

Historically successful launch vehicles have been modified, enlarged, and improved in performance. The newer versions retain most of the old, proven, reliable components, materials, and subsystems. This reduces development effort and cost. Upgrading a vehicle allows an increase in mission energy (more ambitious mission) or payload. Typically, it is done by one or more of these types of improvement: increasing the mass of propellant without an undue increase in tank or case mass; uprating the thrust and strengthening the engine; more specific impulse; or adding successively more or bigger strap-on boosters. It also usually includes a strengthening of the structure to accept higher loads.

Figure 4–15 and Table 4–7 illustrate the growth of payload and mission capability for the early Titan family of space launch vehicles and the effect of the orbit on the payload. The figure shows the evolution of four different multistage configurations of the launch vehicle and their principal propulsion systems; the table defines the increase in payload for the four vehicle configurations and also how the payload is reduced as more ambitious orbits are flown. When each of these vehicles is equipped with an additional third stage, it is able to launch substantial payloads into earth escape or synchronous orbit. The table describes the propulsion for each of the several stages used on those vehicles and the payload for several arbitrarily selected orbits.

Table 4–7 shows the effects of orbit inclination and altitude on the payload. The inclination is the angle between the equatorial plane of the earth and the trajectory. An equatorial orbit has zero inclination and a polar orbit has 90° inclination. Since the earth's rotation gives the vehicle an initial velocity, a

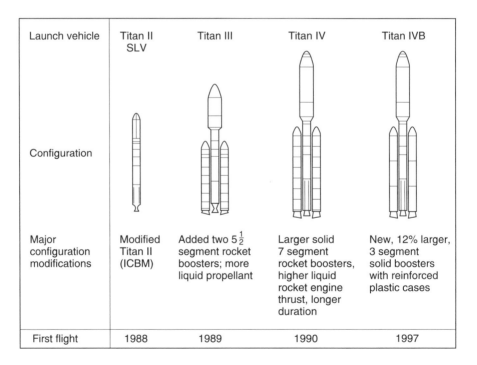

Launch vehicle	Titan II SLV	Titan III	Titan IV	Titan IVB
Configuration				
Major configuration modifications	Modified Titan II (ICBM)	Added two 5 $\frac{1}{2}$ segment rocket boosters; more liquid propellant	Larger solid 7 segment rocket boosters, higher liquid rocket engine thrust, longer duration	New, 12% larger, 3 segment solid boosters with reinforced plastic cases
First flight	1988	1989	1990	1997

FIGURE 4–15. Upgrading methods are illustrated by these four related configurations in the evolution of the Titan Space Launch Vehicle family. Source: Lockheed-Martin Corp.

TABLE 4-7. Payload Capabilities and Rocket Propulsion Systems of Four Titan Space Launch Vehicle Configurations

Space Launch Vehicle	Titan II SLV	Titan III	Titan IV	Titan IV B
	Payloads (lbm) in Low Earth Orbits for 2-Stage Configurations			
100 mi circular orbit, 28.6° inclination from Cape Canaveral	5000	31,000	39,000	47,800
Same, but 99° launch from Vandenberg AFB	4200	26,800	32,000	38,800
Elliptic orbit, 100 mi → 1000 mi, 28.6° inclination	3000	25,000	~ 30,000	~ 34,000
	Payloads (lbm) in Synchronous Earth Orbit, 3-Stage Configurations			
Payload for third-stage propulsion system, optional (see below)	2200	4000	10,000	12,700
	Rocket Propulsion Systems in Titan Launch Vehicles			
Solid rocket boosters (United Technologies/CSD)	None	2 units, each metal case $5\frac{1}{2}$ segments $I_t = 123 \times 10^6$ lbf-sec	Same, but 7 segments $I_t = 159.7 \times 10^6$ lbf-sec	12% more propellant, 3 segments $I_t = 179 \times 10^6$ lbf-sec
Stage I, Aerojet LR 87-AJ-11 engine, N_2O_4 with 50% N_2H_4/50% UDMH	2 thrust chambers 430,000 lbf thrust at SL	Same, 529,000 lbf thrust (vacuum)	Same, but uprated to 550,000 lbf thrust in a vacuum	Same
Stage II, Aerojet LR 91-AJ-11 engine N_2O_4 with 50% N_2H_4 50% UDMH	101,000 lbf thrust in vacuum	Same	Uprated to 106,000 lbf thrust in vacuum	Same
Stage III has several alternative systems for each vehicle; only one is listed here	SSPS with Aerojet liquid storable propellant engine AJ 10-118 K (9800 lbf thrust)	United Technologies/ CSD, Interim Upper Stage (IUS) solid propellant rocket motor (see Table 11-3)	Centaur; 2 Pratt & Whitney RL 10A-3-3A rocket engines, 33,000 lbf thrust, H_2/O_2	Same

Source: Lockheed-Martin Astronautics, Aerojet Propulsion Company, and Pratt & Whitney Division of United Technologies Corp.

launch from the equator in a eastward direction will give the highest payload. For the same orbit altitude other trajectory inclinations have a lower payload. For the same inclination the payload decreases with orbit altitude, since more energy has to be expended to overcome gravitational attraction.

The Space Shuttle has its maximum payload when launched due east into an orbit with 28.5° inclination from Kennedy Space Flight Center in Florida, namely about 56,000 lb (or 25,455 kg) at a 100 nautical mile (185 km) orbit altitude. The payload decreases by about 100 lb (45.4 kg) for every nautical mile increase in altitude. If the inclination is 57°, the payload diminishes to about 42,000 lb (or 19,090 kg). If launched in a southerly direction from Vandenberg Air Force Base on the west coast in a 98° inclination into a circular, nearly polar orbit, the payload will be only about 30,600 lb or 13,909 kg.

The dramatic decrease of payload with circular orbits of increasing altitude and with different inclination is shown for the Pegasus, a relatively small, air-launched, space launch vehicle, in Fig. 4–16. The payload is a maximum when launching from the earth equator in the east direction, that is at 0° inclination.

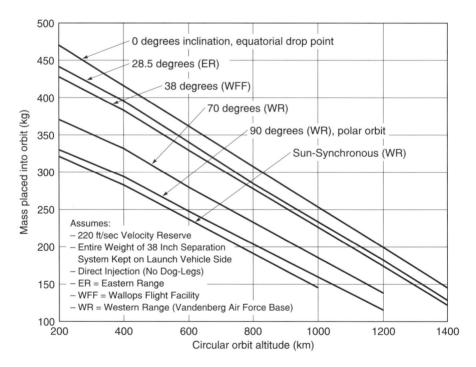

FIGURE 4–16. Decrease of payload with circular orbit altitude and orbit inclination for the Pegasus launch vehicle. This is an air-launched, relatively simple, three-stage launch vehicle of 50 in. diameter driven by a solid propellant rocket motor in each stage. (Courtesy Orbital Sciences Corporation)

The figure shows that a practical payload becomes too small for orbits higher than about 1200 km. To lift heavier payloads and to go to higher orbits requires a larger launch vehicle than the Pegasus. Figure 4–16 is based on the assumption of a particular payload separation mechanism (38 in.) and a specific Δu vehicle velocity reserve (220 ft/sec), for items such as the normal changes in atmospheric density (which can double the drag) or mass tolerances of the propulsion systems. Similar curves can be provided by the makers of all launch vehicles.

4.8. MILITARY MISSILES

The majority of all rocket propulsion systems built today are for military purposes. There is a large variety of missiles and military missions and therefore many different propulsion systems. All are chemical propulsion systems. They range from simple, small, unguided, fin-stabilized single-stage rocket projectiles (used in air-to-surface missions and surface-to-surface bombardment) up to complex, sophisticated, expensive, long-range, multistage ballistic missiles, which are intended for faraway military or strategic targets. The term "surface" means either land surface (ground launch or ground target), ocean surface (ship launched), or below the ocean surface (submarine launched). A *tactical missile* is used for attacking or defending ground troops, nearby military or strategic installations, military aircraft, or war missiles. The armed forces also use *military satellites* for missions such as reconnaissance, early warning of impending attack, secure communication, or navigation.

Strategic missiles with a range of 3000 km or more have been two- or three-stage surface-to-surface rocket-propelled missiles. Early designs used liquid propellant rocket engines and some are still in service. Beginning about 30 years ago, newer strategic missiles have used solid propellant rocket motors. Both types usually also have a liquid propellant reaction control system (RCS) for accurately adjusting the final payload flight velocity (in magnitude, direction, and position in space) at the cut-off of the propulsion system of the last stage. A solid propellant RCS version also exists. The flight analysis and ballistic trajectories of the long-range missiles are similar in many ways to those described for launch vehicles in this chapter. See Fig. 4–9.

Solid propellant rocket motors are preferred for most tactical missile missions, because they allow simple logistics and can be launched quickly (Ref. 4–15). If altitudes are low and flight durations are long, such as with a cruise missile, an air-breathing jet engine and a winged vehicle, which provides lift, will usually be more effective than a long-duration rocket. However, a large solid propellant rocket motor is still needed as a booster to launch the cruise missile and bring it up to speed. There are a variety of different tactical missions, resulting in different sized vehicles with different propulsion needs, as explained later in this section and in Ref. 4–15.

For each of the tactical missile applications, there is an optimum rocket propulsion system and almost all of them use solid propellant rocket motors. For each application there is an optimum total impulse, an optimum thrust–time profile, an optimum nozzle configuration (single or multiple nozzles, with or without thrust vector control, optimum area ratio), optimum chamber pressure, and a favored solid propellant grain configuration. Low exhaust plume gas radiation emissions in the visible, infrared or ultraviolet spectrum and certain safety features (making the system insensitive to energy stimuli) can be very important in some of the tactical missile applications; these are discussed in Chapters 12 and 18.

Short-range, uncontrolled, unguided, single-stage rocket vehicles, such as military rocket projectiles (ground and air launched) and rescue rockets, are usually quite simple in design. Their general equations of motion are derived in Section 4.3, and a detailed analysis is given in Ref. 4–1.

Unguided military rocket-propelled missiles are today produced in larger numbers than any other category of rocket-propelled vehicles. The 2.75 in. diameter, folding fin unguided solid propellant rocket missile has recently been produced in the United States in quantities of almost 250,000 per year. Guided missiles for anti-aircraft, anti-tank, or infantry support have been produced in annual quantities of hundreds and sometimes over a thousand. Table 1–6 lists several guided missiles.

Because these rocket projectiles are essentially unguided missiles, the accuracy of hitting a target depends on the initial aiming and the dispersion induced by uneven drag, wind forces, oscillations, and misalignment of nozzles, body, and fins. Deviations from the intended trajectory are amplified if the projectile is moving at a low initial velocity, because the aerodynamic stability of a projectile with fins is small at low flight speeds. When projectiles are launched from an aircraft at a relatively high initial velocity, or when projectiles are given stability by spinning them on their axis, their accuracy of reaching a target is increased two- to ten-fold, compared to a simple fin-stabilized rocket launched from rest.

In guided *air-to-air* and *surface-to-air rocket-propelled missiles* the time of flight to a given target, usually called the *time to target t_t*, is an important flight-performance parameter. With the aid of Fig. 4–17 it can be derived in a simplified form by considering the distance traversed by the rocket (called the range) to be the integrated area underneath the velocity–time curve. This simplification assumes no drag, no gravity effect, nearly horizontal flight, a relatively small distance traversed during powered flight compared to the total range, and a linear increase in velocity during powered flight.

$$t_t = \frac{S + \frac{1}{2} u_p t_p}{u_0 + u_p} \tag{4-37}$$

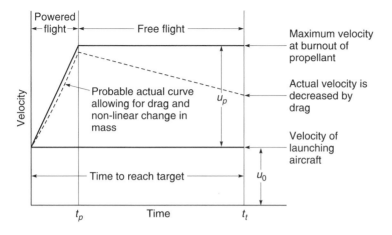

FIGURE 4-17. Simplified trajectory for an unguided, non-maneuvering, air-launched rocket projectile. Solid line shows flight velocity without drag or gravity and dashed curve shows likely actual flight.

Here S is the free-flight (unpowered) range, u_p is the velocity increase of the rocket during powered flight up to the time of burnout, t_p is the time of rocket burning, and u_0 is the initial velocity of the launching aircraft. For more accurate values, the velocity increase u_0 is the initial velocity of the launching aircraft. For more accurate values, the velocity increase u_p is given by Eq. 4–19. More accurate values can only be obtained through a detailed step-to-step trajectory analysis that considers the effects of drag and gravity.

In unguided air-launched air-to-air or air-to-surface projectiles the aiming is done by orienting the launching aircraft. In guided missiles (air-to-air, air-to-ground, ground-to-air, or ground-to-incoming-missile) the rocket's thrust direction, thrust magnitude, or thrust pulse timing can be commanded by an intelligent guidance and control system to chase a maneuvering moving target. The guidance system senses the flight path of the target, calculates a predicted impact point, and then controls the flight path of the guided missile to achieve an impact (or near-impact if a proximity fuse is used) with the target. It can also apply to a ground-launched or a satellite-launched antiballistic missile. In both the unguided projectile and the guided missile the hit probability increases as the time to target t_t is reduced. In one particular air-to-air combat situation, the effectiveness of the rocket projectile varied approximately inversely as the cube of the time to target. The best results (e.g., best hit probability) are usually achieved when the time to target is as small as practically possible.

The analysis of the missile and propulsion configuration that gives the minimum time to target over all the likely flight scenarios can be complex. The following rocket propulsion features and parameters will help to reduce the time to target, but their effectiveness will depend on the specific mission, range, guidance and control system, and the particular flight conditions.

1. *High initial thrust* or *high initial acceleration* for the missile to quickly reach a high-initial-powered flight velocity.

2. Application of additional *lower thrust* to counteract drag and gravity losses and thus maintain a high flight velocity. This can be a single rocket propulsion system that has a short high initial thrust and a smaller (10 to 25%) sustaining thrust of lower duration. It can also be a system that applies discrete pulses of thrust to increase vehicle velocity after drag forces have caused it to diminish, thus maintaining a higher average flight velocity.

3. For higher supersonic flight speeds, a *two-stage missile* can be more effective. Here the first stage is dropped off after its propellant has been consumed, thus reducing the inert mass of the next stage, and improving its mass ratio and thus its flight velocity increase.

4. If the target is highly maneuverable and if the closing velocity between missile and target is large, it may be necessary not only to provide an axial thrust, but also to apply large *side forces* or side accelerations to a tactical missile. This can be accomplished either by aerodynamic forces (lifting surfaces or flying at an angle of attack) or by multiple nozzle propulsion systems with variable or pulsing thrusts; the rocket engine then has an axial thruster and several side thrusters. The thrusters have to be so located that all the thrust forces are essentially directed through the center of gravity of the vehicle. The thrusters that provide the side accelerations have also been called *divert* thrusters, since they divert the vehicle in a direction normal to the axis of flight direction.

5. *Drag losses* can be reduced if the missile has a large L/D ratio (or a small cross-sectional area) and if the propellant density is high, allowing a smaller missile volume. The drag forces can be high if the missile travels at low altitude and high speed.

A unique military application is *rocket assisted gun launched projectiles* for attaining longer artillery ranges. Their small rocket motors withstand very high accelerations in the gun barrel (5000 to 10,000 g_0 is typical). They are in production.

4.9. AERODYNAMIC EFFECT OF EXHAUST PLUMES

The effect of rocket exhaust jets or plumes on the aerodynamic characteristics of a missile is usually to decrease the vehicle drag at supersonic missile speeds and to increase it at subsonic speeds. On *subsonic* vehicles, a supersonic rocket plume acts very much like an ejector and sucks adjacent air into its path. This affects vehicles where the rocket is located on a tapering aft end. The ejector action of the flame accelerates the adjacent air, thereby increasing the skin friction locally and usually reducing the pressure on the vehicle aft body or base plate near the nozzle exit location.

At *supersonic* speeds there often is a turbulent wake area with a low local pressure at the aft end of projectile. With the action of a rocket plume, the void space is filled with rocket gases and the pressure on the aft portion of the body is increased. This increases the pressure thrust and thus reduces the base drag. Exhaust plume effects are discussed in Chapter 18. In fact, some artillery munitions and short-range rockets can achieve increased range (by 10 to 50%) by adding a small rocket-type gas generator; its plume fills the void at the base of the projectile with reaction gas at a finite pressure, thus increasing the base pressure of the projectile and reducing the base drag.

4.10. FLIGHT STABILITY

Stability of a vehicle is achieved when the vehicle does not rotate or oscillate in flight. Unstable flights are undesirable, because pitch or yaw oscillations increase drag (flying at an angle of attack most of the time) and cause problems with instruments and sensors (target seekers, horizon scanners, sun sensors, or radar). Instability often leads to tumbling (uncontrolled turning) of vehicles, which causes missing of orbit insertion, missing targets, or sloshing of liquid propellant in tanks.

Stability can be built in by proper design so that the flying vehicle will be inherently stable, or stability can be obtained by appropriate controls, such as the aerodynamic control surfaces on an airplane, a reaction control system, or hinged multiple rocket nozzles.

Flight stability exists when the overturning moments (e.g., those due to a wind gust, thrust misalignment, or wing misalignment) are smaller than the stabilizing moments induced by thrust vector controls or by aerodynamic control surfaces. When the destabilizing moments exceed the stabilizing moments about the center of gravity, the vehicle turns or tumbles. In unguided vehicles, such as low-altitude rocket projectiles, stability of flight in a rectilinear motion is achieved by giving a large stability margin to the vehicle by using tail fins and by locating the center of gravity ahead of the center of aerodynamic pressure. In a vehicle with an active stability control system, a nearly neutral inherent stability is desired, so that the applied control forces are small, thus requiring small control devices, small RCS thrusters, small actuating mechanisms, and structural mass. Neutral stability is achieved by locating aerodynamic surfaces and the mass distribution of the components within the vehicle in such a manner that the center of gravity is only slightly above the center of aerodynamic pressure. Because the aerodynamic moments change with Mach number, the center of pressure does not stay fixed during accelerating flight but shifts, usually along the vehicle axis. The center of gravity also changes its position as propellant is consumed and the vehicle mass decreases. Thus it is usually very difficult to achieve neutral missile stability at all altitudes, speeds, and flight conditions.

Stability considerations affect rocket propulsion system design in several ways. By careful nozzle design it is possible to minimize thrust misalignment and thus to minimize torques on the vehicle and the reaction control propellant consumption. It is possible to exercise control over the travel of the center of gravity by judicious design. In liquid propellant rockets, special design provisions, special tank shapes, and a careful selection of tank location in the vehicle afford this possibility. The designer generally has less freedom in controlling the travel of the center of gravity of solid propellant rockets. By using nozzles at the end of a blast tube, as shown in Fig. 14–6, it is possible to place the solid propellant mass close to the vehicle's center of gravity. Attitude control liquid propellant engines with multiple thrusters have been used satisfactorily to obtain control moments for turning vehicles in several ways, as described in Section 4.6 and in Chapter 6.

Unguided rocket projectiles and missiles are often given a roll or rotation by inclined aerodynamic fins or inclined multiple rocket exhaust gas nozzles to improve flight stability and accuracy. This is similar to the rotation given to bullets by spiral-grooved rifles. This *spin stability* is achieved by gyroscopic effects, where an inclination of the spin axis is resisted by torques. The centrifugal effects cause problems in emptying liquid propellant tanks and extra stresses on solid propellant grains. In some applications a low-speed roll is applied not for spin stability but to assure that any effects of thrust vector deviations or aerodynamic shape misalignments are minimized and canceled out.

PROBLEMS

1. For a vehicle in gravitationless space, determine the mass ratio necessary to boost the vehicle velocity by 1600 m/sec when the effective exhaust velocity is 2000 m/sec.
   Answer: 0.449.

2. What is the mass ratio m_p/m_0 for a vehicle that has one-fifth its original takeoff mass at the time of the completion of rocket operation?
   Answer: 0.80.

3. Determine the burnout velocity and burnout altitude for a dragless projectile with the following parameters for a simplified vertical trajectory: $\bar{c} = 2209$ m/sec; $m_p/m_0 = 0.57$; $t_p = 5.0$ sec; and $u_0 = h_0 = 0$.
   Answers: $u_p = 1815$ m/sec; $h_p = 3.89 \times 10^3$ m.

4. Assume that this projectile had a drag coefficient essentially similar to the $0°$ curve in Fig. 4–3 and redetermine the answers of Problem 3 and the approximate percentage errors in u_p and h_p. Use a step-by-step method.

5. A research space vehicle in gravity-free and drag-free outer space launches a smaller spacecraft into a meteor shower region. The 2 kg instrument package of this spacecraft (25 kg total mass) limits the maximum acceleration to no more than 50 m/sec². It is launched by a solid propellant rocket motor ($I_s = 260$ sec and $\zeta = 0.88$). Determine

   (a) the maximum allowable burn time, assuming steady propellant mass flow;

   (b) the maximum velocity relative to the launch vehicle.

(c) Solve for (a) and (b) if half of the total impulse is delivered at the previous propellant mass flow rate, with the other half at 20% of this mass flow rate.

6. For a satellite cruising in a circular orbit at an altitude of 500 km, determine the period of revolution, the flight speed, and the energy expended to bring a unit mass into this orbit.
   Answers: 1.58 hr, 7613 m/sec, 33.5 MJ/kg.

7. A large ballistic rocket vehicle has the following characteristics: propellant mass flow rate: 12 slugs/sec (1 slug = 32.2 lbm = 14.6 kg); nozzle exit velocity: 7100 ft/sec; nozzle exit pressure: 5 psia (assume no separation); atmospheric pressure: 14.7 psia (sea level); takeoff weight: 12.0 tons (1 ton = 2000 lbf); burning time: 50 sec; nozzle exit area: 400 in.^{2}. Determine (*a*) the sea-level thrust; (*b*) the sea-level effective exhaust velocity; (*c*) the initial thrust-to-weight ratio; (*d*) the initial acceleration; (*e*) the mass inverse ratio m_0/m_f.
   Answers: 81,320 lbf; 6775 ft/sec; 3.38; 2.38g_0.

8. In Problem 7 compute the altitude and missile velocity at the time of power plant cutoff, neglecting the drag of the atmosphere and assuming a simple vertical trajectory.

9. A spherical satellite has 12 identical monopropellant thrust chambers for attitude control with the following performance characteristics: thrust (each unit): 5 lbf; I_s (steady state or more than 2 sec): 240 sec; I_s (pulsing duration 20 msec): 150 sec; I_s (pulsing duration 100 msec): 200 sec; satellite weight: 3500 lbf; satellite diameter: 8 ft; satellite internal density distribution is essentially uniform; disturbing torques, Y- and Z-axes: 0.00005 ft-lbf average; disturbing torque, for X-axis: 0.001 ft-lbf average; distance between thrust chamber axes: 8 ft; maximum allowable satellite pointing position error: ±1°. Time interval between pulses is 0.030 sec.

   (a) What would be the maximum and minimum vehicle angular drift per hour if no correction torque were applied?
   Answers: 0.466 and 0.093 rad.

   (b) What is the frequency of pulsing action (how often does an engine pair operate?) at 20-msec, 100-msec, and 2-sec pulses in order to correct for angular drift? Discuss which pulsing mode is best and which is impractical.

   (c) If the satellite was to remain in orbit for 1 year with these same disturbances and had to maintain the accurate positions for 24 hr each day, how much propellant would be required? Discuss the practicality of storing and feeding such propellant.

10. For an ideal multistage launch vehicle with several stages, discuss the following: (*a*) the effect on the ideal mission velocity if the second and third stages are not started immediately but are each allowed to coast for a short period after shutoff and separation of the prior stage before rocket engine start of the next stage; (*b*) the effect on the mission velocity if an engine malfunctions and delivers a few percent less than the intended thrust but for a longer duration and essentially the full total impulse of that stage.

11. Given a cylindrically shaped space vehicle ($D = 1$ m, height is 0.7 m, average density is 1.1 g/cm^3) with a flat solar cell panel on an arm (mass of 32 kg, effective moment arm is 1.5 m, effective average area facing normally toward sun is 0.6 m^2) in a set of

essentially frictionless bearings and in a low orbit at 160 km altitude with sunlight being received, on the average, about 60% of the period:

(a) Compute the maximum solar pressure-caused torque and the angular displacement this would cause during 1 day if not corrected.

(b) Using the data from the atmospheric table in Appendix 2 and an arbitrary average drag coefficient of 1.0 for both the body and the flat plate, compute the drag force and torque.

(c) Using stored high-pressure air at 14×10^6 N/m^2 initial pressure as the propellant for attitude control, design an attitude control system to periodically correct for these two disturbances (F, I_s, t, I_t, etc.).

(d) If the vector of the main thrust rocket of the vehicle (total impulse of 67×10^3 N-sec) is misaligned and misses the center of gravity by 2 mm, what correction would be required from the attitude control system? What would need to be done to the attitude control system in c above to correct for this error also?

12. A bullet-shaped toy rocket has a pressurized tank of volume V_0, and is partly filled with water (an incompressible liquid) and partly with compressed air at initial pressure of 50 psia and initial ambient temperature T_0. Assume no water losses during start. Also assume that the ambient air pressure is constant for the altitudes attained by this toy rocket. The empty weight of the toy is 0.30 lbf and it can carry 1.0 lbm of water when the V_0 is half-filled with water. Make other assumptions to suit the calculations.

(a) What type of nozzle is best for this application?
    Answer: Converging nozzle.

(b) What are the desired nozzle dimensions to assure vertical takeoff with about 0.5 g acceleration?

(c) What is the specific impulse of the water at start and near propellant exhaustion?

(d) What happens if only 50 psia air (no water) is ejected?

(e) What is the approximate proportion of water to air volume for maximum altitude?

(f) Sketch a simple rocket release and thrust start device and comment on its design and potential problems.

(g) About how high will it fly vertically?

13. Determine the payload for a single-stage vehicle in Example 4–3. Compare it with the two-stage vehicle.
    Answer: 50.7 kg, which is 18.4% of the payload for a two-stage vehicle.

14. Use the data given in Example 4–3, except that the payload is fixed at 250 kg and the Δu is not given but has to be determined for both cases, namely equal-sized stages and stages of equal mass ratio. What can be concluded from these results and the results in the example?

15. An airplane that is flying horizontally at a 7000 m altitude, at a speed of 700 km/hr over flat country, releases an unguided missile with four small tail fins for flight stability. Determine the impact location (relative to the release point as projected

onto the earth surface), the impact angle, and the time from release to target. Assume that the drag causes an average of about 8% reduction in flight velocities.

16. An earth satellite is in an elliptical orbit with the perigee at 600 km altitude and an eccentricity of $e = 0.866$. Determine the parameters of the new satellite trajectory, if a rocket propulsion system is fired in the direction of flight giving an incremental velocity of 200 m/sec (a) when fired at apogee, (b) when fired at perigee, and (c) when fired at perigee, but in the opposite direction, reducing the velocity.

17. A sounding rocket (75 kg mass, 0.25 m diameter) is speeding vertically upward at an altitude of 5000 m and a velocity of 700 m/sec. What is the deceleration in multiples of g due to gravity and drag? (Use C_D from Fig. 4–3 and use Appendix 2).

18. A single-stage weather sounding rocket has a take-off mass of 1020 kg, a sea-level initial acceleration of 2.00 g, carries 799 kg of useful propellant, has an average specific gravity of 1.20, a burn duration of 42 sec, a vehicle body shaped like a cylinder with an L/D ratio of 5.00 with a nose cone having a half angle of 12 degrees. Assume the center of gravity does not change during the flight. The vehicle tumbled (rotated in an uncontrolled manner) during the flight and failed to reach its objective. Subsequent evaluation of the design and assembly processes showed that the maximum possible thrust misalignment was 1.05 degrees with a maximum lateral off-set of 1.85 mm. Assembly records show it was 0.7 degrees and 1.1 mm for this vehicle. Since the propellant flow rate was essentially constant, the thrust at altitude cutoff was 16.0% larger than at take-off. Determine the maximum torque applied by the thrust at start and at cutoff. Then determine the approximate maximum angle through which the vehicle will rotate during powered flight, assuming no drag. Discuss the result.

SYMBOLS

a	major axis of ellipse, m, or acceleration, m/sec^2 (ft/sec^2)
A	area, m^2
b	minor axis of ellipse, m
B	numerical value of drag integral
c	effective exhaust velocity, m/sec (ft/sec)
$\bar{c}$	average effective exhaust velocity, m/sec
C_D	drag coefficient
C_L	lift coefficient
d	total derivative
D	drag force, N (lbf)
e	eccentricity of ellipse, $e = \sqrt{1 - b^2/a^2}$
e	base of natural logarithm (2.71828)
E	energy, J
F	thrust force, N (lbf)
F_f	final thrust, N
F_g	Gravitational attraction force, N
F_0	initial thrust force, N

g	gravitational acceration, m/sec^2
g_0	gravitational acceleration at sea level, 9.8066 m/sec^2
$\bar{g}$	average gravitational attraction, m/sec^2
G	universal or Newton's gravity constant, 6.6700×10^{11} m^3/kg-sec^2
h	altitude, m
h_p	altitude of rocket at power cutoff, m
I_s	specific impulse, sec
k_d	diffuse coefficient of reflectivity
k_s	specular coefficient of reflectivity
l	distance of moment arm, m
L	lift force, N (lbf)
m	instantaneous mass, kg (lbm)
m_f	final mass after rocket operation, kg
m_p	propellant mass, kg
m_0	initial launching mass, kg
$\dot{m}$	mass flow rate of propellant, kg/sec
M_a	angular moment of inertia, kg-m^2
MR	mass ratio of vehicle $= m_f/m_0$
n	number of stages
p	pressure, N/m^2 or Pa (psi)
r	radius, m, or distance between the centers of two attracting masses, m
R	instantaneous radius from vehicle to center of Earth, m
R_0	Effective earth radius, 6.3742×10^6 m
S	range, m
t	time, sec
t_p	time from launching to power cutoff or time from propulsion start to thrust termination, sec
t_t	time to target, sec
T	torque, N-m (ft-lbf)
u	vehicle flight velocity, m/sec (ft/sec)
u_a	orbital velocity at apogee, m/sec
u_p	velocity at power cutoff, m/sec, or orbital velocity at perigee, m/sec
u_0	initial or launching velocity, m/sec
w	weight, N (in some problems, lbf)

Greek Letters

α	angle of attack, or angular acceleration, angle/sec^2
ζ	propellant mass fraction ($\zeta = m_p/m_0$)
θ	angle between flight direction and horizontal, or angle of incident radiation, deg or rad
μ	gravity constant for earth, 3.98600×10^{14} m^3/sec^2
ρ	mass density, kg/m^3
τ	period of revolution, sec

ψ angle of thrust direction with horizontal
ω angular speed, deg/sec (rad/sec)

Subscripts

e escape condition
f final condition at rocket thrust termination
max   maximum
p power cutoff or propulsion termination
s satellite
z zenith
0   initial condition or takeoff condition

REFERENCES

4–1. J. B. Rosser, R. R. Newton, and G. L. Gross, *Mathematical Theory of Rocket Flight*, McGraw-Hill Book Company, 1947; or F. R. Gantmakher and L. M. Levin, *The Flight of Uncontrolled Rockets*, Macmillan, New York, 1964.

4–2. *Orbital Flight Handbook*, NASA SP33, 1963, Part 1: Basic Techniques and Data. Part 2: Mission Sequencing Problems. Part 3: Requirements.

4–3. V. A. Chobotov (Ed.) *Orbital Mechanics*, Educational Series, AIAA, 1991.

4–4. J. W. Cornelisse, H. F. R. Schöyer, and K. F. Wakker, *Rocket Propulsion and Space Flight Dynamics*, Pitman Publishing, London, 1979.

4–5. R. S. Wolf, "Development of a Handbook for Astrobee F Flight Performance Predictions," *Journal of Spacecraft and Rockets*, Vol. 24, No. 1, January–February 1987, pp. 5–6.

4–6. W. Hohmann, *Die Erreichbarkeit der Himmelskörper (Accessibility of Celestial Bodies)*, Oldenburg, Munich, 1925.

4–7. "Spacecraft Aerodynamic Torques," *NASA SP 8058*, January 1971 (N 71-25935).

4–8. "Spacecraft Radiation Torques," *NASA SP 8027*, October 1969 (N 71-24312).

4–9. "Spacecraft Gravitational Torques," *NASA SP 8024*, May 1964 (N 70–23418).

4–10. "Spacecraft Magnetic Torques," *NASA SP 8018*, March 1969 (N 69-30339).

4–11. W. J. Larson and J. R. Wertz, *Space Mission Analysis and Design*, Second edition, published jointly by Microcosm, Inc. and Kluwer Academic Press, 1992.

4–12. J. J. Pocha, *An Introduction to Mission Design for Geostationary Satellites*, Kluwer Academic Publishers, Hingham, MA 1987, 222 pages.

4–13. M. H. Kaplan, *Orbital Spacecraft Dynamics and Control*, John Wiley & Sons, New York, 1976.

4–14. J. R. Wertz (Ed.), *Spacecraft Attitude Determination and Control*, Kluwer Academic Publishers, Hingham, MA, 1980, 858 pages.

4–15. G. E. Jensen and D. W. Netzer, *Tactical Missile Propulsion*, Vol. 170, Progress in Astronautics and Aeronautics, AIAA, 1996.

CHAPTER 5

CHEMICAL ROCKET PROPELLANT PERFORMANCE ANALYSIS

In Chapter 3, simplified one-dimensional performance relations were developed. They require a knowledge of the composition of the hot rocket gas and the properties of the propellant reaction products, such as their combustion temperature T_1, average molecular mass $\mathfrak{M}$, the specific heat ratio or the enthalpy change $(h_1 - h_2)$. This chapter discusses several theoretical approaches to determine these thermochemical properties for a given composition of propellant, chamber pressure, nozzle shape, and nozzle exit pressure. This then allows the determination of performance parameters, such as theoretical specific impulse or exhaust velocity values for chemical rockets.

By knowing the calculated gas temperature, pressure, and gas composition (e.g., whether reducing or oxidizing species) it is possible to calculate other gas properties. This knowledge also allows a more intelligent analysis and selection of materials for chamber and nozzle structures. Heat transfer analyses require the determination of the specific heats, thermal conductivity, and specific heat ratio for the gas mixture. The calculated exhaust gas composition forms the basis for estimating environmental effects, such as the potential spreading of a toxic cloud near a launch site, as discussed in Chapter 20. The exhaust gas parameters also form the basis for the analysis of exhaust plumes (Chapter 18) or flames external to the nozzle.

With the advent of digital computers it has been possible to solve the set of equations involving mass balance, energy balance, or thermodynamic and chemical equilibria of complex systems with a variety of propellant ingredients. This chapter is intended to introduce the basic approach to this theoretical analysis, so the reader can understand the thermodynamic and chemical basis of the several computer programs that are in use today. This chapter does not describe any specific computer analysis programs. However, it discusses which of the physical phenomena or chemical reactions can or cannot be adequately simulated by computer analysis.

The reader is referred to Refs. 5–1 to 5–5 for general chemical and thermodynamic background and principles. For a detailed description of the properties of each of the possible reactant and reaction products, see Refs. 5–6 to 5–12.

All of these theoretical analyses are only approximations of what really happens in rocket combustion and nozzle flow, and they all require some simplifying assumptions. As more of the different phenomena are understood and mathematically simulated, the analysis approach and the computer implementation become more sophisticated, but also more complex. The 11 assumptions made in Section 3.1 for an ideal rocket are valid here also, but only for a quasi-one-dimensional flow. However, some of the more sophisticated analyses can make one or more of these assumptions unnecessary. The analysis is usually divided into two somewhat separate sets of calculations:

1. The *combustion process* is the first part. It usually occurs in the combustion chamber at essentially constant chamber pressure (isobaric) and the resulting gases follow Dalton's law. The chemical reactions or the combustions occur very rapidly. The chamber volume is assumed to be large enough and the residence time in the chamber long enough for attaining chemical equilibrium in the chamber.

2. The *nozzle gas expansion process* constitutes the second set of calculations. The fully reacted, equilibrated gas combustion products enter the nozzle and undergo an adiabatic expansion in the nozzle. The entropy remains constant during a reversible (isentropic) nozzle expansion, but in real nozzle flows it increases slightly.

The principal chemical reactions occur inside the combustion chamber of a liquid propellant rocket engine or inside the grain cavity of a solid propellant rocket motor, usually within a short distance from the burning surface. These chamber combustion analyses are discussed further in Chapters 9 and 13. However, some chemical reactions also occur in the nozzle as the gases expand; the composition of the reaction products can therefore change in the nozzle, as described in this chapter. A further set of chemical reactions can occur in the exhaust plume outside the nozzle, as described in Chapter 18; many of the same basic thermochemical analysis approaches described in this chapter also apply to exhaust plumes.

5.1. BACKGROUND AND FUNDAMENTALS

The principle of chemical reaction or combustion of one or more fuels with one or more oxidizing reactants is the basis of chemical rocket propulsion. The heat liberated in this reaction transforms the propellants (reactants) into hot gaseous reaction products, which in turn are thermodynamically expanded in a nozzle to produce thrust.

The chemical reactants or propellants can initially be either liquid or solid and occasionally also gaseous. The reaction products are usually gaseous, but with some propellants one or more reactant species remain in the solid or liquid phase. For example, with aluminized solid propellants, the chamber reaction gases contain liquid aluminum oxide and the colder gases in the nozzle exhaust contain solid, condensed aluminum oxide particles. For some of the chemical species, therefore, the analysis must consider as many as all three phases and the energy changes for the phase transitions must be included. If the amount of solid or liquid in the exhaust is small and the particles are small, then to assume a perfect gas introduces only small errors.

It is necessary to accurately know the chemical composition of the propellants and their relative proportion. In liquid propellant this means the mixture ratio and the major propellant impurities; in gelled or slurried liquid propellants it also includes suspended or dissolved solid materials; and in solid propellants it means all the ingredients, their proportions and impurities and phase (some ingredients, such as plasticizers, can be in a liquid state).

Dalton's law applies to the gas resulting from the combustion. It states that a mixture of gases at equilibrium exerts a pressure that is the sum of the partial pressures of the individual gases, all at the same temperature. The subscripts a, b, c, etc. refer to individual gas constituents.

$$p = p_a + p_b + p_c + \cdots \tag{5-1}$$

$$T = T_a = T_b = T_c = \cdots \tag{5-2}$$

The perfect gas equation $pV = RT$ applies very closely to high temperature gases. Here V is the specific volume or the volume per unit mass of gas mixture, and the gas constant R for the mixture is obtained by dividing the universal gas constant R' (8314.3 J/kg-mol-K) by the average molecular mass $\mathfrak{M}$ (often erroneously called the molecular weight) of the gas mixture. Using Dalton's law, Eq. 5-1 can be rewritten

$$p = R_a T / V_a + R_b T / V_b + R_c T / V_c + \cdots = R' T / (\mathfrak{M} V_{\text{mix}}) \tag{5-3}$$

The volumetric proportions of gas species in a gas mixture are determined from the *molar concentration* or *molar fractions*, n_j, expressed as kg-mol for a particular species j per kg of mixture. If n is the total number of kg-mol of species j per kilogram of uniform gas mixture, then

$$n = \sum_{j=1}^{j=m} n_j \tag{5-4}$$

where n_j is the kg-mol of species j per kilogram of mixture, m is the number of different gaseous species present in the equilibrium combustion gas products. The effective average molecular mass $\mathfrak{M}$ of a gas mixture is then

$$\mathfrak{M} = \frac{\sum_{j=1}^{m} n_j \mathfrak{M}_j}{\sum_{j=1}^{m} n_j} \tag{5-5}$$

There are n possible species which enter into the relationship and of these only m are gases, so $n - m$ represents the number of condensed species. The *molar specific heat* for a gas mixture at constant pressure C_p can be determined from the individual gas molar fractions n_j and their molar specific heats as shown by Eq. 5–6. The specific heat ratio k of the mixture can be determined by a similar summation or from Eq. 5–7.

$$(C_p)_{\text{mix}} = \frac{\sum_{j=1}^{m} n_j (C_p)_j}{\sum_{j=1}^{m} n_j} \tag{5-6}$$

$$k_{\text{mix}} = \frac{(C_p)_{\text{mix}}}{(C_p)_{\text{mix}} - R'} \tag{5-7}$$

When a chemical reaction goes to completion, that is, all of the reactants are consumed and transformed into reaction products, the reactants are in *stoichiometric* proportions. For example, consider this reaction:

$$H_2 + \tfrac{1}{2}O_2 \rightarrow H_2O \tag{5-8}$$

All the hydrogen and oxygen are fully consumed to form the single product— water vapor—without any reactant residue of either hydrogen or oxygen. In this case it requires 1 mol of the H_2 and $\frac{1}{2}$ mole of the O_2 to obtain 1 mol of H_2O. On a mass basis this *stoichiometric mixture* requires half of 32.0 kg of O_2 and 2 kg of H_2, which are in the stoichiometric mixture mass ratio of 8:1. The release of energy per unit mass of propellant mixture and the combustion temperature are highest at or near the stoichiometric mixture.

Rocket propulsion systems usually do not operate with the proportion of their oxidizer and fuel in the stoichiometric mixture ratio. Instead, they usually operate fuel-rich because this allows lightweight molecules such as hydrogen to remain unreacted; this reduces the average molecular mass of the reaction products, which in turn increases the specific impulse (see Eq. 3–16). For rockets using H_2 and O_2 propellants the best operating mixture mass ratio for high-performance rocket engines is typically between 4.5 and 6.0, not at the stoichiometric value of 8.0.

Equation 5–8 is a reversible chemical reaction; by adding energy to the H_2O the reaction can be made to go backward to create H_2 and O_2 and the arrow in the equation would be reversed. The decompositions of solid propellants into reaction product gases are irreversible chemical reactions, as is the reaction of liquid propellants burning to create gases. However, reactions among combustion product gases are usually reversible.

Chemical equilibrium exists in reversible chemical reactions when the rate of forming products is exactly equal to the reverse reaction of forming reactants from the products. Once this equilibrium is reached, no further changes in concentration can take place. In Equation 5–8 all three gases would be present and their relative proportions would depend on the pressure, temperature, and initial mixture.

The *heat of formation* $\Delta_f H^0$ is the energy released (or absorbed), or the value of enthalpy change, when 1 mole of a chemical compound is formed from its constituent atoms or elements at 1 bar (100,000 Pa) and isothermally at 298.15 K or 25°C. The Δ implies that it is an energy change. The subscript f refers to formation and the superscript ^{0} means that each product or reactant substance is at its thermodynamic standard state and at the reference pressure and temperature. By convention, the heat of formation of the gaseous elements (e.g., H_2, O_2, Ar, Xe, etc.) is set to zero at these standard conditions of temperature and pressure. Typical values of $\Delta_f H^0$ and other properties are given in Table 5–1 for selected species. When heat is absorbed in the formation of a product, then $\Delta_f H^0$ has a negative value. Earlier analyses have been made with the standard temperature at other values, such as 273.15 K and a slightly higher standard reference pressure of 1 atm (101,325 Pa).

The *heat of reaction* $\Delta_r H^0$ is the energy released or absorbed when products are formed from its reactants at standard reference conditions, namely at 1 bar and 25°C. The heat of reaction can be negative or positive, depending on whether the reaction is *exothermic* or *endothermic*. The heat of reaction at other temperatures or pressures has to be corrected in accordance with the change in enthalpy. When a species changes from one state to another (e.g., liquid becomes gas or vice versa), it may lose or gain energy. In most rocket propulsion the heat of reaction is determined for a constant-pressure combustion process. In general the heat of reaction can be determined from sums of the heats of formation of the products and the reactants, namely

$$\Delta_r H^0 = \sum [n_j (\Delta_f H^0)_j]_{\text{products}} - \sum [n_j (\Delta_f H^0)_j]_{\text{reactants}} \qquad (5\text{–}9)$$

Here n_j is the molar fraction of each particular species j. In a typical rocket propellant there are a number of different chemical reactions going on simultaneously; Equation 5–9 provides the heat of reaction for all of these simultaneous reactions. For data on heats of formation and heats of reaction, see Refs. 5–7 to 5–13.

TABLE 5-1. Chemical Thermodynamic Properties of Selected Substances at 298.15 K (25°C) and 0.1 MPa (1 bar)

Substance	Phase*	Molar Mass (g/mol)	$\Delta_f H^0$ (kJ/mol)	$\Delta_f G^0$ (kJ/mol)	$\log K_f$	S^0 (J/mol-K)	C_p (J/mol-K)
Al (crystal)	s	29.9815	0	0	0	28.275	24.204
Al_2O_3	l	101.9612	−1620.567	−1532.025	268.404	67.298	79.015
C (graphite)	s	12.011	0	0	0	5.740	8.517
CH_4	g	16.0476	−74.873	−50.768	8.894	186.251	35.639
CO	g	28.0106	−110.527	−137.163	24.030	197.653	29.142
CO_2	g	44.010	−393.522	−394.389	69.095	213.795	37.129
H_2	g	2.01583	0	0	0	130.680	28.836
HCl	g	36.4610	−92.312	−95.300	16.696	186.901	29.136
HF	g	20.0063	−272.546	−274.646	48.117	172.780	29.138
H_2O	l	18.01528	−285.830	−237.141	41.546	69.950	75.351
H_2O	g	18.01528	−241.826	−228.582	40.047	188.834	33.590
N_2H_4	l	32.0451	+50.626	149.440	−28.181	121.544	98.840
N_2H_4	g	32.0451	+95.353	+159.232	−27.897	238.719	50.813
NH_4ClO_4	s	117.485	−295.767	−88.607	15.524	184.180	128.072
ClF_5	g	130.4450	−238.488	−146.725	25.706	310.739	97.165
ClF_3	g	92.442	−158.866	−118.877	20.827	281.600	63.845
N_2O_4	l	92.011	−19.564	+97.521	−17.085	209.198	142.509
N_2O_4	g	92.011	9.079	97.787	−17.132	304.376	77.256
NO_2	g	46.0055	33.095	51.258	−8.980	240.034	36.974
HNO_3	g	63.0128	−134.306	−73.941	12.954	266.400	53.326
N_2	g	28.0134	0	0	0	191.609	29.125
O_2	g	31.9988	0	0	0	205.147	29.376
NH_3	g	17.0305	−45.898	−16.367	2.867	192.774	35.652

*s = solid, l = liquid, g = gas. Several species are listed twice, as a liquid and as a gas; the difference is due to evaporation or condensation. The molar mass can be in g/g-mol or kg/kg-mol and C_p can be in J/g-mol-K or kJ/kg-mol-K.

Source: Refs. 5–8 and 5–9.

165

Various thermodynamic criteria that represent the necessary and sufficient conditions for an equilibrium to be stable were first advanced by J. W. Gibbs early in the 20th century; they are based on minimizing the free energy. The Gibbs *free energy* G (often called the *chemical potential*) is a convenient derived function or property of the state of a chemical material describing its thermodynamic potential and is directly related to the internal energy U, the pressure p, molar volume V, enthalpy h, temperature T, and entropy S. For a single species j the free energy is defined as G_j; it can be determined for specific thermodynamic conditions, for mixtures of gas as well as an individual gas species.

$$G = U + pV - TS = h - TS \qquad (5\text{–}10)$$

For most materials used as rocket propellant the free energy has been determined and tabulated as a function of temperature. It can be corrected for pressure. Its units are J/kg-mol. For a series of different species the mixture free energy G is

$$G = \sum_{j=1}^{n} G_j n_j \qquad (5\text{–}11)$$

The *free energy* is a function of temperature and pressure. It is another property of a material, just like enthalpy or density; only two such independent parameters are required to characterize a gas condition. The free energy may be thought of as the tendency or driving force for a chemical material to enter into a chemical (or physical) change. Although it cannot be measured directly, differences in chemical potential can be measured. When the chemical potential of the reactants is higher than that of the likely products, a chemical reaction can occur and the chemical composition can change. The change in free energy ΔG for reactions at constant temperature and pressure is the chemical potential of the products less that of the reactants.

$$\Delta G = \sum_{j=1}^{m} [n_j (\Delta_f G^0)_j]_{\text{products}} - \sum_{j=1}^{n} [n_j (\Delta_f G^0)_j]_{\text{reactants}} \qquad (5\text{–}12)$$

Here the superscript m gives the number of gas species in the combustion products, the superscript n gives the number of gas species in the reactants, and the ΔG represents the maximum energy that can be "freed" to do work on an "open" system where mass enters and leaves the system. At equilibrium the free energy is a minimum; at its minimum a small change in mixture fractions causes almost no change in ΔG and the free energies of the products and the reactants are essentially equal. Then

$$d\,\Delta G/dn = 0 \qquad (5\text{–}13)$$

and a curve of molar concentration n versus ΔG would have a minimum.

If reacting propellants are liquid or solid materials, energy will be needed to change phase, vaporize them, or break them down into other gaseous species. This energy has to be subtracted from the heat or the energy available to heat the gases from the reference temperature to the combustion temperature. Therefore, the values of ΔH^0 and ΔG^0 for liquid and solid species are different from those of the same species in a gaseous state. The standard *free energy of formation* $\Delta_f G^0$ is the increment in free energy associated with the reaction of forming a given compound or species from its elements at their reference state. Table 5–2 gives values of $\Delta_f H^0$ and $\Delta_f G^0$ and other properties of carbon monoxide as a function of temperature. Similar data for other species can be obtained from Refs. 5–7 and 5–13. The *entropy* is another thermodynamic property of matter that is relative, which means that it is determined as a change in entropy. In the analysis of isentropic nozzle flow, it is assumed that the entropy remains constant. It is defined as

$$dS = \frac{dU}{T} + \frac{p\,dV}{T} = C_p \frac{dT}{T} - R \frac{dp}{p} \tag{5–14}$$

and the corresponding integral is

$$S - S_0 = C_p \ln \frac{T}{T_0} - R \ln \frac{p}{p_0} \tag{5–15}$$

where the zero applies to the reference state. In an isentropic process, entropy is constant. For a mixture the entropy is

TABLE 5–2. Variation of Thermochemical Data with Temperature for Carbon Monoxide (CO) as an Ideal Gas

Temp. (K)	C_p^0	S^0	$H^0 - H^0(T)$	$\Delta_f H^0$ (kJ/mol)	$\Delta_f G^0$ (kJ/mol)	$\log K_f$
	(J/mol-K)					
0	0	0	−8.671	−113.805	−113.805	∞
298.15	29.142	197.653	0	−110.527	−137.163	24.030
500	29.794	212.831	5.931	−110.003	−155.414	16.236
1000	33.183	234.538	21.690	−111.983	−200.275	10.461
1500	35.217	248.426	38.850	−115.229	−243.740	8.488
2000	36.250	258.714	56.744	−118.896	−286.034	7.470
2500	36.838	266.854	74.985	−122.994	−327.356	6.840
3000	37.217	273.605	93.504	−127.457	−367.816	6.404
3500	37.493	279.364	112.185	−132.313	−407.497	6.082
4000	37.715	284.386	130.989	−137.537	−446.457	5.830

Source: Refs. 5–8 and 5–9.

$$S = \sum_{j=1}^{n} S_j n_j \qquad (5\text{--}16)$$

Here entropy is in J/kg-mol-K. The entropy for each gaseous species is

$$S_j = (S_T^0)_j - R \ln \frac{n_j}{n} - R \ln p \qquad (5\text{--}17)$$

For solid and liquid species the last two terms are zero. Here (S_T^0) refers to the standard state entropy at temperature T. Typical values for entropy are listed in Tables 5–1 and 5–2.

When a chemical reaction is in equilibrium, an *equilibrium constant* has been devised which relates the partial pressures and the molar fractions of the species. For example, in the general reaction

$$a\text{A} + b\text{B} \rightleftharpoons c\text{C} + d\text{D} \qquad (5\text{--}18)$$

a, b, c, and d are the stoichiometric molar concentration coefficients of the chemical molecules (or atoms) A, B, C, and D. The equilibrium constant K, when expressed as partial pressures, is a function of temperature.

$$K_p = \frac{p_C^c p_D^d}{p_A^a p_B^b} p_0^{-c-d+a+b} \qquad (5\text{--}19)$$

Here p_0 is the reference pressure. All pressures are in bars or 10^5 Pa. When $a + b = c + d$, then K_p is independent of pressure. This condition is not valid for a reaction like Eq. 5–8. In this case the pressure increase will drive the equilibrium reaction into the direction of fewer moles and in the direction of absorbing heat if the temperature is increased. For Eq. 5–8 the hydrogen and oxygen equilibrium relation would be

$$K_p = \frac{p_{\text{H}_2\text{O}}}{p_{\text{H}_2} p_{\text{O}_2}^{0.5}} p_0^{-1+1+0.5} \qquad (5\text{--}20)$$

The equilibrium constant can also be expressed as a function of the molar fractions n_j because each partial pressure p_{n_j} is equal to the actual pressure p at which the reaction occurs multiplied by its molar fraction ($p_j = p n_j$). From Equation 5–19 the equilibrium constant K can also be expressed as

$$K_n = \frac{n_C^c n_D^d}{n_A^a n_B^b} \left(\frac{p}{p_0}\right)^{c+d-a-b} \qquad (5\text{--}21)$$

The equilibrium constant for the chemical formation of a given species from its elements is K_f. Typical values of K_f are shown in Tables 5–1 and 5–2. The free energy and the equilibrium constant for the formation of a particular species at standard conditions from its atomic elements are related, namely

$$\Delta G^0 = -RT \ln K_f \qquad (5\text{--}22)$$

Equations 5–19, 5–20, and 5–22 are often used together with mass balance and energy balance relations to solve the simultaneous equations; the equilibrium constant K is primarily used when chemical compounds are formed from their elements.

5.2. ANALYSIS OF CHAMBER OR MOTOR CASE CONDITIONS

The objectives here are to determine the theoretical combustion temperature and the theoretical composition of the resulting reaction products, which in turn will allow the determination of the physical properties of the combustion gases (C_p, k, or ρ). Before we can make this analysis, some basic data (e.g., propellants, their ingredients, desired chamber pressure, or all likely reaction products) have to be known or postulated. Although the combustion process really consists of a series of different chemical reactions that occur almost simultaneously and includes the breakdown of chemical compounds into intermediate and subsequently into final products, the analysis is only concerned with the initial and final conditions, before and after combustion. We will mention several approaches to the analysis of chamber conditions. In this section we will first give some definitions of key terms and explain some concepts and principles. The first principle concerns the *conservation of energy*. The heat created by the combustion is equal to the heat necessary to raise the resulting gases adiabatically to their final combustion temperature. The heat of reaction of the combustion $\Delta_r H$ has to equal the enthalpy change ΔH of the gases.

The *energy balance* can be thought of as a two-step process. The chemical reaction occurs instantaneously but isothermally at the reference temperature, and the resulting energy release then heats the gases from this reference temperature to the final combustion temperature. The heat of reaction is

$$\Delta_r H = \sum_1^n n_j \int_{T_{\text{ref}}}^{T_1} C_p \, dT = \sum_1^n n_j \Delta h_j \big|_{T_{\text{ref}}}^{T_1} \qquad (5\text{--}23)$$

Here Δh is the increase in enthalpy for each species multiplied by its molar fraction, and C_p is the molar specific heat at constant pressure.

The second principle is the *conservation of mass*. The mass of any of the atomic species present in the reactants before the chemical reaction must be equal to the mass of the same species in the products. This can be illustrated by a more general case of the reaction of Equation 5–8. In this case the reactants are not in stoichiometric proportion.

In the combustion of hydrogen with oxygen it is possible to form six products: water, hydrogen, oxygen, hydroxyl, atomic oxygen, and atomic hydrogen. In this case all the reactants and products are gaseous. Theoretically, there could be two additional products: ozone O_3 and hydrogen peroxide H_2O_2; however, these are unstable materials that do not readily exist at high temperature, and they can be ignored. In chemical notation this can be stated by

$$aH_2 + bO_2 \rightarrow n_{H_2O}H_2O + n_{H_2}H_2 + n_{O_2}O_2 + n_O O + n_H H + n_{OH}OH \quad (5\text{–}24)$$

The left side shows the condition before and the right side after the reaction. Since H_2 and O_2 can be found on both sides, it means that not all of these species are consumed and a portion, namely n_{H_2} and n_{O_2}, will remain unreacted. With chemical equilibrium at a particular temperature and pressure the molar concentrations on the right side will remain fixed. Here a, b, n_{H_2O}, n_{H_2}, n_{O_2}, n_O, n_H, and n_{OH} are the respective molar fractions or molar quantities of these substances before and after the reaction, and they can be expressed in kg-mol per kilogram of propellant reactants or reaction products. The initial proportions of a and b are usually known. The number of kg-mol per kilogram of mixture of each element can be established from this initial mix of oxidizer and fuel ingredients. For the hydrogen–oxygen relation above, the mass balances would be

$$\left.\begin{array}{l} \text{for hydrogen: } 2a = 2n_{H_2O} + 2n_{H_2} + n_H + n_{OH} \\ \text{for oxygen: } \quad 2b = n_{H_2O} + 2n_{O_2} + n_O + n_{OH} \end{array}\right\} \quad (5\text{–}25)$$

The *mass balance* of Eq. 5–25 provides two more equations for this reaction (one for each atomic species) in addition to the energy balance equation. There are six unknown product percentages and an unknown combustion or equilibrium temperature. However, three equations provide a solution for only three unknowns, say the combustion temperature and the molar fractions of two of the species. If, for example, it is known that the initial mass mixture ratio of b/a is fuel rich, so that the combustion temperature will be relatively low, the percentage of remaining O_2 and the percentage of the dissociation products (O, H, and OH) would all be very low and can be neglected. Thus n_O, n_H, n_{OH}, and n_{O_2} are set to be zero. The solution requires knowledge of the enthalpy change of each of the species, and that information can be obtained from existing tables, such as Table 5–2 or Refs. 5–8 and 5–9.

In more general form, the mass for any given element must be the same before and after the reaction. The number of kg-mol of a given element per kilogram of reactants and product is equal, or their difference is zero. For any one atomic species, such as the H or the O in Eq. 5–25,

$$\left[\sum_{j=1}^{m} a_{ij}n_j\right]_{\text{products}} - \left[\sum_{j=1}^{n} a_{ij}n_j\right]_{\text{propellants}} = 0 \qquad (5\text{–}26)$$

Here the atomic coefficients a_{ij} are the number of kilogram atoms of element i per kg-mol of species j, and m and n are as defined above. The average molecular mass of the products in Eq. 5–5 would be

$$\mathfrak{M} = \frac{2n_{H_2} + 32n_{O_2} + 18n_{H_2O} + 16n_O + n_H + 17n_{OH}}{n_{H_2} + n_{O_2} + n_{H_2O} + n_O + n_H + n_{OH}} \qquad (5\text{–}27)$$

Another way to determine the molar fractions for the equilibrium composition is to use a factor λ that represents the degree of advancement of the chemical reaction. This factor λ has the value of zero for the initial conditions before the reaction starts and 1.0 for the final conditions, when the reaction is completed and all the reaction gases are converted to product gases. For the reaction described by Eq. 5–24, λ can be used in this way:

$$\text{Number of moles of A: } n_A = a\lambda \qquad (5\text{–}28)$$
$$\text{Number of moles of B: } n_B = b\lambda$$
$$\text{Number of moles of C: } n_C = c(1 - \lambda) \qquad (5\text{–}29)$$
$$\text{Number of moles of D: } n_D = d(1 - \lambda)$$

By substituting these molar fractions into the Gibbs free energy equation (Eq. 5–12), then differentiating the expression with respect to λ and setting the derivative $dG/d\lambda = 0$, one can determine the value of λ at which G is a minimum for the gas mixture. The degree of advancement λ then determines the values of n_A, n_B, n_C, and n_D at equilibrium.

The approach used in Ref. 5–13 is commonly used today for thermochemical analysis. It relies on the minimization of the Gibbs free energy and on mass balance and energy balance equations. As was explained in Eq. 5–12, the change in the Gibbs free energy function is zero at equilibrium ($\Delta G = 0$): the chemical potential of the gaseous propellants has to equal that of the gaseous reaction products, which is Eq. 5–12:

$$\Delta G = \sum (n_j \Delta G_j)_{\text{products}} - \sum (n_j \Delta G_j)_{\text{reactants}} = 0 \qquad (5\text{–}30)$$

To assist in solving this equation a Lagrangian multiplier or a factor of the degree of the completion of the reaction is often used. An alternative method for solving for the gas composition, temperature, and gas properties is to use the energy balance (Eq. 5–23) together with several mass balances (Eq. 5–26) and equilibrium relationships (Eq. 5–21).

After assuming a chamber pressure and setting up the energy balance, mass balances, and equilibrium relations, one method of solving all the equations is to estimate a combustion temperature and then solve for the various values of n_j. Then a balance has to be achieved between the heat of reaction $\Delta_r H^0$ and the heat absorbed by the gases, $H_T^0 - H_0^0$, to go from the reference temperature to the combustion temperature. If they do not balance, another value of the combustion temperature is chosen until there is convergence and the energy balances.

The *energy release efficiency*, sometimes called the *combustion efficiency*, can now be defined as the ratio of the actual change in enthalpy per unit propellant mixture to the calculated change in enthalpy necessary to transform the propellants from the initial conditions to the products at the chamber temperature and pressure. The actual enthalpy change can be evaluated if the initial propellant condition and the actual composition and the temperature of the combustion gases are measured. Experimental measurements of combustion temperature and gas composition are difficult to perform accurately, and the combustion efficiency is therefore actually evaluated only in rare instances. The combustion efficiency in liquid propellant rocket thrust chambers depends on the method of injection and mixing and increases with increased combustion temperature. In solid propellants the combustion efficiency is a function of the grain design, the propellant, and the degree of mixing between the several solid constituents. Actual measurements on well designed rocket propulsion systems indicate efficiency values of 94 to 99%. These high values indicate that the combustion is essentially complete, that very little, if any, unreacted propellant remains, and that chemical equilibrium is indeed established.

The number of compounds or species in the exhaust can be 50 or more with solid propellants or with liquid propellants that have certain additives. The number of nearly simultaneous chemical reactions that have to be considered can easily exceed 150. Fortunately, many of these chemical species are present only in very small amounts and can usually be neglected.

5.3. ANALYSIS OF NOZZLE EXPANSION PROCESSES

There are several methods for analyzing the nozzle flow, depending on the assumptions made for chemical equilibrium, nozzle expansion, particulates, or energy losses. Several are outlined in Table 5–3.

Once the gases reach the nozzle, they experience an adiabatic, reversible expansion process which is accompanied by a drop in temperature and pressure and a conversion of thermal energy into kinetic energy. Several increasingly more complicated methods have been used for the analysis of the process. For the simple case of frozen equilibrium and one-dimensional flow the state of the gas throughout expansion in the nozzle is fixed by the entropy of the system, which is presumed to be invariant as the pressure is reduced to the value assigned to the nozzle exit plane. All the assumptions listed in Chapter 3 for an ideal rocket are also valid here. Again, the effects of friction, divergence angle, heat exchange, shock waves, or nonequilibrium are neglected in the simple cases, but are considered in the more sophisticated solutions. The condensed (liquid or solid) phases are again assumed to have zero volume and to be in kinetic as well as thermal equilibrium with the gas flow. This implies that particles or droplets are very small in size, move at the same velocity as the gas stream, and have the same temperature as the gas at all places in the nozzle. The chemical equilibrium during expansion in the nozzle can be analytically regarded in the following ways:

1. When the composition is invariant throughout the nozzle, there are no chemical reactions or phase changes and the product composition at the nozzle exit is identical to that of its chamber condition. The results are known as *frozen equilibrium* rocket performance. This method usually is simple, but underestimates the performance, typically by 1 to 4%.

2. Instantaneous chemical equilibrium among all molecular species is maintained under the continuously variable pressure and temperature conditions of the nozzle expansion process. Thus the product composition shifts; similarly, instantaneous chemical reactions, phase changes or equilibria occur between gaseous and condensed phases of all species in the exhaust gas. The results so calculated are called *shifting equilibrium* performance. The gas composition mass percentages are different in the chamber and the nozzle exit. This method usually overstates the performance values, such as c^* or I_s, typically by 1 to 4%. Here the analysis is more complex.

3. The chemical reactions do not occur instantaneously, but even though the reactions occur rapidly they require a finite time. The reaction rates of specific reactions can be estimated; the rates are usually a function of temperature, the magnitude of deviation from the equilibrium molar composition, and the nature of the chemicals or reactions involved. The values of T, c^*, or I_s for these types of equilibrium analysis usually are between those of frozen and instantaneously shifting equilibria. This approach is almost never used, because of the lack of good data on reaction rates with multiple simultaneous chemical reactions.

For an axisymmetric nozzle, both one- and two-dimensional analyses can be used. The simplest nozzle flow analysis is one-dimensional, which means that all velocities and temperatures or pressures are equal at any normal cross section of an axisymmetric nozzle. It is often satisfactory for preliminary estimates. In a two-dimensional analysis the velocity, temperature, density, and/or Mach number do not have a flat profile and vary somewhat over the cross sections. For nozzle shapes that are not bodies of revolution (e.g., rectangular, scarfed, or elliptic) a three-dimensional analysis can be performed.

If solid particles or liquid droplets are present in the nozzle flow and if the particles are larger than about 0.1 µm average diameter, there will be a thermal lag and velocity lag. The solid particles or liquid droplets do not expand like a

TABLE 5–3. Typical Steps and Alternatives in the Analysis of Rocket Thermochemical Processes in Nozzles

Step	Process	Method/Implication/Assumption
Nozzle inlet condition	Same as chamber exit; need to know T_1, p_1, v_1, H, c^*, ρ_1, etc.	For simpler analyses assume the flow to be uniformly mixed and steady.
Nozzle expansion	An adiabatic process, where flow is accelerated and thermal energy is converted into kinetic energy. Temperature and pressure drop drastically. Several different analyses have been used with different specific effects. Can use one-, two-, or three-dimensional flow pattern.	1. Simplest method is inviscid isentropic expansion flow with constant entropy. 2. Include internal weak shock waves; no longer a truly isentropic process. 3. If solid particles are present, they will create drag, thermal lag, and a hotter exhaust gas. Must assume an average particle size and optical surface properties of the particulates. Flow is no longer isentropic. 4. Include viscous boundary layer effects and/or non-uniform velocity profile.

Often a simple single correction factor is used with one-dimensional analyses to correct the nozzle exit condition for items 2, 3, and/or 4 above. Computational fluid dynamic codes with finite element analyses have been used with two- and three-dimensional nozzle flow.

Chemical equilibrium during nozzle expansion	Due to rapid decrease in T and p, the equilibrium composition can change from that in the chamber. The four processes listed in the next column allow progressively more realistic simulation and require more sophisticated techniques.	1. Frozen equilibrium; no change in gas composition; usually gives low performance. 2. Shifting equilibrium or instantaneous change in composition; usually overstates the performance slightly. 3. Use reaction time rate analysis to estimate the time to reach equilibrium for each of the several chemical reactions; some rate constants are not well known; analysis is more complex. 4. Use different equilibrium analysis for boundary layer and main inviscid flow; will have nonuniform gas temperature, composition, and velocity profiles.

TABLE 5–3. (*Continued*)

Step	Process	Method/Implication/Assumption
Heat release in nozzle	Recombination of dissociated molecules (e.g., $H + H = H_2$) and exothermic reactions due to changes in equilibrium composition cause an internal heating of the expanding gases. Particulates release heat to the gas.	Heat released in subsonic portion of nozzle will increase the exit velocity. Heating in the supersonic flow portion of nozzle can increase the exit temperature but reduce the exit Mach number.
Nozzle shape and size	Can use straight cone, bell-shaped, or other nozzle contour; bell can give slightly lower losses. Make correction for divergence losses and nonuniformity of velocity profile.	Must know or assume a particular nozzle configuration. Calculate bell contour by method of characteristics. Use Eq. 3–34 for divergence losses in conical nozzle. Most analysis programs are one- or two-dimensional. Unsymmetrical non-round nozzles may need three-dimensional analysis.
Gas properties	The relationships governing the behavior of the gases apply to both nozzle and chamber conditions. As gases cool in expansion, some species may condense.	Either use perfect gas laws or, if some of the gas species come close to being condensed, use real gas properties.
Nozzle exit conditions	Will depend on the assumptions made above for chemical equilibrium, nozzle expansion, and nozzle shape/contour. Assume no jet separation. Determine velocity profile and the pressure profile at the nozzle exit plane. If pressure is not uniform across a section it will have some cross flow.	Need to know the nozzle area ratio or nozzle pressure ratio. For quasi-one-dimensional and uniform nozzle flow, see Eqs. 3–25 and 3–26. If v_2 is not constant over the exit area, determine effective average values of v_2 and p_2. Then calculate profiles of T, ρ, etc. For nonuniform velocity profile, the solution requires an iterative approach. Can calculate the gas conditions (T, p, etc.) at any point in the nozzle.
Calculate specific impulse	Can be determined for different altitudes, pressure ratios, mixture ratios, nozzle area ratios, etc.	Can be determined for average values of v_2, p_2, and p_3 based on Eqs. 2–6, 3–35, and/or 2–14.

gas; their temperature decrease depends on losing energy by convection or radiation, and their velocity depends on the drag forces exerted on the particle. Larger-diameter droplets or particles are not accelerated as rapidly as the smaller ones and flow at a velocity lower than that of the adjacent accelerating gas. Also, the particulates are hotter than the gas and provide heat to the gas. While these particles contribute to the momentum of the exhaust mass, they are not as efficient as an all-gaseous exhaust flow. For composite solid propellants with aluminum oxide particles in the exhaust gas, the loss due to particles could typically be 1 to 3%. The analysis of a two- or three-phase flow requires knowledge of or an assumption about the nongaseous matter, the sizes (diameters), size distribution, shape (usually assumed to be spherical), optical surface properties (for determining the emission/absorption or scattering of radiant energy), and their condensation or freezing temperatures. Some of these parameters are not well known. Performance estimates of flows with particles are explained in Section 3–5.

The viscous boundary layer next to the nozzle wall has velocities substantially lower than that of the inviscid free stream. The slowing down of the gas flow near the wall due to the viscous drag actually causes the conversion of kinetic energy into thermal energy, and thus some parts of the boundary layer can be hotter than the local free-stream static temperature. A diagram of a two-dimensional boundary layer is shown in Figure 3–16. With turbulence this boundary layer can be relatively thick in large-diameter nozzles. The boundary layer is also dependent on the axial pressure gradient in the nozzle, the nozzle geometry, particularly in the throat region, the surface roughness, or the heat losses to the nozzle walls. Today, theoretical boundary layer analyses with unsteady flow are only approximations, but are expected to improve in the future as our understanding of the phenomena and computational fluid dynamics (CFD) techniques are validated. The net effect is a nonuniform velocity and temperature profile, an irreversible friction process in the viscous layers, and therefore an increase in entropy and a slight reduction (usually less than 5%) of the kinetic exhaust energy. The slower moving layers adjacent to the nozzle walls have laminar and subsonic flow.

At the high combustion temperatures a small portion of the combustion gas molecules dissociate (split into simpler species); in this *dissociation* process some energy is absorbed. When energy is released during reassociation (at lower pressures and temperatures in the nozzle), this reduces the kinetic energy of the exhaust gas at the nozzle exit. This is discussed further in the next section.

For propellants that yield only gaseous products, extra energy is released in the nozzle, primarily from the recombination of free-radical and atomic species, which become unstable as the temperature is decreased in the nozzle expansion process. Some propellant products include species that condense as the temperature drops in the nozzle expansion. If the heat release on condensation is large, the difference between frozen and shifting equilibrium performance can be substantial.

In the simplest method the exit temperature T_2 is determined for an isentropic process (frozen equilibrium) by considering the entropy to be constant. The entropy at the exit is the same as the entropy in the chamber. This determines the temperature at the exit and thus the gas condition at the exit. From the corresponding change in enthalpy it is then possible to obtain the exhaust velocity and the specific impulse. For those analysis methods where the nozzle flow is not really isentropic and the expansion process is only partly reversible, it is necessary to include the losses due to friction, shock waves, turbulence, and so on. The result is a somewhat higher average nozzle exit temperature and a slight loss in I_s. A possible set of steps used for the analysis of nozzle processes is given in Table 5–3.

When the contraction between the combustion chamber (or the port area) and the throat area is small ($A_p/A_t \leq 3$), the acceleration of the gases in the chamber causes a drop in the effective chamber pressure at the nozzle entrance. This pressure loss in the chamber causes a slight reduction of the values of c and I_s. The analysis of this chamber configuration is treated in Ref. 5–14 and some data are briefly shown in Tables 3–2 and 6–4.

Example 5–1. Various experiments have been conducted with a liquid monopropellant called nitromethane (CH_3NO_2), which can be decomposed into gaseous reaction products. Determine the values of T, $\mathfrak{M}$, k, c^*, C_F, and I_s using the water–gas equilibrium conditions. Assume no dissociations and no O_2.

SOLUTION. The chemical reaction for 1 mol of reactant can be described as

$$1.0 \ CH_3NO_2 \rightarrow n_{CO}CO + n_{CO_2}CO_2 + n_{H_2}H_2 + n_{H_2O}H_2O + n_{N_2}N_2$$

Neglect other minor products. The mass balances are obtained for each atomic element.

$$
\begin{aligned}
\text{C} \quad & 1 = n_{CO} + n_{CO_2} \\
\text{H} \quad & 3 = 2n_{H_2} + 2n_{H_2O} \\
\text{O} \quad & 2 = n_{CO} + 2n_{CO_2} + n_{H_2O} \\
\text{N} \quad & 1 = 2n_{N_2} \quad \text{or} \quad n_{N_2} = 0.5
\end{aligned}
$$

The reaction commonly known as the water–gas reaction is

$$H_2 + CO_2 \rightleftharpoons H_2O + CO$$

Its equilibrium constant K, expressed as molar concentrations, is a function of temperature.

$$K = \frac{n_{H_2O}n_{CO}}{n_{H_2}n_{CO_2}}$$

The five equations above have six unknowns: namely, the five molar concentrations and K, which is a function of temperature. Solving for n_{H_2} and K:

$$(K - 1)n_{H_2}^2 + (3 - K/2)n_{H_2} = 2.25$$

K can be obtained from a table of the water–gas reaction as a function of temperature. Try $T = 2500\,K$ and $K = 6.440$ and substitute above.

$$5.440n_{H_2}^2 - 0.220n_{H_2} - 2.25 = 0$$

then

$$n_{H_2} = 0.664$$
$$n_{H_2O} = 1.500 - n_{H_2} = 0.836$$
$$n_{CO_2} = 0.164 \qquad n_{CO} = 0.836$$

The heats of formation $\Delta_f H^0$ for the various species are listed in the table below [from the JANAF thermochemical tables (Refs. 5–7 and 5–9)]. The heat of reaction is obtained from Eq. 5–9 in kilojoules per mole. By definition, the heat of formation of H_2 or N_2 is zero. From Eq. 5–9,

$$\Delta_r H^0 = \sum (n\Delta_f H)_{\text{products}} - (\Delta_f H^0)_{\text{reactant}}$$
$$= 0.836(-241.8) + 0.164(-393.5) + 0.836(-110.5) - 1.0(-113.1)$$
$$= -246\,\text{kJ/mol}$$

The enthalpy change of the gases going from the reference conditions to the combustion temperature can also be obtained from tables in Refs. 5–7 and 5–8 and is again listed below.

Species	$\Delta_f H^0$	Δh_j^{2500}	Molecular Weight	n_j
N_2	0	74.296	28	0.500
H_2O	−241.826	99.108	18	0.836
H_2	0	70.498	2	0.664
CO	−110.53	74.985	28	0.836
CO_2	−393.522	121.917	44	0.164
CH_3NO_2	−113.1		61	1.000

The gas enthalpy change of the hot gas in the combustion chamber is numerically equal to the heat of formation. Using data from the table,

$$\Delta H_{298}^{2500} = \sum n_j \Delta h_j = 249.5\,\text{kJ/mol}$$

This is not identical to the 246 kJ/mol obtained previously, and therefore a lower temperature is to be tried. After one or two iterations the final combustion temperature of 2470 K will be found where the heat of reaction balances the enthalpy rise. The above-mentioned composition will be approximately the same at the new temperature. The molecular weight can then be obtained from Eq. 5–5:

$$\mathfrak{M} = \frac{\sum n_j \mathfrak{M}}{\sum n_j} = \frac{28 \times 0.5 + 18 \times 0.836 + 2 \times 0.664 + 28 \times 0.836 + 44 \times 0.164}{2 \times (0.836) + 0.664 + 0.164 + 0.500} = 20.3$$

The specific heat varies with temperature, and average specific heat values $\bar{c}_p$ can be obtained from each species by integrating

$$\bar{c}_p = \frac{\int_{298}^{2470} c_p \, dT}{\int_{298}^{2470} dT}$$

Values of $\bar{c}_p$ can be obtained from tables in Ref. 5–7 and, if not done by computer, the integration can be done graphically. The result is

$$\bar{c}_p = 41{,}440 \ \text{kJ/K-kg-mol}/20.3$$
$$= 2040 \ \text{kJ/kg-K}$$

The specific heat ratio is, from Eq. 5–7,

$$k = \frac{C_p}{C_p - R'} = \frac{41{,}440}{41440 - 8314} = 1.25$$

With $\mathfrak{M}$, k, and T_1 now determined, the ideal performance of a nitromethane rocket engine can be established from Eqs. 3–16, 3–30, and 3–32 for $p_1 = 69$ atm and $p_2 = 1.0$ atm. The results are

$$c^* = 1525 \ \text{m/sec}$$
$$C_F = 1.57 \ \text{(from Fig. 3–6)}$$
$$c = 1.57 \times 1525 = 2394 \ \text{m/sec}$$
$$I_s = 2394/9.80 = 244 \ \text{sec}$$

5.4. COMPUTER ANALYSIS

All the analysis discussed in this chapter is done today by computer programs. Most are based on minimizing the free energy. This is a simpler approach than relying on equilibrium constants, which was used some years ago. Once the values of n_j and T_1 are determined, it is possible to calculate the molecular mass of the gases (Eq. 5–5), the average molar specific heats C_p by a similar formula, and the specific heat ratio k from Eqs. 3–6 and 5–7. This then characterizes the thermodynamic conditions in the combustion chamber. With these data we can calculate c^*, R, and other parameters of the chamber combustion. The nozzle expansion process simulated by computer gives the performance (such as I_s, c, or A_2/A_t) and the gas conditions in the nozzle; it usually includes several of the corrections mentioned in Chapter 3. Programs exist for one-, two-, and three-dimensional flow patterns.

More sophisticated solutions include a supplementary analysis of combustion chamber conditions where the chamber velocities are high (see Ref. 5–14), a boundary layer analysis, a heat transfer analysis, or a two-dimensional axisymmetric flow with nonuniform flow properties across any cross section of the nozzle. Time-dependent chemical reactions in the chamber are usually neglected, but they can be analyzed by estimating the time rate at which the reaction occurs; one way is to calculate the time derivative of the degree of advancement $d\lambda/dt$ and then to set this derivative to zero. This is described in Ref. 5–3.

An example of a commonly used computer program, based on chemical equilibrium compositions, was developed at the NASA Lewis Laboratory. It is described in Ref. 5–13, Vols. 1 and 2. The key assumptions for this program are one-dimensional forms of the continuity, energy, and momentum equations, zero velocity at the forward end of the chamber, isentropic expansion in the nozzle, using ideal gas laws, and chemical equilibrium in the combustion chamber. It includes options to use frozen equilibrium and narrow chambers (for liquid propellant combustion) or port areas with small cross sections (for solid propellant grains), where the chamber flow velocities are high, causing an extra pressure loss and a slight loss in performance.

Table 5–4 shows calculated data for a liquid oxygen, liquid hydrogen thrust chamber taken from an example of this reference. It has shifting equilibrium in the nozzle flow. The narrow chamber has a cross section that is only a little larger than the throat area. The large pressure drop in the chamber (approximately 126 psi) is due to the energy needed to accelerate the gas, as discussed in Section 3.3 and Table 3–2.

5.5. RESULTS OF THERMOCHEMICAL CALCULATIONS

Voluminous results of these machine calculations are available and only a few samples are indicated here to illustrate typical effects of the variations of various parameters. In general, high specific impulse or high values of c^* can be obtained if the average molecular weight of the reaction products is low (usually this implies a formulation rich in hydrogen) or if the available chemical energy (heat of reaction) is large, which means high combustion temperatures (see Eq. 3–16).

Values of calculated specific impulse will be higher than those obtained from firing actual propellants in rocket units. In practice it has been found that the experimental values are, in general, 3 to 12% lower than those calculated by the method explained in this chapter. Because the nozzle inefficiencies explained in Chapter 3 must be considered, only a portion of this correction (perhaps 1 to 4%) is due to combustion inefficiencies.

Figures 5–1 to 5–6 indicate the results of performance calculations for the liquid propellant combination, liquid oxygen–RP-1. These data are taken from

TABLE 5–4. Calculated Parameters for Liquid Oxygen and Liquid Hydrogen Rocket Engine for Four Different Nozzle Expansions

Chamber pressure at injector 773.3 psia or 53.317 bar; $c^* = 2332.1$ m/sec; shifting equilibrium nozzle flow mixture ratio $O_2/H_2 = 5.551$; chamber to throat area ratio $A_1/A_t = 1.580$.

Parameters

Location	Injector face	Comb. end	Throat	Exit	Exit	Exit	Exit
p_{inj}/p	1.00	1.195	1.886	10.000	100.000	282.15	709.71
T (K)	3389	3346	3184	2569	1786	1468	1219
$\mathfrak{M}$ (molec. mass)	12.7	12.7	12.8	13.1	13.2	13.2	13.2
k (spec. heat ratio)	1.14	1.14	1.15	1.17	1.22	1.24	1.26
C_p (spec. heat, kJ/kg-K)	8.284	8.250	7.530	4.986	3.457	3.224	3.042
M (Mach number)	0.00	0.413	1.000	2.105	3.289	3.848	4.379
A_2/A_t	1.580[a]	1.580[a]	1.000	2.227	11.52	25.00	50.00
c (m/sec)	NA	NA	2879[b]	3485	4150	4348	4487
v_2 (m/sec)	NA	NA	1537[b]	2922	3859	4124	4309

Mole fractions of gas mixture

H	0.03390	0.03336	0.02747	0.00893	0.00024	0.00002	0.00000
HO_2	0.00002	0.00001	0.00001	0.00000	0.00000	0.00000	0.00000
H_2	0.29410	0.29384	0.29358	0.29659	0.30037	0.30050	0.30052
H_2O	0.63643	0.63858	0.65337	0.68952	0.69935	0.69948	0.69948
H_2O_2	0.00001	0.00001	0.00000	0.00000	0.00000	0.00000	0.00000
O	0.00214	0.00204	0.00130	0.00009	0.00000	0.00000	0.00000
OH	0.03162	0.03045	0.02314	0.00477	0.00004	0.00000	0.00000
O_2	0.00179	0.00172	0.00113	0.00009	0.00000	0.00000	0.00000

[a]Chamber contraction ratio A_1/A_t.

[b]If cut off at throat.

c is the effective exhaust velocity in a vacuum.

v_2 is the nozzle exit velocity at optimum nozzle expansion.

NA means not applicable.

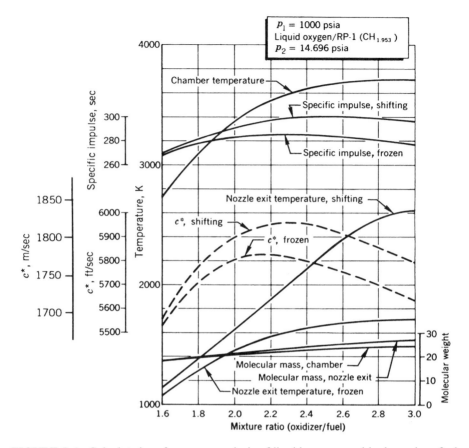

FIGURE 5–1. Calculated performance analysis of liquid oxygen and hydrocarbon fuel as a function of mixture ratio.

Refs. 5–7 and 5–8. The RP-1 fuel is a narrow-cut hydrocarbon similar to kerosene with an average of 1.953 mol of hydrogen for each mole of carbon; thus it has a nominal formula of $CH_{1.953}$. The calculation is limited to a chamber pressure of 1000 psia. Most of the curves are for optimum area ratio expansion to atmospheric pressure, namely, 1 atm or 14.696 psia, and a limited range of oxidizer-to-fuel mixture ratios.

For maximum specific impulse, Figs. 5–1 and 5–4 show an optimum mixture ratio of approximately 2.3 (kg/sec of oxidizer flow divided by kg/sec of fuel flow) for frozen equilibrium expansion and 2.5 for shifting equilibrium with gas expansion to sea level pressure. The maximum values of c^* are at slightly different mixture ratios. This optimum mixture ratio is not the value for highest temperature, which is usually fairly close to the stoichiometric value. The stoichiometric mixture ratio is more than 3.0; much of the carbon is burned to CO_2 and almost all of the hydrogen to H_2O.

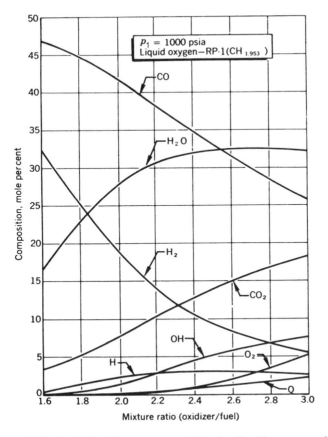

FIGURE 5–2. Calculated chamber gas composition for liquid oxygen and hydrocarbon fuel as a function of mixture ratio. Aggressive gases, such as O_2, O, or OH, can cause oxidation of the wall materials in the chamber and the nozzle.

Because shifting equilibrium makes more enthalpy available for conversion to kinetic energy, it gives higher values of performance (higher I_s or c^*) and higher values of nozzle exit temperature for the same exit pressure (see Fig. 5–1). The influence of mixture ratio on chamber gas composition is evident from Fig. 5–2. A comparison with Fig. 5–3 indicates the marked changes in the gas composition as the gases are expanded under shifting equilibrium conditions. The influence of the degree of expansion, or of the nozzle exit pressure on the gas composition, is shown in Fig. 5–6. As the gases are expanded to higher area ratios and lower exit pressure (or higher pressure ratios) the performance increases; however, the relative increase diminishes as the pressure ratio is further increased (see Figs. 5–5 and 5–6).

Dissociation of molecules requires considerable energy and causes a decrease in the combustion temperature, which in turn can reduce the specific impulse.

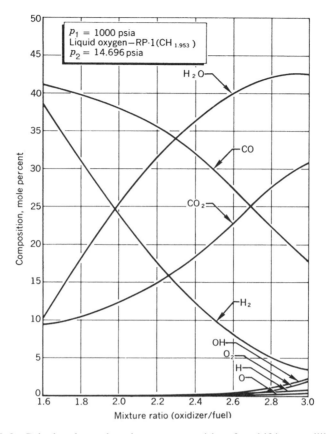

FIGURE 5–3. Calculated nozzle exit gas composition for shifting equilibrium conditions as a function of mixture ratio. Breakdown into O, OH, *or* H *and free* O_2 occurs only at the higher temperatures or higher mixture ratios.

Dissociation of the reaction products increases as the chamber temperature rises, and decreases with increasing chamber pressure. Atoms or radicals such as monatomic O or H and OH are formed, as can be seen from Fig. 5–2; some unreacted O_2 also remains at the higher mixture ratios and very high combustion temperatures. As the gases are cooled in the nozzle expansion, the dissociated species react again to form molecules and release heat into the flowing gases. As can be seen from Fig. 5–3, only a small percentage of dissociated species persists at the nozzle exit and only at the high mixture ratio, where the exit temperature is relatively high. (See Fig. 5–1 for exit temperatures with shifting equilibria). Heat released in a supersonic flow actually reduces the Mach number.

Results of calculations for several different liquid and solid propellant combinations are given in Tables 5–5 and 5–6. For the liquid propellant combinations, the listed mixture ratios are optimum and their performance is a

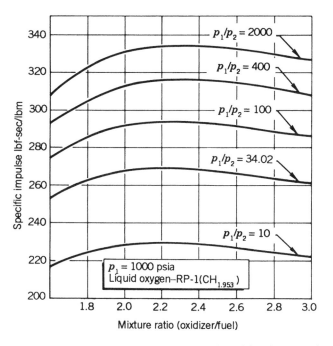

FIGURE 5–4. Variation of theoretical specific impulse with mixture ratio and pressure ratio, calculated for frozen equilibrium.

maximum. For solid propellants, practical considerations (such as propellant physical properties) do not always permit the development of a satisfactory propellant grain when the ingredients are mixed in optimum performance proportions (insufficient binder); therefore the values listed for solid propellants in Table 5–6 correspond in part to practical formulations with reasonable physical and ballistic properties.

Calculated data obtained from Ref. 5–13 are presented in Tables 5–7 to 5–9 for a specific solid propellant to indicate typical variations in performance or gas composition. This particular propellant consists of 60% ammonium perchlorate (NH_4ClO_4), 20% pure aluminum powder, and 20% of an organic polymer of an assumed chemical composition, namely, $C_{3.1}ON_{0.84}H_{5.8}$. Table 5–7 shows the variation of several performance parameters with different chamber pressures expanding to atmospheric exit pressure. The area ratios listed are optimum for this expansion with shifting equilibrium. The exit enthalpy, exit entropy, thrust coefficient, and the specific impulse also consider shifting equilibrium conditions. The characteristic velocity c^* and the chamber molecular mass are functions of chamber conditions only. Table 5–8 shows the variation of gas composition with chamber pressure. Some of the reaction products are in the liquid phase, such as Al_2O_3. Table 5–9 shows the variation of nozzle exit characteristics and composition for shifting equilibria as a func-

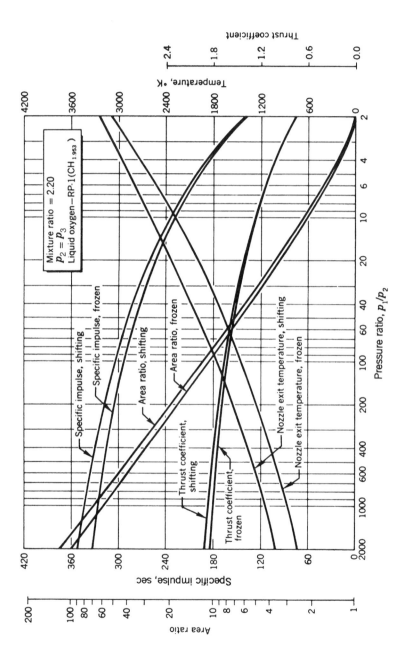
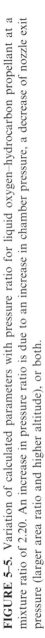

FIGURE 5–5. Variation of calculated parameters with pressure ratio for liquid oxygen–hydrocarbon propellant at a mixture ratio of 2.20. An increase in pressure ratio is due to an increase in chamber pressure, a decrease of nozzle exit pressure (larger area ratio and higher altitude), or both.

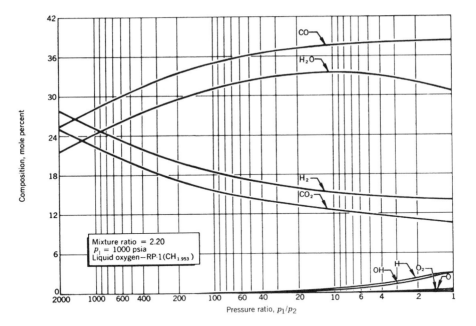

FIGURE 5–6. Variation of exhaust gas composition at nozzle exit with pressure ratio at a fixed mixture ratio and for shifting equilibrium. For frozen equilibrium the composition would be the same as in the chamber, as shown in Fig. 5–2.

tion of exit pressure or pressure ratio for a fixed value of chamber pressure. Table 5–9 shows how the composition is shifted during expansion in the nozzle and how several of the species present in the chamber have disappeared at the nozzle exit. These three tables show theoretical results calculated on a computer; some of the thermodynamic properties of the reactants and reaction products probably do not warrant the indicated high accuracy of five significant figures which are obtained from the computer. In the analysis for chemical ingredients of this solid propellant, approximately 76 additional reaction products were considered in addition to the major product species. This includes, for example, CN, CH, CCl, Cl, NO, and so on. Their calculated mole fractions were very small and therefore they have been neglected and are not included in Table 5–8 or 5–9.

Calculations of this type are useful in estimating performance (I_s, c^*, C_F, ϵ, etc.) for a particular chamber pressure and nozzle exit pressure, and knowledge of the gas composition, as indicated by the previous figures and tables, permits a more detailed estimate of other design parameters, such as gas-film properties for heat transfer determination, radiation characteristics of the flame inside and outside the thrust chambers, and the acoustic characteristics of the gases. Performance data calculated for hybrid propellants are presented briefly in Chapter 15.

TABLE 5–5. Theoretical Performance of Liquid Rocket Propellant Combinations

Oxidizer	Fuel	Mixture Ratio By Mass	Mixture Ratio By Volume	Average Specific Gravity	Chamber Temp. (K)	Chamber c* (m/sec)	𝔐 (kg/mol)	I_s (sec) Shifting	I_s (sec) Frozen	k
Oxygen	Methane	3.20	1.19	0.81	3526	1835	18.3	311	296	1.25
		3.00	1.11	0.80	3526	1853	19.3			
	Hydrazine	0.74	0.66	1.06	3285	1871		313	301	1.26
		0.90	0.80	1.07	3404	1892				
	Hydrogen	3.40	0.21	0.26	2959	2428	8.9	389.5	386	1.24
		4.02	0.25	0.28	2999	2432	10.0			
	RP-1	2.24	1.59	1.01	3571	1774	21.9	285.4	300	1.24
		2.56	1.82	1.02	3677	1800	23.3			
	UDMH	1.39	0.96	0.96	3542	1835	19.8	310	295	1.25
		1.65	1.14	0.98	3594	1864	21.3			
Fluorine	Hydrazine	1.83	1.22	1.29	4553	2128	18.5	334		1.33
		2.30	1.54	1.31	4713	2208	19.4			
	Hydrogen	4.54	0.21	0.33	3080	2534	8.9	410	365	1.33
		7.60	0.35	0.45	3900	2549	11.8		389	
Nitrogen tetroxide	Hydrazine	1.08	0.75	1.20	3258	1765	19.5	292	283	1.26
		1.34	0.93	1.22	3152	1782	20.9			
	50% UDMH–50% hydrazine	1.62	1.01	1.18	3242	1652	21.0	289	278	1.24
		2.00	1.24	1.21	3372	1711	22.6			
	RP-1	3.4	1.05	1.23	3290	1747	24.1			
	MMH	2.15	1.30	1.20	3396		22.3	289	297	1.23
		1.65	1.00	1.16	3200		21.7			
Red fuming nitric acid	RP-1	4.1	2.12	1.35	3175	1591	24.6		278	1.23
		4.8	2.48	1.33	3230	1594	25.8		258	1.22
	50% UDMH–50% hydrazine	1.73	1.00	1.23	2997	1609	20.6	269	272	1.22
		2.20	1.26	1.27	3172	1682	22.4			
Hydrogen peroxide (90%)	RP-1	7.0	4.01	1.29	2760	1701	21.7	279	297	1.19

Notes:

Combustion chamber pressure—1000 psia (6895 kN/m²); nozzle exit pressure—14.7 psia (1 atm); optimum expansion.

Adiabatic combustion and isentropic expansion of ideal gas.

The specific gravity at the boiling point was used for those oxidizers or fuels that boil below 20°C at 1 atm pressure.

Mixture ratios are for approximate maximum value of I_s.

188

TABLE 5–6. Theoretical Performance of Typical Solid Rocket Propellant Combinations

Oxidizer	Fuel	ρ_b (g/cm^3)^a	T_1 (K)	c^* (m/sec)^b	$\mathfrak{M}$ (kg/kg–mol)	I_s (sec)^b	k
Ammonium nitrate	11% binder and 7% additives	1.51	1282	1209	20.1	192	1.26
Ammonium perchlorate 78–66%	18% organic polymer binder and 4–20% aluminum	1.69	2816	1590	25.0	262	1.21
Ammonium perchlorate 84 to 68%	12% polymer binder and 4 to 20% aluminum	1.74	3371	1577	29.3	266	1.17

^{a} Average specific gravity of solid propellant.

^{b} *Conditions for I_s and c^*:*
Combustion chamber pressure: 1000 psia
Nozzle exit pressure: 14.7 psia
Optimum nozzle expansion ratio
Frozen equilibrium

In gas generators and preburners (see Section 10.5), for staged combustion cycle rocket engines (explained in Section 6.5) the gas temperatures are much lower, to avoid damage to the turbine blades. Typically, the combustion reaction gases are at 900 to 1200 K, which is lower than the gas in the thrust chamber (2900 to 3600 K). The thermochemical analysis of this chapter can also be applied to gas generators; the results (such as gas temperature T_1, the specific heat c_p, specific heat ratio k, or composition) are used for estimating turbine inlet conditions or turbine power. Examples are listed in Table 5–10 for a chamber pressure of 1000 psia. Some species in the gases will not be present (such as atomic oxygen or hydroxyl), and often real gas properties will need to be used because some of these gases do not behave as a perfect gas at these temperatures.

TABLE 5-7. Variation of Calculated Performance Parameters for an Aluminized Ammonium Perchlorate Propellant as a Function of Chamber Pressure for Expansion to Sea Level (1 atm) with Shifting Equilibrium

Chamber pressure (psia)	1500	1000	750	500	200
Chamber pressure (atm) or pressure ratio p_1/p_2	102.07	68.046	51.034	34.023	13.609
Chamber temperature (K)	3346.9	3322.7	3304.2	3276.6	3207.7
Nozzle exit temperature (K)	2007.7	2135.6	2226.8	2327.0	2433.6
Chamber enthalpy (cal/g)	−572.17	−572.17	−572.17	−572.17	−572.17
Exit enthalpy (cal/g)	−1382.19	−1325.15	−1282.42	−1219.8	−1071.2
Entropy (cal/g-K)	2.1826	2.2101	2.2297	2.2574	2.320
Chamber molecular mass (kg/mol)	29.303	29.215	29.149	29.050	28.908
Exit molecular mass (kg/mol)	29.879	29.853	29.820	29.763	29.668
Exit Mach number	3.20	3.00	2.86	2.89	2.32
Specific heat ratio—chamber, k	1.1369	1.1351	1.1337	1.1318	1.1272
Specific impulse, vacuum (sec)	287.4	280.1	274.6	265.7	242.4
Specific impulse, sea level expansion (sec)	265.5	256.0	248.6	237.3	208.4
Characteristic velocity, c^* (m/sec)	1532	1529	1527	1525	1517
Nozzle area ratio, A_2/A_t [a]	14.297	10.541	8.507	8.531	6.300
Thrust coefficient, C_F [a]	1.700	1.641	1.596	1.597	1.529

[a] At optimum expansion.

TABLE 5–8. Mole Fraction Variation of Chamber Gas Composition with Combustion Chamber Pressure for a Solid Propellant

Pressure (psia)	1500	1000	750	500	200
Pressure (atm) or pressure ratio	102.07	68.046	51.034	34.023	13.609
Ingredient					
Al	0.00007	0.00009	0.00010	0.00012	0.00018
AlCl	0.00454	0.00499	0.00530	0.00572	0.00655
$AlCl_2$	0.00181	0.00167	0.00157	0.00142	0.00112
$AlCl_3$	0.00029	0.00023	0.00019	0.00015	0.00009
AlH	0.00002	0.00002	0.00002	0.00002	0.00002
AlO	0.00007	0.00009	0.00011	0.00013	0.00019
AlOCl	0.00086	0.00095	0.00102	0.00112	0.00132
AlOH	0.00029	0.00032	0.00034	0.00036	0.00041
AlO_2H	0.00024	0.00026	0.00028	0.00031	0.00036
Al_2O	0.00003	0.00004	0.00004	0.00005	0.00006
Al_2O_3 (solid)	0.00000	0.00000	0.00000	0.00000	0.00000
Al_2O_3 (liquid)	0.09425	0.09378	0.09343	0.09293	0.09178
CO	0.22434	0.22374	0.22328	0.22259	0.22085
COCl	0.00001	0.00001	0.00001	0.00001	0.00000
CO_2	0.00785	0.00790	0.00793	0.00799	0.00810
Cl	0.00541	0.00620	0.00681	0.00772	0.01002
Cl_2	0.00001	0.00001	0.00001	0.00001	0.00001
H	0.02197	0.02525	0.02776	0.03157	0.04125
HCl	0.12021	0.11900	0.11808	0.11668	0.11321
HCN	0.00003	0.00002	0.00001	0.00001	0.00000
HCO	0.00003	0.00002	0.00002	0.00002	0.00001
H_2	0.32599	0.32380	0.32215	0.31968	0.31362
H_2O	0.08960	0.08937	0.08916	0.08886	0.08787
NH_2	0.00001	0.00001	0.00001	0.00000	0.00000
NH_3	0.00004	0.00003	0.00002	0.00001	0.00001
NO	0.00019	0.00021	0.00023	0.00025	0.00030
N_2	0.09910	0.09886	0.09867	0.09839	0.09767
O	0.00010	0.00014	0.00016	0.00021	0.00036
OH	0.00262	0.00297	0.00324	0.00364	0.00458
O_2	0.00001	0.00001	0.00002	0.00002	0.00004

TABLE 5-9. Calculated Variation of Thermodynamic Properties and Exit Gas Composition for an Aluminized Perchlorate Propellant with $p_1 = 1500$ psia and Various Exit Pressures at Shifting Equilibrium and Optimum Expansion

	Chamber	Throat	Nozzle Exit				
Pressure (atm)	102.07	58.860	2.000	1.000	0.5103	0.2552	0.1276
Pressure (MPa)	10.556	5.964	0.2064	0.1032	0.0527	0.0264	0.0132
Nozzle area ratio	>0.2	1.000	3.471	14.297	23.972	41.111	70.888
Temperature (K)	3346.9	3147.3	2228.5	2007.7	1806.9	1616.4	1443.1
Ratio chamber pressure/local pressure	1.000	1.7341	51.034	102.07	200.00	400.00	800.00
Molecular mass (kg/mol)	29.303	29.453	29.843	29.879	29.894	29.899	29.900
Composition (mol %)							
Al	0.00007	0.00003	0.00000	0.00000	0.00000	0.00000	0.00000
AlCl	0.00454	0.00284	0.00014	0.00008	0.00000	0.00000	0.00000
AlCl$_2$	0.00181	0.00120	0.00002	0.00000	0.00000	0.00000	0.00000
AlCl$_3$	0.00029	0.00023	0.00002	0.00000	0.00000	0.00000	0.00000
AlOCl	0.00086	0.00055	0.00001	0.00000	0.00000	0.00000	0.00000
AlOH	0.00029	0.00016	0.00000	0.00000	0.00000	0.00000	0.00000
AlO$_2$H	0.00024	0.00013	0.00000	0.00000	0.00000	0.00000	0.00000
Al$_2$O	0.00003	0.00001	0.00000	0.00000	0.00000	0.00000	0.00000
Al$_2$O$_3$ (solid)	0.00000	0.00000	0.00000	0.00000	0.00000	0.00000	0.00000
Al$_2$O$_3$ (liquid)	0.09425	0.09608	0.09955	0.09969	0.09974	0.09976	0.09976
CO	0.22434	0.22511	0.22553	0.22416	0.22008	0.21824	0.21671
CO$_2$	0.00785	0.00787	0.00994	0.01126	0.01220	0.01548	0.01885
Cl	0.00541	0.00441	0.00074	0.00028	0.00009	0.00002	0.00000
H	0.02197	0.01722	0.00258	0.00095	0.00030	0.00007	0.00001
HCl	0.12021	0.12505	0.13635	0.13707	0.13734	0.13743	0.13746
H$_2$	0.32599	0.33067	0.34403	0.34630	0.34842	0.35288	0.35442
H$_2$O	0.08960	0.08704	0.08091	0.07967	0.07796	0.07551	0.07214
NO	0.00019	0.00011	0.00001	0.00000	0.00000	0.00000	0.00000
N$_2$	0.09910	0.09950	0.10048	0.10058	0.10063	0.10064	0.10065
O	0.00010	0.00005	0.00000	0.00000	0.00000	0.00000	0.00000
OH	0.00262	0.00172	0.00009	0.00005	0.00002	0.00000	0.00000

TABLE 5–10. Typical Gas Characteristics for Fuel-rich Liquid Propellant Gas Generators

Propellant	T_1 (K)	k	Gas Constant R (ft-lbf/lbm-R)	Oxidizer-to-fuel ratio	Specific heat c_p (kcal/kg-K)
Liquid oxygen and liquid	900	1.370	421	0.919	1.99
liquid hydrogen	1050	1.357	375	1.065	1.85
	1200	1.338	347	1.208	1.78
Liquid oxygen and	900	1.101	45.5	0.322	0.639
kerosene	1050	1.127	55.3	0.423	0.654
	1200	1.148	64.0	0.516	0.662
Nitrogen tetroxide and	1050	1.420	87.8	0.126	0.386
dimethyl hydrazine	1200	1.420	99.9	0.274	0.434

PROBLEMS

1. Explain the physical or chemical reasons for a maximum value of specific impulse at a particular mixture ratio of oxidizer to fuel.

2. Explain why, in Table 5–8, the relative proportion of monatomic hydrogen and monatonic oxygen changes markedly with different chamber pressures and exit pressures.

3. This chapter contains several charts for the performance of liquid oxygen and RP-1 hydrocarbon fuel. By mistake the next shipment of cryogenic oxidizer contains at least 15% liquid nitrogen. Explain what general trends should be expected in the results of the next test in the performance values, the composition of the exhaust gas under chamber and nozzle conditions, and the optimum mixture ratio.

4. A mixture of perfect gases consists of 3 kg of carbon monoxide and 1.5 kg of nitrogen at a pressure of 0.1 MPa and a temperature of 298.15 K. Using Table 5–1, find (a) the effective molecular mass of the mixture, (b) its gas constant, (c) specific heat ratio, (d) partial pressures, and (e) density.

   Answers: (a) 28 kg/kg-mol, (b) 297 J/kg-K, (c) 1.40, (d) 0.0666 and 0.0333 MPa, (e) 1.13 kg/m^3.

5. Using information from Table 5–2, plot the value of the specific heat ratio for carbon monoxide (CO) as a function of temperature. Notice the trend of this curve; it is typical of the temperature behavior of other diatomic gases.

   Answers: $k = 1.28$ at 3500 K, 1.30 at 2000 K, 1.39 at 500 K.

6. Modify and tabulate two entries in Table 5–5 for operation in the vacuum of space, namely oxygen/hydrogen and nitrogen tetroxide/hydrazine. Assume the data in the table represents the design condition.

7. The figures in this chapter show several parameters and gas compositions of liquid oxygen burning with RP-1, which is a kerosene-type material. For a mixture ratio of 2.0, use the compositions to verify the molecular mass in the chamber and the specific impulse (frozen equilibrium flow in nozzle) in Fig. 5–1.

SYMBOLS

(Symbols referring to chemical elements, compounds, or mathematical operators are not included in this list.)

a	number of kilogram atoms
A_t	throat area, m^2
c^*	characteristic velocity, m/sec
c_p	specific heat per unit mass, J/kg-K
C_p	molar specific heat at constant pressure of gas mixture, J/kg-mol-K
g_0	acceleration of gravity at sea level, 9.8066 m/sec^2
G	Gibbs free energy for a propellant combustion gas mixture, J/kg
$\Delta_f G^0$	change in free energy of formation at 298.15 K and 1 bar
G_j	free energy for a particular species j, J/kg
ΔH	overall enthalpy change, J/kg or J/kg-mol
ΔH_j	enthalpy change for a particular species j, J/kg
$\Delta_r H^0$	heat of reaction at reference 298.15 K and 1 bar, J/kg
$\Delta_f H^0$	heat of formation at reference 298.15 K and 1 bar, J/kg
h	enthalpy for a particular species, J/kg or J/kg-mol
I_s	specific impulse, N-sec^3/kg;m^2 (lbf-sec/lbm)
k	specific heat ratio
K_f	equilibrium constant when a compound is formed from its elements
K_n	equilibrium constant as a function of molar fractions
K_p	equilibrium constant as a function of partial pressure
m	number of gaseous species
$\dot{m}$	mass flow rate, kg/sec
$\mathfrak{M}$	molecular mass (also called molecular weight) of gas mixture, kg/mol
n	total number of species or moles per unit mass (kg-mol/kg) of mixture
n_j	mole fraction or volume percent of species j, kg-mol/kg-mixture
p	pressure of gas mixture, N/m^2
R	gas constant, J/kg-K
R'	universal gas constant, 8314.3 J/kg mol-K
S	entropy, J/kg mol-K
T	absolute temperature, K
U	internal energy, J/kg-mol
v	gas velocity, m/sec
V	specific volume, m^3/kg

Greek Letters

ϵ nozzle exit area ratio (exit/throat area)

λ Lagrange multiplier, or factor for the degree of advancement of a
chemical reaction

ρ density, kg/m^3

Subscripts

a, b molar fractions of reactant species A or B

c, d molar fractions of product species C or D

i atomic species in a specific propellant

j constituents or species in reactants or products

mix   mixture of gases

ref   at reference condition (also superscript ^{0})

1   chamber condition

2   nozzle exit condition

3   ambient atmospheric condition

REFERENCES

5–1. F. Van Zeggeren and S. H. Storey, *The Computation of Chemical Equilibria*, Cambridge University Press, Cambridge, 1970.

5–2. S. S. Penner, *Thermodynamics for Scientists and Engineers*, Addison-Wesley Publishing Co., Reading, MA, 1968.

5–3. S. I. Sandler, *Chemical and Engineering Thermodynamics*, John Wiley & Sons, 1999, 656 pages.

5–4. M. W. Zemansky and R. H. Dittman, *Heat and Thermodynamics*, McGraw-Hill Book Company, New York, 1981.

5–5. K. Denbigh, *The Principles of Chemical Equilibrium*, 4th ed., Cambridge University Press, Cambridge, 1981.

5–6. K. K. Kuo, *Principles of Combustion*, John Wiley & Sons, 1986.

5–7. *JANAF Thermochemical Tables*, Dow Chemical Company, Midland, MI, Series A (June 1963) through Series E (January 1967).

5–8. M. W. Chase, C. A. Davies, J. R. Downey, D. J. Frurip, R. A. McDonald, and A. N. Syverud, *JANAF Thermochemical Tables*, 3rd ed., Part I, *Journal of Physical and Chemical Reference Data*, Vol. 14, Supplement 1, American Chemical Society, American Institute of Physics, and National Bureau of Standards, 1985.

5–9. D. D. Wagman et al., "The NBS Tables of Chemical Thermodynamic Properties," *Journal of Physical and Chemical Reference Data*, Vol. 11, Supplement 2, American Chemical Society, American Institute of Physics, and National Bureau of Standards, 1982.

5–10. J. B. Pedley, R. D. Naylor, and S. P. Kirby, *Thermochemical Data of Organic Compounds*, 2nd ed., Chapman & Hall, London, 1986.

5–11. B. J. McBride, S. Gordon, and M. Reno, "Thermodynamic Data for Fifty Reference Elements," *NASA Technical Paper 3287*, January 1993.

5–12. B. J. McBride and S. Gordon, "Computer Program for Calculating and Fitting Thermodynamic Functions," *NASA Reference Publication 1271*, November 1992.

5–13. S. Gordon and B. J. McBride, "Computer Program for Calculation of Complex Chemical Equilibrium Compositions and Applications, Vol. 1: Analysis" (October 1994) and "Vol. 2: User Manual and Program Description" (June 1996), *NASA Reference Publication 1311*.

5–14. S. Gordon and B. J. McBride, "Finite Area Combustor Theoretical Rocket Performance," *NASA TM 100785*, April 1988.

CHAPTER 6

LIQUID PROPELLANT ROCKET ENGINE FUNDAMENTALS

This is the first of five chapters devoted to liquid propellant rocket engines. It gives an overview of the engines (a definition of various propellants, engine performance, propellant budget), and of the smaller reaction control engines. It also presents several of their principal subsystems, such as two types of feed systems (including engine cycles), propellant tanks and their pressurization subsystems, valves and piping systems, and engine structures. Chapter 7 covers liquid propellants in more detail, Chapter 8 deals with thrust chambers (and nozzles), Chapter 9 with combustion, and Chapter 10 discusses turbopumps, engine design, engine controls, propellant budgets, engine balance and calibration, overall engine systems.

A liquid propellant rocket propulsion system is commonly called a *rocket engine*. It has all the hardware components and propellants necessary for its operation, that is, for producing thrust. It consists of one or more *thrust chambers*, one or more *tanks** to store the propellants, a *feed mechanism* to force the propellants from the tanks into the thrust chamber(s), a *power source* to furnish the energy for the feed mechanism, suitable *plumbing* or *piping* to transfer the liquids, a *structure** to transmit the thrust force, and *control* devices to initiate and regulate the propellant flow and thus the thrust. In some applications an engine may also include a *thrust vector control system*, various *instrumentation* and *residual propellant* (trapped in pipes, valves, or wetting tank walls). It does not include hardware for non-propulsive purposes, such

---

*The *tanks* and some or all of the *engine structure* and *piping* are sometimes considered to be part of the *vehicle* or the *test facility* and not the engine, depending on the preference of the organizations working on the project.

as aerodynamic surfaces, guidance, or navigation equipment, or the useful payload, such as a scientific space exploration package or a missile warhead.

Figures 1–3 and 1–4 show the basic flow diagrams for simple rocket engines with a *pressurized* and a *turbopump feed system*. Figure 6–1 shows a complex, sophisticated, high-performance liquid propellant rocket engine. References 6–1 and 6–2 give general liquid propellant rocket engine information. Additional data and figures on other rocket engines can be found in Chapter 10.

The design of any propulsion system is tailored to fit a specific application or *mission requirement*. These requirements are usually stated in terms of the application (anti-aircraft rocket, upper stage launch vehicle propulsion, or projectile assist), mission velocity, the desired flight trajectories (surface launch, orbit transfer, altitude–performance profile), vulnerability, attitude control torques and duty cycle, minimum life (during storage or in orbit), or number of units to be built and delivered. They include constraints on cost, schedule, operating conditions (such as temperature limits), storage conditions, or safety rules. Additional criteria, constraints, and the selection process are explained in Chapter 17.

The *mission requirements* can be translated into *rocket engine requirements* in terms of thrust–time profile, propellants, number of thrust chambers, total impulse, number of restarts, minimum reliability, likely propellant, and engine masses and their sizes or envelopes. We can do this only if we select several of the key engine features, such as the feed system, chamber pressure, the method of cooling the thrust chambers, thrust modulation (restart, throttle, thrust vector control), engine cycle (if using turbopump feed), and other key design features. We can arrive at one or more engine concepts and their preliminary or conceptual designs. Tables 1–3 to 1–5 give typical data. Many different types of rocket engines have been built and flown, ranging in thrust size from less than 0.01 lbf to over 1.75 million pounds, with one-time operation or multiple starts (some have over 150,000 restarts), with or without thrust modulation (called throttling), single use or reusable, arranged as single engines or in clusters of multiple units.

One way to categorize liquid propellant rocket engines is described in Table 6–1. There are two categories, namely those used for *boosting* a payload and imparting a significant velocity increase to a payload, and *auxiliary propulsion* for *trajectory adjustments* and *attitude control*. Liquid propellant rocket engine systems can be *classified* in several other ways. They can be *reusable* (like the Space Shuttle main engine or a booster rocket engine for quick ascent or maneuvers of fighter aircraft) or suitable for a *single flight* only (as the engines in the Atlas or Titan launch vehicles) and they can be *restartable*, like a reaction control engine, or *single firing*, as in a space launch vehicle. They can also be categorized by their *propellants, application*, or *stage*, such as an upper stage or booster stage, their *thrust level*, and by the *feed system type* (*pressurized* or *turbopump*).

The *thrust chamber* or *thruster* is the combustion device where the liquid propellants are metered, injected, atomized, mixed, and burned to form hot

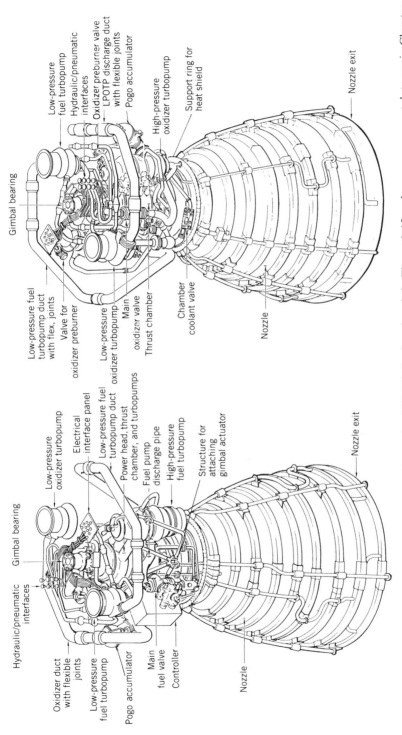

FIGURE 6-1. Two views of the Space Shuttle Main Engine (SSME). Its flowsheet is in Figure 6–12 and some component data are in Chapter 10. (Courtesy of The Boeing Company, Rocketdyne Propulsion and Power.)

Gimbal bearing

Low-pressure fuel turbopump

Hydraulic/pneumatic interfaces

Oxidizer preburner valve

LPOTP discharge duct with flexible joints

Pogo accumulator

High-pressure oxidizer turbopump

Support ring for heat shield

Nozzle exit

Low-pressure fuel turbopump duct with flex, joints

Valve for oxidizer preburner

Low-pressure oxidizer turbopump

Main oxidizer valve

Thrust chamber

Chamber coolant valve

Nozzle

Low-pressure oxidizer turbopump

Electrical interface panel

Low-pressure fuel turbopump duct

Power head, thrust chamber, and turbopumps

Fuel pump discharge pipe

High-pressure fuel turbopump

Structure for attaching gimbal actuator

Nozzle exit

Gimbal bearing

Hydraulic/pneumatic interfaces

Oxidizer duct with flexible joints

Low-pressure fuel turbopump

Pogo accumulator

Main fuel valve

Controller

Nozzle

199

TABLE 6–1. Characteristics of Two Categories of Liquid Propellant Rocket Engines

Purpose	Boost Propulsion	Auxiliary Propulsion
Mission	Impart significant velocity to propel a vehicle along its flight path	Attitude control, minor space maneuvers, trajectory corrections, orbit maintenance
Applications	Booster stage and upper stages of launch vehicles, large missiles	Spacecraft, satellites, top stage of anti-ballistic missile, space rendezvous
Total impulse	High	Low
Number of thrust chambers per engine	Usually 1; sometimes 4, 3, or 2	Between 4 and 24
Thrust level	High; 4500 N up to 7,900,000 N or 1000–1,770,000 lbf	Small; 0.001 up to 4500 N, a few go up to 1000 lbf
Feed system	Mostly turbopump type; occasionally pressurized feed system for smaller thrusts	Pressurized feed system with high-pressure gas supply
Propellants	Cryogenic and storable liquids (see next section)	Storable liquids, monopropellants, and/or stored cold gas
Chamber pressure	2.4–21 MPa or 350–3600 psi	0.14–2.1 MPa or 20–300 psi
Number of starts during a single mission	Usually no restart; sometimes one, but up to four in some cases	Several thousand starts are typical for small thrusters; fewer for larger thrust chambers, perhaps up to 10 starts
Cumulative duration of firing	Up to a few minutes	Up to several hours
Shortest firing duration	Typically 5–40 sec	0.02 sec typical for small thrusters
Time elapsed to reach full thrust	Up to several seconds	Usually very fast, 0.004–0.080 sec
Life in space	Hours, days, or months	10 years or more in space

gaseous reaction products, which in turn are accelerated and ejected at a high velocity to impart a thrust force. A thrust chamber has three major parts: an *injector*, a *combustion chamber*, and a *nozzle*. In a *cooled thrust chamber*, one of the propellants (usually the fuel) is circulated through cooling jackets or a special cooling passage to absorb the heat that is transferred from the hot reaction gases to the thrust chamber walls (see Figs 8–2 and 8–3). A *radiation-cooled* thrust chamber uses a special high-temperature material, such as niobium metal, which can radiate away its excess heat. There are *uncooled* or *heat-absorbing* thrust chambers, such as those using *ablative* materials. Thrust chambers are discussed in Chapter 8.

There are two types of *feed systems* used for liquid propellant rocket engines: those that use pumps for moving the propellants from their flight vehicle tanks to the thrust chamber, and those that use high-pressure gas for expelling or displacing their propellants from their tanks. They are discussed further in Chapter 10 and in Section 6.2 of this chapter.

Tables 17–1 to 17–4 compare the advantages and disadvantages of liquid propellant rocket engines and solid propellant rocket motors.

6.1. PROPELLANTS

The propellants, which are the working substance of rocket engines, constitute the fluid that undergoes chemical and thermodynamic changes. The term *liquid propellant* embraces all the various liquids used and may be one of the following:

1. Oxidizer (liquid oxygen, nitric acid, etc.)
2. Fuel (gasoline, alcohol, liquid hydrogen, etc.).
3. Chemical compound or mixture of oxidizer and fuel ingredients, capable of self-decomposition.
4. Any of the above, but with a gelling agent.

All are described in Chapter 7.

A *bipropellant* rocket unit has two separate liquid propellants, an oxidizer and a fuel. They are stored separately and are not mixed outside the combustion chamber. The majority of liquid propellant rockets have been manufactured for bipropellant applications.

A *monopropellant* contains an oxidizing agent and combustible matter in a single substance. It may be a mixture of several compounds or it may be a homogeneous material, such as hydrogen peroxide or hydrazine. Monopropellants are stable at ordinary atmospheric conditions but decompose and yield hot combustion gases when heated or catalyzed.

A *cold gas propellant* (e.g., nitrogen) is stored at very high pressure, gives a low performance, allows a simple system and is usually very reliable. It has been used for roll control and attitude control.

A *cryogenic propellant* is liquified gas at low temperature, such as liquid oxygen ($-183°C$) or liquid hydrogen ($-253°C$). Provisions for venting the storage tank and minimizing vaporization losses are necessary with this type.

Storable propellants (e.g., nitric acid or gasoline) are liquid at ambient temperature and can be stored for long periods in sealed tanks. *Space storable propellants* are liquid in the environment of space; this storability depends on the specific tank design, thermal conditions, and tank pressure. An example is ammonia.

A *gelled propellant* is a thixotropic liquid with a gelling additive. It behaves like a jelly or thick paint. It will not spill or leak readily, can flow under pressure, will burn, and is safer in some respects. It is described in a separate section of Chapter 7.

The propellant *mixture ratio* for a bipropellant is the ratio at which the oxidizer and fuel are mixed and react to give hot gases. The mixture ratio r is defined as the ratio of the oxidizer mass flow rate $\dot{m}_o$ and the fuel mass flow rate $\dot{m}_f$ or

$$r = \dot{m}_o/\dot{m}_f \tag{6–1}$$

The mixture ratio defines the composition of the reaction products. It is usually chosen to give a maximum value of specific impulse or $T_1/\mathfrak{M}$, where T_1 is the combustion temperature and $\mathfrak{M}$ is the average molecular mass of the reaction gases (see Eq. 3–16 or Fig. 3–2). For a given thrust F and a given effective exhaust velocity c, the total propellant flow is given by Eq. 2–6; namely, $\dot{m} = \dot{w}/g_0 = F/c$. The relationships between $r, \dot{m}, \dot{m}_o$, and $\dot{m}_f$ are

$$\dot{m}_o + \dot{m}_f = \dot{m} \tag{6–2}$$
$$\dot{m}_o = r\dot{m}/(r+1) \tag{6–3}$$
$$\dot{m}_f = \dot{m}/(r+1) \tag{6–4}$$

These same four equations are valid when w and $\dot{w}$ (weight) are substituted for m and $\dot{m}$. Calculated performance values for a number of different propellant combinations are given for specific mixture ratios in Table 5–5. Physical properties and a discussion of several common liquid propellants and their safety concerns are described in Chapter 7.

Example 6–1. A liquid oxygen–liquid hydrogen rocket thrust chamber of 10,000-lbf thrust operates at a chamber pressure of 1000 psia, a mixture ratio of 3.40, has exhaust products with a mean molecular mass of 8.9 lbm/lb-mol, a combustion temperature of 4380°F, and a specific heat ratio of 1.26. Determine the nozzle area, exit area for optimum operation at an altitude where $p_3 = p_2 = 1.58$ psia, the propellant weight and volume flow rates, and the total propellant requirements for 2 min of operation. Assume that the actual specific impulse is 97% of the theoretical value.

SOLUTION. The exhaust velocity for an optimum nozzle is determined from Eq. 3–16, but with a correction factor of g_0 for the foot-pound system.

$$
\begin{aligned}
v_2 &= \sqrt{\frac{2g_0 k}{k-1}\frac{R'T_1}{\mathfrak{M}}\left[1-\left(\frac{p_2}{p_1}\right)^{(k-1)/k}\right]} \\
&= \sqrt{\frac{2\times 32.2 \times 1.26}{0.26}\frac{1544 \times 4840}{8.9}(1-0.00158^{0.205})} = 13{,}900 \text{ ft/sec}
\end{aligned}
$$

The theoretical specific impulse is c/g_0, or in this case v_2/g_0 or $13{,}900/32.2 = 431$ sec. The actual specific impulse is $0.97 \times 431 = 418$ sec. The theoretical or ideal thrust coefficient can be found from Eq. 3–30 or from Fig. 3–6 ($p_2 = p_3$) to be $C_F = 1.76$. The actual thrust coefficient is slightly less, say 98% or $C_F = 1.72$. The throat area required is found from Eq. 3–31.

$$A_t = F/(C_F p_1) = 10,000/(1.72 \times 1000) = 5.80 \text{ in.}^2 \text{ (2.71 in. diameter)}$$

The optimum area ratio can be found from Eq. 3–25 or Fig. 3–5 to be 42. The exit area is $5.80 \times 42 = 244$ in.^{2} (17.6 in. diameter). The weight density of oxygen is 71.1 lbf/ft^3 and of hydrogen is 4.4 lbf/ft^3. The propellant weight flow rate is (Equation 2–5)

$$\dot{w} = F/I_s = 10,000/418 = 24.0 \text{ lbf/ sec}$$

The oxygen and fuel weight flow rates are, from Eqs. 6–3 and 6–4,

$$\dot{w}_o = \dot{w}r/(r+1) = 24.0 \times 3.40/4.40 = 18.55 \text{ lbf/ sec}$$
$$\dot{w}_f = \dot{w}/(r+1) = 24/4.40 = 5.45 \text{ lbf/ sec}$$

The volume flow rates are determined from the densities and the weight flow rates.

$$\dot{V}_o = \dot{w}_o/\rho_o = 18.55/71.1 = 0.261 \text{ ft}^3/ \text{sec}$$
$$\dot{V}_f = \dot{w}_f/\rho_f = 5.45/4.4 = 1.24 \text{ ft}^3/ \text{sec}$$

For 120 sec of operations (arbitrarily allow the equivalent of two additional seconds for start and stop transients and unavailable propellant), the weight and volume of required propellant are

$$w_o = 18.55 \times 122 = 2260 \text{ lbf of oxygen}$$
$$w_f = 5.45 \times 122 = 665 \text{ lbf of hydrogen}$$
$$V_o = 0.261 \times 122 = 31.8 \text{ ft}^3 \text{ of oxygen}$$
$$V_f = 1.24 \times 122 = 151 \text{ ft}^3 \text{ of hydrogen}$$

Note that, with the low-density fuel, the volume flow rate and therefore the tank volume of hydrogen are large compared to that of the oxidizer.

6.2. PROPELLANT FEED SYSTEMS

The propellant feed system has two principal functions: to raise the pressure of the propellants and to feed them to one or more thrust chambers. The energy for these functions comes either from a high-pressure gas, centrifugal pumps, or a combination of the two. The selection of a particular feed system and its components is governed primarily by the application of the rocket, the requirements mentioned at the beginning of this chapter, duration, number or type of thrust chambers, past experience, mission, and by general requirements of simplicity of design, ease of manufacture, low cost, and minimum inert mass. A classification of several of the more important types of feed system is shown in Fig. 6–2 and some are discussed in more detail below. All feed systems have piping, a series of valves, provisions for filling and removing (draining and flushing) the liquid propellants, and control devices to initiate, stop, and regulate their flow and operation.

204

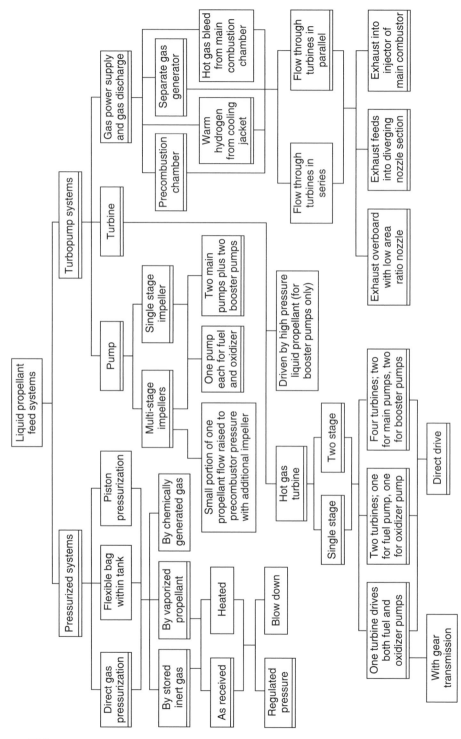

FIGURE 6–2. Design options of fed systems for liquid propellant rocket engines. The more common types are designated with a double line at

In general, a pressure feed system gives a vehicle performance superior to a turbopump system when the total impulse or the mass of propellant is relatively low, the chamber pressure is low, the engine thrust-to-weight ratio is low (usually less than 0.6), and when there are repeated short-duration thrust pulses; the heavy-walled tanks for the propellant and the pressurizing gas usually constitute the major inert mass of the engine system. In a turbopump feed systems the propellant tank pressures are much lower (by a factor of 10 to 40) and thus the tank masses are much lower (again by a factor of 10 to 40). Turbopump systems usually give a superior vehicle performance when the total impulse is large (higher Δu) and the chamber pressure is higher.

The pressurized feed system can be relatively simple, such as for a single-operation, factory-preloaded, simple unit (with burst diaphragms instead of some of the valves), or quite complex, as with multiple restartable thrusters or reusable systems. Table 6–2 shows typical features that have been designed into pressurized feed systems in order to satisfy particular design goals. Figures 1–3, 6–3, 6–4, and 6–13 show some of these features. If the propulsion system is to be reusable or is part of a manned vehicle (where the reliability requirements are very high and the vehicle's crew can monitor and override automatic commands), the feed system becomes more complex (with more safety features and redundancies) and more expensive.

The pneumatic (pressurizing gas) and hydraulic (propellant) flows in a liquid propellant engine can be simulated in a computer analysis that provides for a flow and pressure balance in the oxidizer and the fuel flow paths through the system. One approach is shown in Ref. 6–3. Some of these analyses can provide information on transient conditions (filling up of passages) during start, flow decays at cutoff, possible water hammer, or flow instabilities. The details of such analyses are not described in this book, but the basic mathematical simulation is relatively straightforward.

6.3. GAS PRESSURE FEED SYSTEMS

One of the simplest and most common means of pressurizing the propellants is to force them out of their respective tanks by displacing them with high-pressure gas. This gas is fed into the propellant tanks at a controlled pressure, thereby giving a controlled propellant discharge. Because of their relative simplicity, the rocket engines with pressurized feed systems can be very reliable. Reference 6–3 includes a design guide for pressurized gas systems.

A simple pressurized feed system is shown schematically in Fig. 1–3. It consists of a high-pressure gas tank, a gas starting valve, a pressure regulator, propellant tanks, propellant valves, and feed lines. Additional components, such as filling and draining provisions, check valves, filters, flexible elastic bladders for separating the liquid from the pressurizing gas, and pressure sensors or gauges, are also often incorporated. After all tanks are filled, the high-pressure gas valve in Fig. 1–3 is remotely actuated and admits gas through

TABLE 6–2. Typical Features of Liquid Propellant Feed Systems

---

Enhance Safety

Sniff devices to detect leak of hazardous vapor; used on Space Shuttle orbiter

Check valves to prevent backflow of propellant into the gas tank and inadvertent mixing of propllants inside flow passages

Features that prevent an unsafe condition to occur or persist and shut down engine safely, such as relief valves or relief burst diaphragms to prevent tank overpressurization), or a vibration monitor to shut off operation in the case of combustion instability

Isolation valves to shut off a section of a system that has a leak or malfunction

Burst diaphragms or isolation valves to isolate the propellants in their tanks and positively prevent leakage into the thrust chamber or into the other propellant tank during storage

Inert pressurizing gas

Provide Control

Valves to control pressurization and flow to the thrust chambers (start/stop/throttle)

Sensors to measure temperatures, pressures, valve positions, thrust, etc., and computers to monitor/analyze system status, issue command signals, and correct if sensed condition is outside predetermined limits

Manned vehicle can require system status display and command signal override

Fault detection, identification, and automatic remedy, such as shut-off isolation valves in compartment in case of fire, leak, or disabled thruster

Control thrust (throttle valve) to fit a desired thrust–time profile

Enhance Reliability

Fewest practical number of components/subassemblies

Ability to provide emergency mode engine operation, such as return of Space Shuttle vehicle to landing

Filters to catch dirt in propellant lines, which could prevent valve from closing or small injector holes from being plugged up or bearings from galling.

Duplication of unreliable key components, such as redundant small thrusters, regulators, check valves, or isolation valves

Heaters to prevent freezing of moisture or low-melting-point propellant

Long storage life—use propellants with little or no chemical deterioration and no reaction with wall materials

Provide for Reusability

Provisions to drain remaining propellants or pressurants

Provision for cleaning, purging, flushing, and drying the feed system and refilling propellants and pressurizing gas in field

Devices to check functioning of key components prior to next operation

Features to allow checking of engine calibration and leak testing after operation

Features for access of inspection devices for visual inspection at internal surfaces or components

Enable Effective Propellant Utilization

High tank expulsion efficiency with minimum residual, unavailable propellant

Lowest possible ambient temperature variation or matched propellant property variation with temperature so as to minimize mixture ratio change and residual propellant

Alternatively, measure remaining propellant in tanks (using a special gauge) and automatically adjust mixture ratio (throttling) to minimize residual propellant

Minimize pockets in the piping and valves that cannot be readily drained the pressure regulator at a constant pressure to the propellant tanks. The check valves prevent mixing of the oxidizer with the fuel when the unit is not in an upright position. The propellants are fed to the thrust chamber by opening valves. When the propellants are completely consumed, the pressurizing gas can also scavenge and clean lines and valves of much of the liquid propellant residue. The variations in this system, such as the combination of several valves into one or the elimination and addition of certain components, depend to a large extent on the application. If a unit is to be used over and over, such as space-maneuver rocket, it will include several additional features such as, possibly, a thrust-regulating device and a tank level gauge; they will not be found in an expendable, single-shot unit, which may not even have a tank-drainage provision. Different bipropellant pressurization concepts are evaluated in Refs. 6–3, 6–4, and 6–5. Table 6–2 lists various optional features. Many of these features also apply to pump-fed systems, which are discussed in Section 6..6. With monopropellants the gas pressure feed system becomes simpler, since there is only one propellant and not two, reducing the number of pipes, valves, and tanks.

A complex man-rated pressurized feed system, the combined *Space Shuttle Orbital Maneuver System* (OMS) *and the Reaction Control System* (RCS), is described in Figs 6–3 and 6–4, Ref. 6–6, and Table 6–3. There are three locations for the RCS, as shown in Fig. 1–13: a forward pod and a right and left aft pod. Figures 6–3 and 6–4 refer to one of the aft pods only and show a combined OMS and RCS arrangement. The OMS provides thrust for orbit insertion, orbit circularization, orbit transfer, rendezvous, deorbit, and abort. The RCS provides thrust for attitude control (in pitch, yaw, and roll) and for small-vehicle velocity corrections or changes in almost any direction (translation maneuvers), such as are needed for rendezvous and docking; it can operate simultaneously with or separate from the OMS.

The systems feature various redundancies, an automatic RCS thruster selection system, various safety devices, automatic controls, sensors to allow a display to the Shuttle's crew of the system's status and health, and manual command overrides. The reliability requirements are severe. Several key components, such as all the helium pressure regulators, propellant tanks, some valves, and about half the thrusters are duplicated and redundant; if one fails, another can still complete the mission. It is possible to feed up to 1000 lbm of the liquid from the large OMS propellant tanks to the small RCS ones, in case it is necessary to run one or more of the small reaction control thrusters for a longer period and use more propellant than the smaller tanks allow; it is also possible to feed propellant from the left aft system to the one on the vehicle's right side, and vice versa. These features allow for more than nominal total impulse in a portion of the thrusters, in case it is needed for a particular mission mode or an emergency mode.

The compartmented steel propellant tanks with antislosh and antivortex baffles, sumps, and a surface tension propellant retention device allow propellant to be delivered independent of the propellant load, the orientation, or the

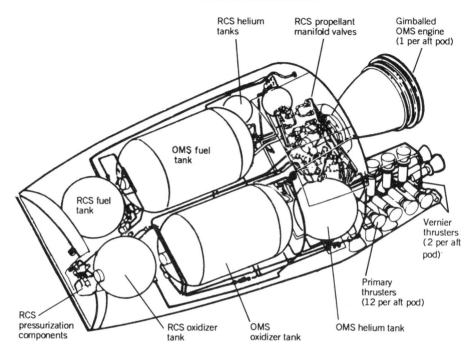

RCS helium tanks

RCS propellant manifold valves

Gimballed OMS engine (1 per aft pod)

OMS fuel tank

RCS fuel tank

Vernier thrusters (2 per aft pod)

Primary thrusters (12 per aft pod)

RCS pressurization components

RCS oxidizer tank

OMS oxidizer tank

OMS helium tank

FIGURE 6–3. Simplified sketch at the left aft pod of the Space Shuttle's Orbiting Maneuvering System (OMS) and the Reaction Control System (RCS). (Source: NASA.)

acceleration environment (some of the time in zero-*g*). Gauges in each tank allow a determination of the amount of propellant remaining, and they also indicate a leak. Safety features include sniff lines at each propellant valve actuator to sense leakage. Electrical heaters are provided at propellant valves, certain lines, and injectors to prevent fuel freezing or moisture forming into ice.

A typical RCS feature that enhances safety and reliability is a self-shutoff device is small thrusters that will cause a shutdown in case they should experience instability and burn through the walls. Electrical lead wires to the propellant valves are wrapped around the chamber and nozzle; a burnout will quickly melt the wire and cut the power to the valve, which will return to the spring-loaded closed position and shut off the propellant flow.

The majority of pressurized feed systems use a pressure regulator to maintain the propellant tank pressure and thus also the thrust at constant values. The required mass of pressurizing gas can be significantly reduced by a *blowdown system* with a "tail-off" pressure decay. The propellants are expelled by the expansion of the gas already in the enlarged propellant tanks. The tank pressure and the chamber pressure decrease or progressively decay during this adiabatic expansion period. The alternatives of either regulating the inert gas pressure or using a blowdown system are compared in Table 6–4; both types

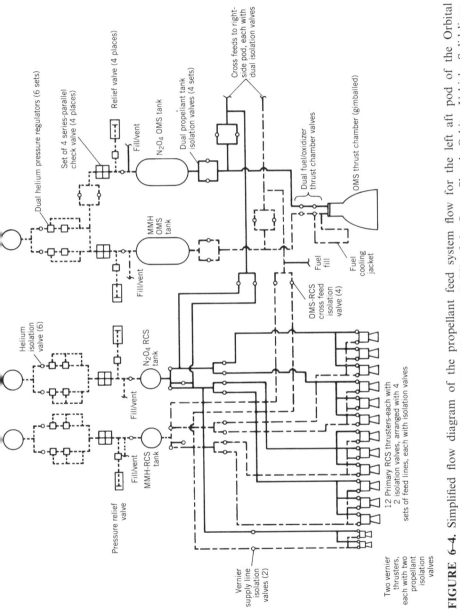

FIGURE 6–4. Simplified flow diagram of the propellant feed system flow for the left aft pod of the Orbital Maneuvering System (OMS) and Reaction Control System (RCS) of the Space Shuttle Orbiter Vehicle. Solid lines: nitrogen tetroxide (N_2O_4); dash-dot lines: monomethylhydrazine (MMH); short dashed lines: high-pressure helium. (Source: NASA.)

209

TABLE 6–3. Characteristics of the Orbital Maneuver System (OMS) and the Reaction Control System (RCS) of the Space Shuttle in One of the Aft Side Podes

Item	OMS	Primary RCS	Vernier RCS
Thrust (per nozzle) (lbf)	6000	870	25
Number of thrusters per pod	1	12	2
Thrust chamber cooling	Regenerative and radiation	Radiation cooling	
Chamber pressure, nominal (psi)	125	152	110
Specific impulse (vacuum nominal) (sec)	313	280[a]	265[a]
Nozzle area ratio	55	22–30[a]	20–50[a]
Mixture ratio (oxide/fuel mass flow)	1.65	1.6	1.6
Burn time, minimum (sec)	2	0.08	0.08
Burn time, maximum (sec)	160	150	125
Burn time, cumulative (sec)	54,000	12,800	125,000
Number of starts, cumulative (sec)	1000	20,000	330,000
Oxidizer (N_2O_4) weight in tank (lb)	14,866	1464	
Fuel (MMH) weight in tank (lb)	9010	923	
Number of oxidizer/fuel tanks	1/1	1/1	
Propellant tank volume, each tank (ft^3)	90	17.9	
Ullage volume, nominal (full tank) (ft^3)	7.8	1.2–1.5	
Tank pressure, nominal (psi)	250	280	
Helium storage tank pressure (psi)	4700	3600	
Number of helium tanks	1	2	
Volume of helium tanks (ft^3)	17	1.76	

[a]Depends on specific vehicle location and scarfing of nozzle.
Sources: NASA, Aerojet Propulsion Company and Kaiser Marquardt Company.

are currently being used. The selection depends on specific application requirements, cost, inert mass, reliability, and safety considerations (see Refs. 6–4 and 6–5).

Some pressure feed systems can be prefilled with propellant and pressurizing agent at the factory and stored in readiness for operation. Compared to a solid propellant rocket unit, these storable prepackaged liquid propellant pressurized feed systems offer advantages in long-term storability and resistance to transportation vibration or shock.

The *thrust* level of a rocket propulsion system with a pressurized gas feed system is determined by the magnitude of the propellant flow which, in turn, is determined by the gas pressure regulator setting. The propellant *mixture ratio* in this type of feed system is controlled by the hydraulic resistance of the liquid propellant lines, cooling jacket, and injector, and can usually be adjusted by means of variable or interchangeable restrictors. Further discussion of the adjusting of thrust and mixture ratio can be found in Section 10.6 and in Example 10–3.

TABLE 6-4. Comparison of Two Types of Gas Pressurization Systems

Type	Regulated Pressure	Blowdown
Pressure/thrust	Stays essentially constant	Decreases as propellant is consumed
Gas storage	In separate high-pressure tanks	Gas is stored inside propellant tank with large ullage volume (30 to 60%)
Required components	Needs regulator, filter, gas valve, and gas tank	Larger, heavier propellant tanks
Advantages	Constant-pressure feed gives essentially constant propellant flow and approximately constant thrust, constant I_s and r Better control of mixture ratio	Simpler system Less gas required Can be less inert mass
Disadvantages	Slightly more complex Regulator introduces a small pressure drop Gas stored under high pressure Shorter burning time	Thrust decreases with burn duration Somewhat higher residue propellant due to less accurate mixture ratio control Thruster must operate and be stable over wide range of thrust values and modest range of mixture ratio Propellants stored under pressure; slightly lower I_s toward end of burning time

6.4. PROPELLANT TANKS

In liquid bipropellant rocket engine systems propellants are stored in one or more oxidizer tanks and one or more fuel tanks; monopropellant rocket engine systems have, of course, only one set of propellant tanks. There are also one or more high-pressure gas tanks, the gas being used to pressurize the propellant tanks. Tanks can be arranged in a variety of ways, and the tank design can be used to exercise some control over the change in the location of the vehicle's center of gravity. Typical arrangements are shown in Fig. 6-5. Because the propellant tank has to fly, its mass is at a premium and the tank material is therefore highly stressed. Common tank materials are aluminum, stainless steel, titanium, alloy steel, and fiber-reinforced plastics with an impervious thin inner liner of metal to prevent leakage through the pores of the fiber-reinforced walls.

The extra volume of gas above the propellant in sealed tanks is called *ullage*. It is necessary space that allows for thermal expansion of the propellant liquids, for the accumulation of gases that were originally dissolved in the propellant, or for gaseous products from slow reactions within the propellant during storage. Depending on the storage temperature range, the propellants' coefficient of thermal expansion, and the particular application, the ullage volume is usually between 3 and 10% of the tank volume. Once propellant is loaded into

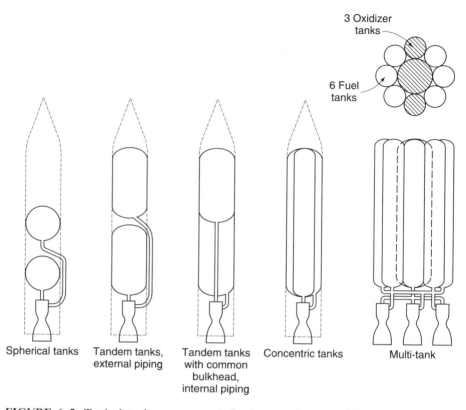

FIGURE 6–5. Typical tank arrangements for large turbopump-fed liquid propellant rocket engines.

a tank, the ullage volume (and, if it is sealed, also its pressure) will change as the bulk temperature of the propellant varies.

The *expulsion efficiency* of a tank and/or propellant piping system is the amount of propellant expelled or available divided by the total amount of propellant initially present. Typical values are 97 to 99.7%. The losses are unavailable propellants that are trapped in grooves or corners of pipes, fittings, and valves, are wetting the walls, retained by surface tension, or caught in instrument taps. This *residual propellant* is not available for combustion and must be treated as inert mass, causing the vehicle mass ratio to decrease slightly. In the design of tanks and piping systems, an effort is made to minimize the residual propellant.

The optimum shape of a propellant tank (and also a gas pressurizing tank) is spherical, because for a given volume it results in a tank with the least weight. Small spherical tanks are often used with reaction control engine systems, where they can be packaged with other vehicle equipment. Unfortunately, the larger spheres, which are needed for the principal propulsion systems, are not very efficient for using the space in a vehicle. These larger tanks are often made integral with the vehicle fuselage or wing. Most are cylindrical with half ellipses at the ends, but they can be irregular in shape. A more detailed discussion of tank pressurization is given in the next section.

Cryogenic propellants cool the tank wall temperature far below the ambient air temperature. This causes condensation of moisture on the outside of the tank and usually also formation of ice during the period prior to launch. The ice is undesirable, because it increases the vehicle inert mass and can cause valves to malfunction. Also, as pieces of ice are shaken off or break off during the initial flight, these pieces can damage the vehicle; for example, the ice from the Shuttle's cryogenic tank can hit the orbiter vehicle.

For an extended storage period, cryogenic tanks are usually thermally insulated; porous external insulation layers have to be sealed to prevent moisture from being condensed inside the insulation layer. With liquid hydrogen it is possible to liquify or solidify the ambient air on the outside of the fuel tank. Even with heavy insulation and low-conductivity structural tank supports, it is not possible to prevent the continuous evaporation of the cryogenic fluid. Even with good thermal insulation, all cryogenic propellants evaporate slowly during storage and therefore cannot be kept in a vehicle for more than perhaps a week without refilling of the tanks. For vehicles that need to be stored or to operate for longer periods, a storable propellant combination must be used.

Prior to loading very cold cryogenic propellant into a flight tank, it is necessary to remove or evacuate the air to avoid forming solid air particles or condensing any moisture as ice. These frozen particles would plug up injection holes, cause valves to freeze shut, or prevent valves from being fully closed. Tanks, piping, and valves need to be chilled or cooled down before they can contain cryogenic liquid without excessive bubbling. This is usually done by letting the initial amount of cryogenic liquid absorb the heat from the relatively warm hardware. This initial propellant is vaporized and vented through appropriate vent valves.

If the tank or any segment of piping containing low-temperature cryogenic liquid is sealed for an extended period of time, heat from ambient-temperature hardware will result in evaporation and this will greatly raise the pressure until it exceeds the strength of the container (see Ref. 6–7). This self-pressurization will cause a failure, usually a major leak or even an explosion. All cryogenic tanks and piping systems are therefore vented during storage on the launch pad, equipped with pressure safety devices (such as burst diaphragms or relief valves), and the evaporated propellant is allowed to escape from its container. For long-term storage of cryogenic propellants in space vacuum (or on the ground) some form of a powered refrigeration system is needed to recondense the vapors and minimize evaporation losses. The tanks are refilled or topped off just before launch to replace the evaporated vented propellant. When the tank is pressurized, just before launch, the boiling point is usually raised slightly and the cryogenic liquid can usually absorb the heat transferred to it during the several minutes of rocket firing.

There are several categories of tanks in liquid propellant propulsion systems:

1. For pressurized feed systems the propellant tanks typically operate at an average pressure between 1.3 and 9 MPa or about 200 to 1800 lbf/in.^{2}. These tanks have thick walls and are heavy.

2. For high-pressure gas (used to expel the propellants) the tank pressures are much higher, typically between 6.9 and 69 MPa or 1000 to 10,000 lbf/in.^{2}. These tanks are usually spherical for minimum inert mass. Several small spherical tanks can be connected together and then they are relatively easy to place within the confined space of a vehicle.

3. For turbopump feed systems it is necessary to pressurize the propellant tanks slightly (to suppress pump cavitation as explained in Section 10.1) to average values of between 0.07 and 0.34 MPa or 10 to 50 lbf/in^2. These low pressures allow thin tank walls, and therefore turbopump feed systems have relatively low tank weights.

Liquid propellant tanks can be difficult to empty under side accelerations, zero-g, or negative-g conditions during flight. Special devices and special types of tanks are needed to operate under these conditions. Some of the effects that have to be overcome are described below.

The oscillations and side accelerations of vehicles in flight can cause *sloshing* of the liquid in the tank, very similar to a glass of water that is being jiggled. In an antiaircraft missile, for example, the side accelerations can be large and can initiate sloshing. Typical analysis of sloshing can be found in Refs. 6–8 and 6–9. When the tank is partly empty, sloshing can uncover the tank outlet and allow gas bubbles to enter into the propellant discharge line. These bubbles can cause major combustion problems in the thrust chambers; the aspirating of bubbles or the uncovering of tank outlets by liquids therefore needs to be avoided. Sloshing also causes shifts in the vehicle's center of gravity and makes flight control difficult.

Vortexing can also allow gas to enter the tank outlet pipe; this phenomenon is similar to the Coriolis force effects in bath tubs being emptied and can be augmented if the vehicle spins or rotates in fight. Typically, a series of internal baffles is often used to reduce the magnitude of sloshing and vortexing in tanks with modest side accelerations. A positive expulsion mechanism can prevent gas from entering the propellant piping under multidirectional major accelerations or spinning (centrifugal) acceleration. Both the vortexing and sloshing can greatly increase the unavailable or residual propellant, and thus cause a reduction in vehicle performance.

In the gravity-free environment of space, the stored liquid will float around in a partly emptied tank and may not always cover the tank outlet, thus allowing gas to enter the tank outlet or discharge pipe. Figure 6–6 shows that gas bubbles have no orientation. Various devices have been developed to solve this problem: namely, *positive expulsion devices* and *surface tension devices*. The positive expulsion tank design include movable pistons, inflatable

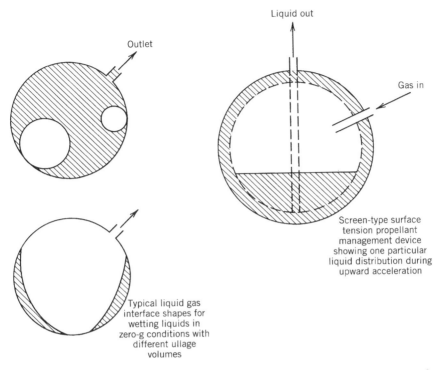

FIGURE 6–6. Ullage bubbles can float around in a zero-gravity environment; surface tension device can keep tank outlet covered with liquid.

flexible bladders, or thin movable, flexible metal diaphragms. Surface tension devices rely on surface tension forces to keep the outlet covered with liquid.

Several basic types of *positive expulsion devices* have been used successfully in propellant tanks of pressurized feed systems. They are compared in Table 6–5 and shown in Fig. 6–7 for simple tanks. These devices mechanically separate the pressurizing gas from the liquid propellant in the propellant tank. Separation is needed for these reasons:

1. It prevents pressurizing gas from dissolving in the propellant. Dissolved pressurizing gas dilutes the propellant, reduces its density as well as its specific impulse, and makes the pressurization inefficient.

2. It allows hot and reactive gases (generated by gas generators) to be used for pressurization, and this permits a reduction in pressurizing system mass and volume. The mechanical separation prevents a chemical reaction between the hot gas and the propellant, prevents gas from being dissolved in the propellant, and reduces the heat transfer to the liquid.

3. In some cases tanks containing toxic propellant must be vented without spilling any toxic liquid propellant or its vapor. For example, in servicing

TABLE 6-5. Comparison of Propellant Expulsion Methods for Spacecraft Hydrazine Tanks

Selection Criteria	Positive Expulsion Devices					
	Single Elastomeric Diaphragm (Hemispherical)	Inflatable Dual Elastomeric Bladder (Spherical)	Foldable Metallic Diaphragm (Hemispherical)	Piston or Bellows	Rolling Diaphragm	Surface Tension Screens
Application history	Extensive	Extensive	Limited	Extensive in high acceleration vehicles	Limited	Extensive
Weight (normalized)	1.0	1.1	1.25	1.2	1.0	0.9
Expulsion efficiency	Excellent	Good	Good	Excellent	Very good	Good or fair
Maximum side acceleration	Low	Low	Medium	High	Medium	Lowest
Control of center of gravity	Poor	Limited	Good	Excellent	Good	Poor
Long service life	Excellent	Excellent	Excellent	Very good	Unproven	Excellent
Preflight check	Leak test	Leak test	Leak test	Leak test	Leak test	None
Disadvantages	Chemical deterioration	Chemical deterioration; fits only into a few tank geometries	High-pressure drop; fits only certain tank geometries; high weight	Potential seal failure; critical tolerances on piston seal; heavy	Weld inspection is difficult; adhesive (for bonding to wall) can deteriorate	Limited to low accelerations

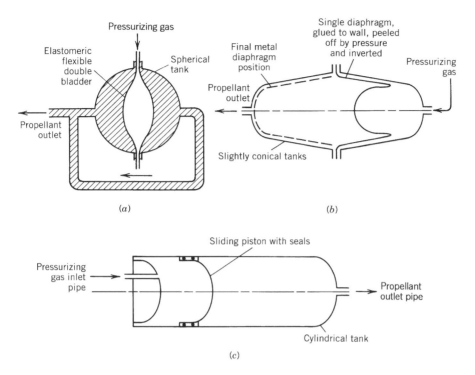

FIGURE 6–7. Three concepts of propellant tanks with positive expulsion: (a) inflatable dual bladder; (b) rolling, peeling diaphragm; (c) sliding piston. As the propellant volume expands or contracts with changes in ambient temperature, the piston or diaphragm will also move slightly and the ullage volume will change during storage.

a reusable rocket, the tank pressure needs to be relieved without venting or spilling potentially hazardous material.

A *piston expulsion* device permits the center of gravity (CG) to be accurately controlled and its location to be known. This is important in rockets with high side accelerations such as antiaircraft missiles or space defense missiles, where the thrust vector needs to go through the CG; if the CG is not well known, unpredictable turning moments may be imposed on the vehicle. A piston also prevents sloshing or vortexing.

Surface tension devices use capillary attraction for supplying liquid propellant to the tank outlet pipe. These devices (see Fig. 6–6) are often made of very fine (300 mesh) stainless steel wire woven into a screen and formed into tunnels or other shapes (see Refs. 6–10 and 6–11). These screens are located near the tank outlet and, in some tanks, the tubular galleries are designed to connect various parts of the tank volume to the outlet pipe sump. These devices work best in a relatively low-acceleration environment, when surface tension forces can overcome the inertia forces.

The combination of surface tension screens, baffles, sumps, and traps is called a *propellant management device*. Although not shown in any detail, they are included inside the propellant tanks of Figs. 6–6 and 6–13.

High forces can be imposed on the tanks and thus on the vehicle by strong sloshing motions of the liquid and also by sudden changes in position of liquid mass in a partly empty tank during a gravity-free flight when suddenly accelerated by a relatively large thrust. These forces can be large and can cause tank failure. The forces will depend on the tank geometry, baffles, ullage volume, and its initial location and the acceleration magnitude and direction.

6.5. TANK PRESSURIZATION

Subsystems for pressurizing tanks are needed for both of the two types of feed systems, namely pressure feed systems and pump feed systems. The tank pressures for the first type are usually between 200 and 1800 psi and for the second between 10 and 50 psig. Refs. 6–1, 6–3 to 6-5 give further descriptions. Inert gases such as helium or nitrogen are the most common method of pressurization. In pump feed systems a small positive pressure in the tank is needed to suppress pump cavitation. For cryogenic propellants this has been accomplished by heating and vaporizing a small portion of the propellant taken from the high-pressure discharge of the pump and feeding it into the propellant tank, as shown in Fig. 1–4. This is a type of low-pressure gas feed system.

The pressurizing gas must not condense, or be soluble in the liquid propellant, for this can greatly increase the mass of required pressurant and the inert mass of its pressurization system hardware. For example, nitrogen pressurizing gas will dissolve in nitrogen tetroxide or in liquid oxygen and reduce the concentration and density of the oxidizer. In general, about $2\frac{1}{2}$ times as much nitrogen mass is needed for pressurizing liquid oxygen if compared to the nitrogen needed for displacing an equivalent volume of water at the same pressure. Oxygen and nitrogen tetroxide are therefore usually pressurized with helium gas, which dissolves only slightly. The pressurizing gas must not react chemically with the liquid propellant. Also, the gas must be dry, since moisture can react with some propellants or dilute them.

The pressurizing gas above a cryogenic liquid is usually warmer than the liquid. The heat transfer to the liquid cools the gas and that increases the density; therefore a larger mass of gas is needed for pressurization even if none of the gas dissolves in the liquid propellant. If there is major sloshing and splashing in the tank during flight, the gas temperature can drop quickly, causing irregularities in the tank pressure.

Chemical pressurization permits the injection of a small amount of fuel or other suitable spontaneously ignitable chemical into the oxidizer tank (or vice versa) which creates the pressurizing gas by combustion inside the propellant tank. While ideally this type of pressurization system is very small and light, in practice it has not usually given reproducible tank pressures, because of irre-

gular combustion the sloshing of propellant in the tank during vehicle maneuvers has caused sudden cooling of the hot pressurizing gas and thus some
erratic tank pressure changes. This problem can be avoided by physically
separating the hot reactive gas from the liquid propellant by a piston or a
flexible bladder. If hot gas from a solid propellant gas generator of from the
decomposition of a monopropellant is used (instead of a high-pressure gas
supply), a substantial reduction in the gas and inert mass of the pressurizing
system can be achieved. For example, the pressurizing of hydrazine monopropellant by warm gas (from the catalytic decomposition of hydrazine) has been
successful for moderate durations.

The prepackaged compact experimental liquid propellant rocket engine
shown in Fig. 6–8 is unique. It uses a gelling agent to improve propellant safety
and density (see Section 7.5 and Ref. 7–11), a solid propellant for pressurization of propellant tanks, two concentric annular pistons (positive expulsion),
and a throttling and multiple restart capability. It allows missiles to lock on to
targets before or after launch, slow down and search for targets, loiter, maneuver, or speed up to a high terminal velocity. This particular experimental
engine, developed by TRW, has been launched from a regular Army mobile
launcher.

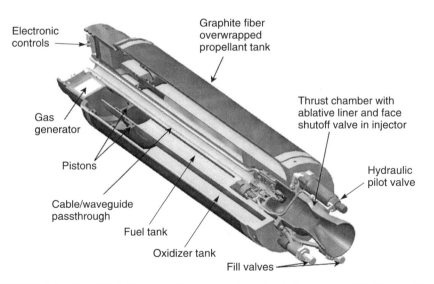

Electronic controls

Graphite fiber overwrapped propellant tank

Gas generator

Thrust chamber with ablative liner and face shutoff valve in injector

Pistons

Hydraulic pilot valve

Cable/waveguide passthrough

Fuel tank

Oxidizer tank

Fill valves

FIGURE 6–8. Simplified diagram of a compact pre-loaded, pressure-fed, bipropellant
experimental rocket engine aimed at propelling smart maneuvering ground-to-ground
missiles. It uses gelled red fuming nitric acid and gelled monomethylhydrazine as propellants. A solid propellant gas generator provides the gas for tank pressurization and
the hot gases are isolated from the propellants by pistons. The concentric spray injector
allows restart, throttling, and flow shut-off at the injector face. The rocket engine is 6 in.
diameter and 23.5 in. long. (Courtesy of Space and Electronics Group, TRW, Inc.)

Estimating the Mass of the Pressurizing Gas

The major function of the pressurizing gas is to expel the propellants from their tanks. In some propulsion system installations, a small amount of the pressurized gas also performs other functions such as the operation of valves and controls. The first part of the gas leaving the high-pressure-gas storage tank is at ambient temperature. If the high-pressure gas expands rapidly, then the gas remaining in the tank undergoes essentially an isentropic expansion, causing the temperature of the gas to decrease steadily; the last portions of the pressurizing gas leaving the tank are very much colder than the ambient temperature and readily absorb heat from the piping and the tank walls. The Joule–Thomson effect causes a further small temperature change.

A *simplified analysis* of the pressurization of a propellant tank can be made on the basis of the conservation of energy principle by assuming an adiabatic process (no heat transfer to or from the walls), an ideal gas, and a negligibly small initial mass of gas in the piping and the propellant tank. Let the initial condition in the gas tank be given by subscript 0 and the instantaneous conditions in the gas tank by subscript g and in the propellant tank by subscript p. The gas energy after and before propellant expulsion is

$$m_g c_v T_g + m_p c_v T_p + p_p V_p = m_0 c_v T_0 \qquad (6\text{--}5)$$

The work done by the gas in displacing the propellants is given by $p_p V_p$. Using Eqs. 3–3 to 3–5, the initial storage gas mass m_0 may be found.

$$c_v p_g V_0 / R + c_v p_p V_p / R + p_p V_p = m_0 c_v T_0$$
$$m_0 = (p_g V_0 + p_p V_p k)/(R T_0) \qquad (6\text{--}6)$$

This may be expressed as

$$m_0 = \frac{p_g m_0}{p_0} + \frac{p_p V_p}{R T_0} k = \frac{p_p V_p}{R T_0} \left(\frac{k}{1 - p_g/p_0} \right) \qquad (6\text{--}7)$$

The first term in this equation expresses the mass of gas required to empty a completely filled propellant tank if the gas temperature is maintained at the initial storage temperature T_0. The second term expresses the availability of the storage gas as a function of the pressure ratio through which the gas expands.

Heating of the pressurizing gas reduces the storage gas and tank mass requirements and can be accomplished by putting a heat exchanger into the gas line. Heat from the rocket thrust chamber, the exhaust gases, or from other devices can be used as the energy source. The reduction of storage gas mass depends largely on the type and design of the heat exchanger and the duration.

If the expansion of the high-pressure gas proceeds slowly (e.g., with an attitude control propulsion system with many short pulses over a long period of time), then the gas expansion comes close to an isothermal process; heat is absorbed from the vehicle and the gas temperature does not decrease appreciably. Here $T_0 = T_g = T_p$. The actual process is between an adiabatic and an isothermal process and may vary from flight to flight.

The heating and cooling effects of the tank and pipe walls, the liquid propellants, and the values on the pressurizing gas require an iterative analysis. The effects of heat transfer from sources in the vehicle, changes in the mission profile, vaporization of the propellant in the tanks, and heat losses from the tank to the atmosphere or space have to be included and the analyses can become quite complex. The design of storage tanks therefore allows a reasonable excess of pressurizing gas to account for these effects, for ambient temperature variations, and for the absorption of gas by the propellant. Equation 6–7 is therefore valid only under ideal conditions.

Example 6–2. What air tank volume is required to pressurize the propellant tanks of a 9000-N thrust rocket thrust chamber using 90% hydrogen peroxide as a monopropellant at a chamber pressure of 2.00 MPa for 30 sec in conjunction with a solid catalyst? The air tank pressure is 14 MPa and the propellant tank pressure is 3.0 MPa. Allow for 1.20% residual propellant.

SOLUTION. The exhaust velocity is 1300 m/sec and the required propellant flow can be found from Eq. 3–42 ($\zeta_d = 1.06$):

$$\dot{m} = \zeta_d F/c = 1.06 \times 9000/1300 - 7.34 \text{ kg/sec}$$

The total propellant required is $m = 7.34$ kg/sec $\times$ 30 sec $\times 1.012 = 222.6$ kg. The density of 90% hydrogen peroxide is 1388 kg/m^3. The propellant volume is $222.6/1388 = 0.160$ m^3. With 5% allowed for ullage and excess propellants, Eq. 6–7 gives the required weight of air ($R = 289$ J/kg-K; $T_0 = 298$ K; $k = 1.40$) for displacing the liquid.

$$m_0 = \frac{p_p V_p}{RT_0} \frac{k}{[1 - (p_g/p_0)]} = \frac{3.0 \times 10^6 \times 0.16 \times 1.05 \times 1.4}{289 \times 298 \times [1 - (3/14)]}$$
$$= 10.4 \text{ kg of compressed air}$$

With an additional 5% allowed for excess gas, the high-pressure tank volume will be

$$V_0 = m_0 RT_0/p_0 = 1.05 \times 10.4 \times 289 \times 298/(14 \times 10^6)$$
$$= 0.067 \text{ m}^3.$$

6.6. TURBOPUMP FEED SYSTEMS AND ENGINE CYCLES

The principal components of a rocket engine with one type of turbopump system are shown in the simplified diagram of Fig. 1–4. Here the propellants are pressurized by means of *pumps*, which in turn are driven by *turbines*. These turbines derive their power from the expansion of hot gases. Engines with turbopumps are preferred for booster and sustainer stages of space launch vehicles, long-range missiles, and in the past also for aircraft performance augmentation. They are usually lighter than other types for these high thrust, long duration applications. The inert hardware mass of the rocket engine (without tanks) is essentially independent of duration. Examples can be seen In Figs. 6–1 and 6–9 and also in Refs. 6–1, 6–2, and 6–6. For aircraft performance augmentation the rocket pump can be driven directly by the jet engine, as in Ref. 6–12. From the turbopump feed system options depicted in Fig. 6–2, the designer can select the most suitable concept for a particular application.

An *engine cycle* for turbopump-fed engines describes the specific propellant flow paths through the major engine components, the method of providing the hot gas to one or more turbines, and the method of handling the turbine exhaust gases. There are *open cycles* and *closed cycles*. *Open* denotes that the working fluid exhausting from the turbine is discharged overboard, after having been expanded in a nozzle of its own, or discharged into the nozzle of the thrust chamber at a point in the expanding section far downstream of the nozle throat. In *closed cycles* or *topping cycles* all the working fluid from the turbine is injected into the engine combustion chamber to make the most efficient use of its remaining energy. In closed cycles the turbine exhaust gas is expanded through the full pressure ratio of the main thrust chamber nozzle, thus giving a little more performance than the open cycles, where these exhaust gases expand only through a relatively small pressure ratio. The overall engine performance difference is typically between 1 and 8% of specific impulse and this is reflected in even larger differences in vehicle performance.

Figure 6–9 shows the three most common cycles in schematic form. Reference 6–13 shows variations of these cycles and also other cycles. The gas generator cycle and the staged combustion cycle can use most of the common liquid propellants. The expander cycle works best with vaporized cryogenic hydrogen as the coolant for the thrust chamber, because it is an excellent heat absorber and does not decompose. The schematic diagrams of Fig. 6–9 show each cycle with a separate turbopump for fuel and for oxidier. However, an arrangement with the fuel and oxdizer pump driven by the same turbine is also feasible and sometimes reduces the hardware mass, volume, and cost. The "best" cycle has to be selected on the basis of the mission, the suitability of existing engines, and the criteria established for the particular vehicle. There is an optimum chamber pressure and an optimum mixture ratio for each application, engine cycle, or optimization criterion, such as maximum range, lowest cost, or highest payload.

In the *gas generator cycle* the turbine inlet gas comes from a separate gas generator. Its propellants can be supplied from separate propellant tanks or can be bled off the main propellant feed system. This cycle is relatively simple; the pressures in the liquid pipes and pumps are relatively low (which reduces inert engine mass). It has less engine-specific impulse than an expander cycle or a staged combustion cycle. The pressure ratio across the turbine is relatively

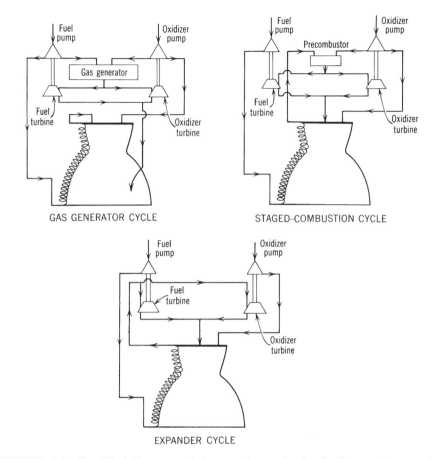

FIGURE 6–9. Simplified diagrams of three engine cycles for liquid propellant rocket engines. The spirals are a symbol for an axisymmetric cooling jacket where heat is absorbed.

high, but the turbine or gas generator flow is small (1 to 4% of total propellant flow) if compared to closed cycles. Some early engines used a separate mono-propellant for creating the generator gas. The German V-2 missile engine used hydrogen peroxide, which was decomposed by a catalyst. Typically, the turbine exhaust gas is discharged overboard through one or two separate small low-area-ratio nozzles (at relatively low specific impulse), as shown schematically in Fig. 1–4 and in the Vulcain engine or RS-68 engine listed in Table 10-3. Alternatively, this turbine exhaust can be aspirated into the main flow through openings in the diverging nozzle section, as shown schematically in Fig. 6–9. This gas then protects the walls near the nozzle exit from high temperatures. Both methods can provide a small amount of additional thrust. The gas generator mixture ratio is usually fuel rich (in some engine it is oxidizer rich) so that the gas temperatures are low enough (typically 900 to 1350 K) to allow the use of uncooled turbine blades and uncooled nozzle exit segments. The RS-68 rocket engine, shown in Fig. 6–10, has a simple gas generator cycle. This engine is the largest liquid hydrogen/liquid oxygen rocket engine built to date. As can be seen from the data in the figure, with a gas generator cycle the specific impulse of the thrust chamber by itself is always a little higher than that of the engine and the thrust of the thrust chamber is always slightly lower than that of the engine.

In the *expander cycle* most of the engine coolant (usually hydrogen fuel) is fed to low-pressure-ratio turbines after having passed through the cooling jacket where it picked up energy. Part of the coolant, perhaps 5 to 15%, bypasses the turbine (not shown in Fig. 6-9) and rejoins the turbine exhaust flow before the entire coolant flow is injected into the engine combustion chamber where it mixes and burns with the oxidizer (see Refs. 6–2 and 6–14). The primary advantages of the expander cycle are good specific impulse, engine simplicity, and relatively low engine mass. In the expander cycle all the propellants are fully burned in the engine combustion chamber and expanded efficiently in the engine exhaust nozzle.

This cycle is used in the RL10 hydrogen/oxygen rocket engine, and different versions of this engine have flown successfully in the upper stages of several space launch vehicles. Data on the RL10-A3-3A are given in Table 10-3. A recent modification of this engine, the RL10B-2 with an extendible nozzle skirt, can be seen in Fig. 8–19 and data on this engine are contained in Table 8–1. It delivers the highest specific impulse of any chemical rocket engine to date. The RL10B-2 flow diagram in Fig. 6–11 shows its expander cycle. Heat absorbed by the thrust chamber cooling jacket gasifies and raises the gas temperature of the hydrogen so that it can be used to drive the turbine, which in turn drives a single-stage liquid oxygen pump (through a gear case) and a two-stage liquid hydrogen pump. The cooling down of the hardware to cryogenic temperatures is accomplished by flowing (prior to engine start) cold propellant through cooldown valves. The pipes for discharging the cooling propellants overboard are not shown here, but can be seen in Fig. 8–19. Thrust is regulated by controlling the flow of hydrogen gas to the turbine, using a bypass to maintain constant chamber pressure. Helium is used as a means of power boost by actuating several of the larger valves through solenoid-operated pilot valves.

In the *staged combustion cycle*, the coolant flow path through the cooling jacket is the same as that of the expander cycle. Here a high-pressure precombustor (gas generator) burns all the fuel with part of the oxidizer to provide high-energy gas to the turbines. The total turbine exhaust gas flow is injected into the main combustion chamber where it burns with the remaining oxidizer. This cycle lends itself to high-chamber-pressure operation, which allows a small thrust chamber size. The extra pressure drop in the precombustor and turbines causes the pump discharge pressures of both the fuel and the oxidizer to be higher than with open cycles, requiring heavier and more

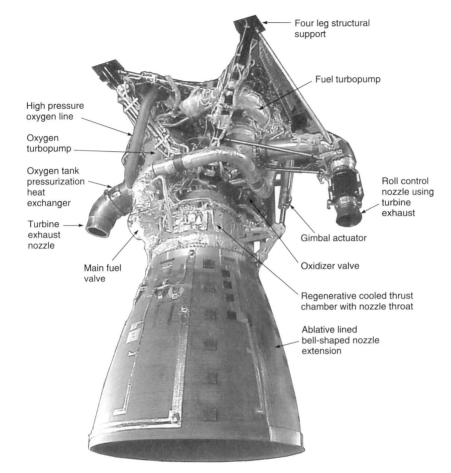

Parameter	Thrust chamber	Engine
Specific impulse at sea level (max.), sec	368	362
Specific impulse in vacuum (max.), sec	421	415
Thrust, at sea level, lbf	640,700	650,000
Thrust in vacuum lbf	732,400	745,000
Mixture ratio	6.74	6.0

FIGURE 6–10. Simplified view of the RS–68 rocket engine with a gas generator cycle. For engine data see Table 10–3. (Courtesy of The Boeing Company, Rocketdyne Propulsion and Power.)

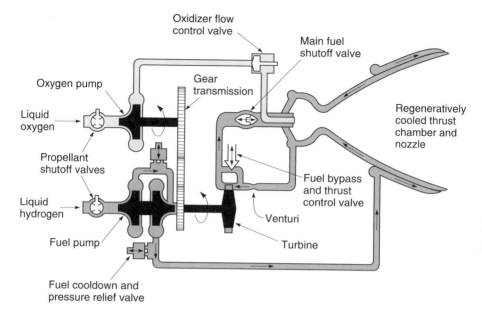

FIGURE 6–11. Schematic flow diagram of the RL10B-2 upper stage rocket engine. For data see Table 8–1. (Courtesy of Pratt & Whitney, a division of United Technologies.)

complex pumps, turbines, and piping. The turbine flow is relatively high and the turbine pressure drop is low, when compared to an open cycle. The staged combustion cycle gives the highest specific impulse, but it is more complex and heavy. In contrast, an open cycle can allow a relatively simple engine, lower pressures, and can have a lower production cost. A variation of the staged combustion cycle is used in the Space Shuttle main engine, as shown in Figs. 6–1 and 6–12. This engine actually uses two separate precombustion chambers, each mounted directly on a separate main turbopump. In addition, there are two more turbopumps for providing a boost pressure to the main pumps, but their turbines are not driven by combustion gases; instead, high-pressure liquid oxygen drives one booster pump and evaporated hydrogen drives the other. The injector of this reusable liquid propellant high-pressure engine is shown in Fig. 9–6 and performance data are given in Tables 10–1 and 10–3. While the space shuttle main engine (burning hydrogen with oxygen) has fuel-rich preburners, oxidizer-rich preburners are used in the RD120 engine (kerosene/oxygen) and other Russian rocket engines. See Table 10–5. Another example of a staged combustion cycle is the Russian engine RD253; all of the nitrogen tetroxide oxidizer and some of the unsymmetrical dimethyl hydrazine fuel are burned in the precombustor, and the remaining fuel is injected directly into the main combustion chamber, as shown in Table 10–5.

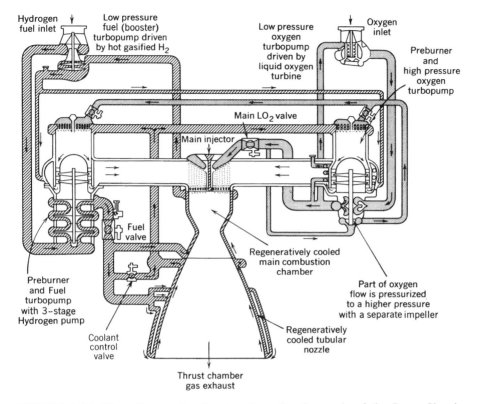

FIGURE 6–12. Flow diagram for the staged combustion cycle of the Space Shuttle Main Engine (SSME) using liquid oxygen and a liquid hydrogen fuel. (Courtesy of The Boeing Company, Rocketdyne Propulsion and Power.)

6.7. FLOW AND PRESSURE BALANCE

From an inspection of the schematic flow diagram of an engine with a gas generator in Fig. 1–4, the following basic feed system relationships are readily deduced. The flow through both pumps $\dot{m}_f$ and $\dot{m}_o$ must equal the respective propellant flow through the gas generator $\dot{m}_{gg}$ and one or more thrust chambers $\dot{m}_c$. With some cycles $\dot{m}_{gg}$ is zero. See equation on Section 10–2.

$$\dot{m}_o = (\dot{m}_o)_{gg} + (\dot{m}_o)_c \tag{6-8}$$

$$\dot{m}_f = (\dot{m}_f)_{gg} + (\dot{m}_f)_c$$

$$\dot{m}_c = (\dot{m}_o)_c + (\dot{m}_f)_c \tag{6-9}$$

$$\dot{m}_{gg} = (\dot{m}_o)_{gg} + (\dot{m}_f)_{gg} \tag{6-10}$$

In the turbopump the torques, powers, and shaft speeds must match. The balance of shaft speeds N can be simply written as

$$N_t = a_o N_o = a_f N_f \tag{6-11}$$

where a_o and a_f are gear ratios. If no gears are used, $a_o = a_f = 1$. The power balance implies that the power of turbine P_T equals the power consumed by pumps and auxiliaries. The power is expressed as the product of torque L and shaft speed N:

$$P_T = L_T N_T = L_o N_o + L_f N_f + P_b \tag{6-12}$$

where P_b represents the bearing, seal, friction, and transmission power losses. If there are no gears in a particular turbopump, then

$$N_T = N_o = N_f \tag{6-13}$$
$$L_T = L_o + L_f + L_b \tag{6-14}$$

The pressure balance equations for the fuel line at a point downstream of the fuel pump can be written as

$$
\begin{aligned}
(p_f)_d &= (p_f)_s + (\Delta p)_{\text{pump}} \\
&= (\Delta p)_{\text{main fuel system}} + p_1 \\
&= (\Delta)_{\text{generator fuel system}} + p_{gg}
\end{aligned}
\tag{6-15}
$$

Here the fuel pump discharge pressure $(p_f)_d$ equals the fuel pump suction pressure $(p_f)_s$ plus the pressure rise across the pump $(\Delta p)_{\text{pump}}$; this in turn equals the chamber pressures p_1 plus all the pressure drops in the main fuel system downstream of the pump, and this is further equal to the chamber pressure in the gas generator combustion chamber p_{gg} augmented by all the pressure losses in the fuel piping between the generator and the downstream side of the fuel pump. The pressure drop in the main fuel system usually includes the losses in the cooling jacket and the pressure decrease in the injector. Equations 6–8 to 6–15 relate to a steady-state condition. A similar pressure balance is needed for the oxidizer flow. The transients and the dynamic change conditions are rather complex but have been analyzed using iterative procedures and digital computers.

6.8. ROCKET ENGINES FOR MANEUVERING, ORBIT ADJUSTMENTS, OR ATTITUDE CONTROL

These engines have usually a set of small thrusters, that are installed at various places in a vehicle, and a common pressurized feed system, similar to Figures

1–3, 4–13, or 6–13. They are called *reaction control systems* or *auxiliary rockets* as contrasted to higher-thrust *primary or boost propulsion systems* in Table 6–1. Most use storable liquid propellants, require a highly accurate repeatability of pulsing, a long life in space, and/or a long-term storage with loaded propellants in flight tanks. Figure 4–13 shows that it requires 12 thrusters for the application of pure torques about three vehicle axes. If a three-degree-of-rotation freedom is not a requrement, or if torques can be combined with some translation maneuvers, fewer thrusters will be needed. These *auxiliary rocket engines* are commonly used in spacecraft or missiles for the accurate *control of flight trajectories*, *orbit adjustments*, or *attitude control* of the vehicle. References 6–1 and 6–2 give information on several of these. Figure 6–13 shows a simplified flow diagram for a post-boost control rocket engine, with one larger rocket thrust chamber for changing the velocity vector and eight small thrusters for attitude control.

Section 4.6 describes various space trajectory correction maneuvers and satellite station-keeping maneuvers that are typically performed by these small auxiliary liquid propellant rocket engines with multiple thrusters. Attitude control can be provided both while a primary propulsion system (of a vehicle or of a stage) is operating and while its auxiliary rocket system operates by itself. For instance, this is done to point satellite's telescope into a specific orientation or to rotate a spacecraft's main thrust chamber into the desired direction for a vehicle turning maneuver.

A good method for achieving accurate velocity corrections or precise angular positions is to use pure modulation, that is, to fire some of the thrusters in a *pulsing mode* (for example, fire repeatedly for 0.020 sec, each time followed by a pause of perhaps 0.020 to 0.100 sec). The guidance system determines the maneuver to be undertaken and the vehicle control system sends command signals to specific thrusters for the number of pulses needed to accomplish this maneuver. Small liquid propellant engine systems are uniquely capable of these pulsing operations. Some thrusters have been tested for more than 300,000 pulses. For very short pulse durations the specific impulse is degraded by 5 to 25%, because the performance during the thrust build-up and thrust decay period (at lower chamber pressure) is inferior to operating only at the rated chamber pressure and the transient time becomes a major portion of the total pulse time.

Ballistic missile defense vehicles usually have highly maneuverable upper stages. These require substantial side forces (200 to 6000 N) during the final closing maneuvers just prior to reaching the target. In concept the system is similar to that of Fig. 6–13, except that the larger thrust chamber would be at right-angles to the vehicle axis. A similar system for terminal maneuvers, but using solid propellants, is shown in Fig. 11-28.

The Space Shuttle performs its reaction control with 38 different thrusters, as shown schematically in Figs. 1–13 and 6–4; this includes several duplicate (spare or redundant) thrusters. Selected thrusters are used for different maneuvers, such as space orbit corrections, station keeping, or positioning the

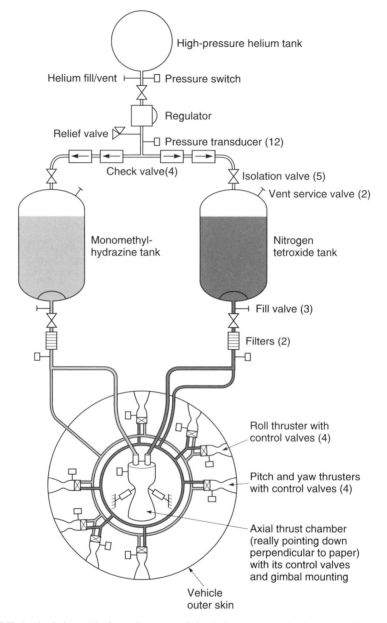

FIGURE 6–13. Schematic flow diagram of the helium-pressurized, bipropellant rocket engine system of the fourth stage of the Peacekeeper ballistic missile, which provides the terminal velocity (in direction and magnitude) to each of several warheads. It has one larger gimballed thrust chamber for trajectory translation maneuvers and eight small thrusters (with scarfed nozzles) for attitude control in pitch, yaw, and roll. (Courtesy of USAF.)

Space Shuttle for reentry or visual observations. These small restartable rocket engines are also used for space *rendezvous* or *docking maneuvers*, where one spacecraft slowly approaches another and locks itself to the other, without causing excessive impact forces during this docking manuever. This docking operation requires rotational and translational maneuvers from a series of rocket engines.

Broadly, the application of pure torque to spacecraft can be divided into two classes, *mass expulsion* types (rockets) and *nonmass expulsion* types. Nonmass expulsion types include momentum storage, gravity gradient, solar radiation, and magnetic systems. Some space satellites are equipped with both the mass and nonmass expulsion types. *Reaction wheels* or flywheels, a momentum storage device, are particularly well suited to obtaining vehicle angular position control with high accuracies of less than $0.01°$ deviation and low vehicle angular rates of less than 10^{-5} degrees/sec with relatively little expenditure of energy. The vehicle angular momentum is changed by accelerating (or decelerating) the wheel. Of course, when the wheel speed reaches the maximum (or minimum) permissible, no further electrical motor torquing is possible; the wheel must be decelerated (or accelerated) to have its momentum removed (or augmented), a function usually accomplished through the simultaneous use of small attitude control rockets, which apply a torque to the vehicle in the opposite direction.

The propellants for *auxiliary rockets* fall into three categories: cold gas jets (also called inert gas jets), warm or heated gas jets, and chemical combustion rockets, such as bipropellant liquid propellant rockets. The specific impulse is typically 50 to 120 sec for cold gas systems and 105 to 250 sec for warm gas systems. Warm gas systems can use inert gas with an electric heater or a monopropellant which is catalytically and/or thermally decomposed. Bipropellant attitude control thrust chambers allow an I_s of 220 to 325 sec and have varied from 5 to 4000 N thrust; the highest thrusts apply to large spacecraft. All basically use pressurized feed systems with multiple thrusters or thrust chambers equipped with fast-acting, positive-closing precision valves. Many systems use small, uncooled, metal-constructed supersonic exhaust nozzles strategically located on the periphery of the spacecraft. Gas jets are used typically for low thrust (up to 10 N) and low total impulse (up to 4000 N-sec). They have been used on smaller satellites and often only for roll control.

Small liquid monopropellant and liquid bipropellant rocket units are common in auxiliary rocket systems for thrust levels typically above 2 N and total impulse values above 3000 N-sec. Hydrazine is the most common monopropellant used in auxiliary control rockets; nitrogen tetroxide and monomethylhydrazine is a common bipropellant combination. The next chapter contains data on all three categories of these propellants, and Chapter 10 shows diagrams of small auxiliary rocket engines and their thrusters.

Combination systems are also in use. Here a bipropellant with a relatively high value of I_s, such as N_2O_4 and N_2H_4, is used in the larger thrusters, which consume most of the propellant; then several simple monopropellant thrusters (with a lower I_s), used for attitude control pulsing, usually consume a relatively small fraction of the total fuel. Another combination system is to employ bipropellant or monopropellant thrusters for adding a velocity increment to a flight vehicle or to bleed or pulse some of the pressurizing gas, such as helium, through small nozzles controlled by electromagnetic valves to provide roll control. The specific mission requirements need to be analyzed to determine which type or combination is most advantageous for a particular application.

Special thruster designs exist which can be used in a bipropellant mode at higher thrust and also in a monopropellant mode for lower thrust. This can offer an advantage in some spacecraft applications. An example is the TRW secondary combustion augmented thruster (SCAT), which uses hydrazine and nitrogen tetroxide, is restartable, vaporizes the propellants prior to injection and therefore has very efficient combustion (over 99%), can operate over a wide range of mixture ratios, and can be throttled from 5 to 15 lbf thrust.

6.9. VALVES AND PIPE LINES

Valves control the flows of liquids and gases and pipes conduct these fluids to the intended components. There are no rocket engines without them. There are many different types of valves. All have to be reliable, lightweight, leakproof, and must withstand intensive vibrations and very loud noises. Table 6–6 gives several key classification categories for rocket engine valves. Any one engine will use only some of the valves listed here.

The art of designing and making valves is based, to a large extent, on experience. A single chapter cannot do justice to it by describing valve design and operation. References 6–1 and 6–2 decribe the design of specific valves, lines, and joints. Often the design details, such as clearance, seat materials, or opening time delay present development difficulties. With many of these valves, any leakage or valve failure can cause a failure of the rocket unit itself. All valves are tested for two qualities prior to installation; they are tested for leaks—through the seat and also through the glands—and for functional soundness or performance.

The propellant valves in high thrust units handle relatively large flows at high service pressures. Therefore, the forces necessary to actuate the valves are large. Hydraulic or pneumatic pressure, controlled by pilot valves, operates the larger valves; these pilot valves are in turn actuated by a solenoid or a mechanical   linkage. Essentially this is a means of power boost.

TABLE 6–6. Classification of Valves Used in Liquid Propellant Rocket Engines

1. *Fluid*: fuel; oxidizer; cold pressurized gas; hot turbine gas.

2. *Application or Use*: main propellant control; thrust chamber valve (dual or single); bleed; drain; fill; by-pass; preliminary stage flow; pilot valve; safety valve; overboard dump; regulator; gas generator control; sequence control; isolation of propellant or high-pressure gas prior to start.

3. *Mode of Actuation*: automatically operated (by solenoid, pilot valve, trip mechanism, pyrotechnic, etc.); manually operated; pressure-operated by air, gas, propellant, or hydraulic fluid (e.g., check valve, tank vent valve, pressure regulator, relief valve), with or without position feedback, rotary or linear actuator.

4. The *flow* magnitude determines the *size* of the valve.

5. *Duty cycle*: single or multiple pulse operation; reusable for other flights; long or short life.

6. *Valve Type*: normally open; normally closed; normally partly open; two-way; three-way, with/without valve position feedback; ball valve, gate valve, butterfly type, spring loaded.

7. *Temperature* and *pressure* allow classification by high, low, or cryogenic temperature fluids, or high or low pressure or vacuum capability.

8. *Accessible or not accessible* to inspection, servicing, or replacement of valve or its seal.

Two valves commonly used in pressurized feed systems are *isolation valves* (when shut, they isolate or shut off a portion of the propulsion system) and *latch valves*; they require power for brief periods during movements, such as to open or shut, but need no power when latched or fastened into position.

A very simple and very light valve is a *burst diaphragm*. It is essentially a circular disk of material which blocks a pipeline and is designed so that it will fail and burst at a predetermined pressure differential. Burst diaphragms are positive seals and prevent leakage, but they can be used only once. The German *Wasserfall* antiaircraft missile used four burst disks; two were in high pressure air lines and two were in the propellant lines.

Figure 6–14 shows a main liquid oxygen valve. It is normally closed, rotary actuated, cryogenic, high pressure, high flow, reusable ball valve, allowing continuous throtting, a controlled rate of opening through a crank and hydraulic piston (not shown), with a position feedback and anti-icing controls.

Pressure regulators are special valves which are used frequently to regulate gas pressures. Usually the discharge pressure is regulated to a predetermined standard pressure value by continuously throttling the flow, using a piston, flexible diaphragm, or electromagnet as the actuating mechanism. Regulators can be seen in Figs. 1–3 and 6–13.

The various fluids in a rocket engine are conveyed by *pipes* or *lines*, usually made of metal and joined by fittings or welds. Their design must provide for

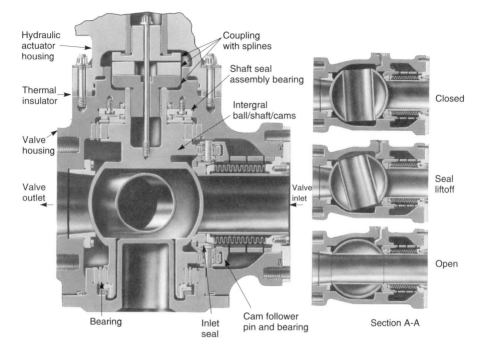

FIGURE 6–14. The SSME main oxidizer valve is a low-pressure drop ball valve representative of high-pessure large valves used in rocket engines. The ball and its integral shaft rotate in two bearings. The seal is a machined plastic ring spring-loaded by a bellows against the inlet side of the ball. Two cams on the shaft lift the seal a short distance off the ball within the first few degrees of ball rotation. The ball is rotated by a precision hydraulic actuator (not shown) through an insulating coupling. (Courtesy of The Boeing Company, Rocketdyne Propulsion and Power.)

thermal expansion and provide support to minimize vibration effects. For gimballed thrust chambers it is necessary to provide flexibility in the piping to allow the thrust axis to be rotated through a small angle, typically ±3 to 10°. This flexibility is provided by flexible pipe joints and/or by allowing pipes to deflect when using two or more right-angle turns in the lines. The high-pressure propellant feed lines of the SSME have both flexible joints and right-angle bends, as shown in Figs 6–1 and 6–15. This joint has flexible bellows as a seal and a universal joint-type mechanical linkage with two sets of bearings for carrying the separating loads imposed by the high pressure.

Sudden closing of valves can cause water hammer in the pipelines, leading to unexpected pressure rises which can be destructive to propellant system components. An analysis of this water hammer phenomenon will allow determination of the approximate maximum pressure (Refs. 6–15 and 6–16). The friction of the pipe and the branching of pipelines reduce this maximum pressure.

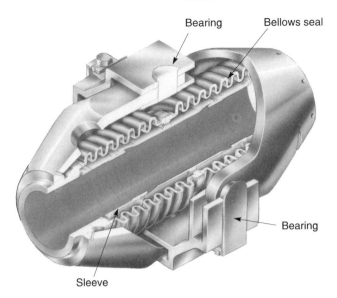

FIGURE 6–15. Flexible high-pressure joint with external gimbal rings for a high-pressure hot turbine exhaust gas. (Courtesy of The Boeing Company, Rocketdyne Propulsion and Power.)

Water hammer can also occur when admitting the initial flow of high-pressure propellant into evacuated pipes. The pipes are under vacuum to remove air and prevent the forming of gas bubbles in the propellant flow, which can cause combustion problems.

Many liquid rocket engines have *filters* in their lines. This is necessary to prevent dirt, particles, or debris, such as small pieces from burst diaphragms, from entering precision valves or regulators (where debris can cause a malfunction) or from plugging small injection holes, which could cause hot streaks in the combustion gases, in turn causing a thrust chamber failure.

Occasionally a convergent–divergent *venturi section*, with a sonic velocity at its throat, is placed into one or both of the liquid propellant lines. The merits are that it maintains constant flow and prevents pressure disturbances from traveling upstream. This can include the propagating of chamber pressure oscillations or coupling with thrust chamber combustion instabilities. The venturi section can also help in minimizing some water hammer effects in a system with multiple banks of thrust chambers.

6.10. ENGINE SUPPORT STRUCTURE

Most of the larger rocket engines have their own mounting structure or support structure. On it the major components are mounted. It also transmits the thrust force to the vehicle. Welded tube structures or metal plate/sheet metal assemblies have been used. In some large engines the thrust chamber is used as a structure and the turbopump, control boxes, or gimbal actuators are attached to it.

In addition to the thrust load, an engine structure has to withstand forces imposed by vehicle maneuvers (in some cases a side acceleration of 10 g_0), vibration forces, actuator forces for thrust vector control motions, and loads from transportation over rough roads.

In low-thrust engines with multiple thrusters there often is no separate engine mounting structure; the major components are in different locations of the vehicle, connected by tubing, wiring, or piping, and each is usually mounted directly to the vehicle or spacecraft structure.

PROBLEMS

1. Enumerate and explain the merits and disadvantages of pressurized and turbopump feed systems.

2. In a turbopump it is necessary to do more work in the pumps if the thrust chamber operating pressure is raised. This of course requires an increase in turbine gas flow which, when exhausted, adds little to the engine specific impulse. If the chamber pressure is raised too much, the decrease in performance due to an excessive portion of the total propellant flow being sent through the turbine and the increased mass of the turbopump will outweigh the gain in specific impulse that can be attained by increased chamber pressure and also by increased thrust chamber nozzle exit area. Outline in detail a method for determining the optimum chamber pressure where the sea level performance will be a maximum for a rocket engine that operates in principle like the one shown in Fig. 1–4.

3. The engine performance data for a turbopump rocket system are as follows:

Engine system specific impulse	272 sec
Engine system mixture ratio	2.52
Engine system thrust	40,000 N
Oxidizer vapor flow to pressurize oxidizer tank	0.003% of total oxidizer flow
Propellant flow through turbine	2.1% of total propellant flow
Gas generator mixture ratio	0.23
Gas generator specific impulse	85 sec

Determine performance of the thrust chamber I_s, r, F (see Sect. 10–2).

4. For a pulsing rocket engine, assume a simplified parabolic pressure rise of 0.005 sec, a steady-state short period of full chamber pressure, and a parabolic decay of 0.007 sec approximately as shown in the sketch. Plot curves of the following ratios as a function of operating time t from $t = 0.013$ to $t = 0.200$ sec; (a) average pressure to

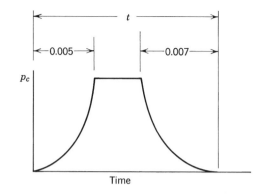

ideal steady-state pressure (with zero rise or decay time); (b) average I_s to ideal steady-state I_s; (c) average F to ideal steady-state F.

5. For a total impulse of 100 lbf-sec compare the volume and system weights of a pulsed propulsion system using different gaseous propellants, each with a single spherical gas storage tank (at 3500 psi and 0°C). A package of small thrust nozzles with piping and controls is provided which weighs 5.2 lb. The gaseous propellants are hydrogen, nitrogen, and argon (see Table 7–3).

6. Compare several systems for a potential roll control application which requires four thrusters of 1 lbf each to operate for a cumulative duration of 2 min each. Include the following:

> Pressurized helium   Cold
> Pressurized nitrogen   Cold
> Pressurized krypton   Cold
> Pressurized helium at 500°F (electrically heated)

The pressurized gas is stored at 5000 psi in a single spherical fiber-reinforced plastic tank; use a tensile strength of 200,000 psi and a density of 0.050 lbm/in.^{3} with a 0.012 in. thick aluminum inner liner as a seal against leaks. Neglect the gas volume in the pipes, valves, and thrusters, but assume the total hardware mass of these to be about 1.3 lbm. Use Table 7–3. Make estimates of the tank volume and total system weight. Discuss the relative merits of these systems.

7. Make tables comparing the merits and disadvantages of engines using the gas generator cycle and engines having the staged combustion cycle.

8. Prepare dimensioned rough sketches of the two propellant tanks needed for operating a single RD253 engine (Table 10–5) for 80 sec at full thrust and an auxiliary rocket system using the same propellants, with eight thrust chambers, each of 100 kg thrust, but operating on the average with only two of the eight firing at any one time, with a duty cycle of 12 percent (fires only 12% of the time), but for a total flight time of 4.00 hours. Describe any assumptions that were made with the propellant budget, the engines, or the vehicle design, as they affect the amount of propellant.

9. Table 10–5 shows that the RD 120 rocket engine can operate at 85% of full thrust and with a mixture ratio variation of ±10.0%. Assume a 1.0% unavailable residual

propellant. The allowance for operational factors, loading uncertainties, off-nominal rocket performance, and a contingency is 1.27% for the fuel and 1.15% for the oxidizer.

(a) In a particular flight the average thrust was 98.0% of nominal and the mixture ratio was off by +2.00% (oxidizer rich). What percent of the total fuel and oxidizer loaded into the vehicle will remain unused at thrust termination?

(b) If we want to run at a fuel-rich mixture in the last 20% of the flight duration (in order to use up all the intended flight propellant), what would the mixture ratio have to be for this last period?

(c) In the worst possible scenario with maximum throttling and extreme mixture ratio excursion (but operating for the nominal duration), what is the largest possible amount of unused oxidizer or unused fuel in the tanks?

SYMBOLS

a	gear ratio
F	thrust, N (lbf)
g_0	acceleration of gravity at sea level, 9.8066 m/sec^2
I_s	specific impulse, sec
k	specific heat ratio
L	shaft torque, m-N (ft-lbf)
m	propellant mass, kg (lbm)
$\dot{m}$	mass flow rate, kg/sec (lb/sec)
N	shaft speed, rpm (rad/sec)
p	pressure, N/m^2 (psi)
Δp	pressure drop, N/m^2 (psi)
P	power, W
r	mixture ratio (oxidizer to fuel mass flow rate)
t	time, sec
T	absolute temperature, K
u	vehicle velocity, m/sec (ft/sec)
V	volume flow rate, m^3/sec (ft^3/sec)
w	total propellant weight, N (lbf)
$\dot{w}$	weight flow rate, N/sec (lbf/sec)
α	nozzle divergence angle

Subscripts

b	bearings, seals
c	chamber or thrust chamber
d	discharge side
f	fuel
gg	gas generator
oa	overall

o	oxidizer
s	suction side
tp	tank pressurization
1	chamber (stagnation condition)
2	nozzle exit
3	ambient atmosphere

REFERENCES

6–1. D. K. Huzel and D. H. Huang. *Design of Liquid Propellant Rocket Engines*, Revised edition, AIAA, 1992, 437 pages.

6–2. G. G. Gakhun, V. I. Baulin, et ala., *Construction and Design of Liquid Propellant Rocket Engines* (in Russian), *Konstruksiya i Proyektirovaniye Zhidkostniyk Raketnykh Dvigateley*, Mashinostroyeniye, Moscow, 1989, 424 pages.

6–3. C. J. G. Dixon and J. G. B. Marshall, "Mathematical Modelling of Bipropellant Combined Propulsion Subsystems," *AIAA Paper 90–2303*, 26th Joint Propulsion Conference, July 1990; and *Design Guide for Pressurized Gas Systems*, Vols. I and II, prepared by IIT Research Institute, NASA Contract NAS7-388, March 1966.

6–4. H. C. Hearn, "Design and Development of a large Bipropellant Blowdown Propulsion System," *Journal of Propulsion and Power*, Vol. 11, No. 5, September–October 1995.

6–5. H. C. Hearn, "Evaluation of Bipropellant Pressurization Concepts for Spacecraft," *Journal of Spacecraft and Rockets*, Vol. 19, July 1982, pp. 320–325.

6–6. *National Space Transportation System Reference*, Vol. 1, National Aeronautics and Space Administration, Washington, DC, June 1988 (description of Space Shuttle system and operation).

6–7. J. I. Hochsten, H.-C. Ji, and J. Ayelott, "Prediction of Self-Pressurization Rate of Cryogenic Propellant Tankage," *Journal of Propulsion and Power*, Vol. 6, No. 1, January–February 1990, pp. 11–17.

6–8. B. Morton, M. Elgersma, and R. Playter, "Analysis of Booster Vehicle Slosh Stability during Ascent to Orbit," *AIAA Paper 90-1876*, July 1990, 7 pages.

6–9. J. J. Pocha, "Propellant Slosh in Spacecraft and How to Live with It," *Aerospace Dynamics*, Vol. 20, Autumn 1986, pp. 26–31.

6–10. G. P. Purohit and L. D. Loudenback, "Application of Etched Disk Stacks in Surface Tension Propellant Management Devices," *Journal of Propulsion and Power*, Vol. 7, No. 1, January–February 1991, pp. 22–30.

6–11. J. R. Rollins, R. K. Grove, and D. R. Walling, Jr. "Design and Qualification of a Surface Tension Propellant Tank for an Advanced Spacecraft," *AIAA Paper 88-2848*, 24th Joint Propulsion Conference, 1988.

6–12. H. Grosdemange and G. Schaeffer. "The SEPR 844 Reuseable Liquid Rocket Engine for Mirage Combat Aircraft", *AIAA Paper 90–1835*, July 1990.

6–13. D. Manski, C. Goertz, H. D. Sassnick, J. R. Hulka, B. D. Goracke, and D. J. H. Levack, "Cycles for Earth to Orbit Propulsion," *Journal of Propulsion and Power*, *AIAA*, Vol. 14, No. 5, September–October 1998.

6–14. J. R. Brown. "Expander Cycle Engines for Shuttle Cryogenic Upper Stages, *AIAA Paper 83–1311*, 1983.

6–15. R. P. Prickett, E. Mayer, and J. Hermel, "Waterhammer in Spacecraft Propellant Feed Systems," *Journal of Propulsion and Power*, Vol. 8, No. 3, May–June 1992.

6–16. Chapter 9 in: I. Karassik, W. C. Krutzsch, W. H. Fraser, and J. P. Messina (eds), *Pump Handbook*, McGraw-Hill Book Company, New York, 1976 (pumps and waterhammer).

CHAPTER 7

LIQUID PROPELLANTS

The classification of liquid propellants has been given in Section 6.1 of the preceding chapter. In this chapter we discuss properties, performance, and characteristics of selected common liquid propellants. These characteristics affect the engine design, test facilities, propellant storage and handling. Today we commonly use three liquid bipropellant combinations. Each of their propellants will be described further in this chapter. They are: (1) the cryogenic *oxygen–hydrogen propellant system*, used in upper stages and sometimes booster stages of space launch vehicles; it gives the highest specific impulse for a non-toxic combination, which makes it best for high vehicle velocity missions; (2) the *liquid oxygen–hydrocarbon propellant combination*, used for booster stages (and a few second stages) of space launch vehicles; its higher average density allows a more compact booster stage, when compared to the first combination; also, historically, it was developed before the first combination and was originally used for ballistic missiles; (3) several *storable propellant combinations*, used in large rocket engines for first and second stages of ballistic missiles and in almost all bipropellant low-thrust, auxiliary or reaction control rocket engines (this term is defined below); they allow long-term storage and almost instant readiness to start without the delays and precautions that come with cryogenic propellants. In Russia the nitric acid–hydrocarbon combination was used in ballistic missiles many years ago. Today Russia and China favor nitrogen tetroxide–unsymmetrical dimethylhydrazine or UDMH for ballistic missiles and auxiliary engines. The USA started with nitrogen tetroxide and a fuel mixture of 50% UDMH with 50% hydrazine in the Titan missile. For auxiliary engines in many satellites and upper stages the USA has used the bipropellant of nitrogen tetroxide with monomethylhydrazine. The orbit maneuvering system of the Space Shuttle uses it. Alternatively, many US satellites have used monopropellant hydrazine for auxiliary engines.

A comparative listing of various performance quantities for a number of propellant combinations is given in Table 5–5 and in Ref. 7–1. Some important physical properties of various propellants are given in Table 7–1. For comparison water is also listed. Specific gravities and vapor pressures are shown in Figs. 7–1 and 7–2.

7.1. PROPELLANT PROPERTIES

It is important to distinguish between the characteristics and properties of the *liquid propellants* (the fuel and oxidizer liquids in their unreacted condition) and those of the *hot gas mixture*, which result from the reaction in the combustion chamber. The chemical nature of the liquid propellants determines the properties and characteristics of both of these types. Unfortunately, none of the practical, known propellants have all the desirable properties, and the selection of the propellant combination is a compromise of various factors, such as those listed below.

Economic Factors

Availability in large quantity and a *low cost* are very important considerations in the selection of a propellant. In military applications, consideration has to be given to *logistics* of production, supply, and other possible military uses. The production process should be simple, requiring only ordinary chemical equipment and available raw materials. It is usually more expensive to use a toxic or cryogenic propellant than a storable, non-toxic one, because it requires additional steps in the operation, more safety provisions, additional design features, longer check-out procedures, and often more trained personnel.

Performance of Propellants

The performance can be compared on the basis of the *specific impulse*, the *effective exhaust velocity*, the *characteristic velocity*, the *specific propellant consumption*, the *ideal exhaust velocity*, or other engine parameters. They have been explained in Chapter 3, 5 and 6. The specific impulse and exhaust velocity are functions of pressure ratio, specific heat ratio, combustion temperature, mixture ratio, and molecular mass. Values of performance parameters for various propellant combinations can be calculated with a high degree of accuracy and several are listed in Table 5–5. Very often the performance is expressed in terms of *flight performance parameters* for a given rocket application, as explained in Chapter 4. Here the average density, the

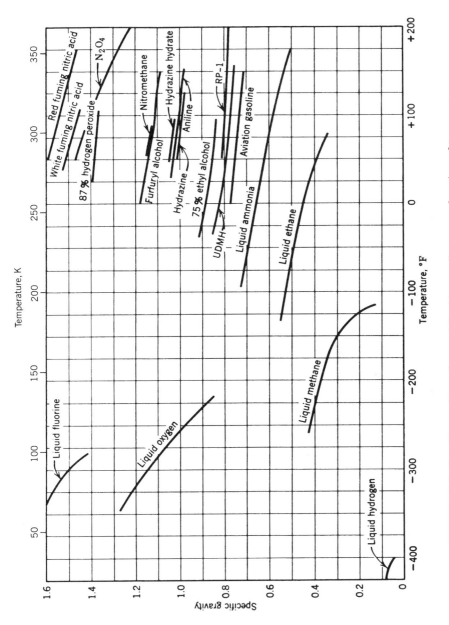

FIGURE 7-1. Specific gravities of several liquid propellants as a function of temperature.

TABLE 7–1. Some Physical Properties of Several Common Liquid Propellants

Propellant	Liquid Fluorine	Hydrazine	Liquid Hydrogen	Methane	Monomethyl-hydrazine
Chemical formula	F_2	N_2H_4	H_2	CH_4	CH_3NHNH_2
Molecular mass	38.0	32.05	2.016	16.03	46.072
Melting or freezing point (K)	53.54	274.69	14.0	90.5	220.7
Boiling point (K)	85.02	386.66	20.4	111.6	360.6
Heat of vaporization (kJ/kg)	166.26[b]	44.7[b] (298.15 K)	446	510[b]	875
Specific heat (kcal/kg-K)	0.368 (85 K)	0.736 (293 K)	1.75[b] (20.4 K)	0.835[b]	0.698 (293 K)
	0.357 (69.3 K)	0.758 (338 K)	—		0.735 (393 K)
Specific gravity[c]	1.636 (66 K)	1.005 (293 K)	0.071 (20.4 K)	0.424 (111.5 K)	0.8788 (293 K)
	1.440 (93 K)	0.952 (350 K)	0.076 (14 K)		0.857 (311 K)
Viscosity (centipoise)	0.305 (77.6 K)	0.97 (298 K)	0.024 (14.3 K)	0.12 (111.6 K)	0.855 (293 K)
	0.397 (70 K)	0.913 (330 K)	0.013 (20.4 K)	0.22 (90.5 K)	0.40 (344 K)
Vapor pressure (MPa)	0.0087 (100 K)	0.0014 (293 K)	0.2026 (23 K)	0.033 (100 K)	0.0073 (300 K)
	0.00012 (66.5 K)	0.016 (340 K)	0.87 (30 K)	0.101 (117 K)	0.638 (428 K)

[a] Red fuming nitric acid (RFNA) has 5 to 20% dissolved NO_2 with an average molecular weight of about 60, and a density and vapor pressure somewhat higher than those of pure nitric acid.
[b] At boiling point.
[c] Reference for specific gravity ratio: 10^3 kg/m^3 or 62.42 lbm/ft^3.

specific impulse, and the engine mass ratio usually enter into a complex flight relation equation.

For high performance a *high content of chemical energy* per unit of propellant mixture is desirable because it permits a high chamber temperature. A *low molecular mass* of the product gases of the propellant combination is also desirable. It can be accomplished by using fuels rich in combined hydrogen, which is liberated during the reaction. A low molecular mass is obtained if a large portion of the hydrogen gas produced does not combine with oxygen. In general, therefore, the best mixture ratio for many bipropellants is not necessarily the stoichiometric one (whch results in complete oxidation and yields a

Nitric Acid[a] (99%) pure	Nitrogen Tetroxide	Liquid Oxygen	Rocket Fuel RP-1	Unsymmetrical Dimethyl-hydrazine (UDMH)	Water
HNO_3	N_2O_4	O_2	Hydrocarbon $CH_{1.97}$	$(CH_3)_2NNH_2$	H_2O
63.016	92.016	32.00	~ 175	60.10	18.02
231.6	261.95	54.4	225	216	273.15
355.7	294.3	90.0	460–540	336	373.15
480	413[b]	213	246[b]	542 (298 K)	2253[b]
0.042	0.374	0.4	0.45	0.672	1.008
(311 K)	(290 K)	(65 K)	(298 K)	(298 K)	(273.15 K)
0.163	0.447			0.71	
(373 K)	(360 K)			(340 K)	
1.549	1.447	1.14	0.58	0.856	1.002
(273.15 K)	(293 K)	(90.4 K)	(422 K)	(228 K)	(373.15 K)
1.476	1.38	1.23	0.807	0.784	1.00
(313.15 K)	(322 K)	(77.6 K)	(289 K)	(244 K)	(293.4 K)
1.45	0.47	0.87	0.75	4.4	0.284
(273 K)	(293 K)	(53.7 K)	(289 K)	(220 K)	(373.15 K)
	0.33	0.19	0.21	0.48	1.000
	(315 K)	(90. 4 K)	(366 K)	(300 K)	(277 K)
0.0027	0.01014	0.0052	0.002	0.0384	0.00689
(273.15 K)	(293 K)	(88.7 K)	(344 K)	(289 K)	(312 K)
0.605	0.2013		0.023	0.1093	0.03447
(343 K)	(328 K)		(422 K)	(339 K)	(345 K)

high flame temperature) but usually a fuel-rich mixture containing a large portion of low-molecular-mass reaction products, as shown in Chapter 5.

If very small metallic fuel particles of beryllium or aluminum are suspended in the liquid fuel, it is theoretically possible to increase the specific impulse by between 9 and 18%, depending on the particular propellant combination, its mixture ratio and the metal powder additive. Gelled propellants with suspended solid particles have been tested successfully with storable fuels. For gelled propellants, see Section 7.5.

The chemical propellant combination that has the highest potential specific impulse (approximately 480 sec at 1000 psia chamber pressure and expansion to sea level atmosphere, and 565 sec in a vacuum with a nozzle area ratio of 50) uses a toxic liquid fluorine oxidizer with hydrogen fuel plus suspended toxic solid particles of beryllium; as yet a practical means for storing these propellants and a practical rocket engine have not been developed.

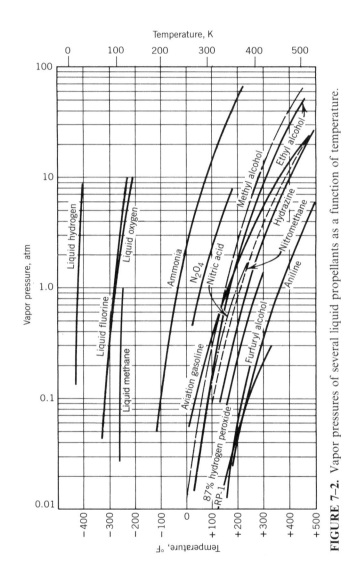

FIGURE 7-2. Vapor pressures of several liquid propellants as a function of temperature.

Common Physical Hazards

Although the several categories of hazards are described below, they do not all apply to every propellant. The hazards are different for each specific propellant and must be carefully understood before working with that propellant. The consequences of unsafe operation or unsafe design are usually also unique to several propellants.

Corrosion. Various propellants, such as nitrogen tetroxide or hydrogen peroxide, have to be handled in containers and pipelines of special materials. If the propellant were permitted to become contaminated with corrosion products, its physical and chemical properties could change sufficiently to make it unsuitable for rocket operation. The corrosion of the gaseous reaction products is important in applications in which the reaction products are likely to damage structure and parts of the vehicle or affect communities and housing near a test facility or launch site.

Explosion Hazard. Some propellants, such as hydrogen peroxide and nitromethane, are unstable and tend to detonate under certain conditions of impurities, temperature, and shock. If liquid oxidizers (e.g., liquid oxygen) and fuels are mixed together they can be detonated. Unusual, rare flight vehicle launch or transport accidents have caused such mixing to occur (see Refs. 7–2 and 7–3).

Fire Hazard. Many oxidizers will start chemical reactions with a large variety of organic compounds. Nitric acid, nitrogen tetroxide, fluorine, or hydrogen peroxide react spontaneously with many organic substances. Most of the fuels are readily ignitable when exposed to air and heat.

Accidental Spills. Unforeseen mishaps during engine operation and traffic accidents on highways or railroads while transporting hazardous materials, including propellants, have on occasion caused spills, which expose people to unexpected fires, or potential health hazards. The U.S. Department of Transportation has rules for marking and containing hazardous materials during transport and also guidelines for emergency action (see Ref. 7–4).

Health Hazards. Many propellants are toxic or poisonous, and special precautions have to be taken to protect personnel. Fluorine, for example, is very poisonous. Toxic propellant chemicals or poisonous exhaust species can enter the human body in several ways. The resulting health disorders are propellant specific. Nitric acid can cause severe skin burn and tissue disintegration. Skin contact with aniline or hydrazine can cause nausea and other adverse health effects. Hydrazine and its derivatives, such as dimethylhydrazine or hydrazine hydrate, are known carcinogens (cancer-causing substances). Many propellant vapors cause eye irritation, even in very small concentration. Inadvertent swallowing of many propellants can also cause severe health degradation.

The inhalation of certain toxic exhaust gases or gaseous or vaporized propellants is perhaps the most common health hazard. It can cause severe damage if the exposure is for long duration or in concentrations that exceed established maximum threshold values. In the United States the Occupational Safety and Health Administration (OSHA) has established limits or thresholds on the allowable exposure and concentration for most propellant chemicals. Several of these propellant gas threshold limits are mentioned later in this chapter. Toxic gases in the exhaust could include hydrofluoric acid (HF) gas; its OSHA 8-hr personnel exposure limit is 3 ppm (volumetric parts per million) and its short-term (typically, 15 min) exposure limit is 6 ppm. A concentration of 3000 ppm or 0.3% can be fatal within a few seconds. Pentaborane, which is very toxic and has been used in experimental engines, has an 8-hr personnel exposure limit at a threshold of 0.005 ppm. References 7–2 and 7–5 give more information on toxic effects.

The corrosion, explosion, and fire hazards of many propellants put severe limitations on the materials, the handling, and the design of rocket-propelled vehicles and their engine compartments. Not only is the rocket system itself exposed to the hazardous propellant, but adjacent personnel, structural parts, electrical and other vehicle equipment, and test and launch facilities have to be properly protected against the effects of possible leaks, fumes, and fires or explosions from propellant accumulations.

Material Compatibility. Many liquid propellants have only a limited number of truly compatible materials, both metals and nonmetals, such as gaskets or O-rings. There have been unfortunate failures (causing fires, leakage, corrosion, or malfunctions) when an improper or incompatible material was used in the hardware of a rocket engine. Depending on the specific component and loading conditions, these structural materials have to withstand high stresses, stress corrosion, high temperatures, or abrasion. Several specific material limitations are mentioned in the next section. Certain materials catalyze a self-decomposition of stored hydrogen peroxide into water and oxygen, making long-term storage difficult and, if confined, causing its container to explode. Many structural materials, when exposed to cold, cryogenic propellants, can become very brittle.

Desirable Physical Properties

Low Freezing Point. This permits operation of rockets in cold weather. The addition of small amounts of special chemicals has been found to help depress the freezing point of some liquid propellants which solidify readily at relatively high temperature.

High Specific Gravity. In order to accommodate a large mass of propellants in a given vehicle tank space, a dense propellant is required. It permits a small vehicle construction and, consequently, a relatively low structural vehicle mass and low aerodynamic drag. Specific gravity, therefore, has an important effect on the maximum flight velocity and range of any rocket-powered vehicle or missile flying within the earth's atmosphere, as explained in Chapter 4. Specific gravities for various propellants are plotted in Fig. 7–1. A variation of the temperature of stored propellant will cause change in liquid level in the tank.

For any given mixture ratio r, the average specific gravity of a propellant combination δ_{av} can be determined from the specific gravities of the fuel δ_f and of the oxidizer δ_o. The average specific gravity is defined as the mass of the fuel and oxidizer, divided by the sum of their volumes. Here the mixture ratio is defined as the oxidizer mass flow rate divided by the fuel mass flow rate.

$$\delta_{av} = \frac{\delta_o \delta_f (1 + r)}{r\delta_f + \delta_o} \tag{7–1}$$

Values of δ_{av} for various propellant combinations are listed in Table 5–5. The value of δ_{av} can be increased by adding heavy materials to the propellants, either by solution or colloidal suspension. The identical type of equation can be written for the average density ρ_{av} in terms of the fuel density and the oxidizer density.

$$\rho_{av} = \frac{\rho_o \rho_f (1 + r)}{\rho_f r + \rho_o} \tag{7–2}$$

In the SI system of units the specific gravity has the same numerical value as the density expressed in units of grams per cubic centimeter or kg/liter. In some performance comparisons the parameter *density specific impulse* I_d is used. It is defined as the product of the average specific gravity δ and the specific impulse I_s:

$$I_d = \delta_{av} I_s \tag{7–3}$$

Stability. No *deterioration* and no *decomposition* with long-term (over 15 years) storage and minimal *reaction with the atmosphere* have been attained with many propellants. Good *chemical stability* means no decomposition of the liquid propellant during operation or storage, even at elevated temperature. A good liquid propellant should also have no *chemical deterioration* when in contact with piping, tank walls, valve seats, and gasket materials, even at relatively high ambient temperatures. No appreciable *absorption of moisture* and no adverse effects of small amounts of *impurities* are desirable properties. There should be no chemical deterioration when liquid flows through the hot cooling jacket passages. Some hydrocarbons (e.g., olefins) decompose and form carbonaceous deposits on the hot inside surfaces of the cooling passage. These deposits can be hard, reduce the heat flow, increase the local metal temperatures, and thus can cause the metal to weaken and fail. About 1% per year of stored concentrated hydrogen peroxide decomposes in clean storage tanks. Between 1 and 20% of a cryogenic propellant (stored in a vehicle) evaporates every day in an insulated tank.

Heat Transfer Properties. *High specific heat, high thermal conductivity,* and a *high boiling or decomposition temperature* are desirable for propellants that are used for thrust chamber cooling (see Section 8.3).

Pumping Properties. A low *vapor pressure* permits not only easier handling of the propellants, but also a more effective pump design in applications where the propellant is pumped. This reduces the potential for cavitation, as explained in Chapter 10. If the *viscosity* of the propellant is too high, then pumping and engine-system calibration become difficult. Propellants with high vapor pressure, such as liquid oxygen, liquid hydrogen, and other liquefied gases, require special design provisions, unusual handling techniques, and special low-temperature materials.

Temperature Variation. The *temperature variation* of the physical properties of the liquid propellant should be small. For example, a wide temperature variation in vapor pressure and density (*thermal coefficient of expansion*) or an unduly high change in viscosity with temperature makes it very difficult to accurately calibrate a rocket engine flow system or predict its performance over any reasonable range of operating temperatures.

Ignition, Combustion, and Flame Properties

If the propellant combination is *spontaneously ignitable*, it does not require an ignition system. This means that burning is initiated as the oxidizer and the fuel come in contact with each other. Spontaneously ignitable propellants are often termed *hypergolic* propellants. Although an ignition system is not a very objectionable feature, its elimination is usually desirable because it simplifies the propulsion system. All rocket propellants should be readily ignitable and have a small ignition time delay in order to reduce the potential explosion hazard during starting. Starting and ignition problems are discussed further in Section 8.4.

Nonspontaneously ignitable propellants have to be heated by external means before ignition can begin. Igniters are devices that accomplish an initial slight pressurization of the chamber and the initial heating of the propellant mixture to the point where steady flow combustion can be self-sustained. The amount of energy added by the igniter to activate the propellants should be small so that low-power ignition systems can be used. The energy required for satisfactory ignition usually diminishes for increasing ambient temperature of the propellant.

Certain propellant combinations burn very smoothly without combustion vibration. Other propellant combinations do not demonstrate this *combustion stability* and, therefore, are less desirable. Combustion is treated in Chapter 9.

Smoke formation is objectionable in many applications because of the smoke deposits on the surrounding equipment and parts. *Smoke* and brilliantly *luminous exhaust flames* are objectionable in certain military applications, because they can be easily detected. In some applications the condensed species in the exhaust gas can cause *surface contamination* on spacecraft windows or optical lenses and the electrons in the flame can cause undesirable *interference* or *attenuation of communications radio signals*. See Chapter 18 for information on exhaust plumes.

Property Variations and Specifications

The propellant properties and quality must not vary, because this can affect engine performance, combustion, and physical or chemical properties. The same propellant must have the same composition, properties, and storage or rocket operating characteristics if manufactured at different times or if made by different manufacturers. For these reasons propellants are purchased against specifications which define ingredients, maximum allowable impurities, packaging methods or compatible materials, allowable tolerances on physical properties (such as density, boiling point, freezing point, viscosity, or vapor pressure), quality control requirements, cleaning procedures for containers, documentation of inspections, laboratory analyses, or test results. A careful chemical analysis of the composition aand impurities is necessary. Reference 7–6 describes some of these methods of analysis.

Additive

Altering and tailoring propellant properties can be achieved with additives. For example, to make a non-hypergolic fuel become hypergolic (readily ignited), a reactive ingredient has been added. To desensitize concentrated hydrogen peroxide and reduce self-decomposition, it is diluted with 3 to 15% water. To increase density or to alleviate certain combustion instabilities, a fine powder of a heavy solid material can be suspended in the propellant.

7.2. LIQUID OXIDIZERS

Many different types of storable and cryogenic liquid oxidizer propellants have been used, synthesized, or proposed. For high specific impulse this includes boron–oxygen–fluorine compounds, oxygen–fluorine compounds, nitrogen–fluorine formulations, and fluorinated hydrocarbons; however, they all have some undesirable characteristics and these synthetic oxidizers have not been proven to be practical. Oxidizer liquids that have been used in experimental liquid rocket engines include mixtures of liquid oxygen and liquid fluorine, oxygen difluoride (OF_2), chlorine trifluoride (ClF_3), or chlorine pentafluoride (ClF_5). All of these are highly toxic and very corrosive. Several commonly used oxidizers are listed below.

Liquid Oxygen (O_2)

Liquid oxygen, often abbreviated as LOX, boils at 90 K at atmospheric pressure; at these conditions it has a specific gravity of 1.14 and a heat of vaporization of 213 kJ/kg. It is widely used as an oxidizer and burns with a bright white-yellow flame with most hydrocarbon fuels. It has been used in combination with alcohols, jet fuels (kerosene-type), gasoline, and hydrogen. As shown in Table 5–5, the attainable performance is relatively high, and liquid oxygen is therefore a desirable and commonly used propellant in large rocket engines. The following missiles and space launch vehicles use oxygen: (1) with jet fuel—Atlas, Thor, Jupiter, Titan I, Saturn booster; (2) with hydrogen—Space Shuttle and Centaur upper stage; (3) with alcohol—V-2 and Redstone. Figures 1–4 and 6–1 show units that use oxygen. Figures 5–1 to 5–6 give theoretical performance data for liquid oxygen with a kerosene-type fuel.

Although it usually does not burn spontaneously with organic materials at ambient pressures, combustion or explosions can occur when a confined mixture of oxygen and organic matter is suddenly pressurized. Impact tests show that mixtures of liquid oxygen with many commercial oils or organic materials will detonate. Liquid oxygen supports and accelerates the combustion of other materials. Handling and storage are safe when contact materials are clean. Liquid oxygen is a noncorrosive and nontoxic liquid and will not cause the deterioration of clean container walls. When in prolonged contact with human skin, the cryogenic propellant causes severe burns. Because liquid oxygen evaporates rapidly, it cannot be stored readily for any great length of time. If liquid oxygen is used in large quantities, it is often produced very close to its geographical point of application. Liquid oxygen can be obtained in several ways, such as by boiling liquid nitrogen out of liquid air.

It is necessary to insulate all lines, tanks, valves, and so on, that contain liquid oxygen in order to reduce the evaporation loss. Rocket propulsion systems which remain filled with liquid oxygen for several hours and liquid oxygen storage systems have to be well insulated against absorbing heat from the surroundings. External drainage provisions have to be made on all liquid oxygen tanks and lines to eliminate the water that condenses on the walls.

Example 7–1. Estimate the approximate temperature and volume change of liquid oxygen if an oxygen tank is pressurized to 8.0 atmospheres for a long time before engine start. Assume the tank is 60% full and the evaporated oxygen is refrigerated and recondensed (constant mass).

SOLUTION. Using Table 7–1 and Figs. 7–1 and 7–2, the vapor pressure goes from 1.0 atm (0.1 MPa) to 8 atm (about 0.8 MPa) and the equilibrium temperature goes from the boiling point of 90 K at 1.0 atm to about 133 K. The corresponding specific gravities are 1.14 and 0.88 respectively. This is an increase of 1.14/0.88 = 1.29 or about 77% full (29% more volume).

In tanks with turbopump feed systems the actual tank pressures are lower (typically 2 to 4 atm) and the evaporated oxygen is vented, causing a cooling effect on the liquid surface. So the numbers calculated above are too large (8 atm was selected to clearly show the effect). The warming occurs when there is a long hold period of a pressurized cryogenic propellant tank and is most pronounced when the final portion of the propellant is being emptied. Nevertheless the higher temperature, higher vapor pressure, and lower density can cause changes in mixture ratio, required tank volume, and pump suction condition (see Section 10.1). Therefore tanks with cryogenic propellant are insulated (to minimize heat transfer and density changes) and are pressurized only shortly before engine start, so as to keep the propellant at its lowest possible temperature.

Hydrogen Peroxide (H_2O_2)

In rocket application, hydrogen peroxide has been used in a highly concentrated form of 70 to 99%; the remainder is mostly water. Commercial peroxide is approximately 30% concentrated. Concentrated hydrogen peroxide was used in gas generator and rocket applications between 1938 and 1965 (X–1 and X–15 research aircraft).

In the combustion chamber, the propellant decomposes according to the following chemical reaction, forming superheated steam and gaseous oxygen:

$$H_2O_2 \rightarrow H_2O + \tfrac{1}{2}O_2 + \text{ heat}$$

This decomposition is brought about by the action of catalysts such as various liquid permanganates, solid manganese dioxide, platinum, and iron oxide. In fact, most impurities act as a catalyst. H_2O_2 is hypergolic with hydrazine and will burn well with kerosene. The theoretical specific impulse of 90% hydrogen peroxide is 154 sec, when used as a monopropellant with a solid catalyst bed.

Even under favorable conditions H_2O_2 will often decompose at a slow rate during storage, about one percent per year for 95%, and gas will bubble out of the liquid. Contaminated liquid peroxide must be disposed of before it reaches a danger point of about 448 K, when an explosion usually occurs. Concentrated peroxide causes severe burns when in contact with human skin and may ignite and cause fires when in contact with wood, oils, and many other organic materials. In the past rocket engines with hydrogen peroxide oxidizer have been used for aircraft boost (German Me 163, and U.S. F 104) and a missile (Britain: Black Knight). It has not been used for a long time, partly because of its long-term storage stability. However, there has been some improvement and some renewed interest in this dense oxidizer, which produces a nontoxic exhaust.

Nitric Acid (HNO₃)

There are several types of nitric acid mixtures that have been used as oxidizers between 1940 and 1965; they are not used extensively today in the United States. The most common type, *red fuming nitric acid* (RFNA), consists of concentrated nitric acid (HNO_3) that contains between 5 and 20% dissolved nitrogen dioxide. The evaporating red-brown fumes are exceedingly annoying and poisonous. Compared to concentrated nitric acid (also called *white fuming nitric acid*), RFNA is more energetic, more stable in storage, and less corrosive to many tank materials.

Nitric acid is highly corrosive. Only certain types of stainless steel, gold, and a few other materials are satisfactory as storage containers or pipeline materials. A small addition of fluorine ion (less than 1% of HF) inhibits the nitric acid, causes a fluoride layer to form on the wall, and greatly reduces the corrosion with many metals. It is called *inhibited red fuming nitric acid* (IRFNA). In case of accident of spilling, the acid should be diluted with water or chemically deactivated. Lime and alkali metal hydroxides and carbonates are common neutralizing agents. However, nitrates formed by the neutralization are also oxidizing agents and must be handled accordingly.

Nitric acid has been used with gasoline, various amines, hydrazine, dimethylhydrazine, and alcohols. It ignites spontaneously with hydrazine, furfuryl alcohol, aniline, and other amines. The specific gravity of nitric acid varies from 1.5 to 1.6, depending on the percentages of nitric oxide, water, and impurities. This high density permits compact vehicle construction.

Vapors from nitric acid or red fuming nitric acid have an OSHA 8-hr personnel exposure limit or a threshold work allowance of 2 ppm (parts per million or about 5 mg/m³) and a short-term exposure limit of 10 ppm. Droplets on the skin cause burns and sores which do not heal readily.

Nitrogen Tetroxide (N₂O₄)

This is a high-density yellow-brown liquid (specific gravity of 1.44). Although it is the most common storable oxidizer used in the United States today, its liquid temperature range is narrow and it is easily frozen or vaporized. It is only mildly corrosive when pure, but forms strong acids when moist or allowed to mix with water. It readily absorbs moisture from the air. It can be stored indefinitely in sealed containers made of compatible material. It is hypergolic with many fuels and can cause spontaneous ignition with many common materials, such as paper, leather, and wood. The fumes are reddish brown and are extremely toxic. Because of its high vapor pressure it must be kept in relatively heavy tanks. The freezing point of N_2O_4 can be lowered (by adding a small amount of nitric oxide or NO) but at the penalty of a higher vapor pressure. This mixture of NO and N_2O_4 is called mixed oxides of nitrogen (MON) and different grades have been 2 and 30% NO content.

Nitrogen tetroxide is a storable propellant oxidizer and is used in the Titan missile together with a fuel mixture consisting of hydrazine and unsymmetrical dimethylhydrazine. It is also used with monomethylhydrazine fuel in the Space Shuttle orbital maneuver system and reaction control system and in many spacecraft propulsion systems. In many of these applications care must be taken to avoid freezing this propellant. The OSHA 8-hr personnel exposure limit is 5 ppm or 9 mg/m^3.

7.3. LIQUID FUELS

Again, many different chemicals have been proposed, investigated, and tested. Only a few have been used in production rocket engines. Liquid fuels other than those listed below have been used in experimental rocket engines, in older experimental designs, and in some older production engines. These include aniline, furfuryl alcohcol, xylidine, gasoline, hydrazine hydrate, borohydrides,- methyl and/or ethyl alcohol, ammonia, and mixtures of some of these with one or more other fuels.

Hydrocarbon Fuels

Petroleum derivatives encompass a large variety of different hydrocarbon chemicals, most of which can be used as a rocket fuel. Most common are those types that are in use with other applications and engines, such as gasoline, kerosene, diesel oil, and turbojet fuel. Their physical properties and chemical composition vary widely with the type of crude oil from which they were refined, with the chemical process used in their production, and with the accuracy of control exercised in their manufacture. Typical values are listed in Table 7–2.

In general, these petroleum fuels form yellow-white, brilliantly radiating flames and give good preformance. They are relatively easy to handle, and there is an ample supply of these fuels available at low cost. A specifically refined petroleum product particularly suitable as a rocket propellant has been designated RP-1. It is basically a kerosene-like mixture of saturated and unsaturated hydrocarbons with a somewhat narrow range of densities and vapor pressure. Several hydrocarbon fuels can form carbon deposits on the inside of cooling passages, impeding the heat transfer and raising wall temperatures. Ref. 7–7 indicates that this carbon formation depends on fuel temperature in the cooling jacket, the particular fuel, the heat transfer, and the chamber wall material. RP-1 is low in olefins and aromatics, which can cause carbonaceous deposits inside fuel cooling passages. RP-1 has been used with liquid oxygen in the Atlas, Thor, Delta, Titan I, and Saturn rocket engines (see Figs. 5–1 to 5–6).

Methane (CH_4) is a cryogenic hydrocarbon fuel. It is denser than liquid hydrogen and relatively low in cost. Compared to petroleum refined hydro-

TABLE 7–2. Properties of Some Typical Hydrocarbon Fuels Made from Petroleum

	Jet Fuel	Kerosene	Aviation Gasoline 100/130	Diesel Fuel	RP-1
Specific gravity at 289 K	0.78	0.81	0.73	0.85	0.80–0.815
Freezing point (K)	213 (max.)	230	213	250	239 (max.)
Viscosity at 289 K (cP)	1.4	1.6	0.5	2.0	16.5 (at 239 K)
Flash point (K) (TCC)	269	331	244	333	316
ASTM distillation (K)					
10% evaporated	347	—	337	—	458–483
50% evaporated	444	—	363	—	—
90% evaporated	511	—	391	617	—
Reid vapor pressure (psia)	2 to 3	Below 1	7	0.1	—
Specific heat (cal/kg-K)	0.50	0.49	0.53	0.47	0.50
Average molecular mass (kg/mol)	130	175	90	—	—

carbons it has highly reproducible properties. With liquid oxygen it is a candidate propellant combination for launch vehicle booster rocket engines and also reaction engines control when oxygen is available from the main engines). Experimental oxygen–methane engines have been tested, but they have not yet flown.

Liquid Hydrogen (H₂)

Liquid hydrogen, when burned with liquid fluorine or liquid oxygen, gives a high performance, as shown in Table 5–5. It also is an excellent regenerative coolant. With oxygen it burns with a colorless flame; however, the shock waves in the plume may be visible. Of all known fuels, liquid hydrogen is the lightest and the coldest, having a specific gravity of 0.07 and a boiling point of about 20 K. The very low fuel density requires bulky fuel tanks, which necessitate very large vehicle volumes. The extremely low temperature makes the problem of choosing suitable tank and piping materials difficult, because many metals become brittle at low temperatures.

Because of its low temperature, liquid hydrogen tanks and lines have to be well insulated to minimize the evaporation of hydrogen or the condensation of moisture or air on the outside with the subsequent formation of liquid or solid air or ice. A vacuum jacket often has been used in addition to insulating materials. All common liquids and gases solidify in liquid hydrogen. These solid particles in turn plug orifices and valves. Therefore, care must be taken to scavenge all lines and tanks of air and moisture (flush with helium or pull vacuum) before introducing the propellant. Mixture of liquid hydrogen and solid oxygen or solid air can be explosive.

Liquid hydrogen has two species, namely, orthohydrogen and parahydrogen, which differ in their nuclear spin state. As hydrogen is liquefied, the relative equilibrium composition of ortho- and parahydrogen changes. The transformation from one species to another is accompanied by a transfer of energy. Liquid hydrogen is manufactured from gaseous hydrogen by successive compression, cooling, and expansion processes.

Hydrogen gas, when mixed with air, is highly flammable and explosive over a wide range of mixture ratios. To avoid this danger, hydrogen gas leakage (a tank vent line) is often intentionally ignited and burned in the air. Liquid hydrogen is used with liquid oxygen in the Centaur upper stage, the Space Shuttle main engine, and upper stage space engines developed in Japan, Russia, Europe, and China.

Hydrogen burning with oxygen forms a nontoxic exhaust gas. This propellant combination has been applied successfully to space launch vehicles because of its high specific impulse. Here the payload capability usually increases greatly for relatively small increases in specific impulse. However, the low density of hydrogen makes for a large vehicle and a relatively high drag.

One method to increase the density of hydrogen is to use a subcooled mixture of liquid hydrogen and suspended frozen small particles of solid hydrogen, which is denser than the liquid. Experiments and studies on this "slush" hydrogen have been performed; it is difficult to produce and maintain a uniform mixture. It has not yet been used in a flight vehicle.

Some studies have shown that, when burned with liquid oxygen, a hydrocarbon (such as methane or RP-1) can give a small advantage in space launch vehicle first stages. Here the higher average propellant density allows a smaller vehicle with lower drag, which compensates for the lower specific impulse of the hydrocarbon when compared to a hydrogen fuel. Also, there are some concepts for operating the booster-stage rocket engine initially with hydrocarbon fuel and then switching during flight to hydrogen fuel. As yet, engines using two fuels, namely methane (or hydrocarbon) and hydrogen, have not yet been fully developed or flown. Some work on an experimental engine was done in Russia.

Hydrazine (N_2H_4)

Reference 7–8 gives a good discussion of this propellant, which is used as a bipropellant fuel as well as a monopropellant. Hydrazine and its related liquid organic compounds, monomethylhydrazine (MMH) and unsymmetrical dimethylhydrazine (UDMH), all have similar physical and thermochemical properties. Hydrazine is a toxic, colorless liquid with a high freezing point (274.3 K). Hydrazine has a short ignition delay and is spontaneously ignitable with nitric acid and nitrogen tetroxide.

Its vapors may form explosive mixtures with air. If hydrazine is spilled on a surface or a cloth, a spontaneous ignition with air can occur.

Pure anhydrous hydrazine is a stable liquid; it has been safely heated above 530 K. It has been stored in sealed tanks for over 15 years. With impurities or at higher temperatures it decomposes and releases energy. Under pressure shock (blast wave) it decomposes at temperatures as low as 367 K. Under some conditions this decomposition can be a violent detonation, and this has caused problems in cooling passages of experimental injectors and thrust chambers. Harmful effects to personnel may result from ingestion, inhalation of vapors, or prolonged contact with skin. The OSHA 8-hr personnel exposure limit is 0.1 ppm or 0.13 mg/m^3. Hydrazine is a known carcinogen.

Hydrazine reacts with many materials, and care must be exercised to avoid storage contact with materials that cause a decomposition (see Ref 7–9). Tanks, pipes, or valves must be cleaned and free of impurities. Compatible materials include stainless steels (303, 304, 321, or 347), nickel, and 1100 and 3003 series of aluminum. Iron, copper and its alloys (such as brass or bronze), monel, magnesium, zinc, and some types of aluminum alloy must be avoided.

Unsymmetrical Dimethylhydrazine [(CH$_3$)$_2$NNH$_2$]

A derivative of hydrazine, namely, unsymmetrical dimethylhydrazine (UDMH), is often used instead of or in mixtures with hydrazine because it forms a more stable liquid, particularly at higher temperatures. Furthermore, it has a lower freezing point (215.9 K) and a higher boiling point (336.5 K) than a hydrazine. When UDMH is burned with an oxidizer it gives only slightly lower values of I_s than pure hydrazine. UDMH is often used when mixed with 30 to 50% hydrazine. This fuel is used in the Titan missile and launch vehicle and spacecraft engines in 50% mixtures and has been used in the lunar landing and take-off engines. UDMH is used in Russian and Chinese rocket engines.

Freezing does not affect UDMH, MMH, or hydrazine, but freezing of a 50:50 mixture of UDMH and hydrazine causes a separation into two distinct layers; a special remixing operation is necessary for reblending if freezing occurs in a space vehicle. The OSHA 8-hr personnel exposure limit for vapor is 0.5 ppm, and UDMH is a carcinogen.

Monomethylhydrazine (CH$_3$NHNH$_2$)

Monomethylhydrazine (MMH) has been used extensively as a fuel in space-craft rocket engines, particularly in small attitude control engines, usually with N$_2$O$_4$ as the oxidizer. It has a better shock resistance to blast waves, better heat transfer properties, and a better liquid temperature range than pure hydrazine. Like hydrazine, its vapors are easily ignited in air; the flammability limits are from 2.5 to 98% by volume at atmospheric sea level pressure and ambient temperature. The materials compatible with hydrazine are also compatible with MMH. The specific impulse with storable oxidizers usually is 1 or 2% lower with MMH than with N_2H_4.

Both MMH an UDMH are soluble in many hydrocarbons; hydrazine is not. All hydrazines are toxic materials, but MMH is the most toxic when inhaled, and UDMH the least toxic. Atmospheric concentrations of all hydrazines should be kept below 0.1 ppm for long periods of exposure.

Monomethylhydrazine, when added in relatively small quantities of 3 to 15% to hydrazine, has a substantial quenching effect on the explosive decomposition of hydrazine. Monomethylhydrazine decomposes at 491 K, whereas hydrazine explodes at 369 K when subjected to pressure shocks of identical intensity. MMH is a suspected carcinogen and the OSHA personnel 8-hour exposure limit is 0.2 ppm.

7.4. LIQUID MONOPROPELLANTS

The propellant-feed and control-system simplicity associated with a monopropellant makes this type of propellant attractive for certain applications. Hydrazine is being used extensively as a monopropellant in small attitude and trajectory control rockets for the control of satellites and other spacecraft and also as a hot gas generator. (It is discussed in the preceding section.) Other monopropellants (ethylene oxide or nitromethane) were tried experimentally, but are no longer used today. Concentrated hydrogen peroxide was used for monopropellant gas generation in the USA, Russia, and Germany in engines designed before 1955.

Ignition of monopropellants can be produced thermally (electrical or flame heat) or by a catalytic material. A monopropellant must be chemically and thermally stable to insure good liquid storage properties, and yet it must be easily decomposed and reactive to give good combustion properties.

Hydrazine as a Monopropellant

Hydrazine is not only an excellent storable fuel, but also an excellent monopropellant when decomposed by a suitable solid or liquid catalyst; this catalyst often needs to be preheated for fast startup. Iridium is an effective catalyst at room temperature. At elevated temperature (about 450 K) many materials decompose hydrazine, including iron, nickel, and cobalt. See Ref. 7–8. Different catalysts and different reaction volumes make the decomposition reaction go to different products, resulting in gases varying in composition or temperature. As a monopropellant, it is used in gas generators or in space engine attitude control rockets.

Hydrazine has been stored in sealed tanks for over 15 years. A typical hydrazine monopropellant thrust chamber, its injection pattern, and its decomposition reaction are described in Chapter 10 and typical design parameters are shown in Fig. 7–3 and a monopropellant structure in Fig. 8–16.

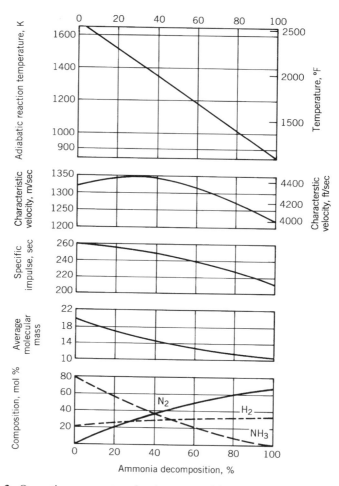

FIGURE 7–3. Operating parameters for decomposed hydrazine at the exit of a catalytic reactor as a function of the ammonia dissociation fraction. Adapted with permission from Ref. 7–8.

The catalytic decomposition of hydrazine can be described ideally as a two-step process; this ignores other steps and intermediate products. First, hydrazine (N_2H_4) decomposes into gaseous ammonia (NH_3) and nitrogen (N_2); this reaction is highly exothermic, i.e., it releases heat. Secondly, the ammonia decomposes further into nitrogen and hydrogen gases, but this reaction is endothermic and absorbs heat. These simplified reactions can be written as

$$3N_2H_4 \rightarrow 4(1-x)NH_3 + (1+2x)N_2 + 6xH_2 \quad (7\text{–}4)$$

Here x is the degree of ammonia dissociation; it is a function of the catalyst type, size, and geometry, the chamber pressure, and the dwell time within the catalyst bed. Figure 7–3 shows several ideal rocket engine parameters for hydrazine monopropellant as a function of x, the fraction of ammonia that is decomposed. The values are for an ideal thruster at 1000 psia chamber pressure with an area ratio of 50 expanding at high altitude. The best specific impulse is attained when little ammonia is allowed to dissociate.

Hydrazine is manufactured in several grades of purity. The standard commercial hydrazine has about 1.5% maximum by weight of water, less than 1.0% aniline, and no more than 10 mg/l of particulates, including carbon. Monopropellant-grade hydrazine has less than 1% water, less than 0.5% aniline (whch is a material commonly used in the manufacture of hydrazine), and traces of ammonia, carbon dioxide, chlorides, and iron- or carbon-containing materials such as UDMH or MMH. Aniline and other organic impurities can poison the catalyst used to decompose monopropellant hydrazine; as mentioned in Chapter 10, this can cause operating problems. There is also a highly purified grade of hydrazine that has less water, less than 0.005% aniline, and less than 0.003% carbon materials; it does not contaminate the catalyst and is used now in many monopropellant applications.

Hydroxyl Ammonium Nitrate ($NH_2OH^+NO_3$)

This is a relatively new, synthetic, propellant material rich in oxygen, but with combined hydrogen and nitrogen (fuel ingredients), it is abbreviated as HAN. It is an opaque hygroscopic solid when pure, and a clear colorless odorless liquid in aqueous solutions. The solid HAN (specific gravity of 1.84) is a potential solid propellant ingredient and the liquid HAN solution is a potential monopropellant (a 13 molar solution has a specific gravity of 1.523). Both can be made to burn smoothly and several catalysts have been effective in obtaining controlled decomposition. The boiling point (110 to 145°C) and the freezing point (−15 to −44°C) vary with the water content. HAN becomes more viscous as the percentage of water is reduced. The liquid is corrosive, toxic, denser than hydrazine monopropellant, and does not seem to be carcinogenic. The liquid is incompatible with alkali materials, many metals, and other materials. Even with relatively very compatible materials HAN solutions decompose slowly in storage; a satisfactory stabilizer has yet to be found. The monopropellant's specific impulse is between 200 and 265 sec, depending on the water content and the mixing of the aqueous HAN with one of several possible compatible organic fuel liquids. The HAN propellant formulation, its rocket engines, and solid motors are still in their research and development phase, as shown in Refs. 7–9 and 7–10.

7.5. GELLED PROPELLANTS

Gelled propellants have additives that make them thixotropic materials. They have the consistency of thick paint or jelly when at rest, but they liquify and flow through pipes, valves, pumps, or injectors when an adequate shear stress is applied. They offer these *advantages*.

Small aluminum particles can be suspended in the fuels where smoky exhaust is not objectionable. Inert solid particles can be suspended in oxidizer liquids. This increases propellant density, density impulse, and thus reduces the size of tanks and vehicles. Smaller vehicles have reduced drag and thus can allow an increase in the range or speed of tactical missiles.

There is no plugging of injector orifices or valve passages and good flow control has been demonstrated.

Individual gelled fuel propellants will be essentially nonflammable and will not usually sustain an open fire.

There is reduced susceptibility of leakage or spill, reduced sloshing of liquids in the tanks, and the boil-off rate is reduced.

Long-term storage without settling or separation is possible; more than 10 years has been demonstrated.

Explosions or detonations, which happen when a vehicle accident causes liquid propellants to become inadvertently premixed, are much less likely with gelled propellants, which are difficult to mix.

Many spilled gelled propellants can be diluted with water and disposed of safely.

Short-duration pulsing is possible.

Most storable oxidizers, a few cryogenic propellants, and most liquid storable fuels can be gelled.

Explosions are much less likely when a propellant tank is penetrated by a bullet or when a missile is exposed to an external fire or a nearby detonation.

These are some of the *disadvantages*:

There is a small decrease in specific impulse due to dilution with a gelling agent, and less efficient atomization or combustion. For example, the characteristic velocity c^* of oxygen–kerosene propellant is decreased by 4 to 6% when the kerosene is gelled and aluminum is suspended in the fuel. When both the fuel and a nitric acid oxidizer are gelled, the performance loss (c^*) can be as high as 8%. Clever injector design and the selection of good gelling agents can reduce this loss.

Loading or unloading of propellants is somewhat more complex.

Residual propellant quantity may be slightly higher, because the thixotropic fluid layer on the walls of the tanks and pipes may be slightly thicker.

Changes in ambient temperature will cause slight changes in propellant density and viscosity and therefore also in mixture ratio; this can result in more leftover or residual propellant and thus in a slight reduction of available total impulse. This can be minimized by careful selection of gelling agents so as to match the rheological property changes of oxidizer and fuel over a particular temperature range.

Suspended metals can make the plume smoky and visible.

Some gelling agents have resulted in unstable gelled propellants; that is, they separated or underwent chemical reactions.

Experimental rocket engines have shown these gelled propellants to be generally safer than ordinary liquid propellants and to have good performance and operational characteristics (see Refs 7–11 and 7–12). This makes them less susceptible to field accidents. A variety of different organic and inorganic gelling agents have been explored with a number of different liquid propellants.

Experimental thrust chambers and rocket engine systems have been satisfactorily demonstrated with several gelled propellant combinations. One experimental engine is shown in Fig. 6–8. As far as is known, no such rocket engine has yet been put into production or flight operation. An effort is underway to demonstrate this technology clearly and to qualify a rocket engine with gelled propellants for an actual flight application.

7.6. GASEOUS PROPELLANTS

Cold gas propellants have been used successfully for reaction control systems (RCS) for perhaps 50 years. The engine system is simple, consisting of one or more high-pressure gas tanks, multiple simple metal nozzles (often aluminum or plastic), an electrical control valve with each nozzle, a pressure regulator, and provisions for filling and venting the gas. The tank size will be smaller if the tank pressures are high. Pressures are typically between 300 and 1000 MPa (about 300 to 10,000 psi). The mass of spherical storage tanks is essentially independent of pressure if they contain the same mass of gas.

Typical cold gas propellants and some of their properties and characteristics are listed in Table 7–3. Nitrogen, argon, dry air, krypton and Freon 14 have been employed in spacecraft RCSs. With high-pressure hydrogen or helium as cold gas, the specific impulse is much higher, but the densities of these gases are much lower. This requires a much larger gas storage volume and heavier high-pressure tanks. In most applications the extra inert mass outweighs the advantage of better performance. In a few applications the gas (and its storage tank) are heated electrically or chemically. This improves the specific impulse and allows a smaller tank, but it also introduces complexity.

The selection of the gas propellant, the storage tanks, and RCS design depend on many factors, such as volume and mass of the storage tanks, the maximum thrust and total impulse, the gas density, required maneuvers, duty cycle, and flight duration. Cold gas systems are used for total impulses of perhaps 1200 N-sec or 5000 lbf-sec. Higher values usually employ liquid propellants.

TABLE 7–3. Properties of Gaseous Propellants Used for Auxiliary Propulsion

Propellant	Molecular Mass	Density[a] (lb/ft^3)	k	Theoretical Specific Impulse[b] (sec)
Hydrogen	2.0	1.77	1.40	284
Helium	4.0	3.54	1.67	179
Methane	16.0	14.1	1.30	114
Nitrogen	28.0	24.7	1.40	76
Air	28.9	25.5	1.40	74
Argon	39.9	35.3	1.67	57
Krypton	83.8	74.1	1.63	50

[a]At 5000 psia and 20°C.
[b]In vacuum with nozzle area ratio of 50:1 and initial temperature of 20°C.

If the operation is short (only a few minutes, while the main engine is running), the gas expansion will be adiabatic (no heat absorption by gas) and often is analyzed as isentropic (constant stagnation presure). The temperature of the gas will drop (the pressure and specific impulse will also drop) as the gas is consumed. For long intermittent operations (months or years in space) the heat from the spacecraft is transfered to the gas and the tank temperature stays essentially constant; the expansion will be nearly isothermal. An analysis of gas expansion is given in Section 6.5.

The advantages and disadvantages of cold gas systems are described on pages 303 and 304.

7.7. SAFETY AND ENVIRONMENTAL CONCERNS

To minimize the hazards and potential damage inherent in reactive propellant materials, it is necessary to be very conscientious about the likely risks and hazards (see Ref. 7–4). This concerns toxicity, explosiveness, fire or spill danger, and others mentioned in Section 7.1. Before an operator, assembler, maintenance mechanic, supervisor, or engineer is allowed to transfer or use a particular propellant, he or she should receive safety training in the particular propellant, its characteristics, its safe handling or transfer, potential damage to equipment or the environment, and the countermeasures for limiting the consequences in case of an accident. They must also understand the potential hazards to the health of personnel, first aid, remedies in case of contact exposure of the skin, ingestion, or inhaling, and the use of safety equipment. Examples of safety equipment are protective clothing, detectors for toxic vapors, remote controls, warning signals, or emergency water deluge. The personnel working with or being close to highly toxic materials usually have to undergo frequent health monitoring. Also rocket engines need to be designed for safety to minimize the occurrence of a leak, an accidental spill, an unexpected fire, or other potentially unsafe conditions. Most organizations have one or more safety specialists who review the safety of the test plans, manufacturing operations, design, procedures, or safety equipment. With the proper training, equipment, precautions, and design safety features, all propellants can be handled safely.

If a safety violation occurs or if an operation, design, procedure, or practice is found to be (or appears to be) unsafe, then a thorough investigation of the particular item or issue should be undertaken, the cause of the lack of safety should be investigated and identified, and an appropriate remedial action should be selected and initiated as soon as possible.

The discharge of toxic exhaust gases to the environment and their dispersion by the wind can cause exposure of operating personnel as well as the people in nearby areas. This is discussed in Section 20.2. The dumping or spilling of toxic liquids can contaminate subterranean aquifers and surface waters, and their vapors can pollute the air. Today the type and amount of gaseous and liquid discharges are regulated and monitored by government authorities. These discharges must be controlled or penalties will be assessed against violators. Obtaining a permit to discharge can be a lengthy and involved procedure.

PROBLEMS

1. Plot the variation of the *density specific impulse* (product of average specific gravity and specific impulse) with mixture ratio and explain the meaning of the curve. Use the theoretical shifting specific impulse values of Figure 5–1 and the specific gravities from Figure 7–1 or Table 7–1 for the liquid oxygen–RP-1 propellant combination. *Answers*: Check point at $r = 2.0$; $I_s = 290$; $I_d = 303$; $\delta_{av} = 1.01$.

2. Prepare a table comparing the relative merits of liquid oxygen and nitric acid as rocket oxidizers.

3. Derive Eq. 7–1 for the average specific gravity.

4. A rocket engine uses liquid oxygen and RP-1 as propellants at a design mass mixture ratio of 2.40. The pumps used in the feed system are basically constant-volume flow devices. The RP-1 hydrocarbon fuel has a nominal temperature of 298 K and it can vary at about $\pm 25°C$. The liquid oxygen is nominally at its boiling point (90 K), but, after the tank is pressurized, this temperature can increase by 30 K. What are the extreme mixture ratios under unfavorable temperature conditions? If this engine has a nominal mass flow rate of 100 kg/sec and a duration of 100 sec, what is the maximum residual propellant mass when the other propellant is fully consumed? Use the curve slopes of Fig. 7–1 to estimate changes in density. Assume that the specific impulse is constant for the relatively small changes in mixture ratio, that vapor pressure changes have no influence on the pump flow, and that the engine has no automatic control for mixture ratio.

5. The vehicle stage propelled by the rocket engine in Problem 4 has a design mass ratio m_f/m_0 of 0.50 (see Eq. 4–6). How much will the worst combined changes in propellant temperatures effect the mass ratio and the ideal gravity-free vacuum velocity?

6. (a) What should be the approximate percent ullage volume for nitrogen tetroxide tank when the vehicle is exposed to ambient temperatures between about 50°F and about 150°F?

   (b) What is maximum tank presure at 150°F.

   (c) What factors should be considered in part (b)?

   Answers: (a) 15 to 17%; the variation is due to the nonuniform temperature distribution in the tank; (b) 6 to 7 atm; (c) vapor pressure, nitrogen monoxide content in the oxidizer, chemical reactions with wall materials, or impurities that result in largely insoluble gas products.

7. An insulated, long vertical, vented liquid oxygen tank has been sitting on the sea level launch stand for a period of time. The surface of the liquid is at atmospheric pressure and is 10.2 m above the closed outlet at the bottom of the tank. If there is no circulation, what will be the temperature, pressure and density of the oxygen at the tank outlet?

SYMBOLS

I_d	density specific impulse, sec
I_s	specific impulse, sec
k	ratio of specific heat
r	mixture ratio (mass flow rate of oxidizer to mass flow rate of fuel)

Greek Letters

δ_{av}	average specific gravity of mixture
δ_f	specific gravity of fuel
δ_o	specific gravity of oxidizer
$\rho_{av}, \rho_f, \rho_o$	densities, kg/m^3 (lbm/ft^3)

REFERENCES

7–1. S. F. Sarner, *Propellant Chemistry*, Reinhold Publishing Company, New York, 1966.

7–2. *Chemical Rocket Propellant Hazards*, Vol. 1, *General Safety Engineering Design Criteria*, Chemical Propulsion Information Agency (CPIA) Publication 194, October 1971.

7–3. L. C. Sutherland, "Scaling Law for Estimating Liquid Propellant Explosive Yields," *Journal of Spacecraft and Rockets*, March–April 1978, pp. 124–125.

7–4. *Hazardous Materials, 1980 Emergency Response Guidebook*, DOT-P 5800.2, U.S. Department of Transportation, Washington, DC, 1980.

7–5. *1990–1991 Threshold Limit Values for Chemical Substances and Physical Agents and Biological Exposure Indices*, American Conference of Government Industrial Hygienists, Cincinnati, OH, 1990 (revised periodically).

7–6. H. E. Malone, *The Analysis of Rocket Propellants*, Academic Press, New York, 1976.

7–7. K. Liang, B.Yang, and Z. Zhang, "Investigation of Heat Transfer and Coking Characteristics of Hydrocarbon Fuels," *Journal of Propulsion and Power*, Vol. 14, No. 5, September–October 1998.

7–8. E. W. Schmidt, *Hydrazine and its Derivatives, Preparation, Properties, Applications*, John Wiley & Sons, New York, 1984.

7–9. O. M. Morgan and D. S. Meinhardt, "Monopropellant Selection Criteria— Hydrazine and other Options," *AIAA Technical Paper 99–2595*, June 1999.

7–10. D. Mittendorf, W. Facinelli, and R. Serpolus, "Experimental Development of a Monopropellant for Space Propulsion Systems," *AIAA Technical Paper 99–2951*, June 1999.

7–11. K. F. Hodge, T. A. Crofoot, and S. Nelson, "*Gelled Technical Propellants for Tactical Missile Application*," AIAA Technical Paper 99–2976, June 1999.

7–12. B. D. Allen, "History, Development and Testing of Thixotropic Gels for Advanced Systems," AIAA Propulsion Conference, July 1985.

CHAPTER 8

THRUST CHAMBERS

The *thrust chamber* is the key subassembly of a rocket engine. Here the liquid propellants are metered, injected, atomized, vaporized, mixed, and burned to form hot reaction gas products, which in turn are accelerated and ejected at high velocity (see Refs. 6–1 and 6–2). This chapter describes thrust chambers, their components, cooling, ignition, and heat transfer. A rocket thrust chamber assembly (Figs. 8–1 and 8–2) has an *injector, a combustion chamber*, a *supersonic nozzle*, and *mounting provisions*. All have to withstand the extreme heat of combustion and the various forces, including the transmission of the thrust force to the vehicle. There also is an ignition system if non-spontaneously ignitable propellants are used. Some thrust chamber assemblies also have integrally mounted propellant valves and sometimes a thrust vector control device, as described in Chapter 16. Table 8–1 (see pages 272–273) gives various data about five different thrust chambers with different kinds of propellants, cooling methods, injectors, feed systems, thrust levels, or nozzle expansions. Some engine parameters are also listed. Some of the terms used in this table will be explained later in this chapter.

The basic analyses for thrust chamber performance (specific impulse, combustion temperature) are given in Chapter 5, the basic design parameters (thrust, flow, chamber pressure, or throat area) are in Chapter 3, and the combustion phenomena in Chapter 9.

Although we use the word *thrust chamber* in this book (for rocket engines generally larger than 1000 lbf thrust), some articles use the term *thrust cylinder* or *rocket combustor*. We will also use the term *thruster* for small thrust units, such as attitude control thrusters, and for electrical propulsion systems.

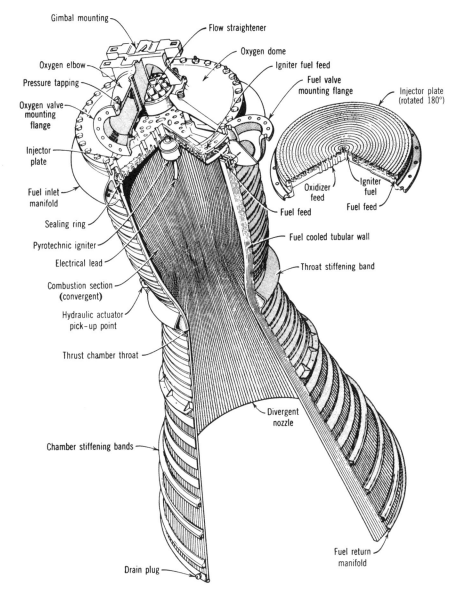

Gimbal mounting
Flow straightener
Oxygen dome
Oxygen elbow
Igniter fuel feed
Pressure tapping
Fuel valve
mounting flange
Injector plate
(rotated 180°)
Oxygen valve
mounting
flange
Injector
plate
Fuel inlet
manifold
Oxidizer
feed
Igniter
fuel
Fuel feed
Sealing ring
Fuel feed
Fuel feed
Pyrotechnic igniter
Fuel cooled tubular wall
Electrical lead
Combustion section
(convergent)
Throat stiffening band
Hydraulic actuator
pick-up point
Thrust chamber throat
Divergent
nozzle
Chamber stiffening bands
Fuel return
manifold
Drain plug

FIGURE 8–1. Construction of a regeneratively cooled tubular thrust chamber using a kerosene-type fuel and liquid oxygen, as originally used in the Thor missile. The nozzle inside diameter is about 15 in. The sea-level thrust was originally 120,000 lbf, but was uprated to 135,000, then 150,000, and finally to 165,000 lbf by increasing the flow and chamber pressure and strengthening and modifying the hardware. The cone-shaped exit cone was replaced by a bell-shaped nozzle. Figure 8–9 shows how the fuel flows down through every other tube and returns through the adjacent tube before flowing into the injector. Figure 8–4 shows the flow passages in a similar injector. (Courtesy of Rolls Royce, England.)

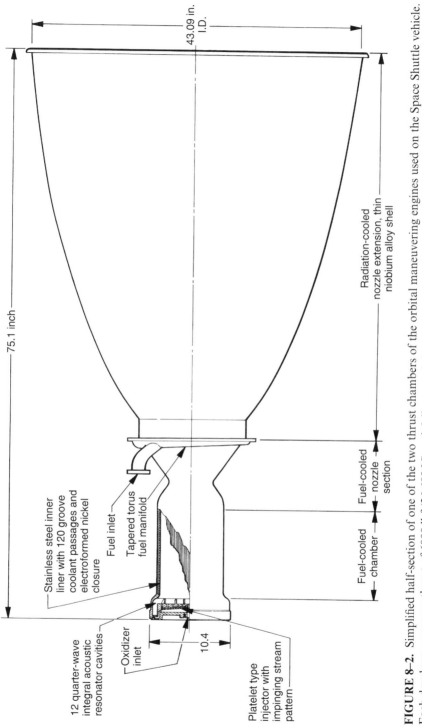

FIGURE 8–2. Simplified half-section of one of the two thrust chambers of the orbital maneuvering engines used on the Space Shuttle vehicle. Each develops a vacuum thrust of 6000 lbf (26,689 N) and delivers a minimum *vacuum* specific impulse of 310 sec, using nitrogen tetroxide and monomethyl hydrazine propellants at a nominal mixture ratio of 1.65 and a nominal chamber pressure of 128 psia. It is designed for 100 flight missions, a service life of 10 years, and a minimum of 500 starts. These engines provide the thrust for final orbit attainment, orbit circularization, orbit transfer, rendezvous, and deorbit maneuvers. The nozzle area ratio of 55:1. (Courtesy of Aerojet Propulsion Company.)

8.1. INJECTORS

The functions of the injector are similar to those of a carburetor of an internal combustion engine. The injector has to introduce and meter the flow of liquid propellants to the combustion chamber, cause the liquids to be broken up into small droplets (a process called atomization), and distribute and mix the propellants in such a manner that a correctly proportioned mixture of fuel and oxidizer will result, with uniform propellant mass flow and composition over the chamber cross section. This has been accomplished with different types of injector designs and elements; several common types are shown in Fig. 8–3 and complete injectors are shown in Figs. 9–6, 8–1, and 8–4.

The *injection hole pattern* on the face of the injector is closely related to the internal manifolds or feed passages within the injector. These provide for the distribution of the propellant from the injector inlet to all the injection holes. A large complex manifold volume allows low passage velocities and good distribution of flow over the cross section of the chamber. A small manifold volume allows for a lighter weight injector and reduces the amount of "dribble" flow after the main valves are shut. The higher passage velocities cause a more uneven flow through different identical injection holes and thus a poorer distribution and wider local gas composition variation. Dribbling results in afterburning, which is an inefficient irregular combustion that gives a little "cutoff" thrust after valve closing. For applications with very accurate terminal vehicle velocity requirements, the cutoff impulse has to be very small and reproducible and often valves are built into the injector to minimize passage volume.

Impinging-stream-type, multiple-hole injectors are commonly used with oxygen–hydrocarbon and storable propellants. For *unlike doublet* patterns the propellants are injected through a number of separate small holes in such a manner that the fuel and oxidizer streams impinge upon each other. Impingement forms thin liquid fans and aids atomization of the liquids into droplets, also aiding distribution. Characteristics of specific injector orifices are given in Table 8–2 (see page 279). Impinging hole injectors are also used for *like-on-like* or *self-impinging patterns* (fuel-on-fuel and oxidizer-on-oxidizer). The two liquid streams then form a fan which breaks up into droplets. Unlike doublets work best when the hole size (more exactly, the volume flow) of the fuel is about equal to that of the oxidizer and the ignition delay is long enough to allow the formation of fans. For uneven volume flow the triplet pattern seems to be more effective.

The *nonimpinging* or *shower head* injector employs nonimpinging streams of propellant usually emerging normal to the face of the injector. It relies on turbulence and diffusion to achieve mixing. The German World War II V-2 rocket used this type of injector. This type is now not used, because it requires a large chamber volume for good combustion. *Sheet* or *spray-type injectors* give cylindrical, conical, or other types of spray sheets; these sprays generally intersect and thereby promote mixing and atomization. By varying the width of the sheet (through an axially moveable sleeve) it is possible to throttle the propel-

TABLE 8-1. Thrust Chamber Characteristics

	Engine Designation				
	RL 10B-2	LE-7 (Japan)	RCS	RS-27	AJ-10-118I
Application	Delta-III and IV upper stage	Booster stage for H-II launcher	Attitude control	Delta II Space Launch booster	Delta II Second stage
Manufacturer	Pratt & Whitney, United Technologies Corporation	Mitsubishi Heavy Industries	Kaiser Marquardt Company	The Boeing Co., Rocketdyne Propulsion & Power	Aerojet Propulsion Company
Thrust Chamber					
Fuel	Liquid H_2	Liquid H_2	MMH	RP-1 (kerosene)	50% N_2H_4/50% UDMH
Oxidizer	Liquid O_2	Liquid O_2	N_2O_4	Liquid oxygen	N_2O_4
Thrust chamber thrust (lbf)					
at sea level (lbf)	No sea level firing	190,400	12	164,700	NA
in vacuum (lbf)	24,750	242,500	18	207,000	9850
Thrust chamber mixture ratio	5.88	6.0	2.0	2.35	1.90
Thrust chamber specific impulse					
at sea level (sec)	NA	349.9	200	257	
in vacuum (sec)	462	445.6	290	294	320
Characteristic exhaust velocity, c^* (ft/sec)	7578	5594.8	5180	5540	5606
Thrust chamber propellant flow (lb/sec)	53.2	346.9	0.062	640	30.63
Injector end chamber pressure (psia)	640	—	70	576	125
Nozzle end stagnation pressure (psia)	NA	1917	68	534	
Thrust chamber sea level weight (lbf)	<150	1560	7	730	137
Gimbal mount sea level weight (lbf)	<10	57.3	NA	70	23
Chamber diameter (in.)	—	15.75	1.09	21	11.7
Nozzle throat diameter (in.)	5.2	9.25	0.427	16.2	7.5
Nozzle exit diameter (in.)	88	68.28	3.018	45.8	60
Nozzle exit area ratio	285	54:1	50:1	8:1	65:1

Chamber contraction area ratio	—	2.8?	6:1	1.6?:1	2.34:1
Characteristic chamber length L^* (in.)	—	30.7	18	38.7	30.5
Thrust chamber overall length (in.)	90	14.8	11.0	86.15	18.7
Fuel jacket and manifold volume (ft³)	—	3.5	—	2.5	—
Nozzle extension	Carbon-carbon	None	None	None	None
Cumul. firing duration (sec)	> 360*	*	*	*	>150
Restart capability	Yes	No	Yes	No	Yes
Cooling system	Stainless steel tubes, 1½ passes regenerative cooled	Regenerative (fuel) cooled, stainless steel tubes	Radiation cooled, niobium	Stainless steel tubes, single pass, regenerative cooling	Ablative layer is partly consumed
Tube diameter/channel width (in.)	NA	0.05 (channel)	NA	0.45	Ablative material: Silica phenolic
Number of tubes	NA	288	0	292	NA
Jacket pressure drop (psi)	253	540	NA	100	NA
Injector type	Concentric annular swirl and resonator cavities	Hollow post/sleeve elements; baffle and acoustic cavities	Drilled holes	Flat plate, drilled rings and baffle	Outer row: shower head; triplets & doublets, with dual tuned resonator
Injector pressure drop—oxidizer (psi)	100	704	50	156	40
Injector pressure drop—fuel (psi)	54	154	50	140	40
Number of oxidizer injector orifices	216	452 (coaxial)	1	1145	1050
Number of fuel injector orifices	216	452 (coaxial)	1	1530	1230

Engine Characteristics

Feed system	Turbopump with expander cycle	Turbopump	Pressure fed tanks	Turbopump with gas generator	Pressure fed tanks
Engine thrust (at sea level) (lb)	NA	190,400	12	165,000	NA
Engine thrust (altitude) (lb)	24,750	242,500	18	207,700	9850
Engine specific impulse at sea level	NA	349.9	200	253	320
Engine specific impulse at altitude	462	445.6	290	288	320
Engine mixture ratio (oxidizer/fuel)	5.88	6.0	2.0	2.27	1.90

*limited only by available propellant.

Sources: Companies listed above and NASA.

The thrust for the thrust chamber is usually slightly less than the thrust of the engine for open cycles, such as a gas generator cycle; the thrust chamber specific impulse is actually slightly higher (about 1%) than the engine specific impulse. For closed cycles such as the staged combustion cycle, the F and I_s values of the engine and thrust chamber are the same.

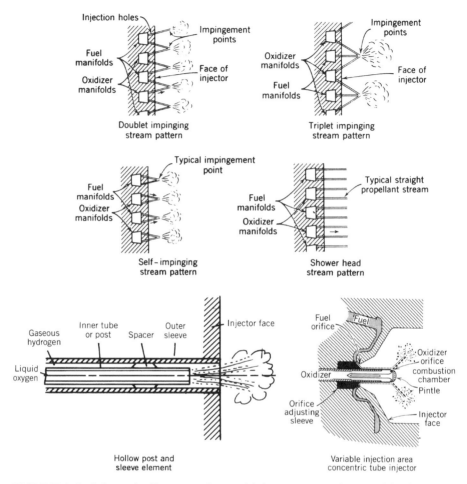

FIGURE 8–3. Schematic diagrams of several injector types. The movable sleeve type variable thrust injector is adapted from Ref. 8–1.

lant flow over a wide range without excessive reduction in injector pressure drop. This type of variable area concentric tube injector was used on the descent engine of the Lunar Excursion Module and throttled over a 10:1 range of flow with only a very small change in mixture ratio.

The *coaxial hollow post injector* has been used for liquid oxygen and gaseous hydrogen injectors by most domestic and foreign rocket designers. It is shown in the lower left of Fig. 8–3. It works well when the liquid hydrogen has absorbed heat from cooling jackets and has been gasified. This gasified hydrogen flows at high speed (typically 330 m/sec or 1000 ft/sec); the liquid oxygen flows far more slowly (usually at less than 33 m/sec or 100 ft/sec) and the differential velocity causes a shear action, which helps to break up the oxygen stream into small droplets. The injector has a multiplicity of these coaxial posts on its face. This

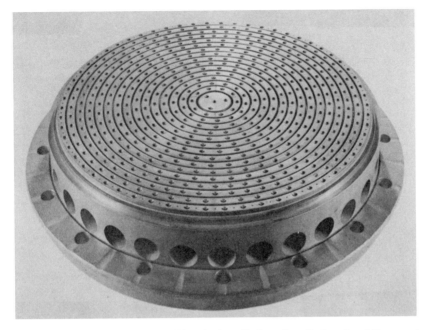

FIGURE 8–4. Injector with 90° self-impinging (fuel-against-fuel and oxidizer-against-oxidizer)-type countersunk doublet injection pattern. Large holes are inlets to fuel manifolds. Pre-drilled rings are brazed alternately over an annular fuel manifold or groove and a similar adjacent oxidizer manifold or groove. A section through a similar but larger injector is shown in Fig. 8–1.

type of injector is not used with liquid storable bipropellants, in part because the pressure drop to achieve high velocity would become too high.

The SSME injector shown in Fig. 9–6 uses 600 of these concentric sleeve injection elements; 75 of them have been lengthened beyond the injector face to form cooled baffles, which reduce the incidence of combustion instability.

The original method of making injection holes was to carefully drill them and round out or chamfer their inlets. This is still being done today. It is difficult to align these holes accurately (for good impingement) and to avoid burrs and surface irregularities. One method that avoids these problems and allows a large number of small accurate injecton orifices is to use multiple etched, very thin plates (often called platelets) that are then stacked and diffusion bonded together to form a monolithic structure as shown in Fig. 8–5. The photo-etched pattern on each of the individual plates or metal sheets then provides not only for many small injection orifices at the injector face, but also for internal distribution or flow passages in the injector and sometimes also for a fine-mesh filter inside the injector body. The platelets can be stacked parallel to or normal to the injector face. The finished injector has been called the platelet injector and has been patented by the Aerojet Propulsion Company.

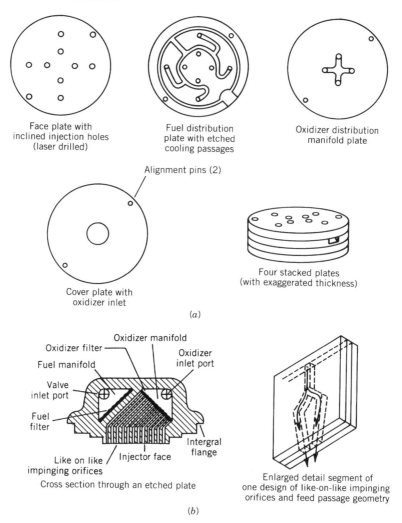

FIGURE 8–5. Simplified diagrams of two types of injector using a bonded platelet construction technique: (a) injector for low thrust with four impinging unlike doublet liquid streams; the individual plates are parallel to the injector face; (b) Like-on-like impinging stream injector with 144 orifices; plates are perpendicular to the injector face. (Courtesy of Aerojet Propulsion Company.)

Injector Flow Characteristics

The differences of the various injector configurations shown in Fig. 8–3 reflect themselves in different hydraulic flow–pressure relationships, different starting characteristics, atomization, resistance to self-induced vibrations, and combustion efficiency.

The *hydraulic injector characteristics* can be evaluated accurately and can be designed for orifices with the desired injection pressures, injection velocities, flows, and mixture ratio. For a given thrust F and a given effective exhaust velocity c, the total propellant mass flow $\dot{m}$ is given by $\dot{m} = F/c$ from Eq. 2–6. The relations between the mixture ratio, the oxidizer, and the fuel flow rates are givenby Eqs. 6–1 to 6–4. For the *flow* of an incompressible fluid through hydraulic orifices,

$$Q = C_d A \sqrt{2\Delta p / \rho} \tag{8–1}$$

$$\dot{m} = Q\rho = C_d A \sqrt{2\rho\Delta p} \tag{8–2}$$

where Q is the volume flow rate, C_d the dimensionless discharge coefficient, ρ the propellant mass density, A the cross-sectional area of the orifice, and Δp the pressure drop. These relationships are general and can be applied to any one section of the propellant feed system, to the injector, or to the overall liquid flow system. A typical variation of injection orifice flow and pressure drop is shown in Fig. 8–6. If the hole has a rounded entrance (top left sketch), it gives the lowest pressure drop or the highest flow. Small differences in chamfers, hole entry radius, or burrs at the edge of a hole can cause significant variations in the discharge coefficient and the jet flow patterns, and these in turn can alter

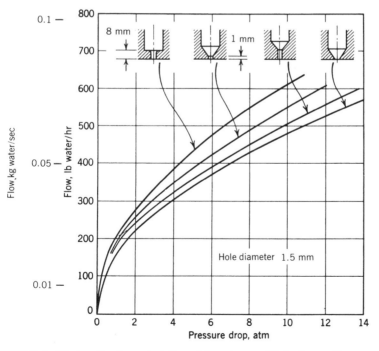

FIGURE 8–6. Hydraulic characteristics of four types of injection orifice.

the quality and distribution of the atomized small droplets, the local mixture ratio, and the local heat transfer rates. An improperly manufactured hole can cause local chamber or injector burnout.

For any given pressure drops the injection orifices determine the *mixture ratio* and the *propellant flows* of the rocket unit. From Eqs. 6–1 and 8–2 the mixture ratio is

$$r = \dot{m}_o/\dot{m}_f = [(C_d)_o/(C_d)_f](A_o/A_f)\sqrt{(\rho_o/\rho_f)(\Delta p_o/\Delta p_f)} \qquad (8\text{–}3)$$

The quantities in the preceding equations have to be chosen so that the correct design mixture ratio is attained, even if the total flow is varied slightly. Orifices whose discharge coefficients are constant over a large range of Reynolds numbers and whose ratio $(C_d)_o/(C_d)_f$ remains invariant should be selected. For a given injector it is usually difficult to maintain the mixture ratio constant at low flows or thrusts, such as in starting.

The quality of the injector is checked by performing cold tests with inert simulant liquids instead of reactive propellant liquids. Often water is used to confirm pressure drops through the fuel or oxidizer side at different flows and this allows determination of the pressure drops with propellants and the discharge coefficients. Nonmixable inert liquids are used with a special apparatus to determine the local cold flow mixture ratio distribution over the chamber cross section. The simulant liquid should be of approximately the same density and viscosity as the actual propellant. All new injectors are hot fired and tested with actual propellants.

The actual mixture ratio can be estimated from cold flow test data, the measured hole areas, and discharge coefficients by correcting by the square root of the density ratio of the simulant liquid and the propellant. When water at the same pressure is fed alternately into both the fuel and the oxidizer sides, $\Delta p_f = \Delta p_o$ and $\rho_f = \rho_o$ and the water mixture ratio will be

$$r = [(C_d)_o/(C_d)_f]A_o/A_f \qquad (8\text{–}4)$$

Therefore, the mixture ratio measured in water tests can be converted into the actual propellant mixture ratio by multiplying it by the square root of the density ratio of the propellant combination and the square root of the pressure drop ratio. The mechanism of propellant atomization with simultaneous vaporization, partial combustion, and mixing is difficult to analyze, and performance of injectors has to be evaluated by experiment within a burning rocket thrust chamber. The *injection velocity* is given by

$$v = Q/A = C_d\sqrt{2\Delta p/\rho} \qquad (8\text{–}5)$$

Values of discharge coefficients for various types of injection orifices are shown in Table 8–2. The velocity is a maximum for a given injection pressure drop

TABLE 8–2. Injector Discharge Coefficients

Orifice Type	Diagram	Diameter (mm)	Discharge Coefficient
Sharp-edged orifice		Above 2.5 Below 2.5	0.61 0.65 approx.
Short-tube with rounded entrance $L/D > 3.0$		1.00 1.57 1.00 (with $L/D \sim 1.0$)	0.88 0.90 0.70
Short tube with conical entrance		0.50 1.00 1.57 2.54 3.18	0.7 0.82 0.76 0.84–0.80 0.84–0.78
Short tube with spiral effect		1.0–6.4	0.2–0.55
Sharp-edged cone		1.00 1.57	0.70–0.69 0.72

when the discharge coefficient equals 1. Smooth and well-rounded entrances to the injection holes and clean bores give high values of the discharge coefficient and this hole entry design is the most common.

When an oxidizer and a fuel jet impinge, the *resultant momentum* can be calculated from the following relation, based on the principle of conservation of momentum. Figure 8–7 illustrates a pair of impinging jets and defines γ_0 as the angle between the chamber axis and the oxidizer stream, γ_f as the angle between the chamber axis and the fuel stream, and δ as the angle between the chamber axis and the average resultant stream. If the total momentum of the two jets before and after impingement is equal,

$$\tan \delta = \frac{\dot{m}_o v_o \sin \gamma_o - \dot{m}_f v_f \sin \gamma_f}{\dot{m}_o v_o \cos \gamma_o + \dot{m}_f v_f \cos \gamma_f} \tag{8–6}$$

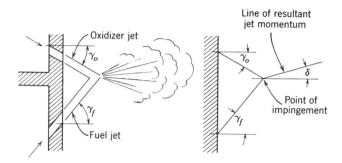

FIGURE 8–7. Angular relation of doublet impinging-stream injection pattern.

Good performance is often obtained when the resultant momentum of impinging streams is approximately axial. If the resultant momentum is along the chamber axis, $\delta = 0$, $\tan \delta = 0$, and the angular relation for an axially directed jet momentum is given by

$$\dot{m}_o v_o \sin \gamma_o = \dot{m}_f v_f \sin \gamma_f \tag{8–7}$$

From these equations the relation between γ_f, γ_0, and δ can be determined. A sample injector analysis is shown in Section 8.6.

Factors Influencing Injector Behavior

A complete theory relating injector design parameters to rocket performance and combustion phenomena has not yet been devised, and therefore the approach to the design and development of liquid propellant rocket injectors has been largely empirical. Yet the available data indicate several important factors that affect the performance and operating characteristics of injectors; some of these are briefly enumerated here.

Propellant Combination. The particular combination of fuel and oxidizer affects such characteristics as the relative chemical reactivity, the ease and speed of vaporization, the ignition temperature, the diffusion of hot gases, the volatility, or the surface tension. Hypergolic (self-igniting) propellants generally require injector designs somewhat different from those required by propellants that must be ignited. Injector designs that perform well with one combination generally do not work too well with a different propellant combination.

Injection Orifice Pattern and Orifice Size. With individual holes in the injector plate, there appears to be an optimum performance and/or heat transfer condition for each of the following parameters: orifice size, angle of impingement, angle of resultant momentum, distance of the impingement locus from the injector face, number of injection orifices per unit of injector face surface, flow per unit of injection orifice, and distribution of orifices over the injector face. These parameters are largely determined experimentally or from similar earlier successful injectors.

Transient Conditions. Starting and stopping may require special provisions (temporary plugging of holes, accurate valve timing, insertion of paper cups over holes to prevent entry of one propellant into the manifold of the other propellant, or check valves) to permit satisfactory transient operation.

Hydraulic Characteristics. The orifice type and the pressure drop across the injection orifice determine the injection velocity. A low pressure drop is desirable to minimize the weight of the feed system or the pumping power and improve the overall rocket efficiency, yet high pressure drops are used often to increase the rocket's resistance to combustion instability and enhance atomization of the liquids.

Heat Transfer. Injectors influence the performance and the heat transferred in rocket thrust chambers. Low heat transfer rates have been obtained when the injection pattern resulted in an intentionally rich mixture near the chamber walls. In general, the higher performance injectors have a higher heat-transfer rate to the walls of the combustion chamber, the nozzle, and the injector face.

Structural Design. The injector is highly loaded by pressure forces from the combustion chamber and the propellant manifolds. During transition (starting or stopping) these pressure conditions can cause stresses which sometimes exceed the steady-state operating conditions. The faces of many modern injectors are flat and must be reinforced by suitable structures which nevertheless provide no obstructions to the hydraulic manifold passages; the structure must also be sufficiently flexible to allow thermal deformations caused by heating the injector face with hot combustion gases or cooling by cryogenic propellants. The injector design must also provide for positive seals between fuel and oxidizer manifolds (an internal leak can cause manifold explosions or internal fires) and a sealed attachment of the injector to the chamber. In large, gimbal-mounted thrust chambers the injector also often carries the main thrust load, and a gimbal mount is often directly attached to the injector, a shown in Figs. 6–1 and 8–1.

Combustion Stability. The injection hole pattern, impingement pattern, hole distribution, and pressure drop have a strong influence on combustion stability; some types are much more resistant to pressure disturbances. As explained in Section 9–3, the resistance to vibration is determined experimentally, and often special antivibration devices, such as baffles or resonance cavities, are designed directly into the injector.

8.2. COMBUSTION CHAMBER AND NOZZLE

The *combustion chamber* is that part of a thrust chamber where the combustion or burning of the propellant takes place. The combustion temperature is much higher than the melting points of most chamber wall materials. Therefore it is necessary either to cool these walls (as described in a later section of this chapter) or to stop rocket operation before the critical wall areas become too hot. If the heat transfer is too high and thus the wall temperatures become locally too high, the thrust chamber will fail. Heat transfer to thrust chambers will be described later in this chapter. Section 8.6 gives a sample analysis of a thrust chamber and Ref. 8–2 describes the design and development of one.

Volume and Shape

Spherical chambers give the least internal surface area and mass per unit chamber volume; they are expensive to build and several have been tried. Today we prefer a cylindrical chamber (or slightly tapered cone frustum) with a flat injector and a converging–diverging nozzle. The chamber volume is defined as the volume up to the nozzle throat section and it includes the cylindrical chamber and the converging cone frustum of the nozzle. Neglecting the effect of the corner radii, the chamber volume V_c is

$$V_c = A_1 L_1 + A_1 L_c (1 + \sqrt{A_t/A_1} + A_t/A_1) \tag{8–8}$$

Here L is the cylinder length, A_1/A_t is the chamber contraction ratio, and L_c is the length of the conical frustum. The approximate surfaces exposed to heat transfer from hot gas comprise the injector face, the inner surface of the cylinder chamber, and the inner surface of the converging cone frustrum. The *volume and shape* are selected after evaluating these parameters:

1. The volume has to be large enough for adequate *mixing, evaporation*, and *complete combustion* of propellants. Chamber volumes vary for different propellants with the time delay necessary to vaporize and activate the propellants and with the speed of reaction of the propellant combination. When the chamber volume is too small, combustion is incomplete and the performance is poor. With higher chamber pressures or with highly reactive propellants, and with injectors that give improved mixing, a smaller chamber volume is usually permissible.

2. The chamber diameter and volume can influence the *cooling requirements*. If the chamber volume and the chamber diameter are large, the

heat transfer rates to the walls will be reduced, the area exposed to heat will be large, and the walls are somewhat thicker. Conversely, if the volume and cross section are small, the inner wall surface area and the inert mass will be smaller, but the chamber gas velocities and the heat transfer rates will be increased. There is an optimum chamber volume and diameter where the total heat absorbed by the walls will be a minimum. This is important when the available cooling capacity of the coolant is limited (for example oxygen–hydrocarbon at high mixture ratios) or if the maximum permissive coolant temperature has to be limited (for safety reasons with hydrazine cooling). The total heat transfer can also be further reduced by going to a rich mixture ratio or by adding film cooling (discussed below).

3. All inert components should have *minimum mass*. The thrust chamber mass is a function of the chamber dimensions, chamber pressure, and nozzle area ratio, and the method of cooling.

4. Manufacturing considerations favor a simple chamber geometry, such as a cylinder with a double cone bow-tie-shaped nozzle, low cost materials, and simple fabrication processes.

5. In some applications the *length* of the chamber and the nozzle relate directly to the overall length of the vehicle. A large-diameter but short chamber can allow a somewhat shorter vehicle with a lower structural inert vehicle mass.

6. The *gas pressure drop* for accelerating the combustion products within the chamber should be a minimum; any pressure reduction at the nozzle inlet reduces the exhaust velocity and the performance of the vehicle. These losses become appreciable when the chamber area is less than three times the throat area.

7. For the same thrust the combustion volume and the nozzle throat area become smaller as the operating chamber pressure is increased. This means that the chamber length and the nozzle length (for the same area ratio) also decrease with increasing chamber pressure. The performance also goes up with chamber pressure.

The preceding chamber considerations conflict with each other. It is, for instance, impossible to have a large chamber that gives complete combustion but has a low mass. Depending on the application, a compromise solution that will satisfy the majority of these considerations is therefore usually selected and verified by experiment.

The *characteristic chamber length* is defined as the length that a chamber of the same volume would have if it were a straight tube and had no converging nozzle section.

$$L^* = V_c/A_t \qquad (8\text{--}9)$$

where L^* (pronounced el star) is the characteristic chamber length, A_t is the nozzle throat area, and V_c is the chamber volume. The chamber includes all the volume up to the throat area. Typical values for L^* are between 0.8 and 3.0 meters (2.6 to 10 ft) for several bipropellants and higher for some monopropellants. Because this parameter does not consider any variables except the throat area, it is useful only for a particular propellant combination and a narrow range of mixture ratio and chamber pressure. The parameter L^* was used about 40 years ago, but today the chamber volume and shape are chosen by using data from successful thrust chambers of prior similar designs and identical propellants.

The *stay time* t_s of the propellant gases is the average value of the time spent by each molecule or atom within the chamber volume. It is defined by

$$t_s = V_c/(\dot{m}V_1) \tag{8–10}$$

where $\dot{m}$ is the propellant mass flow, V_1 is the average specific volume or volume per unit mass of propellant gases in the chamber, and V_c is the chamber volume. The minimum stay time at which a good performance is attained defines the chamber volume that gives essentially complete combustion. The stay time varies for different propellants and has to be experimentally determined. It includes the time necessary for vaporization, activation, and complete burning of the propellant. Stay times have values of 0.001 to 0.040 sec for different types of thrust chambers and propellants.

The *nozzle* dimensions and configuration can be determined from the analyses presented in Chapter 3. The converging section of the supersonic nozzle experiences a much higher internal gas pressure than the diverging section and therefore the design of the converging wall is similar to the design of the cylindrical chamber wall. Most thrust chambers use a shortened bell shape for the diverging nozzle section. Nozzles with area ratios up to 400 have been developed.

In Chapter 3 it was stated that very large nozzle exit area ratios allow a small but significant improvement in specific impulse, particularly at very high altitudes; however, the extra length and extra vehicle mass necessary to house a large nozzle make this unattractive. This disadvantage can be mitigated by a multipiece nozzle, that is stored in annular pieces around the engine during the ascent of the launch vehicle and automatically assembled in space after launch vehicle separation and before firing. This concept, known as extendible nozzle cone, has been successfully employed in solid propellant rocket motors for space applications for about 20 years. The first flight with an extendible nozzle on a liquid propellant engine was performed in 1998 with a modified version of a Pratt & Whitney upper stage engine. Its flight performance is listed in Table 8–1. The engine is shown later in Fig. 8–19 and its carbon–carbon extendible nozzle cone is described in the section on Materials and Fabrication.

Heat Transfer Distribution

Heat is transmitted to all internal hardware surfaces exposed to hot gases, namely the injector face, the chamber and nozzle walls. The *heat transfer rate* or *heat transfer intensity*, that is, local wall temperatures and heat transfer per unit wall area, varies within the rocket. A typical heat transfer rate distribution is shown in Fig. 8–8. Only $\frac{1}{2}$ to 5% of the total energy generated in the gas is transferred as heat to the chamber walls. For a typical rocket of 44,820 N or 10,000 lbf thrust the heat rejection rate to the wall may be between 0.75 and 3.5 MW, depending on the exact conditions and design. See Section 8.3.

The amount of heat transferred by *conduction* from the chamber gas to the walls in a rocket thrust chamber is negligible. By far the largest part of the heat is transferred by means of *convection*. A part (usually 5 to 35%) of the transferred heat is attributable to *radiation*.

For constant chamber pressure, the chamber wall surface increases less rapidly than the volume as the thrust level is raised. Thus the cooling of chambers is generally easier in large thrust sizes, and the capacity of the wall material or the coolant to absorb all the heat rejected by the hot gas is generally more critical in smaller sizes, because of the volume–surface relationship.

Higher chamber pressure leads to higher vehicle performance (higher I_s), but also to higher engine inert mass. However, the resulting *increase of heat transfer with chamber pressure* often imposes design or material limits on the maximum practical chamber pressure for both liquid and solid propellant rockets.

The heat transfer intensity in chemical rocket propulsion can vary from less than 50 W/cm^2 or 0.3 Btu/in.^{2}-sec to over 16 kW/cm^2 or 100 Btu/in.^{2}-sec. The

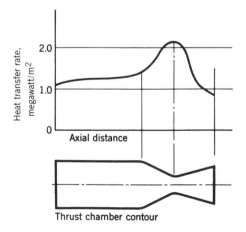

FIGURE 8–8. Typical axial heat transfer rate distribution for liquid propellant thrust chambers and solid propellant rocket motors. The peak is always at the nozzle throat and the lowest value is usually near the nozzle exit.

high values are for the nozzle throat region of large bipropellant thrust chambers and high-pressure solid rocket motors. The lower values are for gas generators, nozzle exit sections, or small thrust chambers at low chamber pressures.

Cooling of Thrust Chambers

The primary objective of cooling is to prevent the chamber and nozzle walls from becoming too hot, so they will no longer be able to withstand the imposed loads or stresses, thus causing the chamber or nozzle to fail. Most wall materials lose strength and become weaker as temperature is increased. These loads and stresses are discussed in the next section. With further heating, the walls would ultimately fail or even melt. Cooling thus reduces the wall temperatures to an acceptable value.

Basically, there are two cooling methods in common use today. The first is the *steady state method*. The heat transfer rate and the temperatures of the chambers reach *thermal equilibrium*. This includes *regenerative cooling* and *radiation cooling*. The duration is limited only by the available supply of propellant.

Regenerative cooling is done by building a cooling jacket around the thrust chamber and circulating one of the liquid propellants (usually the fuel) through it before it is fed to the injector. This cooling technique is used primarily with bipropellant chambers of medium to large thrust. It has been effective in applications with high chamber pressure and high heat transfer rates. Also, most injectors use regenerative cooling.

In *radiation cooling* the chamber and/or nozzle have only a single wall made of high temperature material. When it reaches thermal equilibrium, this wall usually glows red or white hot and radiates heat away to the surroundings or to empty space. Radiation cooling is used with monopropellant thrust chambers, bipropellant and monopropellant gas generators, and for diverging nozzle exhaust sections beyond an area ratio of about 6 to 10 (see Fig. 8–2). A few small bipropellant thrusters are also radiation cooled. This cooling scheme has worked well with lower chamber pressures (less than 250 psi) and moderate heat transfer rates.

The second cooling method relies on *transient heat transfer* or *unsteady heat transfer*. It is also called *heat sink cooling*. The thrust chamber does not reach a thermal equilibrium, and temperatures continue to increase with operating duration. The heat absorbing capacity of the hardware determines its maximum duration. The rocket combustion operation has to be stopped just before any of the exposed walls reaches a critical temperature at which it could fail. This method has mostly been used with low chamber pressures and low heat transfer rates. Heat sink cooling of thrust chambers can be done by absorbing heat in an inner liner made of an ablative material, such as fiber-reinforced plastics. Ablative materials are used extensively in solid propellant rocket motors and will be discussed further in Chapters 11 and 14. The analysis of both of these cooling methods is given in the next section of this chapter.

Film cooling and *special insulation* are supplementary techniques that are used occasionally with both methods to locally augment their cooling capability. All these cooling methods will be described further in this chapter.

Cooling also helps to reduce the oxidation of the wall material and the rate at which walls would be eaten away. The rates of chemical oxidizing reactions between the hot gas and the wall material can increase dramatically with wall temperature. This oxidation problem can be minimized not only by limiting the wall temperature, but also by burning the liquid propellants at a mixture ratio where the percentage of aggressive gases in the hot gas (such as oxygen or hydroxyl) is very small, and by coating certain wall materials with an oxidation-resistant coating; for example iridium has been coated on the inside of rhenium walls.

Cooling with Steady-State Heat Transfer. *Cooled thrust chambers* have provisions for cooling some or all metal parts coming into contact with hot gases, such as chamber walls, nozzle walls, and injector faces. Internal cooling passages, cooling jackets, or cooling coils permit the circulation of a *coolant.* Jackets can consist of separate inner and outer walls or of an assembly of contoured, adjacent tubes (see Figs. 8–1 and 8–9). The inner wall confines the gases, and the spaces between the walls serves as the coolant passage. The *nozzle throat region* is usually the location that has the highest heat-transfer intensity and is therefore the most difficult to cool. For this reason the cooling jacket is often designed so that the coolant velocity is highest at the critical regions by restricting the coolant passage cross section,

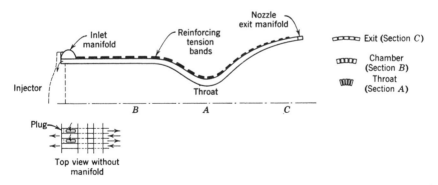

FIGURE 8–9. Diagram of a tubular cooling jacket. The tubes are bent to the chamber and nozzle contours; they are formed hydraulically to give a variable cross section to permit the same number of tubes at the throat and exit diameters. Coolant enters through the inlet manifold into every other tube and proceeds axially to the nozzle exit manifold, where it then enters the alternate tubes and returns axially to go directly to the injector.

and so that the fresh cold coolant enters the jacket at or near the nozzle. While the selection of the coolant velocity and its variation along the wall for any given thrust chamber design depends on heat-transfer considerations, the selection of the coolant passage geometry often depends on pressure loss, stresses, and manufacturing considerations. An *axial flow cooling jacket*, or a *tubular wall*, has a low hydraulic friction loss but is practical only for large coolant flows (above approximately 9 kg/sec). For small coolant flows and small thrust units, the design tolerances of the cooling jacket width between the inner and outer walls or the diameters of the tubes, become too small, or the tolerances become prohibitive. Therefore, most small thrust chambers use radiation cooling or ablative materials.

In *regenerative cooling* the heat absorbed by the coolant is not wasted; it augments the initial energy content of the propellant prior to injection, increasing the exhaust velocity slightly (0.1 to 1.5%). This method is called regenerative cooling because of the similarity to steam regenerators. The design of the tubular chamber and nozzle combines the advantages of a thin wall (good for reducing thermal stresses and high wall temperatures) and a cool, lightweight structure. Tubes are formed to special shapes and contours (see Figs. 8–1 and 8–9), usually by hydraulic means, and then brazed, welded, or soldered together (see Ref. 8–3). In order to take the gas pressure loads in hoop tension, they are reinforced on the outside by high-strength bands or wires. While Fig. 8–9 shows alternate tubes for up and down flow, there are chambers where the fuel inlet manifold is downstream of the nozzle throat area and where the coolant flow is up and down in the nozzle exit region, but unidirectionally up in the throat and chamber regions.

Radiation cooling is another steady-state method of cooling. It is simple and is used extensively in the low heat transfer applications listed previously. Further discussion of radiation cooling is given in the Materials and Fabrication subsection. In order for heat to be radiated into space, it is usually necessary for the bare nozzle and chamber to stick out of the vehicle. Figure 8–18 shows a radiation-cooled thrust chamber. Since the white hot glowing *radiation-cooled* chambers and/or nozzles are potent radiators, they may cause undesirable heating of adjacent vehicle or engine components. Therefore, many have insulation (see Fig. 8–15) or simple external radiation shields to minimize these thermal effects; however, in these cases the actual chamber or nozzle wall temperatures are higher than they would be without the insulation or shielding.

Cooling with Transient Heat Transfer. *Thrust chambers with unsteady heat transfer* are basically of two types. One is a *simple metal chamber* (steel, copper, stainless steel, etc.) made with walls sufficiently thick to absorb plenty of heat energy. For short-duration testing of injectors, testing of new propellants, rating combustion stability, and very-short-duration rocket-propelled missiles, such as an antitank weapon, a heavy-walled simple, short-duration steel chamber is often used. The common method of *ablative cooling* or *heat sink cooling* uses a combination of endothermic reactions

(breakdown or distillation of matrix material into smaller compounds and gases), pyrolysis of organic materials, counter-current heat flow and coolant gas mass flow, charring and localized melting. An ablative material usually consists of a series of strong, oriented fibers (such as glass, Kevlar, or carbon fibers) engulfed by a matrix of an organic material (such as plastics, epoxy resins or phenolic resins). As shown in Fig. 14–11, the gases seep out of the matrix and form a protective film cooling layer on the inner wall surfaces. The fibers and the residues of the matrix form a hard char or porous coke-like material that preserves the wall contour shapes.

The orientation, number and type of fiber determine the ability of the composite ablative material to withstand significant stresses in preferred directions. For example, internal pressure produces longitudinal as well as hoop stresses in the thrust chamber walls and thermal stresses produce compression on the inside of the walls and tensile stresses on the outside. We have learned how to place the fibers in two or three directions, which makes them anisotropic. We then speak of 2-D and 3-D fiber orientation.

A set of strong carbon fibers in a matrix of amorphous carbon is a special, but favorite type of material. It is often abbreviated as C–C or carbon–carbon. The carbon materials lose their ability to carry loads at about 3700 K or 6200 F. Because carbon oxidizes readily to form CO or CO_2, its best applications are with fuel-rich propellant mixtures that have little or no free oxygen or hydroxyl in their exhaust. It is used for nozzle throat inserts. Properties for one type of C–C are given in Table 14–5.

Ablative cooling was first used and is still used extensively with solid propellant rocket motors. It has since been successfully applied to liquid propellant thrust chambers, particularly at low chamber pressure, short duration (including several short-duration firings over a long total time period) and also in nozzle extensions for both large and small thrust chambers, where the static gas temperatures are relatively low. It is not usually effective for cooling if the chamber pressures are high, the exhaust gases contain oxidative species, or the firing durations are long.

Repeatedly starting and stopping (also known as *pulsing*) presents a more severe thermal condition for ablative materials than operating for the same cumulative firing time but without interruption. Figure 8–10 shows that for small pulsing rockets, which fire only 4 to 15% of the time, the consumption or pyrolysis of the ablative liner is a maximum. This curve varies and depends on the specific duty cycle of firings, the design, the materials, and the pauses between the firings. The *duty cycle* for a pulsing thruster was defined in Chapter 6 as the average percent of burning or operating time divided by the total elapsed time. Between pulsed firings there is a heat soak back from the hot parts of the thruster to the cooler outer parts (causing them to become softer) and also a heat loss by radiation to the surroundings. At a duty cycle below 3%, there is sufficient time between firings for cooling by radiation. At long burning times (above 50%) the ablative material's hot layers act as insulators and prevent the stress-bearing portions from becoming too hot.

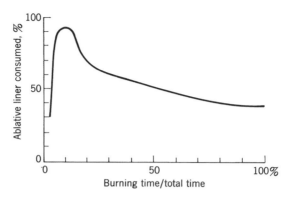

FIGURE 8–10. Relative depth of pyrolysis of ablative material with different duty cycles using many short-duration thrust pulses for a small liquid propellant reaction control thrust chamber of 20 lbf thrust.

Depending on the design, the thrusters with duty cycles between 4 and 25% have the most severe thermal loading.

It is often advantageous to use a different cooling method for the downstream part of the diverging nozzle section, because its heat transfer rate per unit area is usually much lower than in the chamber or the converging nozzle section, particularly with nozzles of large area ratio. There is usually a small saving in inert engine mass, a small increase in performance, and a cost saving, if the chamber and the converging nozzle section and the throat region (up to an area ratio of perhaps 5 to 10) use regenerative cooling and the remainder of the nozzle exit section is radiation cooled (or sometimes ablative cooled). See Fig. 8–2 and Ref. 8–4.

Film Cooling

This is an auxiliary method applied to chambers and/or nozzles, augmenting either a marginal steady-state or a transient cooling method. It can be applied to a complete thrust chamber or just to the nozzle, where heat transfer is the highest. Film cooling is a method of cooling whereby a relatively cool thin fluid film covers and protects exposed wall surfaces from excessive heat transfer. Fig. 8–11 shows film-cooled chambers. The film is introduced by injecting small quantities of fuel or an inet fluid at very low velocity through a large number of orifices along the exposed surfaces in such a manner that a protective relatively cool gas film is formed. A coolant with a high heat of vaporization and a high boiling point is particularly desirable. In liquid propellant rocket engines extra fuel can also be admitted through extra injection holes at the outer layers of the injector; thus a propellant mixture is achieved (at the periphery of the chamber), which has a much lower combustion temperature. This differs from film cooling or transpiration cooling because there does not have to be a chamber

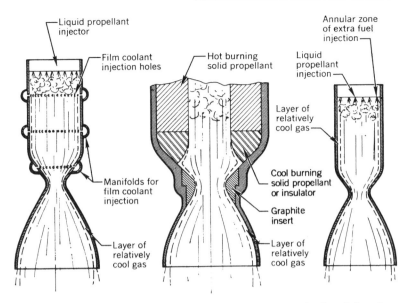

FIGURE 8–11. Simplified diagrams of three different methods of forming a cool boundary layer.

cooling jacket or film-cooling manifolds. In solid propellant rocket engines this can be accomplished by inserting a ring of cool-burning propellant upstream of the nozzle, as shown in Fig. 8–11 or by wall insulation materials, whose ablation and charring will release relatively cool gases into the boundary layer.

Turbine discharge gas (700 to 1100°C) has also been used as a film coolant for uncooled nozzle exit sections of large liquid propellant rocket engines. Of course, the ejection of an annular gas layer at the periphery of the nozzle exit, at a temperature lower than the maximum possible value, causes a decrease in a specific impulse. Therefore, it is desirable to reduce both the thickness of this cooler layer and the mass flow of cooler gas, relative to the total flow, to a practical minimum value.

A special type of film cooling, *sweat cooling* or *transpiration cooling*, uses a porous wall material which admits a coolant through pores uniformly over the surface. This technique has been successfully used to cool injector faces in the upper stage engine (J-2) of the moon launch vehicle and the Space Shuttle main engine (SSME) with hydrogen fuel.

Thermal Insulation

Theoretically, a good thermal insulation layer on the gas side of the chamber wall can be very effective in reducing chamber wall heat transfer and wall temperatures. However, efforts with good insulation materials such as refractory oxides or ceramic carbides have not been successful. They will not withstand differential thermal expansion without cracking or spalling. A sharp edge on the surface (crack or flaked-off piece of insulator) will cause a sudden rise in the stagnation temperature and most likely lead to a local failure. Asbestos is a good insulator and was used several decades ago; because it is a cancer causing agent, it is no longer used. Coating development efforts with rhenium and other materials are continuing.

Insulation or heat shields have been successfully applied on the exterior of radiation-cooled thrust chambers to reduce the heat transfer to adjacent sensitive equipment or structures. With hydrocarbon fuels it is possible to form small carbon particles or soot in the hot gas and that can lead to a carbon deposit on the gas side of the chamber or nozzle walls. If it is a thin, mildly adhesive soot, it can be an insulator, but it is difficult to reproduce such a coating. More likely it forms hard, caked deposits, which can be spalled off in localized flakes and form sharp edges, and then it is undesirable. Most designers have preferred instead to use film cooling or extra high coolant velocities in the cooling jacket with injectors that do not create adhesive soot.

Hydraulic Losses in the Cooling Passage

The cooling coil or cooling jacket should be designed so that the fluid adsorbs all the heat transferred across the inner motor wall, and so that the coolant pressure drop will be small.

A higher pressure drop allows a higher coolant velocity in the cooling jacket, will do a better job of cooling, but requires a heavier feed system, which increases the engine mass slightly and thus also the total inert vehicle mass. For many liquid propellant rockets the coolant velocity in the chamber is approximately 3 to 10 m/sec or 10 to 33 ft/sec and, at the nozzle throat, 6 to 24 m/sec or 20 to 80 ft/sec.

A cooling passage can be considered to be a hydraulic pipe, and the *friction loss* can be calculated accordingly. For a straight pipe,

$$\Delta p / \rho = \tfrac{1}{2} f v^2 (L/D) \tag{8-11}$$

where Δp is the friction pressure loss, ρ the coolant mass density, L the length of coolant passage, D the equivalent diameter, v the average velocity in the cooling passage, and f the friction loss coefficient. In English engineeering units the right side of the equation has to be divided by g_0, the sea-level acceleration of gravity (32.2 ft/sec^2). The friction loss coefficient is a function of Reynolds number and has values betwen 0.02 and 0.05. A *typical pressure loss* of a cooling jacket is between 5 and 25% of the chamber pressure.

A large portion of the pressure drop in a cooling jacket usually occurs in those locations where the flow direction or the flow-passage cross section is changed. Here the sudden expansion or contraction causes a loss, sometimes larger than the velocity head $v^2/2$. This hydraulic situation exists in inlet and outlet chamber manifolds, injector passages, valves, and expansion joints.

The pressure loss in the cooling passages of a thrust chamber can be calculated, but more often it is measured. This pressure loss is usually determined in cold flow tests (with an inert fluid instead of the propellant and without combustion), and then the measured value is corrected for the actual propellant (different physical properties) and the hot firing conditions; a higher temperature will change propellant densities or viscosities and in some designs it changes the cross section of cooling flow passages.

Chamber Wall Loads and Stresses

The analysis of loads and stresses is performed on all propulsion components during their engineering design. Its purpose is to assure the propulsion designer and the flight vehicle user that (1) the components are strong enough to carry all the loads, so that they can fulfill their intended function; (2) potential failures have been identified, together with the possible remedies or redesigns; and (3) their masses have been reduced to a practical minimum. In this section we concentrate on the loads and stresses in the walls of thrust chambers, where high heat fluxes and large thermal stresses complicate the stress analysis. Some of the information on safety factors and stress analysis apply also to all propulsion systems, including solid propellant motors and electric propulsion.

The safety factors (really margins for ignorance) are very small in rocket propulsion when compared to commercial machinery, where these factors can be 2 to 6 times larger. Several *load conditions* are considered for each rocket component; they are:

1. *Maximum expected working load* is the largest likely operating load under all likely operating conditions or transients. Examples are operating at a slightly higher chamber pressure than nominal as set by tolerances in design or fabrication (an example is the tolerance in setting the tank pressure regulator) or the likely transient overpressure from ignition shock.

2. *The design limit load* is typically 1.20 times the maximum expected working load, to provide a margin. If the variation in material composition, material properties, the uncertainties in the method of stress analysis, or predicted loads are significant, a larger factor may be selected.

3. *The damaging load* can be based on the yield load or the ultimate load or the endurance limit load, whichever gives the lowest value. The yield load causes a permanent set or deformation, and it is typically set as 1.10 times the design limit load. The endurance limit may be set by fatigue or creep considerations (such as in pulsing). The ultimate load induces a stress equal to the ultimate strength of the material, where significant elongation and area reduction can lead to failure. Typically it is set as 1.50 times the design limit load.

4. *The proof test load* is applied to engines or their components during development and manufacturing inspection. It is often equal to the design limit load, provided this load condition can be simulated in a laboratory. Thrust chambers and other components, whose high thermal stresses are difficult to simulate, use actual hot firing tests to obtain this proof, often with loads that approach the design limit load (for example, higher than nominal chamber pressure or a mixture ratio that results in hotter gas).

The walls of all thrust chambers are subjected to radial and axial *loads from the chamber pressure, flight accelerations* (axial and transverse), *vibration*, and *thermal stresses*. They also have to withstand a momentary *ignition pressure surge or shock*, often due to excessive propellant accumulation in the chamber. This surge can exceed the nominal chamber pressure. In addition, the chamber walls have to *transmit thrust loads* as well as forces and in some applications also moments, imposed by *thrust vector control* devices described in Chapter 16. Walls also have to survive a "thermal shock", namely, the initial thermal stresses at rapid starting. When walls are cold or at ambient temperature, they experience higher gas heating rates than after the walls have been heated. These loads are different for almost every design, and each unit has to be considered individually in determining the wall strengths.

A heat transfer analysis is usually done only for the most critical wall regions, such as at and near the nozzle throat, at a critical location in the chamber, and sometimes at the nozzle exit. The thermal stresses induced by the temperature difference across the wall are often the most severe stresses and a change in heat transfer or wall temperature distribution will affect the stresses in the wall. Specific failure criteria (wall temperature limit, reaching yield stress, or maximum coolant temperature, etc.) have to be established for these analyses.

The temperature differential introduces a compressive stress on the inside and a tensile stress on the outside of the inner wall; the stress s can be calculated for simple cylindrical chamber walls that are thin in relation to their radius as

$$s = 2\lambda E \ \Delta T/(1 - v) \tag{8-12}$$

where λ is the coefficient of thermal expansion of the wall material, E the modulus of elasticity of the wall material, ΔT the temperature drop across the wall, and v the Poisson ratio of the wall material. Temperature stresses frequently exceed the yield point. The materials experience a change in the yield strength and the modulus of elasticity with temperature. The preceding equation is applicable only to elastic deformations. This yielding of rocket thrust chamber wall materials can be observed by the small and gradual contraction of the throat diameter after each operation (perhaps 0.05% reduction in throat diameter after each firing) and the formation of progressive cracks of the inside wall surface of the chamber and throat after successive runs. These phenomena limit the useful life and number of starts or temperature cycles of a thrust chamber (see Refs. 8–5 and 8–6).

In selecting a working stress for the wall material of a thrust chamber, the variation of strength with temperature and the temperature stresses over the wall thickness have to be considered. The temperature drop across the inner wall is typically between 50 and 550 K, and an average temperature is sometimes used for estimating the material properties. The most severe thermal stresses can occur during the start, when the hot gases cause thermal shock to the hardware. These transient thermal gradients cause severe thermal strain and local yielding.

A picture of a typical stress distribution caused by pressure loads and thermal gradients is shown in Fig. 8–12. Here the inner wall is subjected to a compressive pressure differential caused by a high liquid pressure in the cooling jacket and a relatively large temperature gradient. In a large rocket chamber,

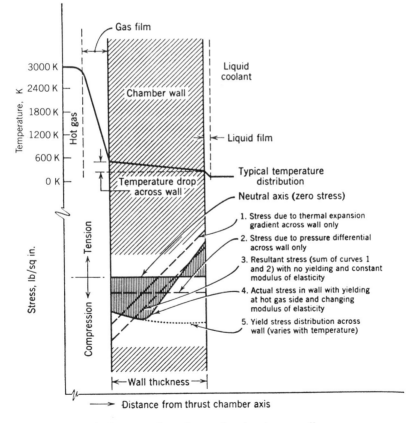

FIGURE 8–12. Typical stresses in a thrust chamber inner wall.

such as is used in the Redstone missile, the wall thickness of the nozzle steel way may be 7 mm and the temperature differential across it may readily exceed several hundred degrees. This temperature gradient causes the hot inner wall surface to expand more than the wall surface on the coolant side and imposes a high compressive thermal stress on the inside surface and a high tensile thermal stress on the coolant side. In these thick walls the stress induced by the pressure load is usually small compared to the thermal stress. The resultant stress distribution in thick inner walls (shown shaded in the sample stress diagram of Fig. 8–12) indicates that the stress in the third of the wall adjacent to the hot gases has exceeded the yield point. Because the modulus of elasticity and the yield point diminish with temperature, the stress distribution is not linear over the yielded portion of the wall. In effect, this inner portion acts as a heat shield for the outer portion which carries the load.

Because of the differential expansion between the hot inner shell and the relatively cold outer shell, it is necessary to provide for axial *expansion joints* to prevent severe temperature stresses. This is particularly critical in larger double-walled thrust chambers. The German V-2 thrust chamber expanded over 5 mm in an axial and 4 mm in a radial direction.

Tubes for tubular wall thrust chambers are subjected to several different stress conditions. Only that portion of an individual cooling tube exposed to hot chamber gases experiences high thermal stresses and deformation as shown in Fig. 8–17. The tubes have to hold the internal coolant pressure, absorb the thermal stresses, and contain the gas pressure in the chamber. The hottest temperature occurs in the center of the outer surface of that portion of each tube exposed to hot gas. The thermal stresses are relatively low, since the temperature gradient is small; copper has a high conductivity and the walls are relatively thin (0.5 to 2 mm). The coolant pressure-induced load on the tubes is relatively high, particularly if the thrust chamber operates at high chamber pressure. The internal coolant pressure tends to separate the tubes. The gas pressure loads in the chamber are usually taken by reinforcing bands which are put over the outside of the tubular jacket assembly (see Fig. 8–1 and 8–9). The joints between the tubes have to be gas tight and this can be accomplished by soldering, welding, or brazing.

When a high-area-ratio nozzle is operated at high ambient pressure, the lower part of the nozzle structure experiences a compression because the pressure in the nozzle near the exit is actually below atmospheric value. Therefore, high-area-ratio nozzles usually have stiffening rings on the outside of the nozzle near the exit to maintain a circular shape and thus prevent buckling, flutter, or thrust misalignment.

Aerospike Thrust Chamber

A separate category comprises thrust chambers using a center body, such as a plug nozzle or aerospike nozzle. They have more surface to cool than ordinary thrust chambers. A circular aerospike thruster is described in Chapter 3 and shown schematically in Fig. 3–12. Here the diameter of the exhaust flow plume increases with altitude. A linear version of a truncated (shortened) aerospike thrust chamber is currently being developed with liquid oxygen and liquid hydrogen as the propellants; see Refs. 8–7 and 8–8. An experimental engine; assembly (XRS-2200) with 20 cells and two hydrogen-cooled, two-dimensional, curved ramps is shown in Figs 8–13 and 8–14. Each individual small (regeneratively cooled) thrust chamber or cell has its own cylindrical combustion chamber, a circular small nozzle throat, but a rectangular nozzle exit of low area ratio. The gas from these 20 rectangular nozzle exits is further expanded (and thus reaches a higher exhaust velocity) along the contour of the spike or ramp. The two fuel-cooled side panels are not shown in these figures. The flat surface at the bottom or base is porous or full of small holes and a low-pressure gas flows through these openings. This causes a back pressure on the flat base surface. This flow can be the exhaust gas from the turbopumps and is typically 1 or 2% of the total flow. The gas region below this base is enclosed by the two gas flows from the ramps and the two side plates and is essentially independent of ambient pressure or altitude. Two of the XRS-2200 engine drive the X-33 wing shaped vehicle aimed at investigating a single stage to orbit concept.

The thrust F of this aerospike thrust chamber consists of (1) the axial component thrusts of each of the little chamber modules, (2) the integral of the pressures acting on the ramps over the axially projected area A_a normal to the axis of the ramps, and (3) the pressure acting over the base area A_{base}.

$$F = [\dot{m}v_2 \cos\theta + (p_2 - p_3)A_2 \cos\theta] + 2\int^{A_a} p\, dA + (p_{\text{base}} - p_3)A_{\text{base}} \qquad (8\text{--}13)$$

Here θ is the angle of the module nozzle axis to the centerline of the spike, $\dot{m}$ is the total propellant flow, v_2 is the exhaust velocity of the module, A_2 is the total exit area of all the modules, p_2 is the exhaust pressure at the exit of the module, and p_3 is the ambient pressure at the nozzle exit level. These expressions are a simplified version of the thrust. Not included, for example, is the negative effect of the slipstream of air around the engine (which causes a low-pressure region) and the friction on the side plates; both actually decrease the thrust slightly. For each application there is an optimum angle θ, an optimum ramp length, an optimum ratio of the projected ramp area to the base area, and an optimum base pressure, which is a function of the base flow.

The local gas pressures on the ramps are influenced by shock wave phenomena and change with altitude. Figure 8–14 shows a typical pressure distribution on a typical ramp surface and the flow patterns for low and high altitude. The hot gas flows coming out of the cell nozzles are turned into a nearly axial direction by multiple expansion waves (shown as dashed lines), which cause a reduction in pressure. At lower altitudes the turning causes compression shock waves (represented as solid lines), which causes local pressures to rise. The compression waves are reflected off the boundary between the hot gas jet

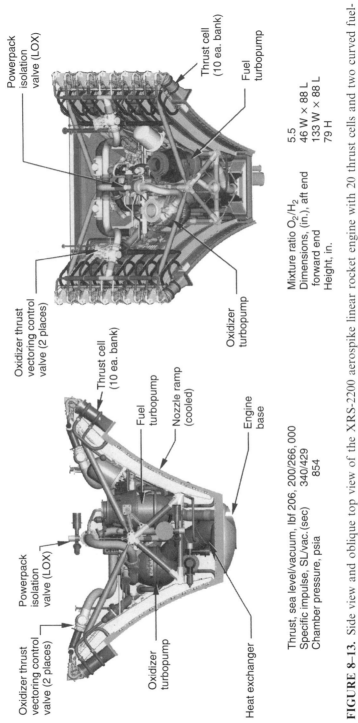

FIGURE 8–13. Side view and oblique top view of the XRS-2200 aerospike linear rocket engine with 20 thrust cells and two curved fuel-cooled ramps. (Courtesy of The Boeing Company, Rocketdyne Propulsion and Power.)

Thrust, sea level/vacuum, lbf 206, 200/266, 000
Specific impulse, SL/vac. (sec)   340/429
Chamber pressure, psia   854

Mixture ratio O$_2$/H$_2$ 5.5
Dimensions, (in.), aft end   46 W × 88 L
forward end   133 W × 88 L
Height, in.   79 H

Powerpack isolation valve (LOX)

Thrust cell (10 ea. bank)

Fuel turbopump

Oxidizer thrust vectoring control valve (2 places)

Oxidizer turbopump

Thrust cell (10 ea. bank)

Fuel turbopump

Nozzle ramp (cooled)

Engine base

Powerpack isolation valve (LOX)

Oxidizer thrust vectoring control valve (2 places)

Oxidizer turbopump

Heat exchanger

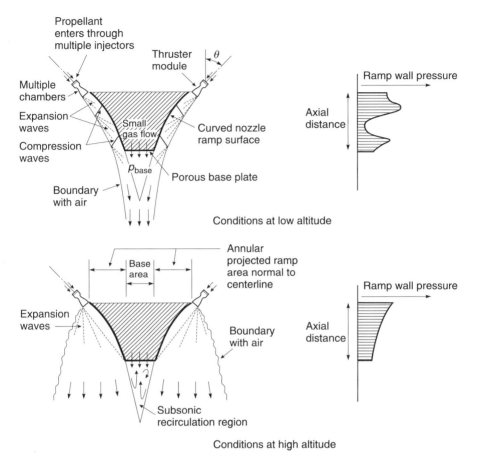

FIGURE 8–14. Pressure profile and flow pattern along the ramp of an aerospike nozzle.

and the ambient air stream, creating further expansion waves. At high altitude there are no compression waves emanating from the ramp and the expanding diverging flow exerts a decreasing pressure on the ramp area and behaves very similarly to the diverging nozzle flow in a bell-shaped conventional nozzle exit section. The contact locations, where the compression waves touch the ramp surface, have higher local heat transfer than other areas on the ramp surface; these locations move downstream as the vehicle gains altitude. The wave patterns and the pressure distribution on the spike or ramp surface can be determined from computerized fluid dynamics programs or a method of characteristics program.

The *advantages* claimed for a linear aerospike engine are these: (1) compared to the axisymmetric rocket engine, it fits well into the trailing edge of a winged or lifting body type vehicle and often has less engine and structural mass; (2) it has altitude compensation and thus operates at optimum nozzle expansion and highest possible performance at every altitude; (3) differential throttling of certain sets of individual thruster modules allows pitch, yaw, and roll control of the vehicle during powered flight, as explained in Chapter 16. There is no gimbal joint, no movement of the nozzle, no actuators, and no actuator power supply or strong structural locations for actuator side loads; (4) the truncated aerospike is short and requires less vehicle volume and structures; and (5) the engine structure can be integrated with the vehicle structure, avoiding a separate vehicle structure at or near the engines. The *disadvantages* include the lack of proven flight experience, proven reliability and performance validation (which are expected to happen soon), and a larger-than-usual surface area subject to high heat transfer.

Low-Thrust Rocket Thrust Chambers or Thrusters

Spacecraft, certain tactical missiles, missile defense vehicles, and upper stages of ballistic missiles often use special, multiple thrusters in their small, liquid propellant rocket engines. They generally have thrust levels between about 0.5 and 10,000 N or 0.1 to 2200 lbf, depending on vehicle size and mission. As mentioned in Chapter 4, they are used for trajectory corrections, attitude control, docking, terminal velocity control in spacecraft or ballistic missiles, divert or side movement, propellant settling, and other functions. Most operate with multiple restarts for relatively short durations during a major part of their duty cycle. As mentioned in Chapter 6, they can be classified into *hot gas thrusters* (high-performance bipropellant with combustion temperatures above 2600 K and I_s of 200 to 325 sec), *warm gas thrusters* such as monopropellant hydrazine (temperatures between 500 and 1600 K and I_s of 18 to 245 sec), and *cold gas thrusters* such as high-pressure stored nitrogen (200 to 320 K) with low specific impules (40 to 120 sec).

A typical small thruster for bipropellant is shown in Fig. 8–15 and for hydrazine monopropellant in Fig. 8–16. For attitude control angular motion these thrust chambers are usually arranged in pairs as explained in Section 4.6. The same control signal activates the valves for both units of such a pair. All these small space rocket systems use a pressurized feed system, some with positive expulsion provisions, as explained in Section 6.3. The vehicle mission and the automatic control system of the vehicle often require irregular and frequent pulses to be applied by pairs of attitude control thrust chambers, which often operate for very short periods (as low as 0.01 to 0.02 sec). This type of frequent and short-duration thrust application is also known as *pulsed thrust operation.* For translation maneuvers a single thruster can be fired (often in a pulsing mode) and the thrust axis usually goes through the center of gravity of the vehicle. The resulting acceleration will depend on the thrust and the location of the thruster on the vehicle; it can be axial or at an angle to the flight velocity vector.

There is a performance degradation with decreasing pulse duration, because propellants are used inefficiently during the buildup of thrust and the decay of

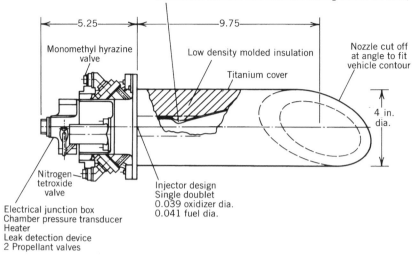

Columbium chamber with disilicide coating (0.003 in. thick)

FIGURE 8–15. This radiation-cooled, insulated vernier thruster is one of several used on the Reaction Control System of the Space Shuttle vehicle for orbit stabilization and orientation, rendezvous or docking maneuvers, station keeping, deorbit, or entry. The nozzle is cut off at an angle to fit the contour of the vehicle. Performance data are given in Table 6–3. Operation can be in a pulse mode (firing durations between 0.08 and 0.32 sec with minimum offtime of 0.08 sec) or a steady-state mode (0.32 to 125 sec). Demonstrated life is 23 hr of operation and more than 300,000 starts. (Courtesy of Kaiser Marquardt Company and NASA.)

thrust, when they operate below full chamber pressure and the nozzle expansion characteristics are not optimum. The specific impulse suffers when the pulse duration becomes very short. In Section 3–5 the actual specific impulse of a rocket operating at a steady state was given at about 92% of theoretical specific impulse. With very short pulses (0.01 sec) this can be lower than 50%, and with pulses of 0.10 sec it can be around 75 to 88%. Also, the reproducibility of the total impulse delivered in a short pulse is not as high after prolonged use. A preheated monopropellant catalyst bed will allow performance improvement in the pressure rise and in short pulses.

One way to minimize the impulse variations in short pulses and to maximize the effective actual specific impulse is to minimize the liquid propellant passage volume between the control valve and the combustion chamber. The propellant flow control valves for pulsing attitude control thrust chambers are therefore often designed as an integral part of the thrust chamber-injector assembly, as shown in Fig. 8–15. Special electrically actuated leakproof, fast-acting valves with response times ranging from 2 to 25 msec for both the opening and closing operation are used. Valves must operate reliably with predictable characteristics for perhaps 40,000 to 80,000 starts. This in turn often requires endurance proof tests of 400,000 to 800,000 cycles.

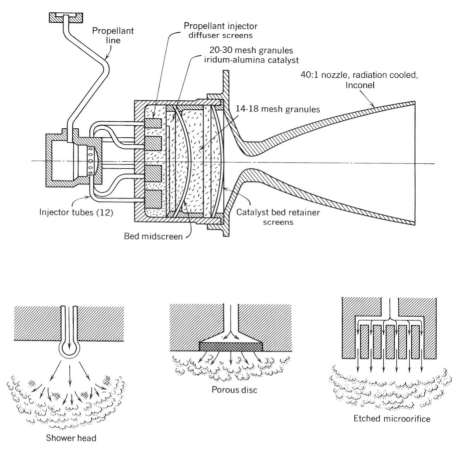

FIGURE 8–16. Typical hydrazine monopropellant small thrust chamber with catalyst bed, showing different methods of injection.

Liquid storable bipropellants such as N_2O_4-monomethylhydrazine are used when high performance is mandatory. Some have used ablative materials for thrust chamber construction, as in the Gemini command module. The Space Shuttle small thrusters use radiation cooling with refractory metals, as shown in Fig. 8–15. A radiation cooled thruster is shown later in Fig. 8–18. Carbon materials made of woven strong carbon fibers in a matrix of carbon have also been used in radiation-cooled bipropellant thrusters.

Hydrazine monopropellant thrusters are used when system simplicity is important and moderate performance is acceptable. They have a nontoxic, clear, clean exhaust plume. Virtually all hydrazine monopropellant attitude control rockets use finely dispersed iridium or cobalt deposited on porous ceramic (aluminum oxide) substrate pellets 1.5 to 3 mm in diameter as a catalyst. Figure 8–16 shows a typical design of the catalyst pellet bed in an attitude control rocket designed for pulse and steady-state operation meeting a specific duty cycle. Each injection hole is covered with a cylindrical screen section which extends into a part of the catalyst bed and distributes the hydrazine propellant. Fig. 8–16 also shows other successful types of hydrazine injector. Several arrangements of catalyst beds are employed; some have spring-loading to keep the pellets firmly packed. Hydrazine monopropellant thrust units range in size from 0.2 to 2500 N of thrust; the scaling procedure is empirical and each size and design requires testing and retesting. The amount of ammonia decomposition, shown in Fig. 7–3, can be controlled by the design of the catalyst bed and its decomposition chamber.

Mechanical, thermal, and chemical problems arise in designing a catalyst bed for igniting hydrazine, the more important of which are catalytic attrition and catalyst poisoning. *Catalytic attrition* or physical loss of catalyst material stems from motion and abrasion of the pellets, with loss of very fine particles. Crushing of pellets can occur because of thermal expansion and momentary overpressure spikes. As explained in Chapter 7, the catalyst activity can also decline because of *poisoning* by trace quantities of contaminants present in commercial hydrazine, such as aniline, monomethylhydrazine, unsymmetrical dimethylhydrazine, sulfur, zinc, sodium, or iron. Some of these contaminants come with the hydrazine and some are added by the tankage, pressurization, and propellant plumbing in the spacecraft. The high-purity grade has less than 0.003% aniline and less than 0.005% carbonaceous material; it does not contaminate catalysts. Catalyst degredation, regardless of cause, produces ignition delays, overpressures, and pressure spikes, decreases the specific impulse, and decreases the impulse duplicate bit per pulse in attitude control engines.

Figure 19–4 shows a combination of chemical monopropellant and electrical propulsion. Electrical post-heating of the reaction gases from catalysis allows an increase of the altitude specific impulse from 240 sec to about 290 or 300 sec. A number of these combination auxiliary thrusters have successfully flown on a variety of satellite applications and are particularly suitable for spacecraft where electrical power is available and extensive short-duration pulsing is needed.

Cold gas thrusters and their performance were mentioned in Section 6.8 and their propellants and specific impulses are listed in Table 7–3. They are simple, low cost, used with pressurized feed systems, used for pulsing operations, and for low thrust and low total impulse. They can use aluminum or plastics for thrusters, valves and piping. The Pegasus air-launched launch vehicle uses them for roll control only. The advantages of cold gas systems are: (a) they are very reliable and have been proven in space flights lasting more than 10 years; (b) the system is simple and relatively inexpensive; (c) the ingredients are nontoxic; (d) no deposit or contamination on sensitive spacecraft surfaces, such as mirrors; (e) they are very safe; and (f) capable of random pulsing. The disadvantages are: (a) engines are relatively heavy with poor propellant mass fractions (0.02 to 0.19); (b) the specific impulses and vehicle velocity increments are low, when compared to mono- or bipropellant systems; and (c) relatively large volumes.

Materials and Fabrication

The choice of the material for the inner chamber wall in the chamber and the throat region, which are the critical locations, is influenced by the hot gases resulting from the propellant combination, the maximum wall temperature, the heat transfer, and the duty cycle. Table 8–3 lists typical materials for several thrust sizes and propellants. For high-performance, high heat transfer, regeneratively cooled thrust chambers a material with high thermal conductivity and a thin wall design will reduce the thermal stresses. Copper is an excellent conductor and it will not really oxidize in fuel-rich non-corrosive gas mixtures, such as are produced by oxygen and hydrogen below a mixture ratio of 6.0. The inner walls are therefore usually made of a copper alloy (with small additions of zirconium, silver, or silicon), which has a conductivity not quite as good as pure (oxygen-free) copper but has improved high temperature strength.

Figure 8–17 shows a cross section of a cooling jacket for a large, regeneratively cooled thrust chamber with formed tapered tubes that are brazed together. The other fabrication technique is to machine nearly rectangular grooves of variable width and depth into the surface of a relatively thick contoured high-conductivity chamber and nozzle wall liner; the grooves are then filled with wax and, by an electrolyte plating technique, a wall of nickel is added to enclose the coolant passages (see Fig. 8–17). The wax is then melted out. As with tubular cooling jackets, a suitable inlet and outlet manifolds are needed to distribute and collect the coolant flow. The figure also shows the locations of maximum wall temperature. For propellant combinations with corrosive or aggressive oxidizers (nitric acid or nitrogen tetroxide) stainless steel is often used as the inner wall material, because copper would chemically react. The depth and width of milled slots (or the area inside formed tubes) vary with the chamber–nozzle profile and its diameters. At the throat region the cooling velocity needs to be at its highest and therefore the cooling passage cross section will be at its lowest.

The failure modes often are bulging on the hot gas side and the opening up of cracks. During hot firing the strain at the hot surface can exceed the local yield point, thus giving it a local permanent compressive deformation. With the cooldown after operation and with each successive firing, some additional yielding and further plastic deformation will occur until cracks form. With successive firings the cracks can become deep enough for a leak and the thrust chamber will then usually fail. The *useful life* of a metal thrust chamber is the maximum number of firings (and sometimes also the cumulative firing duration) without such a failure. The prediction of wall failures is not simple and Refs. 8–5 and 8–6 explain this in more detail. Useful life can also be limited by the storage life of soft components (O-rings, gaskets, valve stem lubricant) and,

TABLE 8–3. Typical Materials used in Several Common Liquid Propellant Thrust Chambers

Application	Propellant	Components	Cooling Method	Typical Materials
Bipropellant TC, cooled, high pressure (Booster or upper stage)	Oxygen–hydrogen	C, N, E	F	Copper alloy
		1	F	Transpiration cooled porous stainless steel face. Structure is stainless steel
		Alternate E	R	Carbon fiber in a carbon matrix, or niobium
		Alternate E	T	Steel shell with ablative inner liner
Same	Oxygen–hycrocarbon or storable propellant*	C, N, E, I	F	Stainless steel with tubes or milled slots
		Alternate E	R	Carbon fiber in a carbon matrix, or niobium
		Alternate E	T	Steel shell with ablative inner liner
Experimental TC (very limited duration—only a few seconds)	All types	C, N, E	U	Low carbon steel
Small bipropellant TC	All types	C, N, E	R	Carbon fiber in carbon matrix, rhenium, niobium
			T	Steel shell with ablative inner linear
		I	F	Stainless steels, titanium
Small monopropellant TC	Hydrazine	C, N, E,	R	Inconel, alloy steels
		I	F	Stainless steel
Cold gas TC	Compressed air, nitrogen	C, N, E, I	U	Aluminum, steel or plastic

*HNO_3 or N_2O_4 oxidizer with N_2H_4, MMH, or UDMH as fuels (see Chapter 7). TC = thrust chamber, C = chamber wall, N = nozzle convering section wall and throat region walls, E = walls at exit region of diverging section of nozzle, I = injector face, F = fuel cooled (regenerative), R = radiation cooled, U = uncooled, T = transient heat transfer or heat sink method (ablative material).

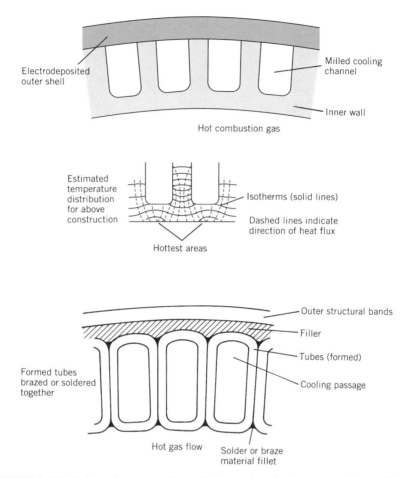

FIGURE 8–17. Enlarged cross section of thrust chamber's regenerative cooling passages for two types of design.

for small thrusters with many pulses, also the fatigue of valve seats. Therefore, there is a maximum limit on the number of firings that such a thrust chamber can withstand safely, and this limits its *useful life* (see Refs. 8–7 and 8–8).

For *radiation cooling*, several different carbon materials have worked in a reducing, fuel-rich rocket atmosphere. At other gas mixtures they can oxidize at the elevated temperatures when they glow red or white. They can be used at wall temperatures up to perhaps 3300 K or 6000 R. Carbon materials and ablative materials are used extensively in solid propellant rocket motors and are discussed further in Chapter 14.

For some small radiation-cooled bipropellant thrusters with storable propellants, such as those of the reaction control thrusters on the Space Shuttle Orbiter, the hot walls are made of niobium coated with disilicide (up to 1120 K or 2050 R). To prevent damage, a fuel-rich mixtures or film cooling is often used. Rhenium walls protected by iridium coatings (oxidation resistant) have come into use more recently and can be used up to about 2300 K or 4100 R (see Ref. 8–9). Other high temperature materials, such as tungsten, molybdenum, alumina, or tantalum, have been tried, but have had problems in manufacture, cracking, hydrogen embrittlement, and excessive oxidation.

A small radiation-cooled monopropellant thruster is shown in Fig. 8–16 and a small radiation cooled bipropellant thruster in Fig. 8–18. This thruster's injection has extra fuel injection holes (not shown in Fig. 8–18) to provide film cooling to keep wall temperatures below their limits. This same thruster will also work with hydrazine as the fuel.

Until recently it has not been possible to make large pieces of carbon–carbon material. This was one of the reasons why large nozzle sections and integral nozzle-exit-cone pieces in solid motors were made from carbon phenolic cloth lay-ups. Progress in manufacturing equipment and technology has now made it possible to build and fly larger c-c pieces. A three-piece extendible c-c nozzle exit cone of 2.3 m (84 in.) diameter and 2.3 to 3 mm thickness has recently flown on an upper-stage engine. This engine with its movable nozzle

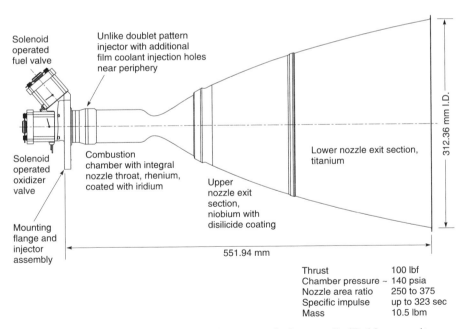

Thrust	100 lbf
Chamber pressure	~ 140 psia
Nozzle area ratio	250 to 375
Specific impulse	up to 323 sec
Mass	10.5 lbm

FIGURE 8–18. Radiation-cooled reaction control thruster R-4D-15 uses nitrogen tetroxide and monomethylhydrazine propellants. The large nozzle area ratio allows good vacuum performance. It has three different nozzle materials, each with a lower allowable temperature (Re 4000°F; Nb 3500°F; Ti 1300°F. (Courtesy of Kaiser-Marquardt Company.)

extension is shown in Fig. 8–19, its parameters are listed in Table 8–1, and its testing is reported in Ref. 8–4.

The material properties have to be evaluated before a material can be selected for a specific thrust chamber application. This evaluation includes physical properties, such as tensile and compressive strengths, yield strength, fracture toughness, modulus of elasticity (for determining deflections under load), thermal conductivity (a high value is best for steady-state heat transfer), coefficient of thermal expansion (some large thrust chambers grow by 3 to 10 mm when they become hot, and that can cause problems with their piping connections or structural supports), specific heat (capacity to absorb thermal energy), reflectivity (for radiation), or density (ablatives require more volume than steel). All these properties change with temperature (they are different when they are hot) and sometimes they change with little changes in composition. The temperature where a material loses perhaps 60 to 75% of its room temperature strength is often selected as the maximum allowable wall temperature, well below its melting point. Since a listing of all the key properties of a single material requires many pages, it is not possible to list them here, but they are usually available from manufacturers and other sources. Other important properties are erosion resistance, little or no chemical reactions with the propellants or the hot gases, reproducible decomposition or vaporization of ablative materials, ease and low cost of fabrication (welding, cutting, forming, etc.), the consistency of the composition (impurities) of different batches of each material (metals, organics, seals, insulators, lubricants, cleaning fluids), and ready availability and low cost of each material.

8.3. HEAT TRANSFER ANALYSIS

In actual rocket development not only is the heat transfer analyzed but the rocket units are almost always tested to assure that heat is transferred satisfactorily under all operating and emergency conditons. Heat transfer calculations are useful to guide the design, testing, and failure investigations. Those rocket combustion devices that are regeneratively cooled or radiation cooled can reach thermal equilibrium and the steady-state heat transfer relationships will apply. Transient heat transfer conditions apply not only during thrust buildup (starting) and shutdown of all rocket propulsion systems, but also with cooling techniques that never reach equilibrirum, such as with ablative materials.

Sophisticated *finite element analysis* (FEA) programs of heat transfer have been available for at least a dozen years and several different FEA computer programs have been used for the analysis of thrust chamber steady-state and transient heat transfer, with different chamber geometries or different materials with temperature variant properties. A detailed description of this powerful analysis is beyond the scope of this book, but can be found in Refs. 8–10 and 8–11. Major rocket propulsion organizations have developed their own ver-

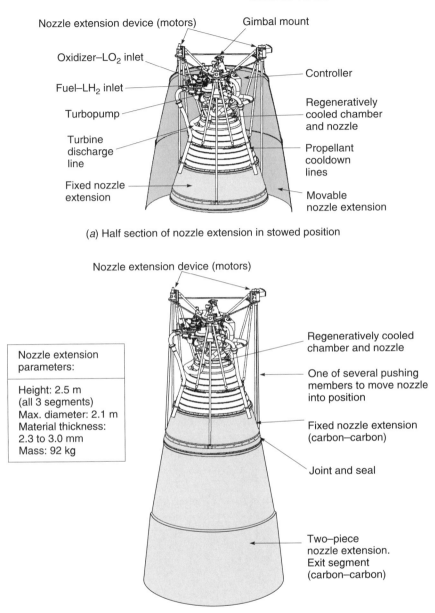

Nozzle extension device (motors)

Gimbal mount

Oxidizer–LO$_2$ inlet

Fuel–LH$_2$ inlet

Turbopump

Turbine discharge line

Fixed nozzle extension

Controller

Regeneratively cooled chamber and nozzle

Propellant cooldown lines

Movable nozzle extension

(a) Half section of nozzle extension in stowed position

Nozzle extension device (motors)

Nozzle extension parameters:

Height: 2.5 m (all 3 segments)
Max. diameter: 2.1 m
Material thickness: 2.3 to 3.0 mm
Mass: 92 kg

Regeneratively cooled chamber and nozzle

One of several pushing members to move nozzle into position

Fixed nozzle extension (carbon–carbon)

Joint and seal

Two–piece nozzle extension. Exit segment (carbon–carbon)

(b) Nozzle extension in deployed position

FIGURE 8–19. The RL-10B-2 rocket engine has an extendible nozzle cone or skirt, which is placed around the engine during the ascent of the Delta III launch vehicle. This extension is lowered into position by electromechanical devices after the launch vehicle has been separated from the upper stage at high altitude and before firing. (Courtesy of Pratt & Whitney, a division of United Technologies.)

sions of suitable computer programs for solving their heat transfer problems. This section gives the basic relationships that are the foundation for FEA programs. They are intended to give some understanding of the phenomena and underlying principles.

General Steady-State Heat Transfer Relations

For heat transfer *conduction* the following general relation applies:

$$\frac{Q}{A} = -\kappa \frac{dT}{dL} = -\kappa \frac{\Delta T}{L} \qquad (8\text{--}14)$$

where Q is the heat transferred per unit across a surface A, dT/dL the temperature gradient with respect to thickness L at the surface A, and κ the thermal conductivity expressed as the amount of heat transferred per unit time through a unit area of surface for $1°$ temperature difference over a unit wall thickness. The negative sign indicates that temperature decreases as thickness increases.

The steady-state heat transfer through the chamber wall of a liqud-cooled rocket chamber can be treated as a series type, steady-state heat transfer problem with a large temperature gradient across the gaseous film on the inside of the chamber wall, a temperature drop across the wall, and, in cases of cooled chambers, a third temperature drop across the film of the moving cooling fluid. It is a combination of convection at the boundaries of the flowing fluids and conduction through the chamber walls. The problem is basically one of heat and mass transport associated with conduction through a wall. It is shown schematically in Fig. 8–20.

The general steady-state heat transfer equations for regeneratively cooled thrust chambers can be expressed as follows:

$$q = h(T_g - T_l) = Q/A \qquad (8\text{--}15)$$

$$= \frac{T_g - T_l}{1/h_g + t_w/\kappa + 1/h_l} \qquad (8\text{--}16)$$

$$= h_g(T_0 - T_{wg}) \qquad (8\text{--}17)$$

$$= (\kappa/t_w)(T_{wg} - T_{wl}) \qquad (8\text{--}18)$$

$$= h_l(T_{wl} - T_l) \qquad (8\text{--}19)$$

where q is heat transferred per unit area per unit time, T_g the absolute chamber gas temperature, T_l the absolute coolant liquid temperature, T_{wl} the absolute wall temperature on the liquid side of the wall, T_{wg} the absolute wall temperature on the gas side of the wall, h the overall film coefficient, h_g the gas film coefficient, h_l the coolant liquid film coefficient, t_w the thickness of the chamber wall, and κ the conductivity of the wall material. The strength and thermal properties of materials are functions of temperature. Any consistent set of units

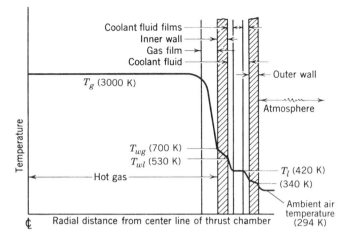

FIGURE 8–20. Temperature gradients in cooled rocket thrust chamber. The listed temperatures are typical.

can be used in these equations. These simple relations assume that the heat flow is radial. The simple quasi-one-dimensional theory also often assumes that the thermal conductivity and the film coefficients are at average values and not functions of temperature or pressure. A two- or three-dimensional finite element model would also need to be used to analyze the heat transfer in the axial directions, which usually occurs in the nozzle throat wall regions; some of the heat from the hot nozzle insert is transferred to wall regions upstream and downstream of the insert.

Because the film coefficients, the gas and liquid coolant temperatures, the wall thickness, and the surface areas usually vary with the axial distance within a combustion chamber (assuming axial heat transfer symmetry), the *total heat transfer per unit time Q* can be found by integrating the local heat transfer over the entire internal surface area of the chamber and the nozzle:

$$Q = \int q \, dA = \pi \int Dq \, dL \tag{8–20}$$

Because both q and D are complicated functions of L, the equation usually has to be solved by dividing the rocket chamber into finite lengths. Assuming that q is given by Eqs. 8–15 to 8–19 and remains constant over the length of each element gives an approximate solution.

The important quantities for controlling the heat transfer across a rocket chamber wall are the fluid film boundaries established by the combustion products on one side of the wall and the coolant flow on the other. The gas film coefficient largely determines the numerical value of the heat transfer rate, and the liquid film largely determines the value of the wall temperatures. The determination of the film coefficients in Eqs. 8–17 and 8–19 is difficult because of the complex geometries, the nonuniform velocity profile, the surface roughness, the boundary layer behavior, and the combustion oscillations.

Conventional heat transfer theory is usually given in terms of several dimensionless parameters (Ref. 8–10):

$$\frac{h_g D}{\kappa} = 0.026 \left(\frac{Dv\rho}{\mu}\right)^{0.8} \left(\frac{\mu c_p}{\kappa}\right)^{0.4} \tag{8–21}$$

where h_g is the film coefficient, D the diameter of the chamber of the nozzle, v the calculated average local gas velocity, κ the conductivity of the gas, μ the absolute gas viscosity, c_p the specific heat of the gas at constant pressure, and ρ the gas density.

In Eq. 8–21 the quantity $h_g D/\kappa$ is known as the Nusselt number, the quantity $Dv\rho/\mu$ as the Reynolds number, and the quantity $c_p\mu/\kappa$ as the Prandtl number Pr. The gas film coefficient h_g can be determined from Eq. 8–21:

$$h_g = 0.026 \frac{(\rho v)^{0.8}}{D^{0.2}} \text{Pr}^{0.4} \kappa/\mu^{0.8} \tag{8–22}$$

where ρv is the local mass velocity, and the constant 0.026 is dimensionless. In order to compensate for some of the boundary layer temperature gradient effects on the various gas properties in rocket combustion, Bartz (Ref. 8–12) has surveyed the agreement between theory and experiment and developed semi-empirical correction factors:

$$h_g = \frac{0.026}{D^{0.2}} \left(\frac{c_p \mu^{0.2}}{\text{Pr}^{0.6}}\right) (\rho v)^{0.8} \left(\frac{\rho_{am}}{\rho'}\right) \left(\frac{\mu_{am}}{\mu_0}\right)^{0.2} \tag{8–23}$$

The subscript 0 refers to properties evaluated at the stagnation or combustion temperature; the subscript am refers to properties at the arithmetic mean temperature of the local free-stream static temperature and the wall temperatures; and ρ' is the free-stream value of the local gas density. Again, the empirical constant 0.026 is dimensionless when compatible dimensions are used for the other terms. The gas velocity v is the local free-stream velocity corresponding to the density ρ'. Since density raised to the 0.8 power is roughly proportional to the pressure and the gas film coefficient is roughly proportional to the heat flux, it follows that the heat transfer rate increases approximately linearly with the chamber pressure. These heat transfer equations have been validated for common propellants, limited chamber pressure ranges, and specific injectors (see Ref. 8–13).

The temperature drop across the inner wall and the maximum temperature are reduced if the wall is thin and is made of material of high thermal conductivity. The wall thickness is determined from strength considerations and thermal stresses, and some designs have as little as 0.025 in. thickness.

Surface roughness can have a large effect on the film coefficients and thus on the heat flux. Measurements have shown that the heat flow can be increased by a factor of up to 2 by surface roughness and to higher factors when designing turbulence-creating obstructions in the cooling channels. Major surface roughness on the gas side will cause the gas locally to come close to stagnation temperature. However, surface roughness on the liquid coolant side of the wall will enhance turbulence and the absorption of heat by the coolant and reduce wall temperatures.

Example 8–1. The effects of varying the film coefficients on the heat transfer and the wall temperatures are to be explored. The following data are given:

Wall thickness	0.445 mm
Wall material	Low-carbon steel
Average conductivity	43.24 W/m^2-K/m
Average gas temperature	3033 K or 2760°C
Average liquid bulk temperature	311.1 K or 37.8°C
Gas-film coefficient	147 W/m^2-°C
Liquid-film coefficient	205,900 W/m^2-°C

Vary h_g (at constant h_l), then vary h_l (at constant h_g), and then determine the changes in heat transfer rate and wall temperatures on the liquid and the gas side of the wall.

SOLUTION. Use Eqs. 8–16 to 8–19 and solve for q, T_{wg}, and T_{wl}. The answers shown in Table 8–4 indicate that variations in the gas-film coefficient have a profound influence on the heat transfer rate but relatively little effect on the wall temperature. The exact opposite is true for variations in the liquid-film coefficient; here, changes in h_l produce little change in q but a fairly substantial change in the wall temperature.

TABLE 8–4. Change in Film Coefficient

Change in Film Coefficient (%)		Change in Heat Transfer (%)	Wall Temperature (K)	
Gas Film	Liquid Film		Gas Side, T_{wg}	Liquid Side, T_{wl}
50	100	50	324.4	321.1
100	100	100	337.2	330.5
200	100	198	362.8	349.4
400	100	389	415.6	386.1
100	50	99	356.1	349.4
100	25	98	393.3	386.7
100	12.5	95	460.0	397.8
100	6.25	91	596.7	590.5

Transient Heat Transfer Analysis

An uncooled (high melting point) metal thrust chamber is the simplest type to analyze, because there is no chemical change. Thermal equilibrium is not reached. The uncooled walls act essentially as a heat sponge and absorb heat from the hot gases. With the air of experimental data to determine some typical coefficients, it is possible in some cases to predict the transient heating of uncooled walls.

Heat is transferred from the hot gases to the wall, and during operation a changing temperature gradient exists across the wall. The heat transferred from the hot wall to the surrounding atmosphere, and by conduction of metal parts to the structure, is negligibly small during this transient heating. Each local point within the wall has its temperature raised as the burning process is extended in time. After the completion of the rocket's operation, the wall temperatures tend to equalize. A typical temperature–time–location history is given in Fig. 8–21. Here the horizontal line at $T = 21°C$ denotes the initial

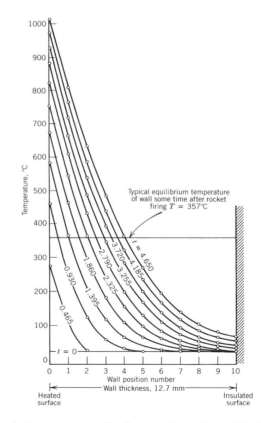

FIGURE 8–21. Typical temperature distributions through a wall of an uncooled metal thrust chamber as a function of heating time.

equilibrium condition of the wall before the rocket operates; the various curves show the temperature profile across the wall at successive time intervals after initiation of combustion. The line at $T = 357°C$ shows an equilibrium temperature of the wall a finite time after cutoff.

The heat transferred across the hot surface of the wall (and distributed within the wall by conduction) must be less than the heat-absorbing capacity of the wall material below the critical temperature. If heat transfer to the outside atmosphere and axially within the metal wall is neglected, this can be expressed in a very simplified form:

$$Q \, \Delta t = -\kappa A (dT/dL) \Delta t = m\bar{c} \Delta T \qquad (8\text{--}24)$$

where Q is the heat per second transferred across area A. Eq. 8–17 shows that Q/A depends on the hot gas temperature, the wall temperature, and the gas film coefficient. The heat conductivity κ depends on the material and its temperature; ΔT denotes the average wall temperature increment; dT/dL the temperature gradient of the heat flow near the hot wall surface in degrees per unit thickness; m the mass of a unit area of wall; $\bar{c}$ the average specific heat of the wall material; and Δt at the time increment. The chamber and nozzle walls can be divided into cylindrical or conical segments, and each wall segment in turn is divided into an arbitrary number of axisymmetric concentric layers, each of a finite thickness. At any given time the heat conducted from any one layer of the wall exceeds the heat conducted into the next outer layer by the amount of heat absorbed in raising the temperature of the particular layer. This iterative approach lends itself readily to two- or three-dimensional computer analysis, resulting in data similar to Fig. 8–21. It is usually sufficient to determine the heat transfer at the critical locations, such as in the nozzle throat region.

A more complex three-dimensional analysis can also be undertaken; here the wall geometry is often more complex than merely cylindrical, heat is conducted also in directions other than normal to the axis, temperature variable properties are used, boundary layer characteristics vary with time and location, and there may be more than one material layer in the wall.

A number of mathematical simulations of transient heat transfer in *ablative materials* have been derived, many with limited success. This approach should include simulation for the pyrolysis, chemical decomposition, char depth, and out-gassing effects on film coefficient, and it requires good material property data. Most simulations require some experimental data.

Steady-State Transfer to Liquids in Cooling Jacket

The term *regenerative cooling* is used for rockets where one of the propellants is circulated through cooling passages around the thrust chamber prior to the injection and burning of this propellant in the chamber. It is really forced convection heat transfer. The term *regenerative* is perhaps not altogether appropriate here, and it bears little relation to the meaning given to it in steam turbine practice. It is intended to convey the fact that the heat absorbed by the coolant propellant is not wasted but augments its initial temperature and raises its energy level before it passes through the injector. This increase in the internal energy of the liquid propellant can be calculated as a correction to the enthalpy of the propellant (see Chapter 5). However, the overall effect on rocket performance is usually very slight. With some propellants the specific impulse can be 1% larger if the propellants are preheated through a temperature differential of 100 to 200°C. In hydrogen-cooled thrust chambers and in small combustion chambers, where the wall-surface-to-chamber volume ratio is relatively large, the temperature rise in the regenerative coolant will be high, and the resulting increase in specific impulse is sometimes more than 1%.

The behavior of the *liquid film* is critical for controlling the wall temperatures in forced convection cooling of rocket devices at high heat fluxes (see Table 8–4 and Refs. 8–14 and 8–15). At least four different types of film appear to exit, as can be interpreted from Fig. 8–22. Here the heat transfer rate per unit of wall surface is shown as a function of the difference between the wall temperature on the liquid side T_{wl} and the bulk temperature of the liquid T_l.

1. The normal *forced convection* region at low heat flux appears to have a liquid boundary layer of predictable characteristics. It is indicated by region A–B in Fig. 8–22. Here the wall temperature is usually below the boiling point of the liquid at the cooling jacket pressure. In steady-state heat transfer analysis the liquid film coefficient can be approximated by the usual equation (see Refs. 8–10 and 8–12):

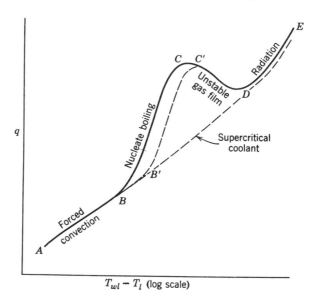

FIGURE 8–22. Regimes in transferring heat from a hot wall to a flowing liquid.

$$h_l = 0.023\bar{c}\frac{\dot{m}}{A}\left(\frac{Dv\rho}{\mu}\right)^{-0.2}\left(\frac{\mu\bar{c}}{\kappa}\right)^{-2/3} \tag{8–25}$$

where $\dot{m}$ is the fluid mass flow rate, $\bar{c}$ its average specific heat, A the cross-sectional flow area, D the equivalent diameter of the coolant passage cross section,* v the fluid velocity, ρ the coolant density, μ its absolute viscosity, and κ its conductivity. Many liquid-cooled rocket devices operate in this regime of heat transfer. Values of the physical properties of several propellants are given in Tables 8–5 and 7–1. In Table 8–5 it can be seen that hydrazine is a good heat absorber, but kerosene is poor.

2. When the wall temperature T_{wl} exceeds the boiling point of the liquid by perhaps 10 to 50 K, small vapor bubbles form at the wall surface. These small, nuclei-like bubbles cause local turbulence, break away from the wall, and collapse in the cooler liquid. This phenomenon is known as *nucleate boiling*. The turbulence induced by the bubbles changes the character of the liquid film and, augmented by the vaporization of some of the propellant, the heat transfer rate is increased without a proportional increase in the temperature drop across the film, as can be seen by the steep slope B–C of the curve in Figure 8–22. If the pressure of the fluid is raised, then the boiling point is also raised and the nucleate

TABLE 8–5. Heat Transfer Characteristics of Several Liquid Propellants

Liquid Coolant	Boiling Characteristics				Nucleate Boiling Characteristics			
	Pressure (MPa)	Boiling Temp. (K)	Critical Temp. (K)	Critical Pressure (MPa)	Temp. (K)	Pressure (MPa)	Velocity (m/sec)	q_{max} (MW/m²)
Hydrazine	0.101	387	652	14.7	322.2	4.13	10	22.1
	0.689	455					20	29.4
	3.45	540			405.6	4.13	10	14.2
	6.89	588					20	21.2
Kerosene	0.101	490	678	2.0	297.2	0.689	1	2.4
	0.689	603					8.5	6.4
	1.38	651			297.2	1.38	1	2.3
	1.38	651					8.5	6.2
Nitrogen tetroxide	0.101	294	431	10.1	288.9	4.13	20	11.4
	0.689	342			322.2			9.3
	4.13	394			366.7			6.2
Unsymmetrical dimethyl hydrazine	0.101	336	522	6.06	300	2.07	10	4.9
	1.01	400					20	7.2
	3.45	489			300	5.52	10	4.7

*The grooves, tubes, or coolant passages in liquid propellant rocket chambers are often of complex cross section. The *equivalent diameter*, needed for fluid-film heat transfer calculations, is usually defined as four times the hydraulic radius of the coolant passage; the hydraulic radius is the cross-sectional flow area divided by the wetted perimeter.

boiling region shifts to the right, to B'–C'. This boiling permits a substantial increase in the heat transfer beyond that predicted by Eq. 8–25. This phenomenon often occurs locally in the nozzle throat area, where the heat flux is high.

3. As the heat transfer is increased further, the rate of bubble formation and the bubble size become so great that the bubbles are unable to escape from the wall rapidly enough. This reaction (shown as C–D in Fig. 8–22) is characterized by an unstable *gas film* and is difficult to obtain reproducibly in tests. When a film consisting largely or completely of gas forms along the hot wall surface, then this film acts as an insulation layer, causing a decrease in heat flux and, usually, a rapid increase in wall temperature, often resulting in a burnout or melting of the wall material. The maximum feasible heat transfer rate (point C) is indicated as q_{max} in Table 8–5 and appears to be a function of the cooling-fluid properties, the presence of dissolved gases, the pressure, and the flow velocity.

4. As the temperature difference across the film is further increased, the wall temperatures reach values in which heat transfer by *radiation* becomes important. Region D–E is not of interest to rocket designers.

Cooling can also be accomplished by a fluid above its critical point with coolants such as hydrogen. In this case there is no nucleate boiling and the heat transfer increases with the temperature difference, as shown by the supercritical (dashed) line in Fig. 8–22. Liquid hydrogen is an excellent coolant, has a high specific heat, and leaves no residues.

Chemical changes in the liquid can seriously influence the heat transfer from hot walls to liquids. Cracking of the fuel, with an attendant formation of insoluble gas, tends to reduce the maximum heat flux and thus promote failure more readily. Hydrocarbon fuel coolants (methane, jet fuel) can break down and form solid, sticky carbon deposits inside the cooling channel, impeding the heat transfer. Other factors influencing steady-state coolant heat transfer are gas radiation to the wall, bends in the coolant passage, improper welds or manufacture, and flow oscillations caused by turbulence or combustion unsteadiness. Some propellants, such as hydrazine, can decompose spontaneously and explode in the cooling passage if they become too hot.

To achieve a good *heat-absorbing capacity of the coolant*, the pressure and the coolant flow velocity are selected so that boiling is permitted locally but the bulk of the coolant does not reach this boiling condition. The total heat rejected by the hot gases to the surface of the hot walls, as given by Eq. 8–15 must be less than that permitted by the temperature rise in the coolant, namely

$$qA = Q = \dot{m}\bar{c}(T_1 - T_2) \tag{8–26}$$

where $\dot{m}$ is the coolant mass flow rate, $\bar{c}$ the average specific heat of the liquid, T_1 the initial temperature of the coolant as it enters the cooling jacket, and T_2

its final temperature. Q is the rate of heat absorption per unit time; q is this same rate per unit heat transfer area. A. T_2 should be below the boiling point prevailing at the cooling jacket pressure.

Radiation

Radiation heat emission is the electromagnetic radiation emitted by a gas, liquid, or solid body by the virtue of its temperature and at the expense of its internal energy. It covers the wavelength range from 10,000 to 0.0001 μm, which includes the visible range of 0.39 to 0.78 μm. Radiation heat transfer occurs most efficiently in a vacuum because there is no absorption by the intervening fluids.

The heat transmitted by the mechanism of radiation depends primarily on the temperature of the radiating body and its surface condition. The second law of thermodynamics can be used to prove that the radiant energy E is a function of the fourth power of the absolute temperature T:

$$E = f \epsilon \sigma A T^4 \qquad (8\text{--}27)$$

The energy E radiated by a body is defined as a function of the emissivity ϵ, which is a dimensionless factor for surface condition and material properties, the Stefan-Boltzmann constant σ (5.67×10^{-8} W/m^2-K^4), the surface area A, the absolute temperature T, and the geometric factor f, which depends on the arrangement of adjacent parts and the shape. At low temperatures (below 800 K) radiation accounts for only a negligible portion of the total heat transfer in a rocket device and can usually be neglected.

In rocket propulsion there are these radiation concerns:

1. Emission of hot gases to the internal walls of a combustion chamber, a solid propellant grain, a hybrid propellant grain or a nozzle.
2. Emission to the surroundings or to space from the external surfaces of hot hardware (radiation-cooled chambers, nozzles, or electrodes in electric propulsion).
3. Radiation from the hot plume downstream of the nozzle exit. This is described in Chapter 18.

In rocket combustion devices gas temperatures are between 1900 and 3900 K or about 3000 to 6600°F; their radiation contributes between 3 and 40% of the heat transfer to the chamber walls, depending on the reaction gas composition, chamber size, geometry, and temperature. It can be a significant portion of the total heat transfer. In solid propellant motors the radiation heating of the grain surfaces can be critical to the burning rate, as discussed in Chapter 13. The absorption of radiation on the wall follows essentially the same laws as those of emission. Metal surfaces and formed tubes reflect much of the radiant energy, whereas ablative materials and solid propellant seem to absorb most of the incident radiation. A highly reflective surface on the inside wall of a combustor tends to reduce absorption and to minimize the temperature increase of the walls.

The hot reaction gases in rocket combustion chambers are potent radiation sources. Gases with symmetrical molecules, such as hydrogen, oxygen, and nitrogen, have been found not to show many strong emission bands in those wavelength regions of importance in radiant heat transfer. Also, they do not really absorb much radiation and do not contribute considerable energy to the heat transfer. Heteropolar gases, such as water vapor, carbon monoxide, carbon dioxide, hydrogen chloride, hydrocarbons, ammonia, oxides of nitrogen, and the alcohols, have strong emission bands of known wavelengths. The radiation of energy of these molecules is associated with the quantum changes in their energy levels of rotation and interatomic vibration. In general, the radiation intensity of all gases increases with their volume, partial pressure, and the fourth power of their absolute temperature. For small thrust chambers and low chamber pressures, radiation contributes only a small amount of energy to the overall heat transfer.

If the hot reaction gases contain small solid particles or liquid droplets, then the radiation heat transfer can increase dramatically by a factor of 2 to 10. The particulates greatly increase the radiant energy as explained in Section 18.1. For example, the reaction gas from some slurry liquid propellants and many solid propellants contains fine aluminum powder. When burned to form aluminum oxide, the heat of combustion and the combustion temperature are increased (raising heat transfer), and the specific impulse is raised somewhat (giving improved performance). The oxide can be in the form of liquid droplets (in the chamber) or solid particles (in the nozzle diverging section), depending on the local gas temperature. Furthermore, the impact of these particulates with the wall will cause an additional increase in heat transfer, particularly to the walls in the nozzle throat and immediately upstream of the nozzle throat region. The particles also cause erosion or abrasion of the walls.

8.4. STARTING AND IGNITION

The starting of a thrust chamber has to be controlled so that a timely and even ignition of propellants is achieved and the flow and thrust are built up smoothly and quickly to their rated value (see Ref. 6–1). The *initial propellant flow* is less than *full flow*, and the *starting mixture ratio* is usually different from the *operating mixture ratio*. A low initial flow prevents an excessive accumulation of unignited propellants in the chamber.

The starting injection velocity is low, the initial vaporization, atomization, and mixing of propellants in a cold combustion chamber is incomplete, and there are local regions of lean and rich mixtures. With cryogenic propellants the initial chamber temperature can be below ambient. The optimum starting mixture is therefore only an average of a range of mixture ratios, all of which should be readily ignited. Mixture ratios near the stoichiometric mixture ratio have a high heat release per unit of propellant mass and therefore permit bringing the chamber and the gases up to equilibrium faster than would be possible with other mixtures. The operating mixture ratio is usually fuel rich and is selected for optimum specific impulse. One method of analytical modeling of the ignition of cryogenic propellants is given in Ref. 8–16.

The *time delay for starting* a thrust chamber ideally consists of the following time periods:

(1) time needed to fully open the propellant valves (typically 0.002 to more than 1.00 sec, depending on valve type and its size and upstream pressure);

(2) time needed to fill the liquid passage volume between the valve seat and the injector face (piping, internal injector feed holes, and cavities);

(3) time for forming discrete streams or jets of liquid propellant (sometimes gaseous propellant, if cryogenic liquid is preheated by heat of ambient temperature cooling jacket) and for initial atomization into small droplets and for mixing these droplets;

(4) time needed for droplets to vaporize and ignite (laboratory tests show this to be very short, 0.02 to 0.05 sec, but this depends on the propellants and the available heat);

(5) once ignition is achieved at a particular location in the chamber, it takes time to spread the flame or to heat all the mixed propellant that has entered into the chamber, to vaporize it, and to raise it to ignition temperature;

(6) time needed to raise the chamber to the point where combustion will be self sustaining, and then to its full pressure.

There are overlaps in these delays and several of them can occur simultaneously. The delays [items (1), (2), (3), (5), and (6) above] are longer with large injectors or large diameter chambers. Small thrusters can usually be started very quickly, in a few milliseconds, while larger units require 1 sec or more.

In starting a thrust chamber one propellant always reaches the chamber a short time ahead of the other; it is almost impossible to synchronize exactly the fuel and oxidizer feed systems so that the propellants reach the chamber simultaneously at all injection holes. Frequently, a more reliable ignition is assured when one of the propellants is intentionally made to reach the chamber first. For example, for a fuel-rich starting mixture the fuel is admitted first. Reference 8–17 describes the control of the propellant lead.

Other factors influencing the starting flows, the propellant lead or lag, and some of the delays mentioned above are the liquid pressures supplied to the injector (e.g., regulated pressure), the temperature of the propellant (some can be close to their vapor point), and the amount of insoluble gas (air bubbles) mixed with the initial quantity of propellants.

The propellant valves (and the flow passages betwen them and the injector face) are often so designed that they operate in a definite sequence, thereby assuring an intentional lead of one of the propellants and a controlled buildup of flow and mixture ratio. Often the valves are only partially opened, avoiding an accumulation of hazardous unburned propellant mixture in the chamber. Once combustion is established, the valves are fully opened and full flow may reach the thrust chamber assembly. The initial reduced flow burning period is called the *preliminary stage*. Section 10.5 describes the starting controls.

Full flow in the larger thrust chambers is not initiated with non-self-igniting propellants until the controller received a signal of successful ignition. The verification of ignition or initial burning is often built into engine controls using visual detection (photocell), heat detection (pyrometer), a fusible wire link, or sensing of a pressure rise. If the starting controls are not designed properly, unburnt propellant may accumulate in the chamber; upon ignition it may then explode, causing sometimes severe damage to the rocket engine. Starting controls and engine flow calibrations are discussed in Section 10.5

Non-spontaneously ignitable propellants need to be activated by absorbing energy prior to combustion initiation. This energy is supplied by the *ignition system*. Once ignition has begun the flame is self-supporting. The igniter has to be located near the injector in such a manner that a satisfactory starting mixture at low initial flow is present at the time of igniter activation, yet it should not hinder or obstruct the steady-state combustion process. At least five different types of successful propellant ignition systems have been used.

Spark plug ignition has been used successfully on liquid oxygen–gasoline and on oxygen–hydrogen thrust chambers, particularly for multiple starts during flight. The spark splug is often built into the injector, as shown in Fig. 9–6.

Ignition by electrically heated wires has been accomplished, but at times has proven to be less reliable than spark ignition for liquid propellants.

Pyrotechnic ignition uses a solid propellant squib or grain of a few seconds' burning duration. The solid propellant charge is electrically ignited and burns with a hot flame within the combustion chamber. Almost all solid propellant rockets and many liquid rocket chambers are ignited in this fashion. The igniter container may be designed to fit directly onto the injector or the chamber (see Fig. 8–1), or may be held in the chamber from outside through the nozzle. This ignition method can only be used once; thereafter the charge has to be replaced.

In *precombustion chamber ignition* a small chamber is built next to the main combustion chamber and connected through an orifice; this is similar to the precombustion chamber used in some internal combustion engines. A small amount of fuel and oxidizer is injected into the precombustion chamber and ignited. The burning mixture enters the main combustion chamber in a torchlike fashion and ignites the larger main propellant flow which is injected into the main chamber. This ignition procedure permits repeated starting of variable-thrust engines and has proved successful with the liquid oxygen–gasoline and oxygen–hydrogen thrust chambers.

Auxiliary fluid ignition is a method whereby some liquid or gas, in addition to the regular fuel and oxidizer, is injected into the combustion chamber for a very short period during the starting operation. This fluid is hypergolic, which means it produces spontaneous combustion with either the fuel or the oxidizer. The combustion of nitric acid and some organic fuels can, for instance, be initiated by the introduction of a small quantity of hydrazine or aniline at the beginning of the rocket operation. Liquids that ignite with air (zinc diethyl or aluminum triethyl), when preloaded in the fuel piping, can accomplish a hypergolic ignition. The flow diagram of the RD 170 Russian rocket engine in Fig. 10–10 shows several cylindrical containers prefilled with a hypergolic liquid, one for each of the high pressure fuel supply lines; this hypergolic liquid is pushed out (by the initial fuel) into the thrust chambers and into the pre-burners to start their ignitions.

In vehicles with multiple engines or thrust chambers it is required to start two or more together. It is often difficult to get exactly simultaneous starts. Usually the passage or manifold volumes of each thrust chamber and their respective values are designed to be the same. The temperature of the initial propellant fed to each thrust chamber and the lead time of the first quantity of propellant entering into the chambers have to be controlled. This is needed, for example, in two small thrusters when used to apply roll torques to a vehicle. It is also one of the reasons why large space launch vehicles are not released from their launch facility until there is assurance that all the thrust chambers are started and operating.

8.5. VARIABLE THRUST

Section 3.8 mentions the equations related to this topic. One of the advantages of liquid propellant rocket engines is the ability to throttle or to randomly vary the thrust over a wide range. Deep throttling over a thrust range of more than 10:1 is required for relatively few applications. Moon landing, interceptor missiles, and gas generators with variable power output are examples. Moderate throttling (a thrust range of up to perhaps 2.5:1) is needed for trajectory velocity control (as in some tactical missiles), space maneuvers, or temporarily limiting the vehicle velocity (to avoid excessive aerodynamic heating during the ascent through the atmosphere), as in the Space Shuttle main engine.

Throttling is accomplished by reducing the propellant flow supply to the thrust chamber and thus reducing the chamber pressure. The pressure drop in the injector is related to the injection velocity by Eq. 8–5. The accompanying reduction of the injector pressure drop can lead to a very low liquid injection velocity and, thus, to poor propellant mixing, improper stream impingement patterns, and poor atomization, which in turn can lead to lower combustion efficiency and thus lower performance and sometimes unstable combustion.

The variation of flow through a given set of injection orifices and of thrust by this method is limited.

There are several throttling methods whereby the injection pressure drop is not decreased unduly. This permits a change in chamber pressure without a major decrease in injector pressure drop. A moving sleeve mechanism for adjusting the fuel and the oxidizer injection circular sheet spray areas is shown in Fig. 8–3.

One way of preventing unstable operation and a drop-off in performance is to use *multiple thrust chambers* or *multiple rocket engines*, each of which operates always at or near rated conditions. The thrust is varied by turning individual thrust chambers on or off and by throttling all of them over a relatively narrow range.

For small reaction control thrusters the average thrust is usually reduced by pulsing. It is accomplished by controlling the number of cycles or pulses (each has one short fixed-duration thrust pulse plus a short fixed-duration zero-thrust pause), by modulating the duration of individual pulses (with short pauses between pulses), or alternatively by lengthening the pause between pulses.

8.6. SAMPLE THRUST CHAMBER DESIGN ANALYSIS

This example shows how a thrust chamber is strongly influenced by the overall vehicle system requirements or the mission parameters and the vehicle design. As outlined in the Design Section of Chapter 10 and in the discussion of the selection of propulsion systems in Chapter 17, each engine goes through a series of rationalizations and requirements that define its key parameters and its design. In this example we describe how the thrust chamber parameters are derived from the vehicle and engine requirements. The overall system requirements relate to the mission, its purpose, environment, trajectories, reusability, reliability, and to restraints such as allowable engine mass, or maximum dimensional envelope. We are listing some, but not all of the requirements. It shows how theory is blended with experience to arrive at the initial choices of the design parameters.

Here we define the application as a new upper stage of an existing multistage space launch vehicle, that will propel a payload into deep space. This means continuous firing (no restart or reuse), operating in the vacuum of space (high nozzle area ratio), modest acceleration (not to exceed 5 g_0), low cost, moderately high performance (specific impulse), and a thrust whose magnitude depends on the payloads, the flight path and acceleration limits. The desired mission velocity increase of the stage is 3400 m/sec. The engine is attached to its own stage, which is subsequently disconnected and dropped from the payload stage. The payload stage (3500 kg) consists of a payload of 1500 kg (for scientific instruments, power supply, or communications and flight control equipment) and its own propulsion systems (including propellant) of 2000 kg (for trajectory changes, station keeping, attitude control, or emergency maneuvers). There are two geometric restraints: the vehicle has an outside diameter of 2.0 m, but when the structure, conduits, certain equipment, thermal insulation, fittings, and assembly are considered, it really is only about 1.90 m. The restraint on the stage length of 4.50 m maximum will affect the length of the thrust chamber. We can summarize the key requirements:

Application	Uppermost stage to an existing multistage launch vehicle
Payload	3500 kg
Desired velocity increase Δu	3400 m/sec in gravity free vacuum
Maximum stage diameter	1.90 m
Maximum stage length	4.50 m
Maximum acceleration	5 g_0

Decisions on Basic Parameters. The following engine design decisions or parameter selection should be made early in the design process:

Propellant combination
Chamber pressure
Nozzle area ratio
Feed system, using pumps or pressurized tanks
Thrust level

From a performance point of view, the best *propellant combination* would be liquid oxygen with liquid hydrogen. However, this bipropellant would have a low average specific gravity (0.36), a very large liquid hydrogen tank, and would cause an increase in vehicle drag during ascent. It would have some potential problems with exceeding the allocated stage volume, hydrogen mass losses, and the vehicle structure. The lower stages of the existing launch vehicle use liquid oxygen with RP-1 fuel with an average specific gravity of about 1.014, and the launch pad is already equipped for supplying these. The new stage is limited in volume and cross section. Because of these factors the propellant combination of liquid oxygen and RP-1 (a type of kerosene) is selected. From Fig. 5–1 we see that the theoretical specific impulse is between 280 and 300 sec, depending on the mixture ratio and whether we use frozen or shifting chemical equilibrium in the nozzle flow expansion. This figure also shows that the maximum value of the characteristic velocity c^* is reached at a mixture ratio of about 2.30, which is a fuel-rich mixture. We select this mixture ratio. Its combustion temperature is lower than the mixture ratios with higher values, and this should make the cooling of the thrust chamber easier. We will see later that cooling may present some problems. Based on universal experience, we select a value of I_s part way (about 40%) between the values for frozen and shifting equilibrium, namely 292 sec at the standard chamber pressure of 1000 psi or 6.895 MPa, and a nozzle big enough for expansion to sea level. From Fig. 5–1 and Table 5–5 we find the molecular mass to be 23 kg/kg-mol and the specific heat ratio k to be about 1.24. Later we will correct this value of I_s from this standard reference condition to the actual vacuum specific impulse of the thrust chamber.

Next we will select a chamber pressure, a nozzle area ratio and a feed system concept. Historically there has been favorable experience with this propellant combination at chamber pressures between 400 and 3400 psia with nozzle area ratios up to about 40 with both gas generator cycles and staged combustion cycles, giving proof that this is feasible. The following considerations enter into this selection:

1. Higher chamber pressures allow a smaller thrust chamber and (for the same nozzle exit pressure) a shorter nozzle cone with a smaller nozzle exit diameter. The thrust chamber is small enough for a toroidal tank to be built around it, and this conserves stage length. This not only saves vehicle space, but usually also some inert mass in the vehicle and the engine. Figure 8–23 shows the relative sizes of thrust chambers for three chamber pressures and two nozzle area ratios (ϵ of 100 and 300). The nozzle length and exit diameter cannot exceed the values given in the requirements, which, as can be seen, rules out low chamber pressure or high area ratio. The dimensions shown are calculated later in this analysis.

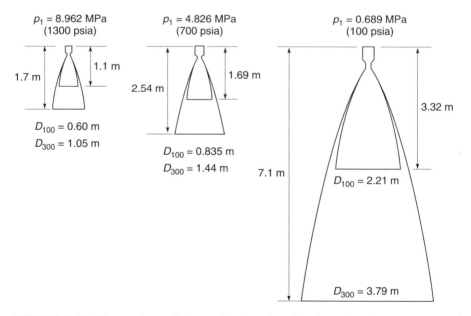

FIGURE 8–23. Comparison of thrust chamber sizes for three chamber pressures and two nozzle area ratios (100 and 300).

2. The heat transfer rate is almost proportional to the gas density, which is proportional to the chamber pressure, as shown by Eq. 8–21 and 8–23. On some prior thrust chambers there have been problems with the formation of solid carbon layer or deposits either inside the cooling jacket (increasing wall temperatures) or on the inner walls of the combustion chamber (the solid can flake off and cause burnout). This favors a lower chamber pressure.

3. Concern over leak-free seals for both static and dynamic seal increases with chamber pressure, which in turn causes all feed presures also to increase.

4. A feed system using pressurized gas is feasible, but its inert masses of tanks and engine are favorable only, if the chamber pressure is very low, perhaps around 100 psia or less. The tanks for propellants and pressurizing gas become very heavy and the thrust chamber will be very large and exceed the dimensional restraints mentioned above. We therefore cannot use this feed system or very low chamber pressures.

5. If we use a pump feed system, the power needed to drive the pumps increases directly with chamber pressure p_1. In a gas generator engine cycle this means a slightly reduced performance as the value of p_1 goes up. For a staged combustion cycle it means high pressures, particularly high pressure hot gas flexible piping, and a more complex, heavier, and expensive engine. We therefore select a gas generator cycle (see Fig. 1–4) at a low enough chamber pressure, so that the thrust chamber (and the other inert hardware) will just fit the geometrical constraints, and the engine inert mass and the heat transfer will be reasonable.

For these reasons we pick a *chamber pressure* of 700 psia or 4.825 MPa and an *area ratio* of 100. With further analysis we could have picked p_1 more precisely; it could be somewhat lower or higher. Next we correct the *specific impulse* to the operating conditions using a ratio of thrust coefficients. We can use Eq. 3–30 or interpolate between Figs. 3–7 and 3–8 for a value of $k = 1.24$. The reference or standard condition (see Fig. 3–6) is for a pressure ratio p_1/p_3 of $1000/14.7 = 68$, which corresponds to an area ratio of about 8. Then $(C_F)_{standard} = 1.58$. For the actual high-altitude operation the pressure ratio is close to infinity and the nozzle has an area ratio of 100; we can determine the thrust coefficient by interpolating $k = 1.24$. The result is $(C_F)_{vacuum} = 1.90$. The new ideal specific impulse value for a chamber threshold of 700 psia and a nozzle area ratio of 100 is therefore $292 \times (1.90/1.58) = 351.1$ sec. In order to correct for losses (divergence, boundary layer, incomplete combustion, some film cooling, etc.) we use a correction factor of 0.96 giving a thrust chamber specific impulse of 337.1 sec. The engine uses a gas generator and this will reduce the engine specific impulse further by a factor of 0.98 or $(I_s)_{engine} = 330.3$ sec or an effective exhaust velocity of 3237 m/sec.

Stage Masses and Thrust Level. An estimate of the stage masses will next be made. We assume that the inert hardware (tanks, gas, generator, turbopumps, etc.) is about 7% of the propellant mass, which is conservative when compared to existing engines. In a full-fledged engine design this number would be verified or corrected once an estimated mass budget becomes available. From Eq. 4–7

$$e^{\Delta u/v} = \frac{m_o}{m_f} = \frac{m_p + 0.07m_p + 3500}{0.07m_p + 3500} = e^{3400/3237}$$

Solve for $m_p = 7639$ kg. The final and initial masses of the stage are then 4023 kg and 11,002 kg respectively.

The maximum *thrust* is limited by the maximum allowed acceleration of $5g_0$. It is $F_{max} = m_0\, a = 11,002 \times 5 \times 9.8 = 539,100$ N. This would become a relatively large and heavy thrust chamber. Considerable saving in inert mass can be obtained if a smaller thrust size (but longer firing duration) is chosen. Since this same thrust chamber is going to be used for another mission where an acceleration of somewhat less than $1.0\,g_0$ is wanted, a thrust level of 50,000 N or 11,240 lbf is chosen. The maximum acceleration of the stage occurs just before cutoff; it is $a = F/m_f = 50,000/4023 = 12.4$ m/sec² or about 1.26 times the acceleration of gravity. This fits the thrust requirements.

The following have now been determined:

Propellant	Liquid oxygen and liquid kerosene (RP-1)
Mixture ratio (O/F)	2.30 (engine)
Thrust	50,000 N or 11,240 lbf
Chamber pressure	700 psia or 4.826 MPa
Nozzle area ratio	100
Specific impulse (engine)	330.3 sec
Specific impulse (thrust chamber)	337.1 sec
Engine cycle	Gas generator
Usable propellant mass	7478 kg

Propellant Flows and Dimensions of Thrust Chamber. From Eq. 2–6 we obtain the propellant mass flow

$$\dot{m} = F/c = 50,000/3200 = 15.625 \text{ kg/sec}$$

When this total flow and the overall mixture ratio are known, then the fuel flow $\dot{m}_f$ and oxidizer flow $\dot{m}_o$ for the engine, its gas generator, and its thrust chamber can be determined from Eqs. 6–3 and 6–4 as shown below.

$$\dot{m}_f = \dot{m}/(r+1) = 15.446/(2.3+1) = 4.680 \text{ kg/sec}$$
$$\dot{m}_o = \dot{m}r/(r+1) = (15.446 \times 2.30)/3.30 = 10.765 \text{ kg/sec}$$

The gas generator flow $\dot{m}_{gg}$ consumes about 2.0% of the total flow and operates at a fuel-rich mixture ratio of 0.055; this results in a gas temperature of about 890 K.

$$(\dot{m}_f)_{gg} = 0.2928 \text{ kg/sec} \qquad (\dot{m}_o)_{gg} = 0.0161 \text{ kg/sec}$$

The flows through the thrust chamber are equal to the total flow diminished by the gas generator flow, which is roughly 98.0% of the total flow or 15.137 kg/sec.

$$(\dot{m}_f)_{tc} = 4.387 \text{ kg/sec} \qquad (\dot{m}_o)_{tc} = 10.749 \text{ kg/sec}$$

The *duration* is the total effective propellant mass divided by the mass flow rate

$$t_b = m_p/\dot{m}_p = 7478/15.446 = 484.1 \text{ sec or a little longer than 8 minutes}$$

The *nozzle throat area* is determined from Eq. 3–31.

$$A_t = F/(p_1 C_F) = 50,000/(4.826 \times 10^6 \times 1.90) = 0.005453 \text{ m}^2 \text{ or } 54.53 \text{ cm}^2$$

The *nozzle throat diameter* is $D_t = 8.326$ cm. The internal diameter of the nozzle at exit A_2 is determined from the area ratio of 100 to be $D_2 = \sqrt{100} \times D_t$ or 83.26 cm. A shortened or *truncated bell nozzle* (as discussed in Section 3.4) will be used with 80% of the length of a 15° conical nozzle, but with the same performance as a 15° cone. The nozzle length (from the throat to the exit) can be determined by an accurate layout or by $L = (D_2 - D_t)/(2 \tan 15)$ as 139.8 cm. For an 80% shortened bell nozzle this length would be about 111.8 cm. The contour or shape of a shortened bell nozzle can be approximated by a parabola (parabola equation is $y^2 = 2px$). Using an analysis (similar to the analysis that resulted in Fig. 3–14) the maximum angle of the diverging section at the inflection point would be about $\theta_i = 34°$ and the nozzle exit angle $\theta_e = 7°$. The approximate contour consists of a short segment of radius $0.4r_t$ of a 34° included angle (between points T and I in Fig. 3–14) and a parabola with two known points at I and E. Knowing the tangent angles (34 and 7°) and the y coordinates $[y_e = r_2$ and $y_i = r_t + 0.382\, r_t\,(1 - \cos\theta_i)]$ allows the determination of the parabola by geometric analysis. Before detail design is undertaken, a more accurate contour, using the method of characteristics, is suggested.

The *chamber diameter* should be about twice the nozzle throat diameter to avoid pressure losses in the combustion chamber ($D_c = 16.64$ cm). Using the approximate length of prior successful smaller chambers and a characteristic length L^* of about 1.1 m, the chamber length (together with the converging nozzle section) is about 11.8 inch or 29.9 cm. The overall length of the thrust chamber (169 cm) is the sum of the nozzle length (111.8 cm), chamber (29.9 cm), injector thickness (estimated at 8 cm), mounted valves (estimated at 10 cm), a support structure, and possibly also a gimbal joint. The middle sketch of the three thrust chambers in Fig. 8–23 corresponds roughly to these numbers.

We have now the stage masses, propellant flows, nozzle and chamber configuration. Since this example is aimed at a thrust chamber, data on other engine components or parameters are given only if they relate directly to the thrust chamber or its parameters.

Next we check if there is enough available vehicle volume (1.90 m diameter and 4.50 m long) to allow making a larger nozzle area ratio and thus gain a little more performance. First we determine how much of this volume is occupied by propellant tanks and how much might be left over or be available for the thrust chamber. This analysis would normally be done by tank design specialists. The average density of the propellant mixture can be determined from Eq. 7–1 to be 1014 kg/m^3 and the total usable propellant of 7478 kg. Using densities from Table 7–1 the fuel volume and the oxidizer volume can be calculated to be 2.797 and 4.571 m^3 respectively. For a diameter of 1.90 m, a nearly spherical fuel tank, a separate oxidizer cylindrical tank with elliptical ends, 6% ullage, and 2% residual propellant, a layout would show an overall tank length of about 3.6 m in a space that is limited to 4.50 m. This would leave only 0.9 m for the length of the thrust chamber, and this is not long enough. Therefore we would need to resort to a more compact tank arrangement, such as using a common bulkhead between the two tanks or building a toroidal tank around the engine. It is not the aim to design the tanks in this example, but the conclusion affects the thrust chamber. Since the available volume of the vehicle is limited, it is not a good idea to try to make the thrust chamber bigger.

This diversion into the tank design shows how a vehicle parameter affects the thrust chamber design. For example, if the tank design would turn out to be difficult or the tanks would become too heavy, then one of these thrust chamber options can be considered: (1) go to a higher chamber pressure (makes the thrust chamber and nozzle smaller, but heavier), (2) go to a lower thrust engine (will be smaller and lighter), (3) store the nozzle of the upper stage thrust chamber in two pieces and assemble them once the lower stages have been used and discarded (see Fig. 8–19; it is more complex and somewhat heavier), or (4) use more than one thrust chamber in the engine (will be heavier, but shorter). We will not pursue these or other options here.

Heat Transfer. The particular computer program for estimating heat transfer and cooling parameters of thrust chambers will depend on the background and experience of specific engineers and rocket organizations. Typical computer programs divide the internal wall surface of the chamber and nozzle into axial incremental axial steps. Usually in a preliminary analysis the heat transfer is estimated only for critical locations such as for the throat and perhaps the chamber.

From Fig. 5–1 and Eq. 3–12 or 3–22 we determine the following gas temperatures for the chamber, nozzle throat region, and a location in the diverging exit section. They are: $T_1 = 3600$ K, $T_t = 3243$ K, and $T_e = 1730$ K at an area ratio of 6.0 in the diverging nozzle section. The chamber and nozzle down to an exit area ratio of 6 will have to be cooled by fuel. For this propellant combination and for the elevated wall temperatures a stainless steel has been successfully used for the inner wall material.

Notice that beyond this area ratio of about 6, the nozzle free stream gas temperatures are relatively low. Uncooled high temperature metals can be used here in this outer nozzle region. Radiation cooling, using a material such as niobium (coated to prevent excessive oxidation) or carbon fibers in a nonporous carbon matrix, is suitable between an area ratio of 6 and about 25. For the final large nozzle exit section, where the temperatures are even lower, a lower cost material such as stainless steel or titanium is suggested. Ablative materials have been ruled out, because of the long duration and the aggressive ingredients in the exhaust gas. The gas compositions of Figs. 5–2 and 5–3 indicate that some free oxygen and hydroxyl is present.

We now have identified the likely materials for key chamber components. The best way to cool the radiation cooled exit segment of the nozzle (beyond area ratio of 6) is to let it stick out of the vehicle structure; the heat can then be freely radiated to space. One way to accomplish this, is to discard the vehicle structure around the nozzle end.

As in Fig. 8–8, the maximum heat transfer rate will be at the nozzle throat region. A variety of heat transfer analysis programs are available for estimating this heat transfer. If a suitable computer program is not available, then an approximate steady-state heat transfer analysis can be made using Eqs. 8–15 to 8–19 and the physical properties (specific heat, thermal conductivity, and density) of RP-1 at elevated temperatures. The film coefficients of Eqs. 8–23 and 8–25 are also needed. This is not done in this example, in part because data tables for the physical properties would take up a lot of space and results are not always reliable. Data from prior thrust chambers with the same propellants indicate a heat transfer rate at the nozzle throat region exceeding 10 Btu/in.^{2}-sec or 1.63×10^7 W/m^2.

The RP-1 fuel is an unusual coolant, since it does not have a distinct boiling point. Its composition is not consistent and depends on the oil stock from which it was refined and the refining process. It is distilled or evaporated gradually over a range of temperatures. The very hot wall can cause the RP-1 to locally break down into carbon-rich material and to partially evaporate or gasify. As long as the small vapor bubbles are recondensed when they are mixed with the cooler portions of the coolant flow, a steady heat transfer process will occur. If the heat transfer is high enough, then these bubbles will not be condensed, may contain noncondensable gases, and the flow will contain substantial gas bubbles and become unsteady, causing local overheating. The recondensing is aided by high cooling passage velocities (more than 10 m/sec at the throat region) and by turbulence in these passages. A coolant flow velocity of 15 m/sec is selected for the throat and 7 m/sec in the chamber and nozzle exit segment.

The material for the cooling jacket will be stainless steel to resist the oxidation and erosion of the fast moving, aggressive hot gas, which contains free oxygen and hydroxyl species. The cooling by fuel will assure that the temperatures of this stainless steel are well below its softening temperature of about 1050 K.

The construction of the cooling jacket can be tubular, as shown in Figs. 8–1 and 8–9, or it can consist of milled channels as shown in Figs. 8–2 and 8–17. The cross section of each tube or cooling channel will be a minimum at the throat region, gradually become larger, and be about two or more times as large at the chamber and diverging nozzle regions. The wall thickness (on the hot gas side) should be as small as possible to reduce the temperature drop across the wall (which reduces the thermal stresses and allows a lower wall temperature) and to minimize the yielding of the material that occurs due to thermal deformation and pressure loads. Figure 8–12 shows this behavior, but for a thick wall. Practical considerations such as manufacturability, the number of test firings before flight, the deformation under pressure loads, the temperature gradient and dimensional tolerances also enter into the selection of the wall thickness. A thickness of 0.5 mm and a cooling velocity of 15 m/sec have been selected for the throat region of the cooling jacket and cooling velocities of 7 m/sec in the chamber and the cooled nozzle segment. Milled slots (rather than tubes) have been selected for this thrust chamber.

The selection of the number of milled slots, their cross sections, and the wall thickness is a function of the coolant mass flow, its pressure, wall stresses, wall material, and the shape of the channel. Figure 8–24 and Table 8–6 describe the channel width and height for different numbers of channels and different locations. The fuel coolant flow is diminished by the gas generator fuel flow (0.293 kg/sec) and is about 4.387 kg/sec. For this flow and a cooling velocity of 15 m/sec in the throat region the cumulative cross-sectional area of all the channels is only about 3.62 cm^2. The cooling velocity is lower in the chamber and nozzle regions and the cumulative channel flow area will be larger there. The variables are the number of channels, the thickness of the hot wall, the rib thickness between channels, the cooling velocity, the gas temperature, and the

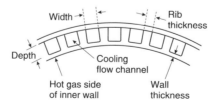

FIGURE 8–24. Segment of cooling jacket with milled channels and an electroformed outer wall.

TABLE 8–6. Alternative Milled Channel Configurations for Fuel (cooling) Flow of 4.387 kg/sec

Throat Section			Chamber Section		
Wall thickness	0.05 cm		Wall thickness	0.06 cm	
Rib thickness	0.08 cm		Rib thickness	0.08	
Total flow area	3.653 cm^2		Total flow area	7.827 cm^2	
Flow velocity	15 m/sec		Flow velocity	7.0 m/sec	
Number of Channels	Channel Width, cm	Channel Depth, cm	Number of Channels	Channel Width, cm	Channel Depth, cm
80	0.257	0.177			
100	0.193	0.189	100	0.456	0.171
120	0.145	0.210	120	0.367	0.179
140	0.113	0.231	140	0.303	0.184
150	0.100	0.243	150	0.277	0.188
160	0.092	0.247	160	0.255	0.192
180	0.070	0.289	180	0.218	0.196

location along the thrust chamber profile. The number of channels or tubes will determine the shape of the cross section, ranging from deep and thin to almost square. The effect of varying the number of channels or channel dimensions and shape is shown in Table 8–6. The minimum inert mass of the cooling jacket and a low friction loss occur, when the shape (which varies axially throughout the jacket) is on the average close to a square. On the basis of analyses, as shown in the table, a 150-channel design has been selected for giving favorable cross section, reasonable dimensions for ease of fabrication, good cooling and often low thermal wall stresses.

Reinforcing bands have to be put on the outside of the tubes or channels to hold the internal gas pressure during operation, to contain the coolant pressures, which cause heated walls wanting to become round, and any surge pressures during the start transient or arising from water hammer in the lines. We assume a surge pressure of 50% above chamber pressure and a steel strength σ of 120,000 psi. In the chamber the inside diameter is 16.7 cm (6.57 in.), the walls and channels are 0.3 cm thick, and the pressure is 700 psia or 4.826 MPa. If one band allows the reinforcing of a length of chamber of 3.0 in. the cross sectional area of that reinforcing band will be $A = pDL/(2\sigma) = [700 \times 1.5 \times (6.57 + 0.3) \times 3]/(2 \times 120,000) = 0.0902$ in. If the band were 1.0 in. wide, its thickness would be 0.09 in. and if it were 3 in. wide it would be 0.3 in. thick. Large nozzle exit sections have been observed to experience flutter or cyclic deformation, and therefore some stiffening rings may be needed near the exit.

The capacity of the fuel to absorb heat is approximately $c_p m_f \Delta T = 0.5 \times 4.81 \times 200 = 278,000$ J/sec. The maximum ΔT is established by keeping the fuel well below its chemical decomposition point. This calculated heat absorption is less than the heat transfer from the hot gases. It is therefore necessary to reduce the gas temperature near the chamber and nozzle walls or to increase the heat absorption. This can be accomplished by (1) introducing film cooling by injection into the chamber just ahead of the nozzle, by (2) modifying the injection patterns, so that a cooler, fuel-rich thick internal boundary layer is formed, or (3) by allowing some nucleate boiling in the throat region. The analysis of these three methods is not given here. Item (2), supplementary cooling, is selected because it is easy to design and build, and can be based on extensive data of prior favorable experience. However it causes a small loss of performance (up to about 1% in specific impulse).

Injector Design. The injector pattern can be any one of the several types shown in Figs. 8–3 and 8–4. For this propellant combination we have used both doublets (like and unlike), and triplets in the USA, and the Russians have used multiple hollow double posts with swirling or rotation of the flow in the outer annulus. Based on good experience and demonstrated combustion stability with similar designs, we select a doublet self impinging type stream pattern and an injector design similar to Fig 8–4. The impinging streams form fans of liquid propellant, which break up into droplets. Oxidizer and fuel fans alternate radially. We could also use a platelet design, like Fig. 8–5.

The pressure drop across the injector is usually set at values between 15 and 25% of the chamber pressure, in part to obtain high injection velocities, which aid in atomization and droplet breakup. In turn this leads to more complete combustion (and thus better performance) and to stable combustion. We will use 20% or 140 psi or 0.965 MPa for the injector pressure drop. There is a small pressure loss in the injector passages. The injection velocities are found from Eqs. 8–1 and 8–5. The equation is solved for the area A, which is the cumulative cross-section area of all the injection holes of one of the propellants in the injector face.

With rounded and clean injection hole entrances the discharge coefficient will be about 0.80 as shown in Table 8–2. Solving for the cumulative injection hole area for the fuel and the oxidizer flow gives 1.98 cm^2 for the fuel and 4.098 cm^2 for the oxidizer. A typical hole diameter in this size of injector would be about 0.5 to 2.5 mm. We will use a hole size of 1.5 mm for the fuel holes (with 90% of the fuel flow) and 2.00 mm for the oxidizer hole size, resulting in 65 doublets of oxidizer holes and 50 doublets of fuel. By using a slightly smaller fuel injection hole diameter, we can match the number of 65 doublets as used with the oxidizer holes. These injection doublets will be arranged on the injector face in concentric patterns similar to Fig. 8–4. We may be able to obtain a slightly higher performance by going to smaller hole sizes and a large number of fuel and oxidizer holes. In addition there will be extra fuel holes on the periphery of the injector face to help in providing the cooler boundary layer, which is needed to reduce heat transfer. They will use 10% of the fuel flow and, for a 0.5 mm hole diameter, the number of holes will be about 100. To make a good set of liquid fans, equal inclination angles of about 25° are used with the doublet impingements. See Fig. 8–7.

Ignition. A pyrotechnic (solid propellant) igniter will be used. It has to have enough energy and run long enough to provide the pressure and temperature in the thrust chamber for good ignition. Its diameter has to be small enough to be inserted through the throat, namely 8.0 cm maximum diameter and 10 to 15 cm long.

Layout Drawings, Masses, Flows, and Pressure Drops. We now have enough of the key design parameters of the selected thrust chamber, so a preliminary layout drawing can be made. Before this can be done well, we will need some analysis or estimates on the manifolds for fuel and oxidizer, valve mounting provisions and their locations, a nozzle closure during storage, a thrust structure, and possibly an actuator and gimbal mount, if gimballing is required by the mission. A detailed layout or CAD (Computer Aided Design) image (not shown in this analysis) would allow a more accurate picture and a good determination of the mass of the thrust chamber and its center of gravity both with and without propellants.

Estimates of gas pressures, liquid pressures (or pressure drops) in the flow passages, injector, cooling jacket, and the valves are needed for the stress analysis, so that various wall thicknesses and component masses can be determined. Material properties will need to be obtained from references or tests. A few of these analyses and designs may actually change some of the data we selected or estimated early in this sample analysis and some of the calculated parameters may have to be re-analyzed and revised. Further changes in the thrust chamber design may become evident in the design of the engine, the tanks, or the interface with the vehicle. The methods, processes and fixtures for manufacturing and testing (and the number and types of tests) will have to be evaluated and the number of thrust chambers to be built has to be decided, before we can arrive at a reasonable manufacturing plan, a schedule and cost estimates.

PROBLEMS

1. How much total heat per second can be absorbed in a thrust chamber with an inside wall surface area of 0.200 m^2 if the coolant is liquid hydrogen and the coolant temperature does not exceed 145 K in the jacket? Coolant flow is 2 kg/sec. What is the average heat transfer rate per second per unit area? Use the data from Table 7–1 and the following:

Heat of vaporization near boiling point	446 kJ/kg
Thermal conductivity (gas at 21 K)	0.013 W/m-K
(gas at 194.75 K)	0.128 W/m-K
(gas at 273.15 K)	0.165 W/m-K

2. During a static test a certain steel thrust chamber is cooled by water. The following data are given:

Average water temperature	100°F
Thermal conductivity of water	1.07×10^{-4} Btu/sec-ft^2-°F/ft
Gas temperature	4500°F
Viscosity of water	2.5×10^{-5} lbf-sec/ft^2
Specific heat of water	1.0 Btu/lb-°F
Cooling passage dimensions	$\frac{1}{4} \times \frac{1}{2}$ in.
Water flow through passage	0.585 lb/sec
Thickness of inner wall	$\frac{1}{8}$ in.
Heat absorbed	1.3 Btu/in.^{2}-sec
Thermal conductivity of wall material	26 Btu/hr-ft^2-°F/ft

Determine (a) the film coefficient of the coolant; (b) the wall temperature on the coolant side; (c) the wall temperature on the gas side.

3. In the example of Problem 2 determine the water flow required to decrease the wall temperature on the gas side by 100°F. What is the percentage increase in coolant velocity? Assume that the various properties of the water and the average water temperature do not change.

4. Express the total temperature drop in Problem 2 in terms of the percentage temperature drops through the coolant film, the wall, and the gas film.

5. Determine the absolute and relative reduction in wall temperatures and heat transfer caused by applying insulation in a liquid-cooled rocket chamber with the following data:

Tube wall thickness	0.381 mm
Gas temperature	2760 K
Gas-side wall temperature	1260 K
Heat transfer rate	15 MW/m^2
Liquid film coefficient	23 kW/m^2-K
Wall material	Stainless steel AISI type 302

A 0.2 mm thick layer of insulating paint is applied on the gas side; the paint consists mostly of magnesia particles. The conductivity of this magnesia is 2.59 W/m^2-K/m. The stainless steel has an average thermal conductivity of 140 Btu/hr-ft^2-°F/in.

6. A small thruster has the following characteristics:

Propellants	Nitrogen tetroxide and monomethyl hydrazine
Injection individual hole size	Between 0.063 and 0.030 in.
Injection hole pattern	Unlike impinging doublet
Thrust chamber type	Ablative liner with a carbon–carbon nozzle throat insert
Specific gravities:	1.446 for oxidizer and 0.876 for fuel
Impingement point	0.25 in. from injector face
Direction of jet momentum	Parallel to chamber axis after impingement
$r = 1.65$ (fuel rich)	$(I_s)_{actual} = 251$ sec
$F = 300$ lbf	$t_b = 25$ sec
$p_1 = 250$ psi	$A_1/A_t = 3.0$
$(\Delta p)_{inj} = 50.0$ psi	$(C_d)_0 = (C_d)_f = 0.86$

Determine the number of oxidizer and fuel holes and their angles. Make a sketch to show the symmetric hole pattern and the feed passages in the injector. To protect the wall, the outermost holes should all be fuel holes.

7. A large, uncooled, uninsulated, low carbon steel thrust chamber burned out in the throat region during a test. The wall (0.375 in. thick) had melted and there were several holes. The test engineer said that he estimated the heat transfer to have been about 15 Btu/in.^{2}. The chamber was repaired and you are responsible for the next test. Someone suggested that a series of water hoses be hooked up to spray plenty of water on the outside of the nozzle wall at the throat region during the next test to prolong the firing duration. The steel's melting point is estimated to be 2550°F. Because of the likely local variation in mixture ratio and possibly imperfect impingement, you anticipate some local gas regions that are oxidizer rich and could start the rapid oxidation of the steel. You therefore decide that 2250°F should be the maximum allowable inner wall temperature. Besides knowing the steel weight density (0.284 lbf/in.^{3}), you have the following data for steel for the temperature range from ambient to 2250°F: the specific heat is 0.143 Btu/lbm-°F and the thermal conductivity is 260 Btu/hr-ft^2-°F/in. Determine the approximate time for running the next test (without burnout) both with and without the water sprays. Justify any assumptions you make about the liquid film coefficient of the water flow. If the water spray seems to be worth while (getting at least 10% more burning time), make sketches with notes on how the mechanic should arrange for this water flow so it will be most effective.

8. The following conditions are given for a double-walled cooling jacket of a rocket thrust chamber assembly:

Rated chamber pressure	210 psi
Rated jacket pressure	290 psi
Chamber diameter	16.5 in.
Nozzle throat diameter	5.0 in.
Nozzle throat gas pressure	112 psi
Average inner wall temperature at throat region	110°F
Average inner wall temperature at chamber region	800°F
Cooling passage height at chamber and nozzle exit	0.375 in.
Cooling passage height at nozzle throat	0.250 in.
Nozzle exit gas pressure	14.7 psi.
Nozzle exit diameter	10 in.
Wall material	1020 carbon steel
Safety factor on yield strength	2.5
Cooling fluid	RP-1
Average thermal conductivity of steel	250 Btu/hr-ft^2-F/in.

Assume other parameters, if needed. Compute the outside diameters and the thickness of the inner and outer walls at the chamber, at the throat, and at the nozzle exit.

9. Determine the hole sizes and the angle setting for a multiple-hole, doublet impinging stream injector that uses alcohol and liquid oxygen as propellants. The resultant momentum should be axial, and the angle between the oxygen and fuel jets ($\gamma_o + \gamma_f$) should be 60°. Assume the following:

$(C_d)_o$	0.87	Chamber pressure	300 psi
$(C_d)_f$	0.91	Fuel pressure	400 psi

ρ_o	71 lb/ft^3	Oxygen pressure	380 psi
ρ_f	51 lb/ft^3	Number of jet pairs	4
r	1.20	Thrust	250 lbf
	Actual specific impulse	218 sec	

Answers: 0.0197 in.; 0.0214 in.; 32.3°; 27.7°.

10. Explain in a rational manner why Fig. 8–10 has a maximum and how this maximum would be affected by the duty cycle, ablative material, heat loss from the thrust chamber, effect of altitude, and so on. Why does this maximum not occur at 90% burn time?

11. Table 10–5 shows that the RD 120 rocket engine can operate down to 85% of full thrust and at a mixture ratio variation of ±10.0%. In a particular static test the average thrust was held at 96% of nominal and the average mixture ratio was 2.0% fuel rich. Assume a 1.0% residual propellant, but neglect other propellant budget allowances. What percentage of the fuel and oxidizer that have been loaded will remain unused at thrust termination? If we want to correct the mixture ratio in the last 20.0% of the test duration and use up all the available propellant, what would be the mixture ratio and propellant flows for this last period?

12. Make a simple sketch to scale of the thrust chamber that was analyzed in Section 8.6. The various dimensions should be close, but need not be accurate. Include or make separate sketches of the cooling jacket and the injector. Also compile a table of all the key characteristics, similar to Table 8–1, but include gas generator flows, and key materials. Make estimates or assumptions for any key data that is not mentioned in Section 8.6

SYMBOLS

A	area, m^2 (ft^2)
A_a	Projected area of linear aerospike ramp, m^2 (ft^2)
c_p	specific heat at constant pressure, J/kg-K (Btu/lbm R)
$\bar{c}$	average liquid specific heat, J/kg-K (Btu/lbm R)
C_d	discharge coefficient
C_F	thrust coefficient (see Eq. 3–31)
D	diameter, m (ft)
E	modulus of elasticity, N/m^2 (lbf/in.^{2}), or radiation energy, kg-m^2/sec^2
f	friction loss coefficient, or geometric factor in radiation
g_0	sea level acceleration of gravity, 9.806 m/sec^2 (32.17 ft/sec^2)
h	film coefficient, W/(m^2-K)
Δh	enthalpy change, J/kg (Btu/lb)
I_s	specific impulse, sec
k	specific heat ratio
L	length, m (ft)
L^*	characteristic chamber length, m (ft)
m	mass, kg

$\dot{m}$	mass flow rate, kg/sec (lb/sec)
p	pressure, N/m^2 or Pa (lbf/in.^{2})
Pr	Prandtl number ($c_p \mu / \kappa$)
q	heat-transfer rate or heat flow per unit area, J/m^2-sec (Btu/ft^2-sec)
Q	volume flow rate, m^3/sec (ft^3/sec), or heat flow rate, J/sec
r	flow mixture ratio (oxidizer to fuel)
s	stress N/m^2 (lbf/in.^{2})
t	time, sec
t_s	stay time, sec
t_w	wall thickness, m (in.)
T	absolute temperature, K (R)
v	velocity, m/sec (ft/sec)
$\mathcal{V}$	specific volume, m^3/kg(ft^3/lb)
V_c	combustion chamber volume (volume up to throat), m^3 (ft^3)

Greek Letters

γ_o	angle between chamber axis and oxidizer stream
γ_f	angle between chamber axis and fuel stream
Δ	finite differential
δ	angle between chamber axis and the resultant stream
ϵ	nozzle area ratio ($\epsilon = A_2 / A_t$), or emissivity of radiating body
θ	angle
κ	thermal conductivity, J/(m^2-sec-K)/m (Btu/in.^{2}-sec^2- R/in.)
λ	coefficient of thermal expansion, m/m-K (in./in.-R)
μ	viscosity, m^2/sec
ν	Poisson ratio
ρ	density, kg/m^3 (lbf/ft^3)
σ	Stefan–Boltzmann constant (5.67×10^{-8} W/m^2-K^4)

Subscripts

am	arithmetic mean
c	chamber
f	fuel or final condition
g	gas
l	liquid
o	oxidizer
t	throat
w	wall
wg	wall on side of gas
wl	wall on side of liquid
0	initial condition

1 inlet or chamber condition
2 nozzle exit condition
3 atmosphere or ambient condition

REFERENCES

8–1. R. Sackheim, D. F. Fritz, and H. Maklis, "The Next Generation of Spacecraft Propulsion Systems," *AIAA Paper 79-1301*, 1979.

8–2. S. Peery and A. Minnick, "Design and Development of an Advanced Expander Combustor," *AIAA Paper 98-3675*, July 1998.

8–3. R. D. McKown, "Brazing the SSME," *Threshold, an Engineering Journal of Power Technology*, No. 1, Rocketdyne Division of Rockwell International (now The Boeing Co.), Canoga Park, CA, March 1987, pp. 8–13.

8–4. R. A. Ellis, J. C. Lee, F. M. Payne, M. Lacoste, A. Lacombe, and P. Joyes, "Testing of the RL10B-2 Carbon–Carbon Nozzle Extension," *AIAA Conference Paper 98-3363*, July 1998.

8–5. M. Niino, A. Kumakawa, T. Hirano, K. Sumiyashi, and R. Watanabe, "Life Prediction of CIP Formed Thrust Chambers," *Acta Astronautica*, Vol. 13, Nos. 6–7, 1986, pp. 363–369 (fatigue life prediction).

8–6. J. S. Porowski, W. J. O'Donnell, M. L. Badlani, B. Kasraie, and H. J. Kasper, "Simplified Design and Life Predictions of Rocket Thrustchambers," *Journal of Spacecraft and Rockets*, Vol. 22, No. 2, March–April 1985, pp. 181–187.

8–7. J. S. Kinkaid, "Aerospike Evolution," *Threshold*, The Boeing Co., Rocketdyne Propulsion and Power, No. 18, Spring 2000, pp. 4–13.

8–8. T. Harmon, "X-33 Linear Aerospike on the Fast Track in System Engineering," *AIAA Paper 99-2181, Joint Propulsion Conference*, June 1999.

8–9. A. J. Fortini and R. H. Tuffias, "Advanced Materials for Chemical Propulsion: Oxide–Iridium/Rhenium Combustion Chambers," *AIAA Paper 99-2894*, June 1999.

8–10. F. P. Incropera and D. P. DeWitt, *Introduction to Heat Transfer*, John Wiley & Sons, New York, 1996.

8–11. A. A. Samarskii and P. N. Vabishchevich, *Computational Heat Transfer, Vol. 1. Mathematical Modeling* and *Vol. 2. The Finite Difference Methodology*, John Wiley & Sons, New York, 1995.

8–12. D. R. Bartz, "Survey of Relationships between Theory and Experiment for Convective Heat Transfer in Rocket Combustion Gases," in *Advances in Rocket Propulsion*, S. S. Penner (Ed.), AGARD, Technivision Services, Manchester, UK, 1968.

8–13. N. Sugathan, K. Srinivathan, and S. Srinivasa Murthy, "Experiments on Heat Transfer in a Cryogenic Engine Thrust Chamber," *Journal of Propulsion and Power*, Vol. 9, No. 2, March–April 1993.

8–14. E. Mayer, "Analysis of Pressure Feasibility Limits in Regenerative Cooling of Combustion Chambers for Large Thrust Rockets," in *Liquid Rockets and*

Propellants, L. E. Bollinger, M. Goldsmith, and A. W. Lemmon, Jr (Eds.), Academic Press, New York, 1969, pp. 543–561.

8–15. J. M. Fowler and C. F. Warner, "Measurements of the Heat-Transfer Coefficients for Hydrogen Flowing in a Heated Tube," *American Rocket Society Journal*, Vol. 30, No. 3, March 1960, pp. 266–267.

8–16. P.-A. Baudart, V. Duthoit, T. Delaporte, and E. Znaty, "Numerical Modeling of the HM 7 B Main Chamber Ignition," *AIAA Paper 89-2397*, 1989.

8–17. A. R. Casillas, J. Eninger, G. Josephs, J. Kenney, and M. Trinidad, "Control of Propellant Lead/Lag to the LEA in the AXAF Propulsion System," *AIAA Paper 98-3204*, July 1998.

CHAPTER 9

COMBUSTION OF LIQUID PROPELLANTS

The design, development, and operation of liquid rocket engines requires efficient stable burning of the propellants and the generation of a high-temperature, uniform gas that is the rocket's working fluid. In this chapter we treat the complex phenomena of the combustion processes in the combustion chamber of a liquid bipropellant thrust chamber. We describe in general terms the combustion behavior, the progress in analysis of combustion, the several types of combustion instability with its undesirable effects, and semiempirical remedies. The objective is to operate at very high combustion efficiencies and to prevent the occurrence of combustion instability. Thrust chambers should operate with stable combustion over a wide range of operating conditions. For a treatment of these subjects see Refs. 9–1 to 9–3.

The combustion of liquid propellants is very efficient in well-designed thrust chambers, precombustion chambers, or gas generators. Efficiencies of 95 to 99.5% are typical compared to turbojets or furnaces, which can range from 50 to 97%. This is due to the very high reaction rates at the high combustion temperatures and the thorough mixing of fuel and oxidizer reaction species by means of good injection distribution and gas turbulence. The losses are due to incomplete burning or inadequate mixing (nonuniform mixing ratio). For very small bipropellant thrust chambers, where the injector has very few injection orifices or elements, the combustion efficiency can be well below 95%.

9.1. COMBUSTION PROCESS

In describing the combustion processes, it is convenient and helpful to the understanding to divide the combustion chamber into a series of discrete zones, as shown in Fig. 9–1 for a typical configuration. It has a flat injector face with many small injection orifices for introducing both fuel and oxidizer liquids as many discrete individual streams, jets, or thin sprays or sheets. The relative thicknesses of these zones, their behavior, and their transitions are influenced by the specific propellant combination, the operating conditions (pressure, mixture ratio, etc.), the design of the injector, and chamber geometry. The boundaries between the zones shown in Fig. 9–1 are really not flat surfaces and do not display steady flow. They are undulating, dynamically movable, irregular boundaries with localized changes in velocity, temporary bulges, locally intense radiation emissions, or variable temperature. Table 9–1 shows the major interacting physical and chemical processes that occur in the chamber. This table is a modification of tables and data in Refs. 9–2 and 9–3.

The combustion behavior is propellant dependent. If the fuel were hydrogen that has been used to cool the thrust chamber, the hydrogen would be gaseous and fairly warm (60 to 500 K); there would be no liquid hydrogen droplets and no evaporation. With hypergolic propellants there is an initial chemical reaction in the liquid phase when a droplet of fuel impinges on a droplet of oxidizer. Experiments show that the contact can create local explosions and enough energy release to suddenly vaporize a thin layer of the fuel and the oxidizer locally at the droplet's contact face; there immediately follows a vapor chemical reaction and a blow-apart and breakup of the droplets, due to the explosion shock wave pressure (Refs. 9–4 and 9–5).

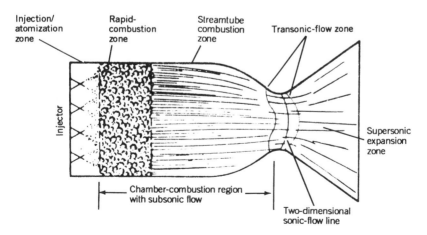

FIGURE 9–1. Division of combustion chamber into zones for analysis. (Reprinted with permission from Ref. 8–1, copyright by AIAA.)

TABLE 9–1. Physical and Chemical Processes in the Combustion of Liquid Propellants

Injection	Atomization	Vaporization
Liquid jets enter chamber at relatively low velocities	Impingement of jets or sheets	Droplet gasification and diffusion
Sometimes gas propellant is injected	Formation of liquid fans	Further heat release from local chemical reactions
Partial evaporation of liquids	Formation of droplets	Low gas velocities and some cross flow
Interaction of jets and high pressure gas	Secondary breakup of drops	Heat absorbed by radiation and conduction from blowback of turbulent gases from the hot reaction zone
	Liquid mixing and some liquid–liquid chemical reaction	Acceleration to higher velocities
	Oscillations of jets or fans as they become unstable during breakup	Vaporization rate influenced by pressure or temperature oscillations and acoustic waves
	Vaporization begins and some vapor reactions occur	

Mixing and Reaction	Expansion in Chamber
Turbulent mixing (three-dimensional)	Chemical kinetics causes attainment of final combustion temperature and final equilibrium reaction gas composition
Multiple chemical reactions and major heat releases	Gas dynamics displays turbulence and increasing axial gas velocities
Interactions of turbulence with droplets and chemical reactions	Formation of a boundary layer
Temperature rise reduces densities	Acceleration to high chamber velocities
Local mixture ratios, reaction rates, or velocities are not uniform across chamber and vary rapidly with time	Streamlined high-velocity axial flow with very little cross flow
Some tangential and radial flows	

Rapid Combustion Zone

In this zone intensive and rapid chemical reactions occur at increasingly higher temperature; any remaining liquid droplets are vaporized by convective heating and gas pockets of fuel-rich and fuel-lean gases are mixed. The mixing is aided by local turbulence and diffusion of the gas species.

The further breakdown of the propellant chemicals into intermediate fractions and smaller, simpler chemicals and the oxidation of fuel fractions occur rapidly in this zone. The rate of heat release increases greatly and this causes the specific volume of the gas mixture to increase and the local axial velocity to increase by a factor of 100 or more. The rapid expansion of the heated gases also forces a series of local transverse gas flows from hot high-burning-rate sites to colder low-burning-rate sites. The liquid droplets that may still persist in the upstream portion of this zone do not follow the gas flow quickly and are difficult to move in a transverse direction. Therefore, zones of fuel-rich or oxidizer-rich gases will persist according to the orifice spray pattern in the upstream injection zone. The gas composition and mixture ratio across the chamber section become more uniform as the gases move through this zone, but the mixture never becomes truly uniform. As the reaction product gases are accelerated, they become hotter (due to further heat releases) and the lateral velocities become relatively small compared to the increasing axial velocities.

The combustion process is not a steady flow process. Some people believe that the combustion is locally so intense that it approches localized explosions that create a series of shock waves. When observing any one specific location in the chamber, one finds that there are rapid fluctuations in pressure, temperature, density, mixture ratio, and radiation emissions with time.

Injection/Atomization Zone

Two different liquids are injected with storable propellants and with liquid oxygen/hydrocarbon combinations. They are injected through orifices at velocities typically between 7 and 60 m/sec or about 20 to 200 ft/sec. The injector design has a profound influence on the combustion behavior and some seemingly minor design changes can have a major effect on instability. The pattern, sizes, number, distribution, and types of orifices influence the combustion behavior, as do the pressure drop, manifold geometry, or surface roughness in the injection orifice walls. The individual jets, streams, or sheets break up into droplets by impingement of one jet with another (or with a surface), by the inherent instabilities of liquid sprays, or by the interaction with gases at a different velocity and temperature. In this first zone the liquids are atomized into a large number of small droplets (see Refs. 9–3 and 9–6). Heat is transferred to the droplets by radiation from the very hot rapid combustion zone and by convection from moderately hot gases in the first zone. The droplets evaporate and create local regions rich either in fuel vapor or oxidizer vapor.

This first zone is heterogeneous; it contains liquids and vaporized propellant as well as some burning hot gases. With the liquid being located at discrete sites, there are large gradients in all directions with respect to fuel and oxidizer mass fluxes, mixture ratio, size and dispersion of droplets, or properties of the gaseous medium. Chemical reactions occur in this zone, but the rate of heat generation is relatively low, in part because the liquids and the gases are still relatively cold and in part because vaporization near the droplets causes fuel-rich and fuel-lean regions which do not burn as quickly. Some hot gases from the combustion zone are recirculated back from the rapid combustion zone, and they can create local gas velocities that flow across the injector face. The hot gases, which can flow in unsteady vortexes or turbulence patterns, are essential to the initial evaporation of the liquids.

The injection, atomization and vaporization processes are different if one of the propellants is a gas. For example, this occurs in liquid oxygen with gaseous hydrogen propellant in thrust chambers or precombustion chambers, where liquid hydrogen has absorbed heat from cooling jackets and has been gasified. Hydrogen gas has no droplets and does not evaporate. The gas usually has a much higher injection velocity (above 120 m/sec) than the liquid propellant. This causes shear forces to be imposed on the liquid jets, with more rapid droplet formation and gasification. The preferred injector design for gaseous hydrogen and liquid oxygen is different from the individual jet streams used with storable propellants, as shown in Chapter 8.

Stream Tube Combustion Zone

In this zone oxidation reactions continue, but at a lower rate, and some additional heat is released. However, chemical reactions continue because the mixture tends to be driven toward an equilibrium composition. Since axial velocities are high (200 to 600 m/sec) the transverse convective flow velocities become relatively small. Streamlines are formed and there is relatively little turbulent mixing across streamline boundaries. Locally the flow velocity and the pressure fluctuate somewhat. The residence time in this zone is very short compared to the residence time in the other two zones. The streamline type, inviscid flow, and the chemical reactions toward achieving chemical equilibrium presist not only throughout the remainder of the combustion chamber, but are also extended into the nozzle.

Actually, the major processes do not take place strictly sequentially, but several seem to occur simultaneously in several parts of the chamber. The flame front is not a simple plane surface across the combustion chamber. There is turbulence in the gas flow in all parts of the combustion chamber.

The residence time of the propellant material in the combustion chamber is very short, usually less than 10 milliseconds. Combustion in a liquid rocket engine is very dynamic, with the volumetric heat release being approximately 370 MJ/m^3-sec, which is much higher than in turbojets. Further, the higher temperature in a rocket causes chemical reaction rates to be several times faster (increasing exponentially with temperature) than in turbojet.

9.2. ANALYSIS AND SIMULATION

For the purpose of analysing the combustion process and its instabilities, it has been convenient to divide the acoustical characteristics into linear and nonlinear behavior. A number of computer simulations with linear analyses have been developed over the last 45 years and have been used to understand the combustion process with liquid propellant combustion devices and to predict combustion oscillation frequencies. The nonlinear behavior (for example, why does a disturbance cause an apparently stable combustion to suddenly become unstable?) is not well understood and not properly simulated. Mathematical simulations require a number of assumptions and simplifications to permit feasible solutions (see Refs. 9–1, 9–3, 9–6, and 9–7). Good models exist for relatively simple phenomena such as droplets of a propellant vaporizing and burning in a gaseous atmosphere or the steady-state flow of gases with heat release from chemical reactions. The thermochemical equilibrium principles mentioned in Chapter 5 also apply here. Some programs who consider some turbulence and film cooling effects.

The following phenomena are usually ignored or greatly simplified: cross flows; nonsymmetrical gradients; unsteadiness of the flow; time variations in the local temperature, local velocity, or local gas composition; thermochemical reactions at local off-design mixture ratios and at different kinetic rates; enhancement of vaporization by acoustic fields (see Ref. 9-8); uncertainties in the spatial as well as the size distribution of droplets from sprays; or drag forces on droplets. It requires skilled, experienced personnel to use, interpret, and modify the more complex programs so that meaningful results and conclusions can be obtained. The outputs of these computer programs can give valuable help and confirmation about the particular design and are useful guides in interpreting actual test results, but by themselves they are not sufficient to determine the designs, select specific injector patterns, or predict the ocurrence of combustion instabilities.

All the existing computer programs known to the authors are suitable for steady-state flow conditions, usually at a predetermined average mixture ratio and chamber pressure. However, during the starting, thrust change, and stopping transients, the mixture ratio and the pressure change drastically. The analysis of these transient conditions is more difficult.

The combustion is strongly influenced by the injector design. The following are some of the injection parameters which influence combustion behavior: injector spray or jet pattern; their impingement; hole sizes or hole distribution; droplet evaporation; injection pressure drop; mixture ratio; pressure or temperature gradients near the injector; chamber/injector geometry; initial propellant temperature, and liquid injection pressure drop. Attempts to analyze these effects have met with only partial success.

Computational fluid dynamics (CFD) is a relatively new analytical tool that can provide a comprehensive description of complex fluid dynamic and thermodynamic behavior. It allows for a time history of all parameters and can even include some nonlinear effects. Numerical approaches are used to evaluate sets of equations and models that represent the behavior of the fluid. For complex geometries the information has been tracked with up to 250,000 discrete locations and can include changes in gas composition, thermodynamic conditions, equilibrium reactions, phase changes, viscous or nonviscous flow, one-, two-, or three-dimensional flow, and steady-state or transient conditions. It has been applied to resonance cavities in injectors or chambers and to the flow of burning gases through turbines. A comprehensive rocket combustion model using CFD is not yet available, but could become useful in the future.

9.3. COMBUSTION INSTABILITY

If the process of rocket combustion is not controlled (by proper design), then combustion instabilities can occur which can very quickly cause excessive pressure vibration forces (which may break engine parts) or excessive heat transfer (which may melt thrust chamber parts). The aim is to prevent occurrence of this instability and to maintain reliable operation (see Ref. 9–8). Although much progress has been made in understanding and avoiding combustion instability, new rocket engines can still be plagued by it.

Table 9–2 lists the principal types of combustion vibrations encountered in liquid rocket thrust chambers (see Refs. 9–3 and 9–9). Admittedly, combustion in a liquid rocket is never perfectly smooth; some fluctuations of pressure, temperature, and velocity are always present. When these fluctuations interact with the natural frequencies of the propellant feed system (with and without vehicle structure) or the chamber acoustics, periodic superimposed oscillations, recognized as instability, occur. In normal rocket practice *smooth combustion* occurs when pressure fluctuations during steady operation do not exceed about ±5% of the mean chamber pressure. Combustion that gives greater pressure fluctuations at a chamber wall location which occur at completely random intervals is called *rough combustion*. Unstable combustion, or *combustion instability*, displays organized oscillations occurring at well-defined intervals with a pressure peak that may be maintained, may increase, or may die out. These periodic peaks, representing fairly large concentrations

TABLE 9–2. Principal Types of Combustion Instability

Type and Word Description	Frequency Range (Hz)	Cause Relationship
Low frequency, called chugging or feed system instability	10–400	Linked with pressure interactions between propellant feed system, if not the entire vehicle, and combustion chamber
Intermediate frequency, called acoustic,[a] buzzing, or entropy waves	400–1000	Linked with mechanical vibrations of propulsion structure, injector manifold, flow eddies, fuel/oxidizer ratio fluctuations, and propellant feed system resonances
High frequency, called screaming, screeching, or squealing	Above 1000	Linked with combustion process forces (pressure waves) and chamber acoustical resonance properties

[a]Use of the word *acoustical* stems from the fact the frequency of the oscillations is related to combustion chamber dimensions and velocity of sound in the combustion gas.

of vibratory energy, can be easily recognized against the random-noise background (see Fig. 9–2).

Chugging, the first type of combustion instability listed in Table 9–2, stems mostly from the elastic nature of the feed systems and structures of vehicles or the imposition of propulsion forces upon the vehicle. Chugging of an engine or thrust chamber assembly can occur in a test facility, especially with low chamber pressure engines (100 to 500 psia), because of propellant pump cavitation, gas entrapment in propellant flow, tank pressurization control fluctuations, and vibration of engine supports and propellant lines. It can be caused by resonances in the engine feed system (such as an oscillating bellows inducing a periodic flow fluctuation) or a coupling of structural and feed system frequencies.

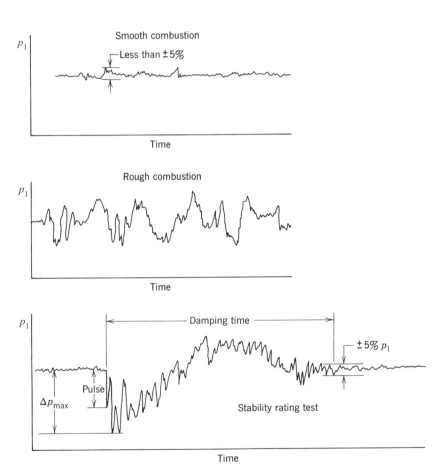

FIGURE 9–2. Typical oscillogrpah traces of chamber pressure p_1 with time for different combustion events.

When both the vehicle structure and the propellant liquid in the feed system have about the same natural frequency, then force coupling can occur, not only to maintain, but also to strongly amplify oscillations. Propellant flow rate disturbances, usually at 10 to 50 Hz, give rise to low-frequency longitudinal combustion instability, producing a longitudinal motion of vibration in the vehicle. This vehicle flight instability phenomenon has been called *pogo instability* since it is similar to pogo jumping stick motion. Pogo instabilities can occur in the propellant feed lines of large vehcles such as space launch vehicles or ballistic missiles.

Avoiding objectionable engine–vehicle coupled oscillation is best accomplished at the time of initial design of the vehicle, as contrasted to applying "fixes" later as has been the case with rocket engines for the Thor, Atlas, and Titan vehicles. Analytical methods exist for understanding the vibration modes and damping tendencies of major vehicle components, including the propellant tanks, tank pressurization systems, propellant flow lines, engines, and basic vehicle structure. Figure 9–3, a simplified spring–mass model of a typical two-stage vehicle, indicates the complexity of the analytical problem. Fortunately, the vibrational characteristics of the assembly can be affected substantially by designing damping into the major components or subassemblies. Techniques for damping pogo instability include the use of energy-absorption devices in fluid flow lines, perforated tank liners, special tank supports, and properly designed engine, interstage, and payload support structures (see Refs. 9–10 and 9–11).

A partially gas-filled pogo accumulator has been an effective damping device; it is attached to the main propellant feed line. Such an accumulator is used in the oxidizer feed line of the Space Shuttle main engine (SSME) betwen the two oxidizer turbopumps; it can be seen in Figs. 6–1 and 6–12. The SSME fuel line does not need such a damping device, because the fuel has a relatively very low density and a lower mass flow.

The dynamic characteristics of a propellant pump can also have an influence on the pogo-type vibrations, as examined in Ref. 9–12. The pogo frequency will change as propellant is consumed and the remaining mass of propellant in the vehicle changes. The bending or flexing of pipes, joints or bellows, or long tanks also has an influence.

Buzzing, the intermediate type of instability, seldom represents pressure perturbations greater than 5% of the mean in the combustion chamber and usually is not accompanied by large vibratory energy. It often is more noisy and annoying than damaging, although the occurrence of buzzing may initiate high-frequency instability. Often it is characteristic of coupling between the combustion process and flow in a portion of the propellant feed system. Initiation is thought to be from the combustion process. Acoustic resonance of the combustion chamber with a critical portion of the propellant flow system, sometimes originating in a pump, promotes continuation of the phenomenon. This type of instability is more prevalent in medium-size engines (2000 to 250,000 N thrust or about 500 to 60,000 lbf) than in large engines.

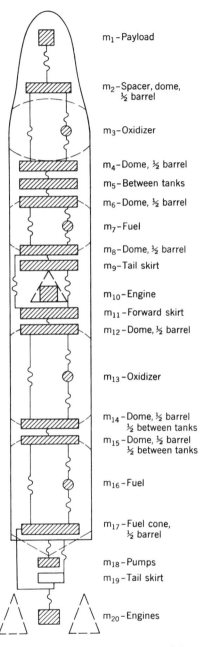

m_1 – Payload

m_2 – Spacer, dome, ½ barrel

m_3 – Oxidizer

m_4 – Dome, ½ barrel
m_5 – Between tanks
m_6 – Dome, ½ barrel
m_7 – Fuel
m_8 – Dome, ½ barrel
m_9 – Tail skirt

m_{10} – Engine
m_{11} – Forward skirt
m_{12} – Dome, ½ barrel

m_{13} – Oxidizer

m_{14} – Dome, ½ barrel
　　　 ½ between tanks
m_{15} – Dome, ½ barrel
　　　 ½ between tanks

m_{16} – Fuel

m_{17} – Fuel cone, ½ barrel

m_{18} – Pumps
m_{19} – Tail skirt

m_{20} – Engines

FIGURE 9–3. Typical two-stage vehicle spring–mass model used in analysis of pogo vibration in the veritcal direction.

The third type, *screeching* or *screaming*, has high frequency and is most perplexing and most common in the development of new engines. Both liquid and solid propellant rockets commonly experience high-frequency instablity during their development phase. Since energy content increases with frequency, this type is the most destructive, capable of destroying an engine in much less than 1 sec. Once encountered, it is the type for which it is most difficult to prove that the incorporated "fixes" or improvements render the engine "stable" under all launch and flight conditions. It can be treated as a phenomenon isolated to the combustion chamber and not generally influenced by feed system or structure.

High-frequency instability occurs in at least two modes, *longitudinal* and *transverse*. The *longitudinal mode* (sometimes called *organ pipe mode* propagates along axial planes of the combustion chamber and the pressure waves are reflected at the injector face and the converging nozzle cone. The *transverse modes* propagate along planes perpendicular to the chamber axis and can be broken down into *tangential* and *radial* modes. Transverse mode instability predominates in large liquid rockets, particularly in the vicinity of the injector. Figure 9–4 shows the distribution of pressure at various time intervals in a cylindrical combustion chamber (cross section) encountering transverse mode instability. Two kinds of wave form have been observed for tangential vibrations. One can be considered a *standing* wave that remains fixed in position while its pressure amplitude fluctuates. The second is a *spinning* or *traveling* tangential wave which has associated with it a rotation of the whole vibratory system. This waveform can be visualized as one in which the amplitude remains constant while the wave rotates. Combinations of transverse and longitudinal modes can also occur and their frequency can also be estimated.

Energy that drives screeching is believed to be predominantly from acoustically stimulated variations in droplet vaporization and/or mixing, local detonations, and acoustic changes in combustion rates. Thus, with favorable acoustic properties, high-frequency combustion instability, once triggered, can rapidly drive itself into a destructive mode. Invariable, a distinct boundary layer seems to disappear and heat transfer rates increase by an order of magnitude, much as with detonation, causing metal melting and wall burnthroughs, sometimes within less than 1 sec. The tangential modes appear to be the most damaging, heat transfer rates during instability often increasing 4 to 10 times. Often the instantaneous pressure peaks are about twice as high as with stable operation.

One possible source of triggering high-frequency instability is a rocket combustion phenomenon called *popping*. Popping is an undesirable random high-amplitude pressure disturbance that occurs during steady-state operation of a rocket engine with hypergolic propellants. It is a possible source for initiation of high-frequency instability. "Pops" exhibit some of the characteristics of a detonation wave. The rise time of the pressure is a few microseconds and the pressure ratio across the wave can be as high as 7:1. The elimination of popping

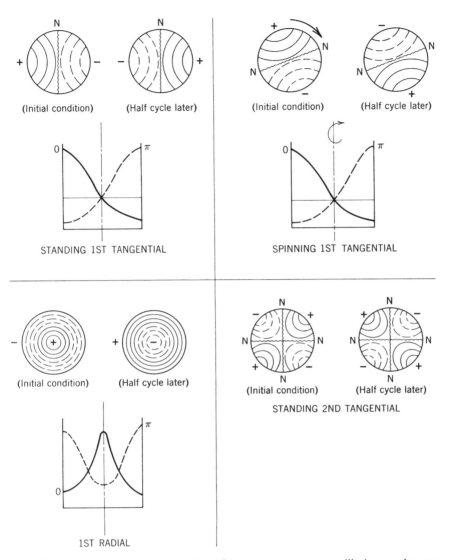

FIGURE 9–4. Simplified representation of transverse pressure oscillation modes at two time intervals in a cylindrical combustion chamber. The solid line curves indicate pressures greater than the normal or mean operating pressure and the dashed lines indicate lower pressures. The N–N lines show the node locations for these wave modes.

is usually achieved by redesign of the injector rather than by the application of baffles or absorbers.

Some combustion instabilities can be induced by *pulsations in the liquid flow* originating in turbopumps. Unsteady liquid flow can be caused by irregular cavitation at the leading edge of the inducer impellers or the main pump impellers. Also, when an impeller's trailing edge passes a rib or stationary vane of the volute, a small pressure perturbation always occurs in the liquid flow that travels downstream to the injector. These two types of pressure fluctuation can be greatly amplified if they coincide with the natural frequency of combustion vibrations in the chamber.

The estimated natural frequencies can be determined from the wavelength l, or the distance traveled per cycle, and the acoustic velocity a (see Eq. 3–10). The frequency, or number of cycles per second, is

$$\text{frequency} = a/l = (1/l)\sqrt{kTR'/\mathfrak{M}} \qquad (9\text{--}1)$$

where k is the specific heat ratio, R' the universal gas constant, $\mathfrak{M}$ the estimated molecular weight of the hot chamber gases, and T the local average absolute temperature. The length of wave travel depends on the vibrational mode, as shown in Fig. 9–4. Smaller chambers give higher frequencies.

Table 9–3 shows a list of estimated vibration frequencies for the Vulcain HM 60 rocket thrust chamber; it operates with liquid hydrogen and liquid oxygen propellants at a vacuum thrust of 1008 kN, a nominal chamber pressure of 10 MPa, and a nominal mixture ratio of 5.6 (see Ref. 9–13). The data in the table are based on acoustic measurements at ambient conditions with corrections for an appropriate sonic velocity correlation; since the chamber has a shallow conical shape and no discrete converging nozzle section, the purely longitudinal vibration modes would be weak; in fact, no pure longitudinal modes were detected.

Figure 9–5 shows a series of time-sequenced diagrams of frequency–pressure–amplitude measurements taken in the oxygen injector manifold of the Vulcain HM 60 engine during the first 8 sec of a static thrust chamber test while operating at off-nominal design conditions. Chugging can be seen at low

TABLE 9–3. Estimated Acoustic Hot Gas Frequencies for Nominal Chamber Operating Conditions for the Vulcain HM-60 Thrust Chamber

Mode[a]	(L, T, R)	Frequency (Hz)	Mode[a]	(L, T, R)	Frequency (Hz)
T1	(0, 1, 0)	2424	L1T3	(1, 3, 0)	6303
L1T1	(1, 1, 0)	3579	T4	(0, 4, 0)	6719
T2	(0, 2, 0)	3856	L2R1	(2, 0, 1)	7088
R1	(0, 0, 1)	4849	T5	(0, 5, 0)	8035
L1T2	(1, 2, 0)	4987	TR21	(0, 2, 1)	8335
T3	(0, 3, 0)	5264	R2	(0, 0, 2)	8774
L1R1	(1, 0, 1)	5934			

Reprinted with AIAA permission from Ref. 9–13.
[a]Modes are classified as L (longitudinal), T (tangential), or R (radial) and the number refers to the first, second, or third natural frequency.

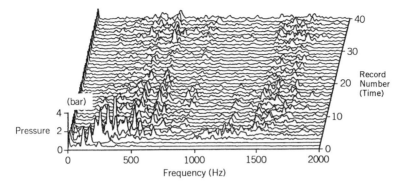

FIGURE 9–5. Graphical representation of a series of 40 superimposed frequency–amplitude diagrams taken 0.200 sec apart during the start phase (for the first 8 sec) of the Vulcain HM 60 thrust chamber. In this static hot-firing test the thrust chamber was operating at 109 bar chamber pressure and an oxidizer-to-fuel mass flow mixture ratio of 6.6. (Copied with permission from Ref. 9–13).

frequency (up to 500 Hz) during the first few seconds and a natural frequency around 1500 Hz is attributed to the natural resonance frequency of the oxygen injector dome structure where the high-frequency pressure transducer was-mounted. The continued oscillations observed at about 500 and 600 Hz are probably resonances associated with the feed system.

Rating Techniques

Semi-empirical techniques exist for artificially disturbing combustion in a rocket thrust chamber during test operation and evaluating its resistance to instability (see Ref. 9–14). These include: (1) nondirectional "bombs" placed within the combustion chamber; (2) oriented explosive pulses from a "pulse gun" directed through the chamber sidewall; and (3) directed flows of inert gas through the sidewall into the chamber. Often heavy prototype thrust chambers are used because they are less expensive and more resistant to damage than flight-weight engines. Other techniques used less widely but which are important, especially for small engines, include: (1) momentary operation at "off-mixture ratio;" (2) introduction of "slugs" of inert gas into a propellant line; and (3) a purposeful "hard start" achieved by introducing a quantity of unreacted propellant at the beginning of the operation.

The objective of these rating techniques is to measure and demonstrate the ability of an engine system to return quickly to normal operation and stable combustion after the combustion process has intentionally been disturbed or perturbed.

All techniques are intended to introduce shock waves into the combustion chamber or to otherwise perturb the combustion process, affording opportunity for measuring recovery time for a predetermined overpressure disturbance, assuming stable combustion resumes. Important to the magnitude and mode of the instability are the type of explosive charge selected, the size of the charge, the location and direction of the charge, and the duration of the exciting pulse. The bottom curve in Fig. 9–2 characterizes the recover of stable operation after a combustion chamber was "bombed." The time interval to recover and the magnitude of explosive or perturbation pressure are then used to rate the resistance of the engine to instability.

The nondirectional bomb method and the explosive pulse-gun method are the two techniques in common use. The bomb that can be used in large flight-weight thrust chambers without modification consists of six 250 grains of explosive powder (PETN,RDX,etc.) encased in a Teflon, nylon, or micarta case. Detonation of the bomb is achieved either electrically or thermally. Although the pulse gun requires modification of a combustion chamber, this technique affords directional control, which is important to tangential modes of high-frequency instability and allows several data points to be observed in a single test run by installing several pulse guns on one combustion chamber. Charges most frequently used are 10, 15, 20, 40, and 80 grains of pistol powder. Pulse guns can be fired in sequence, introducing successive pressure perturbations (approximately 150 msec apart), each of increasing intensity, into the combustion chamber.

Control of Instabilities

The control of instabilities is an important task during the design and development of a rocket engine. The designer usually relies on prior experience with similar engines and tests on new experimental engines. He also has available analytical tools with which to simulate and evaluate the combustion process. The design selection has to be proven in actual experiments to be free of instabilities over a wide range of transient and steady-state operating conditions. Some of the experiments can be accomplished on a subscale rocket thrust chamber that has a similar injector, but most tests have to be done on a full-scale engine.

The design features to control instabilities are different for the three types described in Table 9–2. Chugging is usually avoided if there is no resonance in the propellant feed system and its coupling with the elastic vehicle structure. Increased injection pressure drop and the addition of artificial damping devices in the propellant feed lines have been used successfully. Chugging and acoustical instabilities sometimes relate to the natural frequency of a particular feed system component that is free to oscillate, such as a loop of piping that can vibrate or a bellows whose oscillations cause a pumping effect.

With the choice of the propellant combination usually fixed early in the planning of a new engine, the designer can alter combustion feedback (depressing the driving mechanism) by altering injector details, (such as changing the injector hole pattern, hole sizes or by increasing the injection pressure drop), or alternatively by increasing acoustical damping within the combustion chamber. Of the two methods, the second has been favored in recent years because it is very effective, it is better understood, and theory fits. This leads to the application of *injector face baffles, discrete acoustic energy absorption cavities*, and *combustion chamber liners* or *changes in injector design*, often by using a trial and error approach.

Injector face baffles (see Fig. 9–6) were a widely accepted design practice in the 1960s for overcoming or preventing high-frequency instability. Baffle design is predicated on the assumption that the most severe instability, oscillations, along witht he driving source, are located in or near the injector–atomization zone at the injector end of the combustion chamber. The baffles minimize influential coupling and amplification of gas dynamic forces within the chamber. Obviously, baffles must be strong, have excellent resistance to combustion temperatures (they are usually cooled by propellant), and must protrude into the chamber enough to be effective, yet not so far as to act like an individual combustion chamber with its own acoustical characteristics. The number of baffle compartments is always odd. An even number of compartments enhances the standing modes of instability, with the baffles acting as nodal lines separating regions of relatively high and low pressure. The design and development of baffles remains highly empirical. Generally, baffles are designed to minimize acoustical frequencies below 4000 Hz, since experience has shown damaging instability is rare at frequencies above 4000 Hz.

MAIN INJECTOR ASSEMBLY

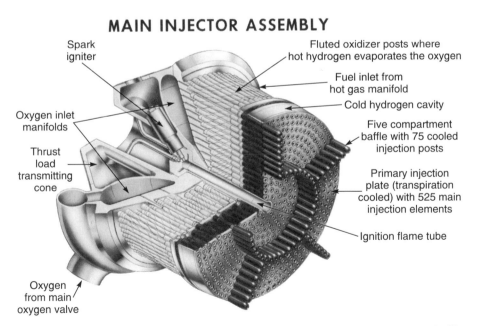

FIGURE 9–6. Main injector assembly of the Space Shuttle main engine showing baffle with five outer compartments. (Courtesy of The Boeing Company, Rocketdyne Propulsion and Power.)

Various mechanisms of *energy absorption* or *vibration damping* exist in a thrust chamber. Damping by well friction in combustion chambers is not significant. The exhaust nozzle produces the main damping of longitudinal mode oscillations; the reflection of waves from the convergent nozzle entrance departs from that of an ideal closed end. The principal damping source affecting propagation in the transverse plane is combustion itself. The great volumetric change in going from liquid to burned gases and the momentum imparted to a particle (solid or liquid) both constitute damping phenomena in that they take energy from high instantaneous local pressures. Unfortunately, the combustion process can generate a great deal more pressure oscillation energy than is absorbed by its inherent damping mechanism.

Acoustical absorbers are applied usually as discrete *cavities* along or in the wall of the combustion chamber near the injector end. Both act as a series of Helmholtz resonators that remove energy from the vibratory system which otherwise would maintain the pressure oscillations. Figure 9–7 shows the application of discrete cavities (interrupted slots) at the "corner" of the injector face. The corner location usually minimizes the fabrication problems, and it is the one location in a combustion chamber where a pressure antinode exists for all

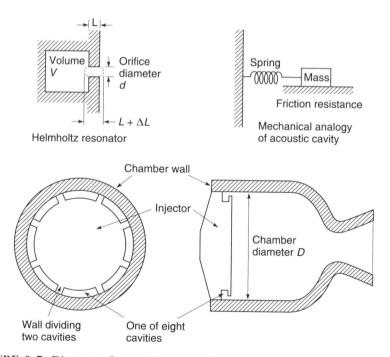

FIGURE 9–7. Diagram of acoustic energy absorber cavities at the periphery of an injector. In this thrust chamber the cavity restriction is a slot (in the shape of sections of a circular arc) and not a hole. Details of the chamber cooling channels, injector holes, or internal feed passages are not shown.

resonant modes of vibration, including longitudinal, tangential, radial, and combinations of these. Velocity oscillations are minimal at this point, which favors absorber effectiveness. Transverse modes of instability are best damped by locating absorbers at the corner location. Figure 9–7 also shows a Helmholtz resonator cavity and its working principles in simple form. Taking one resonator element, the mass of gas in the orifice with the volume of gas behind it forms an oscillatory system analogous to the spring–mass system shown (see Ref. 9–15). Even though Helmholtz resonator theory is well understood, problems exist in applying the theory to conditions of high pressure, temperature, chamber flow, and sound energy levels present when screech occurs, end in properly tuning the cavities to the estimated frequencies.

Absorption cavities designed as Helmholtz resonators placed in or near the injector face offer relatively high absorption bandwidth and energy absorbed per cycle. The Helmholtz resonator (an enclosed cavity with a small passage entry) dissipates energy twice each cycle (jets are formed upon inflow and outflow). Modern design practice favors acoustic absorbers over baffles. The storable propellant rocket engine shown in Fig. 8–2 has acoustic absorption cavities in the chamber wall at a location next to the injector.

The resonance frequency f of a Helmholtz cavity can be estimated as

$$f = \frac{a}{2\pi}\sqrt{\frac{A}{VL}} \tag{9–2}$$

Here a is the local acoustic velocity, A is the restrictor area, $A = (\pi/4)d^2$, and other symbols are as shownin Fig. 9–7. The ΔL is an empirical factor between 0.05 and 0.9 to allow for additional oscillating gas mass. It varies with the L/d ratio and the edge condition of the restricted orifice (sharp edge, rounded, chamfered). Resonators in thrust chambers are tuned or designed to perform their maximum damping at predicted frequencies.

Small changes in injector geometry or design can cause an unstable combustion to become stable and vice versa. New injectors, therefore, use the design and geometry of proven, stable prior designs with the same propellants. For example, the individual pattern of concentric tube injector elements used with gaseous hydrogen and liquid oxygen (shown in Fig. 8–3) are likely to be more stable, if the hydrogen gas is relatively warm and the injection velocity of the hydrogen is at least 10 times larger than that of the liquid oxygen.

In summary, the designer needs to (1) use data from prior successful engines and simulation programs to establish key design features and estimate the likely resonances, (2) design the feed system and structure to avoid these resonances, (3) use a robust injector design that will provide good mixing and dispersion of propellants and be resistant to disturbances, and (4) if needed, include tuned damping devices (cavities) to overcome acoustic oscillations. To validate that a particular thrust chamber is stable, it is necessary to test it over the range of likely operating conditions without encountering instability. An analysis is needed to determine the maximum and minimum likely propellant temperatures, maximum and minimum probable chamber pressures, and the highest and lowest mixture ratios, using a propellant budget as shown in Section 10.3. These limits then establish the variations of test conditions for this test series. Because of our improved understanding, the amount of testing needed to prove stability has been greatly reduced.

PROBLEMS

1. For a particular liquid propellant thrust chamber the following data are given:

Chamber presure	68 MPa
Chamber shape	cylindrical
Internal chamber diameter	0.270 m
Length of cylindrical section	0.500 m
Nozzle convergent section angle	45°
Throat diameter and radius of wall curvature	0.050 m
Injector face	Flat
Average chamber gas temperature	2800 K
Average chamber gas molecular weight	20 kg/kg-mol
Specific heat ratio	1.20

Assume the gas composition and temperature to be uniform in the cylindrical chamber section. State any other assumptions that may be needed. Determine the approximate resonance frequencies in the first longitudinal mode, radial mode, and tangential mode.

2. In Problem 1, explain how these three frequencies will change with combustion temperature, chamber pressure, chamber length, chamber diameter, and throat diameter.

3. Why does heat transfer increase during combustion instability?

4. Prepare a list of steps for undertaking a series of tests to validate the stability of a new pressure-fed liquid bipropellant rocket engine.

5. Estimate the resonant frequency of a set of each of nine cavities similar to Fig. 9–7. Here the chamber diameter $D = 0.200$ m, the slot width is 1.0 mm, and the width and height of the cavity are each 20.0 mm. The walls separating the individual cavities are 10.0 mm thick. Assume $L = 4.00$ mm, $\Delta L = 3.00$ mm, and $a = 1050$ m/sec. *Answer*: approximately 3138 cycles/sec.

REFERENCES

9–1. R. D. Sutton, W. S. Hines, and L. P. Combs, "Development and Application of a Comprehensive Analysis of Liquid Rocket Combustion," *AIAA Journal*, Vol. 10, No. 2, Feburary 1972, pp. 194–203.

9–2. K. K. Kuo, *Principles of Combustion*, John Wiley & Sons, New York, 1986.

9–3. V. Yang and W. Anderson (Eds.) *Liquid Rocket Engine Combustion Instability*, Vol. 169 of *Progress in Astronautics and Aeronautics*, AIAA, 1995, in particular Chapter 1, F.E.C. Culick and V. Yang, "Overview of Combustion Instabilities in Liquid Propellant Rocket Engines."

9–4. B. R. Lawver, "Photographic Observations of Reactive Stream Impingement," *Journal of Spacecraft and Rockets*, Vol. 17, No. 2, March–April 1980, pp. 134–139.

9–5. M. Tanaka and W. Daimon, "ExplosionPhenomena from Contact of Hypergolic Liquids," *Journal of Propulsion and Power*, Vol. 1, No. 4, 1984, pp. 314–316.

9–6. P. Y. Liang, R. J. Jensen, and Y. M. Chang, "Numerical Analysis of the SSME Preburner Injector Atomization and Combustion Process," *Journal of Propulsion and Power,* Vol. 3, No. 6, November–December 1987, pp. 508–513.

9–7. M. Habiballah, D. Lourme, and F. Pit, "PHEDRE—Numerical Model for Combustion Stability Studies Applied to the Ariane Viking Engine," *Journal of Propulsion and Power*, Vol. 7, No. 3, May–June 1991, pp. 322–329.

9–8. R. I. Sujith, G. A. Waldherr, J. I. Jagoda and B. T. Zinn, "Experimental Investigation of the Evaporation of Droplets in Axial Acoustic Fields," *Journal of Propulsion and Power*, AIAA, Vol. 16, No. 2, March–April 2000, pp. 278–285.

9–9. D. T. Hartje (Ed.), "Liquid Propellant Rocket Combustion Instability," *NASA SP-194*, U.S. Government Printing Office, No. 3300–0450, 1972.

9–10. B. W. Oppenheim and S. Rubin," Advanced Pogo Analysis for Liquid Rockets," *Journal of Spacecraft and Rockets*, Vol. 30, No. 3, May–June 1993.

9–11. G. About et al., "A New Approach of POGO Phenomenon Three-Dimensional Studies on the Ariane 4 Launcher," *Acta Astronautica*, Vol. 15, Nos. 6 and 7, 1987, pp. 321–330.

9–12. T Shimura and K. Kamijo, "Dynamic Response of the LE-5 Rocket Engine Liquid Oxygen Pump," *Journal of Spacecraft and Rockets*, Vol. 22, No. 7, March–April 1985.

9–13. E. Kirner, W. Oechslein, D. Thelemann and D. Wolf, "Development Status of the Vulcain (HM 60) Thrust Chamber." *AIAA Paper 90;2255*, July 1990.

9–14. F. H. Reardon, "Combustion Stability Specification and Verification Procedure," *CPIA Publication 247*, October 1973.

9–15. T. L. Acker and C. E. Mitchell, "Combustion Zone–Acoustic Cavity Interactions in Rocket Combustors," *Journal of Propulsion and Power*, Vol. 10, No 2, March–April 1994, pp. 235–243.

CHAPTER 10

TURBOPUMPS, ENGINE DESIGN, ENGINE CONTROLS, CALIBRATION, INTEGRATION, AND OPTIMIZATION

In this chapter we first discuss a complex high-precision, high-speed, rotating subsystem, namely the turbopump. Only some high-thrust engines have turbopumps. This chapter contains an overall engine discussion which applies to all engines. This includes the liquid propellant rocket engine's design, performance, controls, calibration, propellant budget, integration, and optimization.

10.1. TURBOPUMPS

The assembly of a turbine with one or more pumps is called a *turbopump*. Its purpose is to raise the pressure of the flowing propellant. Its principal subsystems are a hot gas powered *turbine* and one or two *propellant pumps*. It is a high precision rotating machine, operating at high shaft speed with severe thermal gradients and large pressure changes, it usually is located next to a thrust chamber, which is a potent source of noise and vibration.

This turbopump feed system and its several cycles have been discussed in Section 6.6 and Fig. 6–2 categorizes the various common turbopump configurations. Turbopumps or installation of turbopumps in rocket engines are shown in Figs. 1–4, 6–1, 6–12, 8–19, 10–1, 10–2, 10–3, and 10–11; they are discussed in Refs. 6–1 and 10–1. A schematic diagram of different design arrangements of pumps and turbines for common turbopump types can be seen in Fig. 10–4. Table 10–1 shows lists parameters of pumps and turbines of two large rocket engines.

Specific nomenclature and terminology used in the next few paragraphs will be explained later in this Chapter. In Fig. 10–1 a simple turbopump with a

TABLE 10-1. Turbopump Characteristics

Engine:	Space Shuttle Main Engine[a]				LE-7[b]	
Feed System Cycle:	Modified Staged Combustion Cycle				Modified Staged Combustion Cycle	
Propellants:	Liquid Oxygen and Liquid Hydrogen				Liquid Oxygen and Liquid Hydrogen	

Pumps

	LPOTP	LPFTP	HPOTP	HPFTP	HPFTP	HPOTP
Designation[c]	LPOTP	LPFTP	HPOTP	HPFTP	HPFTP	HPOTP
Type	Axial flow	Axial flow	Dual inlet	Radial flow	Radial flow	Radial flow
No. of impeller stages	1	1	1+1[d]	3	2	1+1[d]
No. of aux. or inducers	—	—	—	1	1	1
Flow rate (kg/sec)	425	70.4	509 50.9	70.4	35.7	211.5 46.7[d]
Inlet pressure (MPa)	0.6	0.9	2.70 NA	1.63	0.343	0.736 18.2[d]
Discharge pressure (MPa)	2.89	2.09	27.8 47.8[d]	41.0	26.5	18.2 26.7[d]
Pump efficiency (%)	68	75	72 75[d]	75	69.9	76.5 78.4[d]

Turbines

	LPOTP	LPFTP	HPOTP	HPFTP	HPFTP	HPOTP
No. of stages	6	2	3	2	2	2
Type	Hydraulic LOX driven	Reaction-impulse	Reaction-impulse	Reaction-impulse	Reaction-impulse	Reaction-impulse
Flow rate (kg/sec)	105.5	264	27.7	66.8	33.1	15.4
Inlet temperature (K)	26.2	29.0	756	1000	871	863
Inlet pressure (MPa)	NA	1.29	32.9	32.9	20.5	19.6
Pressure ratio			1.54	1.50	1.43	1.37
Turbine efficiency (%)	69	60	74	79	73.2	48.1
Turbine speed (rpm)	5020	15,670	22,300	34,270	41600	18300
Turbine power (kW)	1120	2290	15,650	40,300	25,350	7012
Mixture ratio, O/F	LOX only	H₂ only	~0.62	~0.88	~0.7	~0.7

[a]Data courtesy of The Boeing Company, Rocketdyne Propulsion and Power, at flight power level of 104.5% of design thrust.
[b]Data courtesy of Mitsubishi Heavy Industries, Ltd.
[c]LPOTP, low-pressure oxidizer turbopump; HPFTP, high-pressure fuel turbopump.
[d]Boost impeller stage for oxygen flow to preburners or gas generator.

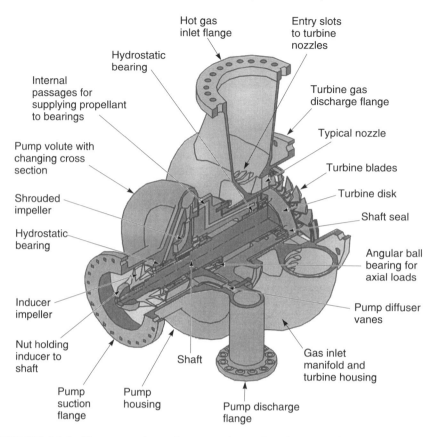

FIGURE 10–1. Cut-away view of an experimental turbopump demonstrator with a single-stage liquid oxygen pump impeller, an inducer impeller, and a single-stage turbine (one row of blades) on the same shaft. (Courtesy of The Boeing Company, Rocketdyne Propulsion and Power.)

single-stage propellant pump (with an inducer impeller ahead of the main impeller) is driven by a single-stage axial-flow turbine. The hot combustion gases, which drive this turbine, are burned in a separate gas generator (or a precombustion chamber or preburner) at a mixture ratio that gives gases between 900 and 1200 K; this is sufficiently cool, so that the hot turbine hardware (blades, nozzles, manifolds, or disks) still have sufficient strength without needing forced cooling. The gases are expanded (accelerated) in an annular set of converging–diverging supersonic turbine nozzles, which are cast into the cast turbine inlet housing. The gases then enter a set of rotating blades, which are mounted on a rotating wheel or turbine disk. The blades remove the tangential energy of the gas flow. The exhaust gas velocity exiting from the blades is relatively low and its direction is essentially parallel to the shaft. The pump is driven by the turbine through an interconnecting solid shaft. The propellant enters the pump through an *inducer*, a special impeller where the pressure of the propellant is raised only slightly (perhaps 5 to 10% of the total pressure rise). This is just enough pressure so that there will be no cavitation as the flow enters the main pump impeller. Most of the kinetic energy given to the flow by the pump impeller is converted into hydrostatic pressure in the diffusers (the diffuser vanes are not clearly visible, since they are inclined) and/or volutes of the pump. The two hydrostatic bearings support the shaft radially. All bearings and shaft seals create heat as they run. They are cooled and lubricated by a small flow of propellant, which is supplied from the pump discharge through drilled passages. One bearing (near the pump) is very cold and the other is hot, since it is close to the hot turbine. The angular ball bearing accepts the axial net loads from the unbalanced hydrodynamic pressures around the shrouded impeller, the inducer, and also the turbine blades or the turbine disk.

A novel, high speed, compact, and light weight liquid hydrogen turbopump is shown in Fig. 10–2 and in Ref 10.2. It is intended to be used with a new upper stage hydrogen/oxygen rocket engine with a thrust of about 50,000 lbf (22.4 kN), a delivered engine specific impulse of 450.6 sec at an engine mixture ratio of 6.0. This engine will run on an expander cycle, with a chamber pressure of 1375 psia (96.7 kg/m^2) and a maximum internal fuel pressure of 4500 psi (323.4 kg/m^2) at the fuel pump discharge. The unique single-piece titanium rotor turns nominally at 166,700 rpm, has two machined sets of pump vanes, a machined inducer impeller, a set of machined radial inflow turbine

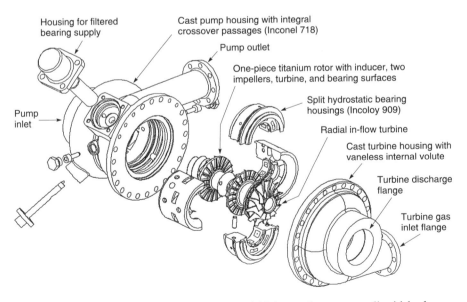

FIGURE 10–2. Exploded view of an advanced high-speed, two-stage liquid hydrogen fuel pump driven by a radial flow turbine. (Copied with permission of Pratt & Whitney, a division of United Technologies; adapted from Ref. 10–2.)

blades, and radial as well as axial bearing surfaces. A small filtered flow of hydrogen lubricates the hydrostatic bearing surfaces. The cast pump housing has internal crossover passages between stages. The unique radial in-flow turbine (3.2 in. dia.) produces about 5900 hp at an efficiency of 78%. The hydrogen pump impellers are only 3.0 in. diameter and produce a pump discharge pressure of about 4500 psi at a fuel flow of 16 lbm/sec and an efficiency of 67%. A high pump inlet pressure of about 100 psi is needed to assure cavitation-free operation. The turbopump can operate at about 50% flow (at 36% discharge pressure and 58% of rated speed). The number of pieces to be assembled is greatly reduced, compared to a more conventional turbopump, thus enhancing its inherent reliability.

The geared turbopump in Fig. 10–3 has a higher turbine and pump efficiencies, because the speed of the two-stage turbine is higher than the pump shaft speeds and the turbine is smaller. The auxiliary power package (e.g., hydraulic pump) was used only in an early application. The precision ball bearings and

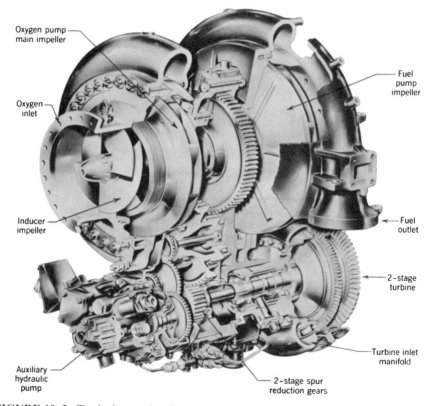

FIGURE 10–3. Typical geared turbopump assembly used on the RS-27 engine (Delta I and II Launch Vehicles) with liquid oxygen and RP-1 propellants. (Courtesy of The Boeing Company, Rocketdyne Propulsion and Power.)

seals on the turbine shaft can be seen, but the pump bearings and seals are not visible in this figure.

Approach to Turbopump Preliminary Design

With all major rocket engine components the principal criteria (high performance or efficiency, minimum mass, high reliability, and low cost) have to be weighted and prioritized for each vehicle mission. For example, high efficiency and low mass usually mean low design margins, and thus lower reliability. A higher shaft speed will allow a lower mass turbopump, but it cavitates more readily and requires a higher tank pressure and heavier vehicle tanks (which often outweigh the mass savings in the turbopump) in order to have acceptable life and reliability.

The engine requirements give the initial basic design goals for the turbopump, namely propellant flow, the pump outlet or discharge pressure (which has to be equal to the chamber pressure plus the pressure drops in the piping, valves, cooling jacket, and injector), the desired best engine cycle (gas generator or staged combustion, as shown in Fig. 6–9), the start delay, and the need for restart or throttling, if any. Also, the propellant properties (density, vapor pressure, viscosity, or boiling point) must be known. Some of the design criteria are explained in Refs. 6–1 and 10–3, and basic texts on turbines and pumps are listed as Refs. 10–4 to 10–8.

There are several design variations or geometrical arrangements for transmitting turbine power to one or more propellant pumps; some are shown schematically in Fig. 10–4. If the engine has propellants of similar density (such as liquid oxygen and RP-1), the fuel and oxidizer pumps will have similar shaft speeds and can usually be placed on a common shaft driven by a single turbine (F-1, RS-27/Delta Fig. 10–3, Atlas, or Redstone engines). If there is a mismatch between the optimum pump speed and the optimum turbine speed (which is usually higher), it may save inert mass and turbine drive gas mass to interpose a gear reduction between their shafts. See Fig. 6–11. For the last two decades designers have preferred to use direct drive, which avoids the complication of a gear case but at a penalty in efficiency and the amount of turbine drive propellant gas required. See Figs. 6–12, 10–1, or 10–2.

If the densities are very different (e.g., liquid hydrogen and liquid oxygen), the pump head rise* (head = $\Delta p / \rho$) is much higher for the lower-density propellant, and the hydrogen pump usually has to have more than one impeller or one stage and will typically operate at a higher shaft speed; in this case separate

---

Pump head means the difference between pump discharge and pump suction head. Its units are meters or feet. The head is the height of a column of liquid with equivalent pressure at its bottom. The conversion from pounds per square inch into feet of head is: (X) psi $= 144(X)/$density (lb/ft^3). To convert pascals (N/m^2) of pressure into column height (m), divide by the density (kg/m^3) and g_0 (9.806 m/sec^2).

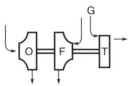

Two pumps on same shaft
with outboard turbine.
Shaft goes through fuel pump inlet.

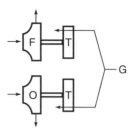

Two turbines, each with one pump.
Gas flow shown in parallel.
(alternate is gas flow in series, first through
one and then the other turbine).

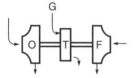

Direct drive with turbine in middle.
Shaft goes through turbine
discharge manifold.

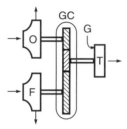

With gear case, turbine can
run faster. The two pumps
have different speeds.

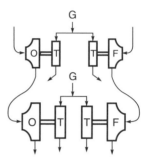

Two main pumps and two booster
pumps, each with its own gas turbine

FIGURE 10–4. Simplified diagrams of different design arrangements of turbopumps. F is fuel pump, O is oxidizer pump, T is turbine, G is hot gas, and GC is gear case.

turbopumps for the fuel and the oxidizer can give the lowest energy and overall mass (J-2, SSME, LE-7, Vulcain 60).

Usually, the preliminary analysis for the pump is done first. Avoiding excessive cavitation sets a key pump parameter, namely the maximum shaft speed. This is the highest possible shaft speed, which in turn allows the lightest turbopump mass, without excessive cavitation in the pump. If excessive cavitation occurs at the leading edge of the first impeller (inducer or main impeller), then the flow will become unsteady and variable, leading to lower thrust and possible combustion instability. The amount of pressure in the vehicle (gas pressure in propellant tank plus the static elevation pressure) that can be made available to the engine (at the pump inlet) for suppressing cavitation has to be larger than the impeller vanes' own pressure limit to cavitate. This allows us then to determine the shaft speed, which in turn can establish the approximate pump efficiencies, impeller tip speed (usually limited by the material strength of the impeller), number of pump stages, key dimensions of the impeller, and the pump power requirements. All this will be discussed further (including key equations) in the pump section of this chapter.

The key turbine parameter can be estimated, because the power output of the turbine essentially has to equal the power demand of the pump. If the pump is driven directly, that is without a gear case, then the pump speed and the turbine speed are equal. From the properties of the turbine drive gas (temperature, specific heat, etc.), the strength limits of the turbine materials, and the likely pressure drop, it is possible to determine the basic dimensions of the blades (pitch line velocity, turbine nozzle outlet velocity, number of rows (stages) of blades, turbine type, or turbine efficiency). The particular arrangement or geometry of the major turbopump components is related to their selection process. Most propellant pumps have a single-stage main impeller. For liquid hydrogen with its low density, a two- or three-stage pump is normally needed. Usually some design limit is reached which requires one or more iterations, each with a new changed approach or parameter. The arrangement of the major turbopump components (Fig. 10–4) is also influenced by the position of the bearings on the shaft. For example, we do not want to place a bearing in front of an impeller inlet because it will cause turbulence, distort the flow distribution, raise the suction pressure requirement, and make cavitation more likely to occur. Also, bearings positioned close to a turbine will experience high temperatures, which influences the lubrication by propellant and may demand more cooling of the bearings.

The use of booster pumps allows lower tank pressure, and thus lower inert vehicle mass, and provides adequate suction pressures to the main pump inlet. Booster pumps are used in the Space Shuttle main engine and the Russian RD-170, as seen in Figs. 6–12 and 10–11. Some booster pumps have been driven by a liquid booster turbine using a small flow of high-pressure liquid propellant that has been tapped off the discharge side of the main pump. The discharged turbine liquid then mixes with the main propellant flow at the discharge of the booster pump.

Later in this section a few of the equations that apply to the steady-state (full thrust) operating condition will be described. However, no detailed discussion will be given of the *transient starting conditions*, such as the filling of pipes, pumps, or manifolds with liquid propellants, or the filling of turbines and their manifolds with high-pressure gas. These dynamic conditions can be complex, are related to the combustion reactions, and are sometimes difficult to analyze, yet they are very significant in the proper and safe operation of the engine. Each major rocket engine manufacturer has developed some methodology, usually analysis and hydraulic models, for these system dynamics that are often peculiar to specific engines and hardware (see Refs. 10–3 and 10–4).

Mass is at a premium in all flying installations, and the feed system is selected to have a minimum combined mass of tubines, pumps, gas generator, valves, tanks, and gas generator propellants. Some of the considerations in the design of turbopumps are the thermal stresses, warpage due to thermal expansion or contraction, axial loads, adequate clearances to prevent rubbing yet minimize leakage, alignment of bearings, provisions for dynamic balancing of rotating parts, mounting on an elastic vehicle frame without inducing external forces, and avoiding undue pressure loads in the liquid and gas pipes.

Vibrations of turbopumps have caused problems during development. The analyses of the various vibrations (shaft, turbine blades, liquid oscillations, gas flow oscillations, or bearing vibrations) are not given here. At the *critical speed* the natural structural resonance frequency of the rotating assembly (shaft, impellers, turbine disk, etc.) coincides with the rotation operating speed. A slight unbalance can be amplified to cause significant shaft deflections (in bending), bearing failure, and other damage. The operating speed therefore is usually lower and sometimes higher than the critical speed. A large diameter stiff shaft, rigid bearings, and stiff bearing supports will increase this critical speed, and damping (such as the liquid lubricant film in the bearing) will reduce the vibration amplitude. Also, this critical shaft frequency or the operating speed should not coincide with and excite other natural vibration frequencies, such as those of various parts (piping, bellows, manifolds, or injector dome). The solving of various internal vibrations problems, such as whirl in bearings and blade vibrations, is reported in Ref. 10–5.

Bearings in most existing turbopumps are high precision, special alloy ball or roller bearings. Some ball bearings can take both radial and axial loads. Ball and roller bearings are limited in the loads and speeds at which they can operate reliably. In some turbopump designs this maximum bearing speed determines the minimum size of turbopump, rather than the cavitation limit of the pump. More recently, we use hydrostatic bearings where the shaft rides on a high-pressure fluid film; they have good radial load capacity, can provide some damping of oscillations and a stiff support. Axial loads (due to pressure unbalance on impellers and turbine blades) can be taken by special hydrostatic bearings. Since there is no direct contact between rolling and stationary assemblies, there is little or no wear and the life expectancy of these hydrostatic bearings is long. However, there is rubbing contact and wear at low speeds, namely during start or shutdown (see Ref. 10–6).

Cooling and lubricating the bearings and seals is essential for preventing bearing problems. A small flow of one of the propellants is used. Hydrocarbon fuels are usually good lubricants and hydrogen is a good coolant, but a marginal lubricant. If an oxidizer is used as the coolant and lubricant, then the materials used for bearings and seals have to be resistant to oxidation when heated during operation.

If the turbopump is part of a reusable rocket engine, it becomes more complex. For example, it can include provision to allow for inspection and automatic condition evaluation after each mission or flight. This can include an inspection of bearings through access holes for boroscope instruments, checking for cracks in highly stressed parts (turbine blade roots or hot-gas high-pressure manifolds), or the measurement of shaft torques (to detect possible binding or warpage).

Pumps

Classification and Description. The *centrifugal pump* is generally considered the most suitable for pumping propellant in large rocket units. For the large flows and high pressures involved, they are efficient as well as economical in terms of mass and space requirement.

Figure 10–5 is a schematic drawing of a centrifugal pump. Fluid entering the *impeller*, which is essentially a wheel with spiral curved vanes rotating within a *casing*, is accelerated within the impeller channels and leaves the impeller periphery with a high velocity to enter the *volute*, or collector, and thereafter the *diffuser*, where conversion from kinetic energy (velocity) to potential energy (pressure) takes place. In some pumps the curved diffuser vanes are upstream of the collector. The three-dimensional hydraulic design of impeller vanes, diffuser vanes, and volute passages can be accomplished by computer programs to give high efficiency and adequate strength. Internal leakage, or circulation between the high-pressure (discharge) side and the low-pressure (suction) side of an impeller, is held to a minimum by maintaining close clearances between the rotating and stationary parts at the *seals* or *wear ring surfaces*. External leakage along the shaft is minimized or prevented by the use of a *shaft seal*. Single-stage pumps (one impeller only) are stress-limited in the pressure rise they can impart to the liquid, and multiple-stage pumps are therefore needed for high pump head,[*] such as with liquid hydrogen. References 10–5 to 10–7 give information on different pumps. There is a free passage of flow through the pump at all times, and no positive means for shutoff are provided. The pump characteristics, that is, the pressure rise, flow, and efficiency, are functions of the pump speed, the impeller, the vane shape, and the casing configuration. Figure 10–6 shows a typical set of curves for centrifugal

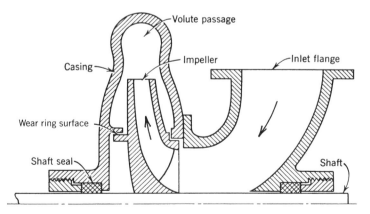

FIGURE 10–5. Simplified schematic half cross section of a typical centrifugal pump.

[*]See footnote on page 367.

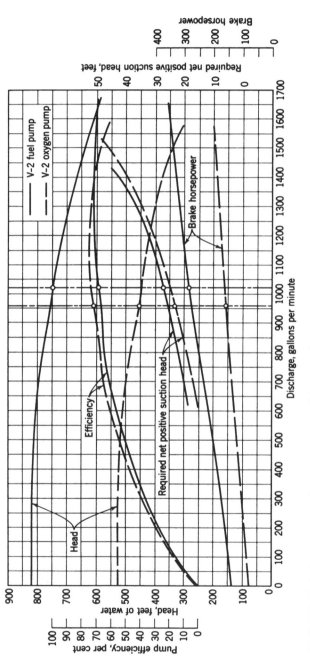

FIGURE 10–6. Water test performance curves of the centrifugal pumps of the German V-2 rocket engine. The propellants are diluted 75% ethyl alcohol and liquid oxygen.

pumps. The negative slope on the head versus flow curve indicates a stable pump behavior. References 10–7 and 10–8 describe the development of a smaller turbopump and the testing of a spiral high-speed first-stage impeller, called an inducer.

A *shrouded impeller* has a shroud or cover (in the shape of a surface of revolution) on top of the vanes as shown in Figs. 10–1, 10–3, and 10–5. This type usually has higher stresses and lower leakage around the impeller. In an unshrouded impeller or turbine the vanes are not covered as seen in the turbine vanes in Fig. 10.2.

Pump Parameters. This section outlines some of the important parameters and features that have to be considered in the design of rocket propellant centrifugal pumps under steady flow conditions.

The *required pump flow* is established by the rocket design for a given thrust, effective exhaust velocity, propellant densities, and mixture ratio. In addition to the flow required by the thrust chamber, the propellant consumption of the gas generator, and in some designs also a bypass around the turbine and auxiliaries have to be considered in determining the pump flows. The required *pump discharge pressure* is determined from the chamber pressure and the hydraulic losses in valves, lines, cooling jacket, and injectors (see Eq. 6–15). To obtain the rated flow at the rated pressure, an additional adjustable pressure drop for a control valve or orifice is usually included which permits a calibration adjustment or change in the required feed pressure. A regulation of the pump speed can also change the required adjustable pressure drop. As described in Section 10.6, this adjustment of head and flow is necessary to allow for hydraulic and performance tolerances on pumps, valves, injectors, propellant density, and so on.

It is possible to predict the *pump performance at various speeds* if the performance is known at any given speed. Because the fluid velocity in a given pump is proportional to the pump speed N, the flow quantity or discharge Q is also proportional to the speed and the head H is proportional to the square of the speed. This gives the following relations:

$$Q \text{ (flow)} \sim N \text{ (rpm or rad/sec)}$$

$$H \text{ (pump head)} \sim N^2 \tag{10–1}$$

$$P \text{ (pump power)} \sim N^3$$

From these relations it is possible to derive a parameter called the *specific speed* N_s. It is a dimensionless number derived from a dimensional analysis of pump parameters as shown in Ref. 10–9.

$$N_s = N\sqrt{Q_e}/(g_0 \Delta H_e)^{3/4} \tag{10–2}$$

Any set of consistent units will satisfy the equation: for example, N in radians per second, Q in m^3/s, g_0 as 9.8 m/sec^2, and H in meters. The subscript e refers to the maximum efficiency condition. In U.S. pump practice it has become the custom to delete g_0, express N in rpm, and Q in gallons per minute or ft^3/sec. Much of the existing U.S. pump data is in these units. This leads to a modified form of Eq. 10–2, where N_s is not dimensionless, namely

$$N_s = 21.2N\sqrt{Q_e}/(\Delta H_e)^{3/4} \tag{10–3}$$

The factor 21.2 applies when N is in rpm, Q is in ft^3/sec, and H is in feet. For each range of specific speed, a certain shape and impeller geometry has proved most efficient, as shown in Table 10–2. Because of the low density, hydrogen can be pumped effectively by axial flow devices.

The *impeller tip speed* in centrifugal pumps is limited by design and material strength considerations to about 60 to 450 m/sec or roughly 200 to 1475 ft/sec. With titanium (lower density than steel) and machined unshrouded impellers a tip speed of over 2150 ft/sec is now possible and used on the pumps shown in Fig. 10–2. For cast impellers this limiting value is lower than for machined impellers. This maximum impeller tip speed determines the maximum head that can be obtained from a single stage. The impeller vane tip speed u is the product of the shaft speed, expressed in radians per second, and the impeller radius and is related to the pump head by

$$u = \psi\sqrt{2g_0\Delta H} \tag{10–4}$$

where ψ has values between 0.90 and 1.10 for different designs. For many pumps, $\psi = 1.0$.

TABLE 10-2. Pump Types

	Impeller type				
	Radial	Francis	Mixed flow	Near axial	Axial
Basic shape (half section)					
Specific speed N_s					
U.S. nomenclature	500–1000	1000–2000	2000–3000	3000–6000	Above 8000
SI consistent units	0.2–0.3	0.4	0.6–0.8	1.0–2.0	Above 2.5
Efficiency %	50–80	60–90	70–92	76–88	75–82

The flow quantity defines the impeller inlet and outlet areas according to the equation of continuity. The diameters obtained from this equation should be in the proportion indicated by the diagrams for a given specific speed in Table 10–2. The continuity equation for an incompressible liquid is

$$Q = A_1 v_1 = A_2 v_2 \qquad (10\text{--}5)$$

where the subscripts refer to the impeller inlet and outlet sections, all areas being measured normal to their respective flow velocity. The inlet velocity v_1 ranges usually between 2 and 6 m/sec or 6.5 to 20 ft/sec and the outlet velocity v_2 between 3 and 15 m/sec or 10 to 70 ft/sec. For a compressible liquid, such as liquid hydrogen, the density will change with pressure. The continuity equation then is:

$$\dot{m} = A_1 v_1 \rho_1 = A_2 v_2 \rho_2 \qquad (10\text{--}6)$$

The head developed by the pump will then also depend on the change in density.

The pump performance is limited by *cavitation*, a phenomenon that occurs when the static pressure at any point in a fluid flow passage becomes less than the fluid's vapor pressure. The formation of vapor bubbles causes cavitation. These bubbles collapse when they reach a region of higher pressure, that is, when the static pressure in the fluid is above the vapor pressure. In centrifugal pumps cavitation is most likely to occur behind the leading edge of the pump impeller vane at the inlet because this is the point at which the lowest absolute pressure is encountered. The excessive formation of vapor causes the pump discharge mass flow to diminish and fluctuate and can reduce the thrust and make the combustion erratic and dangerous (see Ref. 10–10).

When the bubbles travel along the pump impeller surface from the low-pressure region (where they are formed) to the downstream higher-pressure region, the bubbles collapse. The sudden collapses create local high-pressure pulses that have caused excessive stresses in the metal at the impeller surface. In most rocket applications this cavitation erosion is not as serious as in water or chemical pumps, because the cumulative duration is relatively short and the erosion of metal on the impeller is not usually extensive. It has been a concern with test facility transfer pumps.

The *required suction head* $(H_s)_R$ is the limit value of the head at the pump inlet (above the local vapor pressure); below this value cavitation in the impeller will not occur. It is a function of the pump and impeller design and its value increases with flow as can be seen in Fig. 10–6. To avoid cavitation the *suction head* above vapor pressure *required* by the pump $(H_s)_R$ must always be *less* than the *available* or *net positive suction head* furnished by the line up to the pump $(H_s)_A$, that is, $(H_s)_R \leq (H_s)_A$. The required suction head above vapor pressure can be determined from the suction specific speed S:

$$S = 21.2 N \sqrt{Q_e}/(H_s)_R^{3/4} \qquad (10\text{--}7)$$

The suction specific speed S depends on the quality of design and the specific speed N_s, as shown in Table 10–2. The suction specific speed S has a value between 5000 and 60,000 when using ft-lbf units. For pumps with poor suction characteristics it has values near 5000, for the best pump designs without cavitation it has values near 10,000 and 25,000, and for pumps with limited and controllable local cavitation it has values above 40,000. In Eq. 10–7 the required suction head $(H_s)_R$ is usually defined as the critical suction head at which the developed pump discharge head has been diminished arbitrarily by 2% in a pump test with increasing throttling in the suction side. Turbopump development has, over the last several decades, led to impeller designs which can operate successfully with considerably more cavitation than the arbitrary and commonly accepted 2% head loss limit. Inducers are now designed to run stably with extensive vapor bubbles near the leading edge of their vanes, but these bubbles collapse at the trailing end of these vanes. Inducers now can have S values above 80,000. A discussion of the design of impeller blades can be found in Ref. 10–9.

The head that is available at the pump suction flange is called the *net positive suction head* or *available suction head above vapor pressure* $(H_s)_A$. It is an absolute head value determined from the tank pressure (the absolute gas pressure in the tank above the liquid level), the elevation of the propellant level above the pump inlet, the friction losses in the line between tank and pump, and the vapor pressure of the fluid. When the flying vehicle is undergoing accelerations, the head due to elevation must be corrected accordingly. These various heads are defined in Fig. 10–7. The net positive suction head $(H_s)_A$ is the maximum head available for suppressing cavitation at the inlet to the pumps:

$$(H_s)_A = H_{\text{tank}} + H_{\text{elevation}} - H_{\text{friction}} - H_{\text{vapor}} \qquad (10\text{–}8)$$

To avoid pump cavitation, $(H_s)_A$ has to be higher than $(H_s)_R$. If additional head is required by the pump, the propellant may have to be pressurized by external means, such as by the addition of another pump in series (called a booster pump) or by gas pressurization of the propellant tanks. This latter method requires thicker tank walls and, therefore, heavier tanks, and a bigger gas-pressurizing system. For example, the oxygen tank of the German V-2 was pressurized to 2.3 atm, partly to avoid pump cavitation. For a given value of $(H_s)_A$, propellants with high vapor pressure require correspondingly higher tank pressures and heavier inert tank masses. For a given available suction head $(H_s)_A$, a pump with a low required suction pressure usually permits designs with high shaft speeds, small diameter, and low pump inert mass. A small value of $(H_s)_R$ is desirable because it may permit a reduction of the requirements for tank pressurization and, therefore, a lower inert tank mass. The value of $(H_s)_R$ will be small if the impeller and fluid passages are well designed and if the shaft speed N is low. A very low shaft speed, however, requires a large diameter pump, which will be excessively heavy. The trend in

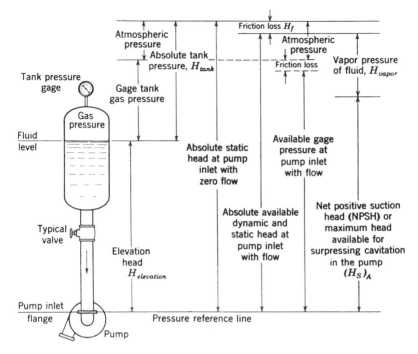

FIGURE 10–7. Definition of pump suction head.

selecting centrifugal pumps for rocket application has been to select the highest shaft speed that gives a pump with a low value of $(H_S)_R$, does not require excessive tank pressurization or other design complications, and thereby permits relatively lightweight pump design. This places a premium on pumps with good suction characteristics.

There have been some low-thrust, low-flow, experimental engines that have used positive displacement pumps, such as diaphragm pumps, piston pumps, or rotary displacement pumps (gear and vane pumps). For low values of N_S these pumps have much better efficiencies, but their discharge pressures fluctuate with each stroke and they are noisy.

One method to provide a lightweight turbopump with minimal tank pressure is to use an *inducer*, which is a special pump impeller usually on the same shaft and rotating at the same speed as the main impeller. It has a low head rise and therefore a relatively high specific speed. Inducer impellers are immediately upstream of the main impeller. They are basically axial flow pumps with a spiral impeller, and many will operate under slightly cavitating conditions. The inducer stage's head rise (typically, 2 to 10% of the total pump head) has to be just large enough to suppress cavitation in the main pump impeller; this allows a smaller, lighter, higher-speed main pump. Figures 10–3 and 10–8 show an inducer and Ref. 10–8 describes the testing of one of them.

FIGURE 10–8. Fuel pump inducer impeller of the Space Shuttle main engine low-pressure fuel turbopump. It has a diameter about 10 in., a nominal hydrogen flow of 148.6 lbm/sec, a suction pressure of 30 psi, a discharge pressure of 280 psi at 15,765 rpm, an efficiency of 77%, and a suction specific speed of 39,000 when tested with water. (Courtesy of The Boeing Company, Rocketdyne Propulsion and Power.)

In some rockets the inert mass of the turbopump and tank system can be further reduced by putting the inducer impeller into a separate low-power, low-speed booster turbopump, driven by its own separate turbine. In the Space Shuttle main engine there are two such low-pressure-rise turbopumps, as shown in the flow diagram of Fig. 6–4 and the engine view of Fig. 6–1. This allows the inducer impeller to be operated at an optimum (lower) shaft speed.

Influence of Propellants. For the same power and mass flow, the pump head is inversely proportional to the propellant density. Since pumps are basically constant-volume flow machines, the propellant with the highest density requires less head, less power and thus allows a smaller pump assembly.

Because many of the propellants are dangerous to handle, special provision has to be made to prevent any leakage through the shaft seals. With spontaneously ignitable propellants the leakages can lead to fires in the pump compartment and may cause explosions. Multiple seals are often used with a drainage provision that safely removes or disposes of any propellants that flow past the first seal. Inert-gas purges of seals have also been used to remove hazardous propellant vapors. The sealing of corrosive propellants puts very severe requirements on the sealing materials and design. With cryogenic propellants the pump bearings are usually lubricated by the propellant, since lubricating oil would freeze at the low pump hardware temperature.

Centrifugal pumps should operate at the highest possible *pump efficiency*. This efficiency increases with the volume flow rate and reaches a maximum value of about 90% for very large flows (above 0.05 m³/sec) and specific speeds above about 2500 (see Refs. 6–1 and 10–9). Most propellant pump efficiencies are between 30 and 70%. The pump efficiency is reduced by surface roughness of casing and impellers, the power consumed by seals, bearings, and stuffing boxes, and by excessive wear ring leakage and poor hydraulic design. The pump efficiency η_P is defined as the fluid power divided by the pump shaft power P_P:

$$\eta_P = \rho Q \, \Delta H / P_P \qquad (10\text{–}9)$$

A correction factor of 550 ft-lbf/hp has to be added if P_P is given in horse-power, H in feet, and Q in ft³/sec. When using propellants, the pump power has to be multiplied by the density ratio if the required power for water tests is to be determined.

Example 10–1. Determine the shaft speed and the overall impeller dimensions for a liquid oxygen pump which delivers 500 lb/sec of propellant at a discharge pressure of 1000 psia and a suction pressure of 14.7 psia. The oxygen tank is pressurized to 35 psia. Neglect the friction in the suction pipe and the suction head changes due to acceleration and propellant consumption. The initial tank level is 15 ft above the pump suction inlet.

SOLUTION. The density of liquid oxygen is 71.2 lbm/ft³ at its boiling point. The volume flow will be $500/71.2 = 7.022$ ft³/sec. The vapor pressure of the oxygen is 1 atm $= 14.7$ psi$= 29.8$ ft. The suction head is $35 \times 144/71.2 = 70.8$ ft. From Eq. 10–8 the available suction head is $70.8 + 14.7 = 85.5$ ft. The available suction head above vapor pressure is $(H_s)_A = 70.8 + 14.7 - 0 - 29.8 = 55.7$ ft. The discharge head is $1000 \times 144/71.2 = 2022$ ft. The head delivered by the pump is then $2022 - 85.5 = 1937$ ft.

The required suction head will be taken as 80% of the available suction head in order to provide a margin of safety for cavitation $(H_s)_R = 0.80 \times 85.5 = 68.4$ ft. Assume a suction specific speed of 15,000, a reasonable value if no test data are available. From Eq. 10–7 solve for the shaft speed N:

$$S = 21.2N\sqrt{Q}/(H_s)_R^{3/4} = 21.2N\sqrt{7.022}/68.4^{0.75} = 15,000$$

Solve for $N = 6350$ rpm   or   664.7 rad/sec.

The specific speed, from Eq. 10–3, is

$$N_s = 21.2N\sqrt{Q}/H^{3/4} = 21.2 \times 6350\sqrt{7.022}/1937^{0.75} = 1222$$

According to Table 10–2, the impeller shape for this value of N_s will be a Francis type. The impeller discharge diameter D_2 can be evaluated from the tip speed by Eq. 10–4:

$$u = \psi\sqrt{2g_0\Delta H} = 1.0\sqrt{2 \times 32.2 \times 1937} = 353 \text{ ft/sec}$$
$$D_2 = 353 \times 2/664.7 = 1.062 \text{ ft} = 12.75 \text{ in.}$$

The impeller inlet diameter D_1 can be found from Eq. 10–5 by assuming a typical inlet velocity of 15 ft/sec and a shaft cross section 5.10 in.^{2} (2.548 in. diameter).

$$A = Q/v_1 = 7.022/15 = 0.468 \text{ ft}^2 = 67.41 \text{ in.}^2$$
$$A = \tfrac{1}{4}\pi D_1^2 + 5.10 = 67.41 + 5.10 = 72.51 \text{ in.}^2$$
$$D_1 = 9.61 \text{ in. (internal flow passage diameter)}$$

This is enough data to draw a preliminary sketch of the impeller.

Turbines

The turbine must provide adequate shaft power for driving the propellant pumps (and sometimes also auxiliaries) at the desired speed and torque. The turbine derives its energy from the expansion of a gaseous working fluid through fixed nozzles and rotating blades. The blades are mounted on disks to the shaft. The gas is expanded to a high, nearly tangential, velocity and through inclined nozzles and then flows through specially shaped *blades*, where the gas energy is converted into tangential forces on each blade. These forces cause the turbine wheel to rotate (see Refs.10–1 and 10–11).

Classification and Description. The majority of turbines have blades at the periphery of a turbine disk and the gas flow is axial, similarly in concept to the axial flow pattern shown for pumps in Table 10–2 and the single-stage turbine of Fig. 10–1. However, there are a few turbines with radial flow (particularly at high shaft speeds), such as the one shown in Fig. 10–2. Ideally there are two types of axial flow turbines of interest to rocket pump drives: impulse turbines and reaction turbines, as sketched in Fig. 10–9. In an *impulse turbine* the enthalpy of the working fluid is converted into kinetic energy within the first set of stationary turbine nozzles and not in the rotating blade elements. High-velocity gases are delivered (in essentially a tangential direction) to the rotating blades, and blade rotation takes place as a result of the impulse imparted by the momentum of the fluid stream of high kinetic energy to the rotating blades which are mounted on the turbine disk. The *velocity-staged impulse turbine* has a stationary set of blades which changes the flow direction after the gas leaves the first set of rototating blades and directs the gas to enter a second set of rotating blades in which the working fluid gives up further energy to the turbine wheel. In a *pressure-staged impulse turbine*, the expansion of the gas takes place in all the stationary rows of blades. In a *reaction turbine* the expansion of the gas is roughly evenly split between the rotating and stationary blade elements. The high pressure drop available for the expansion of the turbine working fluid in a gas generator cycles favors simple, lightweight one- or two-stage impulse turbines for high thrust engines. Many rocket turbines are neither pure impulse nor reaction turbines, but often are fairly close to an impulse turbine with a small reaction in the rotating vanes.

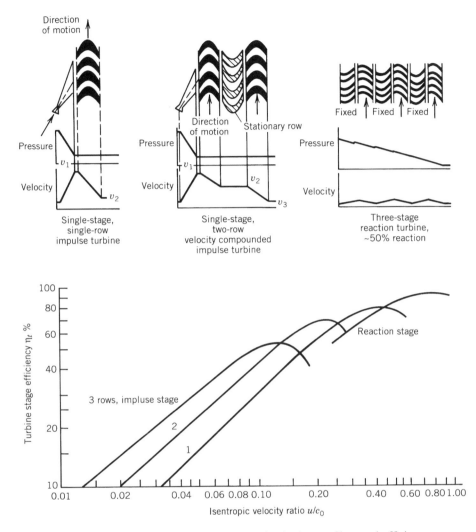

FIGURE 10–9. Top view diagram, pressure and velocity profiles, and efficiency curves for impulse and reaction type turbines. The velocity ratio is the pitch line velocity of the rotating blades u divided by the theoretical gas spouting velocity c_0 derived from the enthalpy drop. Adapted with permission from Refs. 10–1 and 10–12.

With some cycles the turbine exhaust gases pass through a *De Laval nozzle* at the exit of the exhaust pipe (see Fig. 1–4). The high turbine outlet pressure gives critical flow conditions at the venturi throat (particularly at high altitudes) and thereby assures a constant turbine outlet pressure and a constant turbine power which will not vary with altitude. Furthermore, it provides a small additional thrust to the engine.

Turbine Performance and Design Considerations. The power supplied by the turbine is given by a combined version of Eqs. 3–1 and 3–7:

$$P_T = \eta_T \dot{m}_T \Delta h \tag{10–10}$$

$$P_T = \eta_T \dot{m}_T c_p T_1 [1 - (p_2/p_1)^{(k-1)/k}] \tag{10–11}$$

The power delivered by the turbine P_T is proportional to the turbine efficiency η_T, the flow through the turbine $\dot{m}_T$, and the available enthalpy drop per unit of flow Δh. The units in this equation have to be consistent (1 Btu = 778 ft-lbf= 1055 J). This enthalpy is a function of the specific heat c_p, the nozzle inlet temperature T_1, the pressure ratio across the turbine, and the ratio of the specific heats k of the turbine gases. For gas generator cycles the pressure drop between the turbine inlet and outlet is relatively high, but the turbine flow is small (typically 2 to 5% of full propellant flow). For staged combustion cycles this pressure drop is very much lower, but the turbine flow is much larger.

For very large liquid propellant engines with high chamber pressure the turbine power can reach over 250,000 hp, and for small engines this could be perhaps around 35 kW or 50 hp.

According to Eq. 6–12, the power delivered by the turbine P_T has to be equal to the power required by the propellant pumps, the auxiliaries mounted on the turbopump (such as hydraulic pumps, electric generators, tachometers, etc.), and power losses in bearings, gears, seals, and wear rings. Usually these losses are small and can often be neglected. The effect of the turbine gas flow on the specific impulse of the rocket engine system is discussed in Sections 6.2 and 10.2. For gas generator engine cycles, the rocket designer is interested in obtaining a high turbine efficiency and a high turbine inlet temperature T_1 in order to reduce the flow of turbine working fluid, and for gas generator cycles also to raise the overall effective specific impulse, and, therefore, reduce the propellant mass required for driving the turbine. Three-dimensional computer analyses of the gas flow behavior and turbine blade geometry have resulted in efficient blade designs.

Better turbine blade materials (such as single crystals which have been unidirectionally solidified) and specialty alloys can allow turbine inlet temperatures between 1400 K (or about 2050°F) and perhaps 1600 K (or 2420°F); these higher temperatures or higher gas enthalpies reduce the required turbine flow. Reliability and cost considerations have kept actual turbine inlet temperatures at conservative values, such as 1150 to 1250°F or about 900 to 950 K, using lower cost steel alloy as the material. The *efficiency* of turbines for rocket turbopumps is shown in Fig. 10–9. Maximum *blade speeds* with good design and strong high-temperature materials are typically 400 to 700 m/sec or about 1300 to 2300 ft/sec. Higher blade speeds generally allow an improvement in efficiency. For the efficiency to be high the turbine blade and nozzle profiles have to have smooth surfaces. Small clearances at the turbine blade tips are also needed, to minimize leakage.

The low efficiency in many rocket turbines is dictated by centrifugal *pump design considerations*, which *limit the shaft speed* for turbopumps in which the pump and turbine are mounted on a common shaft, as discussed in the next section. A low shaft speed together with minimum mass requirements, which prohibit a very large turbine wheel diameter, give a low blade speed, which in turn reduces the efficiency.

The advantage of increased turbine efficiency (less gas generator propellant requirement) can be realized only if the turbopump design allows high blade speeds. This can be achieved in rockets of medium and low thrust by gearing the turbine to the pumpshaft or by using pumps that permit high shaft speeds; in rockets of very high thrust the pumps have diameters and shaft speeds close to those of the turbines and can be mounted on the same shaft as the turbine. The power input to the turbine can be regulated by controlling the flow to the turbine inlet. This can be accomplished by throttling or by-passing some of the flow of the working fluid to the turbine and varying the turbine inlet pressure.

There is *no warm-up time* available in rocket turbines. The sudden admission of hot gas at full flow causes severe thermal shock and thermal distortion and increases the chances for rubbing between moving metal parts. The most severe stresses of a turbine blade often are thermal stresses; they come during the engine start when the leading edge is very hot but other parts of the blade are still cold. This and other loading conditions can be more severe in rocket turbines than in air-burning gas turbines.

For low-thrust engines the shaft speeds can become very high, such as over 100,000 rpm. Also, the turbine blade height becomes very short and friction losses can become prohibitive. In order to obtain a reasonable blade height we go to partial admission turbine designs. Here a portion of the turbine nozzles are effectively plugged or eliminated.

Gas Generators and Preburners

A gas generator is used as the source of hot gas (from combustion of propellants) for driving many of the turbines of turbopumps in a liquid rocket engine. Depending on the engine cycle, other sources of turbine drive gases are sometimes employed, as described in Section 6.3.

Gas generators can be classified as monopropellant, bipropellant, or solid propellant. Actually the basic design parameters for gas generators are similar to those for engine thrust chambers or solid rocket motors. The combustion temperature is usually kept below 1400 to 1600 K (or 2000 to 2400°F) by intentionally regulating or mixing the propellants in proportions substantially different from stoichiometric mixture, usually fuel rich. These lower gas temperatures allow uncooled chamber construction and prevent melting or limit the erosion or turbine blades. With monopropellants, such as hydrogen peroxide (H_2O_2) or hydrazine (N_2H_4), the flow is easily controlled and the gases are generated at predictable temperatures depending on the details of the catalyst and the gas generator design. In principle a gas generator looks like an uncooled rocket thrust chamber except that the nozzle is replaced by a pipe leading to the turbine nozzles.

Propellants supplied to the liquid propellant gas generators can come from separate pressurized tanks or can be tapped off from the engine propellant pumps. When starting pump-fed gas generators, the turbomachinery needs to be brought up to operating speed. This can be done by a solid propellant gas generator starter, an auxiliary pressurized propellant supply, or by letting the engine "bootstrap" itself into a start using the liquid column head existing in the vehicle tankage and feed system lines—usually called a "tank-head" start.

Gas generators have been used for *other applications* besides supplying power to rocket feed systems. They have a use wherever there is a need for a large amount of power for a relatively short time, because they are simpler and lighter than conventional short-duration power equipment. Typical applications are gas generators for driving torpedo turbines and gas generators for actuating airplane catapults.

In a staged combustion cycle all of one propellant and a small portion of the other propellant (either fuel-rich or oxidizer-rich mixture) are burned to create the turbine drive gases. This combustion device is called a *preburner* and it is usually uncooled. It has a much larger flow than the gas generators mentioned above, its turbines have a much smaller pressure drop, and the maximum pressure of the propellants is higher.

10.2. PERFORMANCE OF COMPLETE OR MULTIPLE ROCKET PROPULSION SYSTEMS

The simplified relations that follow give the basic method for determining the overall specific impulse, the total propellant flow, and the overall mixture ratio as a function of the corresponding component performance terms for complete rocket engine systems. This applies to engine systems consisting of one or more thrust chambers, auxiliaries, gas generators, turbines, and evaporative propellant pressurization systems all operating at the same time.

Refer to Eqs. 2–5 and 6–1 for the specific impulse I_s, propellant flow rate $\dot{w}$ or $\dot{m}$ and mixture ratio r. The overall thrust F_{oa} is the sum of all the thrusts from thrust chambers and turbine exhausts and the overall flow $\dot{m}$ is the sum of their flows. The subscripts oa, o, and f designate the overall engine system, the oxidizer, and the fuel, respectively. Then

$$(I_s)_{oa} = \frac{\sum F}{\sum \dot{w}} = \frac{\sum F}{g_0 \sum \dot{m}} \tag{10-12}$$

$$\dot{w}_{oa} = \sum \dot{w} \quad \text{or} \quad \dot{m}_{oa} = \sum \dot{m} \tag{10-13}$$

$$r_{oa} \sim \frac{\sum \dot{w}_o}{\sum \dot{w}_f} = \frac{\sum \dot{m}_o}{\sum \dot{m}_f} \tag{10-14}$$

These same equations should be used for determining the overall performance when more than one rocket engine is contained in a vehicle propulsion system and they are operating simultaneously. They also apply to multiple solid propellant rocket motors and combinations of liquid propellant rocket engines and solid propellant rocket booster motors, as in the Space Shuttle (see Fig. 1–13).

Example 10–2. For an engine system with a gas generator similar to the one shown in Fig. 1–4, determine a set of equations that will express (1) the overall engine performance and (2) the overall mixture ratio of the propellant flows from the tanks. Let the following subscripts be used: c, thrust chamber; gg, gas generator; and tp, tank pressurization. For a nominal burning time t, a 1% residual propellant, and a 6% overall reserve factor, give a formula for the amount of fuel and oxidizer propellant required with constant propellant flow. Ignore stop and start transients, thrust vector control, and evaporation losses.

SOLUTION. Only the oxidizer tank is pressurized by vaporized propellant. Although this pressurizing propellant must be considered in determining the overall mixture ratio, it should not be considered in determining the overall specific impulse since it stays with the vehicle and is not exhausted overboard.

$$(I_s)_{oa} \sim \frac{F_c + F_{gg}}{(\dot{m}_c + \dot{m}_{gg})g_0} \tag{10–15}$$

$$r_{oa} \sim \frac{(\dot{m}_o)_c + (\dot{m}_o)_{gg} + (\dot{m}_o)_{tp}}{(\dot{m}_f)_c + (\dot{m}_f)_{gg}} \tag{10–16}$$

$$m_f = [(\dot{m}_f)_c + (\dot{m}_f)_{gg}]\, t\, (1.00 + 0.01 + 0.06)$$
$$m_o = [(\dot{m}_o)_c + (\dot{m}_o)_{gg} + (\dot{m}_o)_{tp}]\, t\, (1.00 + 0.01 + 0.06)$$

For this gas generator cycle the engine mixture ratio or r_{oa} is different from the thrust chamber mixture ratio $r_c = (m_o)_c/(m_f)_c$. Similarly, the overall engine specific impulse is slightly lower than the thrust chamber specific impulse. However, for an expander cycle or a staged combustion cycle these two mixture ratios and two specific impulses are the same, provided that there are no gasified propellant used for tank pressurization.

The overall engine specific impulse is influenced by the nozzle area ratio and the chamber pressure, and to a lesser extent by the engine cycle, and the mixture ratio. Table 10–3 describes 11 different rocket engines using liquid oxygen and liquid hydrogen propellants designed by different companies in different countries, and shows the sensitivity of the specific impulse to these parameters. References 10–13 to 10–15 give additional data on several of these engines.

TABLE 10-3. Comparison of Rocket Engines Using Liquid Oxygen and Liquid Hydrogen Propellants

Engine Designation Engine Cycle, Manuf. or Country (Year Qualified)	Vehicle	Thrust in Vacuum, kN (lbf)	Specific Impulse in Vacuum (sec)	Chamber Pressure, bar (psia)	Mixture Ratio	Nozzle Area Ratio	Engine Mass (dry), kg
SSME, staged combustion, Rocketdyne (1998)	Space Shuttle (3 required)	2183 (490,850)	452.5	196 (2870)	6.0	68.8	3400
RS-68, gas generator, Rocketdyne (2000)	Delta	3313 (745,000)	415	97.2 (1410)	6.0	21.5	6800
LE-5A, Expander bleed, MH1, Japan, (1991)	HII	121.5 (27,320)	452	37.2 (540)	5.0	130	255
LE-7, staged combustion, MH1, Japan (1992)	HII	1080 (242,800)	445.6	122 (1769)	6.0	52	1720
Vulcain, gas generator, SEP and other European Co.'s	Ariane 5	1120 (251,840)	433	112 (1624)	5.35	45	1585
HM7, gas generator, SEP France	Ariane 1,2,3,4	62.7 (14,100)	444.2	36.2 (525)	5.1	45	155
RL 10-A3, Expander, Pratt & Whitney (1965)	Various upper stages	73.4 (16,500)	444.4	32.75 (475)	5.0	61	132
RL 10-B2, (1998), same as above	—	110 (24,750)	466.5	44.12 (640)	6.0	375	275
YF 73, China	Long March	44,147 (10,000)	420	26.28 (381)	5.0	40	236
YF 75 (2 required), China		78.45 (17,600)	440	36.7 (532)	5.0	80	550

10.3. PROPELLANT BUDGET

In all liquid propellant rocket engines some of the propellants are used for purposes other than producing thrust or increasing the velocity of the vehicle. This propellant must also be included in the propellant tanks. A propellant budget can include the eleven items listed below, but very few engines have allowances for all these items. Table 10–4 shows an example of a budget for a spacecraft pressure-fed engine system with several small thrusters and one larger thrust chamber.

1. Enough propellant has to be a available for achieving the *required vehicle velocity increase* of the particular application and the particular flight vehicle or stage. The nominal velocity increment is usually defined by systems analysis and mission optimization using an iterative calculation based on Eqs. 4–19 or 4–35. If there are alternative flight paths or missions for the same vehicle, the mission with an unfavorable flight path and the highest total impulse should be selected. This mission-required propellant is the largest portion of the total propellants loaded into the vehicle tanks.

2. In a turbopump system using a gas generator cycle, a small portion of the overall propellant is burned in a separate *gas generator*. It has a lower flame temperature than the thrust chamber and operates at a different mixture ratio; this causes a slight change in the overall mixture ratio of propellants flowing from the tanks, as shown by Eqs. 10–14 and 10–16.

TABLE 10–4. Example of a Propellant Budget for a Spacecraft Propulsion System with a Pressurized Monopropellant Feed System

Budget Element	Typical Value
1. Main thrust chamber (increasing the velocity of stage or vehicle)	70–90% (determined from mission analysis and system engineering)
2. Flight control function (for reaction control thrusters and flight stability)	5–15% (determined by control requirements)
3. Residual propellant (trapped in valves, lines, tanks, etc.)	0.5–2% of total load[a]
4. Loading uncertainty	0.5% of total load[a]
5. Allowance for off-nominal performance	0.5–1.0% of total load[a]
6. Allowance for off-nominal operations	0.25–1.0% of total load[a]
7. Mission margin (reserve for first two items above)	3–10% of items 1 and 2
8. Contingency	2–10% of total load[a]

Source: Adapted from data supplied by TRW, Inc.
[a]Total load is sum of items 1, 2, and 7.

3. In a rocket propulsion system with a *thrust vector control* (TVC) system, such as a swiveling thrust chamber or nozzle, the thrust vector will be rotated by a few degrees. Thrust vector control systems are described in Chapter 16. There is a slight decrease in the axial thrust and that reduces the vehicle velocity increment in item 1. The extra propellant needed to compensate for the small velocity reduction can be determined from the mission requirements and TVC duty cycle. It could be between 0.1 and 4% of the total propellant.

4. In some engines a small portion of cryogenic propellants is heated, vaporized, and used to *pressurize cryogenic propellant tanks*. A heat exchanger is used to heat liquid oxygen from the pump discharge and pressurize the oxygen tank, as shown schematically in Fig. 1–4. This method is used in the hydrogen and oxygen tanks of the Space Shuttle external tank (see Ref. 6–6).

5. Auxiliary rocket engines that provide for *trajectory corrections, station keeping, maneuvers*, or *attitude control* usually have a series of small restartable thrusters (see Chapter 4). The propellants for these auxiliary thrusters have to be included in the propellant budget if they are supplied from the same feed system and tanks as the larger rocket engine. Depending on the mission and the propulsion system concept, this auxiliary propulsion system can consume a significant portion of the available propellants.

6. The *residual propellant* that clings to tank walls or remains trapped in valves, pipes, injector passages, or cooling passages is unavailable for producing thrust. It is typically 0.5 to 2% of the total propellant load. It increases the final vehicle mass at thrust termination and reduces the final vehicle velocity slightly.

7. A *loading uncertainty* exists due to variations in tank volume or changes in propellant density or liquid level in the tank. This is typically 0.25 to 0.75% of the total propellant. It depends, in part, on the accuracy of the method of measuring the propellant mass during loading (weighing the vehicle, flow meters, level gages, etc.).

8. The *off-nominal rocket performance* is due to variations in the manufacture of hardware from one engine to another (such as slightly different pressure losses in a cooling jacket, in injectors and valves, or somewhat different pump characteristics); these cause slight changes in combustion behavior, mixture ratio, or specific impulse. If there are slight variations in *mixture ratio*, one of the two liquid propellants will be consumed fully and an unusable residue will remain in the other propellant's tank. If a minimum total impulse requirement has to be met, extra propellant has to be tanked to allow for these mixture ratio variations. This can amount up to perhaps 2.0% of one of the propellants.

9. *Operational factors* can result in additional propellant requirements, such as filling more propellant than needed into a tank or incorrectly,

adjusting regulators or control valves. It can also include the effect of changes in flight acceleration from the nominal value. For an engine that has been carefully calibrated and tested, this factor can be small, usually betwen 0.1 and 1.0%.

10. When using cryogenic propellants an *allowance for evaporation and cooling down* has to be included. It is the mass of extra propellant that is allowed to evaporate (and be vented overboard while the vehicle is waiting to be launched) or that is fed through the engine to cool it down, before the remaining propellant in the tank becomes less than the minimum needed for the flight mission. Its quantity depends on the amount of time between topping off (partial refilling) of the tank.

11. Finally, an *overall contingency* or ignorance factor is needed to allow for unforeseen propellant needs or inadequate or uncertain estimates of any of the items above. This can also include allowances for vehicle drag uncertainties, variations in the guidance and control system, wind, or leaks.

Only some of the items above provide axial thrust (items 1, 2, and sometimes also 3 and 5), but all the items need to be considered in determining the total propellant mass and volume.

10.4. ENGINE DESIGN

The approach, methods, and resources used for rocket engine preliminary design and final design are usually different for each design organization and for each major type of engine. They also differ by the degree of novelty.

1. *A totally new engine with new major components* and some *novel design concepts* will result in an optimum engine design for a given application, but it is usually the most expensive and longest development approach. One of the major development costs is usually in sufficient testing of components and several engines (under various environmental and performance limit conditions), in order to establish credible reliability data with enough confidence to allow the initial flights and initial production. Since the state of the art is relatively mature today, the design and development of a truly novel engine does not happen very often.

2. *New engine using major components or somewhat modified key components from proven existing engines.* This is a common approach today. The design of such an engine requires working within the capability and limits of existing or slightly modified components. It requires much less testing for proving reliability.

3. *Uprated or improved version of an existing, proven engine.* This approach is quite similar to the second. It is needed when an installed engine for a given mission requires more payload (which really means higher thrust)

and/or longer burning duration (more total impulse). Uprating often means more propellant (larger tanks), higher propellant flows and higher chamber and feed pressures, and more feed system power. The engine usually has an increased inert engine mass (thicker walls).

In a simplified way, we describe here a typical process for designing an engine. At first the basic function and requirements of the new engine must be established. These engine requirements are derived from the vehicle mission and vehicle requirements, usually determined by the customer and/or the vehicle designers, often in cooperation with one or more engine designers. The engine requirements can include key parameters such as thrust level, the desired thrust–time variation, restart or pulsing, altitude flight profile, environmental conditions, engine locations within the vehicle, and limitations or restraints on cost, engine envelope, test location, or schedule. It also includes some of the factors listed later in Table 17–5. If an existing proven engine can be adapted to these requirements, the subsequent design process will be simpler and quite different than the design of a truly new engine.

Usually some early tentative decisions about the engine are made, such as the selection of the propellants, their mixture ratio, or the cooling approach for the hot components. They are based on mission requirements, customer preferences, past experiences, some analysis, and the judgement of the key decision makers. Some additional selection decisions include the engine cycle, having one, two, or more thrust chambers fed from the same feed system, redundancy of auxiliary thrusters, or type of ignition system. Trade-off studies between several options are appropriate at this time. With a modified existing engine these parameters are well established, and require few trade-off studies or analyses. Initial analyses of the pressure balances, power distribution between pumps and turbines, gas generator flow, propellant flows and reserves, or the maximum cooling capacity are appropriate. Sketches and preliminary estimates of inert mass of key components need to be made, such as tanks, thrust chambers, turbopumps, feed and pressurization systems, thrust vector control, or support structure. Alternate arrangements of components (layouts) are usually examined, often to get the most compact configuration. An initial evaluation of combustion stability, stress analysis of critical components, water hammer, engine performance at some off-design conditions, safety features, testing requirements, cost, and schedule are often performed at this time. Participation of appropriate experts from the field of manufacturing, field service, materials, stress analysis, or safety can be critical for selecting the proper engine and the key design features. A design review is usually conducted on the selected engine design and the rationale for new or key features.

Test results of subscale or full-scale components, or related or experimental engines, will have a strong influence on this design process. The key engine selection decisions need to be validated later in the development process by testing new components and new engines.

The *inert mass of the engine and other mass properties (center of gravity or moment of inertia)* are key parameters of interest to the vehicle designer or customer. They are needed during preliminary design and again, in more detail, in the final design. The engine mass is usually determined by summing up the component or subsystem masses, each of which is either weighed or estimated by calculating their volumes and knowing or assuming their densities. Sometimes early estimates are based on known similar parts or subassemblies.

Preliminary engine performance estimates are often based on data from prior similar engines. If these are not available, then theoretical performance values can be calculated (see Chapter 2, 3, and 5) for F, I_s, k, or $\mathfrak{M}$, using appropriate correction factors. Measured static test data are, of course, better than estimates. The final performance values are obtained from flight tests or simulated altitude tests, where airflow and altitude effects can interact with the vehicle or the plume.

If the preliminary design does not meet the engine requirements, then changes need to be made to the initial engine decisions and, if that is not sufficient, sometimes also to the mission requirements themselves. Components, pressure balances, and so forth will be reanalyzed and the results will be a modified version of the engine configuration, its inert mass, and performance. This process is iterated until the requirements are met and a suitable engine has been found. The initial design effort culminates in preliminary layouts of the engine, a preliminary inert mass estimate, an estimated engine performance, a cost estimate, and a tentative schedule. These preliminary design data form the basis for a written proposal to the customer for undertaking the final or detail design, development, testing, and for delivering engines.

Optimization studies are made to select the best engine parameters for meeting the requirements; some of them are done before a suitable engine has been identified, some afterwards. They are described further in Section 10.7. We optimize parameters such as chamber pressure, nozzle area ratio, thrust, mixture ratio, or number of large thrust chambers supplied by the same turbopump. The results of optimization studies indicate the best parameter, which will give a further, usually small, improvement in vehicle performance, propellant fraction, engine volume, or cost.

Once the engine proposal has been favorably evaluated by the vehicle designers, and after the customer has provided authorization and funding to proceed, then the final design can begin. Some of the analyses, layouts, and estimates will be repeated, but in more detail, specifications and manufacturing documents will be written, vendors will be selected, and tooling will be built. The selection of some of the key parameters (particularly those associated with some technical risk) will need to be validated. After another design review, key components and prototype engines are built and ground tested as part of a planned development effort. If proven reliable, one or two sets of engines will be installed in a vehicle and operated during flight. In those programs where a fair number of vehicles are to be built, the engine will then be produced in the required quantity.

Table 10–5 shows some of the characteristics of three different Russian designs staged combustion cycle engine designs, each at a different thrust and with different propellants (see Ref. 10–17). It shows primary engine para-

TABLE 10–5. Data on Three Russian Large Liquid Propellant Rocket Engines Using a Staged Combustion Cycle

Engine Designation	RD-120	RD-170	RD-253
Application (number of engines)	Zenit second stage (1)	Energia launch vehicle booster (4) and Zenit first stage (1)	Proton vehicle booster (1)
Oxidizer	Liquid oxygen	Liquid oxygen	N_2O_4
Fuel	Kerosene	Kerosene	UDMH
Number and types of turbopumps (TP)	One main TP and two boost TPs	One main TP and two boost TPs	Single TP
Thrust control, %	Yes	Yes	±5
Mixture ratio control, %	±10	±7	±12
Throttling (full flow is 100%), %	85	40	None
Engine thrust (vacuum), kg	85,000	806,000	167,000
Engine thrust (SL), kg	—	740,000	150,000
Specific impulse (vacuum), sec	350	337	316
Specific impulse (SL), sec	—	309	285
Propellant flow, kg/sec	242.9	2393	528
Mixture ratio, O/F	2.6	2.63	2.67
Length, mm	3872	4000	2720
Diameter, mm	1954	3780	1500
Dry engine mass, kg	1125	9500	1080
Wet engine mass, kg	1285	10500	1260
Thrust Chamber Characteristics			
Chamber diameter, mm	320	380	430
Characteristic chamber length, mm	1274	1079.6	999.7
Chamber area contraction ratio	1.74	1.61	1.54
Nozzle throat diameter, mm	183.5	235.5	279.7
Nozzle exit diameter, mm	1895	1430	1431
Nozzle area ratio,	106.7	36.9	26.2
Thrust chamber length, mm	2992	2261	2235
Nominal combustion temperature, K	3670	3676	3010
Rated chamber pressure, kg/cm^2	166	250	150
Nozzle exit pressure, kg/cm^2	0.13	0.73	0.7
Thrust coefficient, vacuum	1.95	1.86	1.83
Thrust coefficient, SL	—	1.71	1.65
Gimbal angle, degree	Fixed	8	Fixed
Injector type	Hot, oxidizer-rich precombustor gas plus fuel		

With a staged combustion cycle the thrust, propellant flow, and mixture ratio for the thrust chamber have the same values as for the entire engine.

TABLE 10–5. (*Continued*)

Engine Designation	RD-120		RD-170		RD-253	
	Turbopump Characteristics					
Pumped liquid	*Oxidizer*	*Fuel*	*Oxidizer*	*Fuel*	*Oxidizer*	*Fuel*
Pump discharge pressure, kg/cm^2	347	358	614	516	282	251
Flow rate, kg/sec	173	73	1792	732	384	144
Impeller diameter, mm	216	235	409	405	229	288
Number of stages	1	1	1	$1 + 1^a$	1	$1 + 1^a$
Pump efficiency, %	66	65	74	74	68	69
Pump shaft power, hp	11,210	6145	175,600	77,760	16,150	8850
Required pump NPSH, m	37	23	260	118	45	38
Shaft speed, rpm	19,230		13,850		13,855	
Pump impeller type	Radial flow		Radial flow		Radial flow	
Turbine power, hp	17,588		257,360		25,490	
Turbine inlet pressure, main turbine, kg/cm^2	324		519		239	
Pressure ratio	1.76		1.94		1.42	
Turbine inlet temperature, K	735		772		783	
Turbine efficiency, %	72		79		74	
Number of turbine stages	1		1		1	
	Preburner Characteristics					
Flow rate, kg/sec	177		836		403.5	
Mixture ratio, O/F	53.8		54.3		21.5	
Chamber pressure, kg/cm^2	325		546		243	
Number of preburners	1		2		1	

^{a}Fuel flow to precombustor goes through a small second-stage pump. (Courtesy of NPO Energomash, Moscow.)

meters (chamber pressure, thrust, specific impulse, weight, propellant combination, nozzle area ratio, dimensions, etc.) which influence the vehicle performance and configuration. It also shows secondary parameters, which are internal to the engine but important in component design and engine optimization. The Space Shuttle main engine (see Figs. 6–1 and 6–12) has two fuel-rich preburners, but the Russian engines use oxidizer-rich preburners. Figure 10–10 shows the RD-170 engine with four thrust chambers (and their thrust vector actuators) supplied by a centrally located single large turbopump (257,000 hp; not visible in the photo) and one of the two oxidizer-rich preburners. The flow diagram of Fig. 10–11 shows this turbopump and the two booster turbopumps; one is driven by a turbine using a bleed of oxygen-rich gas from the turbine exhaust (the gas is condensed when it mixes with the liquid oxygen flow) and the other by a liquid turbine using high-pressure liquid fuel.

Much of today's engine design, preliminary design and design optimization can be performed on computer programs. These include infinite element analyses, codes for stress and heat transfer, weight and mass properties, stress and strain analysis of a variety of structures, water hammer, engine performance analyses, feed system analyses (for balance of flow, pressures, and power), gas

FIGURE 10–10. The RD-170 rocket engine, shown here on a transfer cart, can be used as an expendable or reusable engine (up to 10 flights). It has been used on the Zenith, Soyuz booster, and Energiya launch vehicles. The tubular structure supports the four hinged thrust chambers and its control actuators. It is the highest thrust liquid rocket engine in use today. (Courtesy of NPO Energomash, Moscow.)

pressurization, combustion vibrations, and various exhaust plume effects (see Ref. 10–16). Some customers require that certain analyses (e.g., safety, static test performance) be delivered to them prior to engine deliveries.

Many computer programs are specific to a particular design organization, a certain category of application (e.g., interplanetary flight, air-to-air combat, long-range ballistic missile, or ascent to earth orbit), and many are specific to a particular engine cycle. One is called *engine balance program* and it balances the pressure drops in the fuel, oxidizer, and pressurizing gas flow systems; similar programs balance the pump and turbine power, speeds, and torques (see

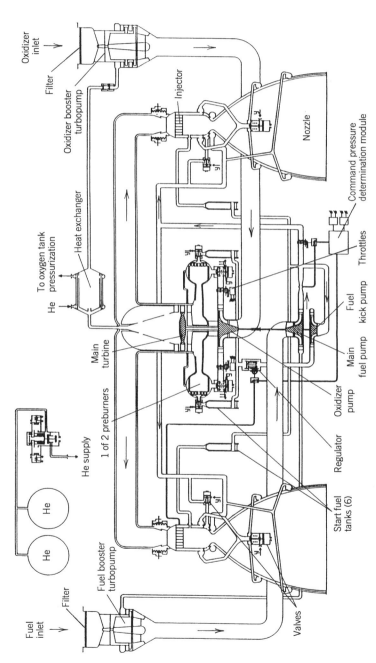

FIGURE 10–11. Simplified flow diagram of the RD-170 high-pressure rocket engine. The single-shaft large turbopump has a single-stage reaction turbine, two fuel pumps, and a single-stage oxygen pump with an inducer impeller. All of the oxygen and a small portion of the fuel flow supply two oxidizer-rich preburners. Only two of the four thrust chambers are shown in this diagram. The two booster pumps prevent cavitation in the main pumps. The pressurized helium subsystem (only shown partially) supplies various actuators and control valves; it is indicated by the symbol y. Ignition is accomplished by injecting a hypergolic fuel into the two preburners and the four thrust chambers. (Courtesy of NPO Energomash, Moscow.)

Section 10.7), compare different turbopump configurations (see Section 10.1); some balance programs also calculate approximate masses for engine, tanks, turbine drive fluids. The program allows iterations of various pressures and pressure drops, mixture ratios, thrust levels, number of thrust chambers, distribution of total velocity increment between different vehicle stages, trades between constant thrust (or propellant flow) and decreasing thrust (throttling) or pulsed (intermittent) thrust.

10.5. ENGINE CONTROLS

All liquid propellant rocket engines have controls to accomplish some or all of the following tasks:

1. Start rocket operation
2. Shut down rocket operation.
3. Restart, if desired.
4. Maintain programmed operation (predetermined constant or randomly varied thrust, preset propellant mixture ratio and flow).
5. When safety devices sense an impending malfunction or a critical condition of the vehicle or the engine, the control system will automatically change the engine operating conditions to remedy the defected defect, or cause a safe emergency engine shutdown. Only some of the likely failure modes can be remedied by sensing a potential problem and initiating a remedial action. Some failures occur so rapidly that there is not enough time to counteract them. Others are too difficult to identify reliably as a failure and others are not well understood.
6. Fill with propellants.
7. Drain excess propellant after operation.
8. With cryogenic propellants the pipes, pumps, cooling jackets, injectors, and valves have to be cooled to the cryogenic fluid temperature prior to start, by bleeding cold propellant through them; this cooling propellant is not used to produce thrust. Its periodic flow has to be controlled.
9. Check out proper functioning of critical components or a group of components without actual hot operation before and/or after flight.
10. For recoverable and reusable rocket engines, also provide built-in self-test features to perform continuous checks in flight and on the ground and recycle the engine to a ready condition within a few minutes after a launch abort without any ground servicing.

The *complexity* of these control elements and the complexity of the engine systems depend very much on the mission of the vehicle. In general, rockets that are used only once (single-shot devices), that are filled with propellants at the factory, and that have to operate over a narrow range of environmental conditions tend to be simpler than rocket systems intended for repeated use, for applications where satisfactory operation must be demonstrated prior to use, and for manned vehicles. Because of the nature of the liquid propellants, most of the control actuation functions are achieved by valves, regulators, pressure switches, and flow controls. The use of special computers for automatic control in large engines is now common. The flow control devices, namely the valves, were discussed in Section 6.9.

Safety controls are intended to protect personnel and equipment in case of malfunction. For example, the control system is usually so designed that a failure of the electrical power supply to the rocket causes a nonhazardous shutdown (all electrical valves automatically returning to their normal position) and no mixing or explosion of unreacted propellant can occur. Another example is an electrical interlock device which prevents the opening of the main propellant valves until the igniter has functioned properly.

Check-out controls permit a simulation of the operation of critical control components without actual hot operation of the rocket unit. For example, many rockets have provisions for permitting actuation of the principal valves without having propellant or pressure in the system.

Control of Engine Starting and Thrust Buildup

In the *starting* and *stopping* process of a rocket engine, it is possible for the mixture ratio to vary considerably from the rated design mixture ratio because of a lead of one of the propellants and because the hydraulic resistances to propellant flow are not the same for the fuel and the oxidizer passages. During this transition period it is possible for the rocket engine to pass through regions of chamber pressure and mixture ratio which can permit combustion instability. The starting and stopping of a rocket engine is very critical in timing, valve sequencing, and transient characteristics. A good control system must be designed to avoid undesirable transient operation. Close *control* of the *flow* of propellant of the *pressure*, and of the *mixture ratio* is necessary to obtain reliable and repeatable rocket performance. The starting and ignition of thrust chambers has been discussed in Section 8.4.

Fortunately, most rocket units operate with a nearly constant propellant consumption and a constant mixture ratio, which simplifies the operating control problem. Stable operation of liquid propellant flows can be accomplished without automatic control devices because the liquid flow system in general tends to be inherently stable. This means that the hydraulic system reacts to any disturbance in the flow of propellant (a sudden flow increase or decrease) in such a manner as to reduce the effect of the disturbance. The system, therefore, usually has a natural tendency to control itself. However, in some cases the natural resonances of the system and its components can have frequency values that tend to destabilize the system.

The *start delay time* for a pressure feed system is usually small. Prior to start, the pressurization system has to be activated and the ullage volume has to be pressurized. This start delay is the time to purge the system if needed, open valves, initiate combustion, and raise the flow and chamber pressure to rated values. A turbopump system usually requires more time to start. In addition to the foregoing starting steps for a pressurized system, it has to allow a time period for starting a gas generator or preburner and for bringing the turbopumps up to a speed at which combustion can be sustained and thereafter up to full flow. If the propellant is nonhypergolic, additional time has to be allowed for the igniter to function and for feedback to confirm that it is working properly. All these events need to be controlled. Table 10–6 describes many of these steps.

Starting of small thrusters with a pressurized feed system can be very fast, as short as 3 to 15 millisec, enough time for a small valve to open, the propellant to flow to the chamber and to ignite, and the small chamber volume to be filled with high-pressure combustion gas. For turbopump-fed systems and larger thrust engines, the time from start signal to full chamber pressure is longer, about 1 to 5 sec, because the pump rotors have inertia, the igniter flame has to heat a relatively large mass of propellants, the propellant line volumes to be filled are large, and the number of events or steps that need to take place is larger.

Large turbopump-fed rocket engines have been started in at least four ways:

1. A *solid propellant start grain* or start cartridge is used to pressurize the gas generator or preburner, and this starts turbine operations. This method is used on Titan III hypergolic propellant rocket engines (first and second stages) and on the H-1 (nonhypergolic), where the start grain flame also ignites the liquid propellants in the gas generator. This is usually the fastest start method, but it does not provide for a restart.

2. This method, known as *tank head start*, is used on the SSME, is slower, does not require a start cartridge, and permits engine restart. The head of liquid from the vehicle tanks (usually in vertically launched large vehicles) plus the tank pressure cause a small initial flow of propellants; then slowly more pressure is built up as the turbine begins to operate and in a couple of seconds the engine "bootstraps" its flows and the pressures then rise to their rated values.

3. A small *auxiliary pressurized propellant feed system* is used to feed the initial quantity of fuel and oxidizer (at essentially full pressure) to the thrust chamber and gas generator. This method was used on the RS-27 engine in the first stage of a Delta II space launch vehicle.

4. The *spinner start* method uses clean high-pressure gas from a separate tank to spin the turbine (usually at less than full speed) until the engine provides enough hot gas to drive the turbine. The high-pressure tank is heavy, the connections add complexity, and this method is seldom used today.

TABLE 10–6. Major Steps in the Starting and Stopping of a Typical Large Liquid Propellant Rocket Engine with a Turbopump Feed System

---

1. *Prior to Start*

Check out functioning of certain components (without propellant flow), such as the thrust vector control or some valve actuators.

Fill tanks with propellants.

Bleed liquid propellants to eliminate pockets of air or gas.

When using propellants that can react with air (e.g., hydrogen can freeze air, small solid air crystals can plug injection holes, and solid air with liquid hydrogen can form an explosive mixture), it is necessary to purge the piping system (including injector, valves and cooling jacket) with an inert, dry gas (e.g., helium) to remove air and moisture. In many cases several successive purges are undertaken.

With cryogenic propellants the piping system needs to be cooled to cryogenic temperatures to prevent vapor pockets. This is done by repeated bleeding of cold propellant through the engine system (valves, pumps, pipes, injectors, etc.) just prior to start. The vented cold gas condenses moisture droplets in the air and this looks like heavy billowing clouds escaping from the engine.

Refill or "top off" tank to replace cryogenic propellant that has evaporated or been used for cooling the engine.

Pressurize vehicle's propellant tanks just before start.

2. *Start: Preliminary Operation*

Provide start electric signal, usually from vehicle control unit or test operator.

With nonhypergolic propellants, start the ignition systems in gas generator or preburner and main chambers; for nonhypergolic propellants a signal has to be received that the igniter is burning before propellants are allowed to flow into the chambers.

Initial operation: opening of valves (in some cases only partial opening or a bypass) to admit fuel and oxidizer at low initial flows to the high pressure piping, cooling jacket, injector manifold, and combustion chamber(s). Valve opening rate and sequencing may be critical to achieve proper propellant lead. Propellants start to burn and turbine shaft begins to rotate.

Using an automated engine control, make checks (e.g., shaft speed, igniter function, feed pressures) to assure proper operation before initiating next step.

In systems with gearboxes the gear lubricant and coolant fluid start to flow.

For safety reasons, one of the propellants must reach the chamber first.

3. *Start: Transition to Full Flow/Full Thrust*

Turbopump power and shaft speed increase.

Propellant flows and thrust levels increase until they reach full-rated values. May use controls to prevent exceeding limits of mixture ratio or rates of increase during transient.

Principal valves are fully opened. Attain full chamber pressure and thrust.

In systems where vaporized propellant is fed into the propellant tanks for tank pressurization, the flow of this heated propellant is initiated.

Systems for controlling thrust or mixture ratio or other parameter are activated.

4. *Stop*

Signal to stop deactivates the critlcal valve(s).

Key valves close in a predetermined sequence. For example, the valve controlling the gas generator or preburner will be closed first. Pressurization of propellant tanks is stopped.

As soon as turbine drive gas supply diminishes the pumps will slow down. Pressure and flow of each propellant will diminish quickly until it stops. The main valves are closed, often by spring forces, as the fluid pressures diminish. Tank pressurization may also be stopped. In some engines the propellant trapped in the lines or cooling jacket may be blown out by vaporization or gas purge.

SSME Start and Stop Sequences. This is an example of the transient start and stop behavior of a complex staged combustion cycle engine with a tank head start. It illustrates the rapid functions of an electronic controller. The SSME flow sheet in Fig. 6–12 identifies the location of the key components mentioned below and Fig. 10–12 shows the sequence and events of these transients. The remainder of this subsection is based on information provided by The Boeing Company, Rockerdyne Propulsion and Power.

For a tank head start, initial energy to start the turbines spinning is all derived from initial propellant tank pressures (fuel and oxidizer) and gravity (head of liquid column). Combining the tank head start with a staged combustion cycle consisting of five pumps, two preburners, and a main combustion chamber (MCC) results in a complicated and sophisticated start sequence, which is very robust and reliable. Prior to test, the SSME turbopumps and ducting (down to the main propellant valves) are chilled with liquid hydrogen and liquid oxygen (LOX) to cryogenic temperature to ensure liquid propellants for proper pump operation. At engine start command, the main fuel valve (MFV) is opened first to provide chilling below the MFV and a fuel lead to the engine. The three oxidizer valves sequence the main events during the crucial first two seconds of start. The fuel preburner oxidizer valve (FPOV) is ramped to 56% to provide LOX for ignition in the fuel preburner (FPB) in order to provide initial turbine torque of the high-pressure fuel turbopump (HPFTP). Fuel system oscillations (FSO), which occur due to heat transfer downstream of the initially chilled system can result in flowrate dips. These fuel flow dips can lead to damaging temperature spikes in the FPB as well as the oxidizer preburner (OFB) at ignition and 2 Hz cycles thereafter until the hydrogen is above critical pressure. The oxidizer preburner oxidizer valve (OPOV) and main oxidizer valve (MOV) are ramped open next to provide LOX for OPB and MCC ignition.

The next key event is FPB prime. Priming is filling of the OX system upstream of the injectors with liquid propellant. This results in increased combustion and higher power. This event occurs around 1.4 sec into start. The high-pressure fuel turbopump (HPFTP) speed is automatically checked at 1.24 sec into start to ensure it will be at a high enough level before the next key event, MCC prime, which is controlled by the MOV. Priming and valve timing are critical. We explain some of the events that could go wrong. At MCC prime, an abrupt rise in backpressure on the fuel pump/turbine occurs. If flowrate through the fuel pump at this time is not high enough (high speed), then the heat imparted to the fluid as it is being pumped can vaporize it, leading to unsatisfactory flow in the engine, and subsequent high mixture ratio with high gas temperature and possible burnout in the hot gas system. This occurs if the MCC primes too early or HPFTP speed is abnormally low. If the MCC primes too late, the HPFTP may accelerate too fast due to low backpressure after FPB prime and exceed its safe speed. The MCC prime normally occurs at 1.5 sec. The OPB is primed last since it controls LOX flow and a strong fuel lead and healthy fuel pump flow are desirable to prevent engine burnout due to

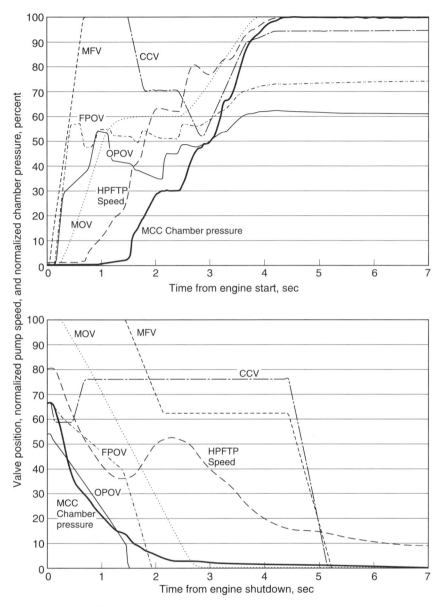

FIGURE 10–12. The sequence and events for starting and shutdown of the SSME (Space Shuttle main engine). This particular start sequence leads to a chamber pressure of 2760 psia (normalized here to 100%), a high-pressure fuel turbopump speed of 33,160 rpm (100%) , at a sea-level thrust of 380,000 lbf (shown as 100%). This shutdown occurs at altitude when the engine has been throttled to 67% of its power level or a vacuum thrust of 312,559 lbf, which is shown as 67% of the MCC chamber pressure. (Courtesy of The Boeing Company, Rocketdyne Propulsion and Power.)

a high mixture ratio. The OPOV provides minimal flowrate during the early part of the start to force the oxidizer to prime last at 1.6 sec into start. Again, the FSO influences temperature spikes in the OPB and must be sequenced around, prior to the MCC prime which raises the fuel pressure above critical in the fuel system. At two seconds into start, the propellant valves are sequenced to provide 25% of rated power level (RPL). During the first 2.4 sec of start, the engine is in an open-loop mode, but proportional control of the OPOV is used, based on MCC pressure. At this point, additional checks are carried out to ensure engine health, and a subsequent ramp to mainstage at 2.4 sec is done using closed-loop MCC–chamber-pressure/OPOV control. At 3.6 sec, closed-loop mixture ratio/FPOV control is activated.

The chamber cooling valve (CCV) is open at engine start and sequenced to provide optimum coolant fuel flow to the nozzle cooling jacket and the chamber and preburners during the ignition and main stage operation. It diverts flow to the cooling passages in the nozzle after MCC prime causes the heat load to increase. The description above is simplified and does not mention several other automatic checks, such as verifying ignition in the MCC or FPB or the fuel or chamber pressure buildup, which are sensed and acted upon at various times during the start sequence. The spark-activated igniters are built into the three injectors (MCC, FPB, OPB) using the same propellants. They are not mentioned above or shown in the flow sheet, but one of them can be seen in Fig. 9–6.

The shutdown sequence is initiated by closing the OPOV, which powers down the engine (reduces oxygen flow, chamber pressure, and thrust); this is followed quickly by closing the FPOV, so the burning will shut down fuel rich. Shortly thereafter the MOV is closed. The MFV stays open for a brief time and then is moved into an intermediate level to balance with the oxygen flow (from trapped oxygen downstream of the valves). The MPV and the CCV are closed after the main oxygen mass has been evaporated or expelled.

Automatic Controls

Automatically monitored controls are frequently used in liquid propellant rockets to accomplish thrust control or mixture ratio control. The automatic control of the thrust vector is discussed in Chapter 16.

Before electronic controls became common for large engines, pneumatic controls were used with helium gas. We still use helium to actuate large valves, but no longer for logic control. A pressure ladder sequence control was used, where pressures (and a few other quantities) were sensed and, if satisfactory, the next step of the start sequence was pneumatically initiated. This was used on the H-1 engine and the Russian RD-170 engine, whose flow sheet is shown in Figure 10–11.

Most automatic controls use a servomechanism. They generally consist of three basic elements: a *sensing mechanism*, which measures or senses the variable quantity to be controlled; a *computing or controlling mechanism*, which compares the output of the sensing mechanism with a reference value and gives a control signal to the third component, the *actuating device*, which manipulates the variable to be controlled. Additional discussion of computer control with automatic data recording and analysis is given in Chapter 20.

Figure 10–13 shows a typical simple thrust control system for a gas generator cycle aimed at regulating the chamber pressure (and therefore also the thrust) during the flight to a predetermined value. A pressure-measuring device with an electric output is used for the sensing element, and an automatic control device compares this gauge output signal with a signal from the reference gauge or a computer voltage and thus computes an error signal. This error signal is amplified, modulated, and fed to the actuator of the throttle valve. By controlling the propellant flow to the gas generator, the generator pressure is regulated and, therefore, also the pump speed and the main propellant flow; indirectly, the chamber pressure in the thrust chamber is regulated and, therefore, also the thrust. These quantities are varied until such time as the error signal approaches zero. This system is vastly simplified here, for the sake of

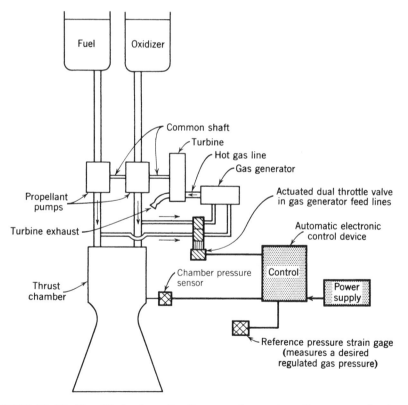

FIGURE 10–13. Simplified schematic diagram of an automatic servomechanism-type chamber pressure control of a liquid propellant rocket engine with a turbopump feed system, a gas generator, and a tank head, boot strap (self-pumping) starting system.

illustration; in actual practice the system may have to be integrated with other automatic controls. In this diagram the mixture of the gas generator is controlled by the pintle shapes of the fuel and oxidizer valves of the gas generator and by yoking these two valves together and having them moved in unison by a single actuator.

In the expander cycle shown schematically in Fig. 6–11, the thrust is regulated by maintaining a desired chamber pressure and controlling the amount of hydrogen gas flowing to the turbine by means of a variable bypass. The flow through this bypass is small (typically 5% of gas flow) and is controlled by the movement of a control valve.

In a *propellant utilization* system the mixture ratio is varied to insure that the fuel and oxidizer propellant tanks are both simultaneously and completely emptied; no undue propellant residue should remain to increase the empty mass of the vehicle, which in turn would detrimentally decrease the vehicle mass ratio and the vehicle's flight performance (see Chapter 4). For example, the oxidizer flow rate may be somewhat larger than normal due to its being slightly denser than normal or due to a lower than normal injector pressure drop; if uncontrolled, a fuel residue would remain at the time of oxidizer exhaustion; however, the control system would cause the engine to operate for a period at a propellant mixture ratio slightly more fuel-rich than normal, to compensate and assure almost simultaneous emptying of both propellant tanks. Such a control system requires accurate measurement of the amount of propellant remaining in the two propellant tanks during the flight.

Any of the three principal components of an automatic control system can have many differerent forms. Typical sensing devices include those that measure chamber pressure, propellant pressures, pump rotational speeds, tank level, or propellant flow. The actuating device can throttle propellant flow or control a bypass device or the gas generator discharge. There are many operating mechanisms for the controller, such as direct electrical devices, electronic analog or digital computers, hydraulic or pneumatic devices, and mechanical devices. The actuators can be driven by electrical motors, hydraulic, pneumatic, or mechanical power. The hydraulic actuators can provide very high forces and quick response. The exact type of component, the nature of the power supply, the control logic, the system type, and the operating mechanisms for the specific control depend on the details of the application and the requirements. Controls are discussed further in Refs. 6–1 and 10–18.

In applications where the final vehicle velocity must be accurately determined, the amount of impulse that is imparted to the vehicle during the cutoff transient may be sufficiently variable to exceed the desired velocity tolerance. Therefore, in these applications close control over the thrust decay curve is necessary and this can be accomplished by automatic control over the sequencing and closing rates of the main propellant valves and the location of the valves in relation to the injector.

Control by Computer

Early rocket engines used simple timers and, later, a pressure ladder sequence to send commands to the engine for actuating valves and other steps in the operation. Pneumatic controllers were also used in some engines for starting and stopping. For the last 20 years we have used *digital computers* in large liquid propellant rocket engines for controlling their operation (see Ref. 10–15). In addition to controlling the start and stop of engines, they can do a lot more and can contribute to making the engine more reliable. Table 10–7 gives a list of typical functions that a modern engine control computer has undertaken in one or more engines. This list covers primarily large turbopump-fed engines and does not include consideration of multiple small thruster attitude control rocket engines.

The design of control computers is beyond this text. In general it has to consider carefully all the possible engine requirements, all the functions that have to be monitored, all the likely potential failure modes and their compensating or ameliorating steps, all the sensed parameters and their scales, the method of control, such as open, closed, or multiple loops, adaptive or self-learning (expert system), the system architecture, the software approach, the interrelation and division of tasks with other computers on board the vehicle or on the ground, and the method of validating the events and operations. It is also convenient to have software that will allow some changes (which become necessary because of engine developments or failures) and allow the control of several parameters simultaneously. While the number of functions performed by the control computer seems to have increased in the last 20 years, the size and mass of the control computer has actually decreased substantially.

The control computer is usually packaged in a waterproof, shockproof black box, which is mounted on the engine. Fire-resistant and waterproof cable harnesses lead from this box to all the instrument sensors, valve position indicators, tachometers, accelerometers, actuators, and other engine components, to the power supply, the vehicle's controller, and an umbilical, severable multi-wire harness leads to the ground support equipment.

10.6. ENGINE SYSTEM CALIBRATION

Although an engine has been designed to deliver a specific performance $(F, I_s, \dot{m}, r)$, a newly manufactured engine will not usually perform precisely at these nominal parameters. If the deviation from the nominal performance values is more than a few percent, the vehicle will probably not complete its intended flight course. There are several reasons for these deviations. Because of unavoidable dimensional tolerances on the hardware, the flow–pressure profile or the injector impingement (combustion efficiency) will deviate slightly from the nominal design value. Even a small change in mixture

TABLE 10–7. Typical Functions to Be Performed by Digital Computers in Monitoring and Controlling the Operation of a Liquid Propellant Rocket Engine

1. *Sample the signals from significant sensors* (e.g., chamber pressure, gas and hardware temperatures, tank pressure, valve position, etc.) at frequent intervals, say once, 10, 100, or 1000 times per second. For parameters that change slowly, e.g., the temperature of the control box, sampling every second or every five seconds may be adequate, but chamber pressure would be sampled at a high frequency.

2. *Keep a record of all the significant signals* received and all the signals generated by the computer and sent out as commands or information.

3. *Control the steps and sequence of the engine start.* Figure 10–12 and Table 10–6 list typical steps that have to be taken, but do not list the measured parameters that will confirm that the commanded step was implemented. For example, if the igniter is activated, a signal change from a properly located temperature sensor or a radiation sensor could verify that the ignition had indeed happened.

4. *Control the shutdown of the engine.* For each of the steps listed at the bottom of Table 10–6 or in Fig. 10–12 there often has to be a sensing of a pressure change or other parameter change to verify that the commanded shutdown step was taken. An *emergency shutdown* may be commanded by the controller, when it senses certain kinds of malfunctions, that allow the engine to be shut down safely before a dramatic failure occurs. This emergency shutdown procedure must be done quickly and safely and may be different from a normal shutdown, and must avoid creating a new hazardous condition.

5. *Limit the duration of full thrust operation.* For example, cutoff is to be initiated just before the vehicle attains the desired mission flight velocity.

6. *Safety monitoring and control.* Detect combustion instability, over-temperatures in precombustors, gas generators, or turbopump bearings, violent turbopump vibration, turbopump overspeed or other parameter known to cause rapid and drastic component malfunction, that can quickly lead to engine failure. Usually, more than one sensor signal will show such a malfunction. If detected by several sensors, the computer may identify it as a possible failure whose in-flight remedy is well known (and preprogrammed into the computer); then a corrective action or a safe shutdown may be automatically commanded by the control computer.

7. *Control propellant tank pressurization.* The tank pressure value has to be within an allowable range during engine operation and also during a coasting flight period prior to a restart. Sensing the activation of relief valves on the tank confirms overpressure. Automatically, the computer can then command stopping or reducing the flow of pressurant.

8. *Perform automatic closed-loop control of thrust and propellant utilization* (described before).

9. *Transmit signals to a flying vehicle's telemetering system*, which in turn can send them to a ground station, thus providing information on the engine status, particularly during experimental or initial flights.

10. *Self-test the computer and software.*

11. *Analyze key sensor signals for deviation from nominal performance* before, during, and after engine operation. Determine whether sensed quantities are outside of predicted limits. If appropriate and feasible, if more than one sensor indicates a possible out-of-limit value, and if the cause and remedy can be predicted (preprogrammed), then the computer can automatically initiate a compensating action.

ratio will cause a significant increase of residual, unused propellant. Also, minor changes in propellant composition or storage temperature (which affects density and viscosity) can cause deviations. Regulator setting tolerances or changes in flight acceleration (which affects static head) are other factors. An engine calibration is the process of adjusting some of its internal parameters so that it will deliver the intended performance within the allowed tolerance bands.

Hydraulic and pneumatic components (valves, pipes, expansion joints) can readily be water flow tested on flow benches and corrected for pressure drops and density (and sometimes also viscosity) to determine their pressure drop at rated flow. Components that operate at elevated temperatures (thrust chambers, turbines, preburners, etc.) have to be hot fired and cryogenic components (pumps, some valves) often have to be tested at the cryogenic propellant temperature. The engine characteristics can be estimated by adding together the corrected values of pressure drops at the desired mass flow. Furthermore, the ratio of the rated flows $\dot{m}_o/\dot{m}_f$ has to equal the desired mixture ratio r. This is shown in the example below. The adjustments include adding pressure drops with judiciously placed orifices, or changing valve positions or regulator setting.

In most pressurized feed systems the pressurizing gas is supplied from its high pressure tank through a regulator to pressurize both the fuel and the oxidizer in their respective tanks. The pressure drop equations for the oxidizer and the fuel (subscripts o and f) are given below for a pressurized feed system at nominal flows.

$$p_{\text{gas}} - (\Delta p_{\text{gas}})_f = p_1 + \Delta p_f + (\Delta p_{\text{inj}})_f + (\Delta p_j)_f + \tfrac{1}{2}\rho_f v_f^2 + La\rho_f \qquad (10\text{--}17)$$

$$p_{\text{gas}} - (\Delta p_{\text{gas}})_o = p_1 + \Delta p_o + (\Delta p_{\text{inj}})_o + \tfrac{1}{2}\rho_o v_o^2 + La\rho_o \qquad (10\text{--}18)$$

The gas pressure in the propellant tank is the regulated pressure p_{gas}, diminished by the pressure losses in the gasline Δp_{gas}. The static head of the liquid $La\rho$ (L is the distance of the liquid level above the thrust chamber, a is the flight acceleration, and ρ is the propellant density) augments the gas pressure. It has to equal the chamber pressure p_1 plus all the pressure drops in the liquid piping or valves Δp, the injector Δp_{inj}, the cooling jacket Δp_j, and the dynamic flow head $\tfrac{1}{2}\rho v^2$. If the required liquid pressures do not equal the gas pressure in the propellant tank at the nominal propellant flow, then an additional pressure drop (calibration orifice) has to be inserted. A good design provides an extra pressure drop margin for this purpose.

Two methods are available for precise control of the engine performance parameters. One uses an automatic system with feedback and a digital computer to control the deviations in real time, while the other relies on an initial static calibration of the engine system. The latter appoach is simpler and is sometimes preferred, and is still quite accurate.

The *pressure balance* is the process of balancing the available pressure supplied to the engine (by pumps and/or pressurized tanks) against the pressure drops plus the chamber pressure. It is necessary to do this balancing in order to calibrate the engine, so it will operate at the desired flows and mixture ratio. Figure 10–14 shows the pressure balance for one of the two branches of propellant systems in a bipropellant engine with a pressurized feed system. It plots the pressure drops (for injector, cooling passages, pressurizing gas passages, valves, propellant feed lines, etc.) and the chamber pressure against the propellant flow, using actual component pressure drop measurements (or estimated data) and correcting them for different flows. The curves are generally plotted in terms of head loss and volumetric flow to eliminate the fluid density as an explicit variable for a particular regulated pressure. The regulated pressure is the same for the fuel and oxidizer pressure balance and it also can be adjusted. This balance of head and flow must be made for both the fuel and oxidizer systems, because the ratio of their flows establishes the actual mixture ratio and the sum of their flows establishes the thrust. The pressure balance between available and required tank pressure, both at the desired flow, is achieved by adding a calibration orifice into one of the lines, as can be seen in Fig. 10–14. Not shown in the figure is the static head provided by the elevation of the liquid level, since it is small for many space launch systems. However, with high acceleration and dense propellants, it can be a significant addition to the available head.

For a pumped feed system of a bipropellant engine, Fig. 10–15 shows a balance diagram for one branch of the two propellants systems. The pump speed is an additional variable. The calibration procedure is usually more complex for a turbopump system, because the pump calibration curves (flow–head–power relation) can not readily be estimated without good test data and cannot easily be approximated by simple analytical relations. The flow of the propellants to a gas generator or preburner also needs to be calibrated. In this case the turbine shaft torque has to equal the torque required by the pumps and the energy losses in bearings, seals or windage. Thus a power balance must be achieved in addition to the matching of pressures and the individual propellant flows. Since these parameters are interdependent, the determination of the calibration adjustments may not always be simple. Many rocket organizations have developed computer programs to carry out this balancing.

Example 10–3. The following component data and design requirements are given for a pressurized liquid propellant rocket system similar to that in Figs. 1–3 and 10–14: fuel, 75% ethyl alcohol; oxidizer, liquid oxygen; desired mixture ratio, 1.30; desired thrust, 5000 lbf at sea level. For this propellant combustion gas $k = 1.22$.

Component test data: Pressure losses in gas systems were found to be negligible. Fuel valve and line losses were 9.15 psi at a flow of 9.63 lbm/sec of water. Oxidizer valve and line losses were 14.2 psi at a flow of 12.8 lbm/sec of liquid oxygen. Fuel cooling jacket prssure loss was 52 psi at a flow of 9.61 lbm/sec of water. Oxidizer side injector pressure

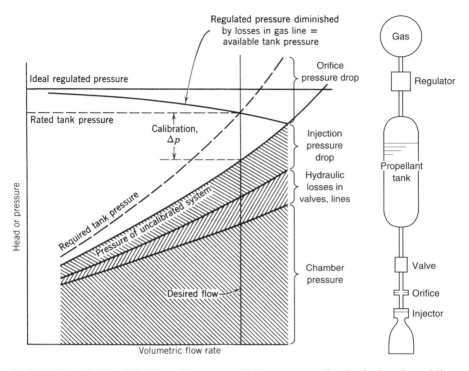

FIGURE 10–14. Simplified flow diagram and balance curves for the fuel or the oxidizer of a typical gas-pressurized bipropellant feed system. This diagram is also the same for a monopropellant feed system, except that it has no calibration orifice; it is calibrated by setting the proper regulated pressure.

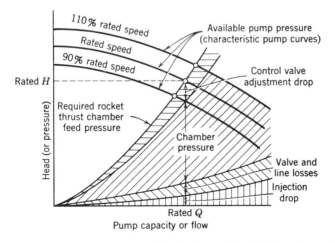

FIGURE 10–15. Simplified diagram of the balance of available and required feed pressures versus flow for one of the propellants in a rocket engine with a turbopump feed system. Chamber pressure is increased by liquid column.

drop was 90.0 psi at 10.2 lb/sec of oxygen flow under thrust chamber operating conditions. Fuel side injector pressure drop was 48.3 psi at 10.2 lb/sec of fuel flow under thrust chamber operating conditions. Average results of several sea-level thrust chamber tests were: thrust = 5410 lbf; mixture ratio = 1.29; specific impulse = 222 sec; chamber pressure = 328 psia; nozzle area ratio = 4.0. Determine regulator setting and size and location of calibration orifices.

SOLUTION. First, the corrections necessary to obtain the desired thrust chamber conditions have to be determined. The experimental thrust chamber data must be adjusted for deviations in mixture ratio, thrust, and specific impulse. The variation of specific impulse with *mixture ratio* is determined from experimental data or (on a relative basis) from theoretical calculations similar to those that are the basis of Fig. 5–1. Because the value of I_s at the desired mixture ratio of 1.30 is within 0.08% of the value of I_s under the actual test conditions ($r = 1.29$), any mixture ratio correction of I_s is neglected here.

The correction of the *specific impulse* for chamber pressure is made next. The specific impulse is essentially proportional to the thrust coefficients as determined from Eq. 3–30. For $k = 1.22$, and the pressure ratios $p_1/p_3 = 328/14.7 = 22.2$ and $300/14.7 = 20.4$, the values of C_F can be calculated as 1.420 and 1.405, respectively. In this calculation p_2 has to be determined for isentropic conditions, such as those in Figs. 3–7 or 3–8 for the given nozzle area ratio. The sea-level specific impulse is therefore corrected to $I_s = 222$ (1.405/1.420) = 220 sec. The *chamber pressure* has to be reduced from 328 psi to a lower value in order to bring the thrust from its test value of 5410 lbf to the design value of 5000 lbf. In accordance with Eq. 3–31, $F = C_F A_t p_1$. The chamber pressure is inversely proportional to the thrust coefficient C_F and proportional to the thrust, and therefore

$$p_1/p_1' = (F_1/F_1')(C_F'/C_F)$$

The primes refer to the component test condition.

$$p_1 = 328(5000/5410)(1.420/1.405) = 306 \text{ psi}$$

The desired total *propellant flow* is, from Eq. 2–5,

$$\dot{w} = F/I_s = 5000/220 = 22.7 \text{ lbf/sec}$$

For a mixture ratio of 1.3, the desired *fuel and oxidizer flows* are obtained from Eqs. 6–3 and 6–4 as $\dot{w}_f = 9.9$ lbf/sec and $\dot{w}_o = 12.8$ lbf/sec. Next, the various component pressure drops are corrected to the desired flow values and to the corrected propellant densities in accordance with Eq. 8–2, which applies to all hydraulic devices. By neglecting variations in discharge coefficients, this equation can be rewritten into a convenient form:

$$\dot{w}/\dot{w}' = \sqrt{\rho/\rho'}\sqrt{\Delta p/\Delta p'}$$

With this equation and the specific gravity values (from Fig. 7–1) of 1.14 for oxygen, 0.85 for diluted ethyl alcohol, and 1.0 for water, the new pressure drops for the corrected flow conditions can be found, and these are tabulated below with flow values given in pounds per second and pressure values in pounds per square inch.

Component	Component Test Data			Design Conditions		
	Fluid	$\dot{w}$	Δp	Fluid	$\dot{w}$	Δp
Fuel injector	Fuel	10.2	48.3	Fuel	9.9	45.3
Oxidizer injector	Oxygen	14.0	90.0	Oxygen	12.8	75.0
Fuel cooling jacket	Water	9.61	52.0	Fuel	9.9	64.9
Fuel valve and line	Water	9.63	9.15	Fuel	9.9	11.4
Oxidizer valve and line	Oxygen	12.8	14.2	Oxygen	12.8	14.2

The total pressure drop in the fuel system is 45.3 + 64.9 + 11.4 = 121.6 psi, and in the oxidizer system it is 75.0 + 14.2 = 89.2 psi.

The tank pressures required to obtain the desired flows are calculated by adding the chamber pressure to these pressure drops; that is, $(p)_o = 306 + 89.2 = 395.2$ psi and $(p)_f = 306 + 121.6 = 427.6$ psi. To equalize the tank pressures so that a single gas pressure regulator can be used, an *additional pressure loss must be introduced into the oxygen system*. The correction to this simple pressurized liquid propellant system is accomplished by means of an orifice, which must be placed in the propellant piping between the oxidizer tank and the thrust chamber. Allowing 10 psi for regulator functioning, the pressure drop in a calibration orifice will be $\Delta p = 427.6 - 395.2 + 10 = 42.4$ psi. The *regulator setting* should be adjusted to give a regulated downstream pressure of 427.6 psi under flow conditions. The orifice area (assume $C_d = 0.60$ for a sharp-edged orifice) can be obtained from Eq. 8–2, but corrected with a g_o for English units.

$$A = \frac{\dot{m}}{C_d \sqrt{2g\rho\,\Delta p}} = \frac{12.8 \times 144}{0.60\sqrt{2 \times 32.2 \times 1.14 \times 62.4 \times 42.4 \times 144}}$$
$$= 0.581 \text{ in.}^2 \text{ (or } 0.738 \text{ in. diameter)}$$

A set of balancing equations can be assembled into a computer program to assist in the calibration of engines. It can also include some of the system's dynamic analogies that enable proper calibration and adjustment of transient performance of the engine as during start. There is a trend to require tighter tolerances on rocket engine parameters (such as thrust, mixture ratio, or specific impulse), and therefore the measurements, calibrations, and adjustments are also being performed to much tighter tolerances than was customary 25 years ago.

10.7. SYSTEM INTEGRATION AND ENGINE OPTIMIZATION

Rocket engines are part of a vehicle and must interact and be integrated with other vehicle subsystems. There are *interfaces* (connections, wires, or pipelines) between the engine and the vehicle's structure, electric power system, flight control system (commands for start or thrust vector control), and ground support system (check-out or propellant supply). The engine also imposes limitations on vehicle components by its heat emissions, noise, and vibrations.

Integration means that the engine and the vehicle are compatible with each other, interfaces are properly designed, and there is no interference or unnecessary duplication of functions with other subsystems. The engine works with other subsystems to enhance the vehicle's performance and reliability, and reduce the cost. In Chapter 17 we describe the process of selecting rocket propulsion systems and it includes a discussion of interfaces and vehicle integration. This discussion in Chapter 17 is supplementary and applies to several different rocket propulsion systems. This section concerns liquid propellant rocket engines.

Since the propulsion system is usually the major mass of the vehicle, its structure (which usually includes the tanks) often becomes a key structural element of the vehicle and has to withstand not only the thrust force but also various vehicle loads, such as aerodynamic forces or vibrations. Several alternate tank geometries and locations (fuel, oxidizer, and pressurizing gas tanks), different tank pressures, and different structural connections have to be evaluated to determine the best arrangement.

The thermal behavior of the vehicle is strongly affected by the heat generation (hot plume, hot engine components, or aerodynamic heating) and the heat absorption (the liquid propellants are usually heat sinks) and by heat rejection to its surroundings. Many vehicle components must operate within narrow temperature limits, and their thermal designs can be critical when evaluated in terms of the heat balance during, after, and before the rocket engine operation.

Optimization studies are conducted to select the best values or to optimize various engine parameters such as chamber pressure (or thrust), mixture ratio (which affects average propellant density and specific impulse), number of thrust chambers, nozzle area ratio, or engine volume. By changing one or more of these parameters, it is usually possible to make some improvement to the vehicle performance (0.1 to 5.0%), its reliability, or to reduce costs. Depending on the mission or application, the studies are aimed at maximizing one or more vehicle parameter such as range, vehicle velocity increment, payload, circular orbit altitude, propellant mass fraction, or minimizing costs. For example, the mixture ratio of hydrogen–oxygen engines for maximum specific impulse is about 3.6, but most engines operate at mixture ratios between 5 and 6 because the total propellant volume is less, and this allows a reduced mass for the propellant tanks and the turbopump (resulting in a higher vehicle velocity increment) and a reduced vehicle drag (more net thrust). The selection of the best nozzle area ratio was mentioned in Chapter 3; it depends on the flight path's altitude–time history; the increase in specific impulse is offset by the extra nozzle weight and length. The best thrust–time profile can also usually be optimized, for a given application, by using trajectory analyses.

PROBLEMS

1. Estimate the mass and volume of nitrogen required to pressurize an N_2O_4–MMH feed system for a 4500 N thrust chamber of 25 sec duration ($\zeta_v = 0.92$, the ideal, $I_s = 285$ sec at 1000 psi or 6894 N/M^2 and expansion to 1 atm). The chamber pressure is 20 atm (abs.) and the mixture ratio is 1.65. The propellant tank pressure is 30 atm, and the initial gas tank pressure is 150 atm. Allow for 3% excess propellant and 50% excess gas to allow some nitrogen to dissolve in the propellant. The nitrogen regulator requires that the gas tank pressure does not fall below 29 atm.

2. What are the specific speeds of the four SSME pumps? (See the data given in Table 10–1.)

3. Compute the turbine power output for a gas consisting of 64% by weight of H_2O and 36% by weight of O_2, if the turbine inlet is at 30 atm and 658 K with the outlet at 1.4 atm and with 1.23 kg flowing each second. The turbine efficiency is 37%.

4. Compare the pump discharge gage pressures and the required pump powers for five different pumps using water, gasoline, alcohol, liquid oxygen, and diluted nitric acid. The respective specific gravities are 1.00, 0.720, 0.810, 1.14, and 1.37. Each pump delivers 100 gal/min, a head of 1000 ft, and arbitrarily has a pump efficiency of 84%. *Answers*: 433, 312, 350, 494, and 594 psi; 30.0, 21.6, 24.3, 34.2, and 41.1 hp.

5. The following data are given on a liquid propellant rocket engine:

Thrust	40,200 lbf
Thrust chamber specific impulse	210.2 sec
Fuel	Gasoline (sp. gr. 0.74)
Oxidizer	Red fuming nitric acid (sp. gr. 1.57)
Thrust chamber mixture ratio	3.25
Turbine efficiency	58%
Required pump power	580 hp
Power to auxiliaries mounted on turbopump gear case	50 hp
Gas generator mixture ratio	0.39
Turbine exhaust pressure	37 psia
Turbine exhaust nozzle area ratio	1.4
Enthalpy available for conversion in turbine per unit of gas	180 Btu/lb
Specific heat ratio of turbine exhaust gas	1.3

Determine the engine system mixture ratio and the system specific impulse. *Answers*: 3.07 and 208.

SYMBOLS

a	acceleration, m/sec^2 (ft/sec^2)
A	area, m^2 (ft^2)
c_p	specific heat at constant pressure, J/kg-K (Btu/lbm-R)

C_F	thrust coefficient (see Eq. 3–30)
D	diameter, m (ft)
F	thrust, N (lbf)
g_0	sea-level acceleration of gravity, 9.806 m/sec^2 (32.17 ft/sec^2)
Δh	enthalpy change, J/kg (Btu/lb)
H	head, m (ft)
$(H_s)_A$	available pump suction head above vapor pressure, often called net positive suction head, m (ft)
$(H_s)_R$	required pump suction head above vapor pressure, m (ft)
I_s	specific impulse, sec (lbf-sec/lbf)
k	specific heat ratio
L	length, m (ft)
$\dot{m}$	mass flow rate, kg/sec
N	shaft speed, rpm (rad/sec)
N_s	specific speed of pump
p	pressure, N/m^2 (lbf/in.^{2})
P	power, W (hp)
Q	volume flow rate, m^3/sec (ft^3/sec)
r	flow mixture ratio (oxidizer to fuel flow)
S	suction specific speed of pump
t	time, sec
T	absolute temperature, K (R)
u	tip speed or mean blade speed, m/sec (ft/sec)
u	velocity, m/sec (ft/sec)

Greek Letters

Δ	finite differential
ζ_d	discharge correction factor
ζ_F	thrust correction factor
η	efficiency
λ	coefficient of thermal expansion, m/m-K (in./in.-R)
ρ	density, kg/m^3 (lb/ft^3)
ψ	constant

Subscripts

c	chamber
e	maximum efficiency
f	fuel
gg	gas generator
o	oxidizer
oa	overall engine system
P	pump
T	turbine

0   initial condition
1   inlet
2   outlet

REFERENCES

10–1. M. L. Strangeland, "Turbopumps for Liquid Rocket Engines," *Threshold, an Engineering Journal of Power Technology*, Rocketdyne Division of Rockwell International, Canoga Park, CA, No. 3, Summer 1988, pp. 34–42.

10–2. A. Minnick and S. Peery, "Design and Development of an Advanced Liquid Hydrogen Turbopump," *AIAA Paper 98-3681*, July 1998, and G. Crease, R. Lyda, J. Park, and A. Minick, "Design and Test Results of an Advanced Liquid Hydrogen Pump," *AIAA Paper 99-2190*, 1999.

10–3. V. M. Kalnin and V. A. Sherstiannikov, "Hydrodynamic Modelling of the Starting Process in Liquid Propellant Engines," *Acta Astronautica*, Vol. 8, 1980, pp. 231–242.

10–4. T. Shimura and K. Kamijo, "Dynamic Response of the LE-5 Rocket Engine Oxygen Pump," *Journal of Spacecraft and Rockets*, Vol. 22, No. 2, March–April 1985.

10–5. M. C. Ek, "Solving Subsynchronous Whirl in the High Pressure Hydrogen Turbomachinery of the Space Shuttle Main Engine," *Journal of Spacecraft and Rockets*, Vol. 17, No. 3, May–June 1980, pp. 208–218, and M. Lalanne and G. Ferraris, *Rotordynamics Prediction in Engineering*, John Wiley & Sons, Inc., New York, 1998, 433 pages.

10–6. R. W. Bursey, Jr., et al., "Advanced Hybrid Rolling Element Bearing for the Space Shuttle Main Engine High Pressure Alternate Turbopump," *AIAA Paper 96-3101*, 1996.

10–7. K. Kamijo, E. Sogame, and A. Okayasu, "Development of Liquid Oxygen and Hydrogen Turbopumps for the LE-5 Rocket Engine," *Journal of Spacecraft and Rockets*, Vol. 19, No. 3, May–June 1982, pp. 226–231.

10–8. H. Yamada, K. Kamijo, and T. Fujita, "Suction Performance of High Speed Cryogenic Inducers," *AIAA Paper 83-1387*, June 1983.

10–9. I. Karassik, W. C. Krutzsch, W. H. Fraser, and J. P. Messina (Eds.), *Pump Handbook*, McGraw-Hill Book Company, New York, 1976 (waterhammer and pumps).

10–10. C. E. Brennan, *Hydrodynamics of Pumps*, Concepts ETI, Inc. and Oxford University Press, 1994.

10–11. S. Andersson and S. Trollheden, "Aerodynamic Design and Development of a Two-Stage Supersonic Turbine for Rocket Engines," *AIAA Paper 99-2192*, 1999.

10–12. "Liquid Rocket Engine Turbines," NASA Space Vehicle Design Criteria (Chemical Propulsion), *NASA SP-8110*, January 1974.

10–13. P. Brossel, S. Eury, P. Signol, H. Laporte, and J. B. Micewicz, "Development Status of the Vulcain Engine," *AIAA Paper 95-2539*, 1995.

10–14. G. Mingchu and L. Guoqui, "The Oxygen/Hydrogen Engine for Long March Vehicle," *AIAA Paper 95-2838*, 1995.

10–15. Y. Fukushima and T. Imoto, "Lessons Learned in the Development of the LE-5 and LE-7 Engines," *AIAA Paper 94-3375*, 1994, and M. Fujita and Y. Fukushima, "Improvement of the LE-5A and LE-7 Engines," *AIAA Paper 96-2847*, 1996.

10–16. R. Iffly, "Performance Model of the Vulcain Ariane 5 Main Engine," *AIAA Paper 96-2609*, 1996.

10–17. V. S. Rachuk, A. V. Shostak, A. I. Dimitrenko, G. I. Goncharov, R. Hernandez, R. G. Starke, and J. Hulka, "Benchmark Testing of an Enhanced Operability LO$_2$/LH$_2$ RD-0120 Engine," *AIAA Paper 96-2609*, 1996.

10–18. R. M. Mattox and J. B. White, "Space Shuttle Main Engine Controller," *NASA Technical Paper 1932*, 1981, p. 19.

CHAPTER 11

SOLID PROPELLANT ROCKET FUNDAMENTALS

This is the first of four chapters on solid propellant rockets. It discusses the burning rates, motor performance, grain configurations, and structural analysis. In solid propellant rocket motors—and the word "motor" is as common to solid rockets as the word "engine" is to liquid rockets—the propellant is contained and stored directly in the combustion chamber, sometimes hermetically sealed in the chamber for long-time storage (5 to 20 years). Motors come in many different types and sizes, varying in thrust from about 2 N to over 4 million N (0.4 to over 1 million lbf). Historically, solid propellant rocket motors have been credited with having no moving parts. This is still true of many, but some motor designs include movable nozzles and actuators for vectoring the line of thrust relative to the motor axis. In comparison to liquid rockets, solid rockets are usually relatively simple, are easy to apply (they often constitute most of the vehicle structure), and require little servicing; they cannot be fully checked out prior to use, and thrust cannot usually be randomly varied in flight.

Figures 1–5 and 11-1 show the principal components and features of relatively simple solid propellant rocket motors. The *grain* is the solid body of the hardened *propellant* and typically accounts for 82 to 94% of the total motor mass. Design and stresses of grains are described later in this chapter. Propellants are described in the next chapter. The *igniter* (electrically activated) provides the energy to start the combustion. The grain starts to burn on its exposed inner surfaces. The combustion and ignition of solid propellants are discussed in Chapter 13. This grain configuration has a central cylindrical cavity with eight tapered slots, forming an 8-pointed star. Many grains have slots, grooves, holes, or other geometric features and they alter the initial

417

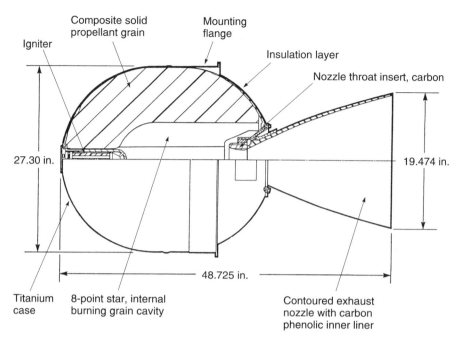

FIGURE 11–1. Cross section of the STAR^{TM} 27 rocket motor, which has been used for orbit and satellite maneuvers. It has an altitude thrust of 6000 lbf, nominally burns for 34.4 sec and has an initial mass of 796 lbm. For more data see Table 11–3. (Courtesy of Thiokol Propulsion, a Division of Cordant Technologies.)

burning surface, which determines the initial mass flow and the initial thrust. The hot reaction gases flow along the *perforation* or *port cavity* toward the nozzle. The inner surfaces of the *case* (really a pressure vessel), which are exposed directly to hot gas, have a thermal protection or *insulation layer* to keep the case from becoming too hot, in which case it could no longer carry its pressure and other loads. The case is either made of metal (such as steel, aluminum or titanium) or a composite fiber-reinforced plastic material.

The *nozzle* accelerates the hot gas; it is made of high temperature materials (usually a graphite and/or an ablative material to absorb the heat) to withstand the high temperatures and the erosion. The majority of all solid rockets have a simple fixed nozzle, as shown here, but some nozzles have provision to rotate it slightly so as to control the direction of the thrust to allow vehicle steering. Chapter 14 describes nozzles, cases, insulators, liners, and the design of solid propellant rocket motors.

Each motor is fastened to its vehicle by a thrust-carrying *structure*. In Fig. 11–1 there is a skirt (with a flange) integral with the case; it is fastened to the vehicle.

The subject of thrust vector control, exhaust plumes, and testing are omitted from these four chapters but are treated for both liquid and solid propellant units in Chapters 16, 18, and 20, respectively. Chapter 17 provides a comparison of the advantages and disadvantages of solid and liquid propellant rocket units. Chapters 3 to 5 are needed as background for these four chapters.

Applications for solid propellant rockets are shown in Tables 1–3, 1–4, and 11–1; each has its own mission requirements and thus propulsion requirements. Figures 11–2, 11–3, and 11–4 illustrate representative designs for some of the major categories of rocket motors listed in Table 11-1: namely, a large booster or second stage, a motor for space flight, and a tactical missile motor. Reference 11–1 is useful for component and design information.

There are several ways for classifying solid propellant rockets. Some are listed in Table 11–2 together with some definitions. Table 11–3 gives characteristics for three specific rocket motors, and from these data one can obtain a feeling for some of the magnitudes of the key parameters. These motors are shown in Figs. 16–5 and 16–9.

Solid propellant rocket motors are being built in approximately 35 different countries today, compared to only three countries about 50 years ago. The technology is well enough understood and disseminated that many companies or government arsenals are now capable of designing developing, and manufacturing solid rockets in several categories.

Almost all rocket motors are used only once. The hardware that remains after all the propellant has been burned and the mission completed—namely, the nozzle, case, or thrust vector control device—is not reusable. In very rare applications, such as the Shuttle solid booster, is the hardware recovered, cleaned, refurbished, and reloaded; reusability makes the design more complex, but if the hardware is reused often enough a major cost saving will result. Unlike some liquid propellant rocket engines, a solid propellant rocket motor and its key components cannot be operationally pretested. As a result, individual motor reliability must be inferred by assuring the structural integrity and verifying manufacturing quality on the entire population of motors.

11.1. PROPELLANT BURNING RATE

The rocket motor's operation and design depend on the combustion characteristics of the propellant, its burning rate, burning surface, and grain geometry. The branch of applied science describing these is known as *internal ballistics*; the effect of grain geometry is treated in Section 11.3.

The burning surface of a propellant grain recedes in a direction essentially perpendicular to the surface. The rate of regression, usually expressed in cm/sec, mm/sec, or in./sec, is the *burning rate r*. In Fig. 11–5 we can visualize the change of the grain geometry by drawing successive burning surfaces with a constant time interval between adjacent surface contours. Figure 11–5 shows this for a two-dimensional grain with a central cylindrical cavity with five slots. Success in rocket motor design and development depends significantly on knowledge of burning rate behavior of the selected propellant under all

(text continues on page 426)

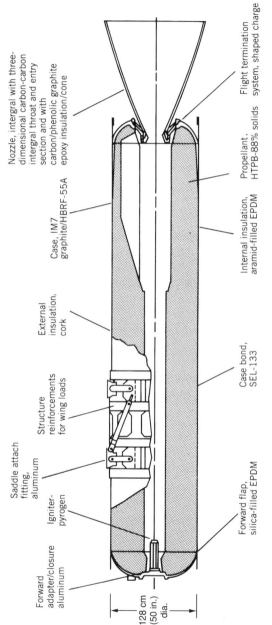

FIGURE 11-2. Booster rocket motor for the Pegasus air-launched three-stage satellite launch vehicle. It has a cylinder grain cavity with fins. The 50 in. diameter case has structural reinforcements to attach the Pegasus vehicle to its launch airplane and also to mount a wing to the case. It produces a maximum vacuum thrust of 726 kN (163,200 lbf) for 68.6 sec, a vacuum specific impulse of 295 sec, with a propellant mass of 15,014 kg and an initial mass of 16,383 kg. (Courtesy of Orbital Sciences, Corp. and Alliant Tech Systems.)

Labels in figure:

Nozzle, intergral with three-dimensional carbon-carbon intergral throat and entry section and with carbon/phenolic graphite epoxy insulation/cone

Case, IM7 graphite/HBRF-55A

External insulation, cork

Structure reinforcements for wing loads

Saddle attach fitting, aluminum

Igniter-pyrogen

Forward adapter/closure aluminum

128 cm (50 in.) dia.

Forward flap, silica-filled EPDM

Case bond, SEL-133

Internal insulation, aramid-filled EPDM

Propellant, HTPB-88% solids

Flight termination system, shaped charge

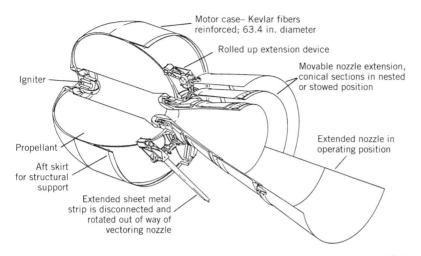

FIGURE 11–3. Inertial upper stage (IUS) rocket motor with an extendible exit cone (EEC). This motor is used for propelling upper launch vehicle stages or spacecraft. The grain is simple (internal tube perforation). With the EEC and a thrust vector control, the motor has a propellant fraction of 0.916. When launched, and while the two lower vehicle stages are operating, the two conical movable nozzle segments are stowed around the smaller inner nozzle segment. Each of the movable segments is deployed in space and moved into its operating position by three identical light-weight, electrically driven actuators. The nozzle area ratio is increased from 49.3 to 181; this improves the specific impulse by about 14 sec. This motor (without the EEC) is described in Table 11–3 and a similar motor is shown in Fig. 16–5. (Courtesy of United Technologies Corp., Chemical Systems.)

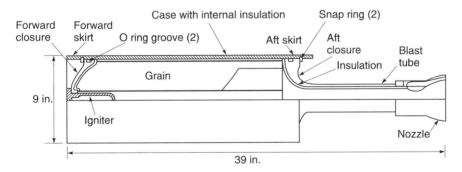

FIGURE 11–4. Simplified cross section through a typical tactical motor. The blast tube allows the grain to be close to the center of gravity of the vehicle; there is very little movement of the center of gravity. The nozzle is at the missile's aft end. The annular space around the blast tube is usually filled with guidance, control, and other non-propulsive equipment. A free-standing grain is loaded before the aft closure is assembled.

TABLE 11–1. Major Application Categories for Solid Propellant Rocket Motors

Category	Application	Typical Characteristics
Large booster and second-stage motors	Space launch vehicles; lower stages of long-range ballistic missiles (see Figs. 11–2 and 14–2)	Large diameter (above 48 in.); L/D of case = 2 to 7; burn time t = 60 to 120 sec; low-altitude operations with low nozzle area ratios (6 to 16)
High-altitude motors	Upper stages of multistage ballistic missiles, space launch vehicles; space maneuvers	High-performance propellant; large nozzle area ratio (20 to 200); L/D of case = 1 to 2; burn time t = 40 to 120 sec (see Fig. 11–3)
Tactical missiles	1. High acceleration: short-range bombardment, antitank missile	Tube launched, L/D = 4 to 13; very short burn time (0.25 to 1 sec); small diameter (2.75 to 18 in.); some are spin stabilized
	2. Modest acceleration: air-to-surface, surface-to-air, short-range guided surface-to-surface, and air-to-air missiles	Small diameter (5 to 18 in.); L/D of case = 5 to 10; usually has fins and/or wings; thrust is high at launch and then is reduced (boost-sustain); many have blast tubes (see Fig. 11–4); wide ambient temperature limits: sometimes minimum temperature −65° F or −53°C, maximum temperature +160°F or +71°C; usually high acceleration; often low-smoke or smokeless propellant
Ballistic missile defense	Defense against long- and medium-range ballistic missiles	Booster rocket and a small upper maneuverable stage with multiple attitude control nozzles and one or more side or divert nozzles
Gas generator	Pilot emergency escape; push missiles from submarine launch tubes or land mobile cannisters; actuators and valves; short-term power supply; jet engine starter; munition dispersion; rocket turbine drive starter; automotive air bags	Usually low gas temperature (< 1300°C); many different configurations, designs, and propellants; purpose is to create high-pressure, energetic gas rather than thrust

TABLE 11–2. Classification of Solid Rocket Motors

Basis of Classification	Examples of Classification
Application	See Table 11–1.
Diameter/Length	0.005–6.6 m or 0.2–260 in.; 0.025 to 45 m or 1 to 1800 in.
Propellant	*Composite*: Heterogeneous (physical) mixture of powdered metal (fuel), crystalline oxidizer and polymer binder
	Double-base: Homogeneous mixture (colloidal) of two explosives (usually nitroglycerin in nitrocellulose)
	Composite-modified double-base: Combines composite and double-base ingredients
	Gas generator and others: See Chapter 12
Case design	*Steel monolithic*: One-piece steel case
	Fiber monolithic: Filament wound (high-strength fibers) with a plastic matrix
	Segmented: Case (usually steel) and grain are in segments which are transported separately and fastened together at launch site
Grain configuration	*Cylindrical*: Cylindrically shaped, usually hollow
	End-burning: Solid cylinder propellant grain
	Other configurations: See Figs. 11–16 and 11–17
Grain installation	*Case-bonded*: Adhesion exists between grain and case or between grain and insulation and case; propellant is usually cast into the case
	Cartridge-loaded: Grain is formed separately from the motor case and then assembled into case
Explosive hazard	*Class 1.3*: Catastrophic failure shows evidence of burning and explosion, not detonation
	Class 1.1: Catastrophic failure shows evidence of detonation
Thrust action	*Neutral grain*: Thrust remains essentially constant during the burn period
	Progressive grain: Thrust increases with time
	Regressive grain: Thrust decreases with time
	Pulse rocket: Two or more independent thrust pulses or burning periods
	Step-thrust rocket: Usually, two distinct levels of thrust
Toxicity	Toxic and nontoxic exhaust gases

TABLE 11–3. Characteristics of Missile Motor and Space Motor

Characteristic	First Stage Minuteman Missile Motor[a]	Orbus-6 Inertial Upper Stage Motor[b]	STAR^{TM} 27 Apogee Motor[a]
Motor Performance (70°F, sea level)			
Maximum thrust (lbf)	201,500	23,800	6,404 (vacuum)
Burn time average thrust (lbf)	194,600	17,175	6,010 (vacuum)
Action time average thrust (lbf)[c]	176,600	17,180	5,177 (vacuum)
Maximum chamber pressure (psia)	850	839	569
Burn time average chamber pressure (psia)[c]	780	611	552
Action time average chamber pressure (psia)[c]	720	604	502
Burn time/action time (sec)[c]	52.6/61.3	101.0/103.5	34.35/36.93
Ignition delay time (sec)	0.130		0.076
Total impulse (lbf-sec)	10,830,000	1,738,000	213,894
Burn time impulse (lbf-sec)	10,240,000	1,737,000	
Altitude specific impulse (sec)	254	289.6 (vacuum)	290.8 (vacuum)
Temperature limits (°F)	60 to 80	45 to 82	20 to 100
Propellant			
Composition:			
NH_4ClO_4 (%)	70	68	72
Aluminum (%)	16	18	16
Binder and additives (%)	14	14	12
Density (lbm/in.³)	0.0636	0.0635	0.0641
Burning rate at 1000 psia (in./sec)	0.349	0.276	0.280
Burning rate exponent	0.21	0.3 to 0.45	0.28
Temperature coeffcient of pressure (%°F)	0.102	0.09	0.10
Adiabatic flame temperature (°F)	5790	6150	5,909
Characteristic velocity (ft/sec)	5180	5200	5,180
Propellant Grain			
Type	Six-point star	Central perforation	8-point star
Propellant volume (in.³)	709,400	94,490	11,480
Web (in.)	17.36	24.2	8.17
Web fraction (%)	53.3	77.7	60
Sliver fraction (%)	5.9	0	2.6
Average burning area (in.²)	38,500	3905	1,378
Volumetric loading (%)	88.7	92.4	
Igniter			
Type	Pyrogen	Pyrogen	Pyrogen
Number of squibs	2	2 through-the bulkhead initiators	2
Minimum firing current (A)	4.9	NA	5.0
Weights (lbf)			
Total	50,550	6515	796.3
Total inert	4719	513	60.6
Burnout	4264	478	53.4

TABLE 11–3. (*Continued*)

Characteristic	First Stage Minuteman Misisle Motor[a]	Orbus-6 Inertial Upper Stage Motor[b]	STAR^{TM} 27 Apogee Motor[a]
Propellant	45,831	6000	735.7
Internal insulation	634	141	12.6
External insulation	309	0	0
Liner	150	Incl. with insulation	0.4
Igniter	26	21	2.9 (empty)
Nozzle	887	143	20.4
Thrust vector control device	Incl. with nozzle	49.4	0
Case	2557	200	23.6
Miscellaneous	156	4	0.7
Propellant mass fraction	0.912	0.921	0.924
	Dimensions		
Overall length (in.)	294.87	72.4	48.725
Outside diameter (in.)	65.69	63.3	27.30
	Case		
Material	Ladish D6AC steel	Kevlar fibers/epoxy	6 A1-4V titanium
Nominal thickness (in.)	0.148	0.35	0.035
Minimum ultimate strength (psi)	225,000	—	165,000
Minimum yield strength (psi)	195,000	—	155,000
Hydrostatic test pressure (psi)	940	~ 1030	725
Hydrostatic yield pressure (psi)	985	NA	—
Minimum burst pressure, psi	—	1225	76.7
Typical burst pressure, psi	—	> 1350	—
	Liner		
Material	Polymeric	HTPB system	TL-H-304
	Insulation		
Type	Hydrocarbon– asbestos	Silica-filled EPDM	NA
Density (lbm/in.^{3})	0.0394	0.044	
	Nozzle		
Number and type	4, movable	Single, flexible	Fixed, contoured
Expansion area ratio	10:1	47.3	48.8/45.94
Throat area (in.^{2})	164.2	4.207	5.900
Expansion cone half angle (deg)	11.4	Initial 27.4, final 17.2	Initial 18.9, exit 15.5
Throat insert material	Forged tungsten	Three-dimensional carbon–carbon	3D carbon–carbon
Shell body material	AISI 4130 steel	NA	NA
Exit cone material	NA	Two-dimensional carbon–carbon	Carbon phenolic

[a]Courtesy of Thiokol Propulsion, a Division of Cordant Technologies, Inc.
[b]Courtesy United Technologies Corp., Chemical Systems; there is also a version Orbus 6-E (see Fig. 11–3) with an extendible, sliding nozzle; it has a specific impulse of 303.8 sec, a total weight of 6604 lb and a burnout weight of 567 lb.
[c]Burn time and action time are defined in Fig. 11–13.
NA: not applicable or not available.

FIGURE 11–5. Diagram of successive burning surface contours, each a fixed small time apart. It shows the growth of the internal cavity. The lengths of these contour lines are roughly the same (within ±15%), which means that the burning area is roughly constant.

motor operating conditions and design limit conditions. Burning rate is a function of the propellant composition. For composite propellants it can be increased by changing the propellant characteristics:

1. Add a burning rate *catalyst*, often called burning rate *modifier* (0.1 to 3.0% of propellent) or increase percentage of existing catalyst.
2. *Decrease* the *oxidizer particle size.*
3. *Increase oxidizer percentage.*
4. Increase the *heat of combustion* of the binder and/or the plasticizer.
5. Imbed *wires* or *metal staples* in the propellant.

Aside from the propellant formulation and propellant manufacturing process, burning rate in a full-scale motor can be increased by the following:

1. Combustion chamber pressure.
2. Initial temperature of the solid propellant prior to start.
3. Combustion gas temperature.
4. Velocity of the gas flow parallel to the burning surface.
5. Motor motion (acceleration and spin-induced grain stress).

Each of these influencing factors will be discussed. The explanation of the behavior of the burning rate with various parameters is largely found in the combustion mechanism of the solid propellant, which is described in Chapter 13. Analytical models of the burning rate and the combustion process exist and are useful for preliminary designs and for extending actual test data; for detail designs and for evaluation of new or modified propellants, engineers need some actual test data. *Burning rate data* are usually obtained in three ways—namely, from testing by:

1. Standard *strand burners*, often called *Crawford burners*.
2. Small-scale *ballistic evaluation motors*.
3. *Full-scale motors* with good instrumentation.

A strand burner is a small pressure vessel (usually with windows) in which a thin strand or bar of propellant is ignited at one end and burned to the other end. The strand can be inhibited with an external coating so that it will burn only on the exposed cross-sectional surface; chamber pressure is simulated by pressurizing the container with inert gas. The burning rate can be measured by electric signals from embedded wires, by ultrasonic waves, or by optical means (Ref. 11–2). The burning rate measured on strand burners is usually lower than that obtained from motor firing (by 4 to 12%) because it does not truly simulate the hot chamber environment. Also small ballistic evaluation motors usually have a slightly lower burning rate than full-scale larger motors, because of scaling factors. The relationship between the three measured burning rates is determined empirically for each propellant category and grain configuration. Strand-burner data are useful in screening propellant formulations and in quality control operations. Data from full-scale motors tested under a variety of conditions constitute the final proof of burning-rate behavior. Obviously, the strand burner and other substitutes for the full-scale motor must be exploited to explore as many variables as practicable.

During development of a new or modified solid propellant, it is tested extensively or *characterized*. This includes the testing of the burn rate (in several different ways) under different temperatures, pressures, impurities, and conditions. It also requires measurements of physical, chemical, and manufacturing properties, ignitability, aging, sensitivity to various energy inputs or stimuli (e.g., shock, friction, fires), moisture absorption, compatibility with other materials (liners, insulators, cases), and other characteristics. It is a lengthy, expensive, often hazardous program with many tests, samples, and analyses.

The burning rate of propellant in a motor is a function of many parameters, and at any instant governs the mass flow rate $\dot{m}$ of hot gas generated and flowing from the motor (stable combustion):

$$\dot{m} = A_b r \rho_b \qquad (11-1)$$

Here A_b is the burning area of the propellant grain, r the burning rate, and ρ_b the solid propellant density prior to motor start. The total mass m of effective propellant burned can be determined by integrating Eq. 11–1:

$$m = \int \dot{m} \, dt = \rho_b \int A_b r \, dt \qquad (11\text{–}2)$$

where A_b and r vary with time and pressure.

Burning Rate Relation with Pressure

Classical equations relating to burning rate are helpful in preliminary design, data extrapolation, and understanding the phenomena; however, analytical modeling and the supportive research have yet to adequately predict the burning rate of a new propellant in a new motor. Elemental laws and equations on burning rate usually deal with the influence of some of the important parameters individually. Unless otherwise stated, burning rate is expressed for 70°F or 294 K propellant (prior to ignition) burning at a reference chamber pressure of 1000 psia or 6.895 MPa.

With many propellants it is possible to approximate the *burning rates* as a function of *chamber pressure*, at least over a limited range of chamber pressures. A log–log plot is shown in Fig. 11–6. For most production-type propellants, this empirical equation is

$$r = a p_1^n \qquad (11\text{–}3)$$

where r, the burn rate, is usually in centimeters per second or inches per second, and the chamber pressure p_1 is in MPa or psia; a is an empirical constant influenced by ambient grain temperature. This equation applies to all the commonly used double-base, composite, or composite double-base propellants and they are described in the next chapter. Also a is known as the *temperature coefficient* and it is not dimensionless. The *burning rate exponent n*, sometimes called the *combustion index*, is independent of the initial grain temperature and describes the influence of chamber pressure on the burning rate. The change in ambient temperature does not change the chemical energy released in combustion; it merely changes the rate of reaction at which energy is released.

The curves shown in Fig. 11–6 are calculated and are straight lines on a log–log plot; however, many actual burning rate plots deviate somewhat and the actual data have some slight bends in parts of the curve, as seen in Fig. 11–7. For a particular propellant and for wide temperature and pressure limits, the burning rate can vary by a factor of 3 or 4. For all propellants they range from about 0.05 to 75 mm/sec or 0.02 to 3 in./sec; the high values are difficult to achieve, even with considerable burning rate catalyst additives, embedded metal wires, or high pressures (above 14 MPa or 2000 psi). A technology that would give a burning rate of more than 250 mm/sec at a chamber pressure of 1000 psia is desired by motor designers for several applications.

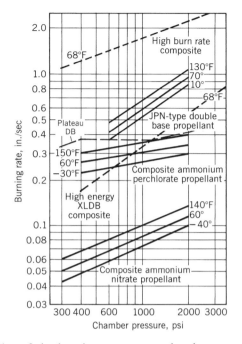

FIGURE 11–6. Plot of the burning rate versus chamber pressure for several typical solid rocket propellants, some at three different temperatures. A particular *double base* plateau propellant shows a constant burning rate over a fairly wide pressure range.

Example 11–1. Tabulate the variation of burning rate with pressure for two propellants with $a_1 = 0.00137$, $n_1 = 0.9$, $a_2 = 0.060$, and $n_2 = 0.4$, with p expressed in pounds per square inch and r in inches per second.

SOLUTION. Use Eq. 11–3 and solve for several conditions, as shown below.

Pressure (psia)	r_1 (in./sec)	r_2 (in./sec)
500	0.367	0.720
1000	0.685	0.95
1500	0.994	1.11
2000	1.28	1.26
2500	1.56	1.33

From inspection of these results and also from Eq. 11–3, it can be seen that the burning rate is very sensitive to the exponent n. For stable operation, n has values greater than 0 and less than 1.0. High values of n give a rapid change of burning rate with pressure. This implies that even a small change in chamber pressure produces substantial changes in the amount of hot gas produced. Most production propellants have a pressure exponent n ranging between 0.2 and 0.6. In practice, as n approaches 1, burning rate and chamber pressure

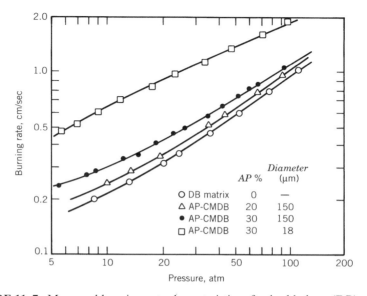

FIGURE 11–7. Measured burning rate characteristics of a double-base (DB) propellant and three composite-modified double-base (CMDB) propellants which contain an increasing percentage of small diameter (159 μm) particles of ammonium perchlorate (AP). When the size of the AP particles is reduced or the percentage of AP is increased, an increase in burning rate is observed. None of these data form straight lines. (Reproduced with permission of the AIAA from Chapter 1 of Ref. 11–3.)

become very sensitive to one another and disastrous rises in chamber pressure can occur in a few milliseconds. When the n value is low and comes closer to zero, burning can become unstable and may even extinguish itself. Some propellants display a negative n which is important for "restartable" motors or gas generators. A propellant having a pressure exponent of zero displays essentially zero change in burning rate over a wide pressure range. *Plateau propellants* are those that exhibit a nearly constant burning rate over a limited pressure range. One is shown with a dashed line in Fig. 11–6; they are relatively insensitive to major changes in chamber pressure for a limited range of pressures. Several double base propellants and a few composite propellants have this desirable plateau characteristic. Table 12–1 lists the nominal burning rate r and the pressure exponent n for several operational (production) propellants.

Burning Rate Relation with Temperature

Temperature affects chemical reaction rates and the initial ambient temperature of a propellant grain prior to combustion influences burning rate, as shown in Figs. 11–6 and 11–8. Common practice in developing and testing larger rocket motors is to "condition" the motor for many hours at a particular temperature before firing to insure that the propellant grain is uniformly at the desired temperature. The motor performance characteristics must stay within specified acceptable limits. For air-launched missile motors the extremes are usually 219 K (−65°F) and 344 K (160°F). Motors using typical composite propellant experience a 20 to 35% variation in chamber pressure and a 20 to 30% variation in operating time over such a range of propellant temperatures (see Fig. 11–8). In large rocket motors an uneven heating of the grain (e.g., by the sun heating one side) can cause a sufficiently large difference in burning rate so that a slight thrust misalignment can be caused (see Ref. 11–4).

The sensitivity of burning rate to propellant temperature can be expressed in the form of temperature coefficients, the two most common being

$$\sigma_p = \left(\frac{\delta \ln r}{\delta T}\right)_p = \frac{1}{r}\left(\frac{\delta r}{\delta T}\right)_p \tag{11–4}$$

$$\pi_K = \left(\frac{\delta \ln p}{\delta T}\right)_K = \frac{1}{p_1}\left(\frac{\delta p}{\delta T}\right)_K \tag{11–5}$$

with σ_p, known as the *temperature sensitivity of burning rate*, expressed as percent change of burning rate per degree change in propellant temperature at a particular value of chamber pressure, and π_K as the *temperature sensitivity of pressure* expressed as percent change of chamber pressure per degree change in propellant temperature at a particular value of K. Here K is a geometric function, namely the ratio of the burning surface A_b to nozzle throat area A_t.

The coefficient σ_p for a new propellant is usually calculated from strand-burner test data, and π_K from small-scale or full-scale motors. Mathematically,

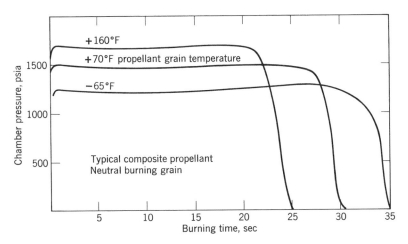

FIGURE 11–8. Effect of propellant temperature on burning time and chamber pressure for a particular motor. The integrated areas under the curves are proportional to the total impulse, which is the same for the three curves.

these coefficients are the partial derivative of the natural logarithm of the burning rate r or the chamber pressure p, respectively, with respect to propellant temperature T. Values for σ_p typically range between 0.001 and 0.009 per degree Kelvin or 0.002 to 0.04 per degree F and for π_K it is 0.067 to 0.278%/°C or 0.12 to 0.50%/°F. With π_K established, the effect of small grain temperature changes on motor chamber pressure is expressed from Eq. 11–5:

$$\Delta p \cong \pi_K p_1 \Delta T \qquad (11\text{--}6)$$

where p_1 is the reference chamber pressure and Δp is the pressure rise (psia) for a value of ΔT or $T - T_0$.

The values of π_K and σ_p depend primarily on the nature of the propellant burning rate, the composition, and the combustion mechanism of the propellant. It is possible to derive a relationship between the two temperature sensitivities, namely

$$\pi_K = \frac{1}{1-n}\sigma_p \qquad (11\text{--}7)$$

This formula is usually valid when the three variables are constant over the chamber pressure and temperature range. When substituting the value of r from Eq. 11–3 into Eq. 11–5, the temperature sensitivity σ_p can be also expressed as

$$\sigma_p = \left[\frac{\delta \ln(ap^n)}{\delta T}\right]_p = \frac{1}{a}\frac{da}{dT} \qquad (11\text{--}8)$$

which then defines σ_p in terms of the changes in the temperature factor a at constant chamber pressure.

It is not simple to predict the motor performance, because of changes in grain temperature and manufacturing tolerances. Reference 11–4 analyses the prediction of burning time.

Example 11–2. For a given propellant with a neutrally burning grain the value of the temperature sensitivity at constant burning area is $\pi_K = 0.005/°\text{F}$ or 0.5%/°F; the value of the pressure exponent n is 0.50. The burning rate r is 0.30 in./sec at 70°F at a chamber pressure of $p_1 = 1500$ psia and an effective nominal burning time of 50 sec. Determine the variation in p_1 and t_b for a change of ±50°F or from +20°F to +120°F assuming that the variation is linear.

SOLUTION. First Eq. 11–5 is modified:

$$\pi_K = \Delta p/(p_1 \Delta T) = \Delta p/[1500(\pm 50)] = 0.005$$

Solving, $\Delta p = \pm 375$ psi or a total excursion of about 750 psi or 50% of nominal chamber pressure.

The total impulse or the chemical energy released in combustion stays essentially constant as the grain ambient temperature is changed; only the rate at which it is released is changed. The thrust at high altitude is approximately proportional to the chamber pressure (with A_t and C_F assumed to be essentially constant in the equation $F = C_F p_1 A_t$) and the thrust will change also, about in proportion to the chamber pressure. Then the burning time is approximately

$$t_1 = 50 \times 1500/(1500 - 375) = 66.7 \text{ sec}$$
$$t_2 = 50 \times 1500/(1500 + 375) = 40.0 \text{ sec}$$

The time change $66.7 - 40.0 = 26.7$ sec is more than 50% of the nominal burning time. The result would be somewhat similar to what is described in Fig. 11–8.

In this example the variation of chamber pressure affects the thrust and burning time of the rocket motor. The thrust can easily vary by a factor of 2, and this can cause significant changes in the vehicle's flight path when operating with a warm or a cold grain. The thrust and chamber pressure increases are more dramatic if the value of n is increased. The least variation in thrust or chamber pressure occurs when n is small (0.2 or less) and the temperature sensitivity is low.

Burning Enhancement by Erosion

Erosive burning refers to the increase in the propellant burning rate caused by the high-velocity flow of combustion gases over the burning propellant surface. It can seriously affect the performance of solid propellant rocket motors. It occurs primarily in the port passages or perforations of the grain as the combustion gases flow toward the nozzle; it is more likely to occur when the port passage cross-sectional area A is small relative to the throat area A_t with a port-to-throat area ratio of 4 or less. An analysis of erosive burning is given in Ref. 11–5. The high velocity near the burning surface and the turbulent mixing in the boundary layers increase the heat transfer to the solid propellant and thus increase the burning rate. Chapter 10 of Ref. 11–3 surveys about 29 different theoretical analytical treatments and a variety of experimental techniques aimed at a better understanding of erosive burning.

Erosive burning increases the mass flow and thus also the chamber pressure and thrust during the early portion of the burning, as shown in Fig. 11–9 for a particular motor. As soon as the burning enlarges the flow passage (without a major increase in burning area), the port area flow velocity is reduced and erosive burning diminishes until normal burning will again occur. Since propellant is consumed more rapidly during the early erosive burning, there usually is also a reduction of flow and thrust at the end of burning. Erosive burning also causes early burnout of the web, usually at the nozzle end, and exposes the insulation and aft closure to hot combustion gas for a longer period of time; this usually requires more insulation layer thickness (and

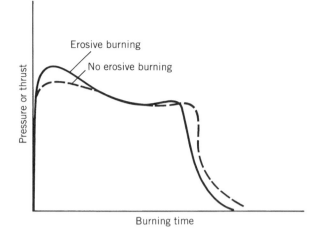

FIGURE 11–9. Typical pressure–time curve with and without erosive burning.

more inert mass) to prevent local thermal failure. In designing motors, erosive burning is either avoided or controlled to be reproducible from one motor to the next.

A relatively simple model for erosive burning, based on heat transfer, was first developed in 1956 by Lenoir and Robillard (Refs. 11–3 and 11–6) and has since been improved and used widely in motor performance calculations. It is based on adding together two burn rates: r_0, which is primarily a function of pressure and ambient grain temperature (basically Eq. 11–3) without erosion, and r_e, the increase in burn rate due to gas velocity or erosion effects.

$$r = r_0 + r_e$$
$$= ap^n + \alpha G^{0.8} D^{-0.2} \exp(-\beta r \rho_b / G) \tag{11–9}$$

Here G is the mass flow velocity per unit area in kg/m²-sec, D is a characteristic dimension of the port passage (usually, $D = 4A_p/S$, where A_p is the port area and S is its perimeter), ρ is the density of the unburned propellant (kg/m³), and α and β are empirically constants. Apparently, β is independent of propellant formulation and has a value of about 53 when r is in m/sec, p_1 is in pascals, and G is in kg/m²-sec. The expression of α was determined from heat transfer considerations to be

$$\alpha = \frac{0.0288 c_p \mu^{0.2} \mathrm{Pr}^{-2/3}}{\rho_b c_s} \frac{T_1 - T_s}{T_2 - T_p} \tag{11–10}$$

Here c_p is the average specific heat of the combustion gases in kcal/kg-K, μ the gas viscosity in kg/m-sec, Pr the dimensionless Prandtl number $(\mu c_p/\kappa)$ based on the molecular properties of the gases, κ the thermal conductivity of the gas, c_s the heat capacity of the solid propellant in kcal/kg-K, T_1 the combustion gas reaction absolute temperature, T_s the solid propellant surface temperature, and T_p the initial ambient temperature within the solid propellant grain.

Figure 11–10 shows the augmentation ratio r/r_0, or the ratio of the burning rate with and without erosive burning, as a function of gas velocity for two similar propellants, one of which has an iron oxide burn rate catalyst. Augmentation ratios up to 3 can be found in some motor designs. There is a pressure drop from the forward end to the aft end of the port passage, because static pressure energy is converted into kinetic gas energy as the flow is accelerated. This pressure differential during erosive burning causes an extra axial load and deformation on the grain, which must be considered in the stress analysis. The erosion or burn rate augmentation is not the same throughout the length of the port passage. The erosion is increased locally by turbulence if there are discontinuities such as protrusions, edges of inhibitors, structural supports, or gaps between segmented grains.

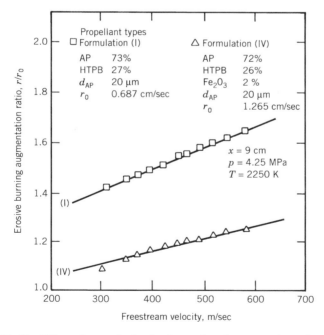

FIGURE 11–10. Effect of gas velocity in the perforation or grain cavity on the erosive burning augmentation factor, which is the burning rate with erosion r divided by the burning rate without erosion r_0. (Reproduced with permission of the AIAA from Chapter 10 of Ref. 11–3.)

Other Burning Rate Enhancements

Enhancement of burning rate can be expected in vehicles that *spin* the rocket motor about its longitudinal axis (necessary for spin-stabilized flight) or have high *lateral* or *longitudinal acceleration*, as occurs typically in antimissile rockets. This phenomenon has been experienced with a variety of propellants, with and without aluminum fuel, and the propellant formulation is one of the controlling variables (see Fig. 11–11). Whether the acceleration is from spin or longitudinal force, burning surfaces that form an angle of 60 to 90° with the acceleration vector are most prone to burning rate enhancement. For example, spinning cylindrical interal burning grains are heavily affected. The effect of spin on a motor with an operational composite propellant internal burning grain is shown in Fig. 11–12. The accelerated burning behavior of candidate propellants for a new motor design is often determined in small-scale motors, or in a test apparatus which subjects burning propellant to acceleration (Ref. 11–8). The stresses induced by rapid acceleration or rapid chamber pressure rise can cause crack formation (see Refs. 11–9 and 11–10), which exposes additional burning surface.

The burning rate of the propellant in an end-burning grain at a location immediately adjacent to or near the propellant-to-insulation bondline along the case wall, can, depending on the propellant formulation and manufacturing process, be higher than that of the propellant elsewhere in the grain.

The *embedding of wires* or other shapes of good metal heat conductors in the propellant grain increases the burning rate. One technique has several silver wires arranged longitudinally in an end-burning grain (see Ref. 11–11). Depending on wire size and the number of wires per grain cross-sectional area, the burning rate can easily be doubled. Aluminum wires are about half as effective as silver wires. Other forms of heat conductors have been wire

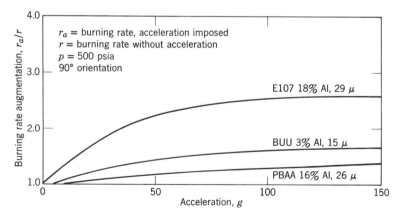

FIGURE 11–11. Acceleration effect on burning rate for three different propellants. (Adapted with permission from Ref. 11–7.)

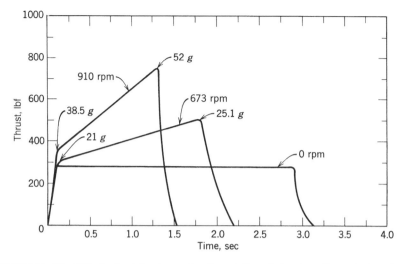

FIGURE 11–12. Effect of axial spin on the thrust–time behavior of a rocket motor with composite propellant using aluminum and PBAN (polybutadiene acrylonitrile) as fuels. (Adapted with permission from Ref. 11–7.)

staples (short bent wires) mixed with the propellant prior to the casting operation.

Intense *radiation emissions* from the hot gases in the grain cavity transfer heat to the burning propellant surfaces. More energetic radiation causes an increase in burning rate. Radiation of the exhaust plume (outside of the nozzle) and the effect of particles in the gas are discussed in Chapter 18.

Combustion instability, also called oscillatory combustion, can affect the burning rate of the propellant because of increased heat-transfer rate, gas velocity, and high pressure. This is discussed in Chapter 13.

11.2. BASIC PERFORMANCE RELATIONS

One basic performance relation is derived from the principle of conservation of matter. The propellant mass burned per unit time has to equal the sum of the change in gas mass per unit time in the combustion chamber grain cavity and the mass flowing out through the exhaust nozzle per unit time.

$$A_b r \rho_b = \frac{d}{dt}(\rho_1 V_1) + A_t p_1 \sqrt{\frac{k}{RT_1}\left(\frac{2}{k+1}\right)^{(k+1)/(k-1)}} \tag{11-11}$$

The term on the left side of the equation gives the mass rate of gas generation from Eq. 11-1. The first term on the right gives the change in propellant mass in the gas volume of the combustion chamber, and the last term gives the nozzle flow according to Eq. 3–24. The burning rate of the propellant is r; A_b is the propellant burning area; ρ_b is the solid propellant density; ρ_1 is the chamber gas density; V_1 is the chamber gas cavity volume, which becomes larger as the propellant is expended; A_t is the throat area; p_1 is the chamber pressure; T_1 is the absolute chamber temperature, which is usually assumed to be constant; and k is the specific heat ratio of the combustion gases. During startup the changing mass of propellant in the grain cavity becomes important. The preceding equation can be simplified and is useful in some numerical solutions of transient conditions, such as during start or shutdown.

The value of the burning surface A_b may change with time and is a function of the grain design, as described in Section 11.3. For preliminary performance calculations the throat area A_t is usually assumed to be constant for the total burning duration. For exact performance predictions, it is necessary also to include the erosion of the nozzle material, which causes a small increase in nozzle throat area as the propellant is burned; this nozzle enlargement is described in Chapter 14. The larger value of A_t causes a slight decrease in chamber pressure, burning rate, and thrust.

The gas volume V_1 will increase greatly with burn time. If the gas mass in the motor cavity is small, and thus if the rate of change in this gas mass is small relative to the mass flow through the nozzle, the term $d(\rho_1 V_1)/dt$ can be neglected. Then a relation for steady burning conditions can be obtained from Eqs. 11–3 and 11–11:

$$
\frac{A_b}{A_t} = \frac{p_1\sqrt{k[2/(k+1)]^{(k+1)/(k-1)}}}{\rho_b r\sqrt{RT_1}} = K
$$
$$
= \frac{(p_1)^{1-n}\sqrt{k[2/(k+1)]^{(k+1)/(k-1)}}}{\rho_b a\sqrt{RT_1}}
\tag{11–12}
$$

As an approximation, the chamber pressure can be expressed as a function of the area ratio of the burning surface to the nozzle throat cross section for a given propellant:

$$
p_1 \sim (A_b/A_t)^{1/(1-n)} = K^{1/(1-n)}
\tag{11–13}
$$

The ratio of the burning area to the nozzle throat area is an important quantity in solid propellant engineering and is given the separate symbol K. Equations 11–12 and 11–13 show the relation between burning area, chamber pressure, throat area, and propellant properties. For example, this relation permits an evaluation of the variation necessary in the throat area if the chamber pressure (and therefore also the thrust) is to be changed. For a propellant with $n = 0.8$, it can be seen that the chamber pressure would vary as the fifth power of the area ratio K. Thus, small variations in burning surface can have large effects on the internal chamber pressure and therefore also on the burning rate. The formation of surface cracks in the grain (due to excessive stress) can cause an unknown increase in A_b. A very low value of n is therefore desirable to minimize the effects of small variations in the propellant characteristics or the grain geometry.

Using this equation and the definition of the characteristic velocity c^* from Eq. 3–32, one can write

$$K = A_b/A_t = p_1^{(1-n)}/(a\rho_b c^*) \qquad (11–14)$$

Here a and ρ_b are constants and c^* does not really vary much. This can be rewritten

$$p_1 = (Ka\rho_b c^*)^{1/(1-n)} \qquad (11–15)$$

The equations above are based on the very simple mathematical dependence of burning rate on chamber pressure. However, for many propellants, this simplification is not sufficiently valid. For accurate evaluation, experimental values must be found.

Those parameters that govern the burning rate and mass discharge rate of motors are called *internal ballistic properties*; they include r, K, σ_p, π_K, and the influences caused by pressure, propellant ingredients, gas velocity, or acceleration. The subsequent solid propellant rocket parameters are *performance parameters*; they include thrust, ideal exhaust velocity, specific impulse, propellant mass fraction, flame temperature, temperature limits, and duration.

The *ideal nozzle exhaust velocity* of a solid propellant rocket is dependent on the thermodynamic theory as given by Eq. 3–15 or 3–16. As explained in Chapter 5, this equation holds only for frozen equilibrium conditions; for shifting equilibrium the exhaust velocity is best defined in terms of the enthalpy drop $(h_1 - h_2)$, which can be computed from $v_2 = \sqrt{2(h_1 - h_2)}$. In deriving the exhaust velocity equation, it was assumed that the approach velocity of gases upstream of the nozzle is small and can be neglected. This is true if the port area A_p (the flow area of gases between and around the propellant grains) is relatively large compared to the nozzle throat area A_t. When the port-to-throat-area ratio A_p/A_t is less than about 4, a pressure drop correction must be made to the effective exhaust velocity.

The thrust for solid propellant rockets is given by the identical definitions developed in Chapters 2 and 3, namely, Eqs. 2–14 and 3–29. The *flame* or *combustion temperature* is a thermochemical property of the propellant formulation and the chamber pressure. It not only affects the exhaust velocity, but also the hardware design, flame radiation emission, materials selection, and the heat transfer to the grain and hardware. In Chapter 5 methods for its calculation are explained. The determination of the nozzle throat area, nozzle expansion area ratio, and nozzle dimensions is discussed in Chapter 3.

The *effective exhaust velocity* c and the *specific impulse* I_s are defined by Eqs. 2–3, 2–4, and 2–6. It is experimentally difficult to measure the instantaneous propellant flow rate or the effective exhaust velocity. However, total impulse and total propellant mass consumed during the test can be measured. The approximate propellant mass is determined by weighing the rocket before and after a test. The effective propellant mass is often slightly less than the total propellant mass, because some grain designs permit small portions of the propellant to remain unburned during combustion, as is explained in a later chapter. Also, a portion of the nozzle and insulation materials erodes and vaporizes during the rocket motor burning and this reduces the final inert mass of the motor and also slightly increases the nozzle mass flow. This explains the difference between the total inert mass and the burnout mass in Table 11–3. It has been found that the *total impulse* can be accurately determined in testing by integrating the area under a thrust time curve. For this reason the average specific impulse is usually calculated from total measured impulse and effective propellant mass. The total impulse I_t is defined by Eq. 2–1 as the integration of thrust F over the operating duration t_b:

$$I_t = \int_0^{t_b} F \, dt = \overline{F} t_b \tag{11–16}$$

where $\overline{F}$ is an average value of thrust over the burning duration t_b.

The *burning time, action time,* and *pressure rise time* at ignition are defined in Fig. 11–13. Time zero is actually when the firing voltage is applied to the ignition squib or prime charge. Visible exhaust gas will actually come out of the rocket nozzle for a period longer than the action time, but the effluent mass flow ahead and behind the action time is actually very small. These definitions are somewhat arbitrary but are commonly in use and documented by standards such as Ref. 2–2.

For flight tests it is possible to derive the instantaneous thrust from the measured flight path acceleration (reduced by an estimated drag) and the estimated instantaneous mass from the chamber pressure measurements, which is essentially proportional to the rocket nozzle mass flow; this gives another way to calculate specific impulse and total impulse.

As explained in Section 3.6, there are at least four values of *specific impulse*: (1) *theoretical specific impulse*, (2) *delivered* or *actual* values as measured from flight tests, static tests, or demonstrations (see Ref. 11–12), (3) delivered specific impulse *at standard or reference conditions*, and (4) the *minimum guaranteed value*. Merely quoting a number for specific impulse without further explanation leaves many questions unanswered. This is similar to the four performance values for liquid propellant engines listed in Section 3.6. Specific impulse as diminished by several losses can be predicted as shown in Ref. 11–13.

Losses include the nozzle inefficiencies due to viscous boundary layer friction and nonaxial flow as described in Chapter 3, thrust vector deflection as described in Chapter 16, residual unburned propellants, heat losses to the walls

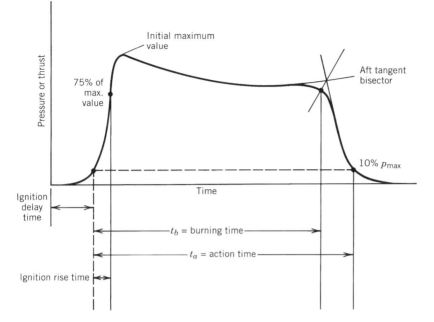

FIGURE 11–13. Definitions of burning time and action time.

or insulators, incomplete combustion, or the presence of solid particles in the gas which need to be accelerated. There are also some performance *gains*; the gases (created by ablation of the ablative nozzle and insulators or the igniter propellants) contribute to an increased mass flow, in many cases also to a somewhat lower average molecular weight of the gas and to a slight reduction of the final inert mass after rocket motor operation.

The *two-phase flow* equations for calculating specific impulse can be solved if the size distribution, shape, and percentage of solid particles in the exhaust gas are known. The assumption of a uniform average spherical particle diameter simplifies the analysis (Ref. 11–13), and this diameter can be estimated from specific impulse measurements on rocket motor tests (Ref. 11–14). Section 3.5 gives a simple theory for two-phase flow of solid particles in a gas flow. Sometimes *density-specific impulse*, the specific gravity of the propellant grain multiplied by specific impulse, is stated as a performance parameter, particularly in rocket motor applications where a compact design is desirable (see Eq. 7–3).

Propellants burn to varying degrees of completeness depending on the fuel, the oxidizer, their ratios, the energy losses, and the environment within the motor. Propellants with nonmetal fuels usually burn with a velocity correction factor of 97 or 98%, as contrasted to 90 to 96% for propellants with aluminum powder as the fuel. The solid particles in the exhaust do not contribute to the gas expansion, require energy to be accelerated, and two-phase flow is less efficient. However, the addition of the aluminum increases the heat of combustion, the chamber gas temperature, and thus the exhaust velocity or specific impulse. This increase usually outweighs the loss for having to accelerate the small solid aluminum oxide particles.

The *propellant mass fraction* ζ was defined in Eq. 2–8 as $\zeta = m_p/m_0$ and it is directly related to the *motor mass ratio* and therefore also to the flight performance of the vehicle. The initial motor mass m_0 is the sum of the useful solid propellant mass m_p and the non-burning, inert hardware mass of the motor. For a vehicle's propellant mass fraction, the payload mass and the nonpropulsion inert mass (vehicle structure, guidance and control, communications equipment, and power supply) have to be added. A high value of ζ indicates a low inert motor mass, an efficient design of the hardware, and high stresses. This parameter has been used to make approximate preliminary design estimates. It is a function of motor size or mass, thrust level, the nozzle area ratio, and the material used for the case. For very small motors (less than 100 lbm) the value of the propellant fraction is between 0.3 and 0.75. Medium-sized motors ($100 < m_0 < 1000$ lbm) have ζ values between 0.8 and 0.91. For larger motors ($1000 < m_0 < 50,000$ lbm) ζ is between 0.88 and 0.945. A range of values is given for each category, because of the influence of the following other variables. Medium- and large-sized motors with steel cases generally have lower ζ values than those with titanium cases, and their values are lower than for cases made of Kevlar fibers in an epoxy matrix. The highest values are for cases made of graphite or carbon fibers in an epoxy matrix. The ζ values are lower for larger area ratio nozzles and motors with thrust vector control. The STAR^{TM} 27 rocket motor, shown in Fig. 11-1 and described in Table 11–3, has a propellant mass fraction of 0.924. This is high for a medium-sized motor with a titanium metal case and a relatively large nozzle exit section.

A number of performance parameters are used to evaluate solid propellant rockets and to compare the quality of design of one rocket with another. The first is the *total-impulse-to-loaded-weight ratio* (I_t/w_G). The loaded weight w_G is the sea-level initial gross weight of propellant and rocket propulsion system hardware. Typical values for I_t/w_G are between 100 and 230 sec, with the higher values representative of high-performance rocket propellants and highly stressed hardware, which means a low inert mass. The total-impulse-to-loaded-weight ratio ideally approaches the value of the specific impulse. When the weight of hardware, metal parts, inhibitors, and so on becomes very small in relation to the propellant weight w_p, then the ratio I_t/w_G approaches I_t/w, which is the definition of the average specific impulse (Eqs. 2–3 and 2–4). The higher the value of I_t/w_G, the better the design of a rocket unit. Another parameter used for comparing propellants is the *volume impulse*; it is defined as the total impulse per unit volume of propellant grain, or I_t/V_b.

The *thrust-to-weight ratio* F/w_G is a dimensionless parameter that is identical to the acceleration of the rocket propulsion system (expressed in multiples of g_0) if it could fly by itself in a gravity-free vacuum; it excludes other vehicle component weights. It is peculiar to the application and can vary from very low values of less than one g_0 to over 1,000 g_0 for high acceleration applications of solid propellant rocket motors. Some rocket assisted gun munitions have accelerations of 20,000 g_0.

The *temperature limits* refer to the maximum and minimum storage temperatures to which a motor can be exposed without risk of damage to the propellant grain. They are discussed further in Section 11.4.

Example 11–3. The following requirements are given for a solid propellant rocket motor:

Sea level thrust	2000 lbf average
Duration	10 sec
Chamber pressure	1000 psia
Operating temperature	Ambient (approx. 70°F)
Propellant	Ammonium nitrate–hydrocarbon

Determine the specific impulse, the throat and exit areas, the flow rate, the total propellant weight, the total impulse, the burning area, and an estimated mass assuming moderately efficient design. Properties for this propellant are: $k = 1.26$; $T_1 = 2700°F = 3160$ R; $r = 0.10$ in./sec at 1000 psia; $c^* = 4000$ ft/sec; $\rho_b = 0.056$ lb/in.^{3}; molecular weight $= 22$ lbm/lb-mol; gas constant $= 1544/22 = 70.2$ ft-lbf/lbm-R.

SOLUTION. From Figs. 3–4 and 3–6, $C_F = 1.57$ (for $k = 1.26$, with optimum expansion at sea level and a pressure ratio of $1000/14.7 = 68$) and $\epsilon = A_2/A_t = 7.8$. The ideal thrust coefficient has to be corrected for nozzle losses. Assume a correction of 0.98; then $C_F = 0.98 \times 1.57 = 1.54$. The *specific impulse* is (Eq. 3–32).

$$I_s = c^* C_F/g_0 = (4000 \times 1.54)/32.2 = 191 \text{ sec}$$

The *required throat* area is obtained from Eq. 3–31:

$$A_t = F/(p_1 C_F) = 2000/(1000 \times 1.54) = 1.30 \text{ in.}^2$$

The *exit area* is $7.8 \times 1.30 = 10.1$ in.^{2} The *nozzle weight flow rate* is obtained from Eq. 2–5, namely $\dot{w} = F/I_s = 2000/191 = 10.47$ lbf/sec. The *effective propellant weight* for a duration of 10 sec is therefore approximately 105 lbf. Allowing for residual propellant and for inefficiencies on thrust buildup, the *total loaded propellant weight* is assumed to be 4% larger, namely, $105 \times 1.04 = 109$ lbf.

The *total impulse* is from Eq. 2–2: $I_t = Ft_b = 2000 \times 10 = 20,000$ lbf-sec. This can also be obtained from $I_t = w \times I_s = 105 \times 191 = 20,000$ lbf-sec. The propellant burning surface can be found by using Eq. 11–12:

$$A_b = \frac{A_t p_1 \sqrt{k[2/(k+1)]^{(k+1)/(k-1)}}}{\rho_b r \sqrt{RT_1}}$$

$$= \frac{1.30 \times 1000}{0.056 \times 0.10} \sqrt{\frac{32.2 \times 1.26}{(1544/22) \times 3160}} (0.885)^{8.7} = 1840 \text{ in.}^2$$

This result can also be obtained from Eq. 11–11 or 11–14. The ratio is given by

$$K = A_b/A_t = 1840/1.30 = 1415$$

The *loaded gross weight* of the rocket motor (not the vehicle) can only be estimated after a detailed design has been made. However, an approximate guess can be made by choosing a total impulse to weight ratio of perhaps 143.

$$w_G = I_t/(I_t/w_G) = 20,000/143 = 140 \text{ lbf}$$

Beause the propellants account for 109 lbf, the hardware parts can be estimated as 140 − 109 = 31 lbf.

11.3. PROPELLANT GRAIN AND GRAIN CONFIGURATION

The grain is the shaped mass of processed solid propellant inside the rocket motor. The propellant material and geometrical configuration of the grain determine the motor performance characteristics. The propellant grain is a cast, molded, or extruded body and its appearance and feel is similar to that of hard rubber or plastic. Once ignited, it will burn on all its exposed surfaces to form hot gases that are then exhausted through a nozzle. A few rocket motors have more than one grain inside a single case or chamber and very few grains have segments made of different propellant composition (e.g., to allow different burning rates). However, most rockets have a single grain.

There are two methods of holding the grain in the case, as seen in Fig. 11–14. *Cartridge-loaded or freestanding grains* are manufactured separately from the case (by extrusion or by casting into a cylindrical mold or cartridge) and then loaded into or assembled into the case. In *case-bonded grains* the case is used as a mold and the propellant is cast directly into the case and is bonded to the case or case insulation. Free-standing grains can more easily be replaced

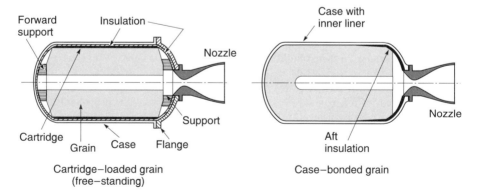

FIGURE 11–14. Simplified schematic diagrams of a free-standing (or cartridge-loaded) and a case-bonded grain.

if the propellant grain has aged excessively. Aging is discussed in the next chapter. Cartridge-loaded grains are used in some small tactical missiles and a few medium-sized motors. They often have a lower cost and are easier to inspect. The case-bonded grains give a somewhat better performance, a little less inert mass (no holding device, support pads, and less insulation), a better volumetric loading fraction, are more highly stressed, and often somewhat more difficult and expensive to manufacture. Today almost all larger motors and many tactical missile motors use case bonding. Stresses in these two types of grains are briefly discussed under structural design in the next section.

Definitions and terminology important to grains include:

Configuration: The shape or geometry of the initial burning surfaces of a grain as it is intended to operate in a motor.

Cylindrical Grain: A grain in which the internal cross section is constant along the axis regardless of perforation shape. (see Fig. 11–3).

Neutral Burning: Motor burn time during which thrust, pressure, and burning surface area remain approximately constant (see Fig. 11–15), typically within about ±15%. Many grains are neutral burning.

Perforation: The central cavity port or flow passage of a propellant grain; its cross section may be a cylinder, a star shape, etc. (see Fig. 11–16).

Progressive Burning: Burn time during which thrust, pressure, and burning surface area increase (see Fig. 11–15).

Regressive Burning: Burn time during which thrust, pressure, and burning surface area decrease (see Fig. 11–15).

Sliver: Unburned propellant remaining (or lost—that is, expelled through the nozzle) at the time of web burnout (see sketch in Problem 11–6).

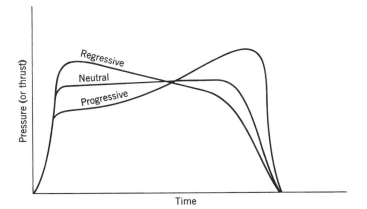

FIGURE 11–15. Classification of grains according to their pressure–time characteristics.

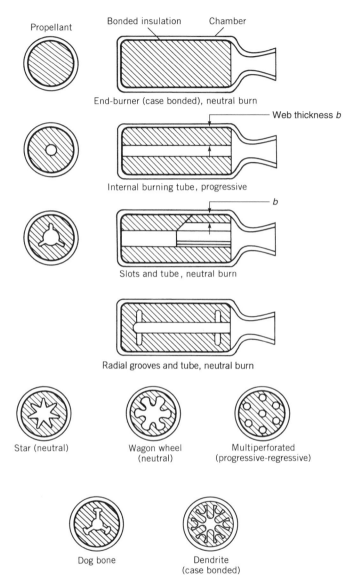

FIGURE 11–16. Simplified diagrams of several grain configurations.

Burning Time, or Effective Burning Time, t_b: Usually, the interval from 10% maximum initial pressure (or thrust) to web burnout, with web burnout usually taken as the aft tangent-bisector point on the pressure–time trace (see Fig. 11–13).

Action Time, t_a: The burning time plus most of the time to burn slivers; typically, the interval between the initial and final 10% pressure (or thrust) points on the pressure–time trace (see Fig. 11–13).

Deflagration Limit: The minimum pressure at which combustion can still be barely self-sustained and maintained without adding energy. Below this pressure the combustion ceases altogether or may be erratic and unsteady with the plume appearing and disappearing periodically.

Inhibitor: A layer or coating of slow- or nonburning material (usually, a polymeric rubber type with filler materials) applied (glued, painted, dipped, or sprayed) to a part of the grain's propellant surface to prevent burning on that surface. By preventing burning on inhibited surfaces the initial burning area can be controlled and reduced. Also called *restrictor*.

Liner: A sticky non-self-burning thin layer of polymeric-type material that is applied to the cases prior to casting the propellant in order to promote good bonding between the propellant and the case or the insulator. It also allows some axial motion between the grain periphery and the case.

Internal Insulator: An internal layer between the case and the propellant grain made of an adhesive, thermally insulating material that will not burn readily. Its purpose is to limit the heat transfer to and the temperature rise of the case during rocket operation. Liners and insulators can be seen in Figs. 11–1, 11–2, 11–4, and 11–14, and are described in Chapter 12.

Web Thickness, b: The minimum thickness of the grain from the initial burning surface to the insulated case wall or to the intersection of another burning surface; for an end-burning grain, b equals the length of the grain (see Fig. 11–16).

Web Fraction, b_f: For a case-bonded internal burning grain, the ratio of the web thickness b to the outer radius of the grain:

$$b_f = b/\text{radius} = 2b/\text{diameter} \tag{11–17}$$

Volumetric Loading Fraction, V_f: The ratio of propellant volume V_b to the chamber volume V_c (excluding nozzle) available for propellant, insulation, and restrictors. Using Eq. 2–4 and $V_b = m/\rho$:

$$V_f = V_b/V_c = I_t/(I_s\rho_b g_0 V_c) \tag{11–18}$$

where I_t is the total impulse, I_s the specific impulse, and ρ_b the propellant density.

A *grain* has to satisfy several *interrelated requirements*:

1. From the *flight mission* one can determine the *rocket motor requirements*. They have to be defined and known before the grain can be designed. They are usually established by the vehicle designers. This can include total impulse, a desired thrust–time curve and a tolerance thereon, motor mass, ambient temperature limits during storage and operation, available

vehicle volume or envelope, and vehicle accelerations caused by vehicle forces (vibration, bending, aerodynamic loads, etc.).

2. The *grain geometry* is selected to fit these requirements; it should be compact and use the available volume efficiently, have an appropriate burn surface versus time profile to match the desired thrust–time curve, and avoid or predictably control possible erosive burning. The remaining unburned propellant slivers, and often also the shift of the center of gravity during burning, should be minimized. This selection of the geo-metry can be complex, and it is discussed in Refs. 11–1 and 11–7 and also below in this section.

3. The *propellant* is usually selected on the basis of its performance cap-ability (e.g., characteristic velocity), mechanical properties (e.g., strength), ballistic properties (e.g., burning rate), manufacturing charac-teristics, exhaust plume characteristics, and aging properties. If neces-sary, the propellant formulation may be slightly altered or "tailored" to fit exactly the required burning time or grain geometry. Propellant selection is discussed in Chapter 12 and in Ref. 11–7.

4. The *structural integrity* of the grain, including its liner and/or insulator, must be analyzed to assure that the grain will not fail in stress or strain under all conditions of loading, acceleration, or thermal stress. The grain geometry can be changed to reduce excessive stresses. This is discussed in the next section of this chapter.

5. The complex *internal cavity volume* of perforations, slots, ports, and fins increases with burning time. These cavities need to be checked for reso-nance, damping, and *combustion stability*. This is discussed in Chapter 13.

6. The *processing* of the grain and the *fabrication* of the propellant should be simple and low cost (see Chapter 12).

The grain configuration is designed to satisfy most requirements, but some-times some of these six categories are satisfied only partially. The geometry is crucial in grain design. For a neutral burning grain (approximately constant thrust), for example, the burning surface A_b has to stay approximately con-stant, and for a regressive burning grain the burning area will diminish during the burning time. From Eqs. 11–3 and 11–14 the trade-off between burning rate and the burning surface area is evident, and the change of burning surface with time has a strong influence on chamber pressure and thrust. Since the density of most modern propellants falls within a narrow range (about 0.066 lbm/in.^{3} or 1830 kg/m^3 +2 to −15%), it has little influence on the grain design.

As a result of motor developments of the past three decades, many *grain configurations* are available to motor designers. As methods evolved for increasing the propellant burning rate, the number of configurations needed decreased. Current designs concentrate on relatively few configurations, since the needs of a wide variety of solid rocket applications can be fulfilled by combining known configurations or by slightly altering a classical configuration. The trend has been to discontinue configurations that give weak grains which can form cracks, produce high sliver losses, have a low volumetric loading fraction, or are expensive to manufacture.

The effect of propellant burning on surface area is readily apparent for simple geometric shapes such as rods, tubes, wedges, and slots, as shown in the top four configurations of Fig. 11–16. Certain other basic surface shapes burn as follows: external burning rod—regressive; external burning wedge—regressive. Most propellant grains combine two or more of these basic surfaces to obtain the desired burning characteristic. The star perforation, for example, combines the wedge and the internal burning tube. Figure 11–17 indicates typical single grains with combinations of two basic shapes. The term *conocyl* is a contraction of the words *cone* and *cylinder*.

Configurations that combine both radial and longitudinal burning, as does the internal–external burning tube without restricted ends, are frequently referred to as "three-dimensional grains" even though all grains are geometrically three-dimensional. Correspondingly, grains that burn only longitudinally

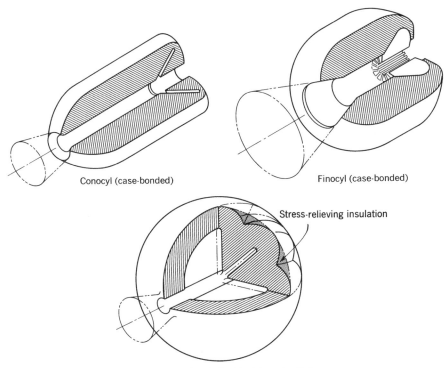

Conocyl (case-bonded)

Finocyl (case-bonded)

Stress-relieving insulation

Spherical (case-bonded) with slots and cylinder

FIGURE 11–17. Typical common grain configurations using combinations of two basic shapes for the grain cavity.

or only radially are "two-dimensional grains." Grain configurations can be classified according to their web fraction b_f, their length-to-diameter ratio L/D, and their volumetric loading fraction V_f. These three dependent variables are often used in selecting a grain configuration in the preliminary design of a motor for a specific application. Obvious overlap of characteristics exists with some of the configurations, as given in Table 11–4 and shown by simple sketches in Fig. 11–16. The configurations listed above the line in the table are common in recent designs. The bottom three were used in earlier designs and usually are more difficult to manufacture or to support in a case. The end burner has the highest volumetric loading fraction, the lowest grain cavity volume for a given total impulse, and a relatively low burning area or thrust with a long duration. The internal burning tube is relatively easy to manufacture and is neutral burning with unrestricted ends of $L/D \simeq 2$. By adding fins or cones (see Fig. 11–17) this configuration works for $2 < L/D < 4$. The star configuration is ideal for web fractions of 0.3 to 0.4; it is progressive above 0.4, but can be neutralized with fins or slots. The wagon wheel is structurally superior to the star shape around 0.3 and is necessary at a web fraction of 0.2 (high thrust and short burn time). Dendrites are used in the lowest web fraction when a relatively large burning area is needed (high thrust and short duration), but stresses may be high. Although the limited number of configurations given in this table may not encompass all the practical possibilities for fulfilling a nearly constant thrust–time performance requirement, combinations of these features should be considered to achieve a neutral pressure–time trace and high volumetric loading before a relatively unproven configuration is accepted. The capabilities of basic configurations listed in these tables can be

TABLE 11–4. Characteristics of Several Grain Configurations

Configuration	Web Fraction	L/D ratio	Volumetric Fraction	Pressure–time Burning Characteristics	C.G. shift
End burner	> 1.0	NA	0.90–0.98	Neutral	Large
Internal burning tube (including slotted tube, trumpet, conocyl, finocyl)	0.5–0.9	1–4	0.80–0.95	Neutral[a]	Small to moderate
Segmented tube (large grains)	0.5–0.9	> 2	0.80–0.95	Neutral	Small
Internal star[b]	0.3–0.6	NA	0.75–0.85	Neutral	Small
Wagon Wheel[b]	0.2–0.3	NA	0.55–0.70	Neutral	Small
Dendrite[b]	0.1–0.2	1–2	0.55–0.70	Neutral	Small
Internal–external burning tube	0.3–0.5	NA	0.75–0.85	Neutral	Small
Rod and tube	0.3–0.5	NA	0.60–0.85	Neutral	Small
Dog bone[b]	0.2–0.3	NA	0.70–0.80	Neutral	Small

[a]Neutral if ends are unrestricted, otherwise progressive.
[b]Has up to 4 or sometimes 8% sliver mass and thus a gradual thrust termination.
NA: not applicable or not available.

extended by alterations. The movement of the center of gravity influences the flight stability of the vehicle. Relative values of this CG shift are also shown in Table 11–4. Most solid propellant manufacturers have specific approaches and sophisticated computer programs for analyzing and optimizing grain geometry alternatives and permitting burn surface and cavity volume analysis. See Refs. 11–15 and 11–16 and Chapters 8 and 9 of Ref. 11–1.

The *end burning grain* (burning like a cigarette) is unique; it burns solely in the axial direction and maximizes the amount of propellant that can be placed in a given cylindrical motor case. In larger motors (over 0.6 m diameter) these end burners show a progressive thrust curve. Figure 11–18 shows that the burning surface soon forms a conical shape, causing a rise in pressure and thrust. Although the phenomenon is not fully understood, two factors contribute to higher burning rate near the bondline: chemical migration of the burning rate catalyst into and towards the bondline, and local high propellant stresses and strains at the bond surface, creating local cracks (Ref. 11–17).

Rockets used in air-launched or certain surface-launched missile applications, weather rockets, certain antiaircraft or antimissile rockets, and other tactical applications actually benefit by reducing the thrust with burn time. A high thrust is desired to apply initial acceleration, but, as propellant is consumed and the vehicle mass is reduced, a decrease in thrust is desirable; this limits the maximum acceleration on the rocket-propelled vehicle or its sensitive payload, often reduces the drag losses, and usually permits a more effective flight path. Therefore, there is a benefit to vehicle mass, flight performance, and cost in having a higher initial thrust during the *boost phase* of the flight, followed by a lower thrust (often 10 to 30% of boost thrust) during the *sustaining phase* of the powered flight. Figure 11–19 shows grains which give two or more discrete thrust periods in a single burn operation. The configurations are actually combinations of the configurations listed in Table 11–4.

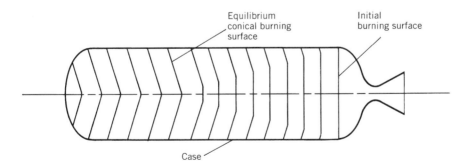

FIGURE 11–18. Schematic diagram of end-burning grain coning effect. In larger sizes (above approximately 0.5 m diameter) the burning surface does not remain flat and perpendicular to the motor axis, but gradually assumes a conical shape. The lines in the grain indicate successively larger-area burning surface contours.

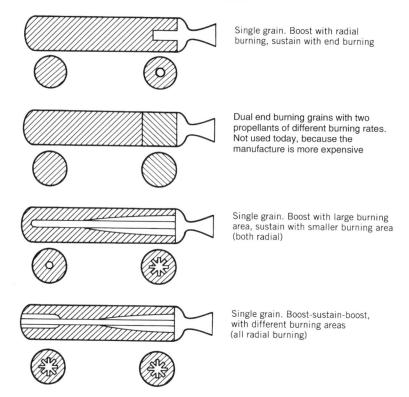

Single grain. Boost with radial burning, sustain with end burning

Dual end burning grains with two propellants of different burning rates. Not used today, because the manufacture is more expensive

Single grain. Boost with large burning area, sustain with smaller burning area (both radial)

Single grain. Boost-sustain-boost, with different burning areas (all radial burning)

FIGURE 11–19. Several simplified schematic diagrams of grain configurations for an initial period of high thrust followed by a lower-thrust period.

In a single-propellant dual-thrust level solid rocket motor, factors relating to the sustain portion usually dominate in the selection of the propellant type and grain configuration if most of the propellant volume is used during the longer sustain portion.

A *restartable rocket motor* has advantages in a number of tactical rocket propulsion systems used for aircraft and missile defense applications. Here two (or sometimes three) grains are contained inside the same case, each with its own igniter. The grains are physically separated typically by a structural bulkhead or by an insulation layer. One method for accomplishing this is shown in Fig. 11–20. The timing between thrust periods (sometimes called *thrust pulses*) can be controlled and commanded by the missile guidance system, so as to change the trajectory in a nearly optimum fashion and minimize the flight time to target. The separation mechanism has to prevent the burning-hot pressurized gas of the first grain from reaching the other grain and causing its inadvertent ignition. When the second grain is ignited the separation devices are automatically removed, fractured, or burned, but in such a manner that the

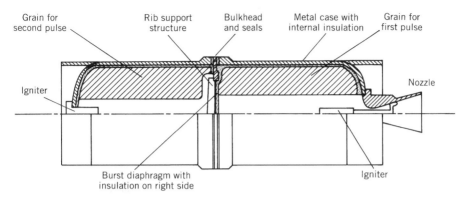

FIGURE 11–20. Simplified diagram of one concept of a two-pulse experimental rocket motor with two grains separated by a bulkhead. During the first pulse operation the metal diaphragm is supported by a spider-web-like structure made of high temperature material. Upon ignition of the second stage, the scored diaphragm is loaded in the other direction; it breaks and its leaves peel back. The bulkhead opening has a much larger area than the nozzle throat.

fragments of hardware pieces will not plug the nozzle or damage the insulation (see Refs. 11–18 and 11–19).

Slivers

Any remaining unburnt propellant is known as slivers. Figure 11–5 and the figure in Problem 11–6 show small slivers or pieces of unburnt propellant remaining at the periphery of the grain, because the pressure went below the deflagration limit (see Ref. 11–20). About 25 years ago grain designs had 2 to 7% propellant slivers; this useless material caused a reduction in propellant mass fraction and vehicle mass ratio. The technology of grain design has advanced so that there are almost no slivers (usually less than 1%). If slivers were to occur in a new unusual grain design, the designer would try to replace the sliver volume with lower-density insulator, which gives less of a mass ratio penalty than the higher-density propellant residue. This is shown in Fig. 11–17.

11.4. PROPELLANT GRAIN STRESS AND STRAIN

The objective of stress analysis of rocket motors is to design the configuration of the grain, the liners, or the grain support in such a way that excessive stresses or excessive strains will not occur and so that there will be no failure. Static and dynamic loads and stresses are imposed on the propellant grains during manufacture, transportation, storage, and operation. Structurally, a rocket motor is a thin shell of revolution (motor case) almost completely filled with a viscoelastic material, the propellant, which usually accounts for 80 to 94% of the motor mass. Propellant has some mechanical properties that are not found in ordinary structural materials and these have received relatively little study. The viscoelastic nature of solid propellant is time–history dependent and the material accumulates damage from repeated stresses; this is known as the *cumulative-damage phenomenon.*

The most common *failure modes* are:

1. Surface cracks are formed when the surface strain is excessive. They open up new additional burning surfaces and this in turn causes the chamber pressure as well as the thrust to be increased. The higher, shorter duration thrust will cause the vehicle to fly a different trajectory and this may cause the mission objective to be missed. With many cracks or deep cracks, the case becomes overpressurized and will fail. The limiting strain depends on the stress level, grain geometry, temperature, propellant age, load history, and the sizes of flaws or voids. At a high strain rate, deeper, more highly branched cracks are more readily formed than at a lower strain rate (see Ref. 11–9).

2. The bond at the grain periphery is broken and an unbonded area or gap can form next to the liner, insulator, or case. As the grain surface regresses, a part of the unbonded area will become exposed to the hot, high-pressure combustion gases, and then suddenly the burning area is increased by the unbonded area.

Other failure modes, such as an excessively high ambient grain temperature causing a large reduction in the physical strength properties, ultimately result in grain cracks and/or debonding. Air bubbles, porosity, or uneven density can locally reduce the propellant strength sufficiently to cause failure, again by cracks or debonds. Other failure modes are excessive deformations of the grain (e.g., slump of large grains can restrict the port area) and involuntary ignition due to the heat absorbed by the viscoelastic propellant from excessive mechanical vibration (e.g., prolonged bouncing during transport).

If the grain has a large number of small cracks or a few deep cracks or large areas of unbonding prior to firing, the burning area will increase, often progressively and unpredictably, and the resulting higher pressure will almost always cause the case to burst. A few small cracks or minor unbonded areas will usually not impede satisfactory motor operation.

Material Characterization

Before a structural analysis can be performed it is necessary to understand the materials and obtain data on their properties. The grain materials (propellant, insulator, and liner) are rubber-like materials that are nearly incompressible. They all have a *bulk modulus in compression* of at least 1400 MPa or about 200,000 psi in their original state (undamaged). Since there are very few voids in a properly made propellant (much less than 1%), its compression strain is low. However, the propellant is easily damaged by applied tension and shear loads. If the strength of propellant in tension and shear (typically betwen 50 and 1000 psi) is exceeded, the grain will be damaged or fail locally. Since grains are three-dimensional, all stresses are combined stresses and not pure compression stresses, and grains are thus easily damaged. This *damage* is due to a "dewetting" of the adhesion between individual solid particles and the binder in the propellant and appears initially as many small voids or porosity. Those very small holes or debonded areas next to or around the solid particles may initially be under vacuum, but they become larger with strain growth.

The propellant, liner, and insulator with a solid filler are viscoelastic materials. They show a nonlinear viscoelastic behavior, not a linear elastic behavior. This means that the maximum stress and maximum elongation or strain diminish each time a significant load is applied. The material becomes weaker and suffers some damage with each loading cycle or thermal stress application. The physical properties also change with the time rate of applying loads; for example, very fast pressurization actually gives a stronger material. Certain binders, such as hydroxyl-terminated polybutadiene (HTPB), give good elongation and a stronger propellant than other polymers used with the same percentage of binder. Therefore HTPB is a preferred binder today. The physical properties are also affected by the manufacturing process. For example, tensile specimens cut from the same conventionally cast grain of composite propellant can show 20 to 40% variation in the strength properties between samples of different orientations relative to the local casting slurry flow direction. Viscoelastic material properties change as a function of prior loading and damage history. They have the capability to reheal and recover partially following damage. Chemical deterioration will in time degrade the properties of many propellants. These phenomena make it difficult to characterize these materials and predict their behavior or physical properties in engineering terms.

Several kinds of laboratory tests on small samples are routinely performed today to determine the physical properties of these materials. (see Refs. 11–21 and 11–22). Simple tests, however, do not properly describe the complex nonlinear behavior. These laboratory tests are conducted under ideal conditions—mostly uniaxial stresses instead of complex three-dimensional stresses—with a uniform temperature instead of a thermal gradient and usually with no prior damage to the material. The application of laboratory test results to real structural analysis therefore involves several assumptions and empirical correction factors. The test data are transformed into derived parameters for determining safety margins and useful life, as described in Chapter 9 of Ref. 11–1. There is no complete agreement on how best to characterize these materials. Nevertheless, laboratory tests provide useful information and several are described below.

The most common test is a simple *uniaxial tensile test* at constant strain rate. One set of results is shown in Fig. 11–21. The test is commonly used for manufacturing quality control, propellant development, and determining fail-

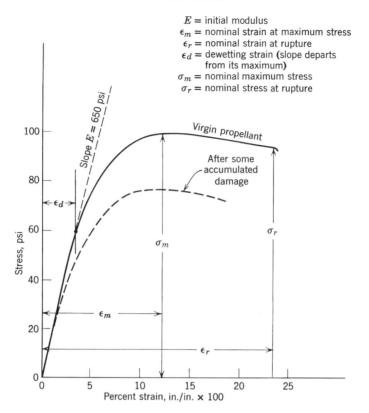

FIGURE 11–21. Stress–strain curves for a typical composite-type solid propellant showing the effect of cumulative damage. The maximum stress σ_m is higher than the rupture stress σ_r, of the tensile test sample.

ure criteria. Once the sample has been loaded, unloaded, and restressed several times, the damage to the material changes its response and properties as shown by the dashed curve in Fig. 11–21.

The dewetting strain is, by definition, the strain (and corresponding maximum stress) where incipient failure of the interface bonds between small solid oxidizer crystals and the rubbery binder occurs. The dewetting stress is analogous to the yield point in elastic materials, because this is when internal material damage begins to happen. The slope E, the modulus at low strain, is not ordinarily used in design, but is often used as a quality control parameter. Data from several such uniaxial tests at different temperatures can then be manipulated to arrive at allowable stresses, permissible safe strains, and a derived artificial modulus, as described later. Once a case-bonded grain has been cooled down from its casting temperature it will have shrunk and be under multidirectional strain. Samples cut from different parts of a temperature-cycled grain will usually give different tensile test results.

Biaxial strength tests are also performed frequently in the laboratory. One type is described in Ref. 11–21. Meaningful three-dimensional stress tests are difficult to perform in the laboratory and are usually not done. There are other sample tests that give information about propellant behavior, such as strain endurance tests to obtain the levels of strain at which the propellant has long endurance and does not suffer significant damage, tests at constant stress levels, fracture tests of samples with known cracks or defects, tensile tests under simulated chamber pressure, or tests to measure the thermal coefficient of expansion. Peel tests of the adhesive bonds of propellants to liners or insulators are very common and their failures are discussed in Ref 11–22. The application and interpretation of all these tests depend on the stress conditions in the grain and company preferences. In addition, strain or stress measurements are made occasionally on full-scale, experimental, flight-weight motors using special embedded sensors. Care must be taken that the implanting of these sensors into the grain will not disturb the local stress–strain distribution, which would lead to erroneous measurements.

The maximum failure stresses of most propellants are relatively low compared to those of plastic materials. Typical values range from about 0.25 to 8 MPa or about 40 to about 1200 psi, with average values between 50 and 300 psi, and elongations range from 4 to 250%, depending on the specific propellant, its temperature, and its stress history. Table 11–5 shows properties for a relatively strong propellant. Some double-base propellants and binder-rich composite propellants can withstand higher stresses (up to about 32 MPa or 4600 psi). The pressure and the strain rate have a major influence on the physical properties. Tensile tests performed at chamber pressure give higher strength than those done at atmospheric pressure, in some cases by a factor of 2 or more. High strain rates (sudden-start pressurization) can also improve the propellant properties temporarily.

The strength properties of the grain material are commonly determined over a range of propellant temperatures. For air-launched missiles these limits are

TABLE 11–5. Range of Tensile Properties of Reduced Smoke Composite Propellant for a Tactical Missile[a]

	Temperature (°F)		
	158	77	−40
Maximum stress (psi)	137–152	198–224	555–633
Modulus (psi)	262–320	420–483	5120–6170
Strain at maximum stress/strain and at ultimate stress (%)	54/55–65/66	56/57–64/66	46/55–59/63

[a]Polybutadiene binder with reduced aluminum and ammonium perchlorate; data are from four different 5-gallon mixes.

Source: Data taken with permission of the AIAA from Ref. 11–23.

wide, with $-65°F$ and $+160°F$ or 219 K and 344 K often being the lower and upper extremes expected during motor exposure. Propellant grains must be strong enough and have elongation capability sufficient to meet the high stress concentrations present during shrinkage at low temperature and also under the dynamic load conditions of ignition and motor operation. The mechanical properties (strength, elongation) can be increased by increasing the percent of binder material in the propellant, but at a reduction in performance.

Structural Design

The structural analysis of a typical case-bonded grain has to consider not only the grain itself but also the liner, insulator, and case, which interact structurally with the propellant grain under various loading conditions (see Chapter 9 or Ref. 11–1). The need to obtain strong bonds between the propellant and the liner, the liner and the insulator, or the insulator and the case is usually satisfied by using properly selected materials and manufacturing procedures to assure a good set of bonds. Liners are usually flexible and can accept large strains without failure, and the vehicle loads can be transmitted from the case (which is usually part of the vehicle structure) into the propellant.

When the propellant is cured (heated in an oven), it is assumed to have uniform internal temperature and to be free of thermal stresses. As the grain cools and shrinks after cure and reaches an equilibrium uniform ambient temperature (say, from -40 to $+75°F$), the propellant experiences internal stresses and strains which can be relatively large at low temperature. The stresses are increased because the case material usually has a thermal coefficient of expansion that is smaller than that of the propellant by an order of magnitude. The stress-free temperature range of a propellant can be changed by curing the motor under pressure. Since this usually reduces the stresses at ambient temperature extremes, this pressure cure is now being used more commonly.

The structural analysis begins when all loads can be identified and quantified. Table 11–6 lists the typical loads that are experienced by a solid propellant motor during its life cycle and some of the failures they can induce. Some of these loads are unique to specific applications. The loads and the timing of these loads during the life cycle of a solid propellant rocket motor have to be analyzed for each application and each motor. They depend on the motor design and use. Although ignition and high accelerations (e.g., impact on a motor that falls off a truck) usually cause high stresses and strains, they may not always be the critical loads. The stresses induced by ambient environmental temperature cycling or gravity slumps are often relatively small; however, they are additive to stresses caused by other loads and thus can be critical. A space motor that is to be fired within a few months after manufacture presents a different problem than a tactical motor that is to be transported, temperature cycled, and vibrated for a long time, and this is different yet from a large-diameter ballistic missile motor that sits in a temperature-conditioned silo for more than 10 years.

TABLE 11–6. Summary of Loads and Likely Failure Modes in Case-Bonded Rocket Motors

Load Source	Description of Load and Critical Stress Area
1. Cool-down during manufacture after hot cure	Temperature differential across case and grain; tension and compression stresses on grain surfaces; hot grain, cool case
2. Thermal cycling during storage or transport	Alternative hot and cold environment; critical condition is with cold grain, hot case; two critical areas: bond-line tensile stress (tearing), inner-bore surface cracking
3. Improper handling and transport vibrations	Shock and vibration, 5 to $30g_0$ forces during road transport at 5 to 300 Hz (5 to 2500 Hz for external aircraft carry) for hours or days; critical failure: grain fracture or grain debonding
4. Ignition shock/pressure loading	Case expands and grain compresses; axial pressure differential is severe with end-burning grains; critical areas; fracture and debonding at grain periphery
5. Friction of internal gas flow in cavity	Axially rearward force on grain
6. Launch and axial flight acceleration	Inertial load mostly axial; shear stress at bond line; slump deformation in large motors can reduce port diameter
7. Flight maneuvers (e.g., antimissile rocket)	High side accelerations cause unsymmetrical stress distribution; can result in debonding or cracks
8. Centrifugal forces in spin-stabilized projectiles/missiles	High strain at inner burning surfaces; cracks will form
9. Gravity slump during storage; only in large motors	Stresses and deformation in perforation can be minimized by rotating the motor periodically; port area can be reduced by slump
10. External air friction when case is also the vehicle's skin	Heating of propellant, liner and insulators will lower their strengths causing premature failure. Induces thermal stresses

Furthermore, the structural analysis requires a knowledge of the material characteristics and *failure criteria*: namely, the maximum stress and strains that can safely be accepted by the propellant under various conditions. The failure criteria are derived from cumulative damage tests, classical failure theories, actual motor failures, and fracture mechanics. This analysis may be an iterative analysis, because the materials and geometry need to be changed if analysis shows that the desired margins of safety are exceeded.

Ideally, the analysis would be based on a nonlinear viscoelastic stress theory; however, such an approach is still being developed and is not yet reliable (see Ref. 11–1). An analysis based on a viscoelastic material behavior is feasible, relatively complex, and requires material property data that are difficult to obtain and uncertain in value. Most structural analyses today are based on an elastic material model; it is relatively simple and many two- and three-dimensional finite element analysis computer programs of this approach are available at rocket motor manufacturing companies. Admittedly, this theory does not fit all the facts, but with some empirical corrections it has given approximate answers to many structural grain design problems. An example of a two-dimensional finite element grid from a computer output is shown in Fig. 11–22 for a segment of a grain using an elastic model (see Refs. 11–24 and 11–25).

With elastic materials the stress is essentially proportional to strain and independent of time; when the load is removed, the material returns to its original condition. Neither of these propositions is valid for grains or their propellant materials. In viscoelastic material a time-related dependency exists between stresses and strains; the relationship is not linear and is influenced by the rate of strain. The stresses are not one-dimensional as many laboratory tests are, but three-dimensional, which are more difficult to visualize. When the load is removed, the grain does not return to its exact original position. References 11–26 and 11–27 and Chapters 9 and 10 of Ref. 11–1 discuss three-dimensional analysis techniques and viscoelastic design. A satisfactory analysis technique has yet to be developed to predict the influence of cumulative damage.

Various techniques have been used to compensate for the nonelastic behavior by using allowable stresses that have been degraded for nonlinear effects and/or an effective modulus that uses a complex approximation based on laboratory strain test data. Many use a modified modulus (maximum stress–strain at maximum stress or σ_m/ϵ_m in Fig. 11–21) called the *stress relaxation modulus* E_R in a master curve against temperature-compensated time to failure, as shown in Fig. 11–23. It is constructed from data collected from a series of uniaxial tests at constant strain rate (typically, 3 to 5%) performed at different temperatures (typically −55 to +43°C). The shifted temperature T_s/T is shown in the inset on the upper right for 3% strain rate and sample tests taken at different temperatures. The factor λ in the ordinate corrects for the necking down of the tension sample during test. The small inset in this figure explains the correction for temperature that is applied to the reduced time to failure. The empirical time–temperature shift factor a_T is set to zero at ambient temperatures (25°C or 77°F) and graphically shifted for higher and lower temperatures. The master curve then provides time-dependent stress–strain data to calculate the response of the propellant for structural analysis (see Ref. 11–21 and Chapter 9 of Ref. 11–1).

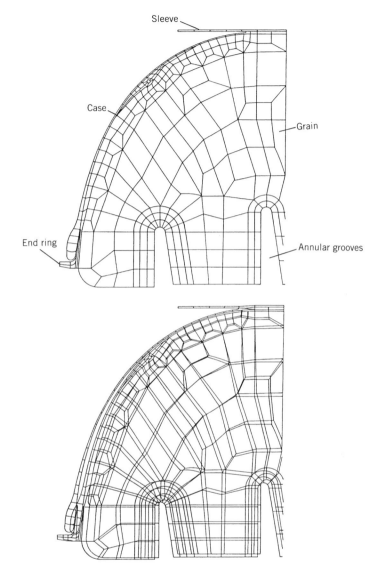

FIGURE 11–22. Finite element analysis grid of the forward end of a cast grain in a filament-wound plastic case. The grain has an internal tube and annular grooves. The top diagram shows the model grid elements and the bottom shows one calculated strain or deformation condition. (Reprinted with permission from A. Turchot, Chapter 10 of Ref. 11–1).

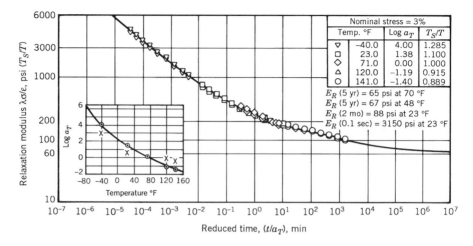

FIGURE 11–23. This stress-relaxation modulus master curve for a particular composite solid propellant is constructed from manipulated data taken from a number of uniaxial tensile tests at constant strain rate but different temperatures. (Reproduced with permission of United Technologies Corp., Chemical Systems from Ref. 11–27.)

Usually, several different grain loading and operating conditions need to be analyzed. Such a structural analysis is useful for identifying locations of maximum stress or strain and to any structural members or grain sectors that are too weak or too heavy, but these analyses have not always been successful. The choice of the best analysis tool and the best pseudo-viscoelastic compensation factors will depend on the experience of the stress analyst, the specific motor design conditions, the complexity of the motor, the geometry, and suitable, available, valid propellant property data.

In a case-bonded motor, special provision is required to reduce the stress concentrations at the grain ends where the case and grain interface, especially for motors expected to operate satisfactorily over a wide range of temperatures. Basically, the high stresses arise from two primary sources. First, the physical properties, including the coefficient of thermal expansion of the case material and the propellant, are grossly dissimilar. The coefficient of expansion of a typical solid propellant is 1.0×10^{-4} m/m-K, which is five times as great as that of a steel motor case. Secondly, the aft-end and head-end geometries at the grain–case juncture often present a discontinuity, with the grain stress theoretically approaching infinity. Actually, finite stresses exist because viscoplastic deformations do occur in the propellant, the liner, and the case insulation. Calculating the stress in a given case–grain termination arrangement is usually impractical, and designers rely on approximations supported by empirical data.

For simple cylindrical grains the highest stresses usually occur at the outer and inner surfaces, at discontinuities such as the bond surface termination point, or at stress concentration locations, such as sharp radii at the roots or tips of star or wagonwheel perforations, as shown in Fig. 11–16. Figure 11–24 shows a *stress relief flap*, sometimes called a *boot*, a device to reduce local stresses. It is usually an area on the outside of the grain near its aft end (and sometimes also its forward end), where the liner material is not sticky but has a non-adhesive coating that permits the grain to shrink away from the wall. It allows for a reduction of the grain at the bond termination point. It moves the location of highest stress into the liner or the insulation at the flap termination or hinge. Normally, the liner and insulation are much stronger and tougher than the propellant.

Parametric studies of propellant and case-bond stresses of a typical grain–case termination design (Fig. 11–24) reveal the following:

1. Flap length is less significant than the thickness of the insulation or the separate flap boot, if one is used, in controlling the local level of stresses at the grain–case termination.

2. The distribution of stresses at the grain–case termination is sensitive to the local geometry; the level of stress at the case bond increases with web fraction and length-to-diameter ratio under loading by internal pressure and thermal shrinkage.

3. As the L/D and web fraction increase, the inner-bore hoop stress and the radial stress at the grain–case bond increase more rapidly than does the

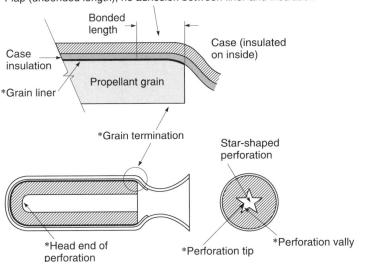

FIGURE 11–24. The asterisks in the bottom simplified diagram denote potentially critical failure areas. The top sketch is an enlargement of the aft termination region of the grain and shows a boot or stress relief flap.

grain–case termination stress under internal pressure and thermal shrink-age loads.

4. The radial case-bond stress level at the grain–case termination is much larger than the case-bond shear stress under axial acceleration loading as well as under internal pressure and thermal shrinkage loading.

Aging of propellants in rocket motors refers to their deterioration in the physical properties with time. It is caused by the *cumulative damage* done to the grain (such as by thermal cycling, and load applications) during storage, hand-ling, or transport. It can also be caused by chemical changes with time, such as the gradual depletion (evaporation) of certain liquid plasticizers or moisture absorption. The ability to carry stress or to allow elongation in propellants diminishes with cumulative damage. The *aging limit* is the estimated time when the motor is no longer able to perform its operation reliably or safely (see Refs. 11–28 and 11–29). Depending on the propellant and the grain design, this age limit or motor life can be betwen 8 and 25 years. Before this limit is reached, the motor should be deactivated and have its propellant removed and replaced. This refurbishing of propellant is routinely done on larger and more expensive rocket motors in the military inventory.

With small tactical rocket motors the aging limit is usually determined by full-scale motor-firing tests at various time periods after manufacture, say 2 or 3 years and with an extrapolation to longer time periods. Accelerated tempera-ture aging (more severe thermal cycles) and accelerated mechanical pulse loads and overstressing are often used to reduce the time needed for these tests. For large rocket motors, which are more expensive, the number of full-scale tests has to be relatively small, and aging criteria are then developed from structural analysis, laboratory tests, and subscale motor tests.

Many of the early grains were *cartridge loaded* and kept the grain isolated from the motor case to minimize the interrelation of the case and the grain stresses and strains resulting from thermal expansion. Also, upon pressuriza-tion the case would expand, but the grain would shrink. The case-bonded grain presents a far more complex problem in stress analysis. With the propellant grain bonded firmly to the case, being a semirubbery and relatively weak material, it is forced to respond to case strains. As a result, several critically stressed areas exist in every case-bonded motor design; some are shown with an asterisk in Fig. 11–24.

The varying nature of the stress analysis problem is brought about by the physical character of propellant; in general terms, solid propellant is relatively weak in tension and shear, is semielastic, grows softer and weaker at elevated temperatures, becomes hard and brittle at low temperatures, readily absorbs and stores energy upon being vibrated, degrades physically during long-term storage because of decomposition and chemical or crystalline changes, and accumulates structural damage under load, including cyclic load. This last phenomenon is shown graphically in Fig. 11–25 and is particularly important in the analysis of motors that are to have a long shelf-life (more than 10 years).

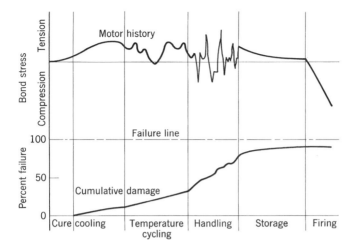

FIGURE 11–25. Representation of the progress in cumulative damage to the bond between the grain and the case in a case-bonded rocket motor experiencing a hypothetical stress history. (Adapted from Ref. 11–30.)

No a priori reason is known for materials to exhibit *cumulative damage*, but propellants and their bond to case material exhibit this trait even under constant load, as shown in Fig. 11–26. Valid theories and analytical methods applicable to cumulative damage include a consideration of both the stress–strain history and the loading path (the material effected). The most important environmental variables affecting the shelf life of a motor are time, temperature

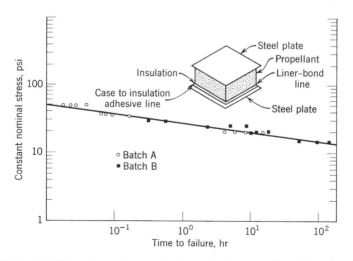

FIGURE 11–26. Time dependent reduction of the propellant–liner–insulator bond strength when subjected to constant load at 77°F. (From Ref. 11–31.)

cycles, propellant mass, stress (gravity forces for large motors), and shock and vibration. Failure due to cumulative damage usually appears as cracks in the face of the perforation or as local "unbonds" in case-bonded motors.

The strength of most propellants is sensitive to the rate of strain; in effect they appear to become more brittle at a given temperature as the strain rate is increased, a physical trait that is important during the ignition process.

11.5. ATTITUDE CONTROL AND SIDE MANEUVERS WITH SOLID PROPELLANT ROCKET MOTORS

A clever attitude control (also called reaction control) system with solid propellants is used on some ballistic missiles. Its hot reaction gas has a low enough temperature so that uncooled hardware can be used for long durations. Ammonium nitrate composite propellant (mentioned as gas generator propellants in Tables 12–1 and 12–2) or a propellant consisting of a nitramine (RDX or HMX, described in Chapter 12) with a polymer binding are suitable. The version shown schematically in Fig. 11–27 provides pitch and yaw control; hot gas flows continuously through insulated manifolds, open hot-gas valves, and all four nozzles. When one of these valves is closed, it causes an unbalance of gas flow and produces a side force. To keep things simple, the four roll-control thrusters have been deleted from this figure.

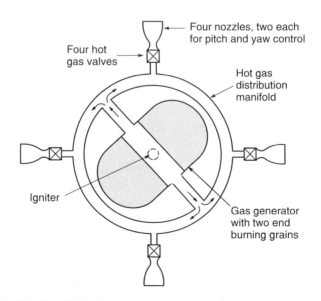

FIGURE 11–27. Simplified diagram of a rocket attitude control system using solid propellant. All four valves are normally open and gas flows equally through all nozzles.

With this type of attitude control system it is possible to achieve variable duration thrust pulsing operations and random pitch, yaw, and roll maneuvers. It is competitive with multi-thruster liquid propellant attitude control systems. The solid propellant versions are usually heavier, because they have heavy insulated hardware and require more propellant (for continuous gas flow), whereas the liquid version is operated only when attitude control motions are required.

A similar approach with hot gas valves applies to upper stages of interceptor vehicles used for missile defense; there is little time available for maneuvers of the upper stage to reach the incoming missile or aircraft and therefore the burning durations are usually short. The solid propellant gas temperatures are higher than with gas generators (typically 1260°C or 2300°F), but lower than with typical composite propellants (3050 K or 5500°F), and this allows the valves and manifolds to be made of high-temperature material (such as rhenium or carbon). In addition to attitude control, the system provides a substantial side force or divert thrust. It displaces the flight path laterally. Figure 11–28 shows such a system. Since all hot-gas valves are normally open, a valve has to be closed to obtain a thrust force as explained in the previous figure. The attitude control system provides pitch, yaw, and roll control to stabilize the vehicle during its flight, to orient the divert nozzle into the desired direction, and sometimes to orient the seeker (at the front of the vehicle) toward the target.

PROBLEMS

1. What is the ratio of the burning area to the nozzle area for a solid propellant motor with these characteristics?

Propellant specific gravity	1.71
Chamber pressure	14 MPa
Burning rate	38 mm/sec
Temperature sensitivity σ_p	0.007 $(K)^{-1}$
Specific heat ratio	1.27
Chamber gas temperature	2220 K
Molecular mass	23 kg/kg-mol
Burning rate exponent n	0.3

2. Plot the burning rate against chamber pressure for the motor in Problem 1 using Eq. 11–3 between chamber pressures of 11 and 20 MPa.

3. What would the area ratio A_b/A_t in Problem 1 be if the pressure were increased by 10%? (Use curve from Problem 2.)

4. Design a simple rocket motor for the conditions given in Problems 1 and 2 for a thrust of 5000 N and a duration of 15 sec. Determine principal dimensions and approximate weight.

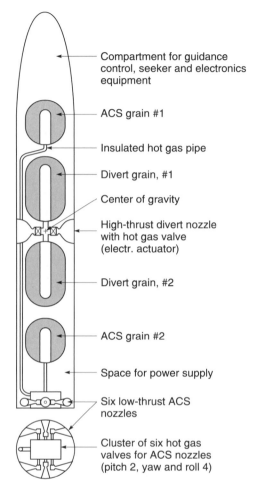

FIGURE 11–28. Simplified schematic diagram of two propulsion systems for one type of maneuverable upper stage of an interceptor missile. The side or divert forces are relatively large and go essentially through the center of gravity (CG) of the upper stage vehicle. To minimize the CG travel two grains are above and two grains are below the CG. Each nozzle has its own hot gas valve, which is normally open and can be pulsed. The attitude control system (ACS) is fed from the reaction gas of two grains and has six small nozzles.

5. For the Orbus-6 rocket motor described in Table 11–3 determine the total impulse-to-weight ratio, the thrust-to-weight ratio, and the acceleration at start and burnout if the vehicle inert mass and the payload come to about 6000 lbm. Use burn time from Table 11-3 and assume $g \approx 32.2 \, \text{ft/sec}^2$.

6. For a cylindrical two-dimensional grain with two slots the burning progresses in finite time intervals approximately as shown by the successive burn surface contours in the drawing on the next page. Draw a similar set of progressive burning surfaces

for any one configuration shown in Figure 11–16 and one shown in Figure 11–17, and draw an approximate thrust–time curve from these plots, indicating the locations where slivers will remain. Assume the propellant has a low value of n and thus the motor experiences little change in burning rate with chamber pressure.

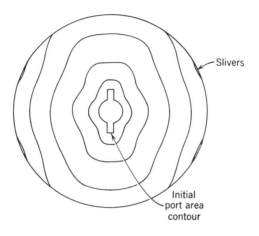

7. Explain the significance of the web fraction, the volumetric loading ratio, and the L/D ratio in terms of vehicle performance and design influence.

8. The partial differential equations 11–4 and 11–5 express the influence of temperature on the burning of a solid propellant. Explain how a set of tests should be set up and exactly what should be measured in order to determine these coefficients over a range of operating conditions.

9. What would be the likely change in r, I_s, p_1, F, t_b, and I_t if the three rocket motors described in Table 11–3 were fired with the grain 100°F cooler than the data shown in the table? Assume typical average temperature effects.

10. A newly designed case-bonded rocket motor with a simple end-burning grain failed and exploded on its first test. The motor worked well for about 20% of its burn time, when the record showed a rapid rise in chamber pressure. It was well conditioned at room temperature before firing and the inspection records did not show any flaws or voids in the grain. Make a list of possible causes for this failure and suggestions on what to do in each case to avoid a repetition of the failure.

11. Derive Eq. 11–7. (*Hint*: First derive π_K by differentiating Eq. 11–3 with respect to temperature.) *Note*: This relation does not fit all the experimental data fully because there are other variables besides n that have a mild influence. For a more complex approach, see Ref. 11–32.

12. What will be the percent change in nominal values of A_t, r, I_s, T_0, t_b, A_b/A_t and the nozzle throat heat transfer rate, if the Orbus-6 rocket motor listed in Table 11–3 is to be downgraded in thrust for a particular flight by 15% by substituting a new nozzle with a larger nozzle throat area but the same nozzle exit area? The propellants, grain, insulation, and igniter will be the same.

13. What would be the new values of I_t, I_s, p_1, F, t_b, and r for the first stage of the Minuteman rocket motor described in Table 11–3, if the motor were fired at sea level with the grain temperature 20°F hotter than the data shown. Use only data from this table.
    Answers: $I_t = 10,240,000$ lbf-sec, $I_s = 224$ sec, $p_1 = 796$ psia, $F = 1.99 \times 10^5$ lbf, $t_b = 51.5$ sec, $r = 0.338$ in./sec.

SYMBOLS

a	burning rate constant, also called temperature coefficient
A_b	solid propellant burning area, m^2 (ft^2)
A_p	port area (flow area of gases inside grain cavity or between and around propellant grains), m^2 (ft^2)
A_t	nozzle throat cross-sectional area, m^2 (ft^2)
b	web thickness, m (in.)
b_f	web fraction, or web thickness-to-radius ratio
c	effective exhaust velocity, m/sec (ft/sec)
c^*	characteristic exhaust velocity, m/sec (ft/sec)
c_p	specific heat of gas, kcal/kg-K
c_s	specific heat of solid, kcal/kg-K
C_F	thrust coefficient
D	diameter, m (ft)
E_R	relaxation modulus, MPa (psi)
F	thrust, N (lbf)
$\overline{F}$	average thrust, N (lbf)
g_0	acceleration due to gravity at sea level, 9.8066 m/sec^2 (32.2 ft/sec^2)
G	mass flow rate, kg-m^2/sec
h	enthalpy per unit mass, J/kg or Btu/lbm
I_s	specific impulse, sec
I_t	total impulse, N-sec (lbf-sec)
k	specific heat ratio
K	ratio of burning surface to throat area, A_b/A_t
L	length, m
m	mass, kg
$\dot{m}$	mass flow rate, kg/sec
n	burning rate exponent
p	pressure, MPa (lbf/in.^{2})
p_1	chamber pressure, MPa (lbf/in.2)
Pr	Prandtl number, $\mu c_p/\kappa$
r	propellant burning rate (velocity of consumption), m/sec or mm/sec or in./sec
R	gas constant, J/kg-K
t	time, sec
t_a	action time, sec

t_b burn time, sec
T absolute temperature, K(R)
v_2 theoretical exhaust velocity, m/sec (ft/sec)
V_b propellant volume, m^3 (ft^3)
V_c chamber volume, m^3 (ft^3)
V_f volumetric loading fraction, %
w total effective propellant weight, N (lbf)
w_G total loaded rocket weight, or gross weight, N (lbf)
$\dot{w}$ weight rate of flow, N/sec (lbf/sec)

Greek Letters

α heat transfer factor
β constant
δ partial derivative
ϵ elongation or strain
κ conductivity
μ viscosity
π_K temperature sensitivity coefficient of pressure, $K^{-1}(R^{-1})$
ρ density, kg/m^3 (lbm/ft^3)
σ stress, N/cm^2 (psi)
σ_p temperature sensitivity coefficient of burning rate, $K^{-1}(R^{-1})$
ζ propellant mass fraction

Subscripts

b solid propellant burning conditions
p pressure or propellant or port cavity
t throat conditions
0 initial or reference condition
1 chamber condition
2 nozzle exit condition

REFERENCES

11–1. P. R. Evans, "Composite Motor Case Design," Chapter 4A; H. Badham and G. P. Thorp, "Considerations for Designers of Cases for Small Solid Propellant Rocket Motors," Chapter 6; B. Zeller, "Solid Propellant Grain Design," Chapter 8; D. I. Thrasher, "State of the Art of Solid Propellant Rocket Motor Grain Design in the United States," Chapter 9; and A. Truchot, "Design and Analysis of Rocket Motor Internal Insulation," Chapter 10; all of *Design Methods in Solid Propellant Rocket Motors*, AGARD Lecture Series 150, Revised Version, 1988.

11–2. N. Eisenreich, H. P. Kugler, and F. Sinn, "An Optical System for Measuring Burning Rates of Solid Propellants," *Propellants, Explosives, Pyrotechnics,* Vol. 12, 1987, pp. 78–80.

11–3. N. Kubota, "Survey of Rocket Propellants and their Combustion Characteristics," Chapter 1; and M. K. Ràzdan and K. K. Kuo, "Erosive Burning of Solid Propellants," Chapter 10; in K. K. Kuo and M. Summerfield (Eds.), *Fundamentals of Solid Propellant Combustion,* Volume 90 in series on *Progress in Astronautics and Aeronautics,* American Institute of Aeronautics and Astronautics, New York, 1984, 891 pages.

11–4. S. D. Heister and R. J. Davis, "Predicting Burning Time Variations in Solid Rocket Motors," *Journal of Propulsion and Power,* Vol. 8, No. 3, May–June 1992.

11–5. M. K. King, "Erosive Burning of Solid Propellants," *Journal of Propulsion and Power,* Vol. 9, No. 6, November–December 1993.

11–6. J. M. Lenoir and G. Robillard, "A Mathematical Method to Predict the Effects of Erosive Burning in Solid-propellant Rocket," *Sixth Symposium (International) on Combustion,* Reinhold, New York, 1957, pp. 663–667.

11–7. "Solid Propellant Selection and Characterization," *NASA SP-8064,* June 19971 (N72-13737).

11–8. M. S. Fuchs, A. Peretz, and Y. M. Timnat, "Parametric Study of Acceleration Effects on Burning Rates of Metallized Solid Propellants," *Journal of Spacecraft and Rockets,* Vol. 19, No. 6, November–December 1982, pp. 539–544.

11–9. K. K. Kuo, J. Moreci, and J. Mantzaras, "Modes of Crack Formation in Burning Solid Propellant," *Journal of Propulsion and Power,* Vol. 3, No. 1, January–February 1987, pp. 19–25.

11–10. M. T. Langhenry, "The Direct Effects of Strain on Burning Rates of Solid Propellants," *AIAA Paper 84-1436,* June 1984.

11–11. M. K. King, "Analytical Modeling of Effects of Wires on Solid Motor Ballistics," *Journal of Propulsion and Power,* Vol. 7, No. 3, May–June 1991, pp. 312–320.

11–12. "Solid Rocket Motor Performance Analysis and Prediction," *NASA SP-8039,* May 1971 (N72-18785).

11–13. E. M. Landsbaum, M. P. Salinas, and J. P. Leavy, "Specific Impulse Predictions of Solid Propellant Motors," *Journal of Spacecraft and Rockets,* Vol. 17, 1980, pp. 400–406.

11–14. R. Akiba and M. Kohno, "Experiments with Solid Rocket Technology in the Development of M-3SII," *Acta Astronautica,* Vol. 13, No. 6–7, 1986, pp. 349–361.

11–15. P. R. Zarda and D. J. Hartman, "Computer-Aided Propulsion Burn Analysis," *AIAA Paper 88-3342,* July 1988 (cavity geometry).

11–16. R. J. Hejl and S. D. Heister, "Solid Rocket Motor Grain Burnback Analysis Using Adaptive Grids," *Journal of Propulsion and Power,* Vol. 11, No. 5, September–October 1995.

11–17. W. H. Jolley, J. F. Hooper, P. R. Holton, and W. A. Bradfield, "Studies on Coning in End-Burning Rocket Motors," *Journal of Propulsion and Power,* Vol. 2, No. 2, May–June 1986, pp. 223–227.

11–18. S. Nishi, K. Fukuda, and N. Kubota, "Combustion Tests of Two-Stage Pulse Rocket Motors," *AIAA Paper 89-2426*, July 1989, 5 pages.

11–19. L. C. Carrier, T. Constantinou, P. G. Harris, and D. L. Smith, "Dual Interrupted Thrust Pulse Motor," *Journal of Propulsion and Power*, Vol. 3, No. 4, July–August 1987, pp. 308–312.

11–20. C. Bruno et al., "Experimental and Theoretical Burning of Rocket Propellant near the Pressure Deflagration Limit," *Acta Astronautica*, Vol. 12, No. 5, 1985, pp. 351–360.

11–21. F. N. Kelley, "Solid Propellant Mechanical Property Testing, Failure Criteria and Aging," Chapter 8 in C. Boyars and K. Klager (Eds.), *Propellant Manufacture Hazards and Testing*, Advances in Chemistry Series 88, American Chemical Society, Washington, DC, 1969.

11–22. T. L. Kuhlmann, R. L. Peeters, K. W. Bills, and D. D. Scheer, "Modified Maximum Principal Stress Criterion for Propellant Liner Bond Failures," *Journal of Propulsion and Power*, Vol. 3, No. 3, May–June 1987.

11–23. R. W. Magness and J. W. Gassaway, "Development of a High Performance Rocket Motor for the Tactical VT-1 Missile," *AIAA Paper 88-3325*, July 1988.

11–24. I-Shih Chang and M. J. Adams, "Three-Dimensional, Adaptive, Unstructured, Mesh Generation for Solid-Propellant Stress Analysis," *AIAA Paper 96-3256*, July 1996.

11–25. W. A. Cook, "Three-Dimensional Grain Stress Analysis Using the Finite Element Method," *AFRPL Report TT-71-51*, Thiokol Corp., April 1971 (AD725043).

11–26. G. Meili, G. Dubroca, M. Pasquier, and J. Thenpenier, "Nonlinear Viscoelastic Design of Case-Bonded Composite Modified Double Base Grains," *AIAA Paper 80-1177R*, July 1980, and S. Y. Ho and G. Care, "Modified Fracture Mechanics Approach in Structural Analysis of Solid-Rocket Motors," *Journal of Propulsion and Power*, Vol. 14, No. 4, July–August 1998.

11–27. P. G. Butts and R. N. Hammond, "IUS Propellant Development and Qualification," Paper presented at the 1983 JANNAF Propulsion Meeting, Monterey, February 1983, 13 pages.

11–28. A. G. Christianson et al., "HTPB Propellant Aging," *Journal of Spacecraft and Rockets*, Vol. 18, No. 3, May–June 1983.

11–29. D. I. Thrasher and J. H. Hildreth, "Structural Service Life Estimates for a Reduced Smoke Rocket Motor," *Journal of Spacecraft and Rockets*, Vol. 19, No. 6, November 1982, pp. 564–570.

11–30. S. W. Tsa (Ed.), *Introduction to Viscoelasticity*, Technomic Publishing Co., Stanford, CT, Conn., 1968.

11–31. J. D. Ferry, *Viscoelastic Properties of Polymers*, John Wiley & Sons, New York, 1970.

11–32. R. E. Hamke, M. T. Gaunce, and J. R. Osborn, "The Effect of Pressure Exponent on Temperature Sensitivity," *Acta Astronautica*, Vol. 15, Nos. 6 and 7, 1987, pp. 377–382.

CHAPTER 12

SOLID PROPELLANTS

In this chapter we describe several common solid rocket propellants, their principal categories, ingredients, hazards, manufacturing processes, and quality control. We also discuss liners and insulators, propellants for igniters, tailoring of propellants, and propellants for gas generators. It is the second of four chapters dealing with solid propellant rocket motors.

Thermochemical analyses are needed to characterize the performance of a given propellant. The analysis methods are described in Chapter 5. Such analyses provide theoretical values of average molecular weight, combustion temperature, average specific heat ratio, and the characteristic velocity; they are functions of the propellant composition and chamber pressure. A specific impulse can also be computed for a particular nozzle configuration.

The term *solid propellant* has several connotations, including: (1) the rubbery or plastic-like mixture of oxidizer, fuel, and other ingredients that have been processed and constitute the finished grain; (2) the processed but uncured product; (3) a single ingredient, such as the fuel or the oxidizer. Acronyms and chemical symbols are used indiscriminately as abbreviations for propellant and ingredient names; only some of these will be used here.

12.1. CLASSIFICATION

Processed modern propellants can be classified in several ways, as described below. This classification is not rigorous or complete. Sometimes the same propellant will fit into two or more of the classifications.

1. Propellants are often tailored to and classified by *specific applications*, such as space launch booster propellants or tactical missile propellants; each has somewhat specific chemical ingredients, different burning rates, different physical properties, and different performance. Table 11–1 shows four kinds of *rocket motor applications* (each has somewhat different propellants) and several *gas generator applications*. Propellants for rocket motors have hot (over 2400 K) gases and are used to produce thrust, but gas generator propellants have lower-temperature combustion gases (800 to 1200 K) and they are used to produce power, not thrust.

   Historically, the early rocket motor propellants used to be grouped into two classes: *double-base* (DB*) propellants were used as the first production propellants, and then the development of polymers as binders made the *composite* propellants feasible.

2. *Double-base* (DB) propellants form a *homogeneous* propellant grain, usually a nitrocellulose (NC*), a solid ingredient which absorbs liquid nitroglycerine (NG) plus minor percentages of additives. Both the major ingredients are explosives and function as a combined fuel and oxidizer. Both *extruded double-base* (EDB) and *cast double-base* (CDB) propellant have found extensive applications, mostly in small tactical missiles of older design. By adding crystalline nitramines (HMX or RDX)* the performance and density can be improved; this is sometimes called *cast-modified double-base* propellant. A further improvement is to add an elastomeric binder (rubber-like, such as crosslinked polybutadiene), which improves the physical properties and allows more nitramine and thus improves the performance slightly. The resulting propellant is called *elastomeric-modified cast double-base* (EMCDB). These four classes of double base have nearly smokeless exhausts. Adding some solid ammonium perchlorate (AP) and aluminum (Al) increases the density and the specific impulse slightly, but the exhaust gas is smoky. The propellant is called *composite-modified double-base propellant* or CMDB.

3. *Composite propellants* form a *heterogeneous* propellant grain with the oxidizer crystals and a powdered fuel (usually aluminum) held together in a matrix of synthetic rubber (or plastic) binder, such as polybutadiene (HTPB)*. Composite propellants are cast from a mix of solid (AP crystals, Al powder)* and liquid (HTPB, PPG)* ingredients. The propellant is hardened by crosslinking or curing the liquid binder polymer with a small amount of curing agent, and curing it in an oven, where it becomes hard and solid. In the past three decades the composite propellants have been the most commonly used class. They can be further subdivided:

   (1) Conventional *composite propellants* usually contain between 60 and 72% ammonium perchlorate (AP) as crystalline oxidizer, up to 22%

---

*Acronyms, symbols, abbreviations, and chemical names of propellant ingredients are explained in Tables 12–6 and 12–7 in Section 12.4.

aluminum powder (Al) as a metal fuel, and 8 to 16% of elastomeric binder (organic polymer) including its plasticizer.

(2) Modified composite propellant where an *energetic nitramine* (HMX or RDX) is added for obtaining a little more performance and also a somewhat higher density.

(3) Modified composite propellant where an *energetic plasticizer* such as nitroglycerine (used in double-base propellant) is added to give a little more performance. Sometimes HMX is also added.

(4) A *high-energy composite solid propellant* (with some aluminum), where the organic elastomeric binder and plasticizer are largely replaced by energetic materials (such as certain explosives) and where some of the AP is replaced by HMX. Some of these are called elastomer-modified cast double-base propellants (EMCDB). Most are experimental propellants. The theoretical specific impulse can be between 270 and 275 sec at standard conditions.

(5) A *lower-energy composite propellant*, where *ammonium nitrate* (AN) is the crystalline oxidizer (no AP). It is used for gas generator propellant. If a large amount of HMX is added, it can become a minimum smoke propellant with fair performance.

Figures 12–1 and 12–2 show the general regions for the specific impulse, burning rate, and density for the more common classes of propellants. Composite propellants give higher densities, specific impulse, and a wider range of burning rates. The ordinate in these figures is an actual or estimated specific impulse at standard conditions (1000 psi and expansion to sea-level atmosphere). It does not include any pressure drops in the chamber, any nozzle erosion, or an assumption about combustion losses and scaling. The composite propellants are shown to have a wide range of burning rates and densities; most of them have specific gravities between 1.75 and 1.81 and burning rates between 7 and 20 mm/sec. Table 12–1 lists performance characteristics for several propellants. The double-base (DB) propellants and the ammonium nitrate (AN) propellants have lower performance and density. Most composite propellants have almost the same performance and density but a wide range of burning rates. The highest performance is for a CMDB propellant whose ingredients are identified as DB/AP-HMX/Al, but it is only four percent higher.

Several of the classifications can be confusing. The term composite-modified double-base propellant (CMDB) has been used for (1) a DB propellant, where some AP, Al, and binder are added; (2) alternatively, the same propellant could be classified as a composite propellant to which some double-base ingredients have been added.

4. Propellants can be classified by the density of the smoke in the exhaust plume as *smoky, reduced smoke,* or *minimum smoke* (essentially smoke-

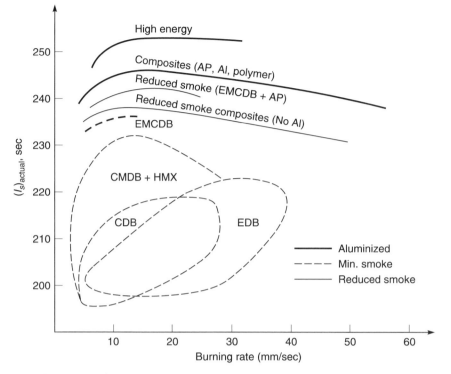

FIGURE 12–1. Estimated actual specific impulse and burning rate for several solid propellant categories. (Adapted and reproduced from Ref. 12–1 with permission of the American Institute of Aeronautics and Astronautics [AIAA].)

less). Aluminum powder, a desirable fuel ingredient, is oxidized to aluminum oxide, which forms visible small solid smoke particles in the exhaust gas. Most composite propellants are smoky. By reducing the aluminum content in composite propellant, the amount of smoke is also reduced. Carbon (soot) particles and metal oxides, such as zirconium oxide or iron oxide, can also be visible if in high enough concentration. This is further discussed in Chapter 18.

5. The *safety rating* for detonation can distinguish propellants as a potentially *detonable* material (class 1.1) or as a *nondetonable* material (class 1.3), as described in Section 11.3. Examples of class 1.1 propellant are a number of double-base propellants and composite propellants containing a significant portion of solid explosive (e.g., HMX or RDX), together with certain other ingredients.

6. Propellants can be classified by some of the principal manufacturing processes that are used. *Cast propellant* is made by mechanical mixing of solid and liquid ingredients, followed by casting and curing; it is the most common process for composite propellants. *Curing* of many cast

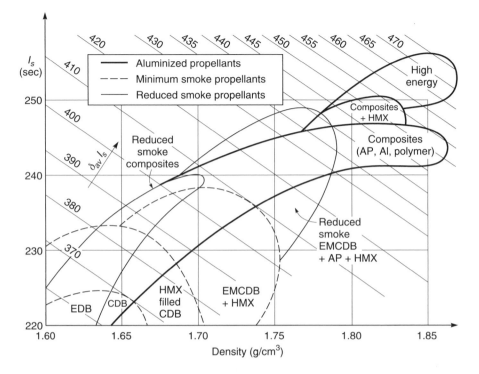

FIGURE 12–2. Estimated actual specific impulse and specific gravity for several solid propellant categories. (Adapted and reproduced from Ref. 12–1 with permission of the AIAA.)

propellants is by chemical reaction between binder and curing agent at elevated temperature (45 to 150°C); however, there are some that can be cured at ambient temperatures (20 to 25°C) or hardened by a nonchemical process such as crystallization. Propellant can also be made by a *solvation* process (dissolving a plasticizer in a solid pelletized matrix, whose volume is expanded). *Extruded propellant* is made by mechanical mixing (rolling into sheets) followed by extrusion (pushing through a die at high pressure). Solvation and extrusion processes apply primarily to double-base propellants.

7. Propellants have also been classified by their *principal ingredient*, such as the *principal oxidizer* (*ammonium perchlorate propellants, ammonium nitrate propellants,* or *azide-type propellants*) or their *principal binder* or *fuel ingredient*, such as *polybutadiene propellants* or *aluminized propellants*. This classification of propellants by ingredients is described in Section 12.4 and Table 12–8.

8. Propellants with *toxic* and *nontoxic* exhaust gases. This is discussed in more detail in Section 12.3.

TABLE 12–1. Characteristics of Some Operational Solid Propellants

Propellant Type[a]	I_s Range (sec)[b]	Flame Temperature[e] (°F)	(°K)	Density or Spec. Gravity[e] (lb/in³)	(sp. gr.)	Metal Content (wt %)	Burning Rate[c,e] (in./sec)	Pressure Exponent[e] n	Hazard Classification[d]	Stress (psi)/Strain (%) −60°F	+150°F	Processing Method
DB	220–230	4100	2550	0.058	1.61	0	0.05–1.2	0.30	1.1	4600/2	490/60	Extruded
DB/AP/Al	260–265	6500	3880	0.065	1.80	20–21	0.2–1.0	0.40	1.3	2750/5	120/50	Extruded
DB/AP–HMX/Al	265–270	6700	4000	0.065	1.80	20	0.2–1.2	0.49	1.1	2375/3	50/33	Solvent cast
PVC/AP/Al	260–265	5600	3380	0.064	1.78	21	0.3–0.9	0.35	1.3	369/150	38/220	Cast or extruded
PU/AP/Al	260–265	5700	3440	0.064	1.78	16–20	0.2–0.9	0.15	1.3	1170/6	75/33	Cast
PBAN/AP/Al	260–263	5800	3500	0.064	1.78	16	0.25–1.0	0.33	1.3	520/16 (at −10°F)	71/28	Cast
CTPB/AP/Al	260–265	5700	3440	0.064	1.78	15–17	0.25–2.0	0.40	1.3	325/26	88/75	Cast
HTPB/AP/Al	260–265	5700	3440	0.067	1.86	4–17	0.25–3.0	0.40	1.3	910/50	90/33	Cast
PBAA/AP/Al	260–265	5700	3440	0.064	1.78	14	0.25–1.3	0.35	1.3	500/13	41/31	Cast
AN/Polymer	180–190	2300	1550	0.053	1.47	0	0.06–0.5	0.60	1.3	200/5	NA	Cast

[a] Al, aluminum; AN, ammonium nitrate; AP, ammonium perchlorate; CTPB, carboxy-terminated polybutadiene; DB, double-base; HMX, cyclotetramethylene tetranitramine; HTPB, hydroxyl-terminatd poly-butadiene; PBAA, polybutadiene–acrylic acid polymer; PBAN, polybutadiene–acrylic acid–acrylonitrile terpolymer; PU, polyurethane; PVC, polyvinyl chloride.

[b] At 1000 psia expanding to 14.7 psia, ideal or theoretical value at reference conditions.

[c] At 1000 psia.

[d] See page 491.

[e] I_s, flame temperature, density, burn rate and pressure exponent will vary slightly with specific composition.

479

A large variety of different *chemical ingredients* and propellant formulations have been synthesized, analyzed, and tested in experimental motors. Later we list many of them. Perhaps only 12 basic types of propellant are in common use today. Other types are still being investigated. Table 12–2 evaluates some of the advantages and disadvantages of several selected propellant classes. A typical propellant has between 4 and 12 different ingredients. Representative formulations for three types of propellant are given in Table 12–3. In actual practice, each manufacturer of a propellant has his own precise formulation and processing procedure. The exact percentages of ingredients, even for a given propellant such as PBAN, not only vary among manufacturers but often vary from motor application to motor application. The practice of adjusting the mass percentage and even adding or deleting one or more of the minor ingredients (additives) is known as *propellant tailoring*. Tailoring is the practice of taking a well-known propellant and changing it slightly to fit a new application, different processing equipment, altered motor ballistics, storage life, temperature limits, or even a change in ingredient source.

New propellant formulations are normally developed using laboratory-size mixers, curing ovens, and related apparatus with the propellant mixers (1 to 5 liters) operated by remote control for safety reasons. Process studies usually accompany the development of the formulation to evaluate the "processibility" of a new propellant and to guide the design of any special production equipment needed in preparing ingredients, mixing, casting, or curing the propellant.

Historically, black powder (a pressed mixture of potassium nitrate, sulfur, and an organic fuel such as ground peach stones) was the first to be used. Other types of ingredients and propellants have been used in experimental motors, including fluorine compounds, propellants containing powdered beryllium, boron, hydrides of boron, lithium, or beryllium, or new synthetic organic plasticizer and binder materials with azide or nitrate groups. Most have not yet been considered satisfactory or practical for production in rocket motors.

12.2. PROPELLANT CHARACTERISTICS

The propellant selection is critical to rocket motor design. The *desirable propellant characteristics* are listed below and are discussed again in other parts of this book. The requirements for any particular motor will influence the priorities of these characteristics:

1. High performance or *high specific impulse*; really this means high gas temperature and/or low molecular mass.
2. Predictable, reproducible, and initially adjustable *burning rate* to fit the need of the grain design and the thrust–time requirement.
3. For minimum variation in thrust or chamber pressure, the *pressure or burning rate exponent* and the *temperature coefficient* should be small.

4. Adequate *physical properties* (including bond strength) over the intended operating temperature range.

5. High *density* (allows a small-volume motor).

6. Predictable, reproducible ignition qualities (such as reasonable ignition overpressure)

7. Good *aging characteristics* and *long life*. Aging and life predictions depend on the propellant's chemical and physical properties, the cumulative damage criteria with load cycling and thermal cycling (see page 461), and actual tests on propellant samples and test data from failed motors.

8. Low absorption of *moisture*, which often causes chemical deterioration.

9. Simple, reproducible, safe, low-cost, controllable, and low-hazard *manufacturing*.

10. Guaranteed availability of all *raw materials* and *purchased components* over the production and operating life of the propellant, and good control over undesirable impurities.

11. *Low technical risk*, such as a favorable history of prior applications.

12. Relative *insensitivity* to certain energy stimuli described in the next section.

13. *Non-toxic exhaust* gases.

14. Not prone to *combustion instability* (see next chapter).

Some of these desirable characteristics will apply also to all materials and purchased components used in solid motors, such as the igniter, insulator, case, or safe and arm device. Several of these characteristics are sometimes in conflict with each other. For example, increasing the physical strength (more binder and or more crosslinker) will reduce the performance and density. So a modification of the propellant for one of these characteristics can often cause changes in several of the others.

Several illustrations will now be given on how the characteristics of a propellant change when the concentration of one of its major ingredients is changed. For composition propellants using a polymer binder [hydroxyl-terminated polybutadiene (HTPB)] and various crystalline oxidizers, Fig. 12–3 shows the calculated variation in combustion or flame temperature, average product gas molecular weight, and specific impulse as a function of oxidizer concentration; this is calculated data taken from Ref. 12–2, based on a thermochemical analysis as explained in Chapter 5. The maximum values of I_s and T_1 occur at approximately the same concentration of oxidizer. In practice the optimum percentage for AP (about 90 to 93%) and AN (about 93%) cannot be achieved, because concentrations greater than about 90% total solids (including the aluminum and solid catalysts) cannot be processed in a mixer. A castable slurry that will flow into a mold requires 10 to 15% liquid content.

TABL-E 12–2. Characteristics of Selected Propellants

Propellant Type	Advantages	Disadvantages
Double-base (extruded)	Modest cost; nontoxic clean exhaust, smokeless; good burn rate control; wide range of burn rates; simple well-known process; good mechanical properties; low temperature coefficient; very low pressure exponent; plateau burning is possible	Free-standing grain requires structural support; low performance, low density; high to intermediate hazard in manufacture; can have storage problems with NG bleeding out; diameter limited by available extrusion presses; class 1.1
Double-base (castable)	Wide range of burn rates; nontoxic smokeless exhaust; relatively safe to handle; simple, well-known process; modest cost; good mechanical properties; good burn rate control; low temperature coefficient; plateau burning can be achieved	NG may bleed out or migrate; high to intermediate manufacture hazard; low performance; low density; higher cost than extruded DB; class 1.1
Composite-modified double-base or CMDB with some AP and Al	Higher performance; good mechanical properties; high density (sp. gr. 1.83–1.86); less likely to have combustion stability problems; intermediate cost; good background experience	Storage stability can be marginal; complex facilities; some smoke in exhaust; high flame temperature; moisture sensitive; moderately toxic exhaust; hazards in manufacture; modest ambient temperature range; the value of n is high (0.8 to 0.9); moderately high temperature coefficient
Composite AP, Al, and PBAN or PU or CTPB binder	Reliable; high density; long experience background; modest cost; good aging; long cure time; good performance; usually stable combustion; low to medium cost; wide temperature range; high density; low to moderate temperature sensitivity; good burn rate control; usually good physical properties; class 1.3	Modest ambient temperature range; high viscosity limits at maximum solid loading; high flame temperature; toxic, smoky exhaust; some are moisture sensitive; some burn-rate modifiers (e.g. aziridines) are carcinogens
Composite AP, Al, and HTPB binder; most common composite propellant today	Slightly better solids loading % and performance than PBAN or CTPB; widest ambient temperature limits; good burn-rate control; usually stable combustion; medium cost; good storage stability; widest range of burn rates; good physical properties; good experience; class 1.3	Complex facilities; moisture sensitive; fairly high flame temperature; toxic, smoky exhaust
Modified composite AP, Al, PB binder plus some HMX or RDX	Higher performance; good burn-rate control; usually stable combustion; high density; moderate temperature sensitivity; can have good mechanical properties	Expensive, complex facilities; hazardous processing; harder-to-control burn rate; high flame temperature; toxic, smoky exhaust; can be impact sensitive; can be class 1.1; high cost; pressure exponent 0.5–0.7

Composite with energetic binder and plasticizer such as NG, AP, HMX	Highest performance; high density (1.8 to 1.86); narrow range of burn rates	Expensive; limited experience; impact sensitive; high pressure exponent
Modified double-base with HMX	Higher performance; high density (1.78 to 1.88); stable combustion; narrow range of burn rates	Same as CMDB above; limited experience; most are class 1.1; high cost
Modified AN propellant with HMX or RDX added	Fair performance; relatively clean; smokeless; nontoxic exhaust	Relatively little experience; can be hazardous to manufacture; need to stabilize AN to limit grain growth; low burn rates; impact sensitive; medium density; class 1.1 or 1.3
Ammonium nitrate plus polymer binder (gas generator)	Clean exhaust; little smoke; essentially nontoxic exhaust; low temperature gas; usually stable combustion; modest cost; low pressure exponent	Low performance; low density; need to stabilize AN to limit grain growth and avoid phase transformations; moisture sensitive; low burn rates
RDX/HMX with polymer	Low smoke; nontoxic exhaust; lower combustion temperature	Low performance; low density; class 1.1

TABLE 12–3. Representative Propellant Formulations

Double-Base (JPN Propellant)		Composite (PBAN Propellant)		Composite Double-Base (CMDB Propellant)	
Ingredient	Wt %	Ingredient	Wt %	Ingredient	Wt %
Nitrocellulose	51.5	Ammonium perchlorate	70.0	Ammonium perchlorate	20.4
Nitroglycerine	43.0	Aluminum powder	16.0	Aluminum powder	21.1
Diethyl phthalate	3.2	Polybutadiene–acrylic acid–acrylonitrile	11.78	Nitrocellulose	21.9
Ethyl centralite	1.0	Epoxy curative	2.22	Nitroglycerine	29.0
Potassium sulfate	1.2			Triacetin	5.1
Carbon black	< 1%			Stabilizers	2.5
Candelilla wax	< 1%				

Source: Courtesy of Air Force Phillips Laboratory, Edwards, California.

A typical composition diagram for a composite propellant is shown in Fig. 12–4. It shows how the specific impulse varies with changes in the composition of the three principal ingredients: the solid AP, solid Al, and viscoelastic polymer binder.

For double-base (DB) propellant the theoretical variations of I_s and T_1 are shown in Figs. 12–1 and 12–5 as a function of the nitroglycerine (NG) or plasticizer percentage. The theoretical maximum specific impulse occurs at about 80% NG. In practice, nitroglycerine, which is a liquid, is seldom found in concentrations over 60%, because the physical properties are poor if NG is high. There need to be other major solid or soluble ingredients to make a usable DB propellant.

For CMDB propellant the addition of either AP or a reactive nitramine such as RDX allows a higher I_s than ordinary DB (where AP or RDX percent is zero), as shown in Fig. 12–6. Both AP and RDX greatly increase the flame temperature and make heat transfer more critical. The maximum values of I_s occur at about 50% AP and at 100% RDX (which is an impractical propellant that cannot be manufactured and will not have reasonable physical properties). At high concentrations of AP or RDX the exhaust gases contain considerable H_2O and O_2 (as shown in Fig. 12–7); these enhance the erosion rate of carbon-containing insulators or nozzle materials. The toxic HCl is present in concentrations between 10 and 20%, but for practical propellants it seldom exceeds 14%.

Nitramines such as RDX or HMX contain relatively few oxidizing radicals, and the binder surrounding the nitramine crystals cannot be fully oxidized. The binder is decomposed at the combustion temperature, forms gases rich in hydrogen and carbon monoxide (which reduces the molecular weight), and

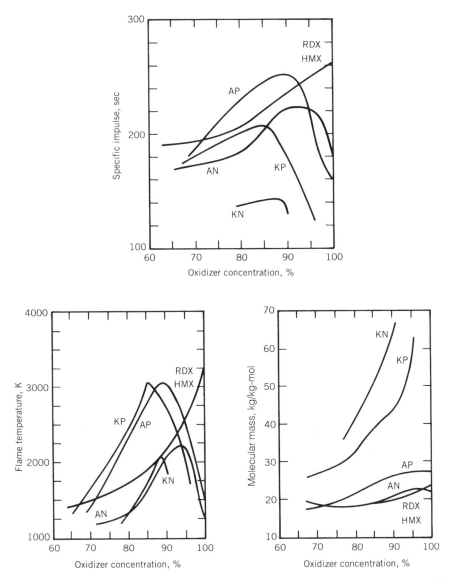

FIGURE 12–3. Variation of combustion temperature, average molecular mass of the combustion gases, and theoretical specific impulse (at frozen equilibrium) as a function of oxidizer concentration for HTPB-based composite propellants. Data are for a chamber pressure of 68 atm and nozzle exit pressure of 1.0 atm. (Reproduced from Ref. 12–2 with permission of the AIAA.)

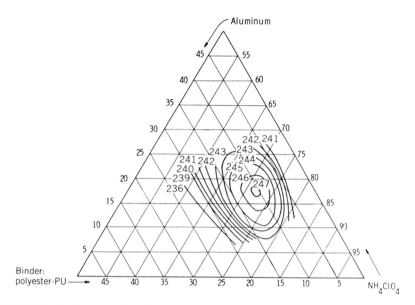

FIGURE 12–4. Composition diagram of calculated specific impulse for an ammonium perchlorate–aluminum–polyurethane (PU is a polyester binder) at standard conditions (1000 psi and expansion to 14.7 psi). The maximum value of specific impulse occurs at about 11% PU, 72% AP, and 17% Al. (Reproduced from Ref. 12–3 with permission of the American Chemical Society.)

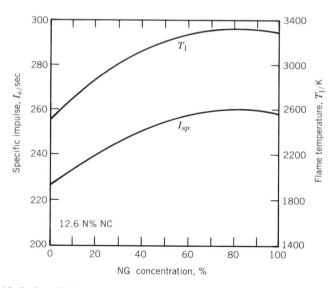

FIGURE 12–5. Specific impulse and flame temperature versus nitroglycerine (NG) concentration of double-base propellants. (Reproduced from Ref. 12–2 with permission of the AIAA.)

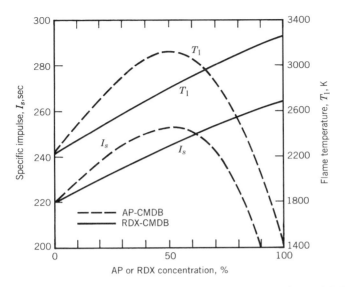

FIGURE 12–6. Specific impulse and flame temperature versus AP or RDX concentration of AP–CMDB propellants. (Reproduced from Ref. 12–2 with permission of the AIAA.)

cools the gases to a lower combustion temperature. The exhaust gases of AP-based and RDX-based CMDB propellant are shown in Fig. 12–7. The solid carbon particles seem to disappear if the RDX content is high.

12.3. HAZARDS

With proper precautions and equipment, all common propellants can be manufactured, handled, and fired safely. It is necessary to fully understand the hazards and the methods for preventing hazardous situations from arising. Each material has its own set of hazards; some of the more common ones are described briefly below and also in Refs. 12–4 and 12–5. Not all apply to each propellant.

Inadvertent Ignition

If a rocket motor is ignited and starts combustion when it is not expected to do so, the consequences can include very hot gases, local fires, or ignition of adjacent rocket motors. Unless the motor is constrained or fastened down, its thrust will suddenly accelerate it to unanticipated high velocities or erratic flight paths that can cause damage. Its exhaust cloud can be toxic and corrosive. Inadvertent ignition can be caused by these effects:

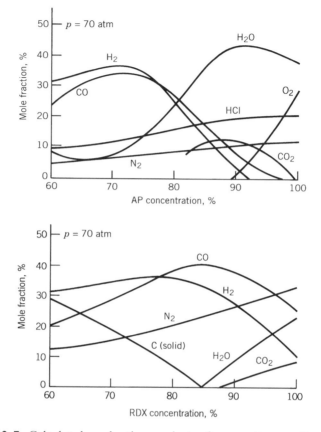

FIGURE 12–7. Calculated combustion products of composite propellant with varying amounts of AP or RDX. (Adapted from Chapter 1 of Ref. 12–2 with permission of the AIAA.)

Stray or induced currents activate the igniter.

Electrostatic discharge causes a spark or arc discharge.

Fires cause excessive heating of motor exterior, which can raise the propellant temperature above the ignition point.

Impact (bullet penetration, or dropping the motor onto a hard surface).

Energy absorption from prolonged mechanical vibration can cause the propellant to overheat.

An electromechanical system is usually provided that prevents stray currents from activating the igniter; it is called *safe and arm system*. It prevents ignition induced by currents in other wires of the vehicle, radar- or radio-frequency-induced currents, electromagnetic surges, or pulses from a nuclear bomb explosion. It prevents electric currents from reaching the igniter circuit during its "unarmed" condition. When put into the "arm" position, it is ready to accept and transmit the start signal to the igniter.

Electrostatic discharges (ESD) can be caused by lightning, friction of insulating materials, or the moving separation of two insulators. The buildup of a high electrostatic potential of thousands of volts can, upon discharge, allow a rapid increase in electric current, which in turn can lead to arcing or exothermic reactions along the current's path. For this reason all propellants, liners, or insulators should have sufficient electric conductivity to prevent the buildup of an electrostatic charge. The inadvertent ignition of a Pershing ground-to-ground missile is believed to have been caused by electrostatic discharge while in the transporter-erector vehicle. ESD is a function of the materials, their surface and volume resistivities, dielectric constants, and the breakdown voltages.

Viscoelastic propellants are excellent absorbers of *vibration* energy and can become locally hot when oscillated for extensive periods at particular frequencies. This can happen in designs where a segment of the grain is not well supported and is free to vibrate at natural frequencies. A propellant can also be accidentally ignited by various other energy inputs, such as mechanical friction or vibration. Standard tests have been developed to measure the propellant's resistance to these energy inputs.

Aging and Useful Life

This topic was discussed briefly in the section on Structural Design in the previous chapter. The *aging* of a propellant can be measured with test motors and propellant sample tests if the loading during the life of the motor can be correctly anticipated. It is then possible to estimate and predict the useful *shelf or storage life* of a rocket motor (see Refs. 12–5 and 12–6). When a reduction in physical properties, caused by estimated thermal or mechanical load cycles (cumulative damage), has reduced the safety margin on the stresses and/or strains to a danger point, the motor is no longer considered to be safe to ignite and operate. Once this age limit or its predicted, weakened condition is reached, the motor has a high probability of failure. It needs to be pulled from the ready inventory, and the old aged propellant needs to be removed and replaced with new, strong propellant.

The *life* of a particular motor depends on the particular propellant, the frequency and magnitude of imposed loads or strains, the design, and other factors. Typical life values range from 5 to 25 years. Shelf life can usually be increased by increasing the physical strength of the propellants (e.g., by increasing the amount of binder), selecting chemically compatible, stable ingredients with minimal long-term degradation, or by minimizing the vibration loads, temperature limits, or number of cycles (controlled storage and transport environment).

Case Overpressure and Failure

The motor case will break or explode if the chamber pressure exceeds the case's burst pressure. The release of high-pressure gas energy can cause an explosion; motor pieces could be thrown out into the adjacent area. The sudden depressurization from chamber pressure to ambient pressure, which is usually below the deflagration limit, would normally cause a class 1.3 propellant to stop burning. Large pieces of unburned propellant can often be found after a violent case burst. This type of motor failure can be caused by one of the following phenomena:

1. The grain is overaged, porous, or severely cracked and/or has major unbonded areas due to severe accumulated damage.
2. There has been a significant chemical change in the propellant due to migration or slow, low-order chemical reactions. This can reduce the allowable physical properties, weakening the grain, so that it will crack or cause unfavorable increases in the burning rate. In some cases chemical reactions create gaseous products which create many small voids and raise the pressure in sealed stored motors.
3. The motor is not properly manufactured. Obviously, careful fabrication and inspection are necessary.
4. The motor has been damaged. For example, a nick or dent in the case caused by improper handling will reduce the case strength. This can be prevented by careful handling and repeated inspections.
5. An obstruction plugs the nozzle (e.g., a loose large piece of insulation) and causes a rapid increase in chamber pressure.
6. Moisture absorption can degrade the strength and strain capabilities by a factor of 3 to 10 in propellants that contain hygroscopic ingredients. Motors are usually sealed to prevent humid air access.

Detonation versus Deflagration. When burning rocket motor propellant is overpressurized, it can either deflagrate (or burn) or detonate (explode violently), as described in Table 12–4. In a detonation the chemical reaction energy of the whole grain can be released in a very short time (microseconds), and in effect it becomes an explosive bomb. This detonation condition can happen with some propellants and some ingredients (e..g, nitroglycerine or HMX, which are described later in this chapter). Detonations can be minimized or avoided by proper design, correct manufacture, and safe handling and operating procedures.

The same material may burn or detonate, depending on the chemical formulation, the type and intensity of the initiation, the degree of confinement, the physical propellant properties (such as density or porosity), and the geometric characteristics of the motor. It is possible for certain propellants to change suddenly from an orderly deflagration to a detonation. A simplified explanation of this transition starts with normal burning at rated chamber pressure;

TABLE 12–4. Comparison of Burning and Detonation

| Characteristic | Burning | | Explosive Detonation |
	With Air	Within Rocket Motors	
Typical material	Coal and air	Propellant, no air	Rocket propellant or explosives
Common means of initiating reaction	Heat	Heat	Shock wave; sudden pressure rise plus heat
Linear reaction rate (m/sec)	10^{-6} (subsonic)	0.2 to 5×10^{-2} (subsonic)	2 to 9×10^3 (supersonic)
Produces shock waves	No	No	Yes
Time for completing reaction (sec)	10^{-1}	10^{-2} to 10^{-3}	10^{-6}
Maximum pressure [MPa (psi)]	0.07–0.14 (10–20)	0.7–100 (100–14,500)	7000–70,000 (10^6–10^7)
Process limitation	By vaporization and heat transfer at burning surface		By physical and chemical properties of material, (e.g., density, composition)
Increase in burning rate can result in:	Potential furnace failure	Overpressure and sudden failure of pressure container	Detonation and violent rapid explosion of all the propellant

the hot gas then penetrates pores or small cracks in the unburned propellant, where the local confinement can cause the pressure to become very high locally, the combustion front speeds up to shock wave speed with a low-pressure differential, and it then accelerates further to a strong, fast, high-pressure shock wave, characteristic of detonations. The degree and rigidity of the geometric confinement and a scale factor (e.g., larger-diameter grain) influence the severity and occurrence of detonations.

Hazard Classification. Propellants that can experience a transition from deflagration to detonation are considered more hazardous and are usually designated as class 1.1-type propellants. Most propellants will burn, the case may burst if chamber pressure becomes too high, but the propellant will not detonate and are class 1.3 propellants. The required tests and rules for determining this hazard category are explained in Ref. 12–7. Propellant samples are subjected to various tests, including impact tests (dropped weight) and card gap tests (which determine the force needed to initiate a propellant detonation when a sample is subjected to a blast from a known booster explosive). If the case should burst violently with a class 1.3 propellant, much of the remaining unburnt propellant would be thrown out, but would then usually stop burning. With a class 1.1 propellant, a powerful detonation can sometimes ensue, which rapidly gasifies all the remaining propellant, and is much more powerful and destructive than the bursting of the case under high pressure. Unfortunately, the term "explosion" has been used to describe both a bursting of a case with its fragmentation of the motor and also the higher rate of energy release of a detonation, which leads to a very rapid and more energetic fragmentation of the motor.

The Department of Defense (DOD) classification of 1.1 or 1.3 determines the method of labeling and the cost of shipping rocket propellants, loaded military missiles, explosives, or ammunition; it also determines the required limits on the amount of that propellant stored or manufactured in any one site and the minimum separation distance of that site to the next building or site. The DOD system (Ref. 12–7) is the same as that used by the United Nations.

Insensitive Munitions

In military operations an accidental ignition and unplanned operation or an explosion of a rocket missile can cause severe damage to equipment and injure or kill personnel. This has to be avoided or minimized by making the motor designs and propellants insensitive to a variety of energy stimuli. The worst scenario is a detonation of the propellant, releasing the explosive energy of all of the propellant mass, and this scenario is to be avoided. The missiles and its motors must undergo a series of prescribed tests to determine their resistance to inadvertent ignition with the most likely energy inputs during a possible battle situation. Table 12–5 describes a series of tests called out in a military specification, which are detailed in Refs. 12–8 and 12–9. A *threat hazard assessment* must be made prior to the tests, to evaluate the logistic and operational threats during the missile's life cycle. The evaluation may cause some modifications to the test setups, changes in the passing criteria, or the skipping of some of these tests.

The missiles, together with their motors, are destroyed in these tests. If the motor should detonate (an unacceptable result), the motor has to be redesigned

TABLE 12–5. Testing for Insensitivity of Rockets and Missiles

Test	Description	Criteria for Passing
Fast cook off	Build a fire (of jet fuel or wood) underneath the missile or its motor	No reaction more severe than burning
Slow cook off	Gradual heating (6°F/hr) to failure	Same as above
Bullet impact	One to three 50 caliber bullets fired at short intervals	Same as above
Fragment impact	Small high-speed steel fragment	Same as above
Sympathetic detonation	Detonation from an adjacent similar motor or a nearby specific munition	No detonation of test motor
Shaped explosive charge impact	Blast from specified shaped charge in specified location	No detonation
Spall impact	Several high-speed spalled fragments from a steel plate which is subjected to a shaped charge	Fire, but no explosion or detonation

and/or have a change in propellant. There are some newer propellants that are more resistant to these stimuli and are therefore preferred for tactical missile applications, even though there is usually a penalty in propulsion performance. If explosions (not detonations) occur, it may be possible to redesign the motor and mitigate the effects of the explosion (make it less violent). For example, the case can have a provision to vent itself prior to an explosion. Changes to the shipping container can also mitigate some of these effects. If the result is a fire (an acceptable result), it should be confined to the particular grain or motor. Under some circumstances a burst failure of the case is acceptable.

Upper Pressure Limit

If the pressure-rise rate and the absolute pressure become extremely high (as in some impact tests or in the high acceleration of a gun barrel), some propellants will detonate. For many propellants these pressures are above approximately 1500 MPa or 225,000 psi, but for others they are lower (as low as 300 MPa or 45,000 psi). They represent an *upper pressure limit* beyond which a propellant should not operate.

Toxicity

A large share of all rockets do not have a significant toxicity problem. A number of propellant ingredients (e.g., some crosslinking agents and burning rate catalysts) and a few of the plastics used in fiber-reinforced cases can be dermatological or respiratory toxins; a few are carcinogens (cancer-causing agents) or suspected carcinogens. They, and the mixed uncured propellant containing these materials, have to be handled carefully to prevent operator exposure. This means using gloves, face shields, good ventilation, and, with some high-vapor-pressure ingredients, gas masks. The finished or cured grain or motor is usually not toxic.

The exhaust plume gases can be very toxic if they contain beryllium or berylium oxide particles, chlorine gas, hydrochloric acid gas, hydrofluoric acid gas, or some other fluorine compounds. When an ammonium perchlorate oxidizer is used, the exhaust gas can contain up to about 14% hydrochloric acid. For large rocket motors this can be many tons of highly toxic gas. Test and launch facilities for rockets with toxic plumes require special precautions and occasionally special decontamination processes, as explained in Chapter 20.

Safety Rules

The most effective way to control hazards and prevent accidents is (1) to train personnel in the hazards of each propellant of concern and to teach them how to avoid hazardous conditions, prevent accidents, and how to recover from an accident; (2) to design the motors, facilities, and the equipment to be safe; and

(3) to institute and enforce rigid safety rules during design, manufacture, and operation. There are many such rules. Examples are no smoking and no matches in areas where there are propellants or loaded motors, wearing spark-proof shoes and using spark-proof tools, shielding all electrical equipment, providing a water-deluge fire extinguishing system in test facilities to cool motors or extinguish burning, or proper grounding of all electrical equipment and items that could build up static electrical charges.

12.4. PROPELLANT INGREDIENTS

A number of relatively common propellant ingredients are listed in Table 12–6 for double-base propellants and in Table 12–7 for composite-type solid propellants. They are categorized by major *function,* such as *oxidizer, fuel, binder, plasticizer, curing agent,* and so on, and each category is described in this section. However, several of the ingredients have more than one function. These lists are not complete and at least 200 other ingredients have been tried in experimental rocket motors.

A classification of modern propellants, including some new types that are still in the experimental phase, is given in Table 12–8, according to their binders, plasticizers, and solid ingredients; these solids may be an oxidizer, a solid fuel, or a combination or compound of both.

The ingredient properties and impurities can have a profound effect on the propellant characteristics. A seemingly minor change in one ingredient can cause measurable changes in ballistic properties, physical properties, migration, aging, or ease of manufacture. When the propellant's performance or ballistic characteristics have tight tolerances, the ingredient purity and properties must also conform to tight tolerances and careful handling (e.g., no exposure to moisture). In the remainder of this section a number of the important ingredients, grouped by function, are briefly, discussed.

Inorganic Oxidizers

Some of the thermochemical properties of several oxidizers and oxygen radical-containing compounds are listed in Table 12–9. Their values depend on the chemical nature of each ingredient.

Ammonium perchlorate (NH_4ClO_4) is the most widely used crystalline oxidizer in solid propellants. Because of its good characteristics, including compatibility with other propellant materials, good performance, quality, uniformity, and availability, it dominates the solid oxidizer field. Other solid oxidizers, particularly ammonium nitrate and potassium perchlorate, were used and occasionally are still being used in production rockets but to a large extent have been replaced by more modern propellants containing ammonium perchlorate. Many oxidizer compounds were investigated during the 1970s, but none reached production status.

TABLE 12–6. Typical Ingredients of Double-Base (DB) Propellants and Composite-Modified Double-Base (CMDB) Propellants

Type	Percent	Acronym	Typical Chemicals
Binder	30–50	NC	Nitrocellulose (solid), usually plasticized with 20 to 50% nitroglycerine
Reactive plasticizer (liquid explosive)	20–50	NG	Nitroglycerine
		DEGDN	Diethylene glycol dinitrate
		TEGDN	Triethylene glycol dinitrate
		PDN	Propanedial-dinitrate
		TMETN	Trimethylolethane trinitrate
Plasticizer (organic liquid fuel)	0–10	DEP	Diethyl phthalate
		TA	Triacetin
		DMP	Dimethyl phthalate
			Dioctile phthalate
		EC	Ethyl centralite
		DBP	Dibutyl phthalate
Burn-rate modifier	up to 3	PbSa	Lead salicylate
		PbSt	Lead stearate
		CuSa	Copper salicylate
		CuSt	Copper stearate
Coolant		OXM	Oxamine
Opacifier		C	Carbon black (powder or graphite powder)
Stabilizer and or antioxidant	> 1	DED	Diethyl diphenyl
		EC	Ethyl centralite
		DPA	Diphenyl amine
Visible flame suppressant	up to 2	KNO₃	Potassium nitrate
		K₂SO₄	Potassium sulphate
Lubricant (for extruded propellant only)	> 0.3	C	Graphite Wax
Metal fuel[a]	0–15	Al	Aluminum, fine powder (solid)
Crystalline oxidizer[a]	0–15	AP	Ammonium perchlorate
		AN	Ammonium nitrate
Solid explosive crystals[a]	0–20	HMX	Cyclotetramethylenetetranitramine
		RDX	Cyclotrimethylenetrinitramine
		NQ	Nitroguanadine

[a] Several of these, but not all, are added to CMDB propellant.

The oxidizing potential of the perchlorates is generally high, which makes this material suited to high specific impulse propellants. Both ammonium and potassium perchlorate are only slightly soluble in water, a favorable trait for propellant use. All the perchlorate oxidizers produce hydrogen chloride (HCl) and other toxic and corrosive chlorine compounds in their reaction with fuels. Care is required in firing rockets, particularly the very large rockets, to safeguard operating personnel or communities in the path of exhaust gas clouds. Ammonium perchlorate (AP) is supplied in the form of small white crystals. Particle size and shape influences the manufacturing process and the propellant burning rate. Therefore, close control of the crystal sizes and the size distribu-

TABLE 12–7. Typical Ingredients of Composite Solid Propellants

Type	Percent	Acronym	Typical Chemicals
Oxidizer (crystalline)	0–70	AP	Ammonium perchlorate
		AN	Ammonium nitrate
		KP	Potassium perchlorate
		KN	Potassium nitrate
		ADN	Ammonium dinitramine
Metal fuel (also acts as a combustion stabilizer)	0–30	Al	Aluminum
		Be	Beryllium (experimental propellant only)
		Zr	Zirconium (also acts as burn-rate modifier)
Fuel/Binder, polybutadiene type	5–18	HTPB	Hydroxyl-terminated polybutadiene
		CTPB	Carboxyl-terminated polybutadiene
		PBAN	Polybutadiene acrylonitrile acrylic acid
		PBAA	Polybutadiene acrylic acid
Fuel/Binder, polyether and polyester type	0–15	PEG	Polyethylene glycol
		PCP	Polycaprolactone polyol
		PGA	Polyglycol adipate
		PPG	Polypropylene glycol
		HTPE	Hydroxyl-terminated polyethylene
		PU	Polyurethane polyester or polyether
Curing agent or crosslinker, which reacts with polymer binder	0.2–3.5	MAPO	Methyl aziridinyl phosphine oxide
		IPDI	Isophorone diisocyanate
		TDI	Toluene-2,4-diisocyanate
		HMDI	Hexamethylene diisocyanide
		DDI	Dimeryl diisocyanate
		TMP	Trimethylol propane
		BITA	Trimesoyl-1(2-ethyl)-aziridine
Burn-rate modifier	0.2–3	FeO	Ferric oxide
		nBF	*n*-Butyl ferrocene
			Oxides of Cu, Pb, Zr, Fe
			Alkaline earth carbonates
			Alkaline earth sulfates
			Metallo-organic compounds
Explosive filler (solid)	0–40	HMX	Cyclotetramethylenetetranitramine
		RDX	Cyclotrimethylenetrinitramine
		NQ	Nitroguanadine
Plasticizer/Pot life control (organic liquid)	0–7	DOP	Dioctyl phthalate
		DOA	Dioctyl adipate
		DOS	Dioctyl sebacate
		DMP	Dimethyl phthalate
		IDP	Isodecyl pelargonate

TABLE 12–7. (*Continued*)

Type	Percent	Acronym	Typical Chemicals
Energetic plasticizer (liquid)	0–14	GAP	Glycidyl azide polymer
		NG	Nitroglycerine
		DEGDN	Diethylene glycol dinitrate
		BTTN	Butanetriol trinitrate
		TEGDN	Triethylene glycol dinitrate
		TMETN	Trimethylolethane trinitrate
		PCP	Polycaprolactone polymer
Energetic fuel/ binder	0–15	GAP	Glycidyl azide polymer
		PGN	Propylglycidyl nitrate
		BAMO/AMMO	Bis-azidomethyloxetane/Azidomethyl-methyloxetane copolymer
		BAMO/NMMO	Bis-azidomethyloxetane/Nitramethyl-methyloxetane copolymer
Bonding agent (improves bond to solid particles)	> 0.1	MT-4	MAPO–tartaric acid–adipic acid condensate
		HX-752	Bis-isophthal-methyl-aziridine
Stabilizer (reduces chemical deterioration)	> 0.5	DPA	Diphenylamine
		—	Phenylnaphthylamine
		NMA	N-methyl-*p*-nitroaniline
		—	Dinitrodiphenylanine
Processing aid	> 0.5	—	Lecithin
		—	Sodium lauryl sulfate

tion present in a given quantity or batch is required. AP crystals are rounded (nearly ball shaped) to allow easier mixing than sharp, fractured crystals. They come in sizes ranging from about 600 μm ($1\mu m = 10^{-6}$ m) diameter to about 80 μm from the factory. Sizes below about 40 μm diameter are considered hazardous (can easily be ignited and sometimes detonated) and are not shipped; instead, the propellant manufacturer takes larger crystals and grinds them (at the motor factory) to the smaller sizes (down to 2 μm) just before they are incorporated into a propellant.

The *inorganic nitrates* are relatively low-performance oxidizers compared with perchlorates. However, *ammonium nitrate* is used in some applications because of its very low cost and smokeless and relatively nontoxic exhaust. Its principal use is with low-burning-rate, low-performance rocket and gas generator applications. Ammonium nitrate (AN) changes its crystal structure at several phase transformation temperatures. These changes cause slight changes in volume. One phase transformation at 32°C causes about a 3.4% change in volume. Repeated temperature cycling through this transition temperature creates tiny voids in the propellant, and causes growth in the grain and a change in physical or ballistic properties. The addition of a small amount

TABLE 12–8. Classification of Solid Rocket Propellants Used in Flying Vehicles According to their Binders, Plasticizers, and Solid Ingredients

Designation	Binder	Plasticizer	Solid Oxidizer and/or Fuel	Propellant Application
Double-base, DB	Plasticized NC	NG, TA, etc.	None	Minimum signature and smoke
CMDB[a]	Plasticized NC	NG, TMETN, TA, BTTN, etc.	Al, AP, KP	Booster, sustainer, and spacecraft
	Same	Same	HMX, RDX, AP	Reduced smoke
	Same	Same	HMX, RDX, azides	Minimum signature, gas generator
EMCDB[a]	Plasticized NC + elastomeric polymer	Same	Like CMDB above, but generally superior mechanical properties with elastomer added as binder	Same
Polybutadiene	HTPB	DOA, IDP, DOP, DOA, etc.	Al, AP, KP, HMX, RDX	Booster, sustainer or spacecraft; used extensively in many applications
	HTPB	Same	AN, HMX, RDX, some AP	Reduced smoke, gas generator
	CTPB, PBAN, PBAA	All like HTPB above, but somewhat lower performance due to higher processing viscosity and consequent lower solids content. Still used in applications with older designs		
TPE[a]	Thermoplastic elastomer	Similar to HTPB, but without chemical curing process. TPEs cure (crosslink) via selective crystallization of certain parts of the binder. Still are experimental propellants		
Polyether and polyesters	PEG, PPG, PCP, PGA, and mixtures	DOA, IDP, TMETN, DEGDN, etc. Al, AP, KP, HMX		Booster, sustainer, or spacecraft
Energetic binder (other than NC)	GAP, PGN, BAMO/ NMMO, BAMO/AMMO	TMETN, BTTN, etc. GAP-azide, GAP-nitrate, NG	Like polyether/polyester propellants above, but with slightly higher performance. Experimental propellant.	

[a] CMDB, composite-modified cast double-base; EMCDB, elastomer-modified cast double-base; TPE, thermoplastic elastomer. For definition of acronyms and abbreviation of propellant ingredients see Tables 12–6 and 12–7.

498

TABLE 12–9. Comparison of Crystalline Oxidizers

Oxidizer	Chemical Symbol	Molecular Mass (kg/kg-mol)	Density (kg/m^3)	Oxygen Content (wt %)	Remarks
Ammonium perchlorate	NH_4ClO_4	117.49	1949	54.5	Low n, low cost, readily available
Potassium perchlorate	$KClO_4$	138.55	2519	46.2	Low burning rate, medium performance
Sodium perchlorate	$NaClO_4$	122.44	2018	52.3	Hygroscopic, high performance
Ammonium nitrate	NH_4NO_3	80.0	1730	60.0	Smokeless, medium performance
Potassium nitrate	KNO_3	101.10	2109	47.5	Low cost, low performance

of stabilizer such as nickel oxide (NiO) or potassium nitrate (KNO_3) seems to change the transition temperature to above 60°C, a high enough value so that normal ambient temperature cycling will no longer cause recrystallization (Refs. 12–10 and 12–11). AN with such an additive is known as *phase-stabilized ammonium nitrate* (PSAN). AN is hygroscopic, and the absorption of moisture will degrade propellant made with AN.

Fuels

This section discusses solid fuels. *Powdered spherical aluminum* is the most common. It consists of small spherical particles (5 to 60 µm diameter) and is used in a wide variety of composite and composite-modified double-base propellant formulations, usually constituting 14 to 20% of the propellant by weight. Small aluminum particles can burn in air and this powder is mildly toxic if inhaled. During rocket combustion this fuel is oxidized into aluminum oxide. These oxide particles tend to agglomerate and form larger particles. The aluminum increases the heat of combustion, the propellant density, the combustion temperature, and thus the specific impulse. The oxide is in liquid droplet form during combustion and solidifies in the nozzle as the gas temperature drops. When in the liquid state the oxide can form a molten slag which can accumulate in pockets (e.g., around an impropely designed submerged nozzle), thus adversely affecting the vehicle's mass ratio. It also can deposit on walls inside the combustion chamber, as described in Refs. 12–12 and 14–13.

Boron is a high-energy fuel that is lighter than aluminum and has a high melting point (2304°C). It is difficult to burn with high efficiency in combustion chambers of reasonable length. However, it can be oxidized at reasonable efficiency if the boron particle size is very small. Boron is used advantageously as a propellant in combination rocket–air-burning engines, where there is adequate combustion volume and oxygen from the air.

Beryllium burns much more easily than boron and improves the specific impulse of a solid propellant motor, usually by about 15 sec, but it and its oxide are highly toxic powders absorbed by animals and humans when inhaled. The technology with composite propellants using powdered beryllium fuel has been experimentally proven, but its severe toxicity makes its application unlikely.

Theoretically, both aluminum hydride (AlH_3) and beryllium hydride (BeH_2) are attractive fuels because of their high heat release and gas-volume contribution. Specific impulse gains are 10 to 15 sec for Al_2H_3 and 25 to 30 sec for BeH_2. Both are difficult to manufacture and both deteriorate chemically during storage, with loss of hydrogen. These compounds are not used today in practical fuels.

Binders

The binder provides the structural glue or matrix in which solid granular ingredients are held together in a composite propellant. The raw materials are liquid prepolymers or monomers. Polyethers, polyesters and poly-butadienes have been used (see Tables 12–6 and 12–7). After they are mixed with the solid ingredients, cast and cured, they form a hard rubber-like material that constitutes the grain. Polyvinylchloride (PVC) and polyurethane (PU) (Table 12–1) were used 40 years ago and are still used in a few motors, mostly of old design. Binder materials are also really fuels for solid propellant rockets and are oxidized in the combustion process. The binding ingredient, usually a polymer of one type or another, has a primary effect on motor reliability, mechanical properties, propellant processing complexity, storability, aging, and costs. Some polymers undergo complex chemical reactions, crosslinking, and branch chaining during curing of the propellant. HTPB has been the favorite binder in recent years, because it allows a somewhat higher solids fraction (88 to 90% of AP and Al) and relatively good physical properties at the temperature limits. Several common binders are listed in Tables 12–1, 12–6 and 12–7. Elastomeric binders have been added to plasticized double-base-type nitrocellulose to improve physical properties. Polymerization occurs when the binder monomer and its crosslinking agent react (beginning in the mixing process) to form long-chain and complex three-dimensional polymers. Other types of binders, such as PVC, cure or plasticize without a molecular reaction (see Refs. 12–2, 12–3, and 12–13). Often called plastisol-type binders, they form a very viscous dispersion of a powdered polymerized resin in nonvolatile liquid. They polymerize slowly by interaction.

Burning-Rate Modifiers

A burning-rate *catalyst* or burning-rate *modifier* helps to accelerate or decelerate the combustion at the burning surface and increases or decreases the value of the propellant burning rate. It permits the tailoring of the burning rate to fit a specific grain design and thrust–time curve. Several are listed in Tables 12–6 and 12–7. Some, like iron oxide or lead stearate, increase the burning rate; however, others, like lithium fluoride, will reduce the burning rate of some composite propellants. The inorganic catalysts do not contribute to the combustion energy, but consume energy when they are heated to the combustion temperature. These modifiers are effective because they change the combustion mechanism, which is described in Chapter 13. Chapter 2 of Ref. 12–2 gives examples of how several modifiers change the burning rate of composite propellants.

Plasticizers

A plasticizer is usually a relatively low-viscosity liquid organic ingredient which is also a fuel. It is added to improve the elongation of the propellant at low temperatures and to improve processing properties, such as lower viscosity for casting or longer pot life of the mixed but uncured propellants. The plasticizers listed in Tables 12–6, 12–7, and 12–8 show several plasticizers.

Curing Agents or Crosslinkers

A curing agent or crosslinker causes the prepolymers to form longer chains of larger molecular mass and interlocks between chains. Even though these materials are present in small amounts (0.2 to 3%), a minor change in the percentage will have a major effect on the propellant physical properties, manufacturability, and aging. It is used only with composite propellants. It is the ingredient that causes the binder to solidify and become hard. Several curing agents are listed in Table 12–7.

Energetic Binders and Plasticizers

Energetic binders and/or plasticizers are used in lieu of the conventional organic materials. They contain oxidizing species (such as azides or organic nitrates) as well as organic species. They add some additional energy to the propellant causing a modest increase in performance. They serve also as a binder to hold other ingredients, or as an energetic plasticizer liquid. They can self-react exothermally and burn without a separate oxidizer. Glycidyl azide polymer (GAP) is an example of an energetic, thermally stable, hydroxyl-terminated prepolymer that can be polymerized. It has been used in experiental propellants. Other energetic binder or plasticizer materials are listed in Tables 12–6, 12–7 and 12–8.

Organic Oxidizers or Explosives

Organic oxidizers are explosive organic compounds with $—NO_2$ radical or other oxidizing fractions incorporated into the molecular structure. References 12–2 and 12–13 describe their properties, manufacture, and application. These are used with high-energy propellants or smokeless propellants. They can be crystalline solids, such as the *nitramines* HMX or RDX, fibrous solids such as NC, or energetic plasticizer liquids such as DEGN or NG. These materials can react or burn by themselves when initiated with enough activating energy, but all of them are explosives and can also be detonated under certain conditions. Both HMX and RDX are stoichiometrically balanced materials and the addition of either fuel or oxidizer only will reduce the T_1 and I_s values. Therefore, when binder fuels are added to hold the HMX or RDX crystals in a viscoelastic matrix, it is also necessary to add an oxidizer such as AP or AN.

RDX and HMX are quite similar in structure and properties. Both are white crystalline solids that can be made in different sizes. For safety, they are shipped in a desensitizing liquid, which has to be removed prior to propellant processing. HMX has a higher density, a higher detonation rate, yields more energy per unit volume, and has a higher melting point. NG, NC, HMX, and RDX are also used extensively in military and commercial explosives. HMX or RDX can be included in DB, CMDB, or composite propellants to achieve higher performance or other characteristics. The percentage added can range up to 60% of the propellant. Processing propellant with these or similar ingredients can be hazardous, and the extra safety precautions make the processing more expensive.

Liquid *nitroglycerine* (NG) by itself is very sensitive to shock, impact, or friction. It is an excellent plasticizer for propellants when desensitized by the addition of other materials (liquids like triacetin or dibutyl phthalate) or by compounding with nitrocellulose. It is readily dissolved in many organic solvents, and in turn it acts as a solvent for NC and other solid ingredients (Ref. 12–13).

Nitrocellulose (NC) is a key ingredient in DB and CMDB propellant. It is made by the acid nitration of natural cellulose fibers from wood or cotton and is a mixture of several organic nitrates. Although crystalline, it retains the fiber structure of the original cellulose (see Ref. 12–13). The nitrogen content is important in defining the significant properties of nitrocellulose and can range from 8 to 14%, but the grades used for propellant are usually between 12.2 and 13.1%. Since it is impossible to make NC from natural products with an exact nitrogen content, the required properties are achieved by careful blending. Since the solid fiber-like NC material is difficult to make into a grain, it is usually mixed with NG, DEGN, or other plasticizer to gelatinize or solvate it when used with DB and CMDB propellant.

Additives

Small amounts of *additives* are used for many purposes, including accelerating or lengthening the *curing time*, improving the *rheological properties* (easier casting of viscous raw mixed propellant), improving the *physical properties*, adding *opaqueness* to a transparent propellant to prevent radiation heating at places other than the burning surface, limiting *migration of chemical species* from the propellant to the binder or vice versa, minimizing the slow oxidation or *chemical deterioration* during storage, and improving the *aging* characteristics or the moisture resistance. *Bonding agents* are additives to enhance adhesion between the solid ingredients (AP or Al) and the binder. *Stabilizers* are intended to minimize the slow chemical or physical reactions that can occur in propellants. *Catalysts* are sometimes added to the crosslinker or curing agent to slow down the curing rate. Lubricants aid the extrusion process. Desensitizing agents help to make a propellant more resistant to inadvertent energy stimulus. These are usually added in very small quantities.

Particle-Size Parameters

The size, shape, and size distribution of the solid particles of AP, Al or HMX in the propellant can have a major influence on the composite propellant characteristics. The particles are spherical in shape, because this allows easier mixing and a higher percentage of solids in the propellant than shapes of sharp-edged natural crystals. Normally, the ground AP oxidizer crystals are graded according to particle size ranges as follows:

Coarse	400 to 600 μm (1 μm = 10^{-6} m)
Medium	50 to 200 μm
Fine	5 to 15 μm
Ultrafine	submicrometer to 5 μm

Coarse and medium-grade AP crystals are handled as class 1.3 materials, whereas the fine and ultrafine grades are considered as class 1.1 high explosives and are usually manufactured on-site from the medium or coarse grades. (See Section 12.3 for a definition of these explosive hazard classifications.) Most propellants use a blend of oxidizer particle sizes, if only to maximize the weight of oxidizer per unit volume of propellant, with the small particles filling part of the voids between the larger particles.

Figure 12–8 shows the influence of varying the ratio of coarse to fine oxidizer particle sizes on propellant burning rate and also the influence of a burning rate additive. Figure 12–9 shows that the influence of particle size of the aluminum fuel on propellant burning rate is much less pronounced than that of oxidizer particle size. Figure 12–8 also shows the effect of particle size. Particle

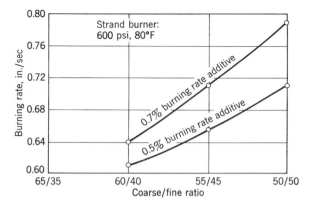

FIGURE 12–8. Typical effect of oxidizer (ammonium perchlorate) particle size mixture and burning rate additive on the burning rate of a composite propellant. (From NASA report SP-72262, Motor Propellant Development, July 1, 1967.)

size range and particle shape of both the oxidizer [usually ammonium perchlorate (AP)] and solid fuel (usually aluminum) have a significant effect on the solid packing fraction and the rheological properties (associated with the flowing or pouring of viscous liquids) of uncured composite propellant. By definition, the *packing fraction* is the volume fraction of all solids when packed to minimum volume (a theoretical condition). High packing fraction makes mixing, casting, and handling during propellant fabrication more difficult. Figure 12–10 shows the distribution of AP particle size using a blend of sizes; the shape of this curve can be altered drastically by controlling the size ranges and ratios. Also, the size range and shape of the solid particles affect the *solids loading ratio*, which is the mass ratio of solid to total ingredients in the uncured propellants. Computer-optimized methods exist for adjusting particle-size distributions for improvement of the solids loading. The solids loading can be as

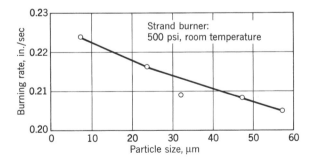

FIGURE 12–9. Typical effect of aluminum particle size on propellant burning rate for a composite propellant. (From NASA Report 8075, Solid Propellant Processing Factors in Rocket Motor Design, October 1971.)

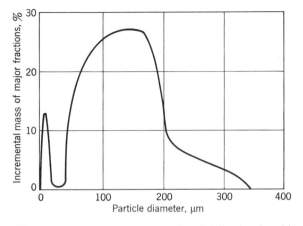

FIGURE 12–10. The oxidizer (AP) particle size distribution is a blend of two or more different particle sizes; this particular composite propellant consists of a narrow cut at about 10 μm and a broad region from 50 to 200 μm.

high as 90% in some composite propellants. High solids loading, desired for high performance, introduces complexity and higher costs into the processing of propellant. Trade-off among ballistic (performance) requirements, processibility, mechanical strength, rejection rates, and facility costs is a continuing problem with many high-specific-impulse composite propellants. References 12–2 and 12–13 give information on the influence of particle size on motor performance.

A *monomodal* propellant has one size of solid oxidizer particles, a *bimodal* has two sizes (say, 20 and 200 μm), and a *trimodal* propellant has three sizes, because this allows a larger mass of solids to be placed into the propellant. Problem 12–1 has a sketch that explains how the voids between the large particles are filled with smaller particles.

12.5. OTHER PROPELLANT CATEGORIES

Gas Generator Propellants

Gas generator propellants produce hot gas but not thrust. They usually have a low combustion temperature (800 to 1600 K), and most do not require insulators when used in metal cases. Typical applications of gas generators were listed in Table 11-1. A large variety of propellants have been used to create hot gas for gas generators, but only a few will be mentioned.

Stabilized *AN-based propellants* have been used for many years with various ingredients or binders. They give a clean, essentially smokeless exhaust and a low combustion temperature. Because of their low burning rate they are useful for long-duration gas generator applications, say 30 to 300 sec. Typical compositions are shown in Ref. 12–11, and a typical propellant is described in Table 12–10.

One method of reducing flame temperature is to burn conventional hot AP propellant and then add water to it to cool the gases to a temperature where uncooled metals can contain them. This is used on the MX missile launcher tube gas generator (Ref. 12–14). Another formulation uses HMX or RDX with an excess of polyether- or polyester-type polyurethane.

For the inflation of automobile collision safety bags the exhaust gas must be nontoxic, smoke free, have a low temperature (will not burn people), be quickly initiated, and be reliably available. One solution is to use alkali azides (e.g., NaN_3 or KN_3) with an oxide and an oxidizer. The resulting nitrates or oxides are solid materials that are removed by filtering and the gas is clean and is largely moderately hot nitrogen. In one model, air can be aspirated into the air

TABLE 12–10. Typical Gas Generator Propellant using Ammonium Nitrate Oxidizer

Ballistic Properties	
Calculated flame temperature (K)	1370
Burning rate at 6.89 MPa and 20°C (mm/sec)	2.1
Pressure exponent n (dimensionless)	0.37
Temperature sensitivity σ_p (%/K)	0.22
Theoretical characteristic velocity, c^* (m/sec)	1205
Ratio of specific heats	1.28
Molecular weight of exhaust gas	19
Composition (Mass Fraction)	
Ammonium nitrate (%)	78
Polymer binder plus curing agent (%)	17
Additives (processing aid, stabilizer, antioxidant) (%)	5
Oxidizer particle size, (μm)	150
Exhaust Gas Composition (Molar %)	
Water	26
Carbon monoxide	19
Carbon dioxide	7
Nitrogen	21
Hydrogen	27
Methane	Trace
Physical Properties at 25°C or 298 K	
Tensile strength (MPa)	1.24
Elongation (%)	5.4
Modulus of elasticity in tension (N/m^2)	34.5
Specific gravity	1.48

bag by the hot, high-pressure gas (see Ref. 12–15). One particular composition uses 65 to 75% NaN_3, 10 to 28% Fe_2O_3, 5 to 16% $NaNO_3$ as an oxidizer, a burn rate modifier, and a small amount of SiO_2 for moisture absorption. The resultant solid nitride slag is caught in a filter.

The power P delivered by a gas generator can be expressed as

$$P = \dot{m}(h_1 - h_2) = [\dot{m}T_1 Rk/(k - 1)][1 - (p_2/p_1)^{(k-1)/k}] \qquad (12–1)$$

where $\dot{m}$ is the mass flow rate, h_1 and h_2 the enthalpies per unit mass, respectively, at the gas generator chamber and exhaust pressure conditions, T_1 is the flame temperature in the gas generator chamber, R the gas constant, p_2/p_1 is the reciprocal of the pressure ratio through which these gases are expanded, and k the specific heat ratio. Because the flame temperature is relatively low there is no appreciable dissociation, and frozen equilibrum calculations are usually adequate.

Smokeless or Low-Smoke Propellant

Certain types of DB propellant, DB modified with HMX, and AN composites can be nearly smokeless. There is no or very little particulate matter in the exhaust gas. These minimum-smoke propellants are not a special class with a peculiar formulation but a variety of one of the classes mentioned previously. Propellants containing Al, Zr, Fe_2O_3 (burn rate modifier), or other metallic species will form visible clouds of small solid metal or metal oxide particles in the exhaust.

For certain military applications a smokeless propellant is needed and the reasons are stated in Chapter 18 (Exhaust Plumes). It is very difficult to make a propellant which has a truly smokeless exhaust gas. We therefore distinguish between *low-smoke* also called *minimum-smoke* (almost smokeless), and *reduced-smoke propellants*, which have a faintly visible plume. A visible smoke trail comes from solid particles in the plume, such as aluminum oxide. With enough of these particles, the exhaust plume will scatter or absorb light and become visible as *primary smoke*. The particles can act as focal points for moisture condensation, which can occur in saturated air or under high humidity, low temperature conditions. Also, vaporized plume molecules, such as water or hydrochloric acid, can condense in cold air and form droplets and thus a cloud trail. These processes create a *vapor trail* or *secondary smoke*.

Several types of DB propellant, DB modified with HMX, nitramine (HMX or RDX) based composites, AN composites, or combinations of these, give very few or no solid particles in their exhaust gas. They do not contain aluminum or AP, generally have lower specific impulse than comparable propellants with AP, and have very little primary smoke, but can have secondary smoke in unfavorable weather. Several of these propellants have been used in tactical missiles.

Reduced-smoke propellants are usually composite propellants with low concentrations of aluminum (1 to 6%); they have a low percentage of aluminum oxide in the exhaust plume, are faintly visible as primary smoke, but can precipitate heavy secondary smoke in unfavorable weather. Their performance is substantially better than that of minimum-smoke propellants, as seen in Fig. 12–1.

Igniter Propellants

The process of propellant ignition is discussed in Section 13.2, and several types of igniter hardware are discussed in Section 14.3. Propellants for igniters, a specialized field of propellant technology, is described here briefly. The requirements for an igniter propellant will include the following:

Fast high heat release and high gas evolution per unit igniter propellant mass to allow rapid filling of grain cavity with hot gas and partial pressurization of the chamber.

Stable initiation and operation over a wide range of pressures (subatmospheric to chamber pressure) and smooth burning at low pressure with no ignition overpressure surge.

Rapid initiation of igniter propellant burning and low ignition delays.

Low sensitivity of burn rate to ambient temperature changes and low burning rate pressure exponent.

Operation over the required ambient temperature range.

Safe and easy to manufacture, safe to ship and handle.

Good aging characteristics and long life.

Minimal moisture absorption or degradation with time.

Low cost of ingredients and fabrication.

Some igniters not only generate hot combustion gas, but also hot solid particles or hot liquid droplets, which radiate heat and impinge on the propellant surface, embed themselves into this surface, and assist in achieving propellant burning on the exposed grain surface.

There have been a large variety of different igniter propellants and their development has been largely empirical. Black powder, which was used in early motors, is no longer favored, because it is difficult to duplicate its properties. Extruded double-base propellants are used frequently, usually as a large number of small cylindrical pellets. In some cases rocket propellants that are used in the main grain are also used for the igniter grain; sometimes they are slightly modified. They are used in the form of a small rocket motor within a large motor that is to be ignited. A common igniter formulation uses 20 to 35% boron and 65 to 80% potassium nitrate with 1 to 5% binder. Binders typically include epoxy resins, graphite, nitrocellulose, vegetable oil, polyisobutylene, and other binders listed in Table 12–7. Another formulation uses magnesium with a fluorocarbon (Teflon); it gives hot particles and hot gas (Refs. 12–16 and 12–17). Other igniter propellants are listed in Ref. 12–18.

12.6. LINERS, INSULATORS, AND INHIBITORS

These three layers at the interface of a grain were defined in Section 11.3. Their materials do not contain any oxidizing ingredients; they will ablate, cook, char, vaporize, or distintegrate in the presence of hot gases. Many will burn if the hot combustion gas contains even a small amount of oxidizing species, but they will not usually burn by themselves. The liner, internal insulator, or inhibitor must be *chemically compatible* with the propellant and each other to avoid migration (described below) or changes in material composition; they must have *good adhesive strength*, so that they stay bonded to the propellant, or to each other. The *temperature* at which they suffer damage or experience a large *surface regression* should be high. They should all have a *low specific gravity*, thus reducing inert mass. Typical materials are neoprene (specific gravity 1.23), butyl rubber (0.93), a synthetic rubber called ethylenepropylene diene or EPDM (0.86), or the binder used in the propellant, such as polybutadiene (0.9 to 1.0); these values are low compared with a propellant specific gravity of 1.6 to 1.8. For low-smoke propellant these three rubber-like materials should give off some gas, but few, if any, solid particles (see Ref. 12–19).

In addition to the desired characteristics listed in the previous paragraph, the *liner* should be a soft stretchable rubber-type thin material (typically 0.02 to 0.04 in. thick with 200 to 450% elongation) to allow relative movement along the bond line between the grain and the case. This differential expansion occurs because the thermal coefficient of expansion of the grain is typically an order of magnitude higher than that of the case. A liner will also seal fiber-wound cases (particularly thin cases), which are often porous, so that high-pressure hot gas cannot escape. A typical liner for a tactical guided missile has been made from polypropylene glycol (about 57%), a titanium oxide filler (about 20%), a di-isocyanate crosslinker (about 20%), and minor ingredients such as an antiox-idant. The motor case had to be preheated to about 82°C prior to application. Ethylenepropylene diene monomer (EPDM) is linked into ethylenepropylene diene terpolymer to form a synthetic rubber which is often used as polymer for liners; it adheres and elongates nicely.

In some motors today the *internal insulator* not only provides for the thermal protection of the case from the hot combustion gases, but also often serves the function of the *liner* for good bonding between propellant and insulator or insulator and case. Most motors still have a separate liner and an insulating layer. The thermal internal insulator should fulfill these additional requirements:

1. It must be erosion resistant, particularly in the insulation of the motor aft end or blast tube. This is achieved in part by using tough elastomeric

materials, such as neoprene or butyl rubber, that are chemically resistant to the hot gas and the impact of particulates. This surface integrity is also achieved by forming a porous black carbon layer on its heated surface called a porous char layer, which remains after some of the interstial materials have been decomposed and vaporized.

2. It must provide good thermal resistance and low thermal conductivity to limit heat transfer to the case and thus keep the case below its maximum allowable temperature, which is usually between 160 and 350°C for the plastic in composite material cases and about 550 and 950°C for most steel cases. This is accomplished by filling the insulator with silicon oxide, graphite, Kevlar, or ceramic particles. Asbestos is an excellent filler material, but is no longer used because of its health hazard.

3. It should allow a large-deformation or strain to accommodate grain deflections upon pressurization or temperature cycling, and transfer loads between the grain and the case.

4. The surface regression should be minimal so as to retain much of its original geometric surface contour and allow a thin insulator.

A simple relationship for the thickness d at any location in the motor depends on the exposure time t_e, the erosion rate r_e (obtained from erosion tests at the likely gas velocity and temperature), and the safety factor f which can range from 1.2 to 2.0:

$$d = t_e r_e f \tag{12-2}$$

Some designers use the simple rule that the insulation depth is twice the charred depth.

The thickness of the insulation is not usually uniform; it can vary by a factor of up to 20. It is thicker at locations such as the aft done, where it is exposed for longer intervals and at higher scrubbing velocities than the insulator layers protected by bonded propellant. Before making a material selection, it is necessary to evaluate the flow field and the thermal environment (combustion temperature, gas composition, pressure, exposure duration, internal ballistics) in order to carry out a thermal analysis (erosion prediction and estimated thickness of insulator). An analysis of loads and the deflections under loads at different locations of the motor are needed to estimate shear and compression stresses. If it involves high stresses or a relief flap, a structural analysis is also needed. Various computer programs, such as the one mentioned in Refs. 12–20 and 12–21, are used for these analyses.

An *inhibitor* is usually made of the same kinds of materials as internal insulators. They are applied (bonded, molded, glued, or sprayed) to grain surfaces that should not burn. In a segmented motor, for example (see Fig. 14–2), where burning is allowed only on the internal port area, the faces of the cylindrical grain sections are inhibited.

Migration is the transfer of mobile (liquid) chemical species from the solid propellant to the liner, insulator, or inhibitor, or vice versa. Liquid plasticizers such as NG or DEGN or unreacted monomers or liquid catalysts are known to migrate. This migratory transfer occurs very slowly; it can cause dramatic changes in physical properties (e.g., the propellant next to the liner becomes brittle or weak) and there are several instances where nitroglycerine migrated into an insulator and made it flammable. Migration can be prevented or inhibited by using (1) propellants without plasticizers, (2) insulators or binders with plasticizers identical to those used in propellants, (3) a thin layer of an impervious material or a migration barrier (such as PU or a thin metal film), and (4) an insulator material that will not allow migration (e.g., PU) (see Ref. 12–22).

The graphite–epoxy motors used to boost the Delta launch vehicle use a three-layer *liner*: EPDM (ethylenepropylene diene terpolymer) as a thin primer to enhance bond strength, a polyurethane barrier to prevent migration of the plasticizer into the EPDM liner, and a plasticized HTPB-rich liner to prevent burning next to the case–bond interface. The composite AP–Al propellant also uses the same HTPB binder.

Liners, insulators, or inhibitors can be applied to the grain in several ways: by painting, coating, dipping, spraying, or by gluing a sheet or strip to the case or the grain. Often an automated, robotic machine is used to achieve uniform thickness and high quality. Reference 12–21 describes the manufacture of particular insulators.

An *external insulation* is often applied to the outside of the motor case, particularly in tactical missiles or high-acceleration launch boosters. This insulation reduces the heat flow from the air boundary layer outside the vehicle surface (which is aerodynamically heated) to the case and then to the propellant. It thus prevents fiber-reinforced plastic cases from becoming weak or the propellant from becoming soft or, in extreme situations, from being ignited. This insulator must withstand the oxidation caused by aerodynamically heated air, have good adhesion, have structural integrity to loads imposed by the flight or launch, and must have a reasonable cure temperature. Materials ordinarily used as internal insulators are unsatisfactory, because they burn in the atmosphere and generate heat. The best is a nonpyrolyzing, low-thermal-conductivity refractory material (Ref. 12–23) such as high-temperature paint. The internal and external insulation also helps to reduce the grain temperature fluctuations and thus the thermal stresses imposed by thermal cycling, such as day–night variations or high- and low-altitude temperature variations for airborne missiles.

12.7. PROPELLANT PROCESSING AND MANUFACTURE

The manufacture of solid propellant involves complex physical and chemical processes. In the past, propellant has been produced by several different processes, including the compaction or pressing of powder charges, extrusion of propellant through dies under pressure using heavy presses, and mixing with a solvent which is later evaporated. Even for the same type of propellant (e.g., double-base, composite, or composite double-base) the fabrication processes are usually not identical for different manufacturers, motor types, sizes, or propellant formulation, and no single simple generalized process flowsheet or fabrication technique is prevalent. Most of the rocket motors in production today use composite-type propellants and therefore some emphasis on this process is given here.

Figure 12–11 shows a representative flowsheet for the manufacture of a complete solid rocket motor with a composite propellant made by batch processes. Processes marked with an asterisk are potentially hazardous, are usually operated or controlled remotely, and are usually performed in buildings designed to withstand potential fires or explosions. The mixing and casting processes are the most complex and are more critical than other processes in determining the quality, performance, burn rate, and physical properties of the resulting propellant.

The rheological properties of the uncured propellant, meaning its flow properties in terms of shear rate, stress, and time, are all-important to the processibility of the propellant, and these properties usually change substantially throughout the length of the processing line. Batch-type processing of propellant, including the casting (pouring) of propellant into motors that serve as their own molds, is the most common method. For very large motors several days are needed for casting perhaps 40 batches into a single case, forming a single grain. Vacuum is almost always imposed on the propellant during the mixing and casting operations to remove air and other dispersed gases and to avoid air bubbles in the grain. Viscosity measurements of the mixed propellant (10,000 to 20,000 poise) are made for quality control. Vacuum, temperature, vibration, energy input of the mixer, and time are some of the factors affecting the viscosity of the uncured propellant. Time is important in terms of *pot life*, that period of time the uncured propellant remains reasonably fluid after mixing before it cures and hardens. Short pot life (a few hours) requires fast operations in emptying mixers, measuring for quality control, transporting, and casting into motors. Some binder systems, such as those using PVC, give a very long pot life and avoid the urgency of haste in the processing line. References 12–3, 12–18, and 12–24 give details on propellant processing techniques and equipment.

Double-base propellants and modified double-base propellants are manufactured by a different set of processes. The key is the diffusion of the liquid nitroglycerine into the fibrous solid matrix or nitrocellulose, thus forming, by means of solvation, a fairly homogeneous, well-dispersed, relatively strong solid material. Several processes for making double-base rocket propellant are in use today, including extrusion and slurry casting. In the slurry casting process the case (or the mold) is filled with solid casting powder (a series of small solid pellets of nitrocellulose with a small amount of nitroglycerine) and the case is then flooded with liquid nitroglycerine, which then solvates the

Chemical ingredients receiving, storage, inspection, weighing and preparation

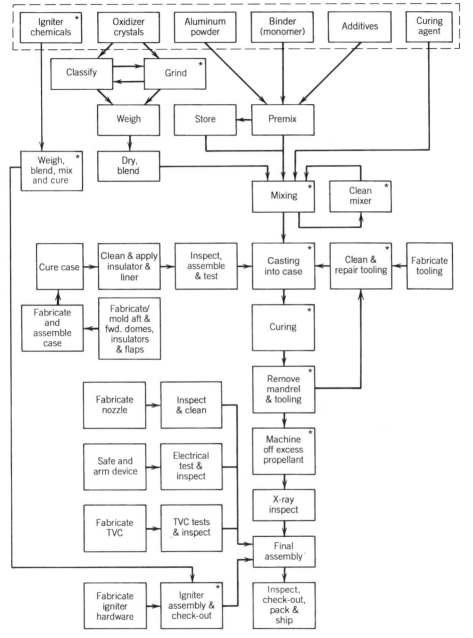

FIGURE 12–11. Simplified manufacturing process flow diagram for a rocket motor and its composite solid propellant.

pellets. Figure 12–12 shows a simplified diagram of a typical setup for a slurry cast process. Double-base propellant manufacturing details are shown in Refs. 12–3 and 12–13.

Mandrels are used during casting and curing to assure a good internal cavity or perforation. They are made of metal in the shape of the internal bore (e.g., star or dogbone) and are often slightly tapered and coated with a nonbonding material, such as Teflon, to facilitate the withdrawal of the mandrel after curing without tearing the grain. For complicated internal passages, such as a conocyl, a complex built-up mandrel is necessary, which can be withdrawn through the nozzle flange opening in smaller pieces or which can be collapsed.

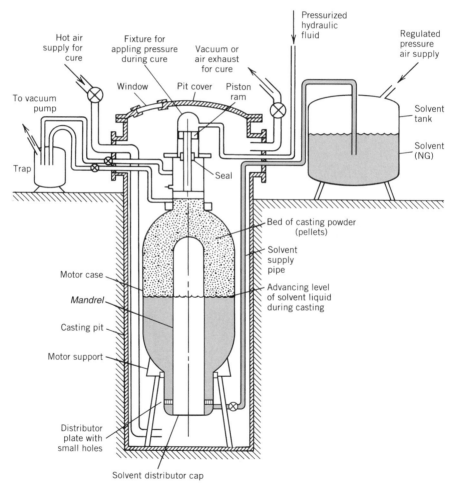

FIGURE 12–12. Simplified diagram of one system for slurry casting and initial curing of a double-base solid propellant.

Some manufacturers have had success in making permanent mandrels (which are not withdrawn but stay with the motor) out of lightweight foamed propellant, which burns very quickly once it is ignited.

An important objective in processing is to produce a propellant grain free of cracks, low-density areas, voids, or other flaws. In general, voids and other flaws degrade the ballistic and mechanical properties of the propellant grain. Even the inclusion of finely dispersed gas in a propellant can result in an abnormally high burning rate, one so high as to cause catastrophic motor failure.

The finished grain (or motor) is usually inspected for defects (cracks, voids, and debonds) using x-ray, ultrasonic, heat conductivity, or other nondestructive inspection techniques. Samples of propellant are taken from each batch, tested for rheological properties, and cast into physical property specimens and/or small motors which are cured and subsequently tested. A determination of the sensitivity of motor performance, including possible failure, to propellant voids and other flaws often requires the test firing of motors with known defects. Data from the tests are important in establishing inspection criteria for accepting and rejecting production motors.

Special process equipment is needed in the manufacture of propellant. For composite propellants this includes mechanical mixers (usually with two or three blades rotating on vertical shafts agitating propellant ingredients in a mixer bowl under vacuum), casting equipment, curing ovens, or machines for automatically applying the liner or insulation to the case. Double-base processing requires equipment for mechanically working the propellant (rollers, presses) or special tooling for allowing a slurry cast process. Computer-aided filament winding machines are used for laying the fibers of fiber-reinforced plastic cases and nozzles.

PROBLEMS

1. Ideally the solid oxidizer particles in a propellant can be considered spheres of uniform size. Three sizes of particles are available: coarse at 500 μm, medium at 50 μm, and fine at 5 μm, all at a specific gravity of 1.95, and a viscoelastic fuel binder at a specific gravity of 1.01. Assume that these materials can be mixed and vibrated so that the solid particles will touch each other, there are no voids in the binder, and the particles occupy a minimum of space similar to the sketch of the cross section shown here. It is desired to put 94 wt % of oxidizer into the propellant mix, for this will give maximum performance. (*a*) Determine the maximum weight percentage of oxidizer if only coarse crystals are used or if only medium-sized crystals are used. (*b*) Determine the maximum weight of oxidizer if both coarse and fine crystals are used, with the fine crystals filling the voids between the coarse particles. What is the optimum relative proportion of coarse and fine particles to give a maximum of oxidizer? (*c*) Same as part (*b*), but use coarse and medium crystals only. Is this better and, if so, why? (*d*) Using all three sizes, what is the ideal weight mixture ratio and what is the maximum oxidizer content possible and the theoretical maximum specific gravity of

the propellant? (*Hint*: The centers of four adjacent coarse crystals form a tetrahedron whose side length is equal to the diameter.)

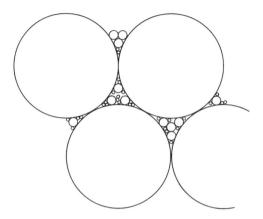

2. Suggest one or two specific applications (intercontinental missile, anti-aircraft, space launch vehicle upper stage, etc.) for each of the propellant categories listed in Table 12–2 and explain why it was selected when compared to other propellants.

3. Prepare a detailed outline of a procedure to be followed by a crew operating a propellant mixer. This 1 m³ vertical solid propellant mixer has two rotating blades, a mixing bowl, a vacuum pump system to allow mix operations under vacuum, feed chutes or pipes with valves to supply the ingredients, and variable-speed electric motor drive, a provision for removing some propellant for laboratory samples, and a double-wall jacket around the mixing bowl to allow heating or cooling. It is known that the composite propellant properties are affected by mix time, small deviations from the exact composition, the temperature of the mix, the mechanical energy added by the blades, the blade speed, and the sequence in which the ingredients are added. It is also known that bad propellant would be produced if there are leaks that destroy the vacuum, if the bowl, mixing blades, feed chutes, and so on, are not clean but contain deposits of old propellant on their walls, if they are not mixed at 80°C, or if the viscosity of the mix becomes excessive. The sequence of loading ingredients shall be: (1) prepolymer binder, (2) plasticizer, (3) minor liquid additives, (4) solid consisting of first powdered aluminum and thereafter mixed bimodal AP crystals, and (5) finally the polymerizing agent or crosslinker. Refer to Fig. 12–11. Samples of the final liquid mix are taken to check viscosity and density. Please list all the sequential steps that the crew should undertake before, during, and after the mixing operation. If it is desired to control to a specific parameter (weight, duration, etc.), that fact should be stated; however, the specific data of ingredient mass, time, power, temperature, and so on, can be left blank. Mention all instruments (e.g., thermometers, wattmeter, etc.) that the crew should have and identify those that they must monitor closely. Assume that all ingredients were found to be of the desired composition, purity, and quality.

4. Determine the longitudinal growth of a 24-in.-long free-standing grain with a linear thermal coefficient of expansion of $7.5 \times 10^{-5}/°F$ for temperature limits of −40 to

140°F.
Answer: 0.32 in.

5. The following data are given for an internally burning solid propellant grain with inhibited end faces and a small initial port area:

Length	40 in.
Port area	27 in.^{2}
Propellant weight	240 lb
Initial pressure at front end of chamber	1608 psi
Initial pressure at nozzle end of chamber	1412 psi
Propellant density	0.060 lb/in.^{3}
Vehicle acceleration	21.2 g_0

Determine the initial forces on the propellant supports produced by pressure differential and vehicle acceleration.
Answers: 19,600 lbf, 5090 lbf.

6. A solid propellant unit with an end-burning grain has a thrust of 4700 N and a duration of 14 sec. Four different burning rate propellants are available, all with approximately the same performance and the same specific gravity, but different AP mix and sizes and different burning rate enhancements. They are 5.0, 7.0, 10, and 13 mm/sec. The preferred L/D is 2.60, but values of 2.2 to 3.5 are acceptable. The impulse-to-initial-weight ratio is 96 at an L/D of 2.5. Assume optimum nozzle expansion. Chamber pressure is 6.894 MPa or 1000 psia and the operating temperature is 20°C or 68°F. Determine grain geometry, propellant mass, hardware mass, and initial mass.

7. For the rocket in Problem 6 determine the approximate chamber pressure, thrust, and duration at 245 and 328 K. Assume the temperature sensitivity (at a constant value of A_b/A_t) of 0.01%/K does not change with temperature.

8. A fuel-rich solid propellant gas generator propellant is required to drive a turbine of a liquid propellant turbopump. Determine its mass flow rate. The following data are given:

Chamber pressure	$p_1 = 5$ MPa
Combustion temperature	$T_1 = 1500$ K
Specific heat ratio	$k = 1.25$
Required pump input power	970 kW
Turbine outlet pressure	0 psia
Turbine efficiency	65%
Molecular weight of gas	22 kg/kg-mol
Pressure drop between gas generator and turbine nozzle inlet	0.10 MPa

Windage and bearing friction is 10 kW. Neglect start transients.
Answer: $\dot{m} = 0.257$ kg/sec.

9. The propellant for this gas generator has these characteristics:

Burn rate at standard conditions	4.0 mm/sec
Burn time	110 sec
Chamber pressure	5.1 MPa

Pressure exponent n 0.55
Propellant specific gravity   1.47

Determine the size of an end-burning cylindrical grain.

Answer: Single end-burning grain 27.2 cm in diameter and 31.9 cm long, or two end-burning opposed grains (each 19.6 cm diameter × 31.9 cm long) in a single chamber with ignition of both grains in the middle of the case.

REFERENCES

12–1. A Davenas, "Solid rocket Motor Design," Chapter 4 of G. E. Jensen and D. W. Netzer (Eds.), *Tactical Missile Propulsion*, Vol. 170, Progress in Astronautics and Aeronautics, AIAA, 1996.

12–2. N. Kubota, "Survey of Rocket Propellants and their Combustion Characteristics," Chapter 1 in K. K. Kuo and M. Summerfield (Eds.), *Fundamentals of Solid-Propellant Combustion*. Progress in Astronautics and Aeronautics, Vol. 90, American Institute of Aeronautics and Astronautics, New York, 1984.

12–3. C. Boyars and K. Klager, *Propellants: Manufacture, Hazards and Testing*, Advances in Chemistry Series 88, American Chemical Society, Washington, DC, 1969.

12–4. Chemical Propulsion Information Agency, *Hazards of Chemical Rockets and Propellants*. Vol. II, *Solid Rocket Propellant Processing, Handling, Storage and Transportation*, NTIS AD-870258, May 1972.

12–5. H. S. Sibdeh and R. A. Heller, "Rocket Motor Service Life Calculations Based on First Passage Method," *Journal of Spacecraft and Rockets*, Vol. 26, No. 4, July–August 1989, pp. 279–284.

12–6. D. I. Thrasher, "State of the Art of Solid Propellant Rocket Motor Grain Design in the United States," Chapter 9 in *Design Methods in Solid Rocket Motors*, Lecture Series LS 150, AGARD/NATO, April 1988.

12–7. "Explosive Hazard Classification Procedures," DOD, U.S. Army Technical Bulletin TB 700-2, updated 1989 (will become a UN specification).

12–8. "Hazards Assessment Tests for Non-Nuclear Ordnance," *Military Standard MIL-STD-2105B* (Government-issued Specification), 1994.

12–9. "Department of Defense—Ammunition and Explosive Safety Standard." U.S. Department of Defense, U.S. Army TB 700-2, U.S. Navy NAVSEAINST 8020.8, U.S. Air Force TO 11A-1-47, Defense Logistics Agency DLAR 8220.1, 1994 rev.

12–10. G. M. Clark and C. A. Zimmerman, "Phase Stabilized Ammonium Nitrate Selection and Development," *JANNAF Publication 435*, October 1985, pp. 65–75.

12–11. J. Li and Y. Xu, "Some Recent Investigations in Solid Propellant Technology for Gas Generators," *AIAA Paper 90-2335*, July 1990.

12–12. S. Boraas, "Modeling Slag Deposition in the Space Shuttle Solid Motor," *Journal of Spacecraft and Rockets*, Vol. 21, No. 1, January–February 1984, pp. 47–54.

12–13. V. Lindner, "Explosives and Propellants," Kirk-Othmer, *Encyclopedia of Chemical Technology*, Vol. 9, pp. 561–671, 1980.

12–14. J. A. McKinnis and A. R. O'Connell, "MX Launch Gas Generator Development," *Journal of Spacecraft and Rockets*, Vol. 20, No. 3, May–June 1983.

12–15. T. H. Vos and G. W. Goetz, "Inflatable Restraint Systems, Helping to Save Lives on the Road," *Quest*, published by TRW, Inc., Redondo Beach, CA, Vol. 12, No. 2, Winter 1989–1990, pp. 2–27.

12–16. A. Peretz, "Investigation of Pyrotechnic MTV Compositions for Rocket Motor Igniters," *Journal of Spacecraft and Rockets*, Vol. 21, No. 2, March–April 1984, pp. 222–224.

12–17. G. Frut, "Mistral Missile Propulsion System," *AIAA Paper 89-2428*, July 1989 ($B–KNO_3$ ignition).

12–18. A. Davenas, *Solid Rocket Propulsion Technology*, Pergamon Press, 1993 (originally published in French, 1988).

12–19. J. L. Laird and R. J. Baker, "A Novel Smokeless Non-flaking Solid Propellant Inhibitor," *Journal of Propulsion and Power*, Vol. 2, No. 4, July–August 1986, pp. 378–379.

12–20. M. Q. Brewster, "Radiation–Stagnation Flow Model of Aluminized Solid Rocket Motor Insulation Heat Transfer," *Journal of Thermophysics*, Vol. 3, No. 2, April 1989, pp. 132–139.

12–21. A. Truchot, "Design of Solid Rocket Motor Internal Insulation," Chapter 10 in *Design Methods in Solid Rocket Motors*, Lecture Series LS 150, AGARD/NATO, April 1988.

12–22. M. Probster and R. H. Schmucker, "Ballistic Anomalies in Solid Propellant Motors Due to Migration Effects," *Acta Astronautica*, Vol. 13, No. 10, 1986, pp. 599–605.

12–23. L. Chow and P. S. Shadlesky, "External Insulation for Tactical Missile Motor Propulsion Systems," *AIAA Paper 89-2425*, July 1989.

12–24. W. W. Sobol, "Low Cost Manufacture of Tactical Rocket Motors," *Proceedings of 1984 JANNAF Propulsion Meeting*, Vol. II, Chemical Propulsion Information Agency, Johns Hopkins University, Columbia, MD, 1984, pp. 219–226.

CHAPTER 13

COMBUSTION OF SOLID PROPELLANTS

This is the third of four chapters on solid propellant motors. We discuss the combustion of solid propellants, the physical and chemical processes of burning, the ignition or startup process, the extinction of burning, and combustion instability.

The combustion process in rocket propulsion systems is very efficient, when compared to other power plants, because the combustion temperatures are very high; this accelerates the rate of chemical reaction, helping to achieve nearly complete combustion. As was mentioned in Chapter 2, the energy released in the combustion is between 95 and 99.5% of the possible maximum. This is difficult to improve. There has been much interesting research on rocket combustion and we have now a better understanding of the phenomena and of the behavior of burning propellants. This combustion area is still the domain of specialists. The rocket motor designers have been concerned not so much with the burning process as with controlling the combustion (start, stop, heat effects) and with preventing the occurrence of combustion instability.

13.1. PHYSICAL AND CHEMICAL PROCESSES

The combustion in a solid propellant motor involves exceedingly complex reactions taking place in the solid, liquid, and gas phases of a heterogeneous mixture. Not only are the physical and chemical processes occurring during solid propellant combustion not fully understood, but the available analytical combustion models remain oversimplified and unreliable. Experimental observations of burning propellants show complicated three-dimensional microstructures, a three-dimensional flame structure, intermediate products in the liquid and gaseous phase, spatially and temporally variant processes, aluminum agglomeration, nonlinear response behavior, formation of carbon particles, and other complexities yet to be adequately reflected in mathematical models.

Some insight into this combustion process can be gained by understanding the behavior of the major ingredients, such a ammonium perchlorate, which is fairly well explored. This oxidizer is capable of self-deflagration with a low-pressure combustion limit at approximately 2 MPa, the existence of at least four distinct "froth" zones of combustion between 2 and 70 MPa, the existence of a liquid froth on the surface of the crystal during deflagration between 2 and 6 MPa, and a change in the energy-transfer mechanism (particularly at about 14 MPa). Its influence on combustion is critically dependent on oxidizer purity. The surface regression rate ranges from 3 mm/sec at 299 K and 2 MPa to 10 mm/sec at 423 K and 1.4 MPa.

The various polymeric binders used in composite propellants are less well characterized and their combustion properties vary, depending on the binder type, heating rate, and combustion chamber pressure.

The addition of powdered aluminum (2 to 40 μm) is known to favorably influence specific impulse and combustion stability. Photography of the burning aluminum particles shows that the particles usually collect into relatively large accumulaties (100 or more particles) during the combustion process. The combustion behavior of this ingredient depends on many variables, including particle size and shape, surface oxides, the binder, and the combustion wave environment. Ref. 13–1 describes solid propellant combustion.

Visual observations and measurements of flames in simple experiments, such as strand burner tests, give an insight into the combustion process. For double-base propellants the combustion flame structure appears to be homogeneous and one-dimensional along the burning direction, as shown in Fig. 13–1. When heat from the combustion melts, decomposes, and vaporizes the solid propellant at the burning surface, the resulting gases seem to be already premixed. One can see a brilliantly radiating bright flame zone where most of the chemical reaction is believed to occur and a dark zone between the bright flame and the burning surface. The brightly radiating hot reaction zone seems to be detached from the combustion surface. The combustion that occurs inside the dark zone does not emit strong radiations in the visible spectrum, but does emit in the infrared spectral region. The dark zone thickness decreases with increasing chamber pressure, and higher heat transfer to the burning surface causes the burning rate to increase. Experiments on strand burners in an inert nitrogen atmosphere, reported in Chapter 1 of Ref. 13–1, show this dramatically: for pressures of 10, 20, and 30 atm the dark zone thickness is 12, 3.3, and 1.4 mm, respectively, and the corresponding burning rates are 2.2, 3.1, and 4.0 mm/sec. The overall length of the visible flame becomes shorter as the chamber pressure increases and the heat release per unit volume near the surface also increases. In the bright, thin fizz zone or combustion zone directly over the burning surface of the DB propellant, some burning and heat release occurs. Beneath

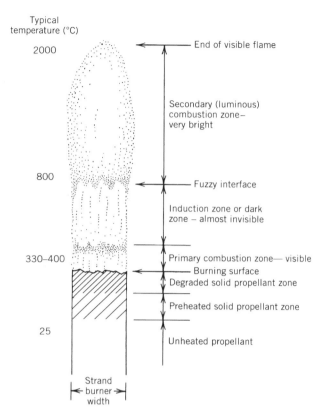

FIGURE 13–1. Schematic diagram of the combustion flame structure of a double-base propellant as seen with a strand burner in an inert atmosphere. (Adapted from Chapter 1 of Ref. 13–1 with permission of the American Institute of Aeronautics and Astronautics, AIAA.)

is a zone of liquefied bubbling propellant which is thought to be very thin (less than 1 μm) and which has been called the foam or degradation zone. Here the temperature becomes high enough for the propellant molecules to vaporize and break up or degrade into smaller molecules, such as NO_2, aldehydes, or NO, which leave the foaming surface. Underneath is the solid propellant, but the layer next to the surface has been heated by conduction within the solid propellant material.

Burn rate catalysts seem to affect the primary combustion zone rather than the processes in the condensed phase. They catalyze the reaction at or near the surface, increase or decrease the heat input into the surface, and change the amount of propellant that is burned.

A typical flame for an AP/Al/HTPB[*] propellant looks very different, as seen in Fig. 13–2. Here the luminous flame seems to be attached to the burning

---

[*]Acronyms are explained in Tables 12–6 and 12–7.

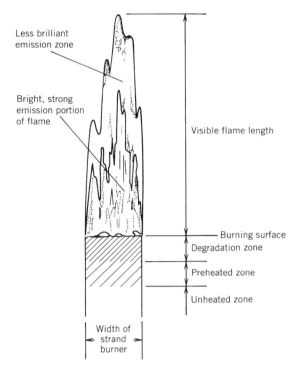

Less brilliant
emission zone

Bright, strong
emission portion
of flame

Visible flame length

Burning surface
Degradation zone

Preheated zone

Unheated zone

Width of
strand
burner

FIGURE 13–2. Diagram of the flickering, irregular combustion flame of a composite propellant (69% AP, 19% Al, plus binder and additives) in a strand burner with a neutral atmosphere. (Adapted from Chapter 1 of Ref. 13–1 with permission of AIAA.)

surface, even at low pressures. There is no dark zone. The oxidizer-rich decomposed gases from the AP diffuse into the fuel-rich decomposed gases from the fuel ingredients, and vice versa. Some solid particles (aluminum, AP crystals, small pieces of binder, or combinations of these) break loose from the surface and the particles continue to react and degrade while in the gas flow. The burning gas contains liquid particles of hot aluminum oxides, which radiate intensively. The propellant material and the burning surface are not homogeneous. The flame structure is unsteady (flicker), three dimensional, not truly axisymmetrical, and complex. The flame structure and the burning rates of composite-modified cast double-base (CMDB) propellant with AP and Al seem to approach those of composite propellant, particularly when the AP content is high. Again there is no dark zone and the flame structure is unsteady and not axisymmetrical. It also has a complex three-dimensional flame structure.

According to Ref. 13–1, the flame structure for double-base propellant with a nitramine addition shows a thin dark zone and a slightly luminous degradation zone on the burning surface. The dark zone decreases in length with increasd pressure. The decomposed gases of RDX or HMX are essentially neutral (not oxidizing) when decomposed as pure ingredients. In this CMDB/RDX propellant the degradation products of RDX solid crystals interdiffuse with the gas from the DB matrix just above the burning surface, before the RDX particles can produce monopropellant flamelets. Thus an essentially homogeneous premixed gas flame is formed, even though the solid propellant itself is heterogeneous. The flame structure appears to be one-dimensional. The burning rate of this propellant decreases when the RDX percentage is increased and seems to be almost unaffected by changes in RDX particle size. Much work has been done to characterize the burning behavior of different propellants. See Chapters 2, 3, and 4 by Kishore and Gayathri, Boggs, and Fifer, respectively, in Ref. 13–1, and Refs. 13–2 to 13–8.

The burning rate of all propellants is influenced by pressure (see Section 11.1 and Eq. 11–3), the initial ambient solid propellant temperature, the burn rate catalyst, the aluminum particle sizes and their size distribution, and to a lesser extent by other ingredients and manufacturing process variables. Erosive burning is basically an accelerated combustion phenomenon stimulated by increased heat transfer and erosion by local high velocities; this was discussed briefly in Chapter 11. Analysis of combustion is treated later in this chapter.

13.2. IGNITION PROCESS

This section is concerned with the mechanism or the process for initiating the combustion of a solid propellant grain. Specific propellants that have been successfully used for igniters have been mentioned in Section 12.5. The hardware, types, design, and integration of igniters into the motor are described in Section 14.4. Chapters 2, 5, and 6 of Ref. 13–1 review the state of the art of ignition, data from experiments, and analytical models, which have been found to be mostly unreliable.

Solid propellant ignition consists of a series of complex rapid events, which start on receipt of a signal (usually electric) and include heat generation, transfer of the heat from the igniter to the motor grain surface, spreading the flame over the entire burning surface area, filling the chamber free volume (cavity) with gas, and elevating the chamber pressure without serious abnormalities such as overpressures, combustion oscillations, damaging shock waves, hangfires (delayed ignition), extinguishment, and chuffing. The *igniter* in a solid rocket motor generates the heat and gas required for motor ignition.

Motor ignition must usually be complete in a fraction of a second for all but the very large motors (see Ref. 13–9). The motor pressure rises to an equilibrium state in a very short time, as shown in Fig. 13–3. Conventionally, the ignition process is divided into three phases for analytical purposes:

Phase I, *Ignition time lag*: the period from the moment the igniter receives a signal until the first bit of grain surface burns.

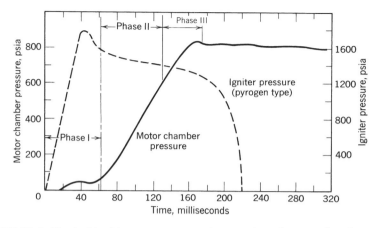

FIGURE 13–3. Typical ignition pressure transient portion of motor chamber pressure–time trace with igniter pressure trace and ignition process phases shown. Electric signal is received a few milliseconds before time zero.

Phase II, *Flame-spreading interval*: the time from first ignition of the grain surface until the complete grain burning area has been ignited.

Phase III, *Chamber-filling interval*: the time for completing the chamber-filling process and for reaching equilibrium chamber pressure and flow.

The ignition will be successful once enough grain surface is ignited and burning, so that the motor will continue to raise its own pressure to the operating chamber pressure. The critical process seems to be a gas-phase reaction above the burning surface, when propellant vapors or decomposition products interact with each other and with the igniter gas products. If the igniter is not powerful enough, some grain surfaces may burn for a short time, but the flame will be extinguished.

Satisfactory attainment of equilibrium chamber pressure with full gas flow is dependent on (1) characteristics of the igniter and the gas temperature, composition and flow issuing from the igniter, (2) motor propellant composition and grain surface ignitability, (3) heat transfer characteristics by radiation and convection between the igniter gas and grain surface, (4) grain flame spreading rate, and (5) the dynamics of filling the motor free volume with hot gas (see Ref. 13–10). The quantity and type of caloric energy needed to ignite a particular motor grain in the prevailing environment has a direct bearing on most of the igniters' design parameters—particularly those affecting the required heat output. The ignitability of a propellant at a given pressure and temperature is normally shown as a plot of ignition time versus heat flux received by the propellant surface, as shown in Fig. 13–4; these data are obtained from laboratory tests. Ignitability of a propellant is affected by many factors, including (1) the propellant formulation, (2) the initial temperature of the propellant grain

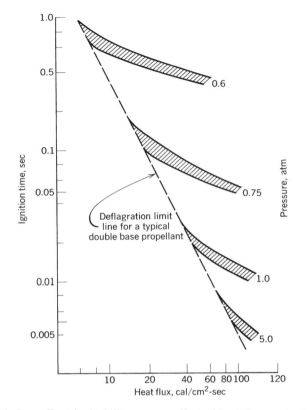

FIGURE 13–4. Propellant ignitability curves: effect of heat flux on ignition time for a specific motor.

surface, (3) the surrounding pressure, (4) the mode of heat transfer, (5) grain surface roughness, (6) age of the propellant, (7) the composition and hot solid particle content of the igniter gases, (8) the igniter propellant and its initial temperature, (9) the velocity of the hot igniter gases relative to the grain surface, and (10) the cavity volume and configuration. Figure 13–4 and data in Chapter 14 show that the ignition time becomes shorter with increases in both heat flux and chamber pressure. If a short ignition delay is required, then a more powerful igniter will be needed. The radiation effects can be significant in the ignition transient as described in Ref. 13–11. In Section 14.3 we describe an analysis and design for igniters.

13.3. EXTINCTION OR THRUST TERMINATION

Sometimes it is necessary to stop or extinguish the burning of a solid motor before all the propellant has been consumed:

1. When a flight vehicle has reached the desired flight velocity (for a ballistic missile to attain a predetermined velocity or for a satellite to achieve an accurate orbit), or a precise total impulse cutoff is needed.

2. As a safety measure, when it appears that a flight test vehicle will unexpectedly fly out of the safe boundaries of a flight test range facility.

3. To avoid collisions of stages during a stage separation maneuver (requiring a thrust reversal) for multistage flight vehicles.

4. During research and development testing, when one wants to examine a partially burned motor.

The common mechanisms for achieving extinction are listed below and described in Chapters 2, 5, and 6 of Ref. 13–1.

1. Very *rapid depressurization*, usually by a sudden, large increase of the nozzle throat area or by fast opening of additional gas escape areas or ports. The most common technique neutralizes the thrust or reverses the net thrust direction by suddenly opening exhaust ports in the forward end of the motor case. Such a thrust reversal using ports located on the forward bulkhead of the case is achieved in the upper stages of Minuteman and Poseidon missiles. This is done by highly predictable and reproducible explosive devices which suddenly open additional gas escape areas (thus causing pressure reduction) and neutralize the thrust by exhausting gases in a direction opposite to that of the motor nozzle. To balance side forces, the thrust termination blow-out devices and their ducts are always designed in symmetrically opposed sets (two or more). In Fig. 1–5 there are four symmetrically placed openings that are blown into the forward dome of the case by circular explosive cords. Two of the sheathed circular cord assemblies are sketched on the outside of the forward dome wall. The ducts that lead the hot gas from these openings to the outside of the vehicle are not shown in this figure. The forward flow of gas occurs only for a very brief period of time, during which the thrust is actually reversed. The rapid depressurization causes a sudden stopping of the combustion at the propellant burning surface. With proper design the explosive cords do not cause a detonation or explosion of the remaining unburned propellant.

2. During some motor development projects it can be helpful to see a partially consumed grain. The motor operation is stopped when the flames are quenched by injecting an inhibiting liquid such as water. Reference 13–12 shows that adding a detergent to the water allows a better contact with the burning surface and reduces the amount of water needed for quenching.

3. *Lowering the combustion pressure* below the pressure deflagration limit. Compared to item 1, this depressurization occurs quite slowly. Many solid propellants have a low-pressure combustion limit of 0.05 to 0.15

MPa. This means that some propellants will not extinguish when vented during a static sea-level test at 1 atm (0.1 MPa) but will stop burning if vented at high altitude.

A sudden depressurization is effective because the primary combustion zone at the propellant surface has a time lag compared to the gaseous combustion zone which, at the lower pressure, quickly adjusts to a lower reaction rate and moves farther away from the burning surface. The gases created by the vaporization and pyrolysis of the hot solid propellant cannot all be consumed in a gas reaction close to the surface, and some will not burn completely. As a result, the heat transfer to the propellant surface will be quickly reduced by several orders of magnitude, and the reaction at the propellant surface will diminish and stop. Experimental results (Chapter 12 of Ref. 13–1) show that a higher initial combustion pressure requires a faster depressurization rate (dp/dt) to achieve extinction.

13.4. COMBUSTION INSTABILITY

There seem to be two types of combustion instability: a set of acoustic resonances or pressure oscillations, which can occur with any rocket motor, and a vortex shedding phenomenon, which occurs only with particular types of grains.

Acoustic Instabilities

When a solid propellant rocket motor experiences unstable combustion, the pressure in the interior gaseous cavities (made up by the volume of the port or perforations, fins, slots, conical or radial groves) oscillates by at least 5% and often by more than 30% of the chamber pressure. When instability occurs, the heat transfer to the burning surfaces, the nozzle, and the insulated case walls is greatly increased; the burning rate, chamber pressure, and thrust usually increase; but the burning duration is thereby decreased. The change in the thrust–time profile causes significant changes in the flight path, and at times this can lead to failure of the mission. If prolonged and if the vibration energy level is high, the instability can cause damage to the hardware, such as overheating the case and causing a nozzle or case failure. Instability is a condition that should be avoided and must be carefully investigated and remedied if it occurs during a motor development program. Final designs of motors must be free of such instability.

There are fundamental differences with liquid propellant combustion behavior. In liquid propellants there is a fixed chamber geometry with a rigid wall; liquids in feed systems and in injectors that are not part of the oscillating gas in the combustion chamber can interact strongly with the pressure fluctuations. In solid propellant motors the geometry of the oscillating cavity increases in size as burning proceeds and there are stronger damping factors, such as solid particles and energy-absorbing viscoelastic materials. In general, combustion instability problems do not occur frequently or in every motor development, and, when they do occur, it is rarely the cause for a drastic sudden motor failure or disintegration. Nevertheless, drastic failures have occurred.

Undesirable oscillations in the combustion cavity propellant rocket motors is a continuing problem in the design, development, production, and even long-term (10 yr) retention of solid rocket missiles. While acoustically "softer" than a liquid rocket combustion chamber, the combustion cavity of a solid propellant rocket is still a low-loss acoustical cavity containing a very large acoustical energy source, the combustion process itself. A small fraction of the energy released by combustion is more than sufficient to drive pressure vibrations to an unacceptable level.

Combustion instability can occur spontaneously, often at some particular time during the motor burn period, and the phenomenon is usually repeatable in identical motors. Both longitudinal and transverse waves (radial and tangential) can occur. Figure 13–5 shows a pressure–time profile with typical instability. The pressure oscillations increase in magnitude, and the thrust and burning rate also increase. The frequency seems to be a function of the cavity geometry, propellant composition, pressure, and internal flame field. As

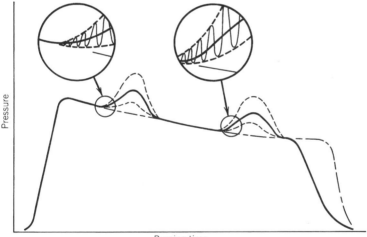

FIGURE 13–5. Simplified diagram showing two periods of combustion instability in the pressure–time history, with enlargements of two sections of this curve. The dashed lines show the upper and lower boundaries of the high-frequency pressure oscillations, and the dot-dash curve is the behavior without instability after a slight change in propellant formulation. The vibration period shows a rise in the mean pressure. With vibration, the effective burning time is reduced and the average thrust is higher. The total impulse is essentially unchanged.

the internal grain cavity is enlarged and local velocities change, the oscillation often abates and disappears. The time and severity of the combustion vibration tend to change with the ambient grain temperature prior to motor operation.

For a simple grain with a cylindrical port area, the resonant transverse mode oscillations (tangential and radial) correspond roughly to those shown in Fig. 9–4 for liquid propellant thrust chambers. The longitudinal or axial modes, usually at a lower frequency, are an acoustic wave traveling parallel to the motor axis between the forward end of the perforation and the convergent nozzle section. Harmonic frequencies of these basic vibration modes can also be excited. The internal cavities can become very complex and can include igniter cases, movable as well as submerged nozzles, fins, cones, slots, star-shaped perforations, or other shapes, as described in the section on grain geometry in Chapter 11; determination of the resonant frequencies of complex cavities is not always easy. Furthermore, the geometry of the internal resonating cavity changes continually as the burning propellant surfaces recede; as the cavity volume becomes larger, the transverse oscillation frequencies are reduced.

The *bulk mode*, also known as the Helmholtz mode, L^* mode, or chuffing mode, is not a wave mode as described above. It occurs at relatively low frequencies (typically below 150 Hz and sometimes below 1 Hz), and the pressure is essentially uniform throughout the volume. The unsteady velocity is close to zero, but the pressure rises and falls. It is the gas motion (in and out of the nozzle) that corresponds to the classical Helmholtz resonator mode, similar to exciting a tone when blowing across the open mouth of a bottle (see Fig. 9–7). It occurs at low values of L^* (see Eq. 8–9), sometimes during the ignition period, and disappears when the motor internal volume becomes larger or the chamber pressure becomes higher. *Chuffing* is the periodic low-frequency discharge of a bushy, unsteady flame of short duration (typically less than 1 sec) followed by periods of no visible flame, during which slow out-gassing and vaporization of the solid propellant accumulates hot gas in the chamber. The motor experiences spurts of combustion and consequent pressure buildup followed by periods of nearly ambient pressure. This dormant period can extend for a fraction of a second to a few seconds (Ref. 13–13 and Chapter 13 by Price in Ref. 13–1).

A useful method of visualizing unstable pressure waves is shown in Figs. 9–5 and 13–6 and Ref. 13–14. It consists of a series of Fourier analyses of the measured pressure vibration spectrum, each taken at a different time in the burning duration and displayed at successive vertical positions on a time scale, providing a map of amplitude versus frequency versus burning time. This figure shows a low-frequency axial mode and two tangential modes, whose frequency is reduced in time by the enlargement of the cavity; it also shows the timing of different vibrations, and their onset and demise.

The initiation or triggering of a particular vibration mode is still not well understood but has to do with energetic combustion at the propellant surface. A sudden change in pressure is known to be a trigger, such as when a piece of

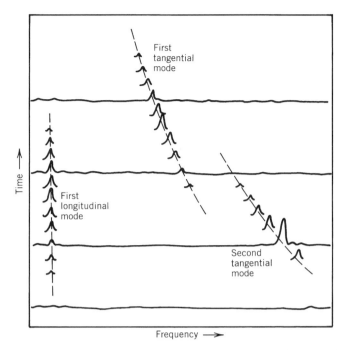

FIGURE 13–6. Example of mode frequency display; also called a "waterfall" diagram of a motor firing. Only four complete time–frequency curves are shown; for easy visualization the other time lines are partly omitted except near the resonating frequencies. The height of the wave is proportional to pressure. As the cavity volume increases, the frequencies of the transverse modes decrease. (Adapted from E. W. Price, Chapter 13 of Ref. 13–1, with permission of AIAA.)

broken-off insulation or unburned propellant flows through the nozzle and temporarily blocks all or a part of the nozzle area (causing a momentary pressure rise).

The shifting balance between amplifying and damping factors changes during the burning operation and this causes the growth and also the abatement of specific modes of vibration. The *response* of a solid propellant describes the change in the gas mass production or energy release at the burning surface when it is stimulated by pressure perturbations. When a momentary high pressure peak occurs on the surface, it increases the instantaneous heat transfer and thus the burning rate, causing the mass flow from that surface to also increase. Velocity perturbations along the burning surface are also believed to cause changes in mass flow. Phenomena that contribute to amplifying the vibrations, or to gains in the acoustic energy (see Ref. 13–1, Chapter 13 by Price), are:

1. The dynamic response of the combustion process to a flow disturbance or the oscillations in the burning rate. This combustion response can be

determined from tests of T-burners as described on pages 533 and 534. The response function depends on the frequency of these perturbations and the propellant formulation. The combustion response may not be in a phase with the disturbance. The effects of boundary layers on velocity perturbations have been investigated in Ref. 13–15.

2. The interactions of flow oscillation with the main flow, similar to the basis for the operation of musical wind instruments or sirens (see Ref. 13–16).

3. The fluid dynamic influence of vortexes.

Phenomena that contribute to a diminishing of vibration or to damping are energy-absorbing processes; they include the following:

1. Viscous damping in the boundary layers at the walls or propellant surfaces.

2. Damping by particles or droplets flowing in an oscillating gas/vapor flow is often substantial. The particles accelerate and decelerate by being "dragged" along by the motion of the gas, a viscous flow process that absorbs energy. The attenuation for each particular vibration frequency is an optimum at a particular size of particles; high damping for low-frequency oscillation (large motors) occurs with relatively large solid particles (8 to 20 μm); for small motors or high-frequency waves the best damping occurs with small particles (2 to 6 μm). The attenuation drops off sharply if the particle size distribution in the combustion gas is not concentrated near the optimum for damping.

3. Energy from longitudinal and mixed transverse/longitudinal waves is lost out through the exhaust nozzle. Energy from purely transverse waves does not seem to be damped by this mechanism.

4. Acoustic energy is absorbed by the viscoelastic solid propellant, insulator, and the motor case; its magnitude is difficult to estimate.

The *propellant characteristics* have a strong effect on the susceptibility to instability. Changes in the binder, particle-size distribution, ratio of oxidizer to fuel, and burn-rate catalysts can all affect stability, often in ways that are not predictable. All solid propellants can experience instability. As a part of characterizing a new or modified propellant (e.g., determining its ballistic, mechanical, aging, and performance characteristics), many companies now also evaluate it for its stability behavior, as described below.

Analytical Models and Simulation of Combustion Stability

Many interesting investigations have been aimed at mathematical models that will simulate the combustion behavior of solid propellants. This was reviewed by T'ien in Chapter 14 of Ref 13–1.

Using complex algorithms and computers it has been possible to successfully simulate the combustion for some limited cases, such as for validating or extrapolating experimental results or making limited predictions of the stability of motor designs. This applies to well-characterized propellants, where empirical constants (such as propellant response or particle-size distribution) have been determined and where the range of operating parameters, internal geometries, or sizes has been narrow. The analytical methods used to date have by themselves not been satisfactory to a motor designer. It is unlikely that a reliable simple analysis will be found for predicting the occurrence, severity, nature, and location of instability for a given propellant and motor design. The physical and chemical phenomena are complex, multidimensional, unsteady, nonlinear, influenced by many variables, and too difficult to emulate mathematically without a good number of simplifying assumptions. However, theoretical analysis gives insight into the physical phenomena, can be a valuable contributor to solving instability problems, and has been used for preliminary design evaluation of grain cavities.

Combustion Stability Assessment, Remedy, and Design

In contrast with liquid rocket technology, an accepted combustion stability rating procedure does not now exist for full-scale solid rockets. Undertaking stability tests on large full-scale flight-hardware rocket motors is expensive, and therefore lower-cost methods, such as subscale motors, T-burners, and other test equipment, have been used to assess motor stability.

The best known and most widely used method of gaining combustion stability-related data is the use of a *T-burner*, an indirect, limited method that does not use a full-scale motor. Figure 13–7 is a sketch of a standard T-burner; it has a 1.5-in. internal diameter double-ended cylindrical burner vented at its midpoint (see Refs. 13–17 to 13–19). Venting can be through a sonic nozzle to the atmosphere or by a pipe connected to a surge tank which maintains a constant level of pressure in the burner cavity. T-burner design and usage usually concentrate on the portion of the frequency spectrum dealing with the transverse oscillations expected in a full-scale motor. The desired acoustical frequency, to be imposed on the propellant charge as it burns, determines the burner length (distance between closed ends).

The nozzle location, midway between the ends of the burner, minimizes attenuation of fundamental longitudinal mode oscillations (in the propellant grain cavity). Theoretically, an acoustic pressure node exists at the center and antinodes occur at the ends of the cavity. Acoustic velocity nodes are out of phase with pressure waves and occur at the ends of the burner. Propellant charges are often in the shape of discs or cups cemented to the end faces of the burner. The gas velocity in the burner cavity is kept intentionally low (Mach 0.2 or less) compared with the velocity in a full-scale motor. This practice minimizes the influence of velocity-coupled energy waves and allows the influence of pressure-coupled waves to be more clearly recognized.

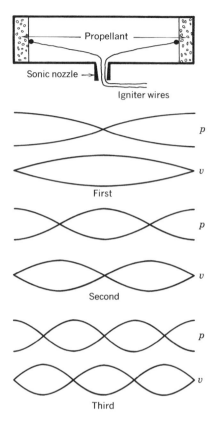

FIGURE 13–7. Standard T-burner and its longitudinal mode standing waves (pressure and velocity).

Use of the T-burner for assessing the stability of a full-scale solid rocket presupposes valid theoretical models of the phenomena occurring in both the T-burner and the actual rocket motor; these theories are still not fully validated. In addition to assessing solid rocket motor combustion stability, the T-burner also is used to evaluate new propellant formulations and the importance of seemingly small changes in ingredients, such as a change in aluminum powder particle size and oxidizer grind method.

Once an instability has been observed or predicted in a given motor, the motor design has to fix the problem. There is no sure method for selecting the right remedy, and none of the cures suggested below may work. The usual alternatives are:

1. Changing the grain geometry to shift the frequencies away from the undesirable values. Sometimes, changing fin locations, port cross-section profile, or number of slots has been successful.

2. Changing the propellant composition. Using aluminum as an additive has been most effective in curing transverse instabilities, provided that the particle-size distribution of the aluminum oxide is favorable to optimum damping at the distributed frequency. Changing size distribution and using other particulates (Zr, Al_2O_3, or carbon particles) has been effective in some cases. Sometimes changes in the binder have worked.

3. Adding some mechanical device for attenuating the unsteady gas motions or changing the natural frequency of cavities. Various inert resonance rods, baffles, or paddles have been added, mostly as a fix to an existing motor with observed instability. They can change the resonance frequencies of the cavities, introduce additional viscous surface losses, but also cause extra inert mass and potential problems with heat transfer or erosion.

Combustion instability has to be addressed during the design process, usually through a combination of some mathematical simulation, understanding similar problems in other motors, studies of possible changes, and supporting experimental work (e.g., T-burners, measuring particle-size distribution). Most solid propellant rocket companies have in-house two- and three-dimensional computer programs to calculate the likely acoustic modes (axial, tangential, radial, and combinations of these) for a given grain/motor, the initial and intermediate cavity geometries, and the combustion gas properties calculated from thermochemical analysis. Data on combustion response (dynamic burn rate behavior) and damping can be obtained from T-burner tests. Data on particle sizes can be estimated from prior experience or plume measurements (Ref. 13–20). Estimates of nozzle losses, friction, or other damping need to be included. Depending on the balance between gain and damping, it may be possible to arrive at conclusions on the grain's propensity to instability for each specific instability mode that is analyzed. If unfavorable, either the grain geometry or the propellant usually have to be modified. If favorable, full-scale motors have to be built and tested to validate the predicted stable burning characteristics. There is always a trade-off between the amount of work spent on extensive analysis, subscale experiments and computer programs (which will not always guarantee a stable motor), and taking a chance that a retrofit will be needed after full-scale motors have been tested. If the instability is not discovered until after the motor is in production, it is often difficult, time consuming, and expensive to fix the problem.

Vortex-Shedding Instability

This instability is associated with burning on the inner surfaces of slots in the grain. Large segmented rocket motors have slots between segments, and some grain configurations have slots that intersect the centerline of the grain. Figure 13–8 shows that hot gases from the burning slot surfaces enter the main flow in

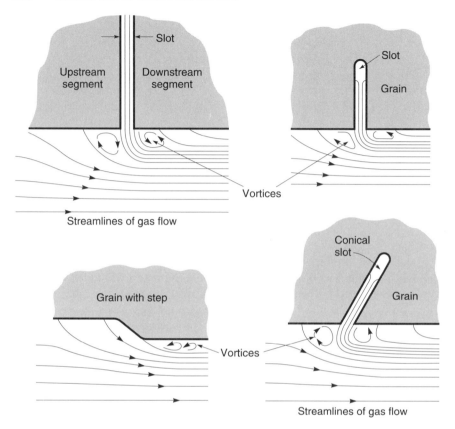

FIGURE 13–8. Simple sketches of four partial grain sections each with a slot or a step. Heavy lines identify the burning surfaces. The flow patterns cause the formation of vortices. The shedding of these vortices can induce flow oscillations and pressure instabilities.

the perforation or central cavity of the grain. The hot gas from the slot is turned into a direction toward the nozzle. The flow from the side stream restricts the flow emanating from the upstream side of the perforation and, in effect, reduces the port area. This restriction causes the upstream port pressure to rise; sometimes there is a substantial pressure rise. The interaction of the two subsonic gas flows causes turbulence. Vortices form and are periodically shed or allowed to flow downstream, thereby causing an unstable flow pattern. The vortex shedding patterns can interact with the acoustic instabilities. Reference 13–21 gives a description and Ref. 13–22 a method for analyzing these vortex-shedding phenomena. The remedy usually is to apply inhibitors to some burning surfaces or to change the grain geometry; for example, by increasing the width of the slot, the local velocities are reduced and the vortices become less pronounced.

PROBLEMS

1. (*a*) Calculate the length of a T-burner to give a first natural oscillation of 2000 Hz using a propellant that has a combustion temperature of 2410 K, a specific heat ratio of 1.25, a molecular weight of 25 kg/kg-mol, and a burning rate of 10.0 mm/ sec at a pressure of 68 atm. The T-burner is connected to a large surge tank and prepressurized with nitrogen gas to 68 atm. The propellant disks are 20 mm thick. Make a sketch to indicate the T-burner dimensions, including the disks.

   (*b*) If the target frequencies are reached when the propellant is 50% burned, what will be the frequency at propellant burnout?

   Answers: (*a*) Length before applying propellant = 0.270 m; (*b*) frequency at burnout = 1854 Hz.

2. An igniter is needed for a rocket motor similar to one shown in Fig. 11–1. Igniters have been designed by various oversimplified design rules such as Fig. 13–3. The motor has an internal grain cavity volume of 0.055 m^3 and an initial burning surface of 0.72 m^2. The proposed igniter propellant has these characteristics: combustion temperature 2500 K and an energy release of about 40 J/kg-sec. Calculate the minimum required igniter propellant mass (*a*) if the cavity has to be pressurized to about 2 atm (ignore heat losses); (*b*) if only 6% of the igniter gas energy is absorbed at the burning surface, and it requires about 20 cal/cm^2-sec to ignite in about 0.13 sec.

3. Using the data from Fig. 13–4, plot the total heat flux absorbed per unit area versus pressure to achieve ignition with the energy needed to ignite being just above the deflagration limit. Then, for 0.75 atm, plot the total energy needed versus ignition time. Give a verbal interpretation of the results and trend for each of the two curves.

REFERENCES

13–1. N. Kubota, "Survey of Rocket Propellants and Their Combustion Characteristics," Chapter 1; K. Kishore and V. Gayathri, "Chemistry of Ignition and Combustion of Ammonium-Perchlorate-Based Propellants," Chapter 2; T. L. Boggs, "The Thermal Behavior of Cyclotrimethylene Trinitrate (RDX) and Cyclotetramethylene Tetranitrate (HMX)," Chapter 3; R. A. Fifer, "Chemistry of Nitrate Ester and Nitramine Propellants," Chapter 4; C. E. Hermance, "Solid Propellant Ignition Theories and Experiments," Chapter 5; M. Kumar and K. K. Kuo, "Flame Spreading and Overall Ignition Transient," Chapter 6; E. W. Price, "Experimental Observations of Combustion Instability," Chapter 13; James S. T'ien, "Theoretical Analysis of Combustion Instability," Chapter 14; all in K. K. Kuo and M. Summerfield (Eds), *Fundamentals of Solid-Propellant Combustion*, Volume 90 of Progress in Astronautics and Aeronautics, American Institute of Aeronautics and Astronautics, New York, 1984.

13–2. V. Duterque and G. Lengelle, "Combustion Mechanism of Nitramine-Based Propellant with Additives," *AIAA Paper 88-3253*, July 1988.

13–3. C. Youfang, "Combustion Mechanism of Double-Base Propellants with Lead Burning Rate Catalyst," *Propellants, Explosives, Pyrotechnics*, Vol. 12, 1987, pp. 209–214.

13–4. N. Kubota et al., "Combustion Wave Structures of Ammonium Perchlorate Composite Propellants," *Journal of Propulsion and Power*, Vol. 2, No. 4, July–August 1986, pp. 296–300.

13–5. T. Boggs, D. E. Zurn, H. F. Cordes, and J. Covino, "Combustion of Ammonium Perchlorate and Various Inorganic Additives," *Journal of Propulsion and Power*, Vol. 4, No. 1, January–February 1988, pp. 27–39.

13–6. T. Kuwahara and N. Kubota, "Combustion of RDX/AP Composite Propellants at Low Pressure," *Journal of Spacecraft and Rockets*, Vol. 21, No. 5, September–October 1984, pp. 502–507.

13–7. P. A. O. G. Korting, F. W. M. Zee, and J. J. Meulenbrugge, "Combustion Characteristics of Low Flame Temperature, Chlorine-Free Composite Propellants," *Journal of Propulsion and Power*, Vol. 6, No. 3, May–June 1990, pp. 250–255.

13–8. N. Kubota and S. Sakamoto, "Combustion Mechanism of HMX," *Propellants, Explosives, Pyrotechnics*, Vol. 14, 1989, pp. 6–11.

13–9. L. H. Caveny, K. K. Kuo, and B. J. Shackleford, "Thrust and Ignition Transients of the Space Shuttle Solid Rocket Booster Motor," *Journal of Spacecraft and Rockets*, Vol. 17, No. 6, November–December 1980, pp. 489–494.

13–10. "Solid Rocket Motor Igniters," *NASA SP-8051*, March 1971 (N71-30346).

13–11. I. H. Cho and S. W. Baek, "Numerical Simulation of Axisymmetric Solid Rocket Motor Ignition with Radiation Effect," *Journal of Propulsion and Power*, Vol. 16, No. 4, July–August 2000, pp. 725–728.

13–12. J.Yin and B. Zhang, "Experimental Study of Liquid Quenching of Solid Rocket Motors," *AIAA Paper 90-2091*.

13–13. B. N. Raghunandam and P. Bhaskariah, "Some New Results of Chuffing in Composite Solid Propellant Rockets," *Journal of Spacecraft and Rockets*, Vol. 22, No. 2, March–April 1985, pp. 218–220.

13–14. P. M. J. Hughes and E. Cerny, "Measurement and Analysis of High Frequency Pressure Oscillations in Solid Rocket Motors," *Journal of Spacecraft and Rockets*, Vol. 21, No. 3, May–June 1984, pp. 261–265.

13–15. R. A. Beddini and T. A. Roberts, "Response of Solid Propellant Combustion to the Presence of a Turbulent Acoustic Boundary Layer," *AIAA Paper 88-2942*, 1988.

13–16. F. Vuillot and G. Avalon, "Acoustic–Mean Flow Interaction in Solid Rocket Motors, Using Navier–Stokes Equations," *AIAA Paper 88-2940*, 1988.

13–17. W. C. Andrepont and R. J. Schoner, "The T-Burner Method for Determining the Combustion Response of Solid Propellants," *AIAA Paper 72-1053*, 1972.

13–18. E. W. Price, H. B. Mathes, O. H. Madden, and B. G. Brown, "Pulsed T-Burner Testing of Combustion Dynamics of Aluminized Solid Propellants," *Aeronautics and Astronautics*, Vol. 10, No. 4, April 1971, pp. 65–69.

13–19. R. L. Coates, "Application of the T-Burner to Ballistic Evaluation of New Propellants," *Journal of Spacecraft and Rockets*, Vol. 3, No. 12, December 1966, pp. 1793–1796.

13–20. E. D. Youngborg, J. E. Pruitt, M. J. Smith, and D. W. Netzer, "Light-Diffraction Particle Size Measurements in Small Solid Propellant Rockets," *Journal of Propulsion and Power*, Vol. 6, No. 3, May–June 1990, pp. 243–249.

13–21. F. Vuillot, "Vortex Shedding Phenomena in Solid Rocket Motors," *Journal of Propulsion and Power*, Vol. 11, No. 4, 1995.

13–22. A. Kourta, "Computation of Vortex Shedding in Solid Rocket Motors using a Time-Dependent Turbulence Model," *Journal of Propulsion and Power*, Vol. 15, No. 3, May–June 1999.

CHAPTER 14

SOLID ROCKET COMPONENTS AND MOTOR DESIGN

This is the last of four chapters on solid propellant rockets. We describe the key inert components of solid propellant rocket motors, namely the motor case, nozzle, and igniter case, and then discuss the design of motors. Although the thrust vector control mechanism is also a component of many rocket motors, it is described separately in Chapter 16. The key to the success of many of these components is new materials which have been developed in recent years.

14.1. MOTOR CASE

The case not only contains the propellant grain, but also serves as a highly loaded pressure vessel. Case design and fabrication technology has progressed to where efficient and reliable motor cases can be produced consistently for any solid rocket application. Most problems arise when established technology is used improperly or from improper design analysis, understating the requirements, or improper material and process control, including the omission of nondestructive tests at critical points in the fabrication process. Case design is usually governed by a combination of motor and vehicle requirements. Besides constituting the structural body of the rocket motor with its nozzle, propellant grain, and so on, the case frequently serves also as the primary structure of the missile or launch vehicle. Thus the optimization of a case design frequently entails trade-offs between case design parameters and vehicle design parameters. Often, case design is influenced by assembly and fabrication requirements.

540

Table 14–1 lists many of the types of loads and their sources; they must be considered at the beginning of a case design. Only some of them apply to any one rocket motor application. In addition, the environmental conditions peculiar to a specific motor and its usage must be carefully considered. Typically, these conditions include the following: (1) temperature (internal heating, aerodynamic heating, temperature cycling during storage, or thermal stresses and strains); (2) corrosion (moisture/chemical, galvanic, stress corrosion, or hydrogen embrittlement); (3) space conditions: vacuum or radiation.

Three classes of materials have been used: high-strength metals (such as steel, aluminum, or titanium alloys), wound-filament reinforced plastics, and a combination of these in which a metal case has externally wound filaments for extra strength. Table 14–2 gives a comparison of several typical materials. For filament-reinforced materials it gives the data not only for the composite material, but also for several strong filaments and a typical binder. The strength-to-density ratio is higher for composite materials, which means that they have less inert mass. Even though there are some important disadvantages, the filament-wound cases with a plastic binder are usually superior on a vehicle performance basis. Metal cases combined with an external filament-wound reinforcement and spiral-wound metal ribbons glued together with plastic have also been successful.

The shape of the case is usually determined from the grain configuration or from geometric vehicle constraints on length or diameter. The case configurations range from long and thin cylinders (L/D of 10) to spherical or near-

TABLE 14–1. Rocket Motor Case Loads

Origin of Load	Type of Load/Stress
Internal pressure	Tension biaxial, vibration
Axial thrust	Axial, vibration
Motor nozzle	Axial, bending, shear
Thrust vector control actuators	Axial, bending, shear
Thrust termination equipment	Biaxial, bending
Aerodynamic control surfaces or wings mounted to case	Tension, compression, bending, shear, torsion
Staging	Bending, shear
Flight maneuvering	Axial, bending, shear, torsion
Vehicle mass and wind forces on launch pad	Axial, bending, shear
Dynamic loads from vehicle oscillations	Axial, bending, shear
Start pressure surge	Biaxial
Ground handling, including lifting	Tension, compression, bending, shear, torsion
Ground transport	Tension, compression, shear, vibration
Earthquakes (large motors)	Axial, bending, shear

TABLE 14–2. Physical Properties of Selected Solid Propellant Motor Case Materials at 20°C

Material	Tensile Strength, N/mm^2 (10^3 psi)	Modulus of Elasticity, N/mm^2 (10^6 psi)	Density, g/cm^3 (lbm/in.^{3})	Strength to Density Ratio (1000)
Filaments				
E-glass	1930–3100 (280–450)	72,000 (10.4)	2.5 (0.090)	1040
Aramid (Kevlar 49)	3050–3760 (370–540)	124,000 (18.0)	1.44 (0.052)	2300
Carbon fiber or graphite fibers	3500–6900 (500–1000)	230,000–300,000 (33–43)	1.53–1.80 (0.055–0.065)	2800
Binder (by itself)				
Epoxy	83 (12)	2800 (0.4)	1.19 (0.043)	70
Filament-Reinforced Composite Material				
E Glass	1030 (150–170)	35,000 (4.6–5.0)	1.94 (0.070)	500
Kevlar 49	1310 (190)	58,000 (8.4)	1.38 (0.050)	950
Graphite IM	2300 (250–340)	102,000 (14.8)	1.55 (0.056)	1400
Metals				
Titanium alloy	1240 (180)	110,000 (16)	4.60 (0.166)	270
Alloy steel (heat treated)	1400–2000 (200–290)	207,000 (30)	7.84 (0.289)	205
Aluminum alloy 2024 (heat treated)	455 (66)	72,000 (10.4)	2.79 (0.101)	165

Source: Data adapted in part from Chapter 4A by Evans and Chapter 7 by Scippa of Ref. 11–1.

spherical geometries (see Figs. 1–5, 11–1 to 11–4, and 11–17). The spherical shape gives the lowest case mass per unit of enclosed volume. The case is often a key structural element of the vehicle and it sometimes has to provide for mounting of other components, such as fins, skirts, electric conduits, or thrust vector control actuators. The propellant mass fractions of the motor are usually strongly influenced by the case mass and typically range from 0.70 to 0.94. The higher values apply to upper stage motors. For small-diameter motors the mass fraction is lower, because of practical wall thicknesses and the fact that the wall surface area (which varies roughly as the square of the diameter) to chamber volume (which varies roughly as the cube of diameter) is less favorable in small sizes. The minimum thickness is higher than would be determined from simple stress analysis; for a a fiber composite case it is two layers of filament strands and the minimum metal thickness is dictated by manufacturing and handling considerations.

Simple membrane theory can be used to predict the approximate stress in solid propellant rocket chamber cases; this assumes no bending in the case walls and that all the loads are taken in tension. For a simple cylinder of radius R and thickness d, with a chamber pressure p, the longitudinal stress σ_l is one-half of the tangential or hoop stress σ_θ:

$$\sigma_\theta = 2\sigma_l = pR/d \tag{14-1}$$

For a cylindrical case with hemispherical ends, the cylinder wall has to be twice as thick as the walls of the end closures.

The combined stress should not exceed the working stress of the wall material. As the rocket engine begins to operate, the internal pressure p causes a growth of the chamber in the longitudinal as well as in the circumferential direction, and these deformations must be considered in designing the support of the motor or propellant grain. Let E be Young's modulus of elasticity, v be Poisson's ratio (0.3 for steel), and d be the wall thickness; then the growth in length L and in diameter D due to pressure can be expressed as

$$\Delta L = \frac{pLD}{4Ed}(1 - 2v) = \frac{\sigma_l L}{E}(1 - 2v) \tag{14-2}$$

$$\Delta D = \frac{pD^2}{4Ed}\left(1 - \frac{v}{2}\right) = \frac{\sigma_\theta D}{2E}\left(1 - \frac{v}{2}\right) \tag{14-3}$$

Details can be found in a text on thin shells or membranes. For a hemispherical chamber end, the stress in each of two directions at right angles to each other is equal to the longitudinal stress of a cylinder of identical radius. For ellipsoidal end-chamber closures, the local stress varies with the position along the surface, and the maximum stress is larger than that of a hemisphere. The radial displacement of a cylinder end is not the same as that of a hemispherical or ellipsoidal closure if computed by thin-shell theory. Thus a discontinuity exists which causes some shearing and bending stresses. Similarly, a boss for the attachment of an igniter, a pressure gauge, or a nozzle can make it necessary for bending and shear stresses to be superimposed on the simple tension stresses of the case. In these locations it is necessary to reinforce or thicken the chamber wall locally.

Finite element computerized stress analysis programs exist and are used in motor design companies today to determine the case design configuration with reasonable stress values. This analysis must be done simultaneously with the stress analysis on the grain (since it imposes loads on the case), and with a finite element thermal analysis to determine thermal stresses and deformations, since these analyses are interdependent on each other.

The fast heating of the inner wall surface produces a temperature gradient and therefore thermal stresses across the wall. The theory of transient heat transfer has been treated by a number of authors, and, by means of a relaxation method, a reasonable approximation of the temperature–time history at any location may be obtained. The inner wall of the case, which is exposed to hot gas, is usually protected by thermal insulation, as described in Section 12.6. Therefore the heat transfer to the case is very low. In fact, for a single operation (not two thrust periods) it is the designer's aim to keep the case temperatures near ambient or at the most 100°C above ambient.

The case design has to provide means for attaching a nozzle (rarely more than one nozzle), for attaching it to the vehicle, igniters, and provisions for loading the grain. Sometimes there are also attached aerodynamic surfaces (fins), sensing instruments, a raceway (external conduit for electrical wires), handling hooks, and thrust vector control actuators with their power supply. For upper stages of ballistic missiles the case can also include blow-out ports or thrust termination devices, as described in Chapter 13. Typical methods for attaching these items include tapered or straight multiple pins, snap rings, or bolts. Gaskets and/or O-ring seals prevent gas leaks.

Metal Cases

Metal cases have several advantages compared to filament-reinforced plastic cases: they are rugged and will take considerable rough handling (required in many tactical missile applications), are usually reasonably ductile and can yield before failure, can be heated to a relatively high temperature (700 to 1000°C or 1292 to 1832°F and higher with some special materials), and thus require less insulation. They will not deteriorate significantly with time or weather exposure and are easily adapted to take concentrated loads, if made thicker at a flange or boss. Since the metal case has much higher density and less insulation, it occupies less volume than does a fiber-reinforced plastic case; therefore, for the same external envelope it can contain somewhat more propellant.

Figure 14–1 shows the various sections of a typical large solid rocket case made of welded steel. The shape of the case, particularly the length-to-diameter ratio for cylindrical cases, influences not only the stresses to be withstood by the case but the amount of case material required to encase a given amount of propellant. For very large and long motors both the propellant grain and the motor case are made in sections; the *case segments* are mechanically attached and sealed to each other at the launch site. The segmented solid rocket booster for the Space Shuttle is shown in Fig. 14–2 and discussed in Ref. 14–1. For the critical seal between the segments a multiple-O-ring joint is often used, as shown in Fig. 14–3 and discussed in Ref. 14–2. Segments are used when an unsegmented motor would be too large and too heavy to be transported over

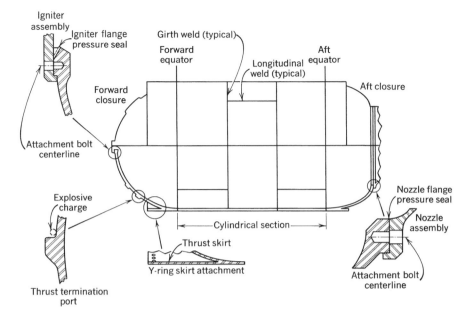

FIGURE 14–1. Typical large solid rocket motor case made of welded alloy steel.

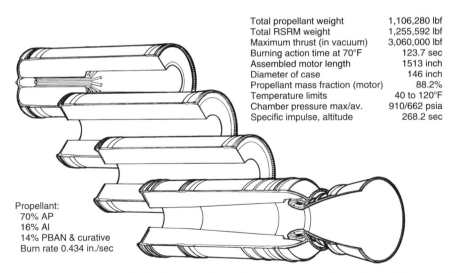

Total propellant weight	1,106,280 lbf
Total RSRM weight	1,255,592 lbf
Maximum thrust (in vacuum)	3,060,000 lbf
Burning action time at 70°F	123.7 sec
Assembled motor length	1513 inch
Diameter of case	146 inch
Propellant mass fraction (motor)	88.2%
Temperature limits	40 to 120°F
Chamber pressure max./av.	910/662 psia
Specific impulse, altitude	268.2 sec

Propellant:
70% AP
16% Al
14% PBAN & curative
Burn rate 0.434 in./sec

FIGURE 14–2. Simplified diagram of the four segments of the Space Shuttle solid rocket motor. Details of the thrust vector actuating mechanism or the ignition system are not shown. (Courtesy of NASA and Thiokol Propulsion, a Division of Cordant Technologies, Inc.)

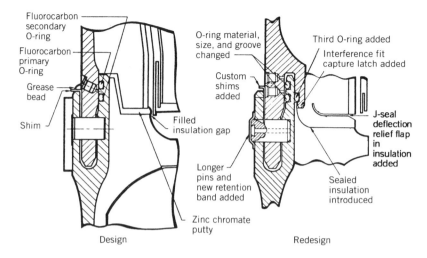

FIGURE 14–3. The joints between segments of the Shuttle solid rocket booster (SRB) were redesigned after a dramatic failure. The improvements were not only in a third O-ring, the mechanical joint, and its locking mechanism, but also featured a redesign of the insulation between propellant segments. (Courtesy of NASA.)

ordinary roads (cannot make turns) or railways (will not go through some tunnels or under some bridges) and are often too difficult to fabricate.

Small metal *cases for tactical missile motors* can be extruded or forged (and subsequently machined), or made in three pieces as shown in Fig. 11–4. This case is designed for loading a free-standing grain and the case, nozzle, and blast tube are sealed by O-rings (see Chapter 6 of Ref. 14–3 and Chapter 7 of Ref. 14–4). Since the mission velocities for most tactical missiles are relatively low (100 to 1500 m/sec), their propellant mass fractions are also relatively low (0.5 to 0.8) and the percentage of inert motor mass is high. Safety factors for tactical missile cases are often higher to allow for rough handling and cumulative damage. The emphasis in selecting motor cases (and other hardware components) for tactical missiles is therefore not on highest performance (lowest inert motor mass), but on reliability, long life, low cost, ruggedness, or survivability.

High-strength alloy steels have been the most common case metals, but others, like aluminum, titanium, and nickel alloys, have also been used. Table 14–2 gives a comparison of motor case material properties. Extensive knowledge exists for designing and fabricating motor cases with low-alloy steels with strength levels to 240,000 psi.

The *maraging steels* have strengths up to approximately 300,000 psi in combination with high fracture toughness. The term *maraging* is derived from the fact that these alloys exist as relative soft low-carbon martensites in the annealed condition and gain high strength from aging at relatively low temperatures.

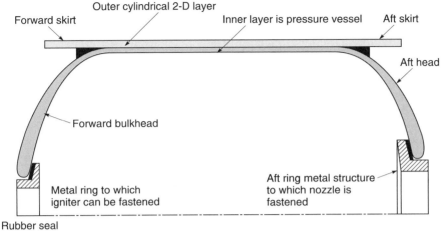

FIGURE 14-4. Simplified half-section of a typical design of a filament-wound composite material case. Elastomeric adhesives are shown in black. The outer layer reinforces the cylinder portion and provides attachment skirts. The thickness of the inner case increases at smaller diameter.

The *HY steels* (newer than the maraging steels) are attractive because of their toughness and resistance to tearing, a property important to motor cases and other pressure vessels because failures are less catastrophic. This toughness characteristic enables a "leak before failure" to occur, at least during hydrostatic proof testing. The HY steels have strengths between 180,000 and 300,000 psi (depending on heat treatment and additives).

Stress-corrosion cracking of certain metals presents a unique problem which can result in spontaneous failure without any visual evidence of impending catastrophe. Emphasis given to lightweight thin metal cases aggravates stress corrosion and crack propagation, often starting from a flaw in the metal, with failure occurring at a stress level below the yield strength of the metal.

Wound-Filament-Reinforced Plastic Cases

Filament-reinforced cases use continuous filaments of strong fibers wound in precise patterns and bonded together with a plastic, usually an epoxy resin. Their principal advantage is their lower weight. Most plastics soften when they are heated above about 180°C or 355°F; they need inserts or reinforcements to allow fastening or assembly of other components and to accept concentrated loads. The thermal expansion of reinforced plastics is often higher than that of metal and the thermal conductivity is much lower, causing a higher temperature gradient. References 14–3 and 14–4 explain the design and winding of these composite cases, and Ref. 14–5 discusses their damage tolerance.

Typical fiber materials are, in the order of increasing strength, glass, aramids (Kevlar), and carbon, as listed in Table 14–2. Typically, the inert mass of a case made of carbon fiber is about 50% of a case made with glass fibers and around 67% of a case mass made with Kevlar fibers.

Individual fibers are very strong in tension (2400 to 6800 MPa or 350,000 to 1,000,000 psi). The fibers are held in place by a plastic binder of relatively low density; it prevents fibers slipping and thus weakening in shear or bending. In a filament-wound composite (with tension, hoop, and bending stresses) the filaments are not always oriented along the direction of maximum stress and the material includes a low-strength plastic; therefore, the composite strength is reduced by a factor of 3 to 5 compared to the strength of the filament itself. The plastic binder is usually a thermosetting epoxy material, which limits the maximum temperature to between 100 to 180°C or about 212 to 355°F. Although resins with higher temperature limits are available (295°C or 563°F), their adhesion to the fibers has not been as strong. The safety factors used (in deterministic structural analysis) are typically for failure to occur at 1.4 to 1.6 times the maximum operating stress, and proof testing is done to 1.15 to 1.25 times the operating pressure.

A typical case design is shown schematically in Fig. 14–4. The forward end, aft end, and cylindrical portion are wound on a preform or mold which already contains the forward and aft rings. The direction in which the bands are laid onto the mold and the tension that is applied to the bands is critical in obtaining a good case. The curing is done in an oven and may be done under pressure to assure high density and minimum voids of the composite material. The preform is then removed. One way is to use sand with a water-soluble binder for the preform; after curing the case, the preform is washed out with water. Since filament-wound case walls can be porous, they must be sealed. The liner between the case and the grain can be the seal that prevents hot gases from seeping through the case walls. Scratches, dents, and moisture absorption can degrade the strength of the case.

In some designs the insulator is placed on the preform before winding and the case is cured simultaneously with the insulator, as seen in Ref. 14–6. In another design the propellant grain with its forward and aft closures is used as the preform. A liner is applied to this grain, then an insulator, and the high-strength fibers of the case are wound in layers directly over the insulated live propellant. Curing has to be done at a relatively low temperature so that the propellant will not be adversely affected. This process works well with extruded cylindrical grains. There are also cylindrical cases made of steel with an over-wrap layer of filament-wound composite material, as described in Ref. 14–7.

The allowable stresses are usually determined from tensile tests of a roving or band and rupture tests on subscale composite cases made by an essentially identical filament winding process. Some companies reduce the allowable strength to account for the degradation due to moisture, manufacturing imperfections, or nonuniform density.

In a motor case the filaments must be oriented in the direction of principal stress and must be proportioned in number to the magnitude of stress. Compromise occurs around parts needed for nozzles, igniters, and so on, and then orientation is kept as close to the ideal as is practicable. Filaments are customarily clustered in *yarns, rovings,* or *bands,* as defined in Fig. 14–5. By using two or more winding angles (i.e., helicals and circumferentials) and calculating the proportion of filaments in each direction, a balanced stress structure is achieved. The ideal in balance is for each fiber in each direction to carry an equal load (tension only). Realistically, the filaments supported by the epoxy resin must absorb stress compression, bending loads, cross-laminar shear, and interlaminar shear. Even though the latter stresses are small compared to the tensile stress, each must be examined by analysis since each can lead to case failure before a filament fails in tension. In a proper design, failure occurs when the filaments reach their ultimate tensile strength, rather than because of stresses in other directions. Figure 16–5 shows a cross section of a Kevlar filament motor case and flexible nozzle made of ablative materials.

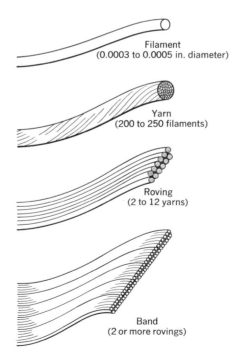

Filament
(0.0003 to 0.0005 in. diameter)

Yarn
(200 to 250 filaments)

Roving
(2 to 12 yarns)

Band
(2 or more rovings)

FIGURE 14–5. Filament winding terminology (each sketch is drawn to a different scale).

14.2. NOZZLES*

The supersonic nozzle provides for the expansion and acceleration of the hot gases and has to withstand the severe environment of high heat transfer and erosion. Advances in material technology have allowed substantial mass reductions and performance improvements. Nozzles range in size from 0.05 in. throat diameter to about 54 in., with operating durations of a fraction of a second to several minutes (see Chapters 2 and 3 of Ref. 14–3 and Chapter 6 in Ref. 14–4).

Classification

Nozzles for solid propellant rocket motors can be classified into five categories as listed below and shown in Fig. 14–6.

1. *Fixed Nozzle.* Simple and used frequently in tactical weapon propulsion systems for short-range air-, ground-, and sea-launched missiles, also as strap-on propulsion for space launch vehicles such as Atlas and Delta, and in spacecraft motors for orbital transfer. Typical throat diameters are between 0.25 and 5 in. for tactical missile nozzles and approximately 10 in. for strap-on motors. Fixed nozzles are generally not submerged (see below) and do not provide thrust vector control (although there are exceptions). See Fig. 14–7.

2. *Movable Nozzle.* Provides thrust vector control for the flight vehicle. As explained in Chapter 16, one movable nozzle can provide pitch and yaw control and two are needed for roll control. Movable nozzles are typically submerged and use a flexible sealed joint or bearing with two actuators 90 degrees apart to achieve omniaxial motion. Movable nozzles are primarily used in long-range strategic propulsion ground- and sea-launched systems (typical throat diameters are 7 to 15 in. for the first stage and 4 to 5 in. for the third stage) and in large space launch boosters such as the Space Shuttle reusable solid rocket motor, Titan boost rocket motor, and Ariane V solid rocket booster, with throat diameters in the 30 to 50 in. range.

3. *Submerged Nozzles.* A significant portion of the nozzle structure is submerged within the combustion chamber or case, as shown in Figs. 14–1 to 14–3. Submerging the nozzle reduces the overall motor length somewhat, which in turn reduces the vehicle length and its inert mass. It is important for length-limited applications such as silo- and submarine-launched strategic missiles as well as their upper stages, and space motor propulsion systems. Reference 14–8 describes the sloshing of trapped molten aluminum oxide that can accumulate in the groove around a submerged noz-

*This section was revised and rewritten by **Terry A. Boardman** of Thiokol Corporation, a Division of Cordant Technologies.

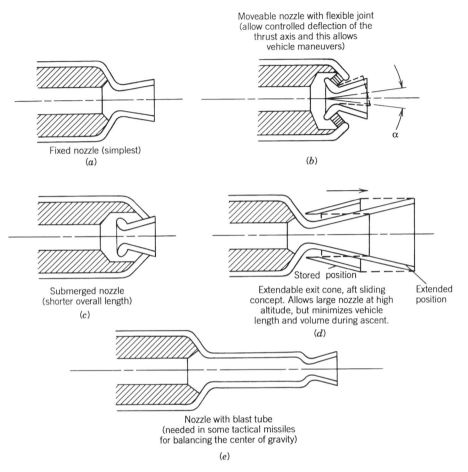

Moveable nozzle with flexible joint
(allow controlled deflection of the
thrust axis and this allows
vehicle maneuvers)

Fixed nozzle (simplest)
(a)

(b)

Submerged nozzle
(shorter overall length)
(c)

Stored position

Extendable exit cone, aft sliding
concept. Allows large nozzle at high
altitude, but minimizes vehicle
length and volume during ascent.
(d)

Extended
position

Nozzle with blast tube
(needed in some tactical missiles
for balancing the center of gravity)
(e)

FIGURE 14–6. Simplified diagrams of five common nozzle configurations.

zle. This accumulation is undesirable, but can be minimized by good design.

4. *Extendible Nozzle.* Commonly referred to as an extendible exit cone, or EEC, although it is not always exactly conical. It is used on strategic missile propulsion upper-stage systems and upper stages for space launch vehicles to maximize motor-delivered specific impulse. As shown in Fig. 11–3, it has a fixed low-area-ratio nozzle section which is enlarged to a higher area ratio by mechanically adding a nozzle cone extension piece. The extended nozzle improves specific impulse by doubling or tripling the initial expansion ratio, thereby significantly increasing the nozzle thrust coefficient. This system thus allows a very high expansion ratio nozzle to be packaged in a relatively short length, thereby reducing vehicle inert mass. The nozzle cone extension is in its retracted position during the

boost phase of the flight and is moved into place before the motor is started but after separation from the lower stage. Typically, electromechanical or turbine-driven ball screw actuators deploy the exit cone extension.

5. *Blast-Tube-Mounted Nozzle.* Used with tactical air- and ground-launched missiles with diameter constraints to allow space for aerodynamic fin actuation or TVC power supply systems. The blast tube also allows the rocket motor's center of gravity (CG) to be close to or ahead of the vehicle CG. This limits the CG travel during motor burn and makes flight stabilization much easier.

Each motor usually has a single nozzle. A few larger motors have had four movable nozzles, which are used for thrust vector control.

Design and Construction

Almost all solid rocket nozzles are ablatively cooled. The general construction of a solid rocket nozzle features steel or aluminum shells (housings) that are designed to carry structural loads (motor operating pressure and nozzle TVC actuator load are the biggest), and composite ablative liners which are bonded to the housings. The ablative liners are designed to insulate the steel or aluminum housings, provide the internal aerodynamic contour necessary to efficiently expand combustion gases to generate thrust, and to ablate and char in a controlled and predictable manner to prevent the buildup of heat which could damage or substantially weaken the structural housings or the bonding materials. Solid rocket nozzles are designed to ensure that the thickness of ablative liners is sufficient to maintain the liner-to-housing adhesive bond line below the temperature that would degrade the adhesive structural properties during motor operation. Nozzle designs are shown in Figs. 1–5, 11–1 to 11–4, and 14–7.

The construction of nozzles ranges from simple single-piece non-movable graphite nozzles to complex multipiece nozzles capable of moving to control the direction of the thrust vector. The simpler, smaller nozzles are typically for applications with low chamber pressure, short durations (perhaps less than 10 sec), low area ratios, and/or low thrust. Typical small, simple built-up nozzles are shown in Fig. 14–7. Complex nozzles are usually necessary to meet more difficult design requirements such as providing thrust vector controls, operating at high chamber pressures (and thus at higher heat transfer rates) and/or higher altitudes (large nozzle expansion ratios), producing very high thrust levels, and surviving longer motor burn durations (above 30 sec).

Figures 14–8 and 14–9 illustrate the design features of the largest and one of the most complex solid rocket nozzles currently in production. This nozzle is used on the Reusable Solid Rocket Booster (RSRM) to provide 71.4% of the lift-off thrust of the Space Shuttle launch vehicle shown in Figs. 1–13 and 14–2. The nozzle is designed to provide large structural and thermal safety margins

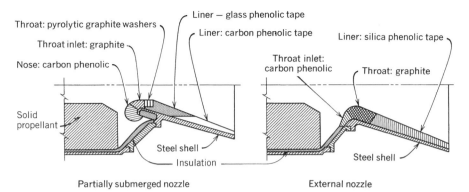

Throat: pyrolytic graphite washers

Liner — glass phenolic tape

Throat inlet: graphite

Liner: carbon phenolic tape

Liner: silica phenolic tape

Nose: carbon phenolic

Throat inlet: carbon phenolic

Throat: graphite

Solid propellant

Steel shell

Insulation

Steel shell

Partially submerged nozzle

External nozzle

FIGURE 14–7. Nozzle designs for small solid propellant motors employ ablative heat sink wall pieces and graphite throat inserts resistant to high temperatures, erosion, and oxidation. The pyrolytic washers or disks are so oriented that their high conductivity direction is perpendicular to the nozzle axis.

during the shuttle booster's 123.7 sec burn time and consists of nine carbon cloth phenolic ablative liners bonded to six steel and aluminum housings. The housings are bolted together to form the structural foundation for the nozzle. A flexible bearing (described further in Chapter 16), made of rubber vulcanized to steel shims, enables the nozzle to vector omniaxially up to eight degrees from centerline to provide thrust vector control. Since the metal housings are recov-

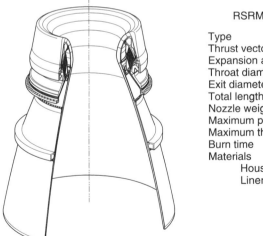

RSRM Nozzle Characteristics

Type	Contoured or bell
Thrust vector control	Flexible bearing
Expansion area ratio	7.72
Throat diameter	53.86 in.
Exit diameter	149.64 in.
Total length	178.75 in.
Nozzle weight	23, 941 lbf
Maximum pressure	1.016 psi
Maximum thrust (vac.)	3, 070, 000 lbf
Burn time	123.7 sec
Materials	
Housings	Steel and aluminum
Liners	Carbon cloth phenolic

FIGURE 14–8. External quarter section view of nozzle configuration of the Space Shuttle reusable solid rocket motor (RSRM). (Courtesy of Thiokol Propulsion, a Division of Cordant Technologies.)

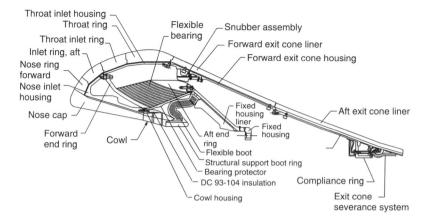

FIGURE 14–9. Section through movable nozzle shown in Fig. 14–8 with component identification. (Courtesy of Thiokol Propulsion, a Division of Cordant Technologies.)

ered and reused after flight, an exit cone severance system (a circumferential linear shaped charge) is used to cut off a major section of the aft exit cone just below the aft exit cone aluminum housing to minimize splashdown loading on the remaining components.

From a performance perspective, the primary nozzle design task is to efficiently expand gas flow from the motor combustion chamber to produce thrust. Simple nozzles with noncontoured conical exit cones can be designed using the basic thermodynamic relationships presented in Chapter 3 to determine throat area, nozzle half angle, and expansion ratio. A more complex contoured (bell-shaped) nozzle is used to reduce the divergence loss, improve the specific impulse slightly, and reduce nozzle length and mass. Section 3.4 gives data on designing bell-shaped nozzles with optimum wall contour (to avoid shock waves) and minimum impact of particulates in the exhaust gas.

Two-dimensional, two-phase, reacting gas method-of-characteristics flow codes are used to analyze the gas-particle flow in the nozzle and determine the optimal nozzle contour which maximizes specific impulse while yielding acceptable erosion characteristics. Such codes provide analytical solutions to all identified specific impulse loss mechanisms which result in less than ideal performance. An example is given in Table 14–3.

Figure 14–10 illustrates the amount of carbon cloth phenolic liner removed by chemical erosion and particle impingement, the liner char depth, and gas temperature and pressure at selected locations in the RSRM nozzle. Erosion on the nose cap (1.73 in.) is high primarily as a result of impingement by Al_2O_3 particles traveling down the motor bore. The impact of the particles mechanically removes the charred liner material. In contrast, the radial throat erosion of 1.07 in. results primarily from the carbon liner material reacting chemically

TABLE 14–3. Calculated Losses in the Space Shuttle Booster RSRM Nozzle

Theoretical specific impulse (vacuum conditions)	278.1 sec
Delivered specific impulse (vacuum conditions)	268.2 sec
Losses (calculated):	*(9.9 sec total)*
Two-dimensional two-phase flow (includes divergence loss)	7.4 sec
Throat erosion (reduces nozzle area ratio)	0.9 sec
Boundary layer (wall friction)	0.7 sec
Submergence (flow turning)	0.7 sec
Finite rate chemistry (chemical equilibrium)	0.2 sec
Impingement (of Al_2O_3 particles on nozzle wall)	0.0 sec
Shock (if turnback angle is too high or nozzle length too low)	0.0 sec
Combustion efficiency (incomplete burning)	0.0 sec

with oxidizing species in the combustion gas flow at the region of greatest heat transfer. At the throat location, impingement erosion is essentially zero because Al_2O_3 particles are traveling parallel to the nozzle surface.

The acronym ITE is often used; it means *integral throat/entrance* and refers to a single-piece nozzle throat insert that also includes a part of the converging entry section. ITE nozzle inserts can be seen in Figs. 1–5, 11–1, 11–3, and 11–4.

Nozzle throat erosion causes the throat diameter to enlarge during operation, and is one of the problems encountered in nozzle design. Usually, a throat area increase larger than 5% is considered unacceptable for most solid rocket appli-

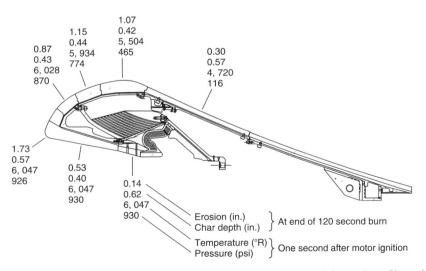

FIGURE 14–10. Erosion measurements and char depth data of the carbon fiber phenolic material of the nozzle of the Space Shuttle reusable solid rocket motor. (Courtesy of Thiokol Propulsion, a Division of Cordant Technologies.)

cations, since it causes a reduction in thrust and chamber pressure. Erosion occurs not only at the throat region (typically, at 0.01 to 0.25 mm/sec or 0.004 to 0.010 in./sec), but also at the sections immediately upstream and downstream of the throat region, as shown in Fig. 14–10. Nozzle assemblies typically lose 3 to 12% of their initial inert mass. Erosion is caused by the complex interaction between the high-temperature, high-velocity gas flow, the chemically aggressive species in the gas, and the mechanical abrasion by particles. The carbon in the nozzle material reacts with species like O_2, O, OH, or H_2O and is oxidized; the cumulative concentration of these species is an indication of the likely erosion. Tables 5–6 and 5–7 give chemical concentrations for the exhaust species from aluminized propellant. Fuel-rich propellants (which contain little free O_2 or O) and propellants where some of the gaseous oxygen is removed by aluminum oxidation show less tendency to cause erosion. Uneven erosion of a nozzle causes thrust misalignment.

Finding the optimum nozzle wall contour requires an analysis (computer codes for bell-shaped nozzles using the method of characteristics are mentioned in Section 3.4) to determine the wall contour which most rapidly turns the gas to near axial flow without introducing shock waves or impinging excessive aluminum oxide (Al_2O_3) particles on the nozzle wall. Figure 3–14 illustrates the key parameters which govern design of the nozzle contour; the initial angle θ_i (angle through which supersonic flow is turned immediately downstream of the nozzle throat), the throat to exit plane length L, the exit plane exit angle θ_e, and the turnback angle $\theta_i - \theta_e$. With solid or liquid particles in the exhaust the impingement can be minimized with an initial angle typically between 20 and 26° and a turnback angle of typically 10 to 15°. The length reduction of a bell-shaped nozzle (with solid particles in the gas) is typically 80 to 90% of the length of an equivalent conical nozzle with 15° half angle. The nozzle throat inlet contour is generally based on a hyperbolic spiral that uniformly accelerates the combustion gas flow to supersonic velocity at the throat plane.

Heat Absorption and Nozzle Materials

Rocket motors never reach thermal equilibrium during their firing. The temperatures of all components exposed to the heat flow increase continuously during operation. In a good thermal design the critical locations reach a maximum allowable temperature a short time after the motor stops running. The nozzle components rely on their heat-absorbing capacity (high specific heat and high energy demand for material decomposition) and slow heat transfer (good insulation with low thermal conductivity) to withstand the stresses and strains imposed by the thermal gradients and loads. The maximum allowable temperature for any of the motor materials is just below the temperature at which excessive degradation occurs (the material loses strength, melts, becomes too soft, cracks, pyrolyses, unglues, oxidizes too rapidly). The operating duration is limited by the design and amount of heat-absorbing and insulating material pieces. Stated in a different way, the objective is to design a nozzle with just sufficient heat-absorbing material mass and insulation mass at the various locations within the nozzle, so that its structures and joints will do the job for the duration of the application under all likely operating conditions.

The selection and application of the proper material is the key to the successful design of a solid rocket nozzle. Table 14–4 groups various typical nozzle materials according to their usage. The high-temperature exhaust of solid rockets presents an unusually severe environment for the nozzle materials, especially when metalized propellants are employed.

About 60 years ago nozzles were made out of a single piece of *molded polycrystalline graphite* and some were supported by metal housing structures. They eroded easily, but were low in cost. We still use them today for short duration, low chamber pressure, low altitude flight applications of low thrust, such as in certain tactical missiles. For more severe conditions a throat insert or ITE was placed into the graphite piece; this insert was a denser, better grade of graphite; later pyrolytic graphite washers and fiber-reinforced carbon materials came into use. For a period of time tungsten inserts were used; they had very good erosion resistance, but were heavy and often cracked. *Pyrolytic graphite* was introduced and is still being used as washers for the throat insert of small nozzles, as shown in Fig. 14–7. The high-strength carbon fiber and the carbon matrix were major advances in high-temperature materials. For small and medium-sized nozzles, ITE pieces were then made of *carbon–carbon*, which is an abbreviation for carbon fibers in a carbon matrix (see Ref. 14–9). The orientation of the fibers can be two-dimensional (2D) or three-dimensional (3D), as described below. Some properties of all these materials are listed in Tables 14–4 and 14–5. For large nozzles the then existing technology did not allow the fabrication of large 3D carbon–carbon ITE pieces, so layups of carbon fiber (or silicon fiber) cloth in a phenolic matrix were used.

The regions immediately upstream and downstream of the throat have less heat transfer, less erosion, and lower temperatures than the throat region, and less expensive materials are usually satisfactory. This includes various grades of graphite, or ablative materials, strong high-temperature fibers (carbon or silica) in a matrix of phenolic or epoxy resins, which are described later in this section. Figure 16–6 shows a movable nozzle with multilayer *insulators* behind the graphite nozzle pieces directly exposed to heat. These insulators (between the very hot throat piece and housing) limit the heat transfer and prevent excessive housing temperatures.

In the *diverging exit section* the heat transfer and temperatures are even lower and similar, but less capable and less expensive materials can be used here. This exit segment can be built integral with the nozzle throat segment (as it is in most small nozzles), or it can be a separate one- or two-piece subassembly which is then fastened to the smaller diameter throat segment. Ablative materials without oriented fibers as in cloth or ribbons, but with short fibers or insulating ceramic particles, can be used here. For large area ratios (upper stages and space transfer), the nozzle will often protrude beyond the vehicle's boat tail surface. This allows radiation cooling, since the exposed exit cone can

TABLE 14-4. Typical Motor Nozzle Materials and Their Functions

Function	Material	Remarks
Structure and pressure container (housing)	Aluminum	Limited to 515°C (959°F)
	Low carbon steel, high-strength steels, and special alloys	Good between 625 and 1200°C (1100 and 2200°F), depending on material; rigid and strong
Heat sink and heat-resistant material at inlet and throat section; severe thermal environment and high-velocity gas, with erosion	Molded graphite	For low chamber temperatures and low pressures only; low cost
	Pyrolitic graphite	Has anisotropic conductivity
	Tungsten, molybdenum, or other heavy metal	Heavy, expensive, subject to cracking; resists erosion
	Carbon or Kevlar fiber cloth with phenolic or plastic resins	Sensitive to fiber orientation. Ablative materials Used with large throats
	Carbon–carbon	Three- or four dimensional interwoven filaments, strong, expensive, limited to 3300°C (6000°F)
Insulator (behind heat sink or flame barrier); not exposed to flowing gas	Ablative plastics, with fillers of silica or Kevlar, phenolic resins	Want low conductivity, good adhesion, ruggedness, erosion resistance; can be filament wound or impregnated cloth layup with subsequent machining
Flame barrier (exposed to hot low-velocity gas)	Ablative plastics (same as insulators but with less filler and tough rubber matrix)	Lower cost than carbon–carbon; better erosion resistance than many insulators
	Carbon, Kevlar or silica fibers with phenolic or epoxy resin	Cloth or ribbon layups; woven and compressed, glued to housng
	Carbon–carbon	Higher temperature than others, three-dimensional weave or layup
Nozzle exit cone	Ablative plastic with metal housing structure	Heavy, limited duration; cloth or woven ribbon lay-ups, glued to housing
	Refractory metal (tantalum, molybdenum)	Radiation cooled, strong, needs coating for oxidation resistance; can be thin, limited to 1650°C (3000°F), unlimited duration
	Carbon–carbon, may need gas seal	Radiation cooled, higher allowable temperature than metals; two- or three-dimensional weave, strong, often porous

TABLE 14-5. Comparison of Properties of Molded and Pyrolytic Graphite, Carbon–Carbon, Carbon Cloth, and Silica Cloth Phenolic

	ATJ Modern Graphite	Pyrolytic Graphite	Three-Dimensional Carbon Fibers in a Carbon Matrix	Carbon Cloth Phenolic	Silica Cloth Phenolic
Density (lbm/in.³)	0.0556	0.079	0.062 to 0.072	0.053	0.062
Thermal expansion (in./in./°F)	0.005 to 0.007	0.00144 (warp) 0.0432 (fill)	$1-9\times10^{-6}$	8.02×10^{-6}	7.6×10^{-6}
Thermal conductivity (Btu/in.-sec/°F) at room temperature	1.2×10^{-3} 1.5×10^{-3}	4.9×10^{-5} (warp)[a] 4.2×10^{-5} (fill)[a]	2 to 21×10^{-5} (warp)[a] 8 to 50×10^{-5} (fill)[a]	2.2 $\times10^{-3}$ (warp)[a]	1.11×10^{-3} (warp)[a]
Modulus of elasticity (psi) at room temperature	1.5×10^{6} (warp)[a] 1.2×10^{6} (fill)[a]	4.5×10^{6} (warp)[a] 1.5×10^{6} (fill)[a]	35 to 80×10^{6}	2.86×10^{-6} (warp)[a] 2.91×10^{-6} (fill)[a]	3.17×10^{-6} (warp)[a] 2.86×10^{-6} (fill)[a]
Shear modulus (psi)	—	0.2×10^{6} (warp)[a] 2.7×10^{6} (fill)[a]	—	0.81×10^{6}	0.80×10^{6}
Erosion rate (typical) (in./sec)	0.004 to 0.006	0.001 to 0.002	0.0005 to 0.001	0.005 to 0.010	0.010 to 0.020

[a]*Warp* is in direction of principal fibers. *Fill* is at right angles to warp.

reject heat by radiation to space. Lightweight thin high temperature metals (niobium, titanium, stainless steel, or a thin carbon–carbon shell) with radiation cooling have been used in a few upper-stage or spacecraft exit cone applications. Since radiation-cooled nozzle exit sections reach thermal equilibrium, their duration is unlimited.

The *housing* or *structural support of the nozzle* uses the same material as the metal case, such as steel or aluminum. The housings are never allowed to become very hot. Some of the simpler, smaller nozzles (with one, two, or three pieces, mostly graphite) do not have a separate housing structure, but use the ITE (integral throat/entry) for the structure.

Estimates of nozzle internal temperatures and temperature distributions with time can be made using two-dimensional finite element difference methods for *transient heat transfer analyses*. These are similar in principle to the transient heat transfer method described in Section 8.3 and shown in Fig. 8–21. After firing, the nozzle temperatures reach an equilibrium value by conducting heat from the hotter inner parts, which were exposed to the hot gas, to the cooler outer pieces. Sometimes the outer pieces will exceed their limit temperatures and suffer damage after firing. The *structural analysis* (stresses and strains) of the key nozzle components is dependent on the heat transfer analysis, which determines the component temperatures. This allows use of the proper material physical properties, which are temperature dependent. The design must also allow for the thermal growth and the differential expansion of adjacent parts.

Typical materials used for the ITE (integral throat and entrance) or nozzle throat insert are listed in Table 14–5. They are exposed to the most severe conditions of heat transfer, thermal stresses, and high temperatures. Their physical properties are often anisotropic; that is, their properties vary with the orientation or direction of the crystal structure or the direction of reinforcing fibers. *Polycrystalline graphites* are extruded or molded. Different grades with different densities and capabilities are available. As already mentioned, they are used extensively for simple nozzles and for ITE parts. *Pyrolytic graphite* is strongly anisotropic and has excellent conductivity in a preferred direction. A nozzle using it is shown in Fig. 14–7. It is fabricated by depositing graphite crystals on a substratum in a furnace containing methane gas. Its use is declining, but it is still installed in current rocket motors of older design.

The *carbon–carbon* material is made from carefully oriented sets of *carbon fibers* (woven, knitted, threaded, or laid up in patterns) in a *carbon matrix*. Two-dimensional (2D) material has fibers in two directions, 3D has fibers oriented in three directions (at right angle to each other), and 4D has an extra set of fibers at about a 45° angle to the other three directions. An organic liquid resin is injected into the spaces between the fibers. The assembly is pressurized, the filler is transformed into a carbon char by heating and is compacted by further injection and densification processes. The graphitization is then performed at temperatures higher than 2000°C. This material is expensive but suited to nozzle applications. Highly densified material is superior in high heat transfer regions, such as the throat. The multidirectional fiber reinforcements allow them to better withstand the high thermal stresses introduced by the steep temperature gradients within the component.

Ablative Materials. These are not only commonly used in the nozzles of rocket motors, but also in some insulation materials. They are usually a composite material of high-temperature organic or inorganic high strength fibers, namely high silica glass, aramids (Kevlar), or carbon fibers, impregnated with organic plastic materials such as phenolic or epoxy resin. The fibers may be individual strands or bands (applied in a geometric pattern on a winding machine), or come as a woven cloth or ribbon, all impregnated with resin.

The ablation process is a combination of surface melting, sublimation, charring, evaporation, decomposition in depth, and film cooling. As shown in Fig. 14–11, progressive layers of the ablative material undergo an endothermic degradation, that is, physical and chemical changes that absorb heat. While some of the ablative material evaporates (and some types also have a viscous liquid phase), enough charred and porous solid material remains on the surface to preserve the basic geometry and surface integrity. Upon rocket start the ablative material acts like any thermal heat sink, but the poor conductivity causes the surface temperature to rise rapidly. At 650 to 800 K some of the resins start to decompose endothermically into a porous carbonaceous char and pyrolysis gases. As the char depth increases, these gases undergo an endothermic cracking process as they percolate through the char in a counter-flow direction to the heat flux. These gases then form an artificial fuel-rich, protective, relatively cool, but flimsy boundary layer over the char.

Since char is almost all carbon and can withstand 3500 K or 6000 R, the porous char layer allows the original surface to be maintained (but with a rough surface texture) and provides geometric integrity. Char is a weak material and can be damaged or abraded by direct impingement of solid particles in the gas. Ablative material construction is used for part or all of the chambers

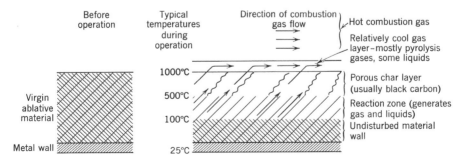

FIGURE 14–11. Zones in an ablative material during rocket operation with fibers at 45° to the flow.

and/or nozzles shown in Figs. 1–5, 6–10, 11–1 to 11–4, and 14–10.

Ablative parts are formed either by high-pressure molding ($\sim$ 55 to 69 MPa or 8000 to 10,000 psi at 149°C or 300°F) or by tapewrapping on a shaped mandrel followed by an autoclave curing process at 1000 to 2000 psi pressure and 300°F temperature. *Tapewrapping* is a common method of forming very large nozzles. The wrapping procedure normally includes heating the shaped mandrel ($\sim$ 54°C or 130°F), heating the tape and resin (66 to 121°C or 150 to 250°F), pressure rolling the tape of fiber material and the injected resin in place while rolling ($\sim$ 35,000 N/m or 200 lbf/in. width), and maintaining the proper rolling speed, tape tension, wrap orientation, and resin flow rate. Experience has proven that as-wrapped density is an important indicator of procedural acceptability, with the desired criterion being near 90% of the autoclaved density. Resin content usually ranges between 25 and 35%, depending on the fabric-reinforcing material and the particular resin and its filler material. Normally, the mechanical properties of the cured ablative material, and also the durability of the material during rocket operation, correlate closely with the cured material density. Within an optimal density range, low density usually means poor bonding of the reinforcing layers, high porosity, low strength, and high erosion rate.

In liquid propellant rockets, ablatives have been effective in very small thrust chambers (where there is insufficient regenerative cooling capacity), in pulsing, restartable spacecraft control rocket engines, and in variable-thrust (throttled) rocket engines. Figure 6–10 shows an ablative nozzle extension for a large liquid propellant rocket engine.

The heat transfer properties of the many available ablative and other fiber-based materials will depend on their design, composition, and construction. Figure 14–12 shows several common fiber orientation and approaches. The *orientation of fibrous reinforcements*, whether in the form of tape, cloth, filaments, or random short fibers, has a marked impact on the erosion resistance of composite nozzles (for erosion data see Figure 14–10). When perpendicular to the gas flow, the heat transfer to the wall interior is high because of the short conducting path. Good results have been obtained when the fibers are at 40 to 60° relative to the gas flow over the surface. Nozzle fabrication variables present wide variations in nozzle life for a given design; the variables include the

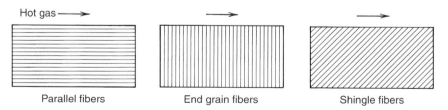

FIGURE 14–12. Simplified sketches of three different types of fiber-reinforced ablative materials.

method of wrapping, molding, and curing, resin batch processes, and resin sources.

14.3. IGNITER HARDWARE

In Section 13.2 the process of ignition was described, and in Section 12.5 some of the propellants used in igniters were mentioned briefly. In this section we discuss specific igniter types, locations, and their hardware (see Ref. 14–10).

Since the igniter propellant mass is small (often less than 1% of the motor propellant) and burns mostly at low chamber pressure (low I_s), it contributes very little to the motor overall total impulse. It is the designer's aim to reduce the igniter propellant mass and the igniter inert hardware mass to a minimum, just big enough to assure ignition under all operating conditions.

Figure 14–13 shows several alternative locations for igniter installations. When mounted on the forward end, the gas flow over the propellant surface helps to achieve ignition. With aft mounting there is little gas motion, particularly near the forward end; here ignition must rely on the temperature, pressure, and heat transfer from the igniter gas. If mounted on the nozzle, the igniter hardware and its support is discarded shortly after the igniter has used all its propellants and there is no inert mass penalty for the igniter case. There are two basic types: pyrotechnic igniters and pyrogen igniters; both are discussed below.

Pyrotechnic Igniters

In industrial practice, pyrotechnic igniters are defined as igniters (other than pyrogen-type igniters as defined further on) using solid explosives or energetic propellant-like chemical formulations (usually small pellets of propellant which

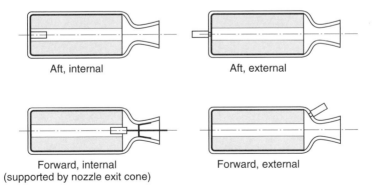

Aft, internal

Aft, external

Forward, internal
(supported by nozzle exit cone)

Forward, external

FIGURE 14–13. Simple diagrams of mounting options for igniters. Grain configurations are not shown.

give a large burning surface and a short burning time) as the heat-producing material. This definition fits a wide variety of designs, known as bag and carbon igniters, powder can, plastic case, pellet basket, perforated tube, combustible case, jellyroll, string, or sheet igniters. The common pellet-basket design in Fig. 14–14 is typical of the pyrotechnic igniters. Ignition of the main charge, in this case pellets consisting of 24% boron–71% potassium perchlorate–5% binder, is accomplished by stages; first, on receipt of an electrical signal the initiator releases the energy of a small amount of sensitive powdered pyrotechnic housed within the initiator, commonly called the squib or the primer charge; next, the booster charge is ignited by heat released from the squib; and finally, the main ignition charge propellants are ignited.

A special form of pyrotechnic igniter is the *surface-bonded* or *grain-mounted igniter*. Such an igniter has its initiator included within a sandwich of flat sheets; the layer touching the grain is the main charge of pyrotechnic. This form of igniter is used with multipulse motors with two or more end-burning grains. The ignition of the second and successive pulses of these motors presents unusual requirements for available space, compatibility with the grain materials, life, and the pressure and temperature resulting from the booster grain operation. Advantages of the sheet igniter include light weight, low volume, and high heat flux at the grain surface. Any inert material employed (such as wires and electric ceramic insulators) is usually blown out of the motor nozzle during ignition and their impacts have caused damage to the nozzle or plugged it, particularly if they are not intentionally broken up into small pieces.

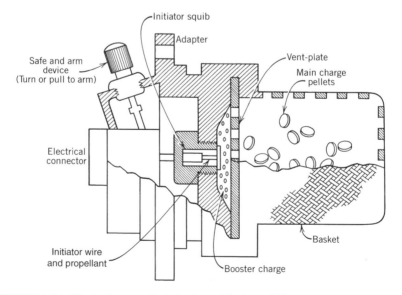

FIGURE 14–14. Typical pyrotechnic igniter with three different propellant charges that ignite in sequence.

Pyrogen Igniters

A pyrogen igniter is basically a small rocket motor that is used to ignite a larger rocket motor. The pyrogen is not designed to produce thrust. All use one or more nozzle orifices, both sonic and supersonic types, and most use conventional rocket motor grain formulations and design technology. Heat transfer from the pyrogen to the motor grain is largely convective, with the hot gases contacting the grain surface as contrasted to a highly radiative energy emitted by pyrotechnic igniters. Figures 11–1, 11–2, and 11–20 illustrate rocket motors with a typical pyrogen igniter. The igniter in Fig. 16–5 has three nozzles and a cylindrical grain with high-burn-rate propellant. For pyrogen igniters the initiator and the booster charge are very similar to the designs used in pyrotechnic igniters. Reaction products from the main charge impinge on the surface of the rocket motor grain, producing motor ignition. Common practice on the very large motors is to mount externally, with the pyrogen igniter pointing its jet up through the large motor nozzle. In this case, the igniter becomes a piece of ground-support equipment.

Two approaches are commonly used to safeguard against motor misfires, or inadvertent motor ignition; one is the use of the classical *safe and arm device* and the second is the *design of safeguards* into the initiator. Energy for unintentional ignition—usually a disaster when it happens—can be (1) static electricity, (2) induced current from electromagnetic radiation, such as radar, (3) induced electrical currents from ground test equipment, communication apparatus, or nearby electrical circuits in the flight vehicle, and (4) heat, vibration, or shock from handling and operations. Functionally, the safe and arm device serves as an electrical switch to keep the igniter circuit grounded when not operating; in some designs it also mechanically misaligns or blocks the ignition train of events so that unwanted ignition is precluded even though the initiator fires. When transposed into the *arm* position, the ignition flame can be reliably propagated to the igniter's booster and main charges.

Electric initiators in motor igniters are also called squibs, glow plugs, primers, and sometimes headers; they always constitute the initial element in the ignition train and, if properly designed, can be a safeguard against unintended ignition of the motor. Three typical designs of initiators are shown in Fig. 14–15. Both (*a*) and (*b*) structurally form a part of the rocket motor case and generically are headers. In the integral diaphragm type (*a*) the initial ignition energy is passed in the form of a shock wave through the diaphragm activating the acceptor charge, with the diaphragm remaining integral. This same principle is also used to transmit a shock wave through a metal case wall or a metal insert in a filament-wound case; the case would not need to be penetrated and sealed. The header type (*b*) resembles a simple glow plug with two high-resistance bridgewires buried in the initiator charge. The exploding bridgewire design (*c*) employs a small bridgewire (0.02 to 0.10 mm) of low-resistance material, usually platinum or gold, that is exploded by application of a high-voltage discharge.

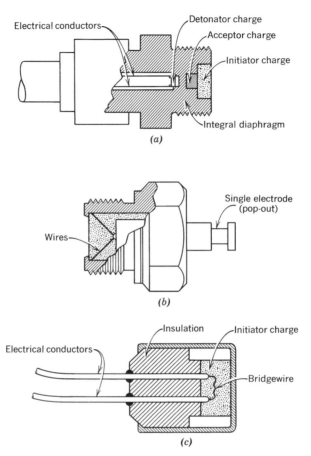

FIGURE 14–15. Typical electric initiators; (a) integral diaphragm type; (b) header type with double bridgewire; (c) exploding bridgewire type.

The safeguard aspect of the initiator appears as a basic design feature in the form of (1) minimum threshold electrical energy required for activation, (2) voltage blockage provisions (usually, air gaps or semiconductors in the electrical circuit), or (3) responsiveness only to a specific energy pulse or frequency band. Invariably, such safeguards compromise to some degree the safety provided by the classical safe and arm device.

A new method of initiating the action of an igniter is to use laser energy to start the combustion of an initiator charge. Here there are no problems with induced currents and other inadvertent electrical initiation. The energy from a small neodymium/YAG laser, external to the motor, travels in fiber-optical glass cables to the pyrotechnic initiator charge (Ref. 14–11). Sometimes an optical window in the case or closure wall allows the initiator charge to be inside the case.

Igniter Analysis and Design

The basic theories of initiating ignition, heat transfer, propellant decomposition, deflagration, flame spreading, and chamber filling are common to the design and application of pyrotechnic and pyrogen igniters. In general, the mathematical models of the physical and chemical processes that must be considered in the design of igniters are far from complete and accurate. See Chapter 5 by Hermance and Chapter 6 by Kumar and Kuo in Ref. 13–1, and Ref. 14–10.

Analysis and design of igniters, regardless of the type, depend heavily on experimental results, including past successes and failures with full-scale motors. The effect of some of the important parameters has become quite predictable, using data from developed motors. For example, Fig. 14–16 is of benefit in estimating the mass of igniter main charge for motors of various sizes (motor free volume). From these data,

$$m = 0.12(V_F)^{0.7} \tag{14–4}$$

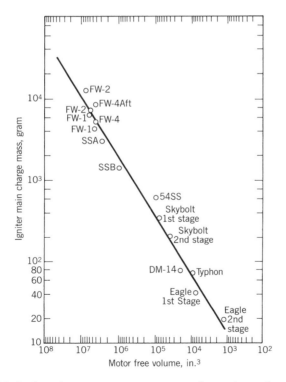

FIGURE 14–16. Igniter charge mass versus motor free volume, based on experience with various-sized rocket motors using AP/Al composite propellant. (Data with permission from Ref. 14–12.)

where m is the igniter charge in grams and V_F is the motor free volume in cubic inches or the void in the case not occupied by propellant. A larger igniter mass flow means a shorter ignition delay. The ignition time events were shown in Fig. 13–3.

14.4. ROCKET MOTOR DESIGN APPROACH

Although there are some common elements in the design of all solid propellant rocket motors, there is no single, well-defined procedure or design method. Each class of application has some different requirements. Individual designers and their organizations have different approaches, background experiences, sequences of steps, or emphasis. The approach also varies with the amount of available data on design issues, propellants, grains, hardware, or materials, with the degree of novelty (many "new" motors are actually modifications of proven existing motors), or the available proven computer programs for analysis.

Usually the following items are part of the preliminary design process. We start with the requirements for the flight vehicle and the motor, such as those listed in Table 14–6. If the motor to be designed has some similarities to proven existing motors, their parameters and flight experience will be helpful in reducing the design effort and enhancing the confidence in the design. The selection of the propellant and the grain configuration are usually made early in the preliminary design; propellant selection was discussed in Chapter 12 and grains in Chapter 11. It is not always easy for the propellant to satisfy its three key requirements, namely the performance (I_s), burning rate to suit the thrust–time curve, and strength (maximum stress and strain). A well-characterized propellant, a proven grain configuration, or a well-tested piece of hardware will usually be preferred and is often modified somewhat to fit the new application. Compared to a new development, the use of proven propellant, grain design, or hardware components avoids many analyses and tests.

An analysis of the structural integrity should be undertaken, at least in a few of the likely places, where stresses or strains might exceed those that can be tolerated by the grain or the other key components at the limits of loading or environmental conditions. An analysis of the nozzle should be done, particularly if the nozzle is complex or includes thrust vector control. Such a nozzle analysis was described briefly in an earlier section of this chapter. If gas flow analysis shows that erosive burning is likely to happen during a portion of the burning duration, it must be decided whether it can be tolerated, or whether it is excessive and a modification of the propellant, the nozzle material, or the grain geometry needs to be made. Usually a preliminary evaluation is also done of the resonances of the grain cavity with the aim of identifying possible combustion instability modes (see Chapter 13). Motor performance analysis, heat transfer, and stress analyses in critical locations will usually be done.

TABLE 14-6. Typical Requirements and Constraints for Solid Rocket Motors

Requirement Category	Examples
Application	Definition of mission, vehicle and propulsion requirements, flight paths, maneuvers, environment
Functional	Total impulse, thrust–time curve, ignition delay, initial motor mass, specific impulse, TVC angles and accelerations, propellant fraction, class 1.1 to 1.3, burn time, and tolerances on all of these parameters
Interfaces	Attachments to vehicle, fins, TVC system, power supply, instruments, lifting and transport features, grain inspection, control signals, shipping container
Operation	Storage, launch, flight environment, temperature limits, transport loads or vibrations, plume characteristics (smoke, toxic gas, radiation), life, reliability, safe and arm device functions, field inspections
Structure	Loads and accelerations imposed by vehicle (flight maneuvers), stiffness to resist vehicle oscillations, safety factors
Insensitive munitions (military application)	Response to slow and fast cook-off, bullet impact, sympathetic detonation, shock tests
Cost and schedule	Stay within the allocated time and money
Deactivation	Method of removing/recycling of propellants, safe disposal of over-age motors
Constraints	Limits on volume, length, or diameter; minimum acceptable performance, maximum cost

There is considerable interdependence and feedback between the propellant formulation, grain geometry/design, stress analysis, thermal analysis, major hardware component designs, and their manufacturing processes. It is difficult to finalize one of these without considering all the others, and there may be several iterations of each. Data from tests of laboratory samples, subscale motors, and full-scale motors have a strong influence on these steps.

Preliminary layout drawings or CAD (computer-aided design) images of the motor with its key components will be made in sufficient detail to provide sizes and reasonably accurate dimensions. For example, a preliminary design of the thermal insulation (often with a heat transfer analysis) will provide preliminary dimensions for that insulator. The layout is used to estimate volumes, inert masses, or propellant masses, and thus the propellant mass fraction.

If any of these analyses or layouts show a potential problem or a possible failure to meet the initial requirements or constraints, then usually a modification of the design, possibly of the propellant, or of the grain configuration may need to be made. The design process needs to be repeated with the changed motor design. If the proposed changes are too complex or not effective, then a change in the motor requirements may be the cure to a particular problem of noncompliance with the requirements. It is common to have several iterations in the preliminary design and the final design. Any major new feature can result in additional development and testing to prove its performance, reliability, operation, or cost; this means a longer program and extra resources.

A simplified diagram of one particular approach to motor preliminary design and development activities for a rocket motor is shown in Fig. 14–17. Not shown in this diagram are many other steps, such as igniter design and tests, liner/insulating selection, thrust vector control design and test, reliability analysis, evaluation of alternative designs, material specifications, inspection/quality control steps, safety provision, special test equipment, special test instrumentation, and so on.

If the performance requirements are narrow and ambitious, it will be necessary to study the cumulative tolerances of the performance or of various other parameters. For example, practical tolerances may be assigned to the propellant density, nozzle throat diameter (erosion), burn rate scale factor, initial burning surface area, propellant mass, or pressure exponent. These, in turn, reflect themselves into tolerances in process specifications, specific inspections, dimensional tolerances, or accuracy of propellant ingredient weighing. Cost is always a major factor and a portion of the design effort will be spent looking for lower-cost materials, simpler manufacturing processes, fewer assembly steps, or lower-cost component designs. For example, tooling for casting, mandrels for case winding, and tooling for insulator molding can be expensive. The time needed for completing a design can be shortened when there is good communication and a cooperative spirit between designers, propellant specialists, analysts, customer representatives, manufacturing people, test personnel, or vendors concerned with this effort. Reference 14–13 deals with some of the uncertainties of a particular booster motor design, and Ref. 14–14 discusses design optimization.

A *preliminary project plan* is usually formulated simultaneously with the preliminary design work. A decade or more ago the project plan was made after the preliminary design was completed. With today's strong emphasis on low cost, the people working on the preliminary designs have also to work on reducing costs on all components and processes. The project plan reflects decisions and defines the number of motors and key components to be built, the availability and lead time of critical materials or components, the type and number of tests (including aging or qualification tests); it identifies the manufacturing, inspection, and test facilities to be used, the number and kind of personnel (and when they will be needed), or any special tooling or fixtures. These decisions and data are needed to make a realistic *estimate of cost* and a *preliminary schedule*. If these exceed the allowable cost or the desired delivery schedule, then some changes have to be made. For example, this may include changes in the number of units to be built, the number and types of tests, or a redesign for easier, less costly assembly. However, such changes must not compromise reliability or performance. It is difficult to

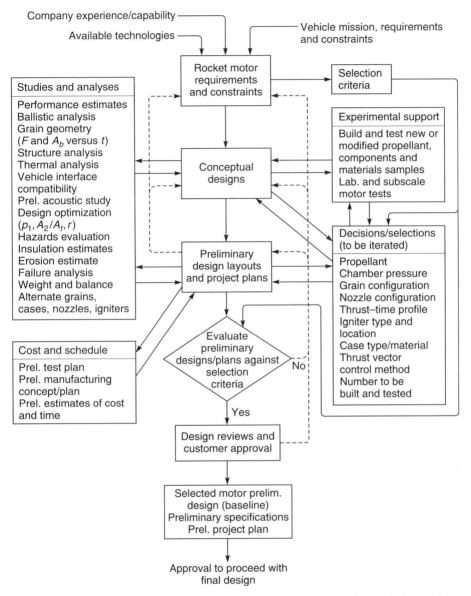

FIGURE 14-17 Simplified diagram of one approach to the preliminary design activity sequences and interrelations. Dashed lines indicate some of the feedback pathes. Some of the specific items listed here apply only to certain types of rocket motors.

make a good plan, and good cost or time estimates, when the rocket motor has not been well defined or designed in sufficient detail. These plans and estimates are therefore largely based on experience with prior successful similar rocket motors.

The final result of a preliminary design will be layout drawings or CAD images of the selected configuration, a prediction of performance, an estimate of motor mass (and, if needed, also the travel of the center of gravity), an identification of the propellant, grain, geometry, insulation, and several of the key materials of the hardware components. An estimate of predicted reliability and motor life would be accompanied by supporting data. All this information would be presented for review of the selected preliminary design. The review would be undertaken by a diverse group of motor experts, the vehicle designer, safety engineers, specialists in manufacturing, assembly, or inspection, customer representatives, analysts, and others. Here the preliminary design team explains why they selected their particular design and how it meets the requirements. With competent reviewers there usually will be suggestions for changes or further improvements. The project plan, preliminary cost estimates, and preliminary schedule are sometimes included with the design review, but more often these are presented to a different group of experts, or just to the customer's experts.

After the design review and the approval of the selected preliminary design, the detail or *final design* of all parts and components and the writing of certain specifications can begin. During manufacture and development testing some design changes may become necessary to improve the manufacture, reduce cost, or remedy a technical problem that became evident. In many organizations the final, detailed design is again submitted to a design review before manufacturing can begin. The new motor will then start its development testing. In some larger, expensive motors, that have a lot of heritage from prior proven motors, the development may consist of a single motor firing. For motors which are built in large quantities and for motors with major new features, the development and qualification may involve the testing of 10 to 30 motors. The final design ends when all detail drawings or CAD images and a final parts list are completed, and specifications for motor testing, certain manufacturing operations, or materials/component acceptance have been prepared. The detail design is considered to be completed when the motor successfully passes its development and qualification tests and begins production for deliveries.

Example 14–1. This example shows one method for making a preliminary determination of the design parameters of a solid rocket using a composite propellant. The rocket is launched at altitude and flies at constant altitude. The following data are given:

Specific impulse (actual)	$I_s = 240$ sec at altitude and 1000 psia
Burning rate	$r = 0.8$ in.,/sec at 1000 psia and 60°F
Propellant density	$\rho_b = 0.066$ lbm/in.^{3} at sea level
Specific heat ratio	$k = 1.25$
Chamber pressure, nominal	$p_1 = 1000$ psi
Desired average thrust	$F = 20,000$ lbf
Maximum vehicle diameter	$D = 16$ in.
Desired duration	$t_b = 5.0$ sec

Ambient pressure   3.0 psi (at altitude)

Vehicle payload   5010 lbm (includes structure)

Approximately neutral burning is desired.

SOLUTION

a. *Basic Design.* The total impulse I_t and propellant weight at sea level w_b are obtained from Eqs. 2–2 and 2–5. $I_t = F t_b = I_s w_b = 20{,}000 \times 5.0 = 100{,}000$ lbf-sec. The propellant weight is $100{,}000/240 = 417$ lbf. Allowing for a loss of 2% for manufacturing tolerances and slivers, the total propellant weight is $1.02 \times 417 = 425$ lbf. The volume required for this propellant V_b is given by $V_b = w_b/\rho_b = 425/0.066 = 6439$ in.^{3}. The web thickness $b = r t_b = 0.80 \times 5 = 4.0$ in.

b. *Case Dimensions.* The outside diameter is fixed at 16.0 in. Heat-treated steel with an ultimate tensile strength of 220,000 psi is to be used. The wall thickness t can be determined from Eq. 14–1 for simple circumferential stress as $t = (p_1/D)/(2\sigma)$. The value of p_1 depends on the safety factor selected, which in turn depends on the heating of the wall, the prior experience with the material, and so on; a safety factor of 2.0 is suggested to allow for surface scratches, combined stresses and welds, and rough field handling. The value of D is the average diameter to the center of the wall. The wall thickness t is

$$t = 2.0 \times 1000 \times 15.83/(2 \times 220{,}000) = 0.086 \text{ in.}$$

A spherical head end and a spherical segment at the nozzle end similar to Fig. 11–1 is assumed.

c. *Grain Configuration.* The grain will be cast into the case but will be thermally isolated from the case with an elastomeric insulator with an average thickness of 0.100 in. inside the case; the actual thickness will be less than 0.10 in. in the cylindrical and forward closure regions, but thicker in the nozzle entry area. The outside diameter D for the grain is determined from the case thickness and liner to be $16.0 - 2 \times 0.086 - 2 \times 0.10 = 15.62$ in. The inside diameter D_i of a simple hollow cylinder grain would be the outside diameter D_o minus twice the web thickness or $D_i = 15.62 - 2 \times 4.0 = 7.62$ in. For a simple cylindrical grain, the volume determines the effective length, which can be determined from the equation

$$V_b = \frac{\pi}{4} L (D_o^2 - D_i^2)$$

$$L = \frac{6439 \times 4}{\pi(15.62^2 - 7.62^2)} = 44.05 \text{ in.}$$

The web fraction would be $2b/D_o = 8/15.62 = 0.512$. The L/D_o is (approximately) $44.05/15.62 = 2.82$.

For grains with this web fraction and this L/D_o ratio, Table 11–4 suggests the use of an internal burning tube with some fins for a cone. A conocyl configuration is selected, although a slotted tube or fins would also be satisfactory. These grain shapes are shown in Figs. 11–1 and 11–17. The initial or average burning area will be found from Eqs. 11–1 and 2–5: namely $F = \dot{w} I_s = \rho_b A_b r I_s$

$$A_b = \frac{F}{\rho_b r I_s} = \frac{20,000}{0.066 \times 0.8 \times 240} = 1578.3 \text{ in.}^2$$

The actual grain now has to be designed into the case with spherical ends, so it will not be a simple cylindrical grain. The approximate volume occupied by the grain is found by subtracting the perforation volume from the chamber volume. There is a full hemisphere at the head end and a partial hemisphere of propellant at the nozzle end (0.6 volume of a full hemisphere).

$$V_b = \tfrac{1}{2}(\pi/6)D_o^3(1 + 0.6) + (\pi/4)D_o^2 L - (\pi/4)D_i^2(L + D_i/2 + 0.3D_i/2) = 6439 \text{ in.}^3$$

This is solved for L, with $D = 15.62$ in. and the inside diameter $D_i = 7.62$ in. The answer is $L = 36.34$ in. The initial internal hollow tube burn area is about

$$\pi D_i(L + D_i/2 + 0.3D_i/2) = 1113 \text{ in.}^2$$

The desired burn area of 1578 in.^{2} is larger by about 465 in.^{2}. Therefore, an additional burn surface area of 465 in.^{2} will have to be designed into the cones of a conocyl configuration or as slots in a slotted tube design. Actually, a detailed geometrical study should be made analyzing the instantaneous burn surface after arbitrary short time intervals and selecting a detailed grain configuration where A_b stays approximately constant. This example does not go through a preliminary stress and elongation analysis, but it should be done.

d. *Nozzle Design.* From Chapter 3 the nozzle parameters can be determined. The thrust coefficient C_F can be found from curves of Figs. 3–6, 3–7, and 3–8 or Eq. 3–30 for $k = 1.25$ and a pressure ratio of $p_1/p_2 = 1000/3 = 333$. Then $C_F = 1.73$. The throat area is from Eq. 3–31:

$$A_t = F/p_1 C_F = 20,000/(1000 \times 1.73) = 11.56 \text{ in.}^2$$

The throat diameter is $D_t = 3.836$ in. The nozzle area ratio for optimum expansion (Fig. 3–6) A_2/A_t is about 27. The exit area and diameter are therefore about $A_e = 312$ in.^{2} and $D_e = 19.93$ in. However, this is larger than the maximum vehicle diameter of 16.0 in. ($A_2 = 201$ in.^{2}), which is the maximum for the outside of the nozzle exit. Allowing for an exit cone thickness of 0.10 in., the internal nozzle exit diameter D_2 is 15.80 in. and A_2 is 196 in.^{2}. This would allow only a maximum area ratio of 196/11.56 or 16.95. Since the C_F values are not changed appreciably for this new area ratio, it can be assumed that the nozzle throat area is unchanged.

This nozzle can have a thin wall in the exit cone, but requires heavy ablative materials, probably in several layers near the throat and convergent nozzle regions. The thermal and structural analysis of the nozzle is not shown here.

c. *Weight Estimate.* The steel case weight (assume a cylinder with two spherical ends and that steel weight density is 0.3 lbf/in.^{3}) is

$$t\pi DL\rho + (\pi/4)tD^2\rho = 0.086\pi \ 15.83 \times 36.34 \times 0.3$$
$$+ 0.785 \times 0.086 \times 15.83^2 \times 0.3$$
$$= 50.9 \text{ lbf}$$

With attachments, flanges, igniter, and pressure tap bosses, this is increased to 57.0 lb. The nozzle weight is composed of the weights of the individual parts, estimated for their densities and geometries. This example does not go through the detailed calculations, but merely gives the result of 30.2 lb. Assume an expended igniter propellant weight of 2.0 lb and a full igniter weight of 5.0 lb. The total weight then is

Case weight at sea level	57.0 lbf
Liner/insulator	14.2 lbf
Nozzle, including fasteners	30.2 lbf
Igniter case and wires	2.0 lbf
Total inert hardware weight	103.4 lbf
Igniter powder	3 lbf
Propellant (effective)	417 lbf
Unuseable propellant (2%)	8 lbf
Total weight	531.4 lbf
Propellant and igniter powder	420.0 lbf

f. *Performance.* The total impulse-to-weight ratio is $100,000/531.4 = 188.2$. Comparison with I_s shows this to be an acceptable value, indicating a good performance. The total launch weight is $5010 + 531.4 = 5541$ lbf, and the weight at burnout or thrust termination is $5541-420 = 5121$ lbf. The initial and final thrust-to-weight ratios and accelerations are

$$F/w = 20,000/5541 = 3.61$$
$$F/w = 20,000/5121 = 3.91$$

The acceleration in the direction of thrust is 3.61 times the gravitational acceleration at start and 3.91 at burnout.

g. *Erosive Burning.* The ratio of the port area to the nozzle throat area at start is $(7.55/3.836)^2 = 3.95$. This is close to the limit of 4.0, and erosive burning is not likely to be significant. A simple analysis of erosive burning in the conical cavity should also be made, but it is not shown here.

PROBLEMS

1. In Figs. 13–4 and 14–16 it can be seen that higher pressures and higher heat transfer rates promote faster ignition. One way to promote more rapid ignition is for the nozzle to remain plugged until a certain minimum pressure has been reached, at which time the nozzle plug will be ejected. Analyze the time saving achieved by such a device, assuming that the igniter gas evolution follows Eqs. 11–3 and

11–11. Under what circumstances is this an effective method? Make assumptions about cavity volume, propellant density, etc.

2. Compare a simple cylindrical case with hemispheric ends (ignore nozzle entry or igniter flanges) for an alloy steel metal and two reinforced fiber (glass and carbon)-wound filament case. Use the properties in Table 14–2 and thin shell structure theory. Given:

Length of cylindrical portion	370 mm
Outside cylinder diameter	200 mm
Internal pressure	6 MPa
Web fraction	0.52
Insulator thickness (average)	
for metal case	1.2 mm
for reinforced plastic case	3.0 mm
Volumetric propellant loading	88%
Propellant specific gravity	1.80
Specific impulse (actual)	248 sec
Nozzle igniter and mounting provisions	0.20 kg

Calculate and compare the theoretical propulsion system flight velocity (without payload) in a gravity-free vacuum for these three cases.

3. The following data are given for a case that can be made of either alloy steel or fiber-reinforced plastic.

Type	Metal	Reinforced Plastic
Material	D6aC	Organic filament composite (Kevlar)
Physical properties		See Table 14–2
Poisson ratio	0.27	0.38
Coefficient of thermal expansions, m/m-K $\times 10^{-6}$	8	45
Outside diameter (m)	0.30	0.30
Length of cylindrical section (m)	0.48	0.48
Hemispherical ends		
Nozzle flange diameter (m)	0.16	0.16
Average temperature rise of case material during operation (°F)	55	45

Determine the growth in diameter and length of the case due to pressurization, heating, and the combined growth, and interpret the results.

4. A high-pressure helium gas tank at 8000 psi maximum storage pressure and 1.5 ft internal diameter is proposed. Use a safety factor of 1.5 on the ultimate strength. The following candidate materials are to be considered:

Kevlar fibers in an epoxy matrix (see Table 14–2)
Carbon fibers in an epoxy matrix
Heat-treated welded titanium alloy with an ultimate strength of 150,000 psi and a weight density of 0.165 lb/in.^{3}

Determine the dimensions and sea level weight of these three tanks and discuss their relative merits. To contain the high-pressure gas in a composite material that is porous, it is also necessary to include a thin metal inner liner (such as 0.016 in.-thick aluminum) to prevent loss of gas; this liner will not really carry structural loads, but its weight and volume need to be considered.

5. Make a simple sketch and determine the mass or sea level weight of a rocket motor case that is made of alloy steel and is cylindrical with hemispherical ends. State any assumptions you make about the method of attachment of the nozzle assembly and the igniter at the forward end.

Outer case and vehicle diameters	20.0 in.
Length of cylinder portion of case	19.30 in.
Ultimate tensile strength	172,000 psi
Yield strength	151,300 psi
Safety factor on ultimate strength	1.65
Safety factor on yield strength	1.40
Nozzle bolt circle diameter	12.0 in.
Igniter case diameter (forward end)	3.00 in.
Chamber pressure, maximum	1520 psi

6. Design a solid propellant rocket motor with insulation and liner. Use the AP/Al-HTPB propellant from Table 11–3 for Orbus 6. The average thrust is 3600 lbf and the average burn time is 25.0 sec. State all the assumptions and rules used in your solution and give your reasons for them. Make simple sketches of a cross section and a half section with overall dimensions (length and diameter), and determine the approximate loaded propellant mass.

7. The STAR 27 rocket motor (Fig. 11–1 and Table 11–3) has an average erosion rate of 0.0011 in./sec. (a) Determine the change in nozzle area, thrust, chamber pressure, burn time, and mass flow at cut-off. (b) Also determine those same parameters for a condition when, somehow, a poor grade of ITE material was used that had three times the usual erosion rate. Comment on the difference and acceptability.
Answer: Nozzle area increases by about (a) 5.3% and (b) 14.7% and chamber pressure at cutoff decreases by approximately the same percentage.

REFERENCES

14–1. NASA, *National Space Transportation System*, Vols. 1 and 2, U.S. Government Printing Office, Washington, DC, June 1988.

14–2. M. Salita, "Simple Finite Element Analysis Model of O-Ring Deformation and Activation during Squeeze and Pressurization," *Journal of Propulsion and Power*, Vol. 4, No. 6, November–December 1988.

14–3. J. H. Hildreth, "Advances in Solid Rocket Motor Nozzle Design and Analysis Technology since 1970," Chapter 2; A. Truchot, "Design and Analysis of Solid Rocket Motor Nozzle," Chapter 3; P. R. Evans, "Composite Motor Case Design," Chapter 4; A. J. P. Denost, "Design of Filament Wound Rocket Cases," Chapter 5; H. Baham and G. P. Thorp, "Consideration for Designers of Cases for small Solid Propellant Rocket Motors," Chapter 6; all in *Design*

Methods in Solid Rocket Motors, AGARD Lecture Series LS 150, Advisory Group for Aerospace Research and Development, NATO, revised 1988.

14–4. B. H. Prescott and M. Macocha, "Nozzle Design," Chapter 6; M. Chase and G. P. Thorp, "Solid Rocket Motor Case Design," Chapter 7; in G. E. Jensen and D. W. Netzer, (Eds), Vol. 170, Progress in Astronautics and Aeronautics, American Institute of Aeronautics and Astronautics, 1996.

14–5. A. de Rouvray, E. Haug, and C. Stavrindis, "Analytical Computations for Damage Tolerance Evaluations of Composite Laminate Structures," *Acta Astronautica*, Vol. 15, No. 11, 1987, pp. 921–930.

14–6. D. Beziers and J. P. Denost, "Composite Curing: A New Process," *AIAA Paper 89-2868*, July 1989.

14–7. A. Groves, J. Margetson, and P. Stanley, "Design Nomograms for Metallic Rocket Cases Reinforced with a Visco-elastic Fiber Over-wind," *Journal of Spacecraft and Rockets*, Vol. 24, No. 5, September–October 1987, pp. 411–415.

14–8. S. Boraas, "Modeling Slag Deposition in Space Shuttle Solid Rocket Motor," *Journal of Spacecraft and Rockets*, Vol. 21, No. 1, January–February 1984.

14–9. P. Gentil, "Design and Development of a New Solid Rocket Motor Nozzle Based on Carbon and Carbon–Ceramic Materials," *AIAA Paper 88-3333*, 1988.

14–10. "Solid Rocket Motor Igniters," *NASA SP-8051*, March 1971 (N71-30346).

14–11. R. Baunchalk, "High Mass Fraction Booster Demonstration," *AIAA Paper 90-2326*, July 1990.

14–12. L. LoFiego, *Practical Aspects of Igniter Design*, Combustion Institute, Western States Section, Menlo Park, CA, 1968 (AD 69-18361).

14–13. R. Fabrizi and A. Annovazzi, "Ariane 5 P230 Booster Grain Design and Performance Study," *AIAA Paper 89-2420*, July 1989.

14–14. A. Truchot, "Overall Optimization of Solid Rocket Motors," Chapter 11 in *Design Methods in Solid Rocket Motors*, AGARD Lecture Series LS 150, Advisory Group for Aerospace Research and Development, NATO, revised 1988.

CHAPTER 15

HYBRID PROPELLANT ROCKETS
Terry A. Boardman[*]

Rocket propulsion concepts in which one component of the propellant is stored in liquid phase while the other is stored in solid phase are called hybrid propulsion systems. Such systems most commonly employ a liquid oxidizer and solid fuel[†]. Various combinations of solid fuels and liquid oxidizers as well as liquid fuels and solid oxidizers have been experimentally evaluated for use in hybrid rocket motors. Most common is the liquid oxidizer–solid fuel concept shown in Fig. 15–1. Illustrated here is a large pressure-fed hybrid booster configuration. The means of pressurizing the liquid oxidizer is not an important element of hybrid technology and a turbopump system could also perform this task.[‡] The oxidizer can be either a noncryogenic (storable) or a cryogenic liquid, depending on the application requirements.

In this hybrid motor concept, oxidizer is injected into a precombustion or vaporization chamber upstream of the primary fuel grain. The fuel grain contains numerous axial combustion ports that generate fuel vapor to react with

---

[*]This is a revision of Chapter 15 in the 6th edition of *Rocket Propulsion Elements* originally authored by Terry A. Boardman, Alan Holzman, and George P. Sutton.
[†]The term hybrid has also been applied to liquid monopropellant systems, electrical propulsion systems where a resistor or electric arc raises the temperature of the reaction gases, solid propulsion systems utilizing separate fuel-rich and oxidizer-rich propellant grains (so called solid–solid hybrid), or the solid fuel ramjet that has a combustion cycle very similar to the hybrid concept discussed in this chapter. The solid fuel ramjet is mentioned briefly in Chapter 1.
[‡]Hybrid technology in this context is construed to encompass combustion physics, fuel grain design, and materials selection for nozzle and internal case insulators. The choice of pressure feeding or turbopump feeding oxidizer to the combustion chamber impacts vehicle performance through differences in mass fraction and specific impulse, but is considered an element of liquid propulsion technology rather than hybrid technology.

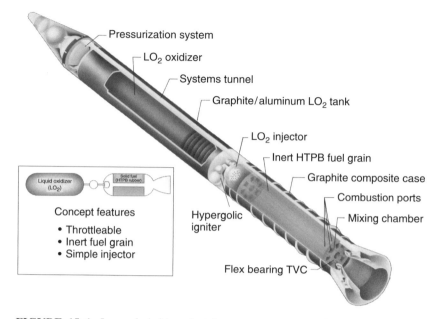

FIGURE 15–1. Large hybrid rocket booster concept capable of boosting the Space Shuttle. It has an inert solid fuel grain, a pressurized liquid oxygen feed system, and can be throttled.

the injected oxidizer. An aft mixing chamber is employed to ensure that all fuel and oxidizer are burned before exiting the nozzle.

The main advantages of a hybrid rocket propulsion system are: (1) safety during fabrication, storage, or operation without any possibility of explosion or detonation; (2) start–stop–restart capabilities; (3) relatively low system cost; (4) higher specific impulse than solid rocket motors and higher density-specific impulse than liquid bipropellant engines; and (5) the ability to smoothly change motor thrust over a wide range on demand.

The disadvantages of hybrid rocket propulsion systems are: (1) mixture ratio and, hence, specific impulse will vary somewhat during steady-state operation and throttling; (2) lower density-specific impulse than solid propellant systems; (3) some fuel sliver must be retained in the combustion chamber at end-of-burn, which slightly reduces motor mass fraction; and (4) unproven propulsion system feasibility at large scale.

15.1. APPLICATIONS AND PROPELLANTS

Hybrid propulsion is well suited to applications or missions requiring throttling, command shutdown and restart, long-duration missions requiring storable nontoxic propellants, or infrastructure operations (manufacturing and launch) that would benefit from a non-self-deflagrating propulsion system. Such applications would include primary boost propulsion for space launch vehicles, upper stages, and satellite maneuvering systems.

Many early hybrid rocket motor developments were aimed at target missiles and low-cost tactical missile applications (Ref. 15–1). Other development efforts focused on high-energy upper-stage motors. In recent years development efforts have concentrated on booster prototypes for space launch applications. Design requirements for one target missile, which entered production in the early 1970s, included a nominal thrust of 2200 N with an 8:1 throttling range, storable liquid oxidizer, and engine shutdown on command. Selected propellants included a nitrogen tetroxide/nitrous oxide oxidizer and a hydrocarbon fuel grain composed of polymethylmethacrylate (plexiglass) and magnesium (Ref. 15–2). Values of vacuum-delivered specific impulse for such storable propellant systems range between 230 and 280 sec. In another program (Ref. 15–3), a hybrid motor was developed for high-performance upperstage applications with design requirements that included a nominal thrust level of 22,240 N and an 8:1 throttling range. Oxygen difluoride was selected as the oxidizer for use with a lithium hydride/polybutadiene fuel grain. Analytical and experimental investigations have been made using other highperformance propellants. High-energy oxidizers include fluorine/liquid oxygen mixtures (FLOX) and chlorine/fluorine compounds such as CIF_3 and CIF_5. Complementary high-energy fuels are typically hydrides of light metals, such as beryllium, lithium, and aluminum, mixed with a suitable polymeric binder (Ref. 15–4). Delivered vacuum-specific impulse levels for these high-energy hybrid propellants are in the 350 to 380 sec range, depending on nozzle expansion ratio. Combustion efficiencies of 95% of theoretical values have been achieved in tests with these propellants; however, none of these exotic formulation systems have seen use on flight vehicles.

A more practical, although lower energy, upper-stage hybrid propellant system is 90 to 95% hydrogen peroxide oxidizer combined with hydroxylterminated polybutadiene (HTPB) fuel. Hydrogen peroxide is considered storable for time periods typical of upper-stage mission cycles (oxidizer tanking to mission completion on the order of several months) and is relatively inexpensive. In solid rocket motors, HTPB is used as the binder to consolidate the aluminum fuel and ammonium perchlorate oxidizer matrix. In a hybrid, HTPB becomes the entire fuel constituent. HTPB is low cost, processes easily, and will not self-deflagrate under any conditions.

The propellant system of choice for large hybrid booster applications is liquid oxygen (LOX) oxidizer and HTPB fuel. Liquid oxygen is a widely used oxidizer in the space launch industry, is relatively safe, and delivers high performance at low cost. This hybrid propellant combination produces a nontoxic, relatively smoke-free exhaust. The LOX/HTPB propellant combination favored for booster applications is chemically and performance-wise equivalent to a LOX–kerosene bipropellant system.

Where a smoky exhaust is not a detriment, hybrid propellants for certain applications may benefit from the addition of powdered aluminum to the fuel. This increases the combustion temperature, reduces the stoichiometric mixture ratio, and increases fuel density as well as overall density-specific impulse. Although density-specific impulse ($\rho_f I_s$) is increased, addition of aluminum to the fuel actually reduces specific impulse. This occurs because the increase in flame temperature gained by adding aluminum does not compensate for the increase in molecular weight of the exhaust products. Figure 15–2 illustrates theoretical vacuum specific impulse levels (calculated at 1000 psia chamber pressure and a 10:1 nozzle expansion ratio) for a variety of cryogenic and storable oxidizers used in conjunction with HTPB fuel. Table 15–1 tabulates the heat of formation for HTPB reacted with various oxidizers.

Large hybrid development work completed to date has focused on motors having a thrust level of approximately 1,112,000 N or 250,000 lbf. The American Rocket Company first tested a 250,000 lbf thrust LOX/HTPB hybrid in 1993 (Ref. 15–5). In 1999, a consortium of aerospace companies also tested several 250,000 lbf thrust LOX/HTPB hybrid prototypes as a candidate strap-on booster for space launch vehicles (see Ref. 15–6 and Fig. 15–3). In these motors, polycyclopentadiene (PCPD) is added to the

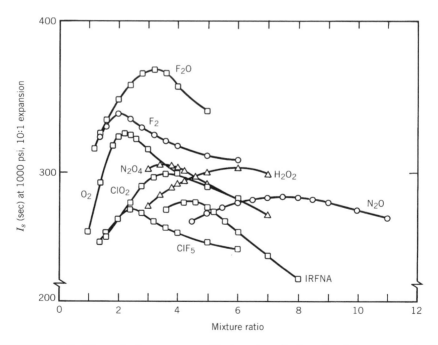

FIGURE 15–2. Theoretical vacuum specific impulse of selected oxidizers reacted with hydroxyl-terminated polybutadiene fuel. The I_s of the O_2/HTPB propellant is comparable to that of a LOX/kerosene bipropellant engine.

TABLE 15–1. Thermochemical Properties of Selected Oxidizers Reacted with HTPB Fuel

Oxidizer	Type	Boiling Point (°C)	Density (g/cm^3)	$\Delta_f H^a$ (kcal/mol)
O_2	Cryogenic	−183	1.149	−3.1
F_2	Cryogenic	−188	1.696	−3.0
O_3	Cryogenic	−112	1.614	+30.9
F_2O	Cryogenic	−145	1.650	+2.5
F_2O_2	Cryogenic	−57	1.450	+4.7
N_2O	Cryogenic	−88	1.226	+15.5
N_2O_4	Storable	+21	1.449	+2.3
IRFNA[b]	Storable	+80 to +120	1.583	−41.0
H_2O_2	Storable	+150	1.463	44.8
ClO_2	Storable	+11	1.640	+24.7
ClF_3	Storable	+11	1.810	−44.4

[a] ΔfH is the heat of formation as defined in Chapter 5.
[b] Inhibited red fuming nitric acid.

HTPB fuel to increase fuel density by about 10% over HTPB alone. The motors were designed to operate for 80 sec at a LOX flow rate of 600 lbm/sec with a maximum chamber pressure of 900 psi. Figure 15–4 illustrates a cross section of one motor configuration. Test results indicated additional work is necessary to develop large hybrid motor configurations that exhibit stable combustion throughout the motor burn, and in understanding fuel regression-rate scale-up factors.

A hybrid fuel grain is ignited by providing a source of heat, which initiates gasification of the solid fuel grain at the head end of the motor. Subsequent initiation of oxidizer flow provides the required flame spreading to fully ignite the motor. Ignition is typically accomplished by injection of a hypergolic fluid into the motor combustion chamber. Using the motor described in Fig. 15–4 as an example, a mixture of triethyl aluminum (TEA) and triethyl borane (TEB) is injected into the vaporization chamber. The TEA/TEB mixture ignites spontaneously on contact with air in the combustion chamber, vaporizing fuel in the dome region. Subsequent injection of liquid oxygen completes ignition of the motor. TEA/TEB mixtures are currently used for motor ignition in the Atlas and Delta commercial launch vehicles. Experimenters (Refs. 15–7 and 15–8) have described solid fuels that will ignite spontaneously at ambient temperature and pressure when sprayed with specific oxidizers other than LOX. Small hybrid motors, such as those used in a laboratory environment with gaseous oxygen oxidizer, are often electrically ignited by passing current through a resistor such as steel wool located in the combustion port, or by use of a propane or hydrogen ignition system.

FIGURE 15–3. Static tests of a 250,000 lbf thrust hybrid motor prototype demonstrated that additional work is needed to understand fuel regression and combustion stability issues at large scale. The fuel case shown here is approximately 6.3 ft diameter.

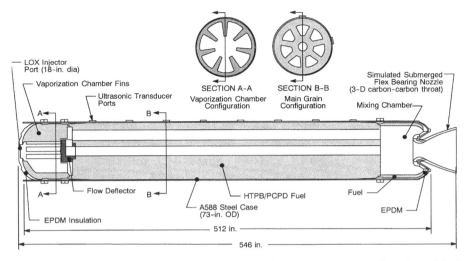

FIGURE 15–4. 250,000 lbf thrust hybrid booster design parameters and section of fuel grain and nozzle. The vaporization chamber fins and flow deflector are designed to promote flame holding in combustion ports.

Maximum operating pressure	900 psia
Maximum vacuum thrust	250,000 lbf
Throat diameter, initial	14.60 in.
Nozzle expansion ratio, initial	12
Liquid oxygen flow rate	420 to 600 lbm/sec (throttlable)
Fuel weight	45,700 lbf
Burn time	80 sec

15.2. PERFORMANCE ANALYSIS AND GRAIN CONFIGURATION

A characteristic operating feature of hybrids is that the fuel regression rate is typically less than one-third that of composite solid rocket propellants. It is very difficult to obtain fuel regression rates comparable to propellant burn rates in solid rocket motors. Consequently, practical high-thrust hybrid motor designs must have multiple perforations (combustion ports) in the fuel grain to produce the required fuel surface area. The performance of a hybrid motor (defined in terms of delivered specific impulse) depends critically on the degree of flow mixing attained in the combustion chamber. High performance stems from high combustion efficiency that is a direct function of the thoroughness with which unburned oxidizer exhausting from the combustion port is mixed with unburned fuel from within sublayers of the boundary layer. Multiple combustion ports serve to promote high combustion efficiency as a result of the turbulent mixing environment for unreacted fuel and oxidizer in the mixing chamber region downstream of the fuel grain.

A cross section of a typical high-thrust hybrid fuel grain is shown in Fig. 15–5. The number of combustion ports required is a motor optimization problem that must account for the desired thrust level, acceptable shifts in mixture ratio during burn, motor length and diameter constraints, and desired oxidizer mass velocity. Hybrid rocket motor design typically begins by specifying a desired thrust level and a propellant system. Subsequently, selection of the desired operating oxidizer-to-fuel mixture ratio (O/F ratio) determines the propellant characteristic velocity. Once the characteristic velocity and mixture ratio are specified, the total propellant flow rate and the subsequent split between oxidizer and fuel flow rates necessary to produce the required thrust level can be computed. The necessary fuel flow rate in a hybrid is determined by the total fuel surface area (perimeter and length of the combustion ports) and the fuel regression rate. As will be shown in subsequent sections, the fuel regression rate is primarily determined by the oxidizer mass velocity, also called oxidizer flux. The oxidizer flux is equal to the mass flow rate of oxidizer in a combustion port divided by the port cross-sectional area. Thus the fuel flow rate is intrinsically linked to the oxidizer flow rate and cannot be independently specified, as in a liquid rocket engine.

Much of the technology from liquid and solid propellant rockets is directly applicable to hybrid rockets; the main differences lie in the driving mechanisms for solid propellant burning and hybrid fuel regression. In a solid system, the oxidizer and fuel ingredients are well mixed during the propellant manufacturing process. Combustion occurs as a result of heterogeneous chemical reactions

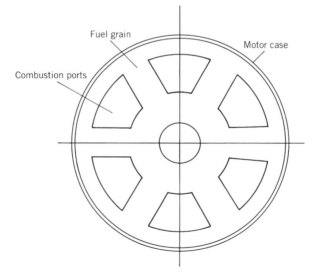

FIGURE 15–5. Cross-sectional sketch of a multi-port fuel grain with web thickness between ports twice that of the outer wall. Multiple ports are required to achieve the large fuel surface area necessary for high fuel flow rates.

on or very near the surface of the solid propellant. The solid propellant burning rate is controlled by chamber pressure and follows the well-established law of Eq. 11–3; it is Eq. 15–1 in this chapter.

$$\dot{r} = ap_1^n \tag{15–1}$$

where a and n are empirical coefficients derived experimentally for specific propellant formulations. Since the rate of propellant gasification per unit area in a solid rocket motor, at a given propellant bulk temperature and in the absence of erosive burning, is determined only by chamber pressure, motor thrust is predetermined by the initial propellant grain surface area and grain geometrical characteristics. Throttling or extinguishment is very difficult to achieve in practical solid rocket motor configurations since the fuel and oxidizer cannot be separated.

As the fuel grain of a hybrid typically contains no oxidizer, the combustion process and hence the regression of the fuel surface is markedly different from that of a solid rocket motor. Because the solid fuel must be vaporized before combustion can occur, the fuel surface regression is intrinsically related to the coupling of combustion port aerodynamics and heat transfer to the fuel grain surface. The primary combustion region over the fuel grain surface has been shown to be limited to a relatively narrow flame zone occurring within the fuel grain boundary layer (Ref. 15–9). Factors affecting the development of the fuel grain boundary layer and, hence, fuel regression characteristics include pressure, gas temperature, grain composition, combustion port oxidizer mass flow rate, and combustion port length. The heat transfer relationships between the gas and solid phase depend on whether the boundary layer is laminar or turbulent. In a typical hybrid using oxygen as the oxidizer, the Reynolds number per unit length is on the order of 1 to 2×10^5 per inch of grain length for flux levels between 0.3 and 0.6 lbm/sec/in.[2] (see Appendix 4 for definitions of non-dimensional parameters used in hybrid boundary layer analyses). Thus, the properties of a turbulent boundary layer govern the convective heat transfer processes to non-metallized fuel grains.

In hybrids with metallized fuel grains, radiation from the metal oxide particle cloud in the combustion port contribues a major portion of the total heat flux to the fuel grain. The local regression rate of the fuel is also quite sensitive to the general turbulence level of the combustion port gas flow (Refs. 15–10 and 15–11). Localized combustion gas eddies or recirculation zones adjacent to the fuel surface act to significantly enhance the regression rate in these areas. Hybrid fuel regression rate is thought to be insensitive to fuel grain bulk temperatures over the range in which solid rocket motors may operate (−65°F to 165°F). This is due to the absence of heterogeneous fuel/oxidizer reactions at the fuel surface (in which the reaction rates are temperature dependent) and because, over the above temperature range, the change in heat content of the solid fuel is small compared to the heat necessary to initiate vaporization of the fuel surface.

Selection of fuel ingredients can also have a significant impact on the grain regression rate, which is largely a function of the energy required to convert the fuel from solid to vapor phase (h_v). This energy is called the heat of gasification and, for polymeric fuels, includes the energy required to break polymer chains (heat of depolymerization) and the heat required to convert polymer fragments to gaseous phase (heat of vaporization). The term "heat of vaporization" is often used as a catchall phrase to include all decomposition mechanisms in hybrid fuels. In non-metallized fuels, low heats of gasification tend to produce higher regression rates. In metallized fuels, the addition of ultra-fine aluminum (UFAl) powder (particle sizes on the order of 0.05 μm to 0.1 μm) to HTPB has been noted to significantly increase the fuel regresion rate relative to a pure HTPB baseline (see Ref. 15–12 and Fig. 15–6). Hybrid propellants containing aluminum particles with diameters typical of those used in solid rocket propellants (40 μm to 400 μm) do not exhibit this effect.

Figure 15–7 depicts a simplified model of the hybrid combustion process for a non-metallized (non-radiating) fuel system. Fuel is vaporized as a result of heat transferred from the flame zone to the fuel mass. Vaporized fuel is convected upward toward the flame zone while oxidizer from the free stream (core flow) is transported to the flame zone by diffusion and flow turbulence. The flame is established at a location within the boundary layer determined by the stoichiometric conditions under which combustion can occur. The thickness of the flame is determined primarily by the rate at which the oxidation reaction

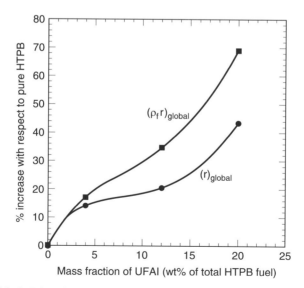

FIGURE 15–6. Ultra-fine aluminum (UFAL) powder mixed with HTPB significantly increases the fuel regression rate.

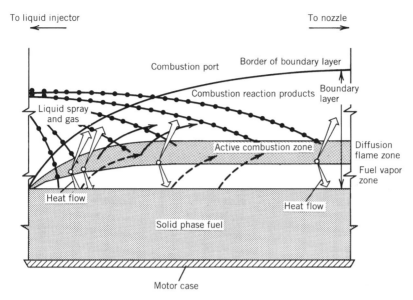

FIGURE 15–7. Simplified model of the diffusion-controlled hybrid combustion process, illustrating the flame zone embedded within the fuel boundary layer.

occurs. This rate is largely dependent on pressure and typically follows an Arrhenius relationship.

The mechanisms of heat transfer to the fuel grain surface in a hybrid are convection and radiation. In a non-metallized fuel grain, at pressures and flux levels of interest for propulsion applications, heat transferred by convection is thought to be much larger than that transferred by gas-phase radiation or radiation from soot particles in the flow. As a result, the basic characteristics of fuel grain regression may be explored via an analysis of convective heat transfer in a turbulent boundary layer (see Appendix 4). Considering an energy balance at the fuel grain surface, one may derive an expression for the fuel surface regression rate as

$$\dot{r} = 0.036 \frac{G^{0.8}}{\rho_f} \left(\frac{\mu}{x} \right)^{0.2} \beta^{0.23} \tag{15–2}$$

where G is the free stream propellant mass velocity (total oxidizer and fuel flow per unit area) in a combustion port at any given axial location x, ρ_f is the solid-phase fuel density, μ is the combustion gas viscosity, and β is the non-dimensionalized fuel mass flux, resulting from fuel vaporization, evaluated at the fuel surface. The parameter β is frequently referred to as a blowing coefficient (see Appendices 4 and 5 for further discussion of β). Equation 15–2 indicates that hybrid fuel regression rate for a non-radiative system is strongly dependent on G and rather weakly dependent on axial location (x) and fuel blowing characteristics (β). One may also note that the regression rate is not explicitly dependent on chamber pressure in this derivation. In fact, experiments have shown that the regression rate for some fuels exhibits little or no dependence on chamber pressure whereas the regression rate for others exhibits a strong dependence. In particular, metallized hybrid fuel systems exhibit a pronounced pressure dependence (Ref. 15–13).

As the combustion port length increases, fuel added to the port mass flow increases the total port mass flux. In ports operating at low mixture ratios, the fuel mass increase may be on the same order as the oxidizer mass flow initially entering the port. Given the weak dependence of regression rate on x in Eq. 15–2, one would therefore expect the fuel regression rate to increase with increasing axial length due to the increase in G. While this generally turns out to be the case, fuel regression rate has been observed to both increase and decrease with increasing x, depending on specifics of the motor configuration. In practice, axial fuel regression characteristics are strongly influenced by oxidizer injection and pre-combustion/vaporization chamber design characteristics. General trends that have been measured in hybrid combustion ports include the following as x increases: total mass flux increases; boundary layer thickness grows; flame standoff from the fuel surface increases; combustion port average gas temperature increases; oxidizer concentration decreases.

Since the blowing coefficient β is not only an aerodynamic parameter but also a thermochemical parameter (see Appendix 5) and the x dependency is of the same order as β in Eq. 15–2, this expression is often simplified for purposes of preliminary engineering design by lumping effects of x, β, fuel density, and gas viscosity into one parameter, a. In practice, deviations from the theoretical 0.8 power mass velocity dependency are also often noted. The result of simplifying Eq. 15–2 is to retain the functional form but fit the free constants a and n using experimental data obtained from characterizing specific fuel and oxidizer combinations. One functional form useful for engineering evaluations is

$$\dot{r} = aG_o^n \tag{15–3}$$

where G_o is the oxidizer mass velocity, which is equal at any time to the oxidizer flow rate divided by the combustion port area. The value of $\dot{r}$ has been observed to vary from 0.05 in./sec to 0.2 in./sec. Likewise, n has been observed to fall in a range between 0.4 and 0.7. An alternative form of Eq. 15–3, to account for an observed pressure and/or port diameter dependency, is given as

$$\dot{r} = aG_o^n p_1^m D_p^l \tag{15–4}$$

where m and l have been observed to vary between zero and 0.25 and zero and 0.7, respectively.

Figure 15–8 illustrates surface regression rate data obtained for the combustion of HTPB fuel grains and gaseous oxygen in rocket motor tests at two

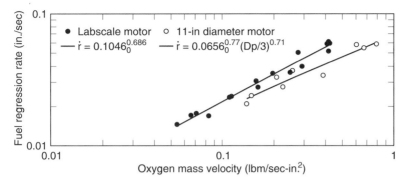

FIGURE 15–8. Hybrid regression rate has been observed to decrease as motor scale (combustion port diameter) increases.

different scales. The first data set were obtained by testing fuel grains in a small laboratory-scale (2-in. motor diameter with a 0.43-in. combustion port diameter) rocket at varying gaseous oxygen flux levels (Ref. 15–14). A least-squares regression analysis, performed to determine the constants in Eq. 15–3, indicates that, at this scale, the following relationship best describes the regression rate characteristics of HTPB as a function of oxygen mass flux:

$$\dot{r}_{\text{HTPB}} = 0.104 G_o^{0.681} \qquad (15\text{–}5)$$

Data obtained with the same propellant system in a larger 11-in. diameter hybrid motor with combustion port diameters ranging between 3 and 6 in. exhibited a relatively strong dependence on combustion port diameter (Ref. 15–15). Data from this testing was best matched with an expression in the form of Eq. 15–4:

$$\dot{r}_{\text{HTPB}} = 0.065 G_o^{0.77} (D_p/3)^{0.71} \qquad (15\text{–}6)$$

The difference in fuel regression characteristics between the two motor scales illustrates one of the central difficulties of hybrid motor design, i.e., that of scaling ballistic performance. Scaling issues in hybrid motors are currently not well understood (in part because of the lack of sufficient valid data for different motor sizes) and the literature abounds with empirical regression rate scaling relationships (Ref. 15–16). Computational fluid dynamic approaches to resolving the hybrid flow field and calculating fuel surface heating appear to offer the best hope of analytically evaluating scale effects.

The dynamic behavior of a hybrid rocket may be analyzed using the continuity equation

$$\frac{\partial(\rho_1 V_1)}{\partial t} = \dot{m}_{\text{in}} - \dot{m}_{\text{out}} \qquad (15\text{–}7)$$

that expresses that the time rate of change of high-pressure gas inside the chamber is equal to the difference between the hot gas generated from inflow of liquid oxidizer, plus that generated from the regressing fuel surface, and the flow through the nozzle. Equation 15–7 may be rewritten as

$$\frac{\partial(\rho_1 V_1)}{\partial t} = \dot{m}_o + \dot{m}_f - \frac{p_1 A_t}{c^*} \tag{15–8}$$

When steady state is reached, Eq. 15–8 reduces to

$$\dot{m} = \dot{m}_o + \dot{m}_f = \frac{p_1 A_t}{c^*} \tag{15–9}$$

The thrust of a hybrid rocket motor can then be expressed as

$$F = \dot{m} I_s g_0 = (\dot{m}_o + \dot{m}_f) I_s g_0 \tag{15–10}$$

Changing the thrust or throttling of a hybrid is achieved by changing the oxidizer flow rate, usually by means of a throttling valve in the oxidizer feed line. The fuel flow is a function of the oxidizer flow but not necessarily a linear function. For circular port geometries with radius R, Eq. 15–3 may be recast as

$$\dot{r} = a\left(\frac{\dot{m}_o}{\pi R^2}\right)^n \tag{15–11}$$

The mass production rate of fuel is given by

$$\dot{m}_f = \rho_f A_b \dot{r} = 2\pi \rho_f R L \dot{r} \tag{15–12}$$

where A_b is the combustion port surface area and L is the port length. Combining Eqs. 15–11 and 15–12, one obtains the fuel production rate in terms of port radius and oxidizer mass flow rate:

$$\dot{m}_f = 2\pi^{1-n} \rho_f L a \dot{m}_o^n R^{1-2n} \tag{15–13}$$

From this expression one will note that, for the particular value of $n = \frac{1}{2}$, the fuel mass flow rate is independent of combustion port radius and varies as the square root of oxidizer mass flow rate. For such a situation, if the oxidizer flow is reduced to one-half of its rated value, then the fuel flow will be reduced by a factor of 0.707 and the motor thrust, which depends on the total propellant flow $(\dot{m}_f + \dot{m}_o)$, will not vary linearly with the change in oxidizer flow. Usually, as the thrust is decreased by reducing the oxizider flow, the mixture ratio $(\dot{m}_o / \dot{m}_f)$ is reduced, becoming increasingly fuel rich. In some hybrid motor concepts, a portion of the oxidizer is injected in a mixing chamber downstream of the fuel grain in order to maintain a more constant mixture ratio. However, for most applications, the system design can be optimized over the range of mixture ratios encountered with very little degradation of average specific impulse due to throttling.

Equation 15–13 also indicates that, for constant oxidizer flow, fuel production will increase with increasing port radius if $n < \frac{1}{2}$. For $n > \frac{1}{2}$, fuel production will decrease with increasing port radius.

For a fuel grain incorporating N circular combustion ports, Eq. 15–11 can be simply integrated to give combustion port radius, instantaneous fuel flow rate, instantaneous mixture ratio, and total fuel consumed as functions of burn time:

Combustion port radius R as a function of time and oxidizer flow rate:

$$R(t) = \left\{ a(2n + 1)\left(\frac{\dot{m}_o}{\pi N}\right)^n t + R_i^{2n+1} \right\}^{\frac{1}{2n+1}} \tag{15–14}$$

Instantaneous fuel flow rate:

$$\dot{m}_f(t) = 2\pi N\, \rho_f La\left(\frac{\dot{m}_o}{\pi N}\right)^n \left\{ a(2n + 1)\left(\frac{\dot{m}_o}{\pi N}\right)^n t + R_i^{2n+1} \right\}^{\frac{1-2n}{1+2n}} \tag{15–15}$$

Instantaneous mixture ratio:

$$\frac{\dot{m}_o}{\dot{m}_f}(t) = \frac{1}{2\rho_f La}\left(\frac{\dot{m}_o}{\pi N}\right)^{(1-n)} \left\{ a(2n + 1)\left(\frac{\dot{m}_o}{\pi N}\right)^n t + R_i^{2n+1} \right\}^{\frac{2n-1}{2n+1}} \tag{15–16}$$

Total fuel consumed:

$$m_f(t) = \pi N\, \rho_f L\left[\left\{ a(2n + 1)\left(\frac{\dot{m}_o}{\pi N}\right)^n t + R_i^{2n+1} \right\}^{\frac{2}{2n+1}} - R_i^2 \right] \tag{15–17}$$

where L is the fuel grain length, R_i is the initial port radius, N is the number of combustion ports of radius R_i in the fuel grain, and $\dot{m}_o$ and $\dot{m}_f$ are the total oxidizer and fuel flow rates, respectively. Although the above equations are strictly valid only for circular combustion ports, they may be used to give a qualitative understanding of hybrid motor behavior which is applicable to the burnout of non-circular ports as well.

15.3. DESIGN EXAMPLE

The preliminary design problem typically posed is to determine the approximate size of a hybrid booster, given numerous system requirements and design assumptions. Suppose that the operating characteristics of a Space Shuttle-class hybrid rocket booster are to be determined, given the following initial design requirements:

Fuel	HTPB
Oxidizer	Liquid oxygen
Required booster initial thrust (vacuum)	3.1×10^6 lbf
Burn time	120 sec
Fuel grain outside diameter	150 in.
Initial chamber pressure	700 psia
Initial mixture ratio	2.0
Initial expansion ratio	7.72

Using the ratio of specific heats from Table 15–2 and the given initial nozzle expansion ratio, the vacuum thrust coefficient is determined from tables or direct calculation to be 1.735. Initial nozzle throat area and throat diameter are determined from

$$A_t = \frac{F_v}{C_{F_v} p_1} = \frac{3.1 \times 10^6 \text{ lbf}}{(1.735)(700 \text{ lbf/in.}^2)} = 2552.5 \text{ in.}^2$$

then $D_t = 57.01$ in. From the data of Table 15–2 for c^* versus mixture ratio, c^* corresponding to an initial mixture ratio of 2.0 is 5912 ft/sec. Theoretical c^* values are typically degraded to account for combustion inefficiency due to incomplete oxidizer/fuel mixing. Using a factor of 95%, the delivered c^* is 5616 ft/sec. Total initial propellant flow rate can now be determined as

TABLE 15–2. Theoretical Characteristic Velocity c^* and Ratio of Specific Heats k for Reaction Gases of Liquid Oxygen–HTPB Fuel

Mass Mixture Ratio	c^* (ft/sec)	k
1.0	4825	1.308
1.2	5180	1.282
1.4	5543	1.239
1.6	5767	1.201
1.8	5882	1.171
2.0	5912	1.152
2.2	5885	1.143
2.4	5831	1.138
2.6	5768	1.135
2.8	5703	1.133
3.0	5639	1.132

$$\dot{m} = \frac{g_o p_1 A_t}{c^*} = \frac{\left(32.174 \frac{\text{lbm-ft}}{\text{lbf-sec}^2}\right)(700 \text{ lbf/in.}^2)(2552.5 \text{ in.}^2)}{(0.95)(5912 \text{ ft/sec}^2)} = 10{,}236 \text{ lbm/sec}$$

Noting that mixture ratio is defined as

$$r = \dot{m}_o / \dot{m}_f$$

initial fuel and oxidizer flow rates follow at the initial mixture ratio of 2.0:

$$\dot{m} = \dot{m}_o + \dot{m}_f = \dot{m}_f(r+1)$$
$$\dot{m}_f = \frac{10{,}236 \text{ lbm/sec}}{3} = 3412 \text{ lbm/sec}$$
$$\dot{m}_o = 10{,}236 - 3412 = 6824 \text{ lbm/sec}$$

Figure 15–9a illustrates a candidate seven-circular-port symmetric fuel grain configuration. The dashed lines represent the diameters to which the combustion ports burn at the end of 120 sec. The problem is to determine the initial port diameter such that, at the end of the specified 120-sec burn time, the grain diameter constraint of 150 in. is satisfied. The unknown quantity in this problem is the initial combustion port radius, R_i, and the fuel burn distance, d_b. In terms of initial port radius, the burn distance can be expressed via Eq. 15–14 as

$$d_b = R(t, R_i)|_{t=120} - R_i$$

The fuel grain diameter requirement of 150 in. is satisfied by the following relation:

$$150 \text{ in.} = 6R_i + 6d_b$$

Sub-scale motor test data indicate that one expression for the fuel surface regression rate can be described by Eq. 15–5. Assuming that these data are valid for the flux levels and port diameters under consideration (ignoring potential regression rate scaling issues), the above two relations can be combined to solve for the initial port radius and distance burned, yielding

$$R_i = 14.32 \text{ in.} \qquad d_b = 10.68 \text{ in.}$$

Knowing the initial port radius, the oxidizer mass velocity can be determined:

$$G_o = \frac{\dot{m}_o}{N \, A_p} = \frac{6824 \text{ lbm/sec}}{7\pi(14.32 \text{ in.})^2} = 1.51 \text{ lbm/in.}^2\text{-sec}$$

The initial fuel regression rate may be explicitly determined from Eq. 15–5:

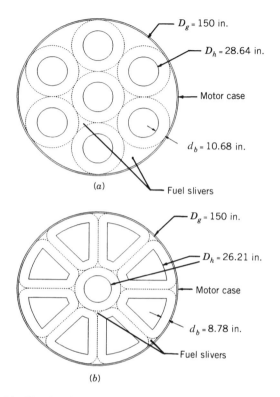

FIGURE 15–9. (a) Circular fuel grain combustion ports are volumetrically inefficient and leave large slivers at burnout. (b) Quadrilateral port hybrid grain configuration minimizes residual fuel sliver at burnout.

$$\dot{r}_i = 0.104G_{oi}^{0.681} = 0.104(1.51 \text{ lbm/ft}^2\text{-sec})^{0.681} = 0.138 \text{ in./sec}$$

From the initial fuel mass flow rate, determined to be 3412 lbm/sec, the fuel grain length required for a seven-circular-port design may be found from Eq. 15–12:

$$L = \frac{\dot{m}_f/N}{2\pi R_i \rho_f \dot{r}_i} = \frac{(3412 \text{ lbm/sec})/7}{\pi(28.65 \text{ in.})(0.033 \text{ lbm/in.}^3)(0.138 \text{ in./sec})} = 1189.6 \text{ in.}$$

Using Eqs. 15–9, 15–10, 15–15, 15–16, and 15–17, while neglecting effects of throat erosion, the general operating characteristics of the booster may be computed with respect to time. The total fuel and oxidizer required for a 120-sec burn time are determined to be 362,577 and 818,880 lbm respectively. The total propellant mass required is therefore 1,181,457 lbm.

Selection of circular fuel ports is not an efficient way of designing a hybrid grain since large fuel slivers will remain at the end of burn. In the preceding example, a sliver fraction (1 minus fuel consumed divided by fuel loaded) of 29.8% can be calculated. Recognizing that uniform burn distances around each port, as well as between combustion ports and the case wall, will minimize residual fuel sliver, the outer ring of circular ports may be replaced with quadrilateral-shaped ports. Such a grain is illustrated in Fig. 15–9b. If, as before, the grain diameter is constrained to be 150 in., the grain geometry is uniquely determined by specification of the initial fuel and oxidizer flow rates, number of ports, burn time, and the requirement that the burn distance around each port be equal. Additionally, the hydraulic diameter D_h (four times port area divided by port perimeter) of all ports should be equal to assure that all ports have the same mass flow rate.

For this example, the nine-port grain configuration results in a theoretical fuel sliver fraction of 4.3%. In reality, the sliver fraction for both designs will be somewhat greater than theoretical values since some web must be designed to remain between ports at the end of the burn duration to prevent slivers from being expelled out of the nozzle. Table 15–3 compares key features of the circular port grain design (Fig. 15–9a) and the quadrilateral grain design (Fig. 15–9b).

In this example, the fuel consumed by the quadrilateral port design is less than that consumed by the circular port design. Therefore, the total impulse of the two designs will be different. If fuel consumed were constrained to be the same in each design, one would find that, as the number of quadrilateral fuel ports would be increased, the grain length would decrease and grain diameter would increase. In practice, the hybrid motor designer must carefully balance

TABLE 15–3. Comparison of Circular Port and Quadrilateral Port Grain Designs

Design Parameter	Circular Port	Quadrilateral Port
Oxidizer flow rate (lbm/sec)	6824	6824
Initial fuel flow rate (lbm/sec)	3412	3412
Burn time (sec)	120	120
Grain diameter (in.)	150	150
Number of combustion ports	7	9
Oxidizer flux (lbm/sec/in.^{2})	1.51	1.07
Fuel regression rate (in./sec)	0.138	0.109
Distance burned (in.)	10.68	8.78
Grain length (in.)	1,189.6	976.1
Combustion port L/D	41.5	37.2
Loaded fuel mass (lbm)	516,664	364,170
Fuel consumed (lbm)	362,577	348,584
Theoretical sliver fraction (%)	29.8	4.28

launch vehicle system requirements, such as total impulse and envelope constraints, with available grain design options to arrive at an optimum motor configuration. Total propellant and propellant contingency necessary to accomplish a specific mission will depend upon such factors as residual fuel and oxidizer allowances at motor cutoff, ascent trajectory throttling requirements, which impact overall mixture ratio and oxidizer utilization, and additional propellant if a Δu (vehicle velocity necessary to achieve mission objectives) contingency reserve is required.

Using Table 15–2 to obtain c^*, the initial vacuum-delivered specific impulse for the circular port booster design may be calculated as

$$I_{s_v} = \frac{(C_F)_v c^*}{g_0} = \frac{(1.735)(0.95)(5912 \text{ ft/sec})}{32.174 \dfrac{\text{lbm-ft}}{\text{lbf-sec}^2}} = 302.87 \text{ sec}$$

At the end of burn, the mixture ratio is determined from Eq. 15–16 to be 2.45. The theortical characteristic velocity corresponding to the mixture ratio is 5815 ft/sec. Assuming the same combustion efficiency factor of 95%, the chamber pressure, neglecting throat erosion, is determined to be

$$p_1 = \frac{\dot{m} c^*}{g_0 A_t} = \frac{(9611 \text{ lbm/sec})(0.95)(5815 \text{ ft/sec})}{\left(32.174 \dfrac{\text{lbm-ft}}{\text{lbf-sec}^2}\right)(2552.5 \text{ in.}^2)} = 646.5 \text{ lbf/in.}^2$$

Using the end-of-burn chamber pressure of 646.5 psia, the end-of-burn specific impulse is calculated to be 299.3 sec.

The throat material erosion rate in a hybrid is generally significantly greater than that of a solid propellant system and is a strong function of chamber pressure and mixture ratio. Erosion of carbonaceous throat materials (carbon cloth phenolic, graphite, etc.) is primarily governed by heterogeneous surface chemical reactions involving the reaction of carbon with oxidizing species present in the flow of combustion gases such as O_2, O, H_2O, OH, and CO_2 to form CO. Hybrid motor operation at oxygen-rich mixture ratios and high pressure will result in very high throat erosion rates. Operation at fuel-rich mixture ratios and pressures below 400 psi will result in very low throat erosion rates.

In general, the effect of throat erosion in ablative nozzles on overall motor performance depends on initial throat diameter. For the booster design under consideration, a 0.010-in./sec erosion rate acting only at the throat will reduce the expansion ratio from 7.72 to 7.11 over the 120-sec burn time. Using the end-of-burn mixture ratio of 2.45 corresponding to a ratio of specific heats of 1.137 (Table 15–2), an end-of-burn chamber pressure and vacuum thrust coefficient of 595.3 psia and 1.730, respectively, may be calculated. Therefore, if throat erosion is accounted for, delivered specific impulse at the end of burn is 297.0 sec, a reduction of only 0.77% compared with the non-eroding throat assumption. As initial throat diameter is reduced, the reduction in expansion ratio due to throat erosion becomes greater, thereby resulting in greater performance losses.

Current practice for preliminary design of hybrid booster concepts is to couple a fuel regression rate model, a grain design model, and booster component design models in an automated preliminary design procedure. Using numerical optimization algorithms, such a computer model can pick the optimum booster design that maximizes selected optimization variables, such as booster ideal velocity or total impulse, while minimizing booster propellant and inert weight.

15.4. COMBUSTION INSTABILITY

The hybrid combustion process tends to produce somewhat rougher pressure versus time characteristics than either liquid or solid rocket engines. However, a well-designed hybrid will typically limit combustion roughness to approximately 2 to 3% of mean chamber pressure. In any combustion device, pressure fluctuations will tend to organize themselves around the natural acoustic frequencies of the combustion chamber or oxidizer feed system. While significant combustion pressure oscillations at chamber natural-mode acoustic frequencies have been observed in numerous hybrid motor tests, such oscillations have not proved to be an insurmountable design problem. When pressure oscillations have occurred in hybrid motors, they have been observed to grow to a limiting amplitude which is dependent on such factors as oxidizer feed system and injector characteristics, fuel grain geometric characteristics, mean chamber pressure level, and oxidizer mass velocity. Unbounded growth of pressure oscillations, such as may occur in solid and liquid rocket motors, has not been observed in hybrid motors.

Hybrid motors have exhibited two basic types of instabilities in static test environments: oxidizer feed system-induced instability (non-acoustic), and flame holding instability (acoustic). Oxidizer feed system instability is essentially a chugging type as described in Chapter 9 and arises when the feed system is sufficiently "soft." In cryogenic systems, this implies a high level of compressibility from sources such as vapor cavities or two-phase flow in feed lines combined with insufficient isolation from motor combustion processes. Figure 15–10a illustrates feed system induced instability in a 24-in. diameter hybrid motor operated at a LOX flow rate of 20 lbm/sec with HTPB fuel. The instability is manifested by high-amplitude, periodic oscillations well below the first longitudinal (1-L) acoustic mode of the combustor. In this example the oscillation frequency is 7.5 Hz whereas the 1-L mode frequency is approximately 60 Hz. Stiffening the feed/injection system can eliminate the oscillation. This is accomplished by increasing the injector pressure drop (thus making propagation of motor pressure disturbances upstream through the feed system more difficult) and eliminating sources of compressibility in the feed system. Chugging-type instabilities in hybrid motors have proven amenable to analysis

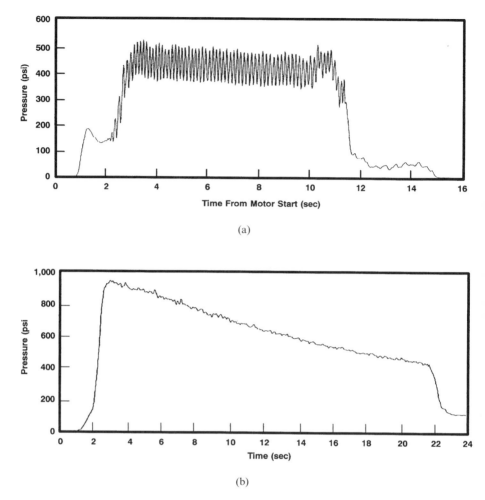

FIGURE 15–10. (a) Periodic, large-amplitude, low-frequency combustion pressure oscillations are an example of oxidizer feed system induced "chug" type combustion instability in a 24-in. diameter LOX/HTBP motor. (b) An example of stable combustion in a 24-in. diameter LOX/HTPB motor, exhibiting an overall combustion roughness level of 1.3%.

in terms of prediction and prevention (Ref. 15–17). For purposes of comparison, Fig. 15–10b shows a pressure–time trace from the same 24-in. diameter hybrid motor exhibiting stable combustion while being operated at a LOX flow rate of 40 lbm/sec at a maximum chamber pressure of 900 psi.

Flame-holding instability relevant to hybrid motors was first observed during the development of solid fuel ramjets (Ref. 15–18). A solid fuel ramjet is essentially a hybrid motor operating on the oxygen available in ram air. Flame-holding instabilities in hybrids are typically manifested at acoustic frequencies and appear in longitudinal modes. No acoustic instabilities in hybrid motors have been observed in higher frequency tangential or radial modes such as in solid rocket motors or liquid engines. Flame-holding instabilities arise due to inadequate flame stabilization in the boundary layer (Ref. 15–19) and are not associated with feed system flow perturbations. Figure 15–11a illustrates flame-holding instability in an 11-in. diameter hybrid motor operated with gaseous oxygen (GOX) oxidizer and HTPB fuel, using an injector producing a conical flow field. In this test, oxygen flow was initiated through the motor at a pressure of 90 psi for two seconds prior to motor ignition. The motor was ignited using a hydrogen torch that continued to operate for approximately one second following motor ignition. During the first second of motor operation, the hydrogen igniter flame stabilizes the motor. When the igniter flame is extinguished, the motor becomes unstable. Figure 15–11b illustrates operation of the same 11-in. diameter motor in which the flame-holding instability has been suppressed without the use of a hydrogen flame. In this case stable combustion was achieved by changing the flow field within the motor, using an injector producing an axial flow field. Figure 15–12 shows the result of decomposing the pressure versus time signal for the unstable example of Fig. 15–11a into its frequency components via fast Fourier transform techniques. The 1-L acoustic oscillation mode is clearly visible at approximately 150 Hz.

It is apparent that flame-holding instability can be eliminated by several means, all of which act to stabilize combustion in the boundary layer. The first method is to use a pilot flame derived from injection of a combustible fluid such as hydrogen or propane to provide sufficient oxidizer preheating in the leading edge region of the boundary layer flame zone. With this technique, motor stability characteristics are relatively insensitive to the nature of the injector flow field. In the previous example, the hydrogen torch igniter acted as a pilot during its period of operation. A second method involves changing the injector flow field to ensure that a sufficiently large hot gas recirculation zone is present at the head end of the fuel grain. Such a zone can be created by forcing the upstream flow over a rearward-facing step or by strong axial injection of oxidizer (see Fig. 15–13). Axial injection in the correct configuration produces a strong counter-flowing hot gas recirculation zone, similar to that of a rearward-facing step, at the head end of the diffusion flame (conical injection produces a much smaller and usually ineffective recirculation zone). These techniques produce a flow field result very similar to that produced by bluff body flame stabilizers used in jet engine afterburners and solid fuel ramjets to prevent flame blowoff. The recirculation zone acts to entrain hot gas from the core flow, which provides sufficient oxidizer preheating for the leading edge of the boundary layer diffusion flame to stabilize combustion.

Comparison of the average pressure levels in Figs. 15–11a and 15–11b illustrates an interesting phenomenon. For the same motor operating conditions (oxidizer flow rate, grain geometry and composition, and throat diameter) the average pressure in the unstable motor is significantly greater than that in the stable motor. This same phenomenon has been noted in solid propellant

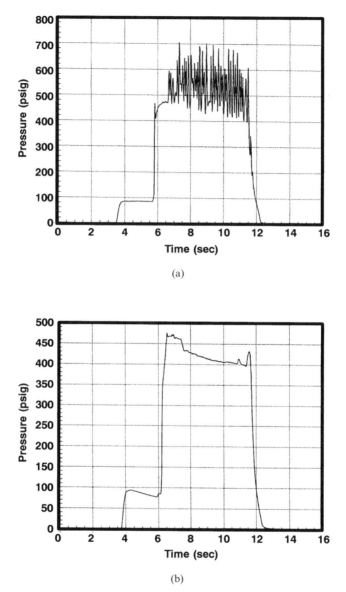

(a)

(b)

FIGURE 15–11. (a) An example of large-amplitude, high-frequency combustion pressure oscillations due to flame-holding instability in an 11-in. diameter GOX/HTPB motor. Instability during the initial one second of burn has been suppressed by the use of a pilot flame. (b) Suppression of flame-holding instability in an 11-in. diameter GOX/HTPB motor by means of strong axial injection of oxidizer.

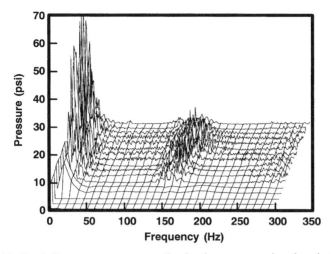

FIGURE 15–12. A frequency-versus-amplitude plot at successive time intervals for an 11-in. diameter GOX/HTPB motor test shows pressure oscillations in the motor 1-L acoustic mode at 150 Hz due to flame-holding instability.

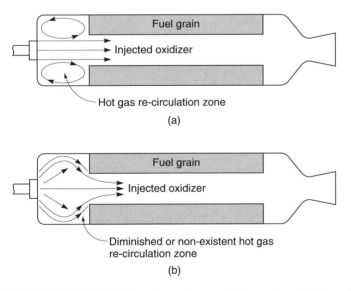

FIGURE 15–13. (a) Axial injection of oxidizer results in a strong hot gas flow recirculation zone at the fuel grain leading edge, producing stable combustion. (b) Conical injection of oxidizer can produce a weak or nonexistent hot gas flow recirculation zone at the fuel grain leading edge, resulting in unstable combustion.

motors and the results from intensification of heat transfer to the fuel surface due to the gas velocity at the fuel surface oscillating at high frequency. The high heating rate results in the vaporization of more fuel than would otherwise occur in equilibrium conditions, thus producing a higher average chamber pressure.

Despite recent advances in understanding causes of and solutions for combustion instability in hybrid motors, development of a comprehensive, predictive theory of combustion stability remains one of the major challenges in hybrid technology development.

SYMBOLS (includes symbols used in Appendices 4 and 5)

a	burning or regression rate coefficient (units of a depend on value of oxidizer flux exponent)	variable
A	particle cloud attenuation coefficient	$m^2 (ft^2)$/particle
A_p	combustion port area	m^2 (in.^{2})
A_s	fuel grain surface area	m^2 (in.^{2})
A_t	nozzle throat area	m^2 (ft^2)
c^*	characteristic velocity	m/sec (ft/sec)
C	particle cloud concentration	particles/unit volume
C_f	skin friction coefficient (blowing)	dimensionless
C_{f_0}	skin friction coefficient (no blowing)	dimensionless
C_{F_v}	vacuum thrust coefficient	dimensionless
C_h	Stanton number	dimensionless
c_p	heat capacity	J/kg-K (Btu/lbm-R)
d_b	fuel grain burn distance	m (in.)
D_h	hydraulic diameter ($4A_p/P$)	m (in.)
D_p	combustion port diameter	m (in.)
D_t	nozzle throat diameter	m (in.)
F_v	vacuum thrust	N (lbf)
G	mass velocity	kg/m^2-sec (lbm/ft^2-sec)
G_o	oxidizer mass velocity	kg/m^2-sec (lbm/ft^2-sec)
g_0	conversion factor—acceleration of gravity	m/sec^2 (lbm-ft/lbf/sec^2)
h	convective heat transfer coefficient	J/m^2-sec/K (Btu/ft^2-sec/R)
h_v	heat of gasification	J/kg (Btu/lbm)
Δh	flame zone–fuel surface enthalpy difference	J/kg (Btu/lbm)
H_f	heat of formation	J/kg-mol (kcal/mol)
I_s	specific impulse	sec
k	specific heat ratio	dimensionless
L	combustion port length	m (in.)
$\dot{m}$	propellant flow rate	kg/sec (lbm/sec)

$\dot{m}_f$	fuel flow rate	kg/sec (lbm/sec)
$\dot{m}_o$	oxidizer flow rate	kg/sec (lbm/sec)
n, m, l	burning or regression rate pressure exponent	dimensionless
P	combustion port perimeter	m (in.)
p_1	chamber pressure	MPa (lbf/in.^{2})
Pr	Prandtl number	dimensionless
Q_c	heat input to fuel surface due to convection	J/m^2-sec (Btu/ft^2-sec)
$Q_{\rm rad}$	heat input to fuel surface due to radiation	J/m^2-sec (Btu/ft^2-sec)
Q_s	total heat input to fuel surface	J/m^2-sec (Btu/ft^2-sec)
R	combustion port radius	m (in.)
R_i	initial combustion port radius	m (in.)
Re	Reynolds number	dimensionless
$\dot{r}$	fuel regression rate	mm/sec (in./sec)
r	oxidizer to fuel mixture ratio	dimensionless
T	temperature	K (F)
u_e	gas free stream velocity in axial direction	m/sec (ft/sec)
v	gas velocity normal to fuel surface	m/sec (ft/sec)
V_1	chamber volume	m^3 (in.^{3})
x	axial distance from leading edge of fuel grain	m (in.)
y	length coordinate normal to fuel surface	m (in.)
z	radiation path length	m (in.)

Greek Letters

α	fuel surface absorptivity	dimensionless
β	boundary layer blowing coefficient	dimensionless
ε_g	emissivity of particle-laden gas	dimensionless
κ_g	gas conductivity	J/m^2-sec/K (Btu/ft-sec-R)
μ	gas viscosity	N-sec/m^2 (lbf-sec/ft^2)
ρ_1	combustion chamber gas density	kg/m^3 (lbm/in.^{3})
ρ_e	free stream gas density	kg/m^3 (lbm/in.^{3})
ρ_f	fuel density	kg/m^3 (lbm/in.^{3})
σ	Stefan–Boltzmann constant	J/m^2-sec/K^4 (Btu/ft^2-sec/R^4)

Subscripts

e	boundary layer edge conditions
f	fuel
i	initial conditions

o	oxidizer	
s	surface conditions	
x	axial distance from leading edge of fuel grain	m (in.)
ref	reference conditions	

REFERENCES

15–1. D. Altman, "Hybrid Rocket Development History," *AIAA Paper 91-2515*, June 1991.

15–2. F. B. Mead and B. R. Bornhorst, "Certification Tests of a Hybrid Propulsion System for the Sandpiper Target Missile," *AFRPL-TR-69–73*, June 1969.

15–3. P. D. Laforce et al., "Technological Development of a Throttling Hybrid Propulsion System," *UTC 2215-FR*, January 1967.

15–4. H. R. Lips, "Experimental Investigation of Hybrid Rocket Engines Using Highly Aluminized Fuels," *Journal of Spacecraft and Rockets*, Vol. 14, No. 9, September 1977, pp. 539–545.

15–5. J. S. McFarlane et al., "Design and Testing of AMROC's 250,000 lbf Thrust Hybrid Motor," *AIAA Paper 93-2551*, June 1993.

15–6. T. A. Boardman, T. M. Abel, S. E. Claflin, and C. W. Shaeffer, "Design and Test Planning for a 250-klbf-Thrust Hybrid Rocket Motor under the Hybrid Propulsion Demonstration Program," *AIAA Paper 97-2804*, July 1997.

15–7. S. R. Jain and G. Rajencran, "Performance Parameters of some New Hybrid Hypergols," *Journal of Propulsion and Power*, Vol. 1, No. 6, November–December 1985, pp. 500–501.

15–8. U. C. Durgapal and A. K. Chakrabarti, "Regression Rate Studies of Aniline–Formaldehyde–Red Fuming Nitric Acid Hybrid System," *Journal of Spacecraft and Rockets*, Vol. 2, No. 6, 1974, pp. 447–448.

15–9. G. A. Marxman, "Combustion in the Turbulent Boundary Layer on a Vaporizing Surface," *Tenth Symposium on Combustion*, The Combustion Institute, 1965, pp. 1337–1349.

15–10. P. A. O. G. Korting, H. F. R. Schoyer, and Y. M. Timnat, "Advanced Hybrid Rocket Motor Experiments," *Acta Astronautica*, Vol. 15, No. 2, 1987, pp. 97–104.

15–11. W. Waidmann, "Thrust Modulation in Hybrid Rocket Engines," *Journal of Propulsion and Power*, Vol. 4, No. 5, September–October 1988, pp. 421–427.

15–12. M. J. Chiaverini et al., "Thermal Pyrolysis and Combustion of HTPB-based Solid Fuels for Hybrid Rocket Motor Applications," *AIAA Paper 96-2845*, July 1996.

15–13. L. D. Smoot and C. F. Price, "Regression Rates of Metalized Hybrid Fuel Systems," *AIAA Journal*, Vol. 4, No. 5, September 1965, pp. 910–915.

15–14. Laboratory data obtained in 2-in. diameter test motors, Thiokol Corporation, 1989.

15–15. T. A. Boardman, R. L. Carpenter, et al., "Development and Testing of 11- and 24-inch Hybrid Motors under the Joint Government/Industry IR&D Program," *AIAA Paper 93-2552*, June 1993.

15–16. P. Estey, D. Altman, and J. McFarlane, "An Evaluation of Scaling Effects for Hybrid Rocket Motors," *AIAA Paper 91-2517*, June 1991.

15–17. T. A. Boardman, K. K. Hawkins, S. R. Wassom, and S. E. Claflin, "Non-Acoustic Feed System Coupled Combustion Instability in Hybrid Rocket Motors," Hybrid Rocket Technical Committee Combustion Stability Workshop, *31st AIAA/ASME/SAE/ASEE Joint Propulsion Conference and Exhibit*, July 1995.

15–18. B. L. Iwanciow, A. L. Holzman, and R. Dunlap, "Combustion Stabilization in a Solid Fuel Ramjet," *10th JANNAF Combustion Meeting*, 1973.

15–19. T. A. Boardman, D. H. Brinton, R. L. Carpenter, and T. F. Zoladz, "An Experimental Investigation of Pressure Oscillations and their Suppression in Suscale Hybrid Rocket Motors," *AIAA Paper 95-2689*, July 1995.

CHAPTER 16

THRUST VECTOR CONTROL

In addition to providing a propulsive force to a flying vehicle, a rocket propulsion system can provide moments to rotate the flying vehicle and thus provide control of the vehicle's attitude and flight path. By controlling the direction of the thrust vectors through the mechanisms described later in the chapter, it is possible to control a vehicle's pitch, yaw, and roll motions.

All chemical propulsion systems can be provided with one of several types of thrust vector control (TVC) mechanisms. Some of these apply either to solid, hybrid, or to liquid propellant rocket propulsion systems, but most are specific to only one of these propulsion categories. We will describe two types of thrust vector control concept: (1) for an engine or a motor with a single nozzle; and (2) for those that have two or more nozzles.

Thrust vector control is effective only while the propulsion system is operating and creating an exhaust jet. For the flight period, when a rocket propulsion system is not firing and therefore its TVC is inoperative, a separate mechanism needs to be provided to the flying vehicle for achieving control over its attitude or flight path.

Aerodynamic fins (fixed and movable) continue to be very effective for controlling vehicle flight within the earth's atmosphere, and almost all weather rockets, antiaircraft missiles, and air-to-surface missiles use them. Even though aerodynamic control surfaces provide some additional drag, their effectiveness in terms of vehicle weight, turning moment, and actuating power consumption is difficult to surpass with any other flight control method. Vehicle flight control can also be achieved by a separate attitude control propulsion system as described in Sections 4.6, 6.8, and 11.3. Here six or more small liquid propellant thrusters (with a separate feed system and a separate control) provide

608

small moments to the vehicle in flight during, before, or after the operation of the main rocket propulsion system.

The reasons for TVC are: (1) to willfully change a flight path or trajectory (e.g., changing the direction of the flight path of a target-seeking missile); (2) to rotate the vehicle or change its attitude during powered flight; (3) to correct for deviation from the intended trajectory or the attitude during powered flight; or (4) to correct for thrust misalignment of a fixed nozzle in the main propulsion system during its operation, when the main thrust vector misses the vehicle's center of gravity.

Pitch moments are those that raise or lower the nose of a vehicle; *yaw moments* turn the nose sideways; and *roll moments* are applied about the main axis of the flying vehicle (Fig. 16–1). Usually, the thrust vector of the main rocket nozzle is in the direction of the vehicle axis and goes through the vehicle's center of gravity. Thus it is possible to obtain pitch and yaw control moments by the simple deflection of the main rocket thrust vector; however, roll control usually requires the use of two or more rotary vanes or two or more separately hinged propulsion system nozzles. Figure 16–2 explains the pitch moment obtained by a hinged thrust chamber or nozzle. The side force and the pitch moment vary as the sine of the effective angle of thrust vector deflection.

16.1. TVC MECHANISMS WITH A SINGLE NOZZLE

Many different mechanisms have been used successfully. Several are illustrated in Refs. 16–1 and 16–2. They can be classified into four categories:

1. Mechanical deflection of the nozzle or thrust chamber.
2. Insertion of heat-resistant movable bodies into the exhaust jet; these experience aerodynamic forces and cause a deflection of a part of the exhaust gas flow.
3. Injection of fluid into the side of the diverging nozzle section, causing an asymmetrical distortion of the supersonic exhaust flow.

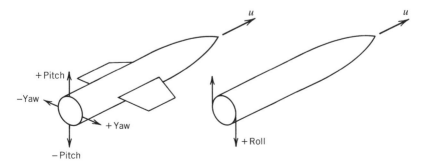

FIGURE 16–1. Moments applied to a flying vehicle.

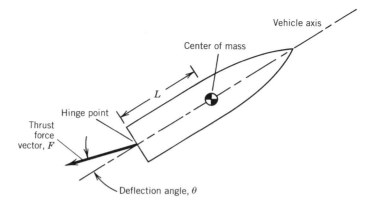

FIGURE 16–2. The pitch moment applied to the vehicle is $FL \sin \theta$.

    4. Separate thrust-producing devices that are not part of the main flow through the nozzle.

Each category is described briefly below and in Table 16–1, where the four categories are separated by horizontal lines. Figure 16–3 illustrates several TVC mechanisms. All of the TVC schemes shown here have been used in production vehicles.

    In the *hinge* or *gimbal* scheme (a hinge permits rotation about one axis only, whereas a gimbal is essentially a universal joint), the whole engine is pivoted on a bearing and thus the thrust vector is rotated. For small angles this scheme has negligible losses in specific impulse and is used in many vehicles. It requires a flexible set of propellant piping (bellows) to allow the propellant to flow from the tanks of the vehicle to the movable engine. The Space Shuttle (Fig. 1–13) has two gimballed orbit maneuver engines, and three gimballed main engines. Figures 6–1, 6–3, and 8–19 show gimballed engines. Some Soviet launch vehicles use multiple thrusters and hinges (Fig. 10–10 shows 4 hinges), while many U.S. vehicles use gimbals.

    Jet vanes are pairs of heat-resistant, aerodynamic wing-shaped surfaces submerged in the exhaust jet of a fixed rocket nozzle. They were first used about 55 years ago. They cause extra drag (2 to 5% less I_s; drag increases with larger vane deflections) and erosion of the vane material. Graphite jet vanes were used in the German V-2 missile in World War II and in the Scud missiles fired by Iraq in 1991. The advantage of having roll control with a single nozzle often outweighs the performance penalties.

    Small auxiliary thrust chambers were used in the Thor and early version of Atlas missiles. They provide roll control while the principal rocket engine operates. They are fed from the same feed system as the main rocket engine. This scheme is still used on some Russian booster rocket vehicles.

    The *injection of secondary fluid* through the wall of the nozzle into the main gas stream has the effect of forming oblique shocks in the nozzle diverging

TABLE 16–1. Thrust Vector Control Mechanisms

Type	L/S[a]	Advantages	Disadvantages
Gimbal or hinge	L	Simple, proven technology; low torques, low power; ±12° duration limited only by propellant supply; very small thrust loss	Requires flexible piping; high inertia; large actuators for high slew rate
Movable nozzle (flexible bearing)	S	Proven technology; no sliding, moving seals; predictable actuation power; up to ±12°	High actuation forces; high torque at low temperatures; variable actuation force
Movable nozzle (rotary ball with gas seal)	S	Proven technology; no thrust loss if entire nozzle is moved; ±20° possible	Sliding, moving hot gas spherical seal; highly variable actuation power; limited duration; needs continuous load to maintain seal
Jet vanes	L/S	Proven technology; low actuation power; high slew rate; roll control with single nozzle; ±9°	Thrust loss of 0.5 to 3%; erosion of jet vanes; limited duration; extends missile length
Jet tabs	S	Proven technology; high slew rate; low actuation power; compact package	Erosion of tabs; thrust loss, but only when tab is in the jet; limited duration
Jetavator	S	Proven on Polaris missile; low actuation power; can be lightweight	Erosion and thrust loss; induces vehicle base hot gas recirculation; limited duration
Liquid-side injection	S/L	Proven technology; specific impulse of injectant nearly offsets weight penalty; high slew rate; easy to adapt to various motors; can check out before flight; components are reusable; duration limited by liquid supply; ±6°	Toxic liquids are needed for high performance; often difficult packaging for tanks and feed system; sometimes requires excessive maintenance; potential spills and toxic fumes with some propellants; limited to low vector angle applications
Hot-gas-side injection	S/L	Lightweight; low actuation power; high slew rate; low volume/compact; low performance loss	Multiple hot sliding contacts and seals in hot gas valve; hot piping expansion; limited duration; requires special hot gas valves; technology is not yet proven
Hinged auxiliary thrust chambers for high thrust engine	L	Proven technology; feed from main turbopump; low performance loss; compact; low actuation power; no hot moving surfaces; unlimited duration	Additional components and complexity; moments applied to vehicle are small; not used for 15 years in USA
Turbine exhaust gas swivel for large engine	L	Swivel joint is at low pressure; low performance loss; lightweight; proven technology	Limited side forces; moderately hot swivel joint; used for roll control only

[a]L, used with liquid propellant engines; S, used with solid propellant motors.

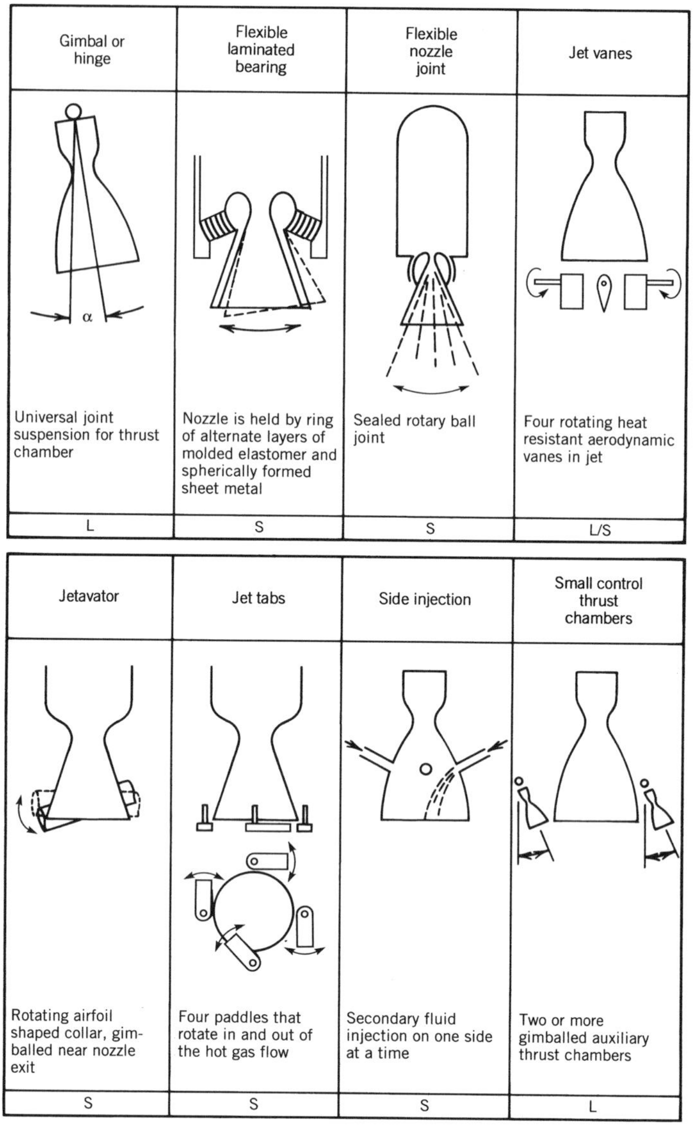

Gimbal or hinge	Flexible laminated bearing	Flexible nozzle joint	Jet vanes
Universal joint suspension for thrust chamber	Nozzle is held by ring of alternate layers of molded elastomer and spherically formed sheet metal	Sealed rotary ball joint	Four rotating heat resistant aerodynamic vanes in jet
L	S	S	L/S

Jetavator	Jet tabs	Side injection	Small control thrust chambers
Rotating airfoil shaped collar, gimballed near nozzle exit	Four paddles that rotate in and out of the hot gas flow	Secondary fluid injection on one side at a time	Two or more gimballed auxiliary thrust chambers
S	S	S	L

FIGURE 16–3. Simple schematic diagrams of eight different TVC mechanisms. Actuators and structural details are not shown. The letter L means it is used with liquid propellant rocket engines and S means it is used with solid propellant motors.

section, thus causing an unsymmetrical distribution of the main gas flow, which produces a side force. The secondary fluid can be stored liquid or gas from a separate hot gas generator (the gas would then still be sufficiently cool to be piped), a direct bleed from the chamber, or the injection of a catalyzed mono-propellant. When the deflections are small, this is a low-loss scheme, but for large moments (large side forces) the amount of secondary fluid becomes excessive. This scheme has found application in a few large solid propellant rockets, such as Titan IIIC and one version of Minuteman.

Of all the mechanical deflection types, the *movable nozzles* are the most efficient. They do not significantly reduce the thrust or the specific impulse and are weight-competitive with the other mechanical types. The flexible nozzle, shown in Figs. 16–3 and 16–4, is a common type of TVC used with solid propellant motors. The molded, multilayer bearing pack acts as a seal, a load transfer bearing, and a viscoelastic flexure. It uses the deformation of a stacked set of doubly curved elastomeric (rubbery) layers between spherical metal sheets to carry the loads and allow an angular deflection of the nozzle axis. The flexible seal nozzle has been used in launch vehicles and large strategic missiles, where the environmental temperature extremes are modest. At low temperature the elastomer becomes stiff and the actuation torques increase substantially, requiring a much larger actuation system. Figure 16–5 describes a different type of flexible nozzle. It uses a movable joint with a toroidal hydraulic bag to transfer loads. There are double seals to prevent leaks of hot gas and various insulators to keep the structure below 200°F or 93°C.

Two of the gimbals will now be described in more detail. Figure 16–6 shows the gimbal bearing assembly of the Space Shuttle main engine. It supports the

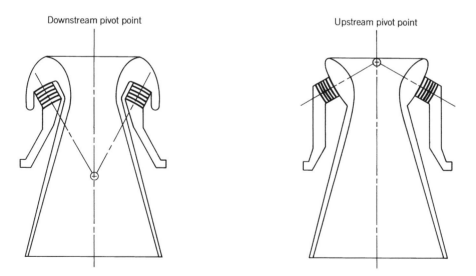

Downstream pivot point   Upstream pivot point

FIGURE 16–4. Two methods of using flexible nozzle bearings with different locations for the center of rotation. The bearing support ring is made of metal or plastic sheet shims formed into rings with spherical contours (white) bonded together by layers of molded elastomer or rubber (black stripes). Although only five elastomeric layers are shown for clarity, many flexible bearings have 10 to 20 layers. (Copied with permission from Ref. 16–1.)

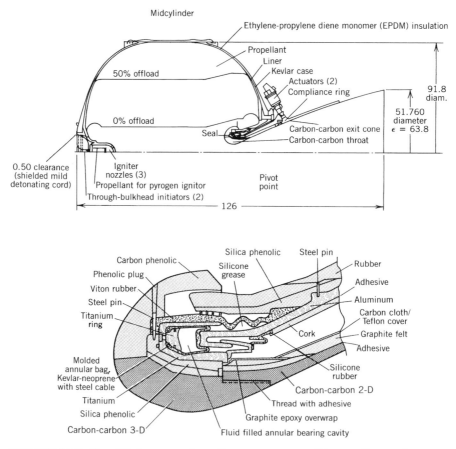

FIGURE 16–5. Simplified cross section of an upper-stage solid propellant rocket motor (IUS) using an insulated carbon-fiber/carbon-matrix nozzle, an insulated Kevlar filament-wound case, a pyrogen igniter, forward and aft stress-relieving boots, a fluid-filled bearing, and an elastomeric seal assembly in the nozzle to allow $4\frac{1}{2}°$ of thrust vector deflection. This motor has a loaded weight of 22,874 lbf, a propellant with hydroxyl-terminated polybutadiene binder, a weight of 21,400 lbf, a burnout weight of 1360 lbf, a motor mass fraction of 0.941, a nozzle throat diameter of 6.48 in., and a nozzle exit area ratio of 63.8. The motor burns 146 sec at an average pressure of 651 psi (886 psi maximum) and an average thrust of 44,000 lbf (60,200 lbf maximum), with an effective altitude specific impulse of 295 sec. Top drawing is cross section of motor; bottom drawing is enlarged cross section of nozzle package assembly. The motor is an enlarged version of Orbus-6 described in Fig. 11–3. (From C. A. Chase, "IUS Solid Motor Overview," *JANNAF Conference*, Monterey, Calif, 1983; courtesy of United Technologies Corp./Chemical Systems.)

weight of the engine and transmits the thrust force. It is a ball-and-socket universal joint with contact and intermeshing spherical (concave and convex) surfaces. Sliding occurs on these surfaces as the gimbal assembly is rotated. When assembling the engine to the vehicle, some offset bushings are used to align the thrust vector. Some of the design features and performance requirements of this gimbal are listed in Table 16–2. The maximum angular motion is actually larger than the deflection angle during operation so as to allow for various tolerances and alignments. The actual deflections, alignment tolerances, friction coefficients, angular speeds, and accelerations during operation are usually much smaller than the maximum values listed in the table.

Table 16–3 and Ref. 16–3 give the design requirements for the actuator system for the TVC for a flexible bearing in the IUS solid rocket motor nozzle. This system is shown in Figs. 11–3 and 16–5 and in Table 11–3. One version of this nozzle can deflect 4° maximum plus 0.5° for margin and another is rated at 7.5°. It has two electrically redundant electromechanical actuators using ball screws, two potentiometers for position indication, and one controller that provides both the power drive and the signal control electronics for each actuator. A variable-frequency, pulse-width-modulated (PWM) electric motor drive is used to allow small size and low weight for the power and forces

TABLE 16–2. Characteristics and Performance Requirements of the Gimbal Bearing Assembly of the Space Shuttle Main Engine

Engine weight to be supported (lbf)	Approx. 7000
Thrust to be transmitted, (lbf)	512,000
Gimbal asembly weight (lbf)	105
Material is titanium alloy	6Al-6V-2Sn
Dimensions (approximate) (in.)	11 dia. × 14
Angular motion (deg)	
Operational requirement (max.)	±10.5
Snubbing allowance in actuators	0.5
Angular alignment	0.5
Gimbal attach point tolerance	0.7
Overtravel vector adjustment	0.1
Maximum angular capability	±12.5
Angular acceleration (max.) (rad/sec^2)	30
Angular velocity (max.) (deg/sec)	20
Angular velocity (min.) (deg/sec)	10
Lateral adjustment (in.)	±0.25
Gimbal duty cycle about each axis	
Number of operational cycles to 10.5°	200
Nonoperational cycles to 10.5°	1400
Coefficient of friction (over a temperature range of 88 to 340 K)	0.01–0.2

Source: Courtesy of Rocketdyne, a Division of Rockwell International.

TABLE 16–3. Design Requirements for TVC Actuation System of an IUS Solid Rocket Motor

Item	Requirement
Performance parameter	
Input power	31 A/axis maximum at 24 to 32 V dc; >900 W (peak)
Stroke	10.2 cm (4.140 in.) minimum
Stall force	1.9 kN (430 lbf) minimum
Accuracy	±1.6 mm (±0.063 in.) maximum
Frequency response	>3.2 Hz at 100° phase lag
No load speed	8.13 cm/sec (3.2 in./sec) minimum
Stiffness	28.9 kN/cm (16,600 lbf/in.) minimum
Backlash	±0.18 mm (0.007 in.) maximum
Reliability	>0.99988 redundant drive train, >0.999972 single thread element
Weight	
Controller	5.9 kg (13 lbf) maximum, each
Actuator	7.04 kg (15.5 lbf) maximum, each
Potentiometer	1.23 kg (2.7 lbf) maximum, each
System	22.44 kg (49.4 lbf) maximum

Source: Reproduced from Ref. 16–3 with permission of United Technologies Corp./Chemical Systems.

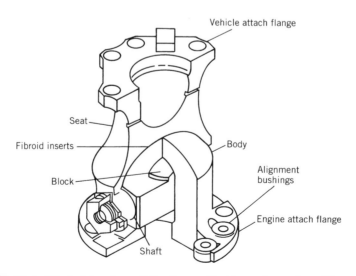

FIGURE 16–6. Gimbal bearing of the Space Shuttle main engine. (Courtesy of the Boeing Company, Rocketdyne Propulsion & Power.)

involved. Also, it has a pair of locking mechanisms that will lock the nozzle in a fixed pitch-and-yaw position as a fail-safe device.

The *alignment* of the thrust vector is a necessary activity during assembly. The thrust vector in the neutral position (no deflection or, in many vehicles, the thrust axis coincides with the vehicle axis) should usually go through the center of gravity of the vehicle. The TVC mechanism has to allow for alignment or adjustments in angle as well as position of the TVC center point with the intended vehicle axis. The geometric centerline of the diverging section of the nozzle is generally considered to be the thrust direction. One alignment provision is shown in Fig. 16–6. An alignment accuracy of one-quarter of a degree and an axis offset of 0.020 in. have been achieved with good measuring fixtures for small-sized nozzles.

The *jet tab TVC system* has low torque, and is simple for flight vehicles with low-area-ratio nozzles. Its thrust loss is high when tabs are rotated at full angle into the jet, but is zero when the tabs are in their neutral position outside of the jet. On most flights the time-averaged position of the tab is a very small angle and the average thrust loss is small. Jet tabs can form a very compact mechanism and have been used successfully on tactical missiles. An example is the jet tab assembly for the booster rocket motor of the Tomahawk cruise missile, shown in Fig. 16–7. Four tabs, independently actuated, are rotated in and out of the motor's exhaust jet during the 15 sec duration of rocket operation. A tab that blocks 16% of the nozzle exit area is equivalent to a thrust vector angle deflection of 9°. The maximum angle is 12° and the slew rate is fast (100°/sec). The vanes are driven by four linear small push–pull hydraulic actuators with two servo valves and an automatic integral controller. The power is supplied by compressed nitrogen stored at 3000 psi. An explosive valve releases the gas to pressurize an oil accumulator in a blowdown mode. The vanes are made of tungsten to minimize the erosion from the solid particles in the exhaust gas.

The *jetavator* was used on submarine-launched missiles. The thrust loss is roughly proportional to the vector angle. This mechanism is shown in Fig. 16–3 and mentioned in Table 16–1.

The concept of TVC by *secondary fluid injection* into the exhaust stream dates back to 1949 and can be credited to A. E. Wetherbee, Jr. (U.S. Patent 2,943,821). Application of *liquid injection thrust vector control* (LITVC) to production vehicles began in the early 1960s. Both inert (water) and reactive fluids (such as hydrazine or nitrogen tetroxide) have been used. Although side injection of reactive liquids is still used on some of the older vehicles, it requires a pressurized propellant tank and a feed system. A high-density injection liquid is preferred because its tank will be relatively small and its pressurization will require less mass. Because other schemes have better preformance, liquid injection TVC will probably not be selected for new applications.

Hot gas injection (HGITVC) of solid rocket propellant or liquid propellant combustion products is inherently attractive from a performance and packaging viewpoint. In the past there has not been a production application of HGITVC because of erosion of materials in hot gas valves. However, two

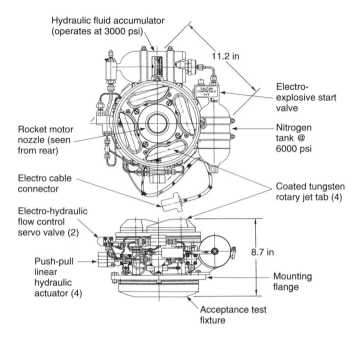

FIGURE 16–7. Two views of the jet tab assembly, packaged in a doughnut shape volume around the nozzle of the Tomahawk cruise missile's solid propellant booster rocket motor. Hydraulic actuators rotate the tabs in and out of the nozzle exhaust jet and are located just beyond the nozzle exit. (Courtesy of Space and Electronics Group, TRW, Inc.)

factors now make hot-gas-side injection feasible: first, hot gas valves can be made with the newer carbon–carbon structural parts and modern insulators. A hot gas system with a limited duration hot gas carbon valve is described in Ref. 16–4. Also, advances in metallurgy have made possible the development of hot valves made of rhenium alloy, a high-temperature metal suitable for hot gas valve applications. The second factor is the development of solid propellants that are less aggressive (less AP, Al_2O_3, and/or fewer oxidizing gas ingredients) and reduce the erosion in nozzles and valves; this helps the hot gas valves and insulated hot gas plumbing to better survive for limited durations but often at the expense of propulsion system performance. Experimental hot gas systems have had difficulties with thermal distortions and in keeping key components cool enough to prevent failure.

With either liquid or solid propellants, the hot gas can be bled off the main combustion chamber or generated in a separate gas generator. The hot gas valves can be used to (1) control side injection of hot gas into a large nozzle, or (2) control a pulsing flow through a series of small fixed nozzles similar to small attitude control thrusters described in Chapters 4, 6, and 11. In liquid propellant engines it is feasible to tap or withdraw gas from the thrust chamber at a location where there is an intentional fuel-rich mixture ratio; the gas temperature would then be low enough (about 1100°C or 2000°F) so that uncooled metal hardware can be used for HGITVC valves and piping.

The total side force resulting from secondary injection of a fluid into the main stream of the supersonic nozzle can be expressed as two force components: (1) the force associated with the momentum of the injectant; and (2) the pressure unbalance acting over areas of the internal nozzle wall. The second term results from the unbalanced wall pressures within the nozzle caused by shock formation, boundary layer separation, difference between injectant and undisturbed nozzle stream pressures, and primary–secondary combustion reactions (for chemically active injectants). The strength of the shock pattern and the pressure unbalance created between opposite walls in the nozzle is dependent on many variables, including the properties of the injectant and whether it is liquid or gas. In the case of injecting a reactive fluid, the combustion occurring downstream of the injection port(s) usually produces a larger pressure unbalance effect than is obtained by liquid vaporization only. However, benefit from combustion is dependent on a chemical reaction rate high enough to keep the reaction zone close to the injection port. The TVC performance that is typical of inert and reactive liquids and hot gas (solid propellant combustion products) is indicated in Fig. 16–8. This plot of force ratios to mass flow ratios is a parametric representation commonly used in performance comparisons.

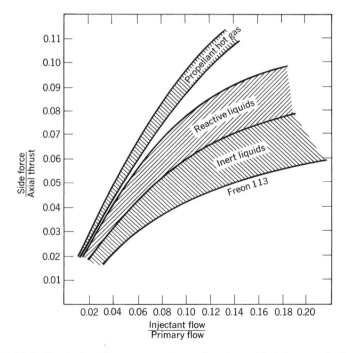

FIGURE 16–8. Typical performance regions of various side injectants in TVC nozzles.

16.2. TVC WITH MULTIPLE THRUST CHAMBERS OR NOZZLES

All the various concepts shown in Fig. 16–3 can provide pitch and yaw moments to a vehicle. Roll control can be obtained only if there are at least two separate vectorable nozzles, four fixed pulsing or throttled flow nozzles, or two jet vanes submerged in the exhaust gas from a single nozzle.

Several concepts have been developed and flown that use two or more rocket engines or a single engine or motor with two or more actuated nozzles. Two fully gimballed thrust chambers or motor nozzles can provide roll control with very slight differential angular deflections. For pitch and yaw control, the deflection would be larger, be of the same angle and direction for both nozzles, and the deflection magnitude would be the same for both nozzles. This can also be achieved with four hinged (see Figure 10–10) or gimballed nozzles. Figure 16–9 shows the rocket motor of an early version of the Minuteman missile booster (first stage) with four movable nozzles. This motor is described in Table 11–3.

The differential throttling concept shown in Fig. 16–10 has no gimbal and does not use any of the methods used with single nozzles as described in Fig. 16–3. It has four fixed thrust chambers and their axes are almost parallel to and set off from the vehicle's centerline. Two of the four thrust chambers are

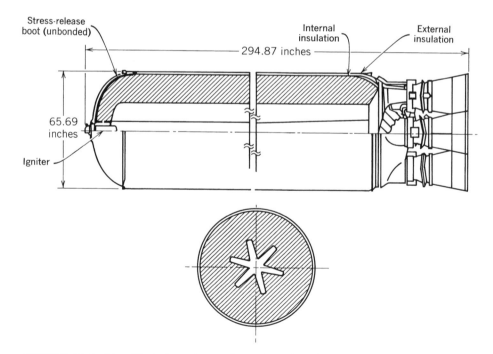

FIGURE 16–9. Simplified view of an early version of the first-stage Minuteman missile motor using composite-type propellant bonded to the motor case. Four movable nozzles provide pitch, yaw, and roll control. (Source: U.S. Air Force.)

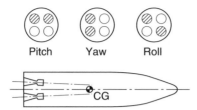

FIGURE 16–10. Differential throttling with four fixed-position thrust chambers can provide flight maneuvers. In this simple diagram the shaded nozzle exits indicate a throttled condition or reduced thrust. The larger forces from the unthrottled engines impose turning moments on the vehicle. For roll control the nozzles are slightly inclined and their individual thrust vectors do not go through the center of gravity of the vehicle.

selectively throttled (typically the thrust is reduced by only 2 to 15%). The four nozzles may be supplied from the same feed system or they may belong to four separate but identical rocket engines. This differential throttling system is used on the Aerospike rocket engine described in Chapters 3 and 8 and on a Russian launch vehicle.

16.3. TESTING

Testing of thrust vector control systems often includes actuation of the system when assembled on the propulsion system and the vehicle. For example, the Space Shuttle main engine can be put through some gimbal motions (without rocket firing) prior to a flight. A typical acceptance test series of the TVC system (prior to the delivery to an engine manufacturer) may include the determination of input power, accuracy of deflected positions, angular speeds or accelerations, signal response characteristics, or validation of overtravel stops. The ability to operate under extreme thermal environment, operation under various vehicle or propulsion system generated vibrations, temperature cycling, and ignition shock (high momentary acceleration) would probably be a part of the qualification tests.

Side forces and roll torques are usually relatively small compared to the main thrust and the pitch or yaw torques. Their accurate static test measurement can be difficult, particularly at low vector angles. Elaborate, multicomponent test stands employing multiple load cells and isolation flextures are needed to assure valid measurements.

16.4. INTEGRATION WITH VEHICLE

The actuations or movements of the TVC system are directed by the vehicle's guidance and control system (see Ref. 16–5). This system measures the three-dimensional position, velocity vectors, and rotational rates of the vehicle and compares them with the desired position, velocity, and rates. The error signals between these two sets of parameters are transformed by computers in TVC controllers into control commands for actuating the TVC system until the error signals are reduced to zero. The vehicle's computer control system determines the timing of the actuation, the direction, and magnitude of the deflection. With servomechanisms, power supplies, monitoring/failure detection devices, actuators with their controllers, and kinetic compensation, the systems tend to become complex.

The criteria governing the selection and design of a TVC system stem from vehicle needs and include the steering-force moments, force rates of change, flight accelerations, duration, performance losses, dimensional and weight limitations, available vehicle power, reliability, delivery schedules, and cost. For the TVC designer these translate into such factors as duty cycle, deflection angle, angle slew rate, power requirement, kinematic position errors, and many vehicle–TVC and motor–TVC interface details, besides the program aspects of costs and delivery schedules.

Interface details include electrical connections to and from the vehicle flight controller, the power supply, mechanical attachment with fasteners for actuators, and sensors to measure the position of the thrust axis or the actuators. Design features to facilitate the testing of the TVC system, easy access for checkout or repair, or to facilitate resistance to a high-vibration environment, are usually included. The TVC subsystem is usually physically connected to the vehicle and mounted to the rocket's nozzle. The designs of these components must be coordinated and integrated. Nozzle–TVC interfaces are discussed in Refs. 6–1 (TVC of liquid rocket engines and their control architecture) and 16–5.

The actuators can be hydraulic, pneumatic, or electromechemical (lead screw), and usually include a position sensor to allow feedback to the controller. The proven power supplies include high-pressure cold stored gas, batteries, warm gas from a gas generator, hydraulic fluid pressurized by cold gas or a warm gas generator, electric or hydraulic power from the vehicle's power supply, and electric or hydraulic power from a separate turbogenerator (in turn driven by a gas generator). The last type is used for relatively long-duration high-power applications, such as the power package used in the Space Shuttle solid rocket booster TVC, explained in Ref. 16–6. The selection of the actuation scheme and its power supply depends on the minimum weight, minimum performance loss, simple controls, ruggedness, reliability, ease of integration, linearity between actuating force and vehicle moments, cost, and other factors. The required frequency response is higher if the vehicle is small, such as with small tactical missiles. The response listed in Table 16–3 is more typical of larger spacecraft applications. Sometimes the TVC system is integrated with a movable aerodynamic fin system, as shown in Ref. 16–7.

REFERENCES

16–1. A. Truchot, "Design and Analysis of Solid Rocket Motor Nozzles," Chapter 3 in *Design Methods in Solid Rocket Motors*, AGARD Lecture Series 150, Advisory Group for Aerospace Research and Development, NATO, Revised Version, 1988.

16–2. B. H. Prescott and M. Macocha, "Nozzle Design," pp. 177–186 in Chapter 6 of G. E. Jensen and D. W. Netzer (Eds.), *Tactical Missile Propulsion*, Vol. 170 in Progress in Astronautics and Aeronautics, American Institute of Aeronautics and Astronautics, 1996.

16–3. G. E. Conner, R. L. Pollock, and M. R. Riola, "IUS Thrust Vector Control Servo System," paper presented at *1983 JANNAF Propulsion Meeting*, Monterey, CA, February 1983.

16–4. M. Berdoyes, "Thrust Vector Control by Injection of Hot Gas Bleed from the Chamber Hot Gas Valve," *AIAA Paper 89-2867*, July 1989.

16–5. J. H. Blakelock, *Automatic Control of Aircraft and Missiles*, 2nd ed., John Wiley & Sons, New York, 1991, 656 pages.

16–6. A. A. McCool, A. J. Verble, Jr., and J. H. Potter, "Space Transportation System's Rocket Booster Thrust Vector Control System," *Journal of Spacecraft and Rockets*, Vol. 17, No. 5, September–October 1980, pp. 407–412.

16–7. S. R. Wassom, L. C. Faupel, and T. Perley, "Integrated Aerofin/Thrust Vector Control for Tactical Missiles," *Journal of Propulsion and Power*, Vol. 7, No. 3, May–June 1991, pp. 374–381.

CHAPTER 17

SELECTION OF ROCKET PROPULSION SYSTEMS

With few exceptions, design problems have several possible engineering solutions from which to select. In this chapter we discuss in general terms the process of selecting propulsion systems for a given mission. Three specific aspects are covered in some detail:

1. A comparison of the merits and disadvantages of liquid propellant rocket engines with solid propellant rocket motors.
2. Some key factors used in evaluating particular propulsion systems and selecting from several competing candidate rocket propulsion systems.
3. The interfaces between the propulsion system and the flight vehicle and/ or the overall system.

A propulsion system is really a subsystem of a flight vehicle. The vehicle, in turn, can be part of an overall system. An example of an overall system would be a communications network with ground stations, computers, transmitters, and several satellites; each satellite is a flight vehicle and has an attitude-control propulsion system with specific propulsion requirements. The length of time in orbit is a system parameter that affects the satellite size and the total impulse requirement of its propulsion system.

Subsystems of a vehicle system (such as the structure, power supply, propulsion, guidance, control, communications, ground support, or thermal control) often pose conflicting requirements. *Only through careful analyses and system engineering studies is it possible to find compromises that allow all subsystems to operate satisfactorily and be in harmony with each other.* The subject of engineering design has advanced considerably in recent times and general references such as Ref. 17–1 should be consulted for details. Other works address the design of space systems (e.g., Refs. 17–2, 17–3) and the design of liquid propellant engines (e.g., Ref. 17–4).

All mission (overall system), vehicle, and propulsion system *requirements* can be related to either performance, cost, or reliability. For a given mission, one of these criteria is usually more important than the other two. There is a strong interdependence between the three levels of requirements and the three categories of criteria mentioned above. Some of the characteristics of the propulsion system (which is usually a second-tier subsystem) can have a strong influence on the vehicle and vice versa. An improvement in the propulsion performance, for example, can have a direct influence on the vehicle size, overall system cost, or life (which can be translated into reliability and cost).

17.1. SELECTION PROCESS

The selection process is a part of the overall design effort for the vehicle system and its rocket propulsion system. The selection is based on a *series of criteria*, which are based on the requirements and which will be used to evaluate and compare alternate propulsion systems. This process for determining the most suitable rocket propulsion system depends on the application, the ability to express many of the characteristics of the propulsion systems quantitatively, the amount of applicable data that are available, the experience of those responsible for making the selection, and the available time and resources to examine the alternate propulsion systems. What is described here is one somewhat idealized selection process as depicted in Fig. 17–1, but there are alternate sequences and other ways to do this job.

All propulsion selections start with a definition of the overall system and its mission. The mission's objectives, payload, flight regime, trajectory options, launch scenarios, probability of mission success, and other requirements have to be defined, usually by the organization responsible for the overall system. Next, the vehicle has to be defined in conformance with the stated flight application. Only then can the propulsion system requirements be derived for the specific mission and/or vehicle. For example, from the mission requirements it is possible to determine the required mass fraction, the minimum specific impulse, and the approximate total propellant mass, as shown in Chapter 4. Furthermore, this can include propulsion parameters such as thrust–time profile, propellant mass fraction, allowable volume or envelope, typical pulsing duty cycle, ambient temperature limits, thrust vector control needs, vehicle interfaces, likely number of units to be built, prior applicable experience, time schedule requirements, and cost limits.

Since the total vehicle's performance, flight control, operation, or maintenance are usually critically dependent on the performance, control, operation, or maintenance of the rocket propulsion system (and vice versa), the process will usually go through several iterations in defining both the vehicle and propulsion requirements, which are then documented. This iterative process

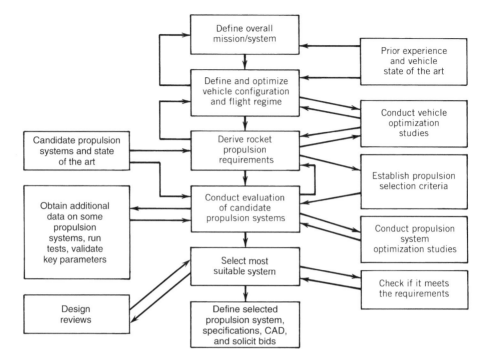

FIGURE 17–1. Idealized process for selecting propulsion systems.

involves both the system organization (or the vehicle/system contractor) and one or more propulsion organizations (or rocket propulsion contractors). Documentation can take many forms; electronic computers have expanded their capability to network, record, and retrieve documents.

A number of competing candidate systems are usually evaluated. They may be proposed by different rocket propulsion organizations, perhaps on the basis of modifications of some existing rocket propulsion system, or may include some novel technology, or may be new types of systems specifically configured to fit the vehicle or mission needs. In making these evaluations it will be necessary to compare several candidate propulsion systems with each other and to rank-order them (in accordance with the selection criteria) on how well they meet each requirement. This requires analysis of each candidate system and also, often, some additional testing. For example, statistical analyses of the functions, failure modes, and safety factors of all key components can lead to quantitative reliability estimates. For some criteria, such as safety or prior related experience, it may not be possible to compare candidate systems quantitatively but only somewhat subjectively.

Various rocket parameters for a particular mission need to be optimized. Trade-off studies are used to determine the number of thrust chambers, engines or motors, optimum chamber pressure, best packaging of the propulsion system(s), optimum mixture ratio, optimum number of stages in a multistage vehicle, best trajectory, optimum nozzle area ratio, number of nozzles, TVC (thrust vector control) concept, optimum propellant mixture ratio or solid propellant formulation, and so on. These trade-off studies are usually aimed at achieving the highest performance, highest reliability, or lowest cost for a given vehicle and mission. Some of these optimizations are needed early in the process to establish propulsion criteria, and some are needed in evaluating competing candidate propulsion systems.

Early in the selection process a tentative recommendation is usually made as to whether the propulsion system should be a solid propellant motor, a liquid propellant engine, an electrical propulsion system, or some other type. Each type has its own regime of thrust, specific impulse, thrust-to-weight ratio (acceleration), or likely duration, as shown in Table 2–1 and Fig. 2–5; these factors are listed for several chemical rocket engines and several types of non-chemical engines. Liquid engines and solid motors are covered in Chapters 6 to 14, hybrids in Chapter 15.

If an existing vehicle is to be upgraded or modified, its propulsion system is usually also improved or modified (e.g., higher thrust, more total impulse, or faster thrust vector control). While there might still be some trade-off studies and optimization of the propulsion parameters that can be modified, one normally does not consider an entirely different propulsion system as is done in an entirely new vehicle or mission. Also, it is rare that an identical rocket propulsion system is selected for two different applications; usually, some design changes and interface modifications are necessary to adapt an existing rocket propulsion system to another application. Proven existing and qualified propulsion systems, that fit the desired requirements, usually have an advantage in cost and reliability.

Electric propulsion systems have a set of unique applications with low thrusts, low accelerations, trajectories exclusively in space, high specific impulse, long operating times, and generally a relatively massive power supply system. They perform well in certain space transfer and orbit maintenance missions. With more electric propulsion systems flying than ever before, the choice of proven electric propulsion thruster types is becoming larger. These systems, together with design approaches, are described in Chapter 19 and Ref. 17–3.

When a chemical rocket is deemed most suitable for a particular application, the selection has to be made between a liquid propellant engine, a solid propellant motor, or a hybrid propulsion system. Some of the major advantages and disadvantages of liquid propellant engines and solid propellant motors are given in Tables 17–1 to 17–4. These lists are general in nature; some items can be controversial, and a number are restricted to particular applications. Items from this list can be transformed into evaluation criteria. For a specific mission, the relevant items on these lists would be rank-ordered in accordance with their relative importance. A quantification of many of the items would be

TABLE 17–1. Solid Propellant Rocket Advantages

Simple design (few or no moving parts).

Easy to operate (little preflight checkout).

Ready to operate quickly.

Will not leak, spill, or slosh.

Sometimes less overall weight for low total impulse application.

Can be throttled or stopped and restarted (a few times) if preprogrammed.

Can provide TVC, but at increased complexity.

Can be stored for 5 to 25 years.

Usually, higher overall density; this allows a more compact package, a smaller vehicle (less drag).

Some propellants have nontoxic, clean exhaust gases, but at a performance penalty.

Some grain and case designs can be used with several nozzles.

Thrust termination devices permit control over total impulse.

Ablation and gasification of insulator, nozzle, and liner materials contribute to mass flow and thus to total impulse.

Some tactical missile motors have been produced in large quantities (over 200,000 per year).

Can be designed for recovery, refurbishing, and reuse (Space Shuttle solid rocket motor).

TABLE 17–2. Liquid Propellant Rocket Advantages

Usually highest specific impulse; for a fixed propellant mass, this increases the vehicle velocity increment and the attainable mission velocity.

Can be randomly throttled and randomly stopped and restarted; can be efficiently pulsed (some small thrust sizes over 250,000 times). Thrust–time profile can be randomly controlled; this allows a reproducible flight trajectory.

Cutoff impulse can be controllable wth thrust termination device (better control of vehicle terminal velocity).

Can be largely checked out just prior to operation. Can be tested at full thrust on ground or launch pad prior to flight.

Can be designed for reuse after field services and checkout.

Thrust chamber (or some part of the vehicle) can be cooled and made lightweight.

Storable liquid propellants have been kept in vehicle for more than 20 years and engine can be ready to operate quickly.

With pumped propulsion feed systems and large total impulse, the inert propulsion system mass (including tanks) can be very low (thin tank walls and low tank pressure), allowing a high propellant mass fraction.

Most propellants have nontoxic exhaust, which is environmentally acceptable.

Same propellant feed system can supply several thrust chambers in different parts of the vehicle.

Can modify operating conditions during firing to prevent some failures that would otherwise result in the loss of the mission or vehicle.

Can provide component redundancy (e.g., dual check valves or extra thrust chamber) to enhance reliability.

With multiple engines, can design for operation with one or more shutoff (engine out capability).

The geometry of low-pressure tanks can be designed to fit most vehicles' space constraints (i.e., mounted inside wing or nose cone).

The placement of propellant tanks within the vehicle can minimize the travel of the center of gravity during powered flight. This enhances the vehicle's flight stability and reduces control forces.

Plume radiation and smoke are usually low.

TABLE 17–3. Solid Propellant Rocket Disadvantages

Explosion and fire potential is larger; failure can be catastrophic; most cannot accept bullet impact or being dropped onto a hard surface.

Many require environmental permit and safety features for transport on public conveyances.

Under certain conditions some propellants and grains can detonate.

Cumulative grain damage occurs through temperature cycling or rough handling; this limits the useful life.

If designed for reuse, it requires extensive factory rework and new propellants.

Requires an ignition system.

Each restart requires a separate ignition system and additional insulation—in practice, one or two restarts.

Exhaust gases are usually toxic for composite propellants containing ammonium perchlorate.

Some propellants or propellant ingredients can deteriorate (self-decompose) in storage.

Most solid propellant plumes cause more radio frequency attenuation than liquid propellant plumes.

Only some motors can be stopped at random, but motor becomes disabled (not reusable).

Once ignited, cannot change predetermined thrust or duration. A moving pintle design with a variety throat area will allow random thrust changes, but experience is limited.

If propellant contains more than a few percent particulate carbon, aluminum, or other metal, the exhaust will be smoky and the plume radiation will be intense.

Integrity of grain (cracks, unbonded areas) is difficult to determine in the field.

Thrust and operating duration will vary with initial ambient grain temperature and cannot be easily controlled. Thus the flight path, velocity, altitude, and range of a motor will vary with the grain temperature.

Large boosters take a few seconds to start.

Thermal insulation is required in almost all rocket motors.

Cannot be tested prior to use.

Needs a safety provision to prevent inadvertent ignition, which would lead to an unplanned motor firing. Can cause a disaster.

TABLE 17–4. Liquid Propellant Rocket Disadvantages

Relatively complex design, more parts or components, more things to go wrong.

Cryogenic propellants cannot be stored for long periods except when tanks are well insulated and escaping vapors are recondensed. Propellant loading occurs at the launch stand and requires cryogenic propellant storage facilities.

Spills or leaks of several propellants can be hazardous, corrosive, toxic, and cause fires, but this can be minimized with gelled propellants.

More overall weight for most short-duration, low-total-impulse applications (low propellant mass fraction).

Nonhypergolic propellants require an ignition system.

Tanks need to be pressurized by a separate pressurization subsystem. This can require high-pressure inert gas storage (2000 to 10,000 psi) for long periods of time.

More difficult to control combustion instability.

Bullet impact will cause leaks, sometimes a fire, but usually no detonations; gelled propellants can minimize or eliminate these hazards.

A few propellants (e.g., red fuming nitric acid) give toxic vapors or fumes.

Usually requires more volume due to lower average propellant density and the relatively inefficient packaging of engine components.

If vehicle breaks up and fuel and oxidizer are intimately mixed, it is possible (but not likely) for an explosive mixture to be created.

Sloshing in tank can cause a flight stability problem, but it can be minimized with baffles.

If tank outlet is uncovered, aspirated gas can cause combustion interruption or combustion vibration.

Smoky exhaust (soot) plume can occur with some hydrocarbon fuels.

Needs special design provisions for start in zero gravity.

With cryogenic liquid propellants there is a start delay caused by the time needed to cool the system flow passage hardware to cryogenic temperatures.

Life of cooled large thrust chambers may be limited to perhaps 100 or more starts.

High-thrust unit requires several seconds to start.

needed. These tables apply to generic rocket propulsion systems; they do not cover systems that use liquid–solid propellant combinations.

A favorite student question has been: Which are better, solid or liquid propellant rockets? A clear statement of strongly favoring one or the other can only be made when referring to a specific set of flight vehicle missions. Today, solid propellant motors seem to be preferred for tactical missiles (air-to-air, air-to-surface, surface-to-air, or short-range surface-to-surface) and ballistic missiles (long- and short-range surface-to-surface) because instant readiness, compactness, and their lack of spills or leaks of hazardous liquids are important criteria for these applications. Liquid propellant engines seem to be preferred for space-launched main propulsion units and upper stages, because of their higher specific impulse, relatively clean exhaust gases, and random throttling capability. They are favored for post-boost control systems and attitude control systems, because of their random multiple pulsing capability with precise cutoff impulse, and for pulsed axial and lateral thrust propulsion on hit-to-kill defensive missiles. However, there are always some exceptions to these preferences.

When selecting the rocket propulsion system for a major new multiyear high-cost project, considerable time and effort are spent in evaluation and in developing rational methods for quantitative comparison. In part this is in response to government policy as well as international competition. Multiple studies are done by competing system organizations and competing rocket propulsion organizations; formal reviews are used to assist in considering all the factors, quantitatively comparing important criteria, and arriving at a proper selection.

17.2. CRITERIA FOR SELECTION

Many criteria used in selecting a particular rocket propulsion system are peculiar to the particular mission or vehicle application. However, some of these selection factors apply to a number of applications, such as those listed in Table 17–5. Again, this list is incomplete and not all the criteria in this table apply to every application. The table can be used as a checklist to see that none of the criteria listed here are omitted.

Here are some examples of important criteria in a few specific applications. For a spacecraft that contains optical instruments (e.g., telescope, horizon seeker, star tracker, or infrared radiation seeker) the exhaust plume must be free of possible contaminants that may deposit or condense on photovoltaic cells, radiators, optical windows, mirrors, or lenses and degrade their performance, and free of particulates that could scatter sunlight into the instrument aperture, which could cause erroneous signals. Conventional composite solid propellants and pulsing storable bipropellants are usually not satisfactory, but cold or heated clean gas jets (H_2, Ar, N_2, etc.) and monopropellant hydrazine reaction gases are usually acceptable. Another example is an emphasis on smokeless propellant exhaust plumes, so as to make visual detection of a smoke or vapor trail very difficult. This applies particularly to tactical missile applications. Only a few solid propellants and several liquid propellants would be truly smokeless and free of a vapor trail under all weather conditions.

Several selection criteria may be in conflict with each other. For example, some propellants with a very high specific impulse are more likely to experience combustion instabilities. In liquid propellant systems, where the oxidizer tank is pressurized by a solid propellant gas generator and where the fuel-rich hot gases are separated by a thin flexible diaphragm from the oxidizer liquid, there is a trade-off between a very compact system and the potential for a damaging system failure (fire, possible explosion, and malfunction of system) if the diaphragm leaks or tears. In electric propulsion, high specific impulse is usually accompanied by heavy power generating and conditioning equipment.

Actual selection will depend on the balancing of the various selection factors in accordance with their importance, benefits, or potential impact on the system, and on quantifying as many of these selection factors as possible through analysis, extrapolation of prior experience/data, cost estimates, weights, and/or separate tests. Design philosophies such as the *Taguchi methodology* and *TQM* (total quality management) can be inferred (Refs. 17–1 and 17–2). Layouts, weight estimates, center-of-gravity analyses, vendor cost estimates, preliminary stress or thermal analysis, and other preliminary design efforts are usually necessary to put numerical values on some of the selection parameters. A comparative examination of the interfaces of alternate propulsion systems is also a part of the process. Some propulsion requirements are incompatible with each other and a compromise has to be made. For example, the monitoring of extra sensors can prevent the occurrence of certain types of failure and thus enhance the propulsion system reliability, yet the extra sensors and control components contribute to the system complexity and their possible failures will reduce the overall reliability. The selection process may also include feedback when the stated propulsion requirements cannot be met or do not make sense, and this can lead to a revision of the initial mission requirements or definition.

Once the cost, performance, and reliability drivers have been identified and quantified, the selection of the best propulsion system for a specified mission proceeds. The final propulsion requirement may come as a result of several iterations and will usually be documented, for example in a *propulsion requirement specification*. A substantial number of records is required (such as engine or motor acceptance documents, CAD (computer-aided design) images, parts lists, inspection records, laboratory test data, etc.). There are many specifications associated with design and manufacturing as well as with vendors, models, and so on. There must also be a disciplined procedure for approving and making design and manufacturing changes. This now becomes the starting point for the design and development of the propulsion system.

TABLE 17.5. Typical Criteria Used in the Selection of a Particular Rocket Propulsion System

---

Mission Definition

Purpose, function, and final objective of the mission of an overall system are well defined and their implications well understood. There is an expressed need for the mission, and the benefits are evident. The mission requirements are well defined. The payload, flight regime, vehicle, launch environment, and operating conditions are established. The risks, as perceived, appear acceptable. The project implementing the mission must have political, economic, and institutional support with assured funding. The propulsion system requirements, which are derived from mission definition, must be reasonable and must result in a viable propulsion system.

Affordability (Cost)

Life cycle costs are low. They are the sum of R&D costs, production costs, facility costs, operating costs, and decommissioning costs, from inception to the retirement of the system (see Ref. 17–5). Benefits of achieving the mission should appear to justify costs. Investment in new facilities should be low. Few, if any, components should require expensive materials. For commercial applications, such as communications satellites, the return on investment must look attractive. No need to hire new, inexperienced personnel, who need to be trained and are more likely to make expensive errors.

System Performance

The propulsion system is designed to optimize vehicle and system performance, using the most appropriate and proven technology. Inert mass is reduced to a practical minimum, using improved materials and better understanding of loads and stresses. Residual (unused) propellant is minimal. Propellants have the highest practical specific impulse without undue hazards, without excessive inert propulsion system mass, and with simple loading, storing, and handling (the specific impulse of the propulsion system is defined in Section 2.1 and is further discussed in Section 19.1). Thrust–time profiles and number of restarts must be selected to optimize the vehicle mission. Vehicles must operate with adequate performance for all the possible conditions (pulsing, throttling, temperature excursions, etc.). Vehicles should be storable over a specified lifetime. Will meet or exceed operational life. Performance parameters (e.g., chamber pressure, ignition time, or nozzle area ratio) should be near optimum for the selected mission. Vehicle should have adequate TVC. Plume characteristics are satisfactory.

Survivability (Safety)

All hazards are well understood and known in detail. If failure occurs, the risk of personnel injury, damage to equipment, facilities, or the environment is minimal. Certain mishaps or failures will result in a change in the operating condition or the safe shutdown of the propulsion system. Applicable safety standards must be obeyed. Inadvertent energy input to the propulsion system (e.g., bullet impact, external fire) should not result in a detonation. The probability for any such drastic failures should be very low. Safety monitoring and inspections must have proven effective in identifying and preventing a significant share of possible incipient failures (see Ref. 17–6). Adequate safety factors must be included in the design. Spilled liquid propellants should cause no undue hazards. All systems and procedures must conform to the safety standards. Launch test range has accepted the system as being safe enough to launch.

Reliability

Statistical analyses of test results indicate a satisfactory high-reliability level. Technical risks, manufacturing risks, and failure risks are very low, well understood, and the impact on the overall system is known. There are few complex components. Adequate storage and operating life of components (including propellants) have been demonstrated. Proven ability to check out major part of propulsion system prior to use or launch. If certain likely failures occur, the system must shut down safely. Redundancy of key components should be provided, where effective. High probability that all propulsion functions must be performed within the desired tolerances. Risk of combustion vibration or mechanical vibration should be minimal.

TABLE 17–5 (*Continued*)

Controllability
Thrust buildup and decay are within specified limits. Combustion process is stable. The time responses to control or command signals are within acceptable tolerances. Controls need to be foolproof and not inadvertently create a hazardous condition. Thrust vector control response must be satisfactory. Mixture ratio control must assure nearly simultaneous emptying of the fuel and oxidizer tanks. Thrust from and duration of afterburning should be negligible. Accurate thrust termination feature must allow selection of final velocity of flight. Changing to an alternate mission profile should be feasible. Liquid propellant sloshing and pipe oscillations need to be adequately controlled. In a zero-gravity environment, a propellant tank should be essentially fully emptied.

Maintainability
Simple servicing, foolproof adjustments, easy parts replacement, and fast, reliable diagnosis of internal failures or problems. Minimal hazard to service personnel. There must be easy access to all components that need to be checked, inspected, or replaced. Trained maintenance personnel are available. Good access to items which need maintenance.

Geometric Constraints
Propulsion system fits into vehicle, can meet available volume, specified length, or vehicle diameter. There is usually an advantage for the propulsion system that has the smallest volume or the highest average density. If the travel of the center of gravity has to be controlled, as is necessary in some missions, the propulsion system that can do so with minimum weight and complexity will be preferred.

Prior Related Experience
There is a favorable history and valid, available, relevant data of similar propulsion systems supporting the practicality of the technologies, manufacturability, performance, and reliability. Experience and data validating computer simulation programs are available. Experienced, skilled personnel are available.

Operability
Simple to operate. Validated operating manuals exist. Procedures for loading propellants, arming the power supply, launching, igniter checkout, and so on, must be simple. If applicable, a reliable automatic status monitoring and check-out system should be available. Crew training needs to be minimal. Should be able to ship the loaded vehicle on public roads or railroads without need for environmental permits and without the need for a decontamination unit and crew to accompany the shipment. Supply of spare parts must be assured. Should be able to operate under certain emergency and overload conditions.

Producibility
Easy to manufacture, inspect, and assemble. All key manufacturing processes are well understood. All materials are well characterized, critical material properties are well known, and the system can be readily inspected. Proven vendors for key components have been qualified. Uses standard manufacturing machinery and relatively simple tooling. Hardware quality and propellant properties must be repeatable. Scrap should be minimal. Designs must make good use of standard materials, parts, common fasteners, and off-the-shelf components. There should be maximum use of existing manufacturing facilities and equipment. Excellent reproducibility, i.e., minimal operational variation between identical propulsion units. Validated specifications should be available for major manufacturing processes, inspection, parts fabrication, and assembly.

Schedule
The overall mission can be accomplished on a time schedule that allows the system benefits to be realized. R&D, qualification, flight testing, and/or initial operating capability are completed on a preplanned schedule. No unforeseen delays. Critical materials and qualified suppliers must be readily available.

TABLE 17–5 (*Continued*)

Environmental Acceptability
No unacceptable damage to personnel, equipment, or the surrounding countryside. No toxic species in the exhaust plume. No serious damage (e.g., corrosion) due to propellant spills or escaping vapors. Noise in communities close to a test or launch site should remain within tolerable levels. Minimal risk of exposure to cancer-causing chemicals. Hazards must be sufficiently low, so that issues on environmental impact statements are not contentious and approvals by environmental authorities become routine. There should be compliance with applicable laws and regulations. No unfavorable effects from currents generated by an electromagnetic pulse, static electricity, or electromagnetic radiation.

Reusability
Some applications (e.g., Shuttle main engine, Shuttle solid rocket booster, or aircraft rocket-assisted altitude boost) require a reusable rocket engine. The number of flights, serviceability, and the total cumulative firing time then become key requirements that will need to be demonstrated. Fatigue failure and cumulative thermal stress cycles can be critical in some of the system components. The critical components have been properly identified; methods, instruments, and equipment exist for careful check-out and inspection after a flight or test (e.g., certain leak tests, inspections for cracks, bearing clearances, etc.). Replacement and/or repair of unsatisfactory parts should be readily possible. Number of firings before disassembly should be large, and time interval between overhauls should be long.

Other Criteria
Radio signal attenuation by exhaust plume to be low. A complete propulsion system, loaded with propellants and pressurizing fluids, can be storable for a required number of years without deterioration or subsequent performance decrease. Interface problems are minimal. Provisions for safe packaging and shipment are available. The system includes features that allow decommissioning (such as to deorbit a spent satellite) or disposal (such as the safe removal and disposal of over-age propellant from a refurbishable rocket motor).

17.3. INTERFACES

In Section 2 of this chapter the interfaces between the propulsion system and the vehicle and/or overall system were identified as some of the criteria to be considered in the selection of a propulsion system. A few rocket propulsion systems are easy to integrate and interface with the vehicles. Furthermore, these interfaces are an important aspect of a disciplined design and development effort. Table 17–6 gives a partial listing of typical interfaces that have been considered in the propulsion system selection, design, and development. It too may be a useful checklist. The interfaces assure system functionality and compatibility between the propulsion system and the vehicle with its other subsystems under all likely operating conditions and mission options. Usually, an interface document or specification is prepared and it is useful to designers, operating personnel, or maintenance people.

Besides cold gas systems, a simple solid propellant rocket motor has the fewest and the least complex set of interfaces. A monopropellant liquid rocket engine also has relatively few and simple interfaces. A solid propellant motor with TVC and a thrust termination capability has additional interfaces, compared to a simple motor. Bipropellant rocket engines are more complex and the

TABLE 17–6. Typical Interfaces between Rocket Propulsion Systems and Flight Vehicle

Interface Category	Typical Detailed Interfaces
Structural	Interface (geometry/location/fastening mechanism) for mounting propulsion system
	Restraints on masses, moments of inertia, or the location of the center of gravity
	Type and degree of damping to minimize vibrations
	Attachment of vehicle components to propulsion system structure, such as wings, electrical components, TVC, or skirts
	Loads (aerodynamic, acceleration, vibrations, thrust, sloshing, dynamic interactions) from vehicle to propulsion system, and vice versa
	Dimensional changes due to loads and/or heating and means for allowing expansions or deflections to occur without overstress
	Interactions from vibration excitation
Mechanical	Interfaces for electric connectors; for pneumatic, hydraulic, propellant pipe connections
	Volume/space available and geometric interference with other subsystems
	Access for assembly, part replacement, inspection, maintenance, repair
	Lifting or handling devices, and lifting attachment locations
	Measurement and adjustment of alignment of fixed nozzles
	Matching of thrust levels when two or more units are fired simultaneously
	Sealing or other closure devices to minimize air breathing and moisture condensation in vented tanks, cases, nozzles, porous insulation, or open pipes
Power	Source and availability of power (usually electric, but sometimes hydraulic or pneumatic) and their connection interfaces
	Identification of all users of power (solenoids, instruments, TVC, igniter, sensors) and their duty cycles. Power distribution to the various users
	Conversion of power to needed voltages, dc/ac, frequencies, or power level
	Electric grounding connections of rocket motors, certain electric equipment or pyrotechnic devices, to minimize voltage buildup and prevent electrostatic discharges
	Shielding of sensitive wires and/or high-voltage components
	Telemetry and radio communications interface

TABLE 17–6 (*Continued*)

Interface Category	Typical Detailed Interfaces
	Heaters (e.g., to keep hydrazine from freezing or to prevent ice formation and accumulation with cryogenic propellants)
	Interfaces with antennas, wiring, sensors, and electronic packages located in the propulsion section of the vehicle
	Thermal management of heat generated in electric components
Propellants	Sharing of propellants between two or more propulsion systems (main thrust chambers and attitude control thrusters)
	Control of sloshing to prevent center of gravity (CG) excursions or to prevent gas from entering the liquid propellant tank outlet
	Design of solid propellant grain or liquid propellant tanks to limit CG travel
	Loading/unloading provisions for liquid propellants
	Access for X-ray inspection of grain for cracks or unbonded areas, while installed
	Access to visually inspect grain cavity for cracks
	Access to inspect cleanliness of tanks, pipes, valves
	Connection of drain pipes for turbopump seal leakage
Vehicle flight control and communications	Command signals (start/stop/throttle, etc.) interface
	Feedback signals (monitoring the status of the propulsion system, e.g., valve positions, thrust level, remaining propellant, pump speed, pressures, temperatures); telemetering devices
	Range safety destruct system
	Attitude control: command actuation in pitch, yaw, or roll; feedback of TVC angle position and slew rate, duty cycle, safety limits
	Division of control logic, computer capability, or data processing and databases between propulsion system controller, vehicle controller, test stand controller, or ground-based computer/controller system
	Number and type of fault detection devices and their connection methods
Thermal	Heat from rocket gas/exhaust plume or aerodynamic airflow will not overheat critical exposed components
	Transfer of heat between propulsion system and the vehicle
	Provisions for venting cryogenic propellant tanks overboard
	Radiators for heat rejection
	Interfaces for cooling, if any

TABLE 17–6. (*Continued*)

Interface Category	Typical Detailed Interfaces
Plume	Radiative and convective heating of vehicle by plume
	Impingement (forces and heating) of plume from attitude control nozzle with vehicle components
	Noise effects on equipment and surrounding areas
	Contamination or condensation of plume species on vehicle or payload parts, such as solar panels, optical components of instruments, or radiation surfaces
	Attenuation of radio signals
Safety	Condition monitoring and sensing of potential imminent failure and automatic remedial actions to prevent or remedy impending failure (e.g., reduce thrust or shut off one of several redundant propulsion systems)
	Arming and disarming of igniter. Access to safe & arm device
	Safe disposal of hazardous liquid propellant leaking through pump shaft seal, valve stem seal, or vented from tanks
	Designed to avoid electrostatic buildup and discharge
Ground support equipment	Interface with standby power system
	Interfaces with heating/cooling devices on ground at launch or test site
	Supply and loading method for liquid propellant, pressurizing gases, and other fluids. Also, interface with method for unloading these
	Electromechanical checkout
	Interface with ground systems for flushing, cleaning, drying the tanks and piping
	Transportation vehicles/boxes/vehicle erection devices
	Lifting devices and handling equipment
	Interface with fire extinguishing equipment on ground

number and difficulty of interfaces increase if they have a turbopump feed system, throttling features, TVC, or pulsing capability. In electric propulsion systems the number and complexity of interfaces is highest for an electrostatic thruster with pulsing capability, when compared to electrothermal systems. More complex electrical propulsion systems generally give higher values of specific impulse. If the mission includes the recovery and reuse of the propulsion system or a manned vehicle (where the crew can monitor and override the propulsion system commands), this will introduce additional interfaces, safety features, and requirements.

REFERENCES

17–1. A. Ertas and J. C. Jones, *The Engineering Design Process*, 2nd Edition, John Wiley & Sons, New York, 1996.

17–2. J. C. Blair and R. S. Ryan, "Role of Criteria in Design and Management of Space Systems," *Journal of Spacecraft and Rockets*, Vol. 31, No. 2, March–April 1994, pp. 323–329.

17–3. R. W. Humble, G. N. Henry, and W. J. Larson, *Space Propulsion Analysis and Design*, McGraw-Hill, New York, 1995.

17–4. D. K. Huzel and D. H. Huang, *Modern Engineering for Design of Liquid Propellant Rocket Engines*, Progress in Astronautics and Aeronautics, Vol. 147, AIAA, Washington, DC, 1992.

17–5. C. J. Meisl, "Life Cycle Cost Considerations for Launch Vehicle Liquid Propellant Engine," *Journal of Propulsion and Power*, Vol. 4, No. 2, March–April 1988, pp. 117–119.

17–6. A. Norman, I. Cannon, and L. Asch, "The History and Future Safety Monitoring in Liquid Rocket Engines," *AIAA Paper 89-2410*, presented at the 25th Joint Propulsion Conference, July 1989.

CHAPTER 18

ROCKET EXHAUST PLUMES

The behavior of rocket exhaust plumes is included in this book because it has gained importance in recent years. In this chapter we provide an introduction to the subject, general background, a description of various plume phenomena and their effects, and references for further study.

The plume is the moving formation of hot rocket exhaust gases (and sometimes also entrained small particles) outside the rocket nozzle. This gas formation is not uniform in structure, velocity, or composition. It contains several different flow regions and supersonic shock waves. It is usually visible as a brilliant flame, emits intense radiation energy in the infrared, visible, and ultraviolet segments of the spectrum, and is a strong source of noise. Many plumes leave a trail of smoke or vapor or toxic exhaust gases. At higher altitudes some of the plume gases can flow backward around the nozzle and reach components of the flight vehicle.

The plume characteristics (size, shape, structure, emission intensity of photons or sound pressure waves, visibility, electrical interference, or smokiness) depend not only on the characteristics of the particular rocket propulsion system or its propellants, but also on the flight path, flight velocity, altitude, weather conditions, such as winds, humidity, or clouds, and the particular vehicle configuration. Progress has been steady in recent decades in gaining understanding of the complex, interacting physical, chemical, optical, aerodynamic, and combustion phenomena within plumes by means of laboratory experiments, computer simulation, measurements on plumes during static firing tests, flight tests, or simulated altitude tests in vacuum test chambers. Yet much is not fully understood or predictable. As shown in Table 18–1, there are

TABLE 18–1. Applications of Plume Technology

Design/develop/operate Flight Vehicles, their Propulsion
Systems, and Launch Stands or Launch Equipment

For a given propulsion system and operating conditions (altitudes, weather, speed, afterburning, with atmospheric oxygen, etc.) determine or predict the plume dimensions, temperature profiles, emissions, or other plume parameters.

Determine likely heat transfer to components of vehicle, test facility, propulsion system or launcher, and prevent damage by design changes. Include afterburning and recirculation.

Estimate the ability of vehicle and test facilities to withstand intensive plume noise.

Determine the aerodynamic interaction of the plume with the airflow around the vehicle, which can cause changes in drag.

Reduce impingement on vehicle components (e.g., plumes from attitude control thrusters hitting a solar panel); this can cause excessive heating or impingement forces that may turn the vehicle.

Estimate and minimize erosion effects on vehicle or launcher components.

Prevent deposits of condensed species on spacecraft windows, optical surfaces, solar panels, or radiating heat emission surfaces.

Determine the backscatter of sunlight by plume particulates or condensed species, and minimize the scattered radiation that can reach into optical instruments on the vehicle, because this can give erroneous signals.

Protect personnel using a shoulder-fired rocket launcher from heat, blast, noise, smoke, and toxic gas.

Detect and Track Flight of Vehicles

Analysis and/or measurement of plume emission spectrum or signature.

Identify plumes of launch vehicles from a distance when observing from spacecraft, aircraft, or ground stations, using IR, UV, or visible radiations and/or radar reflections.

Distinguish their emissions from background signals.

Detect and identify smoke and vapor trails.

Track and predict the flight path.

Alter the propellant or the nozzle to minimize the radiation, radar signature, or smoke emissions.

Estimate weather conditions for appearance of secondary smoke.

Develop Sensors for Measuring Plume Phenomena

Improve calibration and data interpretation.

Develop improved and novel instruments for plume measurements, for both remote and close by locations.

TABLE 18-1. (*Continued*)

Improve Understanding of Plume Behavior

Improve theoretical approaches to plume phenomena.

Improve or create novel computer simulations.

Provide further validation of theory by experimental results from flight tests, laboratory investigations, static tests, or tests in simulated altitude facilities.

Understand and minimize the generation of high-energy noise.

Understand the mechanisms of smoke, soot, or vapor formation, thus learning how to control them.

Provide a better understanding of emission, absorption, and scatter within plume.

Provide a better prediction of chemiluminescence.

Understand the effect of shock waves, combustion vibration, or flight maneuvers on plume phenomena.

Understand the effects of plume remains on the stratosphere or ozone layer.

Develop a better algorithm for simulating turbulence in different parts of the plume.

Minimize Radio-Frequency Interference

Determine the plume attenuation for specific antennas and antenna locations on the vehicle.

Reduce the attenuation of radio signals that have to pass through the plume, typically between an antenna on the vehicle and an antenna on the ground or on another vehicle.

Reduce radar reflections from plumes.

Reduce the electron density and electron collision frequency in the plume; for example, by reducing certain impurities in the gas, such as sodium.

---

many applications or situations where a prediction or a quantitative understanding of plume behavior is needed.

18.1. PLUME APPEARANCE AND FLOW BEHAVIOR

The size, shape, and internal structure of a plume changes dramatically with altitude. Figure 18–1 shows the construction of a low-altitude plume at heights typically between 3 and 10 km. The plume diameter and length are often several times larger than the vehicle diameter and length. In the near field there is an inviscid inner core (exhaust gases that have not yet mixed with air) and a relatively thin outer mixing layer where oxygen from the air burns turbulently with the fuel-rich species in the exhaust gases. In the far field the ambient air and exhaust gases are well mixed throughout a cross section of the plume, and the local pressure is essentially that of the ambient air. In the intermediate field the shock wave intensities diminish and more of the mass flow is mixed with ambient air. The radiation emissions come from all parts of

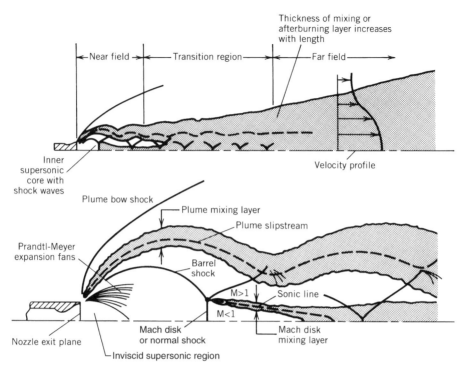

FIGURE 18–1. Half sections of schematic diagrams of a rocket exhaust plume at low altitude. Upper sketch shows full plume and lower sketch is an enlargement of the near field. (Reprinted with permission from Ref. 18–1.)

the plume, whereas the interactions with the vehicle occur only as a result of near-field phenomena.

Figure 18–2 shows sketches of the variation of the plume configuration with altitude. When the nozzle exit pressure is approximately equal to the ambient pressure (condition for optimum nozzle expansion), the plume has a long, nearly cylindrical shape. With increasing altitude the plume shape becomes more of a cone and the plume length and diameter increase. The core of the plume emerges supersonically from the nozzle exit and goes through an oblique compression shock wave, known as the barrel wave, which originates near the nozzle exit lip and has the approximate shape of an inverted but somewhat curved cone. The central part of the plume then goes through the Mach disk, which is a strong normal compression shock wave; here the gases suddenly slow down in velocity and are raised to a higher pressure and temperature. The flow immediately behind the Mach disk is subsonic for a short distance, but downstream again becomes supersonic. This pattern of normal shock waves and short subsonic zones is repeated several times in the core of the plume, but the strength of the shock and the rises in temperature or pressure are reduced in each sequence.

Plume configuration	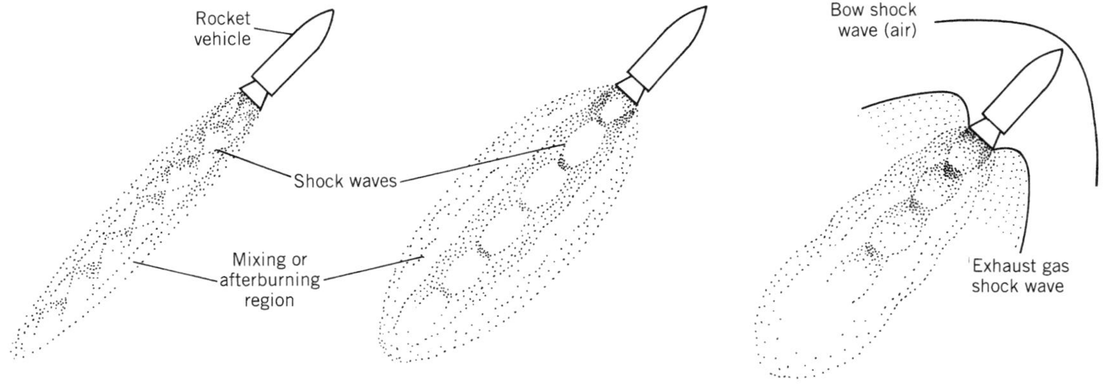		
Inner supersonic core	Narrow, can see several shock diamonds	Larger diameter, some shockwave pattern, fewer visible shock waves	Only one or two sets of shock waves are visible
Mixing layer (afterburning)	Narrow	Wider, unsteady, turbulent	Very wide, irregular
Nozzle exit pressure p_2 and ambient pressure p_3	$p_2 \cong p_3$	$p_2 > p_3$	$p_2 >> p_3$
Flight velocity	Very low, subsonic	Subsonic, transonic and slightly supersonic	Supersonic
Altitude, km	0 to 5	10 to 25	Above 35

FIGURE 18-2. The visible plume grows in length and diameter as the rocket vehicle gains altitude. The afterburning of the fuel-rich combustion products with the oxygen from the air occurs in the mixing layer. At very high altitude, above perhaps 200 km, there is no air and therefore no afterburning.

643

The ambient air mixes with the hot exhaust gases and secondary combustion or afterburning occurs in the mixing layer. It is a turbulent layer surrounding the core and its thickness increases with distance from the nozzle as well as with altitude. The incompletely oxidized fuel species in the exhaust gases, such as H_2, CO, NO, or CH_2, react chemically with the oxygen from the atmosphere and are largely burned to H_2O, CO_2, or NO_2, and the heat of this secondary combustion raises the temperature and the specific volume in this afterburning layer. As explained in Chapter 5, most propellants are fuel rich to achieve optimum specific impulse or optimum flight performance, so additional oxidative heat release is possible.

As the altitude increases, the ambient local air pressure decreases by several orders of magnitude and the pressure ratio in the gases between the nozzle exit and the local ambient pressure is increased greatly, approaching infinity when the rocket operates in a vacuum in space. With higher altitudes, further expansion (increase in specific volume) occurs and this causes a further reduction of gas temperatures and an expansion in both diameter and length; for the principal propulsion systems these usually exceed the dimensions of the vehicle. Some species in the plume will condense and become liquid; they will freeze as the temperature drops and gases like H_2O or CO_2 will form clouds or a vapor trail.

As the vehicle attains supersonic velocity (relative to the ambient air) two shock waves form. One is an oblique compression shock wave in the air ahead of the vehicle and the other is a trailing wave originating at the vehicle's tail, where the air meets the exhaust plume gases. These wave fronts are usually luminescent and highly visible and can reach diameters of several kilometers.

As the supersonic exhaust gas flow emerges from the nozzle, it experiences Prandtl–Meyer-type expansion waves, which attach themselves to the nozzle lip. This expansion allows the outer streamlines just outside the nozzle to be bent and an increase in the Mach number of the gases in the outer layers of the plume. This expansion can, at higher altitudes, cause some portion of the supersonic plume to be bent by more than 90° from the nozzle axis. The theoretical limit of a Prandtl–Meyer expansion is about 129° for gases with $k = 1.4$ (air) and about 160° for gases with $k = 1.3$ (typical for a rocket exhaust mixture; see Ref. 18–2). This backward flow needs to be analyzed to estimate the heat and impingement effects and possible contamination of vehicle components (see Ref. 18–3).

The boundary layer next to the nozzle wall is a region of viscous flow, and the flow velocity is lower than in the main nozzle inviscid flow. The velocity decreases to zero right next to the wall. For large nozzles this boundary layer can be quite thick, say 2 cm or more. Figure 3–16 shows a subsonic and a supersonic region within the boundary layer inside the nozzle divergent section; it also shows a temperature and a velocity profile. While the supersonic flow layer is restricted in the angle through which it can be deflected, the subsonic boundary layer flow at the nozzle lip is in a continuum regime and may be deflected up to 180°. Although the subsonic boundary layer represents only a small portion of the mass flow, it nevertheless lets its exhaust gases flow backward on the outside of the nozzle. This backflow has caused heating of and sometimes chemical damage to the vehicle and propulsion system parts.

The mass distribution or relative density is not uniform, as can be seen in Fig. 18–3, which is based on a calculated set of data for a high-altitude plume. Here 90% of the flow is within $\pm 44°$ of the nozzle axis and only one hundred thousandth or 10^{-5} of the total mass flow is bent by more than 90°. The flow near the center contains most of the heavier molecules, such as CO_2, NO_2, or CO, and the outer regions, which are deflected the most, consist largely of the lighter species, such as mostly H_2 and perhaps some H_2O.

Figure 18–4 shows the drastic change (log scale) in the overall radiation emission intensity as a function of altitude for a typical three-stage satellite launch to a 300- to 500-km orbit or a long-range ballistic missile with a booster stage, a sustainer stage, and a payload velocity adjustment stage. The booster-stage rocket propulsion system gives the largest intensity because it has the

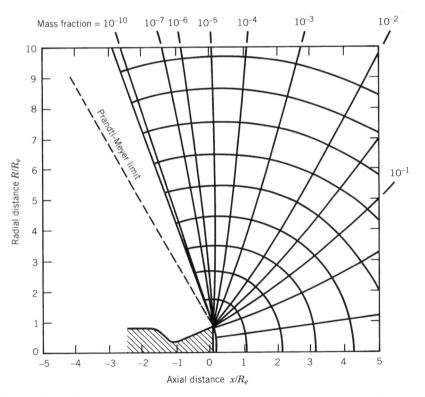

FIGURE 18–3. Density profile for vacuum plume expansion using a one-dimensional flow model for a small storable bipropellant thruster. The axial distance x and the plume radius R have been normalized with the nozzle exit radius R_e. Here $k = 1.25$, the Mach number of the nozzle exit is 4.0, and the nozzle cone half angle is 19°.

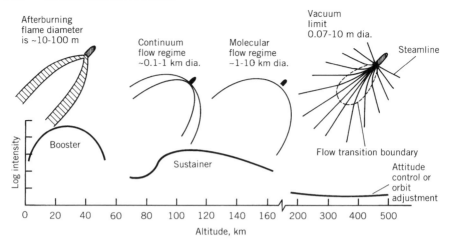

FIGURE 18–4. For a multistage ascending vehicle the plume radiation intensity will vary with the altitude, thrust or mass flow, propellant combination, and plume temperature. The four sketches describing the plume are not drawn to the same scale.

highest rocket gas mass flow or the highest thrust, a relatively dense plume, and its radiation is enhanced by afterburning of the fuel-rich gas with oxygen from the air. The rise in the intensity of the sustainer stage is due to the large increase in plume volume caused by the expansion of the exhaust gases. Both operate in that part of the atmosphere where *continuum flow* prevails; that is, the mean free paths of the molecular motions are relatively small, frequent collisions between molecules occur, the gases follow the basic gas laws, and they can experience compression or expansion waves.

As higher altitude is reached the continuum regime changes into a *free molecular flow* regime, where there are fewer molecules per unit volume and the mean free path of the molecules between collisions becomes larger than the key dimension of the vehicle (e.g., length). Here the plume spreads out even more, reaching diameters in excess of 10 km. Only the exhaust gases close to the nozzle exit experience continuum flow conditions, which allows the streamlines in the flow to spread out by means of successive Prandtl–Meyer expansion waves; once the gas reaches the boundary shown by the elliptical dashed line in the last sketch on the right in Fig. 18–4, the flow will be in the free molecular flow regime and molecules will continue to spread out in straight lines. The regions of free molecular flow and the transition from continuum flow can be analyzed as shown as Ref. 18–4. The third or upper stage, which operates at very high altitudes, has very low emission intensity, because it has a relatively very low gas flow or thrust and because only the inviscid portion of the exhaust gas flow near the nozzle is hot enough to radiate significant energy. This makes it difficult to detect and identify from a distance. The phenomenology of rocket exhaust plumes as seen from a space-based surveillance system is described in Ref. 18–5.

Spectral Distribution of Radiation

The primary radiation emissions from most of the plume gases are usually in the infrared spectrum, to a lesser extent in the ultraviolet spectrum, with relatively little energy in the visible spectrum. The emissions depend on the particular propellants and their respective exhaust gas compositions. For example, the exhaust from the liquid hydrogen–liquid oxygen propellant combination contains mostly water vapor and hydrogen, and with a minor percentage of oxygen and dissociated species. Its radiation is strong in specific wavelength bands characteristic of the emissions from these hot gases (such as 2.7 and 6.3 μm, water—infrared region) and 122 nanometers (hydrogen—ultraviolet region). As shown in Fig. 18–5, the hydrogen–oxygen plume is essentially transparent or colorless, since there are no strong emissions in the visible segment of the spectrum. The propellant combination of nitrogen tetroxide with methylhydrazine fuel gives strong emissions in the infrared region; in addition to the strong emissions for H_2O and hydrogen mentioned previously, there are strong emissions for CO_2 at 4.7 μm, CO at 4.3 μm, and weaker

FIGURE 18–5. Visible plume created by the oxygen–hydrogen propellants of the Vulcain 60 thrust chamber, with a specific impulse of 439 sec at altitude, a nozzle expansion area ratio of 45, and a mixture ratio of 5.6. Multiple shockwave patterns are visible in the core of the plume because of emissions from luminescent minor species. (Courtesy of ESA/CNES/SEP/Daimler-Benz, Europe.)

emission in the ultraviolet (UV) and visible ranges (due to bands of CN, CO, N_2, NH_3, and other intermediate and final gaseous reaction products). This gives it a pink orange-yellow color, but the plume is still partly transparent.

The exhausts of many solid propellants and some liquid propellants contain also solid particles. In Tables 5–8 and 5–9 examples of solid propellant were given that had about 10% of small particles as aluminum oxide (Al_2O_3) in their incandescent white exhaust plumes; some kerosene-burning liquid propellants and most solid propellants have a small percentage of soot or small carbon particles in their exhaust. The radiation spectrum from hot solids is a continuous one, which peaks usually in the infrared (IR) region, but it also has strong emissions in the visible or UV region; this continuous spectrum is usually stronger in the visible range than the narrow-band emissions from the gaseous species in the plume. Afterburning increases the temperature of the particles by several hundred degrees and intensifies their radiation emissions. With 2 to 5% solid particles, the plumes radiate brilliantly and are therefore very visible to the eye. However, these particles in the outer layers of the plume obscure the central core and the shock wave patterns can no longer be observed.

The visible radiation of plumes from double-base propellant can be reduced or suppressed by adding a small amount (1 to 3%) of potassium compound. With composite propellants the control of visible emissions by additives has not been as effective.

The radiation (which is a function of the absolute temperature to the fourth power) cools the plume gases, but it also heats adjacent vehicle or propulsion system components. The prediction of radiative emission requires an understanding of the plume composition, the temperature and density distribution in the plume and the absorption of radiation by intervening atmospheric or plume gases (see Refs. 18–5 to 18–7). The heat transferred from the plume to vehicle components will depend on the propellant combination, the nozzle configuration, the vehicle geometry, the number of nozzles, the trajectory, altitude, and the secondary turbulent flow around the nozzles and the tail section of the vehicle.

The observed or measured values of the radiation emissions have to be corrected. The signal strength diminishes as the square of the distance between the plume and the observation station, and its observed magnitude can change tremendously during a flight. The attenuation is a function of wavelength, rain, fog or clouds, the mass of air and plume gas between the hot part of the plume and the observing location, and depends on the flight vector direction relative to the line of sight. The total emission is a maximum when seen at right angles to the plume (see Refs. 18–5 to 18–7). Radiation measurements can be biased by background radiations (important in satellite observation) or Doppler shift.

Multiple Nozzles

It is common to have more than one propulsion system operating at the same time, or more than one nozzle sending out hot exhaust gas plumes. For example, the Space Shuttle has three main engines and two solid rocket boosters running simultaneously. The interference and impingement of these plumes with one another will cause regions of high temperature in the combined plumes and therefore larger emissions, but the emissions will no longer be axisymmetrical. Also, the multiple nozzles can cause distortions in the airflow near the rear end of the vehicle and influence the vehicle drag and augment the hot backflow from the plume locally.

Plume Signature

This is the term used for all the characteristics of the plume in the infrared, visible, and ultraviolet spectrum, its electron density, smoke or vapor, for a particular vehicle, mission, rocket propulsion system, and propellant (see Refs. 18–8 to 18–10). In many military applications it is desirable to reduce the plume signature in order to minimize being detected or tracked. The initial stagnation temperature of the nozzle exit gas is perhaps the most significant factor influencing plume signature. As plume temperatures increase, higher levels of radiation and radio-frequency interaction will occur. Emissions can be reduced if a propellant combination or mixture ratio with a lower combustion temperature is selected; unfortunately, this usually gives a lower performance. One way to reduce smoke is to choose a reduced-smoke or minimum-smoke solid propellant; they are described in Chapter 12. The plume signature is today often specified as a requirement for a new or modified rocket-propelled vehicle, and it imposes limits on spectral emissions in certain regions of the spectrum and on the amount of acceptable smoke.

The atmosphere absorbs energy in certain regions of the spectrum. For example, the air contains some carbon dioxide and water vapor. These molecules absorb and attenuate the radiation in the frequency bands peculiar to these two species. Since many plume gases contain a lot of CO_2 or H_2O, the attenuation within the plume itself can be significant. The plume energy or intensity, as measured by spectrographic instruments, has to be corrected for the attenuation of the intervening air or plume gas.

Vehicle Base Geometry and Recirculation

The geometry of the nozzle exit(s) and the flight vehicle's tail or aft base have an influence on the plume. Figure 18–6 shows a single nozzle exit whose diameter is almost the same as the base or tail diameter of the vehicle body. If these two diameters are not close to each other, then a flat doughnut-shaped base or tail surface will exist. In this region the high-speed combustion gas velocity is larger than the air speed of the vehicle (which is about the same as the local air velocity relative to the vehicle) and an unsteady flow vortex type recirculation will occur. This greatly augments the afterburning, the heat release to the base, and usually creates a low pressure on this base. This lower pressure in effect increases the vehicle's drag.

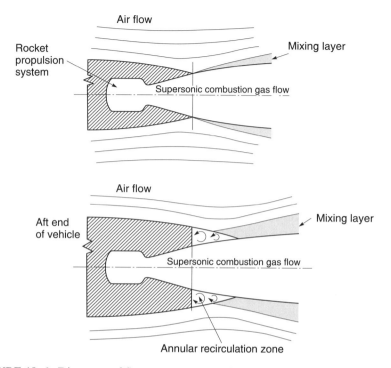

FIGURE 18–6. Diagrams of flow patterns around two different boat tails or vehicle aft configurations, with and without hot gas recirculation.

The air flow pattern at the vehicle tail can be different with different tail geometries, such as cylinder (straight), a diminishing diameter, or an increasing diameter conical shape, which helps to maintain the vehicle's aerodynamic stability. The air flow pattern and the mixing layer change dramatically with angle of attack, causing an unsymmetrical plume shape. Flow separation of the air flow can also occur. In some cases the recirculation of fuel-rich exhaust gas mixed with air will ignite and burn; this dramatically increases the heat transfer to the base surfaces and causes some changes in plume characteristics.

Compression and Expansion Waves

A *shock wave* is a surface of discontinuity in a supersonic flow. In rocket plumes it is the very rapid change of kinetic energy to potential and thermal energy within that very thin wave surface. Fluid crossing a stationary shock wave rises suddenly and irreversibly in pressure and decreases in velocity. When it passes through a shock wave surface that is perpendicular or normal to the incoming supersonic flow, then there is no change in flow direction. Such a normal shock produces the largest increase in pressure (and local downstream temperature) and is known as a *normal shock wave*. The flow velocity behind a normal shock wave is subsonic. When the incoming flow is at an angle less than 90° to the shock wave surface, it is known as a *weak compression wave* or as an *oblique shock wave*. Figure 18–7 illustrates the flow relationships and shows the angle of flow change. The temperature of the gas immediately behind a normal shock wave approaches the stagnation temperature. Here the radiation increases greatly. Also, here (and in other hot plume regions) dissociation of gas species and chemical luminescence (emission of visible light) can occur, as can be seen (downstream of strong shock waves) in Fig. 18–5.

The behavior of gas expansion in the supersonic flow has a fairly similar geometric relationship. It occurs at a surface where the flow undergoes a Prandtl–Meyer *expansion wave*, which is a surface where pressure is reduced and velocities are increased. Often there is a series of weak expansion waves next to each other; this occurs outside the lip of the diverging nozzle exit section when the nozzle exit gas pressure is higher than the ambient pressure, as shown in Fig. 18–1.

The plume from hydrogen–oxygen liquid propellant combustion consists mostly of superheated water vapor and hydrogen gas and is basically invisible. However, it is faintly visible because of the chemically generated luminescence that is believed to be responsible for the pale pink orange and white skeletal wave pattern, particularly in its hot regions. The patterns are shown in Figs. 18–2 and 18–5.

The supersonic gas flow out of the nozzle exit is undisturbed until it changes direction in a wave front or goes through a normal shock. Diamond-shaped

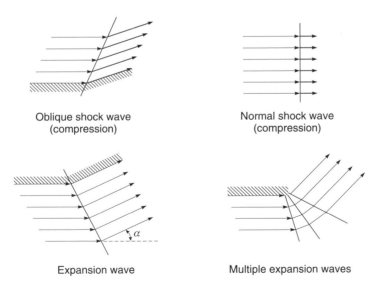

FIGURE 18–7. Simplified diagrams of oblique shock wave or compression wave, normal shock wave, and expansion wave. The change in the length of the arrows is an indication of the change in gas velocity as the flow crosses the wave front.

patterns are formed by compression and expansion wave surfaces. These patterns (shown in Figs. 18–2 and 18–5) then repeat themselves and are clearly visible in largely transparent plumes, such as those from hydrogen–oxygen or alcohol–oxygen propellant combinations. The pattern becomes weaker with each succeeding wave. The mixing layer acts as a reflector, because an expansion wave is reflected as a compression wave.

The inter-face surface between the rocket exhaust plume gas and the air flowing over the vehicle (or the air aspirated by the high velocity plume) acts as a free boundary. Oblique shock waves are reflected at a free boundary as an opposite wave. For example, an oblique compression wave is reflected as an oblique expansion wave. This boundary is not usually a simple surface of revolution, but an annular layer, sometimes called a slip stream or mixing layer. See Figs. 8–1, 8–2, and 8–5.

18.2. PLUME EFFECTS

Smoke and Vapor Trails

Smoke is objectionable in a number of military missiles. It interferes with the transmission of optical signals, such as with a line-of-sight or electro-optical guidance system. Smoke would also hamper the vision of a soldier who guides a wire-controlled anti-tank weapon. Smoke, or a vapor trail, allows easy and rapid detection of a missile being fired, and visually tracking the flight path can reveal a covert launch site. Smoke is produced not only during rocket operation, but also by chuffing, the irregular combustion of propellant remainders after rocket cutoff. Chuffing, described in Chapter 13, produces small puffs of flame and smoke at frequencies of perhaps 10 to 150 Hz.

Primary smoke is a suspension of many very small solid particles in a gas, whereas secondary smoke is a set of condensed small liquid droplets suspended in a gas, such as condensed moisture-forming clouds, fog, or mist. Many propellants leave visible trails of smoke and/or vapor from their plumes during powered rocket flight (see Refs. 18–8 to 18–10). These trails are shifted by local winds after the vehicle has passed. They are most visible in the daytime, because they depend on reflected or scattered sunlight. The solid particles that form the primary smoke are mainly aluminum oxide (Al_2O_3, typically 0.1 to 3 μm diameter) in composite propellants. Other solid particles in the exhaust of solid propellant are unburned aluminum, zirconium or zirconium oxide (from combustion stabilizer), or iron or lead oxides (in burn-rate catalyst). Carbon particles or soot can be formed from various solid propellants and liquid propellants using hydrocarbon fuels.

During the external expansion of rocket exhaust plume gases the gas mixture is cooled by radiation, gas expansion, and convection with colder ambient air to below its dew-point temperature, where the water vapor condenses. Of course, this depends on local weather conditions. If the ambient temperature is low (e.g., at high altitude) and/or if the gas expands to low temperatures, the water droplets can freeze to form small ice crystals or snow. At high altitude, CO_2, HCl, and other gases can also condense. Many rocket exhaust gases contain between 5 and 35% water, but the exhaust from the liquid hydrogen/liquid oxygen propellant combination can contain as much as 80%. If the exhaust contains tiny solid particles, these then act as nuclei upon which water vapor can condense, thus increasing the amount of nongaseous material in the plume, making the plume even more visible.

If reducing smoke in the plume is important to the mission, then a *reduced-smoke solid propellant or a minimum-smoke propellant* is often used. They are described in Chapter 12. Even then, a secondary smoke trail can be formed under cold-weather and high-humidity conditions. However, under most weather conditions it will be difficult to see a trail containing vapor only.

Toxicity

The exhaust gases from many rocket propulsion systems contain toxic and/or corrosive gas species, which can cause severe health hazards and potential environmental damage near launch or test sites. Accidental spills of some liquid oxidizers, such as nitrogen tetroxide or red fuming nitric acid, can create toxic, corrosive gas clouds, which have higher density than air and will stay close to the ground. Exhaust gases such as CO or CO_2 present a health hazard if inhaled in concentrated doses. Gases such as hydrogen chloride (HCl) from solid propellants using a perchlorate oxidizer (see Ref. 18–11), nitrogen dioxide (NO_2), nitrogen tetroxide (N_2O_4), or vapors of nitric acid (HNO_3) have relatively low levels of allowable inhalation concentration before health damage can occur. Chapter 7 lists some of the safe exposure limits. The potential damage increases with the concentration of the toxic species in the exhaust, the mass flow or thrust level, and the duration of the rocket firing at or near the test/launch site.

Dispersion by wind and diffusion and dilution with air can reduce the concentrations of toxic materials to tolerable levels within a few minutes, but this depends on local weather conditions, as explained in Chapter 20. Careful attention is therefore given to scheduling the launch or test operations at times when the wind will carry these gases to nearby uninhabited areas. For very highly toxic exhaust gases (e.g., those containing beryllium oxide or certain fluorine compounds), and usually for relatively small thrust levels, the exhaust gases in static test facilities are captured, chemically treated, and purified before release into the atmosphere.

Noise

Acoustical noise is an unavoidable by-product of thrust; it is particularly important in large launch vehicles and is a primary design consideration in the vehicle and in much of the ground-support equipment, particularly electronic components. Besides being an operational hazard to personnel in and around rocket-propelled vehicles, it can be a severe annoyance to communities near rocket test sites. The acoustic power emitted by the Saturn V vehicle at launch is about 2×10^8 W, enough to light up about 200,000 average homes if it were available as electricity.

Acoustic energy emission is mainly a function of exhaust velocity, mass of gas flow, exhaust gas density, and the velocity of sound in the quiescent medium. In these terms, the chemical rocket is the noisiest of all aircraft and missile propulsion devices. Sound intensity is highest near the nozzle exit and diminishes with the square of the distance from the source. Analytical models of noise from a rocket exhaust usually divide the plume into two primary areas, one being upstream of the shock waves and one being downstream (subsonic), with high-frequency sound coming from the first and low-frequency from the second. The shock wave itself is a generator of sound, as is the highly turbulent mixing of the high-velocity exhaust with its reltively quiescent surroundings. Sound emission is normally measured in terms of microbars (μbar) of *sound pressure*, but is also expressed as *sound power* (W), *sound intensity* (W/ft^2), or *sound level* (decibels, dB). The relationship that exists among a decibel scale, the power, and intensity scales is difficult to estimate intuitively since the decibel is a logarithm of a ratio of two power quantities or two intensity quantities. Further, expression of a decibel quantity must also be accompanied by a decibel scale reference, for example, the quantity of watts corresponding to 0 dB. In the United States, the most common decibel scale references 10^{-13} W power, whereas the European scale references 10^{-12} W.

A large rocket can generate a sound level of about 200 dB (reference 10^{-13} W), corresponding to 10^7 W of sound power, compared to 140 dB for a 75-piece orchestra generating 10 W. Reducing the sound power by 50% reduces the value by only about 3 dB. In terms of human sensitivity, a 10-dB change usually doubles or halves the noise for the average person. Sound levels above 140 dB frequently introduce pain to the ear and levels above 160 become intolerable (see Ref. 18–12).

Spacecraft Surface Contamination

Contamination of sensitive surfaces of a spacecraft by rocket exhaust products can be a problem to vehicle designers and users. It can degrade functional surfaces, such as solar cells, optical lenses, radiators, windows, thermal-control coatings, and mirrored surfaces. Propellants that have condensed liquids or solid particles in their exhaust appear to be more troublesome than propellants with mostly gaseous products, such as oxygen–hydrogen. Plumes from most solid propellant contain contaminating species. Practically all the investigative work has been concerned with small storable liquid propellant attitude control pulse motors in the thrust range 1.0 to 500 N, the type commonly used for controlling vehicle attitude and orbit positioning over long periods of time. Deposits of hydrazinium nitrate and other material have been found. The accumulation of exhaust products on surfaces in the vicinity is a function of many variables, including the propellants, combustion efficiency, combustion pressure, nozzle expansion ratio, surface temperature, and rocket–vehicle interface geometries. The prediction of exhaust contamination of spacecraft surfaces is only partly possible through analytical calculations. Reference 18–13 provides a comprehensive analytical model and computer program for liquid bipropellant rockets.

Another effect of clouds of condensed species (either tiny liquid droplets or solid particles) is to scatter sunlight and cause solar radiation to be diverted to optical instruments on the spacecraft, such as cameras, telescopes, IR trackers, or star trackers; this effect can cause erroneous instrument measurements. The scatter depends on the plume location relative to the instruments, the propellant, the density and size of particulates, the sensitive optical frequency, and the surface temperature of the instrument.

Radio Signal Attenuation

All rocket exhaust plumes generally interfere with the transmission of radio-frequency signals that must pass through the plume in the process of guiding the vehicle, radar detecting, or communicating with it. Solid propellant exhaust plumes usually cause more interference than liquid rocket engine plumes. Signal attenuation is a function of free electron density and electron collision frequency. Given these two parameters for the entire plume, the amount of attenuation a signal experiences in passing through the plume can be calculated. Figure 18–8 shows the minimum plume model sufficient for predicting signal attenuation that contains contours of constant electron density and electron collision frequency for momentum transfer. Free-electron density

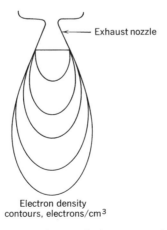

Electron density
contours, electrons/cm^3

FIGURE 18–8. Exhaust plume model for predicting attenuation of radio communications signals. The contours shown are for either equal electron density or electron collision frequency; the highest values are near the nozzle exit.

and activity in the exhaust plume are influenced by many factors, including the propellant formulation, propellant alkali impurities, exhaust temperature, motor size, chamber pressure, flight speed and altitude, and the distance downstream from the nozzle exit. Methods have been developed for analyzing (with computers) the physical and chemical composition, including electron density, and the attenuation characteristics of exhaust plumes (Refs. 18–14 and 18–15).

The relationship between several influential motor and vehicle design factors can be summarized from experience with typical solid propellant rockets as follows:

1. The presence of alkali metal impurities increases attenuation; changing the impurity level of potassium from 10 to 100 ppm increases the relative attenuation some 10-fold at low altitude. Both potassium (~ 150 ppm) and sodium (~ 50 ppm) are impurities in commercial grade nitrocellulose and ammonium perchlorate.

2. The percentage of aluminum fuel is a major influence; increasing the percentage from 10 to 20% increases the attenuation fivefold at sea level and three- to fourfold at 7500 m altitude.

3. Increasing the chamber pressure for a given aluminized propellant from 100 to 2000 psi reduces the relative attenuation by about 50%.

4. Attenuation varies with the distance downstream from the nozzle exit plane and can be four to five times as great as at the nozzle exit plane, depending on the flight altitude, nozzle geometry, oxidizer-to-fuel ratio, flight velocity, altitude, and other parameters.

For many solid rocket applications, attenuation of radio or radar signal strength by the exhaust plume is no problem. When it is, attenuation can usually be kept at acceptable levels by controlling the level of alkali impurities in propellant ingredients and by using nonmetal fuels or a low percentage ($< 5\%$) of aluminum. Motors with high expansion ratio nozzles help, since electrons combine with the positive ions as the exhaust temperature falls.

The electrons in the plume greatly increase its radar cross section, and hot plumes can readily be picked up with radar. The plume is usually a stronger radar reflector than the flight vehicle. A radar homing missile seeker would focus on the plume and not the vehicle. A reduction of the plume cross section is desirable (lower gas temperature, less sodium impurities).

Plume Impingement

In most reaction control systems there are many small thrusters and they are pointed in different directions. There have been cases where the plumes of some of these thrusters have impinged upon a space vehicle surface, such as extended solar cell panels, radiation heat rejection surfaces, or aerodynamic control surfaces. This is more likely to happen at high altitude, where the plume diameters are large. This can lead to the overheating of these surfaces and to unexpected turning moments.

Also, during stage separation, there have been occasions where the plume of the upper stage impinges on the lower vehicle stage. This can cause overheating and impact damage not only to the lower stage (being separated), but by reflection also to the upper stage. Other examples are docking maneuvers or the launching of multiple rockets (nearly simultaneously) from a military barrage launcher. The plume of one of the rocket missiles impinges on another flying missile and causes it to experience a change in flight path, often not hitting the intended target.

18.3. ANALYSIS AND MATHEMATICAL SIMULATION

The complicated structure, the behavior, and many of the physical phenomena of plumes have been simulated by mathematical algorithms, and a number of relatively complex computer programs exist (see Refs. 18–16 to 18–20). Although there has been remarkable progress in using mathematical simulations of plume phenomena, the results of such computer analyses are not always reliable or useful for making predictions of many of the plume characteristics; however, the models help in understanding the plumes and in extrapolating test results to different conditions within narrow limits. There are some physical phenomena in plumes that are not yet fully understood.

All simulations are really approximations, to various degrees; they require simplifying assumptions to make a reasonable mathematical solution possible, and their field of application is usually limited. They are aimed at predicting different plume parameters, such as temperature or velocity or pressure profiles, radar cross section, heat transfer, radiations, condensation, deposits on optical surfaces, impact forces, or chemical species. The analyses are usually limited to separate spatial segments of the plume (e.g., core, mantle, supersonic versus subsonic regions, continuum versus free molecular flow, near or far field), and many have different assumptions about the dynamics or steadiness of the flow (many neglect turbulence effects or the interaction between boundary layers and shock waves). The algorithms are also different in the treatment of chemical reactions, solids content, energy releases, composition changes within the plume, different flight and altitude regimes, the interactions with the airflow and the vehicle, or selected regions of the spectrum (e.g., IR only). Many require assumptions about particle sizes, their amounts, spatial and size distribution, gas velocity distribution, the geometry and boundaries of the mixing layer, or the turbulence behavior. The mathematical models are complex and can use one-, two-, or three-dimensional mesh models. The analysis of a plume often requires using more than one model to solve for different predictions. Many solutions are based in part on extrapolating measured data from actual plumes to guide the analyses. The specific analytical approaches are beyond the scope of this book and their mathematical resolutions are the domain of experts in this field.

The actual measurements on plumes during static and flight tests are used to verify the theories and they require highly specialized instrumentation, careful calibrations and characterization, skilled personnel, and an intelligent application of various correction factors. Extrapolating the computer programs to regions or parameters that have not been validated has often given poor results.

PROBLEMS

1. List at least two parameters that are likely to increase total radiation emission from plumes, and explain how they accomplish this. For example, increasing the thrust increases the radiating mass of the plume.

2. Look up the term *chemiluminescence* in a technical dictionary or chemical encyclopedia; provide a definition and explain how it can affect plume radiation.

3. If a high-altitude plume is seen from a high-altitude balloon, its apparent radiation intensity diminishes with the square of the distance between the plume and the observation platform and as the cosine of the angle of the flight path tangent with the line to the observation station. Establish your own trajectory and its relative location to the observation station. For a plume of an ascending launch vehicle, make a rough estimate of the change in the relative intensity received by the observing sensor during flight. Neglect atmospheric absorption of plume radiation and assume that the intensity of emitted radiation stays constant.

REFERENCES

18–1. S. M. Dash, "Analysis of Exhaust Plumes and their Interaction with Missile Airframes," in M. J. Hemsch and J. N. Nielson (Eds.), *Tactical Missile Aerodynamics*, Progress in Astronautics and Aeronautics, Vol. 104, AIAA, Washington, DC, 1986.

18–2. S. M. Yahya, *Fundamentals of Compressible Flow*, 2nd revised printing, Wiley Eastern Limited, New Delhi, 1986.

18–3. R. D. McGregor, P. D. Lohn, and D. E. Haflinger, "Plume Impingement Study for Reaction Control Systems of the Orbital Maneuvering Vehicle," *AIAA Paper 90-1708*, June 1990.

18–4. P. D. Lohn, D. E. Halfinger, R. D. McGregor, and H. W. Behrens, "Modeling of Near-Continuum Flows using Direct Simulation Monte Carlo Method," *AIAA Paper 90-1663*, June 1990.

18–5. F. S. Simmons, *Rocket Exhaust Plume Phenomenology*, Aerospace Press, The Aerospace Corporation, 2000.

18–6. A. V. Rodionov, Yu A. Plastinin, J. A. Drakes, M. A. Simmons, and R. S. Hiers III, "Modeling of Multiphase Alumina-Loaded Jet Flow Fields," *AIAA Paper 98-3462*, July 1998.

18–7. R. B. Lyons, J. Wormhoudt, and C. E. Kolb, "Calculation of Visible Radiations from Missile Plumes," in *Spacecraft Radiative Heat Transfer and Temperature Control*, Progress in Astronautics and Aeronautics, Vol. 83, AIAA, Washington, DC, June 1981, pp. 128–148.

18–8. A. C. Victor, "Solid Rocket Plumes," Chapter 8 of: G. E. Jensen and D. W. Netzer (Eds.), *Tactical Missile Propulsion*, Progress in Astronautics and Aeronautics, Vol. 170, AIAA, 1996.

18–9. *Rocket Motor Plume Technology*, AGARD Lecture Series 188, NATO, June 1993.

18–10. *Terminology and Assessment Methods of Solid Propellant Rocket Exhaust Signatures*, AGARD Advisory Report 287, NATO, February 1993.

18–11. D. I. Sebacher, R. J. Bendura, and G. L. Gregory, "Hydrogen Chloride Measurements in the Space Shuttle Exhaust Cloud," *Journal of Spacecraft and Rockets*, Vol. 19, No. 4, July–August 1982.

18–12. J. M. Seiner, S. M. Dash, and D. E. Wolf, "Analysis of Turbulent Underexpanded Jets, Part II: Shock Noise Features Using SCIPVIS," *AIAA Journal*, Vol. 23, No. 5, May 1985, pp. 669–677.

18–13. R. J. Hoffman, W. D. English, R. G. Oeding, and W. T. Webber, "Plume Contamination Effects Prediction," Air Force Rocket Propulsion Laboratory, December 1971.

18–14. L. D. Smoot and D. L. Underwood, "Prediction of Microwave Attenuation Characteristics of Rocket Exhausts," *Journal of Spacecraft and Rockets*, Vol. 3, No. 3, March 1966, pp. 302–309.

18–15. W. A. Wood and J. R. De More, "Microwave Attenuation Characteristics of Solid Propellant Exhaust Products," *AIAA Paper 65-183*, February 1965.

18–16. I. Boyd, "Modeling of Satellite Control Thruster Plumes," PhD thesis, Southampton University, England, 1988.

18–17. S. M. Dash and D. E. Wolf, "Interactive Phenomena in Supersonic Jet Mixing Plumes, Part I: Phenomenology and Numerical Modeling Technique," *AIAA Journal*, Vol. 22, No. 7, July 1984, pp. 905–913.

18–18. S. M. Dash, D. E. Wolf, R. A. Beddini, and H. S. Pergament, "Analysis of Two Phase Flow Processes in Rocket Exhaust Plumes," *Journal of Spacecraft*, Vol. 22, No. 3, May–June 1985, pp. 367–380.

18–19. C. B. Ludwig, W. Malkmus, G. N. Freemen, M. Slack, and R. Reed, "A Theoretical Model for Absorbing, Emitting and Scattering Plume Radiations," in *Spacecraft Radiative Transfer and Temperature Control*, Progress in Astronautics and Aeronautics, Vol. 83, AIAA, Washington, DC, 1981, pp. 111–127.

18–20. S. M. Dash, "Recent Developments in the Modeling of High Speed Jets, Plumes and Wakes," *AIAA Paper 85-1616*, presented at AIAA 18th Fluid Dynamics Plasma-Dynamics and Laser Conference, July 1985.

CHAPTER 19

ELECTRIC PROPULSION

As mentioned in Chapters 1 and 2, electric rocket propulsion devices use electrical energy for heating and/or directly ejecting propellant, utilizing an energy source that is independent of the propellant itself. The purpose of this chapter is to provide an introduction to this field. Vector notation is used in several of the background equations presented.

The basic subsystems of a typical electric propulsion thruster are: (1) a raw *energy source* such as solar or nuclear energy with its auxiliaries such as concentrators, heat conductors, pumps, panels, radiators, and/or controls; (2) *conversion devices* to transform this energy into electrical form at the proper voltage, frequency, pulse rate, and current suitable for the electrical propulsion system; (3) a *propellant system* for storing, metering, and delivering the propellant; and (4) one or more *thrusters* to convert the electric energy into kinetic energy of the exhaust. The term *thruster* is commonly used here, as thrust chamber is in liquid propellant rockets.

Electric propulsion is unique in that it includes both thermal and non-thermal systems as classified in Chapter 1. Also, since the energy source is divorced from the propellant, the choice of propellant is guided by factors much different to those in chemical propulsion. In Chapter 3, ideal relations that apply to all thermal thrusters are developed which are also relevant to thermal-electric (or electrothermal) systems. Concepts and equations for non-thermal-electric systems are defined in this chapter. From among the many ideas and designs of electric propulsion devices reported to date, one can distinguish the following three fundamental types:

1. *Electrothermal.* Propellant is heated electrically and expanded thermodynamically; i.e., the gas is accelerated to supersonic speeds through a nozzle, as in the chemical rocket.

2. *Electrostatic.* Acceleration is achieved by the interaction of electrostatic fields on non-neutral or charged propellant particles such as atomic ions, droplets, or colloids.

3. *Electromagnetic.* Acceleration is achieved by the interaction of electric and magnetic fields within a plasma. Moderately dense plasmas are high-temperature or nonequilibrium gases, electrically neutral and reasonably good conductors of electricity.

A general description of these three types was given in Chapter 1, Figs. 1–8 to 1–10. Figure 19–1 and Tables 2–1 and 19–1 show power and performance values for several types of electric propulsion units. Note that the thrust levels are small relative to those of chemical and nuclear rockets, but that values of specific impulse can be substantially higher; the latter may translate into a longer operational life for satellites whose life is presently propellant limited. Inherently, electric thrusters give accelerations too low for overcoming the high-gravity field of earth launches. They function best in space, which also

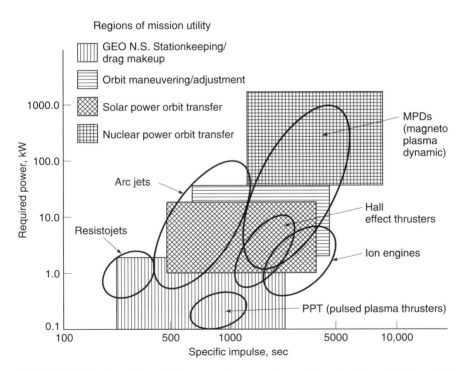

FIGURE 19–1. Overview of the approximate regions of application of different electrical propulsion systems in terms of power and specific impulse.

TABLE 19-1. Typical Performance Parameters of Various Types of Electrical Propulsion Systems

Type	Thrust Range (mN)	Specific Impulse (sec)	Thruster Efficiency[b] (%)	Thrust Duration	Typical Propellants	Kinetic Power per Unit Thrust (W/mN)
Resistojet (thermal)	200–300	200–350	65–90	Months	NH_3, N_2H_4, H_2	0.5–6
Arcjet (thermal)	200–1000	400–1000	30–50	Months	H_2, N_2, N_2H_4, NH_3	2–3
Ion engine	0.01–200	1500–5000	60–80	Months	Xe,Kr,Ar	10–70
Solid pulsed plasma (PPT)	0.05–10	600–2000	10	Years	Teflon	10–50
Magnetoplasma dynamic (MPD)	0.001–2000	2000–5000	30–50	Weeks	Ar, Xe, H_2, Li	100
Hall thruster	0.01–2000	1500–2000	30–50	Months	Xe,Ar	100
Monopropellant rocket[a]	30–100,000	200–250	87–97	Hours or minutes	N_2H_4	

[a] Listed for comparison only.
[b] See Eq. 19-3.

matches the near-vacuum exhaust pressures required of electrostatic and electromagnetic systems. All flight missions envisioned with electric propulsion operate in a reduced-gravity or gravity-free space and, therefore, must be launched from earth by chemical rocket systems.

The many advantages of electric propulsion had been offset by their required use of substantial quantities of electricity which, at certain power levels, had been an expensive commodity in space until recently. All types of electric propulsion presently depend on a vehicle-borne power source—based on solar, chemical, or nuclear energy—and power conversion and conditioning equipment. The mass of the electric generating equipment, even when solar energy is employed, can become much larger than that of the thrusters, particularly when thruster efficiency is low. This causes appreciable increases in inert-vehicle mass (or dry-mass). Modern satellites and other spacecraft have substantial communications requirements. Typically these satellites can share their electrical power sources, thus avoiding the penalty to the propulsion system. What remains to be tagged to the propulsion system is the power-conditioning equipment, except in instances where it is also shared with other spacecraft components.

Electric propulsion has been considered for space applications since the inception of the space program in the 1950s but has only begun to make widespread impact since the mid-1990s. This is a result of the availability of sufficiently large amounts of electrical power in spacecraft. References 19–1 to 19–3 are devoted to electric propulsion. Basic principles on electric propulsion devices are given in these references, along with applications, although the information relates to older versions of such devices. Table 19–2 gives a comparison of advantages and disadvantages of several types of electric propulsion. Pulsed devices differ from continuous or steady-state in that startup and shutdown transients may degrade their effective performance. Pulsed devices, however, are of practical importance, as is detailed later in this chapter.

The *applications* for electric propulsion fall into several broad mission categories (these have already been introduced in Chapter 4):

(1) *Overcoming translational and rotational perturbations* in satellite orbits, such as north–south station keeping (NSSK) of satellites in geosynchronous orbits (GEO) or aligning telescopes or antennas or drag compensation of satellites in low (LEO) and medium earth orbits (MEO). For a typical north–south station-keeping task in a 350-km orbit, a velocity increment of about 50 m/sec every year or 500 m/sec for 10 years might be needed. Several different electric propulsion systems have actually flown in this type of mission.

(2) Increasing satellite speed while overcoming the relatively weak gravitational field some distance away from the earth, such as *orbit raising* from a low earth orbit (LEO) to a higher orbit or even to a geosynchronous orbit (GEO). Circularizing an elliptical orbit may require 2000 m/sec and going from LEO to GEO typically might require a velocity increase

TABLE 19–2. Comparison of Electrical Propulsion Systems

Type	Advantages	Disadvantages	Comments
Resistojet (electrothermal)	Simple device; easy to control; simple power conditioning; low cost; relatively high thrust and efficiency; can use many propellants, including hydrazine augmentation	Lowest I_s; heat loss; gas dissociation; indirect heating of gas; erosion	Operational
Arcjet (electrothermal & electromagnetic)	Direct heating of gas; low voltage; relatively simple device; relatively high thrust; can use catalytic hydrazine augmentation; inert propellant	Low efficiency; erosion at high power; low I_s; high current; heavy wiring; heat loss; more complex power conditioning	High-thrust units need P_e of 100 kW or more. Operational
Ion propulsion (electrostatic)	High specific impulse; high efficiency; inert propellant (xenon)	Complex power conditioning; high voltages; single propellant only; low thrust per unit area; heavy power supply	Flown in satellites (DS1)
Pulsed plasma (PPT) (electromagnetic)	Simple device; low power; solid propellant; no gas or liquid feed system; no zero-g effects on propellant	Low thrust; Teflon reaction products are toxic, may be corrosive or condensable; inefficient	Operational
MPD Steady-state plasma (electromagnetic)	Can be relatively simple; high I_s; high thrust per unit area	Difficult to simulate analytically; high specific power; heavy power supply	Several have flown
Hall thruster (electromagnetic)	Desirable I_s range; compact, relatively simple power conditioning; inert propellant (Xe)	Single propellant; high beam divergence; erosion	Operational

of up to 6000 m/sec. Several electric propulsion units are being developed for these types of mission.

(3) Potential missions such as *interplanetary travel* and *deep space probes* are also candidates for electric propulsion. A return to the moon, missions to Mars, Jupiter, and missions to comets and asteroids are of present interest. These all require relatively high thrust and power. A few electric thrusters for this category of missions (100 kW) are being investigated. The power supply for these missions may require other than solar power; nuclear sources need to be considered.

As an illustration of the benefit in applying electric propulsion, consider a typical geosynchronous communications satellite with a 15-year lifetime and with a mass of 2600 kg. For north–south station-keeping (NSSK) the satellite might need an annual velocity increase of some 50 m/sec; this requires about 750 kg of chemical propellant for the entire period, which is more than one-quarter of the satellite mass. Using an electric propulsion system can increase the specific impulse to 2800 sec (about nines times higher than a chemical rocket), and the propellant mass can be reduced to perhaps less than 100 kg. A power supply and electric thrusters would have to be added, but the inert mass of the chemical system can be deleted. Such an electric system would save perhaps 450 kg or about 18% of the satellite mass. At launch costs of $30,000 per kilogram delivered to GEO, this is a potential saving of some $13,500,000 per satellite. Alternatively, more propellant could be stored in the satellite, thus extending its useful life. Additional savings could materialize if electric propulsion were also used for orbit raising.

The power output (kinetic energy of jet per unit time, P or P_{jet}) is really the basic energy rate supplied by the power source, principally diminished by (1) the losses of the power conversion, such as from solar into electrical energy; (2) conversion into the forms of electric energy suitable for the thrusters; and (3) the losses of the conversion of electric energy into propulsive jet energy. The kinetic power of the jet P per unit thrust F can be expressed by the simple relation (assuming no significant pressure thrust)

$$P/F = \tfrac{1}{2}\dot{m}v^2/\dot{m}v = \tfrac{1}{2}v = \tfrac{1}{2}g_0 I_s \tag{19-1}$$

where $\dot{m}$ is the mass flow rate, v the average jet discharge velocity (v_2 or c in Chapters 2 and 3), and I_s the specific impulse. The power-to-thrust ratio of the jet is therefore proportional to the exhaust velocity or the specific impulse. Thus, electrical propulsion units with very substantial values of I_s require more power per unit of thrust and incidentally a more massive power supply.

Thruster efficiency η_t is defined as the ratio of the thrust-producing kinetic energy (axial component) rate of the exhaust beam to the total electrical power supplied to the thruster, including any used in evaporating or ionizing the propellant, or

$$\eta_t = \frac{\text{power of the jet}}{\text{electrical power input}} = \frac{P_{\text{jet}}}{\Sigma(IV)} \tag{19-2}$$

Then, from the fundamentals in Chapter 2 (Eqs. 2–19 and 2–22),

$$\eta_t = \frac{\tfrac{1}{2}\dot{m}v^2}{P_e} = \frac{FI_s\, g_0}{2P_e} = \frac{FI_s\, g_0}{2\Sigma(IV)} \tag{19-3}$$

with P_e the electric power input to the thruster in watts, usually the product of the electrical current and all associated voltages (hence the Σ-sign).

Thruster efficiency accounts for all the energy losses that do not result in kinetic energy, including (1) the wasted electrical power (stay currents, ohmic resistance, etc.); (2) unaffected or improperly activated propellant particles (propellant utilization); (3) loss of thrust resulting from dispersion of the exhaust (direction and magnitude); and (4) heat losses. It is a measure of how effectively electric power and propellant are used in the production of thrust.

When electrical energy is not the only input energy, Eq. 19–2 has to be modified; for example, the propellant may release energy (chemical monopropellant), as in hydrazine decomposition with a resistojet.

19.1. IDEAL FLIGHT PERFORMANCE

For the low thrust of electric propulsion with its relatively massive power generating systems, the flight regimes of space vehicles propelled by electric rockets are quite different from those using chemical rockets. Accelerations tend to be very low (10^{-4} to 10^{-6} g_0), thrusting times are typically long (several months), and spiral trajectories were originally suggested for spacecraft accelerated by these low thrusts. Figure 19–2 shows schemes for going from LEO to GEO including a spiral, a Hohman ellipse (see Section 4.5 on the *Hohman*

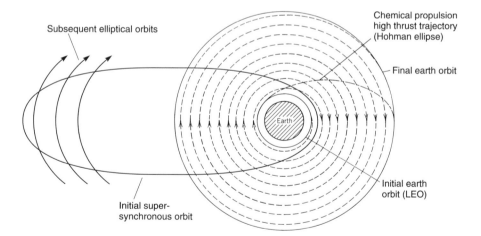

FIGURE 19–2. Simplified diagram of trajectories going from a low earth orbit (LEO) to a high earth orbit using chemical propulsion (short duration), electric propulsion with a multiple spiral trajectory (long duration), and a supersynchronous chemical orbit approach as an alternate to LEO (intermediate duration). From the supersynchronous orbit continuous thrusting with electric propulsion at a fixed inertial attitude lowers the apogee and raises the perigee in each orbit until it reaches the final high circular orbit. See Ref. 19–4.

orbit, which is optimum for chemical propulsion) as well as a "supersynchronous" orbit transfer (Ref. 19–4). Because of the long transfer orbit durations, trajectories other than spiral are presently being considered where one utilizes chemical propulsion to arrive at a very eccentric, supersynchronous elliptical orbit; from there electric propulsion can continuously and effectively be fired to attain a GEO orbit.

The performance of an electrical rocket can be conveniently analyzed in terms of the power and the relevant masses (Ref. 19–5). Let m_0 be the total initial mass of the vehicle stage, m_p the total mass of the propellant to be expelled, m_{pl} the payload mass to be carried by the particular stage under consideration, and m_{pp} the mass of the power plant consisting of the empty propulsion system including the thruster, propellant storage and feed system, the energy source with its conversion system and auxiliaries, and the associated structure. Then

$$m_0 = m_p + m_{pl} + m_{pp} \tag{19–4}$$

The energy source input to the power supply (i.e., solar or nuclear) has to be larger than its electrical power output; they are related by the power conversion efficiency (about 10 to 15% for photovoltaic and up to 30% for rotating machinery) for converting the raw energy into electrical power at the desired voltages, frequencies, and power levels. The converted electrical output P_e is then supplied to the propulsion system. The ratio of the electrical power P_e to the mass of the power plant m_{pp} is defined as α, which is often referred to as the *specific power* of the power plant or of the entire propulsion system. This specific power must be defined for each design, because even for the same type of engine, α is somewhat dependent on the engine–module configuration (this includes the number of engines that share the same power conditioner, redundancies, valving, etc.):

$$\alpha = P_e/m_{pp} \tag{19–5}$$

The specific power is considered to be proportional to engine-power and reasonably independent of m_p. Its value hinges on technological advances and the electric–propulsion engine module configuration. Presently, typical values of α range between 100 and 200 W/kg. In the future it is hoped that α will attain values of 500 to 2000 W/kg pending some breakthrough in power conditioning equipment. Electrical power is converted by the thruster into kinetic energy of the exhaust. Allowing for losses by using the thruster efficiency η_t, defined in Eqs. 19–2 and 19–3, the electric power input is

$$P_e = \alpha m_{pp} = \tfrac{1}{2}\dot{m}v^2/\eta_t = m_p v^2/(2t_p\eta_t) \tag{19–6}$$

where m_p is the propellant mass, v the effective exhaust velocity, and t_p the time of operation or propulsive time when the propellant is being ejected at a uniform rate.

Using Eqs. 19–4, 19–5, and 19–6 together with 4–7, one can obtain a relation for the reciprocal payload mass fraction (see Problem 19–4)

$$\frac{m_0}{m_{pl}} = \frac{e^{\Delta u/v}}{1 - (e^{\Delta u/v} - 1)v^2/(2\alpha t_p \eta_t)} \qquad (19\text{–}7)$$

This assumes a gravity-free and drag-free flight. The change of vehicle velocity Δu which results from the propellant being exhausted at a speed v is plotted in Fig. 19–3 as a function of propellant mass fraction. The specific power α and the thruster efficiency η_t as well as the propulsive time t_p can be combined into a *characteristic speed* (Ref. 19–5)

$$v_c = \sqrt{2\alpha \, t_p \eta_t} \qquad (19\text{–}8)$$

This characteristic speed is not a physical speed but rather a defined grouping of parameters that has units of speed; it can be visualized as the speed the power plant would have if its full power output were converted into the form of kinetic energy of its own inert mass m_{pp}. Equation 19–8 includes the propulsive time t_p which is the actual mission time (certainly, mission time cannot be smaller than the thrusting time). From Fig. 19–3 it can be seen that, for a given payload fraction (m_{pl}/m_0) and characteristic speed (v_c), there is an opti-

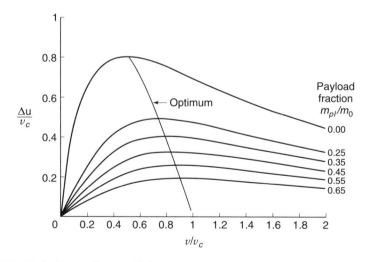

FIGURE 19–3. Normalized vehicle velocity increment as a function of normalized exhaust velocity for various payload fractions with zero inert mass of the propellant tank. The optima of each curve are connected by a line that represents Eq. 19–9.

mum value of v corresponding to the peak vehicle velocity increment; this is later shown to signify that there exists a particular set of most desirable operating conditions.

The peak for the curves in Fig. 19–3 exists because the inert mass of the power plant m_{pp} increases with the specific impulse while the propellant mass decreases with specific impulse. For a constant flow rate, other components are fixed in mass so that they only displace the curves by a constant amount. As indicated in Chapter 17 and elsewhere, this trend is generally true for all propulsion systems and leads to the statement that, *for a given mission, there is theoretically an optimum range of specific impulse and thus an optimum propulsion system design.* The peak of each curve in Fig. 19–3 is nicely bracketed by the ranges $\Delta u/v_c \le 0.805$ and $0.505 \le v/v_c \le 1.0$. This means that for any given electric engine any optimum operating time t_p^* will be proportional to the square of the total required change in vehicle velocity and thus large Δu's would correspond to very long mission times. Similarly, any optimum specific impulse I_s^* will be (nearly) proportional to the change in vehicle velocity and large changes here would necessitate correspondingly high specific impulses. These conclusions will be refined in Section 19.4.

The optimum of the curves in Fig. 19–3 can be found by differentiating Eq. 19–7

$$\left(\frac{v}{\Delta u}\right)\left(e^{\Delta u/v} - 1\right) - \frac{1}{2}\left(\frac{v_c}{v}\right)^2 - \frac{1}{2} = 0 \qquad (19\text{–}9)$$

This relates Δu, v, and v_c for maximum payload fraction (see Ref. 19–1).

All the equations quoted so far apply to all three fundamental types of electric rocket systems. No engine parameters are necessary except for the overall efficiency, which ranges from 0.4 to 0.8 in well-designed electric propulsion units, and α, which varies more broadly.

The problem with the above formulation is that the equations are underconstrained in that, given a velocity increment, mission time and specific impulse can be independently assigned. We will return to this topic in Section 19.4.

Example 19–1. Determine the flight characteristics of an electrical propulsion rocket for raising a low satellite orbit. Data given:

$$I_s = 2000 \text{ sec}$$
$$F = 0.20 \text{ N}$$
$$\text{Duration} = 4 \text{ weeks} = 2.42 \times 10^6 \text{ sec}$$
$$\text{Payload mass} = 100 \text{ kg}$$
$$\alpha = 100 \text{ W/kg}$$
$$\eta_t = 0.5$$

SOLUTION. The propellant flow is, from Eq. 2–13,

$$\dot{m} = F/(I_s g_0) = 0.2/(2000 \times 9.81) = 1.02 \times 10^{-5} \text{kg/sec}$$

The total required propellant is

$$m_p = \dot{m}t = 1.02 \times 10^{-5} \times 2.42 \times 10^6 = 24.69 \text{ kg}$$

The required electrical power is, from Eq. 19–6,

$$P_e = \tfrac{1}{2}\dot{m}v^2/\eta_t = \tfrac{1}{2}(1.02 \times 10^{-5} \times 2000^2 \times 9.81^2)/0.5 = 3.92 \text{ kW}$$

The mass of the propulsion system and energy supply system is, from Eq. 19–5,

$$m_{pp} = P_e/\alpha = 3.92/0.1 = 39.2 \text{ kg}$$

The mass before and after engine operation (see Eq. 19–4) is

$$m_1 = 100 + 24.7 + 39.2 = 163.9 \text{ kg}$$
$$m_2 = 139.2 \text{ kg}$$

The velocity increase of the stage under ideal vacuum and zero-g conditions (Eq. 4–6) is

$$\Delta u = v \, \ln[m_0/(m_0 - m_p)]$$
$$= 2000 \times 9.8 \, \ln(163.9/139.2) = 3200 \text{ m/sec}$$

The average acceleration of the vehicle is

$$a = \Delta u/t = 3200/2.42 \times 10^6 = 1.32 \times 10^{-3} \text{m/sec}^2$$
$$= 1.35 \times 10^{-4} g_0$$

The flight's energy increase after 4 weeks of continuous thrust-producing operation is not enough to get from LEO to GEO (which would have required a change of vehicle velocity of about 4700 m/sec with continuous low thrust). During its travel the satellite will have made about 158 revolutions around the earth and raised the orbit by about 13,000 km. Moreover, this does not represent an optimum. In order to satisfy Eq. 19–9 it would be necessary to increase the burn duration (operating time) or change the thrust, or both.

19.2. ELECTROTHERMAL THRUSTERS

In this category, the electric energy is used to heat the propellant, which is then thermodynamically expanded through a nozzle. There are two basic types in use today:

1. The *resistojet*, in which components with high electrical resistance dissipate power and in turn heat the propellant, largely by convection.

2. The *arcjet*, in which current flows through the bulk of the propellant gas which has been ionized in an electrical discharge. Being relatively devoid of material limitations, this method introduces more heat directly into the gas (it can reach local temperatures of 20,000 K or more). The electrothermal arcjet is a unit where magnetic fields (either external or self-induced by the current) are not as essential for producing thrust as is the nozzle. As shown in Section 19.4, arcjets can also operate as electromagnetic thrusters, but here the magnetic fields are essential for acceleration and propellant densities are much lower. Thus, there are some arc–thruster configurations that could be classified as both electrothermal and electromagnetic.

Resistojets

These devices are the simplest type of electrical thruster because the technology is based on conventional conduction, convection, and radiation heat exchange. The propellant is heated by flowing over an ohmically heated refractory-metal surface, such as (1) coils of heated wire, (2) through heated hollow tubes, (3) over heated knife blades, and (4) over heated cyclinders. Power requirements range between 1 W and several kilowatts; a broad range of terminal voltages, AC or DC, can be designed for, and there are no special requirements for power conditioning. Thrust can be steady or intermittent as programmed in the propellant flow.

Material limitations presently cap the operating temperatures to under 2700 K, yielding maximum specific impulses of about 300 sec. The highest specific impulse has been achieved with hydrogen (because of its lowest molecular mass), but its low density causes propellant storage to be bulky (cryogenic storage being unrealistic for space missions). Since virtually any propellant is appropriate, a large variety of different gases has been used, such as O_2, H_2O, CO_2, NH_3, CH_4, and N_2. Also, hot gases resulting from the catalytic decomposition of hydrazine (which produces approximately 1 volume of NH_3 and 2 volumes of H_2 [see Chapter 7]) have been successfully operated. The system using liquid hydrazine (Ref. 19–6) has the advantage of being compact and the catalytic decomposition preheats the mixed gases to about 700°C (1400°F) prior to their being heated electrically to an even higher temperature; this reduces the required electric power while taking advantage of a well-proven space chemical propulsion concept. Figure 19–4 shows details of such a hybrid resistojet which is fed downstream from a catalyst bed where hydrazine is decomposed into hot gases.

Resistojets have been proposed for manned long-duration deep space missions, where the spacecraft's waste products (e.g., H_2O or CO_2) could then be

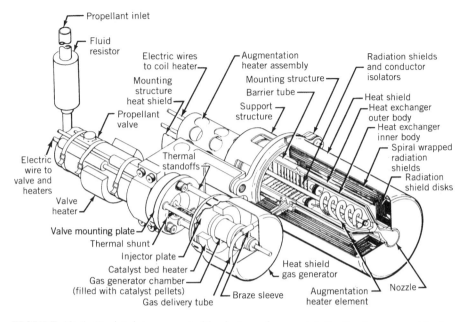

FIGURE 19–4. Resistojet augmented by hot gas from catalytically decomposed hydrazine; two main assemblies are present: (1) a small catalyst bed with its electromagnetically operated propellant valve and heaters to prevent hydrazine from freezing, and (2) an electrical resistance spiral-shaped heater surrounded by thin radiation shields, a refractory metal exhaust nozzle, and high-temperature electrical insulation supporting the power leads. (Courtesy of PRIMEX Aerospace Company.)

used as propellants. Unlike the ion engine and the Hall thruster, the same resistojet design can be used with different propellants.

In common with nearly all electric propulsion systems, resistojets have a propellant feed system that has to supply either gas from high-pressure storage tank or liquid under zero gravity conditions. Liquids require positive tank expulsion mechanisms, which are discussed in Chapter 6, and pure hydrazine needs heaters to keep it from freezing.

Engineering considerations in the development of these rockets include intermittent heat transfer from the heating element to the propellant, conduction and radiation losses from the chamber, the capability of materials to withstand the hot environment, and the heat capacity of the propellant. Procedures have been developed to account for specific heat, thermal conductivity, dissociation, and gas density variations with temperature. The gas flow in the heating chamber is typically considered to be either laminar or vortex flow, and the heat transfer to the stream is by convection.

Available materials limit the maximum gas temperature of a resistojet. High-temperature materials used for the resistance element include rhenium and refractory metals and their alloys (e.g., tungsten, tantalum, molybdenum), platinum (stabilized with yttrium and zirconia), as well as cermets. For high-temperature electrical (but not thermal) insulation, boron nitride has been used effectively.

A design objective is to keep heat losses in the chamber at a low level relative to the power consumed. This can be done by (1) the use of external insulation, (2) internally located radiation shields, and (3) entrant flow layers or cascades. Within reason, the mass of insulation and radiation shields should be small compared to that of the thruster and of the total propulsion system.

The choice of chamber pressure is influenced by several factors. High pressures reduce molecular gas dissociation losses in the chamber, increase the rate of recombination in the exhaust nozzle, improve the heat exchanger performance, and reduce the size of both the chamber and the nozzle for a given mass flow rate. However, high pressures cause higher heat transfer losses, higher stresses on the chamber walls, and can accelerate the rate of nozzle throat erosion. The lifetime of a resistojet is often dictated by the nozzle throat life. Good design practice, admittedly a compromise, sets the chamber pressure in the range of 15 to 200 psi.

Thruster efficiencies of resistojets range between 65 and 85%, depending on the propellant and the exhaust gas temperature, among other things. The specific impulse delivered by any given electrothermal design depends primarily on (1) the molecular mass of the propellant, and (2) the maximum temperature that the chamber and the nozzle surfaces can tolerate.

Table 19–3 gives typical performance values for a resistojet augmented by chemical energy release. The specific impulse and thrust increase as the electric power of the heater is increased. An increase in flow rate (at constant specific power) results in an actual decrease in performance. The highest specific power (power over mass flow rate) is achieved at relatively low flow rates, low thrusts, and modest heater augmentation. At the higher temperatures the dissociation of molecular gases noticeably reduces the energy that is available for thermodynamic expansion.

Even with its comparatively lower value of specific impulse, the resistojet's superior efficiency contributes to far higher values of thrust/power than any of its nearest competitors. Additionally, these engines possess the lowest overall system empty mass since they do not require a power processor and their plumes are uncharged (thus avoiding the additional equipment that ion engines require). Resistojets have been recently employed in Intelsat V, Satcom 1-R, GOMS, Meteor 3-1, Gstar-3, and Iridium spacecraft. They are most attractive for low to modest levels of mission velocity increments, where power limits, thrusting times, and plume effects are mission drivers.

Arcjets

The basic elements of an arcjet thruster are shown in Fig. 1–8 where the relative simplicity of the physical design masks its rather complicated phenomenology. The arcjet overcomes the gas temperature limitations of the resistojet by the use

TABLE 19–3. Selected Performance Values of a Typical Resistojet with Augmentation

Propellant for resistojet	Hydrazine liquid, decomposed by catalysis
Inlet pressure (MPa)	0.689–2.41
Catalyst outlet temperature (K)	1144
Resistojet outlet temperature (K)	1922
Thrust (N)	0.18–0.33
Flow rate (kg/sec)	5.9×10^{-5}–1.3×10^{-4}
Specific impulse in vacuum (sec)	280–304
Power for heater (W)	350–510
Power for valve (max.) (W)	9
Thruster mass (kg)	0.816
Total impulse (N-sec)	311,000
Number of pulses	500,000
Minimum off-pulse bit (N-sec)	0.002
Status	Operational

Source: Data sheet for model MR-501, Primex Aerospace Company.

of an electric arc for direct heating of the propellant stream to temperatures much higher than the wall temperatures. The arc stretches between the tip of a central cathode and an anode, which is part of the coaxial nozzle that accelerates the heated propellant. These electrodes must be electrically insulated from each other and be able to withstand high temperatures. At the nozzle it is desirable for the arc to attach itself as a diffuse annulus in the divergent portion just downstream of the throat. The region of attachment is known to move up or down depending on the magnitude of the arc voltage and on the mass flow rate. In reality, arcs are highly filamentary and tend to heat only a small portion of the flowing gas unless the throat dimension is sufficiently small; bulk heating is done by mixing, often with the aid of vortex flow and turbulence. Since not all the heat is released prior to expansion in the nozzle, there is some loss in that heat released in the divergent portion of the nozzle is not effective in increasing the Mach number of the flow velocity in the exit divergent section.

Arcs are inherently unstable, often forming pinches and wiggles; they can be somewhat stabilized by an external electric field or by swirling vortex motion in the outer layers of the gas flow. The flow structure at the nozzle throat is quite nonuniform and arc instabilities and erosion at the throat are very limiting. The mixing of cooler outer gas with the arc-heated inner gas tends to stabilize the arc while lowering its conductivity, which in turn requires higher voltages of operation. In some designs the arc is made longer by lengthening the throat.

The analysis of arcjets is based on plasma physics, as it applies to a moving ionized fluid. The conduction of electricity through a gas requires that a certain level of ionization be present. This ionization must be obtained from an electrical discharge, i.e., the breakdown of the cold propellant resembling a lightning discharge in the atmosphere (but, unlike lightning, a power supply may feed the current in a continuous or pulsed fashion). Gaseous conductors of electricity follow a modified version of *Ohm's law*. In an ordinary *uniform medium* where an electrical current I is flowing across an area A through a distance d by virtue of a voltage drop V, we can write Ohm's law as

$$V = IR = (I/A)(AR/d)(d) \qquad (19\text{--}10)$$

As given, the medium is uniform and thus we may define the electric field as $E = V/d$, the current density as $j = I/A$, and we introduce the *electrical conductivity* as $\sigma = d/AR$. We can now rewrite the basic Ohm's law as simply $j = \sigma E$. The scalar electrical conductivity is directly proportional to the density of unattached or *free electrons* that, under equilibrium, may be found from Saha's equation (Ref. 19–7). Strictly speaking, Saha's equation applies to thermal ionization only (and not necessarily to electrical discharges). For most gases, either high temperatures or low ionization energies or both are required for plentiful ionization. However, since only about one in a million electrons is sufficient for good conductivity, an inert gas can be seeded with alkali-metal vapors, as is amply demonstrated in plasmas for power generation. The value of plasma electrical conductivity σ may be calculated from

$$\sigma = e^2 n_e \tau / \mu_e \qquad (19\text{--}11)$$

Here e is the electron charge, n_e the electron number density, τ the mean time between collisions, and μ_e the electron mass.

Actually, arc currents are nearly always influenced by magnetic fields, external or self-induced, and a *generalized Ohm's law* (Ref. 19–8) in a moving gas is needed such as the following vector form (this equation is given in scalar forms in the section on electromagnetic devices):

$$j = \sigma[E + v \times B - (\beta/\sigma B)(j \times B)] \qquad (19\text{--}12)$$

The motion of the gas containing charged particles is represented by the velocity v; the magnetic induction field is given as B (a scalar B in the above equation is required in the last term) and the electric field as E. In Eq. 19–12, both the current density j and the conductivity are understood to relate to the free electrons as does β, the Hall parameter. This Hall parameter is made up from the electron cyclotron frequency (ω) multiplied by the mean time it takes an electron to lose its momentum by collisions with the heavier particles (τ). The second term in Eq. 19–12 is the induced electric field due to the motion of the plasma normal to the magnetic field, and the last term represents the Hall electric field which is perpendicular to both the current vector and the applied magnetic field vector as the crossproduct (i.e., the "×") implies (ion-slip and the electron pressure gradient have been omitted above, for simplicity).

Magnetic fields are responsible for most of the peculiarities observed in arc behavior, such as pinching (a constriction arising from the current interacting with its own magnetic field), and play a central role in non-thermal electromagnetic forms of thrusting, as discussed in a following section.

Analytical descriptions of arcjets, based on the configuration shown in Fig. 19–5, may include the following:

1. The energy input occurs largely in the small-diameter laminar flow arc region within the throat of the nozzle. As a first approximation, the power can be computed from Joule heating $[j \cdot E]$; here the current density and the voltage gradient across the arc have to be determined.

2. The cathode tip needs to be hot for thermionic emission of the arc electrons. It is heated by the arc and cooled by the propellant flow. The cathode, typically a coaxial pointed rod, is located in the plenum region.

3. The nozzle inner walls are heated by the arc, which may be at a temperature of 10,000 to 20,000 K. Typically the nozzle is cooled only by conduction and by the boundary layers.

4. The hot gas in the arc proper must mix quickly with the rest of the propellant; this is done by vortexing and turbulence.

5. Portions of the anode are heated to extreme temperatures in a section of the divergent nozzle at the arc footpoint (the arc attachment region of the electrode). The heating of the propellant is not all contained in the plenum chamber, and heating of a supersonic flow is a source of losses.

To start an arcjet, a much higher voltage than necessary for operation has to be applied momentarily in order to break down the cold gas. Some arcjets require an extended initial burn-in period before stable consistent running ensues. Because the conduction of electricity through a gas is inherently unstable, arcs require an external ballast resistance to allow steady-state operation. The cathode must run hot and is usually made of tungsten with 1 or 2% thorium (suitable up to about 3000 K). Boron nitride, an easily shaped high-temperature electrical insulator, is commonly used. Carbon sheets are often used between flanges.

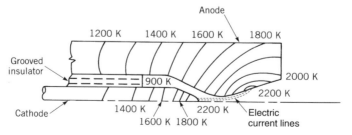

FIGURE 19–5. Typical estimated temperature distribution in the electrodes of an arcjet.

Presently, most arcjets are rather inefficient since less than half of the electrical energy goes into kinetic energy of the jet; the nonkinetic part of the exhaust plume (residual internal energy and ionization) is the largest loss. About 10 to 20% of the electric power input is usually dissipated and radiated as heat to space or transferred by conduction from the hot nozzle to other parts of the system. Arcjets, however, are potentially more scalable to large thrust levels than other electric propulsion systems. Generally, arcjets exhibit about six times the thrust-to-power ratio of a resistojet because of their increased specific impulse coupled with relatively low values of efficiency. Arcjets have another disadvantage in that the required power processing units are somewhat more complex than those for resistojets, due to the complexity of arc phenomena.

The life of an arcjet can be severely limited by local electrode erosion and vaporization, which is specifically due to action of the arc attachment point and of the high operating temperatures in general. The rate of erosion is influenced by the particular propellant in combination with the electrode materials (argon and nitrogen give higher erosion rates than hydrogen), and by pressure gradients, which are usually higher during start or pulsing transients (sometimes by a factor of 100) than during steady-state operation. A variety of propellants has been used in arcjet devices, including N_2, He, H_2, Ne, NH_3, Ar, and the catalytic decomposition products of N_2H_4. Lithium metal, which is a liquid at 180°C, has been considered because of its low molecular mass, ease of ionization, and its potential for transpiration cooling. Also lithium deposits on the cathode tend to reduce cathode erosion. Lithium is very reactive and requires special handling. Specific impulses for H_2 are 1200 to 1500 sec, which, along with other desirable heat-transfer properties, make both hydrogen and lithium the propellants of choice for high performance. There are, however, problems in the handling and storage of these propellants that have been difficult to resolve.

An arcjet downstream of a catalytic hydrazine decomposition chamber looks similar to the resistojet of Fig. 19–4, except that the resistor is replaced by a smaller diameter arc heater. Also, larger cabling is needed to supply the relatively much larger currents. Decomposed hydrazine would enter the arc at a temperature of about 760°C. Liquid hydrazine is easier to store and provides a low-volume, lighter-weight propellant supply system when compared to gaseous propellants. Table 19–4 shows on-orbit performance of a system of 2-kW hydrazine arcjets. Specific impulses from 400 to nearly 600 sec are typical for hydrazine arcjets (Ref. 19–9). A 26-kW ammonia arcjet program (ESEX) is presently undergoing space testing (Refs. 19–10, 19–11) with 787 sec specific impulse and 1.93 N thrust.

19.3. NON-THERMAL ELECTRICAL THRUSTERS

The acceleration of a hot propellant through the use of a supersonic nozzle is the most conspicuous feature of thermal thrusting. Now we turn our study to

TABLE 19–4. On-Orbit 2 kW Hydrazine Arcjet System (PRIMEX, Ref. 19–9)

Propellant	Hydrazine
Steady thrust	222–258 mN
Mass flow rate	36–47 mg/sec
Feed pressure	185–330 psia
Power control unit (PCU) input	4.4 kW (two thrusters)
System input voltage	68–71 V DC
PCU efficiency	93%
Specific impulse	570–600 sec
Dimensions	
Arcjet	$237 \times 125 \times 91$ mm^3
PCU	$632 \times 361 \times 109$ mm^3
Mass	
Arcjet (4) and cable	6.3 kg
PCU	15.8 kg
Total impulse	1,450,000 N-sec

acceleration of a propellant by electrical forces where no area changes are essential for direct gas acceleration. The electrostatic (or Coulomb) force and the electromagnetic (or Lorentz) force can be used to accelerate a suitable propellant to speeds ultimately limited by the speed of light (note that thermal thrusting is essentially limited by the speed of sound in the plenum chamber). The microscopic vector force f_e on a *singly charged particle* can be written as

$$f_e = eE + ev_e \times B \qquad (19\text{–}13)$$

where e is the electron charge magnitude, E the electric field vector, v_e the velocity of the charged particle, and B the magnetic field vector. The sum of the electromagnetic forces on all the charges gives the total force per unit volume vector $\tilde{F}_e$ (scalar forms of this equation follow)

$$\tilde{F}_e = \rho_e E + j \times B \qquad (19\text{–}14)$$

Here ρ_e is the *net charge density* and j the electric current vector density. With plasmas, which by definition have an equal mixture of positively and negatively charged particles within a volume of interest, this net charge density vanishes. On the other hand, the current due to an electric field does not vanish because positive ions move opposite to electrons, thus adding to the current (but in plasmas with free electrons this ion current can be very small). From Eq. 19–14, we see that an electrostatic accelerator must have a nonzero net charge density that is commonly referred to as a *space-charge density*. An example of an electrostatic accelerator is the ion engine, which operates with positive ions; here magnetic fields are unimportant in the accelerator region. Electromagnetic accelerators operate only with plasmas and rely solely on the Lorentz force to accelerate the propellant. The Hall accelerator may be thought of as a crosslink between an ion engine and an electromagnetic engine. These three types of accelerator are discussed next. Research and development efforts in the field of non-thermal thrusters have been extensive and truly international.

Electrostatic and electromagnetic devices require an understanding of the basic laws of electricity and magnetism which are most elegantly summarized in Maxwell's equations complemented by the force relation and Ohm's law, both previously introduced. Moreover, various processes in ionization and gaseous conduction need to be considered. This subject forms the basis of the discipline of magnetohydrodynamics or MHD; however, a proper treatment of this subject is beyond the scope of this book.

Electrostatic Devices

Electrostatic thrusters rely on Coulomb forces to accelerate a propellant composed of non-neutral charged particles. They can operate only in a near vacuum. The electric force depends only on the charge, and all charged particles must be of the same "sign" if they are to move in the same direction. Electrons are easy to produce and are readily accelerated, but they are so extremely light in mass as to be impractical for electric propulsion. From thermal propulsion fundamentals one might deduce that "the lighter the exhaust particle the better." However, the momentum carried by electrons is relatively negligible even at velocities near the speed of light. Thus, the thrust per unit area that can be imparted to such an electron flow remains negligible even when the effective exhaust velocity or specific impulse gets to be very high. Accordingly, electrostatic thrusters use charged heavy-molecular-mass atoms as *positive ions* (a proton is 1840 times heavier than the electron and a typical ion of interest contains hundreds of protons). There has been some research work with small liquid droplets or *charged colloid* which can in turn be some 10,000 times heavier than atomic particles. In terms of power sources and transmission equipment, the use of the heavier particles contributes to more desirable characteristics for electrostatic thrusters—for example, high voltages and low currents in contrast to low voltages and high currents with their associated massive wiring and switching.

Electrostatic thrusters can be categorized by their source of charged particles as follows:

1. *Electron bombardment thrusters.* Positive ions from a monatomic gas are produced by bombarding the gas or vapor, such as xenon or mercury, with electrons emitted from a heated cathode. Ionization can be either DC or RF.

2. *Ion contact thrusters.* Positive ions are produced by passing the propellant vapor, usually cesium, through a hot (about $1100°C$ or $2000°F$) porous tungsten contact ionizer. Cesium vapor was used extensively in the original ion engines.

3. *Field emission or colloid thrusters.* Tiny droplets of propellant are charged either positively or negatively as these droplets pass through an intense electric field discharge. The stability of large, charged particles remains a challenge.

Names such as xenon ion propulsion system (XIPS, Ref. 19–12, and NSTAR/DS1, Refs. 19–10 and 19–13), radio-frequency field ionization (RITA), cesium ion contact rockets, and colloid propulsion have been used to identify electrostatic thrusters. The following general design criteria are desirable for electrostatic thrusters, regardless of the charged particle source:

1. Minimum expenditure of energy per charged particle produced (this energy is an irrecoverable loss).
2. Minimum ion-collision damage to the accelerating electrodes (sputtering) and deterioration of component characteristics over thrust lifetime.
3. Maximum supply of ionized particles (related to propellant utilization factor).
4. Stabilized uniform operation near the space-charge limitations of the thruster (represented by the saturation current density within the accelerator electrodes).
5. Production of particles of uniform mass and charge so that they can be effectively accelerated by the electric field.
6. No reaction of the exhaust plume gases with spacecraft materials (Hg vapor can react with many materials).
7. Nonhazardous propellants with good tankage properties (Hg and Cs are poisonous and Xe is nontoxic but requires extra devices to conserve it). Good tankage means propellant of high density, that is noncorrosive, with stable storage over time.
8. No deposits of condensed species on spacecraft optical components (windows, lenses, mirrors, photovoltaic cell surfaces, or sensitive heat rejection surfaces).
9. Specific impulse near optimum for a given mission (the specific impulse is shown to be a function of accelerating voltage and the particle mass).

Basic Relationships for Electrostatic Thrusters

An electrostatic thruster, regardless of type, consists of the same series of basic ingredients, namely, a propellant source, several forms of electric power, an ionizing chamber, an accelerator region, and a means of neutralizing the exhaust. While Coulomb accelerators require a net charge density of one polarity, the exhaust beam must be neutralized to avoid a space-charge buildup outside of the craft which could easily nullify the operation of the thruster. Neutralization is achieved by the injection of electrons downstream (see the device descriptions that follow). The exhaust velocity is a function of the voltage V_{acc} imposed across the accelerating chamber or grids, the mass of the charged particle μ, and its electrical charge e. In the conservation of energy equation the kinetic energy of a charged particle must equal the electrical energy gained in the field, provided that there are no collisional losses. In its simplest form,

$$\tfrac{1}{2}\mu v^2 = eV_{acc} \tag{19–15}$$

Now, solving for the speed gained in the accelerator,

$$v = \sqrt{2eV_{acc}/\mu} \tag{19–16}$$

When e is in coulombs, μ in kilograms, and V_{acc} is in volts, then v is in meters per second. Using $\mathfrak{M}$ to represent the molecular mass of the ion ($\mathfrak{M} = 1$ for a proton) then, for singly charged ions, the equation above becomes v (m/sec) $= 13,800 \sqrt{V_{acc}/\mathfrak{M}}$. References 19–2 and 19–3 contain a detailed treatment of the applicable theory.

In an ideal ion thruster, the current I across the accelerator represents the sum of all the propellant mass (100% singly ionized) carried per second by the particles accelerated:

$$I = \dot{m}(e/\mu) \tag{19–17}$$

The total ideal thrust from the accelerated particles is given by Eq. 2–14 (without the pressure thrust term, as pressures are extremely low):

$$F = \dot{m}v = I\sqrt{2\mu V_{acc}/e} \tag{19–18}$$

As can be seen, for a given current and accelerator voltage the thrust is proportional to the mass-to-charge ratio of the charged particles. The thrust and power absorbed by the neutralizing electrons are both small (about 1%) and can easily be neglected.

The current density j that can be obtained with a charged particle beam has a *saturation value* depending on the geometry and the electrical field (see Ref. 19–14). This fundamental limit is caused by the internal electric field associated with the ion cloud opposing the electric field from the accelerator when too many charges of the same sign try to pass simultaneously through the accelerator. The saturation current can be derived for a plane-geometry electrode configuration from basic principles. A definition of the current density in terms of the space charge density follows:

$$j = \rho_e v \tag{19–19}$$

The voltage in a one-dimensional space-charge region is found from Poisson's equation, where x represents distance and ε_0 is the *permittivity of free space* which, in SI units, has the value of 8.854×10^{-12} farads per meter:

$$d^2 V / dx^2 = \rho_e / \varepsilon_0 \qquad (19\text{--}20)$$

By solving Eqs. 19–16, 19–19, and 19–20 simultaneously and applying the proper boundary conditions, we obtain the following relation known as the *Child–Langmuir law*:

$$j = \frac{4\varepsilon_0}{9} \sqrt{\frac{2e}{\mu}} \frac{(V_{acc})^{3/2}}{d^2} \qquad (19\text{--}21)$$

In this equation, d is the accelerator interelectrode distance. In SI units the equation for the saturation current density can be expressed (for atomic or molecular ions) as

$$j = 5.44 \times 10^{-8} V_{acc}^{3/2} / (\mathfrak{M}^{1/2} d^2) \qquad (19\text{--}22)$$

Here the current density is in A/m², the voltage is in volts, and the distance in meters. For xenon with electron bombardment schemes, values of j vary from 2 to about 10 mA/cm². The current density and the area are very sensitive to the accelerator voltage as well as to the electrode configuration and spacing.

Using Eqs. 19–18 and 19–22 and letting the cross section be circular so that $I = (\pi D^2 / 4) j$, the thrust can be rewritten as

$$F = (2/9) \pi \varepsilon_0 D^2 V_{acc}^2 / d^2 \qquad (19\text{--}23)$$

In SI units, for molecular ions, this becomes

$$F = 6.18 \times 10^{-12} V_{acc}^2 (D/d)^2 \qquad (19\text{--}24)$$

The ratio of the exhaust beam emitter diameter D to the accelerator-electrode grid spacing d can be regarded as an *aspect ratio* of the ion accelerator region. For multiple grids with many holes (see Figs. 19–6 and 19–7) the diameter D is that of the individual perforation hole and the distance d is the mean spacing between grids. Because of space-charge limitations, D/d can have values no higher than about one for simple, single-ion beams. This implies a rather stubby engine design with many perforations and the need for multiple parallel ion engines for larger thrust values.

Using Eqs. 19–1, 19–2, and 19–17, and assuming η_t conversion of potential energy to kinetic energy, the power of the electrostatic accelerator region is

TABLE 19–5. Ionization Potentials for Various Gases

Gas	Ionization Potential (eV)	Molecular/Atomic Mass (kg/kg-mol)
Cesium vapor	3.9	132.9
Potassium vapor	4.3	39.2
Mercury vapor	10.4	200.59
Xenon	12.08	131.30
Krypton	14.0	83.80
Hydrogen, molecular	15.4	2.014
Argon	15.8	39.948
Neon	21.6	20.183

$$P_e = IV_{acc} = (1/2)\dot{m}v^2/\eta_t \qquad (19\text{–}25)$$

The overall efficiency of an electrostatic thruster will be a function of the thruster efficiency η_t as well as of other loss factors. One loss of energy which is intrinsic to the thruster is the energy expended in charging the propellant, which is related to the ionization energy; it is similar to the dissociation energy in electrothermal devices. Ionization represents an input necessary to make the propellant respond to the electrostatic force and is non-recoverable. The ionization energy is found from the ionization potential (ε_I) of the atom or molecule times the current flow, as the example below shows. Historically, in the development of the ion engine, propellant charging has been of primary concern; the first engine designs used cesium because of its high vapor pressure and ease of ionization, but cesium has many undesirable tankage properties (its high reactivity is very difficult to isolate); then came mercury, with its well-known ionization behavior from fluorescent lamps, but mercury also proved to be unworkable because of its poor tankage characteristics; finally, xenon emerged with its reliable tankage properties and its relative ease of ionization. Table 19–5 shows the molecular mass and first ionization potential for different propellants. In actual practice, considerably higher voltages than the ionization potential are required to operate the ionization chamber.

Example 19–2. For an electron-bombardment ion rocket the following data are given:

Working fluid	xenon (131.3 kg/kg-mol)
Net accelerator voltage	700 V
Distance d between grids	2.5 mm
Diameter D of each grid opening	2.0 mm
Number of holes in the grid	2200
Ionization potential for xenon	12.08 eV

Determine the thrust, exhaust velocity, specific impulse, mass flow rate, propellant needed for 91 days' operation, the power of the exhaust jets, and the thruster efficiency including ionization losses.

SOLUTION. The ideal thrust is obtained from Eq. 19–24:

$$F = 6.18 \times 10^{-12} \times (700)^2 \times (2/2.5)^2 = 1.94 \times 10^{-6} \text{ N per grid opening}$$

The total ideal thrust is then obtained by multiplying by the number of holes

$$F = 2200 \times 1.94 \times 10^{-6} = 4.26 \text{ milliN}$$

The exhaust velocity and specific impulse are obtained from Eq. 19–16:

$$v = 13,800\sqrt{700/131.3} = 31,860 \text{ m/sec}$$
$$I_s = 31,860/9.81 = 3248 \text{ sec}$$

The mass flow rate, obtained from Eq. 2–6, is

$$\dot{m} = F/v = 4.26 \times 10^{-3}/31,860 = 1.34 \times 10^{-7} \text{ kg/sec}$$

For a cumulative period of 91 days of operation, the amount of xenon propellant needed (assuming no losses) is

$$m = \dot{m} \, t_p = 1.34 \times 10^{-7} \times 91 \times 24 \times 3600 = 1.05 \text{ kg}$$

The kinetic energy rate in the jet is

$$\tfrac{1}{2}\dot{m}v^2 = 0.5 \times 1.34 \times 10^{-7} \times (31,860)^2 = 67.9 \text{ W}$$

The ionization losses (l_1) represent the nonrecoverable ionization energy which is related to the ionization potential of the atom (ε_I) times the number of coulombs produced per second (see Table 19–5 and Eq. 19–17):

$$l_I = (12.08) \times (1.34 \times 10^{-7} \times 1.602 \times 10^{-19})/(1.67 \times 10^{-27} \times 131.3) = 1.18 \text{ W}$$

As can be seen, the ionization energy in this ideal case is about 2% of the accelerator energy rate. An equivalent way of calculating the ionization energy is to multiply the ionization potential by the total ion current. The current is found from Eq. 19–17 to be just under 10 mA. Of course, other losses would detract from the high ideal efficiency of this device, which is 98.3%.

Ionization Schemes. Even though all ion acceleration schemes are the same, there are several ionization schemes for electrostatic engines. Most devices are DC but some are RF. To a great extent, the ionization chamber is responsible for most of the size, mass, and perhaps efficiency of these devices. We discuss some of these next.

Ionization of a gas by electron bombardment is a well-established technology (Ref. 19–14). Electrons are emitted from a thermionic (hot) cathode or the more efficient hollow cathode and are forced to interact with the gaseous propellant flow in a suitable ionization chamber. The chamber pressures are low, typically 10^{-3} torr or 0.134 Pa. Figure 19–6 depicts a typical electron-bombardment ionizer which contains neutral atoms, positive ions, and electrons. Emitted electrons are attracted toward the cylindrical anode but are forced by the axial magnetic field to spiral in the chamber, causing numerous collisions with propellant atoms which lead to ionization. A radial electric field removes the electrons from the chamber and an axial electric field moves the ions toward the accelerator grids. These grids act as porous electrodes, which

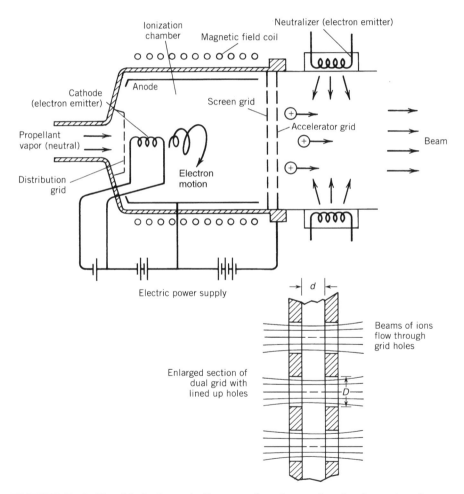

FIGURE 19–6. Simplified schematic diagram of an electron bombardment ion thruster, showing an enlarged section of the double grid.

electrostatically accelerate the positive ions. Loss of electrons is prevented by maintaining the cathode potential negatively biased on both the inner grid electrode and the opposite wall of the chamber. Electrons are routed from the cylindrical anode through an external circuit to another hot cathode at the exhaust beam in order to neutralize the exit beam.

Figure 19–7 shows a cross section of an ion propulsion thruster using xenon as a propellant. It has three perforated electrically charged grids: the inner one keeps the electrons in the ionizer, the middle one has a high voltage (1000 V or more) and accelerates the ions, and the outer one keeps the neutralizing electrons from entering the accelerator region. Each grid hole is lined up with a similar opening in the other grids and the ion beam flows through these holes. If the grids are properly designed, only a few ions are lost by collision with the surface; however, these collisions cause sputtering and greatly diminish the life of the grids. Heavy metals such as molybdenum have been used, with graphite composites being recently introduced. The neutralizer electron source is positioned outside the beam.

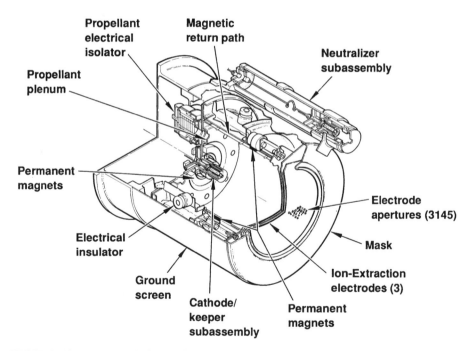

FIGURE 19–7. External view and section of a 500-watt ion propulsion system (XIPS), rated at 18 mN and 2800 sec. Permanent magnets are used on the outside of the ionization chamber; also shown are cathodes for ionization and for beam neutralization. Xenon gas is delivered to the ionizer, then accelerated through the three sheet electrodes, and then the ion beam is neutralized. (Drawing courtesy of Hughes Space and Communications and the American Physical Society.)

Other key components are (1) the heaters for the ionizer and neutralizer cathode, (2) propellant feed and electrical isolator, (3) electrical insulators, and (4) permanent magnets. Reference 19–12 describes a 500 W xenon thruster. Hollow cathodes represent an advancement in the state-of-the-art in electron emission; this cathode consists of a high-temperature metal tube with a flow-limiting orifice and a porous tungsten cylinder impregnated with a barium–oxygen compound located next to the orifice. At about 1370 K the cathode is a good thermionic emitter and thus the hot cylinder produces enough electrons at a relatively low temperature. Xenon, the stable inert gas with the highest molecular mass, is the propellant of choice. Xenon is a minor component of air, in a concentration of about 9 parts in 100 million, so it is a relatively rare and expensive propellant whose availability is currently limited. Its critical point is 289.7 K and 5.84 MPa (the critical density is 1100 kg/m^3). It is easily stored below its critical temperature as a liquid and it does not pose any problems of condensation or toxicity. Pressure regulators for xenon need to be more sophisticated, because no leakages can be tolerated and because flows are very small.

In general, losses can be reduced by (1) decreasing the electron energy and ion density near the walls, (2) increasing the electron energy and ion density near the grid, and (3) optimizing the screen-grid open area. Practical limitations and trade-offs exist for each feature. For example, a reduction in electron energy to reduce the electron flux to the walls also increases collision losses. In practice a small portion of the accelerated ion current impinges on these grids, causing some power loss and some sputtering. Two aspects of the exhaust beam are non-thrust producing: one is the aforementioned *ionization energy* contained in the beam and the other one is any *vector divergence* present which results in beam spreading. Beam spreading, or the radial velocity component of beams, can result from causes both upstream and downstream of the exit electrode. Much of the divergence produced upstream is linked directly with internal geometrical details or "ion optics." Divergence downstream arises from forces within the beam or space charge spreading. Once outside of the accelerator chamber, repelling electrostatic forces between ions rapidly spread the beam radially. Proper neutralization of the beam reduces this spreading, allowing nearly axial velocities.

Other electrical charging schemes include surface or ion contact ionization, field emission ionization, and radio-frequency ionization. These are fairly compact and effective ionizers when compared to electron bombardment. In field emission charging, positive or negative particles are generated when tiny liquid droplets (colloids) pass through a corona discharge. The radio-frequency ion thruster consists of an RF electrodeless discharge that can be compact and produce a high specific impulse; work on this technology is being done primarily in Germany.

The *ion contact thruster* produces ions by surface ionization. The criterion that must be met is that the work function of the metal must be higher than the ionization potential of the propellant. As the propellant atoms "adsorb" on the surface they lose their valence electron to the metal and are re-emitted as a positive ion. The requirement of a high work function and a hot metallic surface restricts this surface to the refractory metals, notably tungsten. Moreover, the requirement of low ionization potential and high atomic mass restricts the propellant to cesium. The operating principle is the same as in the so-called thermionic energy converter. Designs of the cesium/tungsten combination have not yielded high reliability over long lifetimes. Cesium as a propellant is extremely difficult to handle and has proven to be impractical for spacecraft.

Electromagnetic Thrusters

This third major type of electric propulsion device accelerates propellant gas that has been heated to a plasma state. Plasmas are mixtures of electrons, positive ions, and neutrals that readily conduct electricity at temperatures usually above 5000 K or 9000 R. According to electromagnetic theory, whenever a conductor carries a current perpendicular to a magnetic field, a body force is exerted on the conductor in a direction at right angles to both the current and the magnetic field. Unlike the ion engine, this acceleration process yields a neutral exhaust beam. Another advantage is the relatively high thrust density, or thrust per unit area, which is normally about 10 to 100 times that of the ion engines.

Many conceptual arrangements have undergone laboratory study, some with external and some with self-generated magnetic fields, some suited to continuous thrusting and some limited to pulsed thrusting. Table 19–6 shows ways in which electromagnetic thrusters can be categorized. There is a wide variety of devices with a correspondingly wide array of names. We will use the term Lorentz-force accelerators when referring to the principle of operation. For all of these devices the plasma is part of the current-carrying electrical circuit and most are accelerated without the need for area changes. Motion of the propellant, a moderate-density plasma or in some cases a combination of plasma and cooler gas particles, is due to a complex set of interactions. This is particularly true of short duration (3 to 10 μsec) pulsed-plasma thrusters where nothing reaches an equilibrium state. Basically, the designer of an electromagnetic thruster tries to (1) create a body of electrically conductive gas, (2) establish a high current within by means of an applied electric field, and (3) accelerate the propellant to a high velocity in the thrust vector direction with a significantly intense magnetic field (often self-induced).

Conventional Thrusters—MPD and PPT. The description of *magneto-plasma-dynamic (MPD)* and *pulsed-plasma (PPT)* electromagnetic thrusters is based on the Faraday accelerator (Ref. 19–8). In its simplest form, a plasma conductor carries a current in the direction of an applied electric field but perpendicular to a magnetic field, with both of these vectors in turn normal to

TABLE 19–6. Categories of Electromagnetic Thrusters

Thrust Mode	Steady State	Pulsed (Transient)
Magnetic field source	External coils or permanent magnets	Self-induced
Electric current source	Direct-current supply	Capacitor bank and fast switches
Working fluid	Pure gas, as mixture, seeded gas, or vaporized liquid	Pure gas or vaporized liquid or stored as solid
Geometry of path of working fluid	Axisymmetric (coaxial) rectangular, cylindrical, constant or variable cross section	Ablating plug, axisymmetric, other
Special features	Using Hall current or Faraday current	Simple requirement for propellant stage

the direction of plasma acceleration. Equation 19–12 can be specialized to a Cartesian coordinate system where the plasma's "mass-mean velocity" is in the x-direction, the external electric field is in the y-direction, and the magnetic field acts in the z-direction. A simple manipulation of Eq. 19–12, with negligible Hall parameter β, yields a scalar equation for the current,

$$j_y = \sigma(E_y - v_x B_z) \tag{19–26}$$

and the Lorentz force becomes

$$\tilde{F}_x = j_y B_z = \sigma(E_y - v_x B_z)B_z = \sigma B_z^2(E_y/B_z - v_x) \tag{19–27}$$

Here $\tilde{F}_x$ represents the *force "density" within the accelerator* and should not be confused with F the thrust force; $\tilde{F}_x$ has units of force per unit volume (e.g., N/m^3). The axial velocity v_x is a mass-mean velocity that increases internally along the accelerator length; the thrust equals the exit value (v_{max} or v_2) multiplied by the mass flow rate. It is noteworthy that, as long as E_y and B_z (or E/B) remain constant, both the current and the force decrease along the accelerator length due to the *induced field* $v_x B_z$ which subtracts from the impressed value E_y. This increase in plasma velocity translates into a diminishing force along such Faraday accelerators, which limits the final axial velocity. Although not practical it would seem desirable to design for increasing E/B along the channel in order to maintain a substantial accelerating force throughout. But it is not necessarily of interest to design for peak exit velocity because this might translate into unrealistic accelerator lengths (see Problem 19–8). It can be shown that practical considerations might restrict the exit velocity to below one-tenth of the maximum value of E_y/B_z.

A "gasdynamic approximation" (essentially an extension of the classical concepts of Chapter 3 to plasmas in an electromagnetic field) by Resler and Sears (Ref. 19–15) indicates that further complications are possible, namely, that a constant area accelerator channel would *choke* if the plasma velocity does not have the very specific value of $[(k-1)/k](E/B)$ at the sonic location of the accelerator. This *plasma tunnel velocity* would have to be equal to 40% of the value of E/B for inert gases, since k (the ratio of specific heats) equals 1.67. Thus constant area, constant E/B accelerators could be severely constrained because Mach one corresponds only to about 1000 m/sec in typical inert gas plasmas. Constant-area choking in real systems, where the properties E, B, and σ are actually quite variable, is likely to manifest itself as one or more instabilities. Another problem is that values of the conductivity and electric field are usually difficult to determine and a combination of analysis and measurement is required to evaluate, for example, Eq. 19–12. Fortunately, most plasmas are reasonably good conductors when less than 10% of the particles are ionized.

Figure 19–8 shows the simplest plasma accelerator, employing a self-induced magnetic field. This is a *pulsed plasma thruster* (PPT), accelerating plasmas "struck" between two rail electrodes and fed by a capacitor, which is in turn charged by a power supply. The current flow through the plasma quickly discharges the capacitor and hence the mass flow rate must be pulsed according to the discharge schedule. The discharge current forms a current loop, which induces a strong magnetic field perpendicular to the plane of the rails. Analogous to a metal conductor in an electric motor, the Lorentz force acts on the plasma, accelerating it along the rails. For a rail width s, the *total internal accelerating force* has the value $F = sIB$, where I is the total current and B the magnitude of the self-induced field. Hence no area changes are required to accelerate the propellant. Some electrical energy is lost to the electrodes and the ionization energy is never recovered; moreover, this particular plasma does not exit well collimated, and propellant utilization tends to be poor.

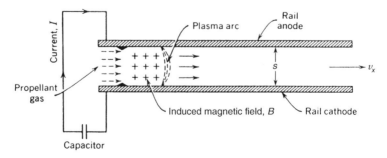

FIGURE 19–8. Simple rail accelerator for self-induced magnetic field acceleration of current-carrying plasma. The concept illustrates the basic physical interactions but suffers from loss of propellant, resulting in low efficiency.

A practical version of the PPT was first put in operation in 1968 and is shown in Chapter 1 as Fig. 1–10. It was used reliably in the USAF's LES-6 communication satellite, which had four PPTs producing approximately 12 million pulses over the life of the thruster. The propellant is stored in a solid Teflon bar that is pushed against the rails by a suitable spring. The rechargeable capacitor discharges across the Teflon surface, momentarily ablating it, and the current flow through the ionized vapor creates its own accelerating magnetic field. About 10^{-5} g of Teflon and 5000 A peak current flow during a 0.6 μsec pulse. In the LES-6 electric propulsion system, one-third of the energy is lost because of capacitor resistance. Other losses occur in ablation, dissociation, ionization, plasma and electrode heating. Teflon stores well in space, is easy to handle, and ablates with insignificant charring. In addition to the overall simplicity of the device, there are no tanks, valves, synchronizing controls, or zero-gravity feed problems. Another advantage is that pulsed thrusting is very compatible with precise control and positioning where the mean thrust is varied by changing the pulsing rate. Besides its very low efficiency, the big disadvantage of this thruster is the size and mass of the power conditioning equipment, which is presently the subject of technology programs toward improvement. Better PPTs are under development (Ref. 19–16).

Figure 19–9 shows a hybrid electrothermal–electromagnetic concept. It produces continuous thrust and Russians claim to have flown several versions. Compared to an electrothermal arcjet, these devices operate at relatively lower pressures and much higher electric and magnetic fields. Hydrogen and argon are common propellants for such MPD arcjets. As with other electromagnetic thrusters, exhaust beam neutralization is unnecessary. Problems of electrode

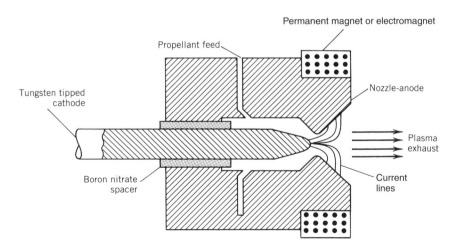

FIGURE 19–9. Simplified diagram of a magnetoplasma dynamic (MPD) arcjet thruster. It is similar in construction to the thermal arcjet shown in Fig. 1–8 but it has a stronger magnetic field to enhance the propellant acceleration.

erosion, massive electrical components, and low efficiencies (with their associated heat dissipation) have slowed implementation of these devices.

Hall-Effect Thrusters. When plasma densities are low enough and/or magnetic fields are high enough, the Hall-effect electric field becomes quite significant. This is the same phenomenon that is observed in the semiconductor Hall effect where a voltage arises transverse to the applied electric field. The Hall current can be understood to represent the motion of the electron "guiding center" (Ref. 19–7) in a crossed electric and magnetic field arrangement where collisions must be relatively insignificant. The Hall thruster is of interest because it represents a practical operating region for space propulsion, which Russian scientists were the first to successfully exploit in a design originally called the *stationary plasma thruster* or SPT, since a portion of the electron current "swirls in place" (Ref. 19–17).

In order to understand the principle of the Hall thruster it is necessary to rewrite in scalar form the generalized Ohm's law, Eq. 19–12. Because the *electron Hall parameter* $\beta = \omega\tau$ is no longer negligible, we arrive at two equations, which are (in Cartesian form):

$$j_x = \frac{\sigma}{1+\beta^2}[E_x - \beta(E_y - v_x B_z)] \tag{19–28}$$

$$j_y = \frac{\sigma}{1+\beta^2}[(E_y - v_x B_z) + \beta E_x] \tag{19–29}$$

For a typical design, the application of a longitudinal electric field E_x causes a current density j_x to flow in the applied field direction together with a Hall current density j_y which flows in the direction transverse to E_x. The Hall electric field E_y is externally shorted to maximize that current and the electrodes are "segmented" in order not to short out the axial electric field E_x. Note that $\beta E_x > v_x B_z$. This arrangement results in a reasonably complicated design (see Fig. 19–10a), one which was deemed impractical. As will be discussed next, for space propulsion, engineers prefer the cylindrical geometry over the rectangular. It yields a simpler, more practical design; here the applied magnetic field is radial and the applied electric field is axial; the thrust-producing Hall current j_θ is azimuthal and counterclockwise and, because it closes on itself, it automatically shorts out its associated Hall electric field. The relevant geometry is shown in Fig. 19–10b, and the equations now become

$$j_x = \frac{\sigma}{1+\beta^2}[E_x + \beta v_x B_r] \tag{19–30}$$

$$j_\theta = \frac{\sigma}{1+\beta^2}[\beta E_x - v_x B_r] \tag{19–31}$$

where, for an accelerator, $\beta E_x > v_x B_r$.

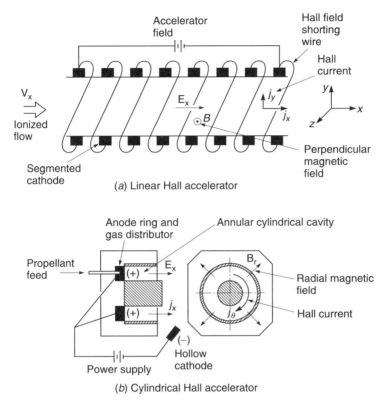

FIGURE 19–10. Linear and cylindrical Hall accelerator configurations showing how an applied axial field results in a transverse current that accelerates the plasma. The Hall current peaks when the external resistance is absent (i.e., shorted). The presence of any significant axial current density j_x represents an inefficiency in Hall devices.

The current j_x is needed for ionization (by electron bombardment) because here the discharge chamber coincides with a portion of the accelerator region. The Hall current j_θ performs the acceleration through the Lorentz force $j_\theta B_r$. The Hall parameter is calculated from the product of the *electron cyclotron frequency* (Ref. 19–7) $\omega = eB/\mu_e$ and the *collision time* τ of the electrons with the heavier particles, which is part of the electrical conductivity in Eq. 9–11. In order for a Hall generator to be of interest, the electron Hall parameter must be much greater than one (in fact, Ref. 19–18 indicates that it should be at least 100), whereas ion motion must proceed relatively unaffected by magnetic effects. Large electron Hall parameters are obtained most readily with low plasma densities which translate into large times between collisions. Figure 19–11 shows a cutout of an SPT design with a redundant set of hollow cathodes and the solenoid magnetic pair responsible for the magnetic field. In Hall thrusters, the propellant gas, xenon or argon, is fed in the vicinity of the anode;

some gas is also provided through the cathode for more efficient cathode operation. While the discharge chamber is not physically separated from the accelerator region, the absence of ions in the first portion of the chamber effectively differentiates the ionization region from the rest of the accelerator. The local charge mass and density of the ions and electrons, together with the magnetic field profiles, need to be tailored such that the ion motion is mostly axial and the electron motion mostly spiral; this makes any given physical design inflexible to changes of propellant. A variation of the original nonconducting accelerator wall SPT design is a smaller channel with metallic walls;

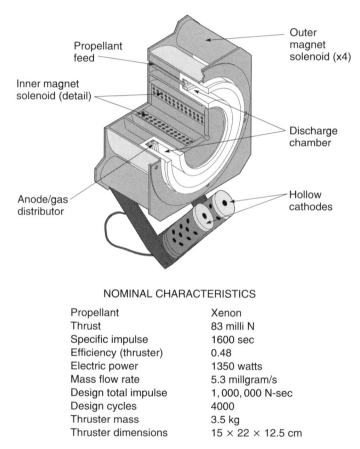

NOMINAL CHARACTERISTICS

Propellant	Xenon
Thrust	83 milli N
Specific impulse	1600 sec
Efficiency (thruster)	0.48
Electric power	1350 watts
Mass flow rate	5.3 millgram/s
Design total impulse	1,000,000 N-sec
Design cycles	4000
Thruster mass	3.5 kg
Thruster dimensions	15 × 22 × 12.5 cm

FIGURE 19–11. External view and quarter section of a 1350-watt Hall accelerator (SPT). It is rated at a thrust of 83 mN and a specific impulse of 1600 sec. The radial magnetic field is produced by an inner solenoid and four external solenoids. Ionization takes place at the beginning of the insulated annular channel. The accompanying table lists the nominal characteristics of SPT-100. (Drawing courtesy of Atlantic Research Corporation and FAKEL.)

this "thruster with an anode layer" (TAL) has comparable performance with a higher thrust density.

The Hall thruster may be classified as either an electromagnetic device (as above) or an electrostatic device where the space charge in the ion acceleration region is neutralized by an electron current transverse to the ion flow (Refs. 19–17, 19–18). If we can mentally separate the process of ionization from that of aceleration, then it is easy to see that electrons swirling within the accelerator neutralize the ionic space charge as it moves from anode to cathode. This, in effect, decreases the magnitude of the accelerating fields and removes most of the beam-focusing requirements. In reality, there is some small interaction between the azimuthal electron current and the ion current, but it diminishes in proportion to the magnitude of the Hall parameter β.

The Hall thruster yields the best β-efficiency (η_H as defined below) when β is very large. The high β-limit is found, from Eqs. 19–30 and 19–31 and the definition of the plasma conductivity σ (Eq. 19–11), as

$$j_x \rightarrow \sigma v_x B_r / \beta = \rho_e v_x \qquad \text{and} \qquad j_\theta \rightarrow \sigma E_x / \beta \qquad (19\text{--}32)$$

$$\tilde{F} = j_\theta B_r \rightarrow \rho_e E_x \qquad (19\text{--}33)$$

$$\eta_H = \tilde{F} v_x / j_x E_x \rightarrow 1.0 \qquad (19\text{--}34)$$

As can be seen, the accelerating force at this high Hall parameter limit is the electrostatic force and, since the exit ionization levels are about 90%, this corresponds in principle to the ion engine without any of its severe space-charge current limitations. Even though electron densities are in the order of 10^{15} to $10^{17}/\text{m}^3$, the effective space-charge densities (ρ_e) are considerably lower because of positive ion neutralization and approach zero at the exit. Furthermore, the Hall β-efficiency η_H as defined in the equations above reflects strictly the influence of β; this efficiency is ideal, being an internal parameter that represents the loss that arises from the total current vector not being perfectly normal to the flow direction. The overall efficiency is still given by Eq. 19–2.

Example 19–3. (a) For the SPT-100 information given in Fig. 19–11, verify the values of thrust and efficiency. (b) Using the definition of the Hall efficiency above, calculate its value for $\beta = 200$ and for a representative value of the parameter $B_r v_x / E_x = 2.5 \times 10^{-2}$ (this grouping of variables can be shown to be intrinsic in the Hall thruster).

SOLUTION. (a) The mass flow rate is 5.3×10^{-6} kg/sec, the specific impulse is 1600 sec, and the input power is 1350 W. Hence

$$F = \dot{m} \, I_s g_0 = (5.3 \times 10^{-6})(1600)(9.81) = 83.2 \text{ mN}$$

$$\eta_t = F I_s g_0 / 2 P_e = (8.3 \times 10^{-2})(1600)(9.81) / 2(1350) = 48.4\%$$

Both of these answers compare very well with the information in Fig. 19–11.

(b) With some manipulation, and defining the Hall local efficiency parameter $\xi = B_r v_x / E_x$, Eq. 19–34 can be written as

$$\eta_H = (-\xi + \beta)\xi/(1 + \beta\xi) = 5/6 = 83.3\%$$

Since the parameter ξ can be highly variable across real accelerator channels, this Hall β-efficiency is not necessarily representative of the overall efficiency, only of the maximum efficiency. Clearly, even the ideal Hall accelerator is not as good as the ideal Faraday or MPD accelerator. Nevertheless, for very large values of β, it can be seen that this efficiency will approach one for any value of ξ.

19.4. OPTIMUM FLIGHT PERFORMANCE

Now that we have discussed the various propulsion devices available, we return to the discussion of flight performance. In Section 19.1 the fundamental background for an optimum propulsion system design was introduced. The discussion remained incomplete because the specific power and the efficiency of individual thrusters, among other things, need to be known for further analysis. In a given mission, the payload m_{pl} and velocity increment Δu are specified along with upper limits on electric power available (Ref. 19–19). In the analysis of Section 19.1, for any desired $\Delta u/v_c$, one can find an optimum v/v_c given a payload ratio; however, even when the choice of an electric propulsion system has been made, thrust time t_p is unspecified and thus the total mass also remains unspecified. Thrust time or "burn time" is the smallest for zero payload and continuously increases with increasing payload ratio. Concurrently, the specific impulse changes, making the problem underconstrained.

Given the payload mass and the vehicle velocity increment, a spacecraft design procedure might be followed using the optimum results of Section 19.1, e.g.:

1. Pick a payload mass fraction—from Fig. 19–3 this yields an optimum $\Delta u/v_c$.
2. From the given Δu, deduce the value of the characteristic speed v_c.
3. From the optimum value of v/v_c in Fig. 19–3 at the given mass fraction, or Eq. 19–9, calculate the corresponding value of v or I_s.
4. Select an engine that can deliver this optimum I_s and from its characteristics (i.e., α and η_t) find the thrusting time t_p from Eq. 19–8.
5. Calculate m_p from Section 19.1, including Eq. 4–7 and the given payload ratio.
6. Check that the available vehicle electrical power (from Eq. 19–6) together with vehicle volume plus the desirable mission time and total cost are not exceeded.

As may be evident, a unique criterion for the choice of the assumed pay-load mass fraction is missing above. One possible solution to this problem is to look for a "dual optimum", namely, to seek the shortest burn time consistent with the highest payload mass fraction. A maximum for the product of m_{pl}/m_0 with $\Delta u/v_c$ does exist as a function of v/v_c (as shown in Ref. 19–20). In other words, this dual optimum defines a minimum overall mass for a specified payload consistent with minimum transfer time. Table 19–7 gives estimated values of α along with the corresponding range of specific impulse and the efficiency for electric propulsion systems in present engine inventories.

The optimum formulation in Section 19.1, however, needs to be modified to account for the portion of tankage mass which results from propellant loading; with few exceptions, an additional 10% of the propellant mass shows up as tank or container mass (this could be further refined to include reserve propellant). Reference 19–16 includes information on this *tankage fraction* for various thrusters. Fortunately, the analysis presented earlier is little modified and it turns out that the optima are driven toward higher specific impulses and longer times of operation. For an arbitrary tankage fraction allowance φ,

$$\frac{\Delta u}{v_c} = \frac{v}{v_c} \ln \left[\frac{(1 + \varphi) + (v/v_c)^2}{(m_{pl}/m_0 + \varphi) + (v/v_c)^2} \right] \qquad (19\text{–}35)$$

When $\varphi = 0.1$ the actual value for the joint-optimized payload ratio can be shown to be 0.46, with corresponding ratios of vehicle velocity increment as 0.299 and propellant exhaust velocity as 0.892. It turns out that this peak is rather broad and that payload ratios between 0.34 and 0.58 are within 6% of the mathematical optimum. Since engine parameters are rather "inelastic", and since spacecraft designers have to deal with numerous constraints which are not propulsion related, this wider range of optima is deemed a practical necessity.

Given the desirable 0.34 to 0.58 optimum payload-ratio range, we may first select one or more engines within the range $0.2268 \leq (I_s^*/\Delta u) \leq 0.4263$, where the optimized specific impulse (I_s^*) is in sec and the velocity change in m/sec. Since the vehicle's change in velocity is known, this condition yields the required limits in specific impulse. We can then proceed to use the following joint-optimized, approximate polynomial relations to find m_{pl}/m_0 and t_p^* as follows:

TABLE 19–7. Summary of Current Technology in Typical Electric Propulsion Engines

Engine Type	Identification (Reference)	Specific power, α (W/kg) (estimated)	Thruster Efficiency, η_t	Specific Impulse, I_s (sec)	Power (W)	Thrust (N)	Lifetime (hr)	Status
Resistojet	N₂H₄ (16, 21) (19–16, 19–21)	333–500	0.8–0.9	280–310	500–1500	0.2–0.8	>390	Operational
	NH₃ (19–16)		0.8	350	500			Operational
	Primex MR-501B (19–21)			303–294	350–510	0.369–0.182	>389	Operational
Arcjet	N₂H₄ (19–21)	313	0.33–0.35	450–600	300–2000	0.2–0.25	>830–1000	Operational
	H₂ (19–16, 19–21)	333	0.4	1000	5–100 K	0.2–0.25	>1000	R&D
	NH₃ (19–16)	270–320	0.27–0.36	500–800	500–30 K	0.2–0.25	1500	Qualified
	Primex Mr-509 (19–21) (c)	115.3	>0.31	>502 (545)	1800	0.213–0.254	>1575	Qualified
	Primex MR-510 (19–21) (c)	150	>0.31	>570–600	2170	0.222–0.258	>2595	Qualified
Ion Propulsion	XIPS (19–21)	100	0.75	2800–3500	200–4000	0.015–0.014	>8000	Operational
	Hughes XIPS-13 (19–21)		0.46, 0.54	2585, 2720	427, 439	0.0178, 0.018	12,000	Qualified
	Hughes XIPS-25 (19–21)		0.65, 0.67	2800	1400	0.0635	>4350	Qualified
	NSTAR/DS1 (19–13)	45	0.6	3100	2300–2500	0.093	>10,000	Operational
	RITA 15 (a)	9.61		3000–4000	540	0.015	>20,000	Qualified
	UK-10/T5 (UK) (19–21)		0.55–0.64	3090–3300	278–636	0.010–0.025	10,700	Qualified
	ETS-VI IES (Jap.) (19–21)		0.4	3000	730	0.02		Operational
	DASA RIT-10 (Ger.) (19–21)		0.38	3000–3150	585	0.015		Operational
Hall	Hall (XE) (19–16)	150	0.5	1500–1600	300–6000	0.04	>7000	Operational
	SPT (XE) (19–21)		0.48	1600	150–1500	0.04–0.2	>4000	Operational
	ARC/Fakel SPT-100 (19–16)	169.8	0.48	1600	1350	0.083	>7424	Operational
	Fakel SPT-70 (19–3)		0.46, 0.50	1510, 1600	640–660	0.04	9000	Operational
	TAL D-55 (Russia) (19–21)	~50.9	0.48, 0.50–0.60	950–1950	600–1500	0.082	>5000	Operational
	Primex BPT Hall (c)		0.5	1500–1800	500–6000			Development
MPD—Steady	Applied Field (19–16)		0.5	2000–5000	1–100 K			R&D
	Self-field (19–16)		0.3	2000–5000	200–4000 K			R&D
MPD-Pulsed	Teflon PPT (19–16)	1	0.07	1000	1–200	4000 N-sec	>10^7 pulses	Operational
	LES 8/9 PPT (19–21)		0.0068, 0.009	836, 1000	25, 30	0.0003	>10^7 pulses	Operational
	NASA/Primex EO-1 (c)	~20	0.098	1150	up to 100	3000 N-sec		Operational
	Primex PRS-101 (c)			1150		1.4 mN, 2 Hz		Operational
	EPEX arcjet (Jap.) (19–21)		0.16	600	430	0.023		Operational

Manufacturers: (*a*): Daimler-Chrysler Aerospace, AG., (*b*): Atlantic Research Corporation, USA Fakel (Russia), (*c*): Primex Aerospace Company

$$\frac{m_{pl}}{m_0} \approx \left[-0.1947 + 2.972\left(\frac{I_s^*}{\Delta u}\right) - 2.7093\left(\frac{I_s^*}{\Delta u}\right)^2 \right] \qquad (19\text{--}36)$$

$$t_p^* \approx \left[67.72 - 39.67\left(\frac{m_{pl}}{m_0}\right) + 20.04\left(\frac{m_{pl}}{m_0}\right)^2 \right] \frac{(I_s^*)^2}{\alpha\eta_t} \text{(sec)} \qquad (19\text{--}37)$$

The success of this approach hinges on the validity of the engine information employed. In particular, the specific power should represent all the inert components of the engine, which can be reasonably assumed to depend on the power level. The payload mass must reflect all mass that is neither proportional to the electrical power nor propellant related. The tankage fraction allowance must reflect the total propellant mass and thus the use of Eq. 19–35 is necessary. It is assumed that there is available a source of electricity (typically from 28 to 110 V DC for solar-powered craft) which is not tagged to the propulsion system. The analysis also assumes that the efficiency is not a function of specific impulse (in contrast to Ref. 19–22); this implies that an average or effective value can be used. Since each individual engine type spans a somewhat limited range of specific impulse, this assumption is not deemed to be too restrictive. For the continuous thrust schedules required by electric engines, thrust time is equal to mission time.

Example 19–4. List the performance of three electric propulsion engines within the dual-optimum criteria to carry a 100 kg payload through a change of velocity of 7000 m/sec. Calculate total mass, burn time, thrust, and power requirements.

SOLUTION. We first calculate the dual-optimum specific-impulse range, which turns out to be between 1590 sec and 2980 sec. Then, we pick engines from the inventory (see Table 19–7). Results are tabulated below for three thrusters.

$$0.2268\Delta u \leq I_s^*[\text{sec}] \leq 0.4263\Delta u$$
$$m_0 = 100/(m_{pl}/m_0) = 100 + 1.1m_p + m_{pp} = 100 + m_p(1.1 + (v/v_c)^2)$$
$$(v/v_c)^* = 0.6953 + 0.5139(m_{pl}/m_0) - 0.1736(m_{pl}/m_0)^2$$

Hall Effect Thruster	Xenon Ion Propulsion System	Magnetoplasma Dynamic
$I_s^* = 1600$ sec	$I_s^* = 2585$ sec	$I_s^* = 3000$ sec
$\alpha\eta_t = 93.5$ W/kg	$\alpha\eta_t = 46$ W/kg	$\alpha\eta_t = 30$ W/kg
(Demonstrated)	(Demonstrated)	(Experimental)
$m_{pl}/m_0 = 0.343$, $m_0 = 291$ kg	$m_{pl}/m_0 = 0.533$, $m_0 = 187$ kg	$m_{pl}/m_0 = 0.581$, $m_0 = 172$ kg
$t_p^* = 17.9$ days	$t_p^* = 87.9$ days	$t_p^* = 178$ days
$F = 1.06$ N	$F = 0.149$ N	$F = 69.7$ mN
$P_e = 15.4$ kW	$P_e = 4.12$ kW	$P_e = 2.05$ kW

As can be seen, total mass, along with thrust, decreases with increasing specific impulse, whereas thrust time increases. The power variation P_e also decreases, reflecting the individual choice of engines and the engine data from Table 19–7. Any engine can be eliminated when the required power exceeds the power available in the spacecraft or when the burn time exceeds some specified mission time constraint. Most often, cost is the ultimate selection criterion and is largely dependent on m_0.

19.5. MISSION APPLICATIONS

Three principal application areas have been described in the introduction to this chapter. The selection of a particular electric propulsion system for a given flight application depends not only on the characteristics of the propulsion system (which are described in this chapter) but also on the propulsive requirements of the particular flight mission, the proven performance of the specific candidate propulsion system, along with vehicle interfaces and the power conversion and storage systems. In general, the following criteria can be enumerated:

1. For very precise low-thrust station-keeping and attitude control applications, *pulsed thrusters* are generally best suited.

2. For deep space missions where the vehicle velocity increment is appreciably high, systems with *very high specific impulse* will give better performance. As shown in Section 19.1, the optimum specific impulse is proportional to the square root of the thrust operating time.

3. The higher the specific impulse, the more electrical power is required for a given thrust level. This translates into larger size and mass requirements for the power conditioning and generating equipment. However, for a given payload and vehicle velocity increase, the total mass and the thrust vary in nontrivial ways with respect to the specific impulse (see Example 19–4).

4. Since most missions of interest require long life, *system reliability* is a key selection criterion. *Extensive testing* under all likely environmental conditions (temperatures, pressures, accelerations, vibration, and radiation conditions) is required for high reliability. Ground testing and qualification of electric engines should be no different from that of their chemical counterparts, where large resources have made it possible to develop the present inventory of reliable engines. Simulation of the low pressures in space requires large vacuum test chambers.

5. There is a premium on *high thruster efficiency* and *high power-conversion efficiency*. This will reduce the inert mass of the power supply system and reduce thermal control requirements, all of which usually results in lower total mass and higher vehicle performance.

6. For every propulsion mission there is a theoretically *optimum range of specific impulse* (see Fig. 19–3) and thus an optimum electrical propulsion system design. While this optimum may be blurred by some conflicting system constraints (e.g., flight time or maximum power or size con-

straints or cost), the present variety in the engine inventory can meet most goals.

7. The present *state of the art in electrical power sources* appears to limit the type and size of electric propulsion systems that can be integrated, particularly for missions to the outer planets, unless nuclear energy power generation on board the spacecraft becomes more developed and acceptable.

8. Practical factors, such as the storing and feeding of liquids in zero gravity, the availability of propellant (in the case of xenon), the conditioning of power to the desired voltage, frequency, and pulse duration, as well as the redundancy in key system elements, the survival of sensors or controllers in long flights, and the inclusion of automatic self-checking devices along with cost, will all influence the selection and application of specific types of electric rockets.

9. In addition to tankage considerations, *propellant selection* will also be influenced by certain interface criteria such as plume noninterference with communication signals. Plumes must also be thermally benign and noncondensing on sensitive surfaces of the spacecraft such as optical windows, mirrors, and solar cells.

Synchronous or geostationary satellites are extremely attractive for communications and earth observation; their long life requires extensive station-keeping propulsion requirements. Until recently, the main limitation to any such life increase had been the propellant requirement. There is also a propulsion need for orbit raising from LEO to GEO. Earth satellites in inclined orbits with precise time–trajectory position requirements need propulsion units to maintain such orbits, counteracting certain perturbing natural forces described in Chapter 4.

The increasing life trend in earth-orbit satellites from a minimum of 8 years to at least 15 years significantly increases their total impulse and durability requirements of the propulsion system. For example, the north–south station-keeping (NSSK) function of a typical geosynchronous satellite requires about 40,000 to 45,000 N-sec or 9000 to 10,000 lbf-sec of impulse per year. Table 19–8 shows some of the characteristics required of small and large electric thrusters for various propulsion functions in space.

19.6. ELECTRIC SPACE-POWER SUPPLIES AND POWER-CONDITIONING SYSTEMS

The availability of substantial amounts of electrical power in space is considered to be one of the most significant factors in electrical propulsion. Several combinations of energy sources and conversion methods have reached prototype stages, but only solar cells (photovoltaic), isotope thermoelectric genera-

TABLE 19–8. Space Propulsion Application and Characteristics for Three Thrust Levels of Electric Propulsion Thrusters

Thrust Class	Application (Life)	Characteristics	Status
Micronewtons (μN)	E–W station keeping Attitude control Momentum wheel unloading (15–20 years)	10–500 W power Precise impulse bits of $\sim 2 \times 10^{-5}$ N-sec	Operational
Millinewtons (mN)	N–S station keeping Orbit changes Drag cancellation Vector positioning (20 years)	Kilowatts of power Impulse bits $\sim 2 \times 10^{-3}$ N-sec for N–S, impulse/year of 46,000 N-sec/100 kg spacecraft mass	Operational
0.2 to 10 N	Orbit raising Interplanetary travel Solar system exploration (1–3 years)	Long duration 10–300 kW of power Intermittent and continuous operation	In development

tion units (nuclear), and fuel cells (chemical) have advanced to the point of routine space-flight operation. Power output capacity of operational systems has been increasing from the low one-kilowatt range to the medium tens of kilowatts required for some missions. The high end of a hundred or more kilowatts is still pending some technological (and political) breakthroughs.

Space power level requirements have been increasing with the increased capacity of earth-orbit communications satellites and with planned missions, manned and robotic, to the moon and nearby planets. Payload requirements and thrust duration dictate the power level. Commercial communications satellites can temporarily reduce the communications volume during orbit maintenance so that the electric power does not require a dedicated power supply for the propulsion system, but larger power demands require enhanced solar cell capabilities. Some communications satellites actually share part or all of the power-conditioning equipment with their electric thrusters.

Power Generation Units

Electric power-generation units are classified as either direct (no moving mechanical parts) or dynamic. When the primary driver is reliability and the total power is small, direct conversion has been preferred but, with the advent of the Space Shuttle and with the proposed manned space station, dynamic systems are being reconsidered. Many diverse concepts have been evaluated for meeting the electrical power demands of spacecraft, including electric propulsion needs. Direct energy conversion methods considered include photovoltaic, thermoelectric, thermionic, and electrochemical, while indirect methods (with moving parts) include the Brayton, Rankine, and Stirling cycles.

Batteries. Batteries can basically be classified as either primary or secondary. *Primary batteries* consume their active materials and convert chemical energy into electrical. *Secondary batteries* store electricity by utilizing a reversible chemical reaction and are designed to be recharged many times. There are both dry-cell and wet-cell primary batteries. The importance of primary batteries passed with the short-lived satellites of the early 1960s. Secondary batteries with recharging provisions provide electrical power at output levels and lifetimes longer than primary batteries. Batteries must be sealed against the space vacuum or housed inside pressurized compartments. Secondary batteries are a critical component of solar cell systems for power augmentation and emergency backup and the periods when the satellite is in the earth's shadow.

Fuel Cells. *Chemical fuel cells* are conversion devices used to supply space-power needs for 2 to 4 weeks and for power levels up to 40 kW in manned missions. A catalyzer controls the reaction to yield electricity directly from the chemical reaction; there is also some heat evolved, which must be removed to maintain a desirable fuel cell temperature. They are too massive for both robotic and long-duration missions, having also had some reliability problems. Recent improvements in fuel cell technology have considerably advanced their performance.

Solar Cell Arrays. Solar cells rely on the photovoltaic effect to convert electromagnetic radiation. They have supplied electrical power for most of the long-duration space missions. The first solar cell was launched in March 1958 on Vanguard I and successfully energized data transmission for 6 years. Solar arrays exist in sizes up to 10 kW and could potentially grow to 100 kW sizes in earth orbits.

An *individual cell* is essentially one-half of a p–n junction in a transistor, except that the surface area has been suitably enlarged. When exposed to sunlight, the p–n junction converts photon energy to electrical energy. Typically, solar cell arrays are designed for 20% over-capacity to allow for material degradation toward the end of life. Loss in performance is due to radiation and particle impact damage, particularly in the radiation belts around the earth. There has been some improvement in efficiency, reliability, and power per unit mass. For example, standard silicon cells deliver 180 W/m^2 and arrays have 40 W/kg. Newer gallium arsenide cells produce 220 W/m^2 and are more radiation resistant than silicon cells; gallium arsenide cells are presently space qualified and integrated; together with parabolic concentrators, their arrays can reach 100 W/kg (Ref. 19–16). In the near future, multi-junction solar cells designed to utilize a greater portion of the solar spectrum will be used; they have already demonstrated 24% efficiency.

Factors that affect the specific mass of a solar array, besides conversion efficiency, include the solar constant (which varies inversely as the square of the distance from the sun) and the manufactured thinness of the cell. Orientation to the sun is a more critical factor when solar concentrators are being used. Cell output is a function of cell temperature; performance can suffer as much as 20% for a 100°F increase in operating temperature so that thermal control is critical. Solar cell panels can be (1) fixed and body mounted to the spacecraft, (2) rigid and deployable (protected during launch and positioned in space), (3) flexible panels that are deployed (rolled out or unfolded), and (4) deployed with concentrator assist.

In addition to the solar arrays, their structure, deployment and orientation equipment, other items including batteries, plus power conditioning and distribution systems are assigned to the power source. Despite their apparent bulkiness and battery dependence, solar-cell electrical systems have emerged as the dominant generating power system for unmanned spacecraft.

Nuclear Thermoelectric and Thermionic Systems. Nuclear energy from long-decay radioisotopes and from fission reactors has played a role in the production of electricity in space. Both thermoelectric (based on the Seebeck effect) and thermionic (based on the Edison effect) devices have been investigated. These generators have no moving parts and can be made of materials reasonably resistant to the radioactive environment. But their specific power is relatively low and cost, availability, and efficiency have been marginal.

Throughout the 1950s and 1960s nuclear fission reactors were regarded as the most promising way to meet the high power demands of space missions, particularly trips to the outer planets involving months and perhaps years of travel. Radioisotope thermoelectric power has been embodied in a series of SNAP (Systems for Nuclear Auxiliary Power) electrical generating units which were designed and tested, ranging from 50 W to 300 kW of electrical output. Fission reactors were included in the SPAR (Space Power Advance Reactor) program, later renamed SP-100, which was to feature a nuclear-thermoelectric generator with an electrical output of 100 kW; this program was discontinued in 1994. The most recent space nuclear reactor generator is the Russian TOPAZ that has been space tested up to nearly 6 kW. It consists of sets of nuclear rods each surrounded by a thermionic generator. TOPAZ technology was obtained by the USA from the Russians and efforts to upgrade and flight-qualify the system were underway in the mid-1990s.

Thermionic converters have a significant mass advantage over thermoelectric ones, based on their higher effective radiator temperatures. Since thermal efficiencies for both thermoelectric and thermionic conversion are below 10% and since all unconverted heat must be radiated, at higher temperatures thermionic radiators are less massive. Moreover, cooling must be present at times when no electricity is generated since the heat source cannot be "turned off." Depending on the location of the waste heat, clever designs are needed, involving heat pipes or recirculating cooling fluids.

Long-Duration High-Output Dynamic Systems. Designs of electric power generation with outputs of 10 to 1000 kW here on earth have been based on Stirling or Rankine heat engine cycles with nuclear, chemical, and even solar power sources. Overall efficiencies can be between 10 and 40%, but the hardware is complex, including bearings, pumps, reactors, control rods, shielding, compressors, turbines, valves, and heat exchangers. Superconducting magnets together with advances in the state-of-the-art of seals, bearings, and flywheel energy storage have made some dynamic units relatively more attractive. There remain development issues about high-temperature materials that will withstand intense nuclear radiation fluxes over several years; there are still some concerns about achieving the required reliability in such complex systems in the space environment. While limited small-scale experiments have been conducted, the development of these systems remains a challenge. The potential of flight accidents, i.e., the unwanted spreading of nuclear materials, remains a concern for the launch and in manned space missions.

Power-Conditioning Equipment

Power-conditioning equipment is a necessary part of electric propulsion systems because of inevitable mismatches in voltage, frequency, power rate, and other electrical properties between the space-power generating unit and the electric thruster. In some earlier systems, the power-conditioning equipment has been more expensive, more massive, and more difficult to qualify than the thruster itself. If the thrust is pulsed, as in the PPT, the power-conditioning unit has to provide pulse-forming networks for momentary high currents, exact timing of different outputs, control and recharging of condensers. Ion engines typically require from 1000 to 3000 V DC; the output of solar-cell arrays is 28 to 300 V DC so there is a need for DC-to-DC inverters and step-up transformers to accomplish the task. Often this equipment is housed in a single "black box," termed the *power conditioner*. Modern conditioning equipment contains all the internal logic required to start, safely operate, and stop the thruster; it is controlled by on-off commands sent by the spacecraft-control processor. Besides the above functions that are specific to each engine, power-conditioning equipment may provide circuit protection and propellant flow control as well as necessary redundancies.

As may be apparent from Table 19–7, one of the largest contributors to the specific mass of the system (α) is the power-conditioning equipment. Here, electrothermal units have the simplest and lightest conditioning equipment. Ion engines, on the other hand, have the heaviest equipment, with Hall thrusters somewhere in between (Ref. 19–17). PPTs tend to be heavy, but advances in energy storage capacitors can improve this situation. In fact, advances in solid-state electronic pulse circuits together with lighter, more efficient, and higher temperature power-conditioning hardware are an area of great interest to the implementation of electric propulsion units. The efficiency of the equipment tends to be high, about 90% or more, but the heat generated is at a low temperature and must be removed to maintain the required moderately low temperatures of operation. When feasible, the elimination of conditioning equipment is desirable, the so-called direct drive, but a low-pass filter would still be necessary for electromagnetic interference (EMI) control (more information in Ref. 19–21).

PROBLEMS

1. The characteristic velocity $v_c = \sqrt{2t_p\alpha\eta}$ was used to achieve a dimensionless representation of flight performance analysis. Derive Eq. 19–35 without any tankage fraction allowance. Also, plot the payload fraction against v/v_c for several values of $\Delta u/v_c$. Discuss your results with respect to optimum performance.

2. For the special case of zero payload, determine the maximized values of $\Delta u/v_c$, v/v_c, m_p/m_0, and m_{pp}/m_0 in terms of this characteristic velocity as defined in Problem 1. *Answer*: $\Delta u/v_c = 0.805$, $v/v_c = 0.505$, $m_p/m_0 = 0.796$, $m_{pp}/m_0 = 0.204$.

3. For a space mission with an incremental vehicle velocity of 85,000 ft/sec and a specific power of $\alpha = 100$ W/kg, determine the optimum values of I_s and t_p for two maximum payload fractions, namely 0.35 and 0.55. Take the thruster efficiency as 100% and $\varphi = 0$.
Answer: for 0.35: $I_s = 5.11 \times 10^3$ sec; $t_p = 2.06 \times 10^7$ sec; for 0.55: $I_s = 8.88 \times 10^3$ sec; $t_p = 5.08 \times 10^7$ sec.

4. Derive Eq. 19–7 showing v_c explicitly.

5. An electric rocket uses heavy charged particles with a charge-to-mass ratio of 500 coulombs per kilogram producing a specific impulse of 3000 sec. (*a*) What acceleration voltage would be required for this specific impulse? (*b*) If the accelerator spacing is 6 mm, what would be the diameter of an ion beam producing 0.5 N of thrust at this accelerator voltage?
Answers: (*a*) 8.66×10^5 V; (*b*) $D = 1.97$ mm.

6. An argon ion engine has the following characteristics and operating conditions:

Voltage across ionizer = 400 V   Voltage across accelerator = 3×10^4 V
Diameter of ion source = 5 cm   Accelerator electrode spacing = 1.2 cm

Calculate the mass flow rate of the propellant, the thrust, and the thruster overall efficiency (including ionizer and accelerator). Assume singly charged ions.
Answer: $\dot{m} = 2.56 \times 10^{-7}$ kg/sec; $F = 9.65 \times 10^{-2}$ N; $\eta_t = 98.7\%$

7. For a given power source of 300 kW electrical output, a propellant mass of 6000 lbm, $\alpha = 450$ W/kg, and a payload of 4000 lbm, determine the thrust, ideal velocity increment, and duration of powered flight for the following three cases:
(*a*) Arcjet $I_s = 500$ sec $\eta_t = 0.35$

(b) Ion engine $I_s = 3000$ sec $\eta_t = 0.75$
(c) Hall engine $I_s = 1500$ sec $\eta_t = 0.50$

Answers:
(a) $t_p = 3.12 \times 10^5$ sec; $\Delta u = 3.63 \times 10^3$ m/sec; $F = 42.8$ N
(b) $t_p = 5.24 \times 10^6$ sec; $\Delta u = 2.18 \times 10^4$ m/sec; $F = 15.29$ N
(c) $t_p = 1.69 \times 10^6$ sec; $\Delta u = 1.09 \times 10^4$ m/sec; $F = 20.4$ N

8. A formulation for the exit velocity that allows for a simple estimate of the accelerator length is shown below; these equations relate the accelerator distance to the velocity implicitly through the acceleration time t. Considering a flow at a constant plasma of density ρ_m (which does not choke), solve Newton's second law first for the speed $v(t)$ and then for the distance $x(t)$ and show that

$$v(t) = (E_y/B_z)[1 - e^{-t/\tau}] + v(0)e^{-t/\tau}$$
$$x(t) = (E_y/B_z)[t + \tau e^{-t/\tau} - \tau] + x(0)$$

where $\tau = \rho_m/\sigma B_z^2$ and has units of sec. For this simplified plasma model of an MPD accelerator, calculate the distance needed to accelerate the plasma from rest up to $v = 0.01(E/B)$ and the time involved. Take $\sigma = 100$ mho/m, $B_z = 10^{-3}$ tesla (Wb/m²), $\rho_m = 10^{-3}$ kg/m³, and $E_y = 1000$ V/m.
Answer: 503 m, 0.1005 sec

9. Assume that a materials breakthrough makes it possible to increase the operating temperature in the plenum chamber of an *electrothermal engine* from 3000 K to 4000 K. Nitrogen gas is the propellant which is available from tanks at 250 K. Neglecting dissociation, and taking $\alpha = 200$ W/kg and $\dot{m} = 3 \times 10^{-4}$ kg/sec, calculate the old and new Δu corresponding to the two temperatures. Operating or thrust time is 10 days, payload mass is 1000 kg, and $k = 1.3$ for the hot diatomic molecule.
Answer: 697 m/sec old, 815 new.

10. An arcjet delivers 0.26 N of thrust. Calculate the vehicle velocity increase under gravitationless, dragless flight for a 28-day thrust duration with a payload mass of 100 kg. Take thruster efficiency as 50%, specific impulse as 2600 sec, and specific power as 200 W/kg. This is not an optimum payload fraction; estimate an I_s which would maximize the payload fraction with all other factors remaining the same.
Answer: $\Delta u = 4.34 \times 10^3$ m/sec; $I_s = 2020$ sec (decrease).

SYMBOLS

a	acceleration, m/sec² (ft/sec²)
A	area, cm² or m²
B	magnetic flux density, Web/m² or tesla
c_p	specific heat, J/kg-K
d	accelerator grid spacing, cm (in.)
D	hole or beam diameter, cm (in.)
e	electronic charge, 1.602×10^{-19} coulomb
E	electric field, V/m

f	microscopic force on a particle
F	thrust force, N or mN (lbf or mlbf)
$\tilde{F}_x$	accelerating force density inside channel, N/m^3 (lbf/ft^3)
g_0	constant converting propellant ejection velocity units to sec, 9.81 m/sec^2 or 32.2 ft/sec^2
I	total current, A
I_s	specific impulse, sec [I_s^* optimum]
j	current density, A/m^2
j_x, j_y	orthogonal current density components
j_θ	Hall current density, A/m^2
k	specific heat ratio
l_I	ionization loss, W
m_p	propellant mass, kg (lbm)
m_{pp}	power plant mass, kg (lbm)
m_{pl}	payload mass, kg (lbm)
m_0	initial total vehicle mass, kg (lbm)
$\dot{m}$	mass flow rate, kg/sec (lbm/sec)
$\mathfrak{M}$	atomic or molecular mass, kg/kg-mol (lbm/lb-mol)
n_e	electron number density, m^{-3} (ft^{-3})
P	power, W
P_e	electrical power, W
P_{jet}	kinetic power of jet, W
R	plasma resistance, ohms
s	distance, cm (in.)
t	time, sec
t_p	propulsive time, sec [t_p^* optimum]
T	absolute temperature, K (R)
Δu	vehicle velocity change, m/sec (ft/sec)
v	propellant ejection velocity, m/sec (ft/sec)
v_x	plasma velocity along accelerator, m/sec
v_c	characteristic speed
V	voltage, V
V_{acc}	accelerator voltage, V
x	linear dimension, m (ft)

Greek Letters

α	specific power, W/kg (W/lbm)
β	electron Hall parameter (dimensionless)
ε_0	permittivity of free space, 8.85×10^{-12} farad/m
ε_I	ionization energy, eV
η_H	Hall thruster β-efficiency
η_t	thruster efficiency
μ	ion mass, kg
μ_e	electron mass, 9.11×10^{-31} kg

ν_e charge particle velocity, m/sec
ξ Hall thruster local efficiency parameter
ρ_e space charge, coulomb/m^3
σ plasma electrical conductivity, mho/m
τ mean collision time, sec (also characteristic time, sec)
φ propellant mass tankage allowance
ω electron cyclotron frequency, $(\text{sec})^{-1}$

REFERENCES

19–1. R. G. Jahn, *Physics of Electric Propulsion*, McGraw-Hill Book Company, New York, 1968, pp. 103–110.

19–2. E. Stuhlinger, *Ion Propulsion for Space Flight*, McGraw-Hill Book Company, New York, 1964.

19–3. P. J. Turchi, "Electric Rocket Propulsion Systems," Chapter 9 in R. W. Humble, G. N. Henry, and W. J. Larson (Eds.), *Space Propulsion Analysis and Design*, McGraw-Hill, New York, 1995, pp. 509–598.

19–4. A. Spitzer, "Near Optimal Transfer Orbit Trajectory Using Electric Propulsion," *AAS Paper 95-215*, American Astronautical Society Meeting, Albuquerque, NM, 13–16 February 1995.

19–5. D. B. Langmuir, "Low-Thrust Flight: Constant Exhaust Velocity in Field-Free Space," in H. Seifert (Ed.), *Space Technology*, John Wiley & Sons, New York, 1959, Chapter 9.

19–6. C. D. Brown, *Spacecraft Propulsion*, AIAA Education Series, Washington, DC, 1996.

19–7. F. F. Chen, *Introduction to Plasma Physics*, Plenum Press, New York, 1974.

19–8. G. W. Sutton and A. Sherman, *Engineering Magnetohydrodynamics*, McGraw-Hill Book Company, New York, 1965.

19–9. D. M. Zube, P. G. Lichon, D. Cohen, D. A. Lichtin, J. A. Bailey, and N. V. Chilelli, "Initial On-Orbit Performance of Hydrazine Arcjets on A2100 Satellites," *AIAA Paper 99-2272*, June 1999.

19–10. T. Randolph, "Overview of Major U.S. Industrial Programs in Electric Propulsion," *AIAA Paper 99-2160*, June 1999.

19–11. R. L. Sackeim and D. C. Byers, "Status and Issues Related to In-Space Propulsion Systems," *Journal of Propulsion and Power*, Vol. 14, No. 5, September–October 1998.

19–12. J. R. Beattie, "XIPS Keeps Satellites on Track," *The Industrial Physicist*, Vol. 4, No. 2, June 1998.

19–13. J. Wang, D. Brinza, R. Goldstein, J. Polk, M. Henry, D. T. Young, J. J. Hanley, J. Nodholt, D. Lawrence, and M. Shappirio, "Deep Space One Investigations of Ion Propulsion Plasma Interactions: Overview and Initial Results," *AIAA Paper 99-2971*, June 1999.

19–14. P. G. Hill and C. R. Peterson, *Mechanics and Thermodynamics of Propulsion*, Addison-Wesley Publishing Company, Reading, MA, 1992.

19–15. E. L. Resler, Jr., and W. R. Sears, "Prospects of Magneto-Aerodynamics," *Journal of Aeronautical Sciences*, Vol. 24, No. 4, April 1958, pp. 235–246.

19–16. M. Martinez-Sanchez and J. E. Pollard, "Spacecraft Electric Propulsion—An Overview," *Journal of Propulsion and Power*, Vol. 14, No. 5, September–October 1998, pp. 688–699.

19–17. C. H. McLean, J. B. McVey, and D. T. Schappell, "Testing of a U.S.-built HET System for Orbit Transfer Applications," *AIAA Paper 99-2574*, June 1999.

19–18. V. Kim, "Main Physical Features and Processes Determining the Performance of Stationary Plasma Thrusters," *Journal of Propulsion and Power*, Vol. 14, No. 5, September–October 1998, pp. 736–743.

19–19. D. Baker, "Mission Design Case Study," in R. W. Humble, G. N. Henry, and W. J. Larson (Eds.), *Space Propulsion Analysis and Design*, McGraw-Hill, New York, 1995, Chapter 10.

19–20. J. J. DeBellis, "Optimization Procedure for Electric Propulsion Engines," MS thesis, Naval Postgraduate School, Monterey, CA, December 1999, 75 pages.

19–21. J. D. Filliben, "Electric Propulsion for Spacecraft Applications," *Chemical Propulsion Information Agency Report CPTR 96-64*, The Johns Hopkins University, December 1996.

19–22. M. A. Kurtz, H. L. Kurtz, and H. O. Schrade, "Optimization of Electric Propulsion Systems Considering Specific Power as a Function of Specific Impulse," *Journal of Propulsion and Power*, Vol. 4, No. 2, 1988, pp. 512–519.

CHAPTER 20

ROCKET TESTING

20.1. TYPES OF TESTS

Before rocket propulsion systems are put into operational use, they are subjected to several different types of tests, some of which are outlined below in the sequence in which they are normally performed.

1. Manufacturing inspection and fabrication tests on individual parts (dimensional inspection, pressure tests, x-rays, leak checks, electric continuity, electromechanical checks, etc.).

2. Component tests (functional and operational tests on igniters, valves, thrusters, controls, injectors, structures, etc.).

3. Static rocket system tests (with complete propulsion system on test stand): (*a*) partial or simulated rocket operation (for proper function, calibration, ignition, operation—often without establishing full thrust or operating for the full duration); (*b*) complete propulsion system tests (under rated conditions, off-design conditions, with intentional variations in environment or calibration). For a reusable or restartable rocket propulsion system this can include many starts, long-duration endurance tests, and postoperational inspections and reconditioning.

4. Static vehicle tests (when rocket propulsion system is installed in a restrained, nonflying vehicle or stage).

5. Flight tests: (*a*) with a specially instrumented propulsion system in a developmental flight test vehicle; (*b*) with a production vehicle.

Each of these five types of tests can be performed on at least three basic types of programs:

1. Research on and development or improvement of a new (or modified) rocket engine or motor or their propellants or components.
2. Evaluation of the suitability of a new (or modified) rocket engine or motor for a specified application or for flight readiness.
3. Production and quality assurance of a rocket propulsion system.

The first two types of programs are concerned with a novel or modified device and often involve the testing and measurement of new concepts or phenomena using experimental rockets. The testing of a new solid propellant grain, the development of a novel control valve assembly, and the measurement of the thermal expansion of a nozzle exhaust cone during firing operation are examples.

Production tests concern themselves with the measurement of a few basic parameters on production propulsion systems to assure that the performance, reliability, and operation are within specified tolerance limits. If the number of units is large, the test equipment and instrumentation used for these tests are usually partly or fully automated and designed to permit the testing, measurement, recording, and evaluation in a minimum amount of time.

During the early development phases of a program, many special and unusual tests are performed on components and complete rockets to prove specific design features and performance characteristics. Special facilities and instrumentation or modification of existing test equipment are used. During the second type of program, some special tests are usually conducted to determine the statistical performance and reliability of a rocket device by operating a number of units of the same design. During this phase tests are also made to demonstrate the ability of the rocket to withstand extreme limits of the operating conditions, such as high and low ambient temperature, variations in fuel composition, changes in the vibration environment, or exposure to moisture, rain, vacuum, or rough handling during storage. To demonstrate safety, sometimes, intentional malfunctions, spurious signals, or manufacturing flaws are introduced into the propulsion system, to determine the capability of the control system or the safety devices to handle and prevent a potential failure.

Before an experimental rocket can be flown in a vehicle it usually has to pass a set of *preliminary flight rating tests* aimed at demonstrating the rocket's safety, reliability, and performance. It is not a single test, but a series of tests under various specified conditions operating limits, and performance tolerances, simulated environments, and intentional malfunctions. Thereafter the rocket may be used in experimental flights. However, before it can be put into production, it usually has to pass another specified series of tests under a variety of rigorous specified conditions, known as the *qualification test* or *preproduction test*. Once a particular propulsion system has been *qualified*, or passed a qualification test, it is usually forbidden to make any changes in design, fabrication processes, or materials without going through a careful review, extensive documentation, and often also a requalification test.

The amount and expense of testing of components and complete propulsion systems has decreased greatly in the last few decades. The reasons are more experience with prior similar systems and more confidence in predicting a number of failure modes and their locations. Validated computer programs have removed many uncertainties and obviated needs for tests. In some applications the number of firing tests has decreased by a factor of 10 or more.

20.2. TEST FACILITIES AND SAFEGUARDS

For chemical rocket propulsion systems, each test facility usually has the following major systems or components:

1. A test cell or test bay where the article to be tested is mounted, usually in a special test fixture. If the test is hazardous, the test facility must have provisions to protect operating personnel and to limit damage in case of an accident.

2. An instrumentation system with associated computers for sensing, maintaining, measuring, analyzing, correcting, and recording various physical and chemical parameters. It usually includes calibration systems and timers to accurately synchronize the measurements.

3. A control system for starting, stopping, and changing the operating conditions.

4. Systems for handling heavy or awkward assemblies, supplying liquid propellant, and providing maintenance, security, and safety.

5. For highly toxic propellants and toxic plume gases it has been required to capture the hazardous gas or vapor (firing inside a closed duct system), remove almost all of the hazardous ingredients (e.g., by wet scrubbing and/or chemical treatment), allow the release of the nontoxic portion of the cleaned gases, and safely dispose of any toxic solid or liquid residues from the chemical treatment. With an exhaust gas containing fluorine, for example, the removal of much of this toxic gas can be achieved by scrubbing it with water that contains dissolved calcium; it will then form calcium fluoride, which can be precipitated and removed.

6. In some tests specialized test equipment and unique facilities are needed to conduct static testing under different environmental conditions or under simulated emergency conditions. For example, high and low ambient temperature tests of large motors may require a temperature-controlled enclosure around the motor; a rugged explosion-resistant facility is needed for bullet impact tests of propellant-loaded missile systems and also for cook-off tests, where gasoline or rocket fuel is burned with air below a stored missile. Similarly, special equipment is needed for

vibration testing, measuring thrust vector forces and moments in three dimensions, or determining total impulse for very short pulse durations at low thrust.

Most rocket propulsion testing is now accomplished in sophisticated facilities under closely controlled conditions. Modern rocket test facilities are frequently located several miles from the nearest community to prevent or minimize effects of excessive noise, vibrations, explosions, and toxic exhaust clouds. Figure 20–1 shows one type of an open-air test stand for vertically down-firing large liquid propellant thrust chambers (100,000 to 2 million pounds thrust). It is best to fire the propulsion system in a direction (vertical

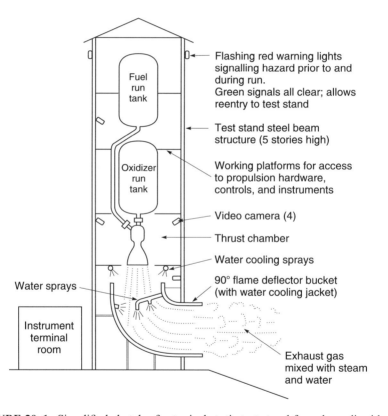

FIGURE 20–1. Simplified sketch of a typical static test stand for a large liquid propellant thrust chamber firing vertically downward. Only a small part of the exhaust plume (between the nozzle exit and flame bucket entrance) is visible. The flame bucket turns the exhaust gas plume by 90° (horizontal) and prevents the flame from digging a hole in the ground. Not shown here are cranes, equipment for installing or removing a thrust chamber, safety railings, high pressure gas tank, the propellant tank pressurization system, separate storage tanks for fuel, oxidizer, or cooling water with their feed systems, or a small workshop.

or horizontal) similar to the actual flight condition. Figure 20–2 shows a simulated altitude test facility for rockets of about 10.5 metric tons thrust force (46,000 lbf). It requires a vacuum chamber in which to mount the engine, a set of steam ejectors to create a vacuum, water to reduce the gas temperature, and a cooled diffuser. With the flow of chemical rocket propellant combustion gases it is impossible to maintain a high vacuum in these kinds of facilities; typically, between 15 to 4 torr (20 to 35 km altitude) can be maintained. This type of test facility allows the operation of rocket propulsion systems with high-nozzle-area ratios that would normally experience flow separation at sea-level ambient pressures.

Prior to performing any test, it is common practice to train the test crew and go through repeated dry runs, to familiarize each person with his or her responsibilities and procedures, including the emergency procedures.

Typical personnel and plant security or safety provisions in a modern test facility include the following:

1. Concrete-walled blockhouse or control stations for the protection of personnel and instruments (see Fig. 20–3) remote from the actual rocket propulsion location.

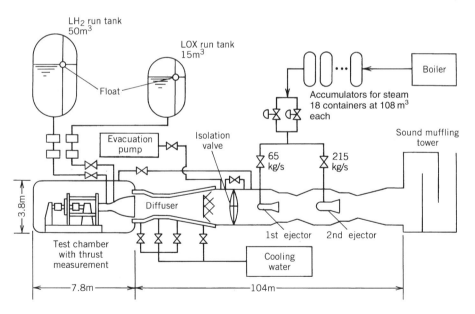

FIGURE 20–2. Simplified diagram of a simulated altitude, horizontal firing test facility for the LE-5 Japanese-designed thrust chamber (liquid oxygen–liquid hydrogen propellants) showing the method of creating a vacuum (6 torr during operation and 13 torr prior to start). The operating duration is limited to about 10 min by the capacity of the steam storage. (Reproduced from Ref. 20–1 with permission of the AIAA.)

FIGURE 20–3. Control room (inside a reinforced concrete blockhouse) for test operators, instrument recorders, and controls. Note the control console, closed-circuit television, radio and telephone, direct read-out meters, strip charts, high-speed tape recorders, oscilloscope, air-quality alarm, and emergency lights. (Courtesy of U.S. Air Force Phillips Laboratory.)

2. Remote control, indication, and recording of all hazardous operations and measurements; isolation of propellants from the instrumentation and control room.

3. Automatic or manual water deluge and fire-extinguishing systems.

4. Closed circuit television systems for remotely viewing the test.

5. Warning signals (siren, bells, horns, lights, speakers) to notify personnel to clear the test area prior to a test, and an all-clear signal when the conditions are no longer hazardous.

6. Quantity and distance restrictions on liquid propellant tankage and solid propellant storage to minimize damage in the event of explosions; separation of liquid fuels and oxidizers.

7. Barricades around hazardous test articles to reduce shrapnel damage in the event of a blast.

8. Explosion-proof electrical systems, spark-proof shoes, and nonspark hand tools to prvent ignition of flammable materials.

9. For certain propellants also safety clothing (see Fig. 20–4), including propellant- and fire-resistant suits, face masks and shields, gloves, special shoes, and hard hats.

FIGURE 20–4. Plastic safety suit, gloves, boots, and hood used by test personnel in handling hazardous or corrosive liquid propellants. Safety shower, which starts automatically when a person steps onto the platform, washes away splashed or spilled propellant. (Official U.S. Air Force photograph.)

10. Rigid enforcement of rules governing area access, smoking, safety inspections, and so forth.
11. Limitations on the number of personnel that may be in a hazardous area at any time.

Monitoring and Control of Toxic Materials

Open-air testing of chemical rockets frequently requires measurement and control of exhaust cloud concentrations and gas movement in the surrounding areas for safeguarding personnel, animals, and plants. A toxic cloud of gas and particles can result from the exhaust gas of normal rocket operation, vapors or reaction gases from unintentional propellant spills, and gases from fires, explosions, or from the intentional destruction of vehicles in flight or rockets on the launch stand. Environmental regulations usually limit the maximum local concentration or the total quantity of toxic gas or particulates released to the atmosphere. The toxic nature of some of these liquids, vapors, and gases has been mentioned in Chapters 7 and 12. One method of control is for tests with discharges of moderately toxic gases or products to be postponed until favorable weather conditions are present.

In ground tests, the toxic cloud source is treated as a *point source*, and in flight tests it is a *ribbon source*. The rate of exhaust cloud diffusion is influenced by many propulsion variables, including propellant, rocket size, exhaust temperature, and thrust duration; by many atmospheric variables, including wind velocity, direction, turbulence, humidity, and vertical stability or lapse rate, and by the surrounding terrain. Extensive analytical studies and measurements of the environmental exposure from explosions, industrial smoke, and gases, and exhausts from missile and space vehicle launchings give background useful for predicting the atmospheric diffusion and downwind concentrations of rocket exhaust clouds. Reference 20–2 describes hazards and toxic gas cloud dispersals and concentrations. Reference 20–3 evaluates the environmental impact of rocket exhausts from large units on the ozone in the stratosphere and on the ground weather near the test site; it concludes that the impacts are generally small and temporary. Reference 20–4 describes a test-area atmospheric measuring network.

A widely used relationship for predicting atmospheric diffusion of gas clouds has been formulated by O. G. Sutton (Ref. 20–5). Many of the most modern equations and models relating to downwind concentrations of toxic clouds are extensions of Sutton's theory. Given below are the Sutton equations of primary interest to rocket and missile operators.

For instantaneous ground-level point source nonisotropic conditions,

$$\chi_{(x,y,z,t)} = \frac{Q}{\pi^{3/2} C_x C_y C_z (\bar{u}t)^{3(2-n)/2}} \exp\left[(\bar{u}t)^{n-2} \left(\frac{x^2}{C_x^2} + \frac{y^2}{C_y^2} + \frac{z^2}{C_z^2} \right) \right] \qquad (20\text{--}1)$$

For continuous ground-level point source nonisotropic conditions,

$$\chi = \frac{2Q}{\pi C_y C_z \bar{u} x^{2-n}} \exp\left[-x^{n-2} \left(\frac{y^2}{C_y^2} + \frac{z^2}{C_z^2} \right) \right] \qquad (20\text{--}2)$$

where χ is the concentration in grams per cubic meter, Q is the source strength (grams for intantaneous, grams per second for continuous); $C_{x,y,z}$ are diffusion coefficients in the x, y, z planes, respectively; $\bar{u}$ is the average wind velocity in meters per second, t is the time in seconds, and the coordinates x, y, z are in meters measured from the center of the moving cloud in the instantaneous case and from a ground point beneath the plume axis in the continuous case. The exponent n is a stability or turbulence coefficient, ranging from almost zero for highly turbulent conditions to 1 as a limit for extremely stable conditions, and usually falling between 0.10 and 0.50.

A few definitions basic to the study of atmospheric diffusion of exhaust clouds are as follows:

1. *Micrometeorology.* Study and forecasting of atmospheric phenomena restricted to a region approximately 300 m above the earth's surface and a horizontal distance of approximately 5 miles.

2. *Lapse Rate.* The rate of decrease in temperature with increasing height above the earth's surface. The United States Standard Atmosphere has a lapse rate of about 6.4°C per 1000 m. Lapse rate is also affected by altitude, wind, and humidity.

3. *Inversion, or Inversion Layer.* Condition of negative lapse rate (temperature increases with increasing height). Usually formed near the ground at night.

The following are a few general rules and observations derived from experience with the atmospheric diffusion of rocket exhaust clouds:

1. Inversion presents a very stable layer and greatly reduces the vertical dispersion (the higher the lapse rate, the greater the vertical dispersion).

2. A highly stable atmospheric condition tends to keep the exhaust plume or cloud intact and away from the earth's surface except when the exhaust products are much heavier than the surrounding air.

3. High wind increases the rate of diffusion and reduces the thermal effects.

4. For short firings (< 500 sec) the approximate dosages downwind are about the same as from an instantaneous point source.

5. When the plume reaches about one-fourth the distance to a given point before emission is stopped, peak concentration will be about three-fourths of that from a continuous source of equal strength.

6. The presence of an inversion layer significantly restricts the mixing or diffusion capacity of the atmosphere in that the effective air mass is that mass existing between the earth's surface and the inversion layer.

7. Penetration of the inversion layer due to the buoyance force of the hot exhaust cloud seldom occurs.

8. Earth surface dosage drops rapidly when missiles or space launch vehicles are destroyed in flight above a height of 1500 m as compared to lower altitudes of 600 to 1000 m.

Interpretation of the hazard that exists once the concentration of the toxic agent is known also requires knowledge of its effects on the human body, plants, and animals. Tolerance limits for humans are given in Chapter 7 and in Ref. 8–5. There are usually three limits of interest: one for the short-time exposure of the general public, one for an 8-hr exposure limit, and an evacuation concentration. Depending on the toxic chemical, the 8-hr limit may vary from 5000 ppm for a gas such as carbon dioxide, to less than 1 ppm for an extremely toxic substance such as fluorine. Poisoning of the human body by exhaust products usually occurs from inhalation of the gases and fine solid particles, but the solid residuals that sometimes remain around a test facility for weeks or months following a test firing can enter the body through cuts and other avenues. Also, certain liquid propellants cause burns and skin rash or are poisonous when ingested, as explained in Chapter 7.

20.3. INSTRUMENTATION AND DATA MANAGEMENT

This section gives only a very brief discussion of this subject. For further study the reader is referred to standard textbooks on instruments and computers used in testing, such as Ref. 20–6. Some of the physical quantities measured in rocket testing are as follows:

1. Forces (thrust, thrust vector control side forces, short thrust pulses).
2. Flows (hot and cold gases, liquid fuel, liquid oxidizer, leakage).
3. Pressures (chamber, propellant, pump, tank, etc.).
4. Temperatures (chamber walls, propellant, structure, nozzle).
5. Timing and command sequencing of valves, switches, igniters, etc.
6. Stresses, strains, and vibrations (combustion chamber, structures, propellant lines, accelerations of vibrating parts) (Ref. 20–7).
7. Time sequence of events (ignition, attainments of full pressure).
8. Movement and position of parts (valve stems, gimbal position, deflection of parts under load or heat).
9. Voltages, frequencies, and currents in electrical or control subsystems.
10. Visual observations (flame configuration, test article failures, explosions) using high-speed cameras or video cameras.
11. Special quantities such as turbopump shaft speed, liquid levels in propellant tanks, burning rates, flame luminosity, or exhaust gas composition.

Reference 20–8 gives a description of specialized diagnostic techniques used in propulsion systems, such as using nonintrusive optical methods, microwaves, and ultrasound for measurements of temperatures, velocities, particle sizes, or burn rates in solid propellant grains. Many of these sensors incorporate specialized technologies and, often, unique software. Each of the measured parameters can be obtained by different types of instruments, sensors, and analyzers, as indicated in Ref. 20–9.

Measurement System Terminology

Each measurement or each measuring system usually requires one or more *sensing elements* (often called transducers or pickups), a device for *recording, displaying,* and/or *indicating* the sensed information, and often also another device for *conditioning, amplifying, correcting,* or *transforming* the sensed signal into the form suitable for *recording, indicating, display,* or *analysis.* Recording of rocket test data has been performed in several ways, such as on *chart recorders* or in digital form on *memory* devices, such as on magnetic tapes or disks. Definitions of several significant terms are given below and in Ref. 20–6.

Range refers to the region extending from the minimum to the maximum rated value over which the measurement system will give a true and linear response. Usually an additional margin is provided to permit temporary overloads without damage to the instrument or need for recalibration.

Errors in measurements are usually of two types: (1) *human errors* of improperly reading the instrument, chart, or record and of improperly interpreting or correcting these data, and (2) *instrument or system errors*, which usually fall into four classifications: static errors, dynamic response errors, drift errors, and hysteresis errors (see Ref. 20–10). *Static errors* are usually fixed errors due to fabrication and installation variations; these errors can usually be detected by careful calibration, and an appropriate correction can then be applied to the reading. *Drift error* is the change in output over a period of time, usually caused by random wander and environmental conditions. To avoid drift error the measuring system has to be calibrated at frequent intervals at standard environmental conditions against known standard reference values over its whole range. *Dynamic response errors* occur when the measuring system fails to register the true value of the measured quantity while this quantity is changing, particularly when it is changing rapidly. For example, the thrust force has a dynamic component due to vibrations, combustion oscillations, interactions with the support structure, etc. These dynamic changes can distort or amplify the thrust reading unless the test strand structure, the rocket mounting structure, and the thrust measuring and recording system are properly designed to avoid harmonic excitation or excessive energy damping. To obtain a good dynamic response requires a careful analysis and design of the total system.

A *maximum frequency response* refers to the maximum frequency (usually in cycles per second) at which the instrument system will measure true values. The natural frequency of the measuring system is usually above the limiting response frequency. Generally, a high-frequency response requires more complex and expensive instrumentation. All of the instrument system (sensing elements, modulators, and recorders) must be capable of a fast response. Most of the measurements in rocket testing are made with one of two types of instruments: those made under nearly steady static conditions, where only relatively gradual changes in the quantities occur, and those made with fast transient conditions, such as rocket starting, stopping, or vibrations (see Ref. 20–11). This latter type of instrument has frequency responses above 200 Hz, sometimes as high as 20,000 Hz. These fast measurements are necessary to evaluate the physical phenomena of rapid transients.

Linearity of the instrument refers to the ratio of the input (usually pressure, temperature, force, etc.) to the output (usually voltage, output display change, etc.) over the range of the instrument. Very often the static calibration error indicates a deviation from a truly linear response. A nonlinear response can cause appreciable errors in dynamic measurements. *Resolution* refers to the minimum change in the measured quantity that can be detected with a given instrument. *Dead zone* or *hysteresis* errors are often caused by energy absorption within the instrument system or play in the instrument mechanism; in part, they limit the resolution of the instrument.

Sensitivity refers to the change in response or reading caused by special influences. For example, the *temperature sensitivity* and the *acceleration sensitivity* refer to the change in measured value caused by temperature and acceleration. These are usually expressed in percent change of measured value per unit of temperature or acceleration. This information can serve to correct readings to reference or standard conditions.

Errors in measurement can arise from many sources. Reference 20–12 gives a standardized method, including mathematical models, for estimating the error, component by component, as well as the cumulative effect in the instrumentation and recording systems. Graphic recordings (error ranges ±0.2 to ±0.5% of strip chart span) and oscillographs (error ranges ±2.0 to ±3.0% of full scale), two of the analog-type recording devices, are used for giving quick-look data and to record high-frequency data or transient conditions; these transients are beyond the capability of digital recorders, which are usually limited to 100 Hz or lower as compared to 5000 Hz or higher for oscillographs.

Electrical interference or "noise" within an instrumentation system, including the power supply, transmission lines, amplifiers, and recorders, can affect the accuracy of the recorded data, especially when low-output transducers are in use. Methods for measuring and eliminating objectionable electrical noise are given in Ref. 20–13.

Use of Computers

Computers have become commonplace in the testing and handling of data in rocket propulsion. They are usually coupled with *sensors* (e.g., pressure transducers, actuator position indicators, temperature sensors, liquid level gauges, etc.), which provide the data inputs, with *controllers* (valve actuators, thrust vector controllers, thrust termination devices), which receive commands resulting from the computer outputs causing a change in the sensed quantity, and with *auxiliaries* such as terminals, data storage devices, or printers. Computers are used in one or more of the following ways:

1. The *analysis of test data* becomes a time-consuming difficult job without computers, simply because of the huge volume of data that is generated in many typical rocket propulsion system tests. All the pertinent data need to be reviewed and evaluated. The computer will permit *automated data reduction*, including *data correction* (e.g., for known instrument error, calibration, or changes in atmospheric pressure), *conversion* of analog data into digital form, and *filtering* of data to eliminate signals

outside the range of interest. It can also include *data manipulation* to put the test information into graphic displays or summary hard-copy read-outs of selected, specific performance parameters.

On the basis of a careful evaluation of the test data the responsible engineers have to decide whether the test objectives were met and what changes to make or what objectives to set for the next test or the remainder of the current test. Reference 20–14 describes a software system that allows *automated test analysis* and *decision support* in evaluating the 50 million bytes of test data that are generated in a typical SSME test; it is based in part on the use of an expert knowledge system.

2. Modern testing systems use digital data bases for *recording and documenting* test records. Often only a portion of the recorded data is actually analyzed and reviewed during or after the test. In complex rocket propulsion system tests, sometimes between 100 or 400 different instrument measurements are made and recorded. Some data need to be sampled frequently (e.g., some transients may be sampled at rates higher than 1000 times per second), whereas other data need to be taken at lower frequencies (e.g., temperature of mounting structure may be needed only every 1 to 10 sec). Multiplexing of data is commonly practiced to simplify data transmission. Most rocket test computer systems contain a configuration file to indicate data characteristics for each channel, such as range, gain, the references, the type of averaging, the parameter characteristics, or the data correction algorithms. Most of the data are not analyzed or printed out as hard copy; a detailed analysis occurs only if there is reason for understanding particular test events in more detail. This analysis may occur months after the actual tests and may not even be done on the same computer.

3. *Sensing and evaluating failures* or *overlimit conditions* (excessive local temperature, vibration, or limiting local pressure) is aimed at detecting an impending malfunction and at deciding whether it is a serious problem. If serious, it can cause either an automatic correction or an automatic and safe shutdown of operation. Sensing of undesirable operating conditions can be accomplished much more rapidly on a computer than would be possible if a human operator were in the control loop. In some engine designs a critical failure is sensed by several sensors and the computer rapidly evaluates the signals from these sensors and causes a correction (or shutdown) only if the majority of sensors indicate an unsafe or undesirable condition, thus eliminating the occasional failure of an individual sensor as a cause for shutdown.

4. *Simulation* of tests can be accomplished by devising algorithms that allow a computer to respond in a manner similar to a rocket propulsion unit. The computer receives inputs from various sensors (valve position, thrust vector control position, unsafe temperatures, etc.), processes the data in a simulation algorithm, and then provides output of control signals (e.g.,

thrust change, shutdown) and also of simulated rocket performance (e.g., chamber pressure, specific impulse, side force, etc.). This computer simulation can be very economical compared to running additional tests. This can be a full off-line simulation (in a separate computer with simulated inputs) or a partial on-line simulation where the computer is coupled to an actual rocket engine or its components; this second type can be used to check out an engine just prior to, or in the first second of, a test run or test flight.

5. *Control of test operation* by computer allows the attainment of the desired test conditions in a minimum amount of time. This could entail a pre-programmed set of pulses for an attitude control thruster, a desired set of different mixture ratios to be achieved for a short time (say, 1 sec each) in a single test, or a planned variation of thrust vector control conditions. It can provide a closed loop control to attain desired operating conditions, including the paths along which these conditions should be achieved. It also makes it possible to control several variables at the same time (e.g., thrust, mixture ratio, and several turbine inlet temperatures). For some component tests programmable logic controllers are used to control the test operation instead of a computer, which usually requires some software development.

In a multiple-static-test facility there can be a group of network-connected computers and databases to achieve some or all of the functions above. Some of the computer hardware would be part of the test article, some part of the test facility, and some can be located remotely and linked by a communications network. Reference 20–15 describes the engine control and computer system for the Space Shuttle main engine.

20.4. FLIGHT TESTING

Flight testing of rocket propulsion systems is always conducted in conjunction with tests of vehicles and other systems such as guidance, vehicle controls, or ground support. These flights usually occur along missile and space launch ranges, sometimes over the ocean. If a flight test vehicle deviates from its intended path and appears to be headed for a populated area, a range safety official (or a computer) will have to either cause a destruction of the vehicle, abort the flight, or cause it to correct its course. Many propulsion systems therefore include devices that will either terminate the operation (shut off the rocket engine or open thrust termination openings into rocket motor cases as described in Chapter 13) or trigger explosive devices that will cause the vehicle (and therefore also the propulsion system) to disintegrate in flight.

Flight testing requires special launch support equipment, means for observing, monitoring, and recording data (cameras, radar, telemetering, etc.), equipment for assuring range safety and for reducing data and evaluating flight test performance, and specially trained personnel. Different launch equipment is needed for different kinds of vehicles. This includes launch tubes for shoulder-held infantry support missile launchers, movable turret-type mounted multiple launchers installed on an army truck or a navy ship, a transporter for larger missiles, and a track-propelled launch platform or fixed complex launch pads for spacecraft launch vehicles. The launch equipment has to have provisions for loading or placing the vehicle into a launch position, for allowing access of various equipment and connections to launch support equipment (checkout, monitoring, fueling, etc.), for aligning or aiming the vehicle, or for withstanding the exposure to the hot rocket plume at launch.

During experimental flights extensive measurements are often made on the behavior of the various vehicle subsystems; for example, rocket propulsion parameters, such as chamber pressure, feed pressures, temperatures, and so on, are measured and the data are telemetered and transmitted to a ground receiving station for recording and monitoring. Some flight tests rely on salvaging and examining the test vehicle.

20.5. POSTACCIDENT PROCEDURES

In the testing of any rocket propulsion system there will invariably be failures, particularly when some of the operating parameters are close to their limit. With each failure comes an opportunity to learn more about the design, the materials, the propulsion performance, the fabrication methods, or the test procedures. A careful and thorough investigation of each failure is needed to learn the likely causes and identify the remedies or fixes to prevent a similar failure in the future. The lessons to be learned from these failures are perhaps the most important benefits of testing. A formalized postaccident approach is often used, particularly if the failure had a major impact, such as high cost, major damage, or personnel injury. A major failure (e.g., the loss of a space launch vehicle or severe damage to a test facility) often causes the program to be stopped and further testing or flights put on hold until the cause of the failure is determined and remedial action has been taken to prevent a recurrence.

Of utmost concern immediately after a major failure are the steps that need to be taken to respond to the emergency. This includes giving first aid to injured personnel, bringing the propulsion system and/or the test facilities to a safe, stable condition, limiting further damage from chemical hazards to the facility or the environment, working with local fire departments, medical or emergency maintenance staff or ambulance personnel, and debris clearing crews, and quickly providing factual statements to the management, the employees, the news media, and the public. It also includes controlling access to the facility where the failure has occurred and preserving evidence for the subsequent investigation. All test personnel, particularly the supervisory people, need to be trained not only in preventing accidents and minimizing the impact of a potential failure, but also how to best respond to the emergency. Reference 20–16 suggests postaccident procedures involving rocket propellants.

REFERENCES

20–1. K. Yanagawa, T. Fujita, H. Miyajima, and K. Kishimoto, "High Altitude Simulation Tests of LOX-LH2 Engine LE-5," *Journal of Propulsion and Power*, Vol. 1, No. 3, May–June 1985, pp. 180–186.

20–2. "Handbook for Estimating Toxic Fuel Hazards," *NASA Report CR-61326*, April 1970.

20–3. R. R. Bennett and A. J. McDonald, "Recent Activities and Studies on the Environmental Impact of Rocket Effluents," *AIAA Paper 98-3850*, July 1998.

20–4. R. J. Grosch, "Micro-Meteorological System," *Report TR-68-37*, Air Force Rocket Propulsion Laboratory, November 1968 (AD 678856).

20–5. O. G. Sutton, *Micrometeorology*, McGraw-Hill Book Company, New York, 1973, Chapter 8.

20–6. D. Ramsey, *Principles of Engineering Instrumentation*, John Wiley & Sons, New York, 1996.

20–7. K. G. McConnell, *Vibration Testing: Theory and Practice*, Wiley Interscience, New York, 1995.

20–8. Y. M. Timnat, "Diagnostic Techniques for Propulsion Systems," *Progress in Aerospace Sciences*, Vol. 26, No. 2, 1989, pp. 153–168.

20–9. R. S. Figliola and D. B. Beasley, *Theory and Design for Mechanical Measurements*, John Wiley & Sons, New York, 1991, 516 pages.

20–10. R. Cerri, "Sources of Measurement Error in Instrumentation Systems," *Preprint 19-LA-61*, Instrument Society of America, Research Triangle Park, NC.

20–11. P. M. J. Hughes and E. Cerny, "Measurement and Analysis of High-Frequency Pressure Oscillations in Solid Rocket Motors," *Journal of Spacecraft and Rockets*, Vol. 21, No. 3, May–June 1984, pp. 261–265.

20–12. *Handbook for Estimating the Uncertainty in Measurements Made with Liquid Propellant Rocket Engine Systems*, Handbook 180, Chemical Propulsion Information Agency, April 30, 1969 (AD 855130).

20–13. "Grounding Techniques for the Minimization of Instrumentation Noise Problems," *Report TR-65-8*, Air Force Rocket Propulsion Laboratory, January 1965 (AD 458129).

20–14. R. C. Heim and K. J. Slusser, "The Measure of Engine Performance," *Threshold*, The Boeing Company, Rocketdyne Propulsion & Power, Summer 1994, pp. 40–48.

20–15. R. M. Mattox and J. B. White, "Space Shuttle Main Engine Controller," *NASA TP-1932*, 1981.

20–16. D. K. Shaver and R. L. Berkowitz, *Post-accident Procedures for Chemicals and Propellants*, Noyes Publications, Park Ridge, NJ, 1984.

APPENDIX 1

CONVERSION FACTORS AND CONSTANTS

Conversion Factors (arranged alphabetically)

Acceleration $(L\ t^{-2})^{*}$

$1\ \text{m/sec}^2 = 3.2808\ \text{ft/sec}^2 = 39.3701\ \text{in./sec}^2$
$1\ \text{ft/sec}^2 = 0.3048\ \text{m/sec}^2 = 12.0\ \text{in./sec}^2$
$g_0 = 9.80665\ \text{m/sec}^2 = 32.174\ \text{ft/sec}^2$ (standard gravity pull at earth's surface)

Area (L^2)

$1\ \text{ft}^2 = 144.0\ \text{in.}^2 = 0.092903\ \text{m}^2$
$1\ \text{m}^2 = 1550.0\ \text{in.}^2 = 10.7639\ \text{ft}^2$
$1\ \text{in.}^2 = 6.4516 \times 10^{-4}\ \text{m}^2$

Density $(M\ L^3)$

Specific gravity is dimensionless, but has the same numerical value as *density* expressed in g/cm^3 or kg/m^3
$1\ \text{kg/m}^3 = 6.24279 \times 10^{-2}\ \text{lbm/ft}^3 = 3.61273 \times 10^{-5}\ \text{lbm/in.}^3$
$1\ \text{lbm/ft}^3 = 16.0184\ \text{kg/m}^3$
$1\ \text{lbm/in.}^3 = 2.76799 \times 10^4\ \text{kg/m}^3$

---

*The letters in parentheses after each heading indicate the dimensional parameters (L = length, M = mass, t = time, and T = temperature).

Energy, also Work or Heat $(M\ L^2\ t^{-2})$

1.0 Btu = 1055.056 J (joule)
1.0 kW-hr = 3.60059×10^6 J
1.0 ft-lbf = 1.355817 J
1.0 cal = 4.1868 J
1.0 kcal = 4186.8 J

Force $(M\ L\ t^{-2})$

1.0 lbf = 4.448221 N
1 dyne = 10^{-5} N
1.0 kg (force) [used in Europe] = 9.80665 N
1.0 ton (force) [used in Europe] = 1000 kg (force)
1.0 N = 0.2248089 lbf
1.0 millinewton (mN) = 10^{-3} N
Weight is the *force* on a *mass* being accelerated by gravity (g_0 applies at the surface of the earth)

Length (L)

1 m = 3.2808 ft = 39.3701 in.
1 ft = 0.3048 m = 12.0 in.
1 in. = 2.540 cm = 0.0254 m
1 mile = 1.609344 km = 1609.344 m = 5280.0 ft
1 nautical mile = 1852.00 m
1 mil = 0.0000254 m = 1.00×10^{-3} in.
1 micron (μm) = 10^{-6} m
1 astronomical unit (au) = 1.49600×10^{11} m

Mass (M)

1 slug = 32.174 lbm
1 kg = 2.205 lbm = 1000 g
1 lbm = 16 ounces = 0.4536 kg

Power $(M\ L^2\ t^{-3})$

1 Btu/sec = 0.2924 W (watt)
1 J/sec = 1.0 W = 0.001 kW
1 cal/sec = 4.186 W
1 horsepower = 550 ft-lbf/sec = 745.6998 W
1 ft-lbf/sec = 1.35581 W

Pressure $(M\ L^{-1}\ t^{-2})$

1 bar = 10^5 N/m^2 = 0.10 MPa
1 atm = 0.101325 MPa = 14.696 psia

1 mm of mercury $= 13.3322$ N/m^2
1 MPa $= 10^6$ N/m^2
1 psi or lbf/in.^{2} $= 6894.757$ N/m^2

Speed (or linear velocity) (L t^{-1})

1 ft/sec $= 0.3048$ m/sec $= 12.00$ in./sec
1 m/sec $= 3.2808$ ft/sec $= 39.3701$ in./sec
1 knot $= 0.5144$ m/sec
1 mile/hr $= 0.4770$ m/sec

Specific Heat (L^2 t^{-2} T^{-1})

1 g-cal/g-°C $= 1$ kg-cal/kg-K $= 1$ Btu/lbm-°F $= 4.186$ J/g-°C $=$
1.163×10^{-3} kW-hr/kg-K

Temperature (T)

1 K $= 9/5$ R $= 1.80$ R
0°C $= 273.15$ K
0°F $= 459.67$ R
$C = (5/9)(F - 32)$ $F = (9/5)C + 32$

Time (t)

1 mean solar day $= 24$ hr $= 1440$ min $= 86,400$ sec
1 calendar year $= 365$ days $= 3.1536 \times 10^7$ sec

Viscosity (M L^{-1} t^{-1})

1 centistoke $= 1.00 \times 10^{-6}$ m^2/sec
1 centipoise $= 1.00 \times 10^{-3}$ kg/m sec
1 lbf-sec/ft^2 $= 47.88025$ kg/m sec

Constants

J	Mechanical equivalent of heat $= 4.186$ joule/cal $= 777.9$ ft-lbf/Btu $= 1055$ joule/Btu	
R'	Universal gas constant $= 8314.3$ J/kg-mole-K $=$ 1545 ft-lbf/lbm-mole-R	
V_{mole}	Molecular volume of an ideal gas $= 22.41$ liter/kg-mole at standard conditions	
e	Electron charge $= 1.6021176 \times 10^{-19}$ coulomb	
ε_0	Permittivity of vacuum $= 8.854187 \times 10^{-12}$ farad/m	
	Gravitational constant $= 6.673 \times 10^{-11}$ m^3/kg-sec	
	Boltzmann's constant	$1.38065003 \times 10^{-23}$ J/°K
	Electron mass	9.109381×10^{-31} kg
	Avogadro's number	6.022142×10^{26}/kg-mol
σ	Stefan–Boltzman constant	5.6696×10^{-8} W/m^2-K^{-4}

APPENDIX 2

PROPERTIES OF THE EARTH'S STANDARD ATMOSPHERE

Sea level pressure is 0.101325 MPa (or 14.696 psia or 1.000 atm).

Altitude (m)	Temperature (K)	Pressure Ratio	Density (kg/m^3)
0 (sea level)	288.150	1.0000	1.2250
1,000	281.651	8.8700×10^{-1}	1.1117
3,000	268.650	6.6919×10^{-1}	9.0912×10^{-1}
5,000	255.650	5.3313×10^{-1}	7.6312×10^{-1}
10,000	223.252	2.6151×10^{-1}	4.1351×10^{-1}
25,000	221.552	2.5158×10^{-2}	4.0084×10^{-2}
50,000	270.650	7.8735×10^{-4}	1.0269×10^{-3}
75,000	206.650	2.0408×10^{-5}	3.4861×10^{-5}
100,000	195.08	3.1593×10^{-7}	5.604×10^{-7}
130,000	469.27	1.2341×10^{-8}	8.152×10^{-9}
160,000	696.29	2.9997×10^{-9}	1.233×10^{-9}
200,000	845.56	8.3628×10^{-10}	2.541×10^{-10}
300,000	976.01	8.6557×10^{-11}	1.916×10^{-11}
400,000	995.83	1.4328×10^{-11}	2.803×10^{-12}
600,000	999.85	8.1056×10^{-13}	2.137×10^{-13}
1,000,000	1000.00	7.4155×10^{-14}	3.561×10^{-15}

Source: U.S. Standard Atmosphere, National Oceanic and Atmospheric Administration, National Aeronautics and Space Administration, and U.S. Air Force, Washington, DC, 1976 (NOAA-S/T-1562).

APPENDIX 3

SUMMARY OF KEY EQUATIONS FOR IDEAL CHEMICAL ROCKETS

Parameter	Equations	Equation Numbers
Average exhaust velocity, v_2 (m/sec or ft/sec) (assume that $v_1 = 0$)	$v_2 = c - (p_2 - p_3)A_2/\dot{m}$ When $p_2 = p_3$, $v_2 = c$ $v_2 = \sqrt{[2k/(k-1)]RT_1[1 - (p_2/p_1)^{(k-1)/k}]}$ $= \sqrt{2(h_1 - h_2)}$	2–16 3–16 3–15
Effective exhaust velocity, c (m/sec or ft/sec)	$c = c^* C_F = F/\dot{m} = I_s g_0$ $c = v_2 + (p_2 - p_3)A_2/\dot{m}$	3–32 2–16
Thrust, F (N or lbf)	$F = c\dot{m} = cm_p/t_p$ $F = C_F p_1 A_t$ $F = \dot{m}v_2 + (p_2 - p_3)A_2$ $F = \dot{m}I_s g_0 = I_s \dot{w}$	2–17 3–31 2–14
Characteristic velocity, c^* (m/sec or ft/sec)	$c^* = c/C_F = p_1 A_t/\dot{m}$ $c^* = \dfrac{\sqrt{kRT_1}}{k\sqrt{[2/(k+1)]^{(k+1)/(k-1)}}}$ $c^* = I_s g_0/C_F = F/(\dot{m}C_F)$	3–32 3–32 3–32, 3–33
Thrust coefficient, C_F (dimensionless)	$C_F = c/c^* = F/(p_1 A_t)$ $C_F = \sqrt{\dfrac{2k^2}{k-1}\left(\dfrac{2}{k+1}\right)^{(k+1)/(k-1)}\left[1 - \left(\dfrac{p_2}{p_1}\right)^{(k-1)/k}\right]}$ $\quad + \dfrac{p_2 - p_3}{p_1}\dfrac{A_2}{A_t}$	3–31, 3–32 3–30
Total impulse	$I_t = \int F\, dt = Ft = I_s w$	2–1, 2–2, 2–5
Specific impulse, I_s (sec)	$I_s = c/g_0 = c^* C_F/g_0$ $I_s = F/\dot{m}g_0 = F/\dot{w}$ $I_s = v_2/g_0 + (p_2 - p_3)A_2/(\dot{m}g_0)$ $I_s = I_t/(m_p g_0) = I_t/w$	 2–5 2–16 2–4, 2–5

Parameter	Equations	Equation Numbers
Propellant mass fraction, ζ (dimensionless)	$\zeta = m_p/m_0 = \dfrac{m_0 - m_f}{m_0}$	2–8, 2-9
	$\zeta = 1 - \mathbf{MR}$	4–4
Mass ratio of vehicle or stage, $\mathbf{MR}$ (dimensionless)	$\mathbf{MR} = \dfrac{m_f}{m_0} = \dfrac{m_0 - m_p}{m_0}$ $= m_f/(m_f + m_p)$	2–7
	$m_0 = m_f + m_p$	2–10
Vehicle velocity increase in gravity-free vacuum, Δv (m/sec or ft/sec) (assume that $v_o = 0$)	$\Delta u = -c \ln \mathbf{MR} = c \ln \dfrac{m_0}{m_f}$ $= c \ln m_0/(m_0 - m_p)$ $= c \ln(m_p + m_f)/m_f$	4–6 4–5, 4-6
Propellant mass flow rate, $\dot{m}$ (kg/sec or lb/sec)	$\dot{m} = Av/V = A_1 v_1/V_1$ $= A_t v_t/V_t = A_2 v_2/V_2$	3–24
	$\dot{m} = F/c = p_1 A_t/c^*$	2–17, 3–31
	$\dot{m} = p_1 A_t k \dfrac{\sqrt{[2/(k+1)]^{(k+1)/(k-1)}}}{\sqrt{kRT_1}}$	3–24
	$\dot{m} = m_p/t_p$	
Mach number, M (dimensionless)	$M = v/a$ $= v/\sqrt{kRT}$ At throat, $v = a$ and $M = 1.0$	3–11
Nozzle area rate, ϵ	$\epsilon = A_2/A_t$	3–19
	$\epsilon = \dfrac{1}{M_2} \sqrt{\left[\dfrac{1 + \dfrac{k-1}{2}M_2^2}{1 + \dfrac{k-1}{2}} \right]^{(k+1)/(k-1)}}$	3–14
Isentropic flow relationships for stagnation and free-stream conditions	$T_0/T = (p_0/p)^{(k-1)/k} = (V/V_0)^{(k-1)}$ $T_x/T_y = (p_x/p_y)^{(k-1)/k} = (V_y/V_x)^{k-1}$	3–7
Satellite velocity, u_s, in circular orbit (m/sec or ft/sec)	$v_s = R_0 \sqrt{g_0/(R_0 + h)}$	4–26
Escape velocity, v_e (m/sec or ft/sec)	$v_e = R_0 \sqrt{2g_0/(R_0 + h)}$	4–25
Liquid propellant engine mixture ratio r and propellant flow $\dot{m}$	$r = \dot{m}_0/\dot{m}_f$	6–1
	$\dot{m} = \dot{m}_0 + \dot{m}_f$	6–2
	$m_f = \dot{m}/(r + 1)$	6–4
	$m_0 = r\dot{m}/(r + 1)$	6–3
Average density ρ_{av} for (or average specific gravity)	$\rho_{av} = \dfrac{\rho_0 \rho_f (r + 1)}{r\rho_f + \rho_0}$	7–2
Characteristic chamber length L^*	$L^* = V_c/A_t$	8–9
Solid propellant mass flow rate $\dot{m}$	$\dot{m} = A_b r \rho_0$	11—1
Solid propellant burning rate r	$r = ap_1^n$	11–3
Ratio of burning area A_b to throat area A_t	$K = A_b/A_t$	11–14
Temperature sensitivity of burning rate at constant pressure	$\sigma_p = \dfrac{1}{r}\left(\dfrac{\delta r}{\delta T}\right)_p$	11–4
Temperature sensitivity of pressure at constant K	$\pi_K = \dfrac{1}{p_1}\left(\dfrac{\delta p}{\delta T}\right)_K$	11–5

APPENDIX 4

DERIVATION OF HYBRID FUEL REGRESSION RATE EQUATION IN CHAPTER 15

Terry A. Boardman

Listed below is an approach for analyzing hybrid fuel regression, based on a simplified model of heat transfer in a turbulent boundary layer. This approach, first developed by Marxman and Gilbert (see Ref. 15–9), assumes that the combustion port boundary layer is divided into two regions separated by a thin flame zone. Above the flame zone the flow is oxidizer rich, while below the flame zone the flow is fuel rich (see Fig. 15–7). An expression is developed to relate fuel regression rate to heat transfer from the flame to the fuel surface. For the definition of the symbols in this appendix, please see the list of symbols in Chapter 15.

Figure A4–1 illustrates a simplified picture of the energy balance at the fuel grain surface. Neglecting radiation and in-depth conduction in the fuel mass, the steady-sate surface energy balance becomes

$$\dot{Q}_c = \rho_f \dot{r} h_v \tag{A4-1}$$

where $\dot{Q}_c$ is the energy transferred to the fuel surface by convection, and ρ_f, $\dot{r}$, and h_v are respectively the solid fuel density, surface regression rate, and overall fuel heat of vaporization or decomposition. At the fuel surface the heat transferred by convection equals that transferred by conduction, so that

$$\dot{Q}_s = h\Delta T = \kappa_g \left.\frac{\partial T}{\partial y}\right|_{y=0} \tag{A4-2}$$

where h is the convective heat transfer film coefficient, ΔT is the temperature difference between the flame zone and the fuel surface, κ_g is the gas phase conductivity, and $\partial T/\partial y|_{y=0}$ is the local boundary layer temperature gradient evaluated at the fuel surface. The central problem in determining the hybrid fuel regression rate is thereby reduced to determining the basic aerothermal

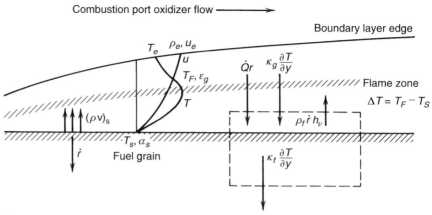

General steady-state energy balance:

Energy input fuel surface = Energy out of fuel surface

$Q_{convection} + Q_{radiation\ in} = Q_{conduction\ out} + Q_{phase\ change} + Q_{radiation\ out}$

$$h\Delta T \text{ or } \kappa_g \frac{\partial T}{\partial y}\bigg|_{y=0} + \alpha \varepsilon_g \sigma T_F^4 = \kappa_f \frac{\partial T}{\partial y} + \rho_f \dot{r} h_v + \varepsilon_s \sigma T_s^4$$

Neglecting radiation and solid phase heat conduction

$$\kappa_g \frac{\partial T}{\partial y}\bigg|_{y=0} = \rho_f \dot{r} h_v$$

FIGURE A4–1. Energy Balance at Fuel Grain Surface.

properties of the boundary layer. Approximate solutions to the flat plate boundary layer problem are well established (Ref. A4–1) and show that the heat transfer coefficient at the wall (in this case, the fuel surface) is related to the skin friction coefficient via the following relationship (called Reynolds' analogy)

$$C_h = \frac{C_f}{2} \text{Pr}^{-2/3} \tag{A4–3}$$

where C_f is the skin friction coefficient with blowing (defined in this case as the evolution of vaporized fuel from the fuel surface and proportional to ρv evaluated at the fuel surface), C_h is the Stanton number, and Pr is the Prandtl number (Stanton, Prandtl, and Reynolds number definitions are summarized in Table A4–1). Furthermore, the Stanton number can be written in terms of the heat flux to the fuel surface as

$$C_h = \frac{\dot{Q}_s}{\Delta h \rho_e u_e} \tag{A4–4}$$

TABLE A4–1. Dimensionless Numbers Used in Hybrid Boundary Layer Analysis

Parameter	Definition	Comment
Stanton number, C_h	$\dfrac{h}{\rho_e u_e c_p}$	Dimensionless heat transfer coefficient
Prandtl number, Pr	$\dfrac{c_p \mu_e g_0}{\kappa_g}$	Ratio of momentum transport via molecular diffusion to energy transport by diffusion
Reynolds number, Re_x	$\dfrac{\rho_e u_e x}{g_0 \mu_e}$	Ratio of gas inertial forces to viscous forces (x is distance from leading edge of fuel grain)

where Δh is the enthalpy difference between the flame zone and the fuel surface, and ρ_e, u_e are the density and velocity of oxidizer at the edge of the boundary layer. Combining Equations A4–1, A4–3, and A4–4, the regression rate of the fuel surface can be written as

$$\dot{r} = \frac{C_f}{2} \frac{\Delta h}{h_v} \frac{\rho_e u_e}{\rho_f} \mathrm{Pr}^{-2/3} \tag{A4–5}$$

From boundary layer theory, one can show that the skin friction coefficient without blowing (C_{f_0}) is related to the local Reynolds number by the relation

$$\frac{C_{f_0}}{2} = 0.0296 \mathrm{Re}_x^{-0.2} \qquad (5 \times 10^5 \leq \mathrm{Re}_x \leq 1 \times 10^7) \tag{A4–6}$$

Experiments (Ref. A4–2) conducted to determine the effect of blowing on skin friction coefficients have shown that C_f is related to C_{f_0} by the following

$$\frac{C_f}{C_{f_0}} = 1.27 \beta^{-0.77} \qquad (5 \leq \beta \leq 100) \tag{A4–7}$$

where the blowing coefficient β is defined as

$$\beta = \frac{(\rho v)_s}{\rho_e u_e C_f / 2} \tag{A4–8}$$

In a turbulent boundary layer, the Prandtl number is very nearly equal to 1. It can be shown that for $\mathrm{Pr} = 1$, β, as defined in Eq. A4–8, is also equal to $\Delta h / h_v$ (see Appendix 5). Noting that $\rho_e u_e$ is the definition of oxidizer mass velocity (G), Eq. A4–5 can be written in the final form as

$$\dot{r} = 0.036 \frac{G^{0.8}}{\rho_f} \left(\frac{\mu}{x}\right)^{0.2} \beta^{0.23} \tag{A4–9}$$

The coefficient 0.036 applies when the quantities are expressed in the English Engineering system of units as given in the list of symbols at the end of Chapter 15. In some hybrid motors, radiation may be a significant contributor to the total fuel surface heat flux. Such motors include those with metal additives to the fuel grain (such as aluminum) or motors in which soot may be present in significant concentrations in the combustion chamber. In these instances, Eq. A4–1 must be modified to account for heat flux from a radiating particle cloud. The radiative contribution affects surface blowing, and hence the convective heat flux as well, so that one cannot simply add the radiative term to Eq. A4–1. Instead, one can show (Ref. A4–3) that the total heat flux to the fuel surface (and hence the fuel regression rate) is expressed by

$$\dot{Q}_s = \rho_f \dot{r} h_v = \dot{Q}_c e^{-Q_{rad}/Q_c} + Q_{rad} \tag{A4–10}$$

which reduces to Eq. A4–1 if $\dot{Q}_{rad} = 0$. The radiation heat flux has been hypothesized to have the following form

$$\dot{Q}_{rad} = \sigma \alpha T_F^4 (1 - e^{-ACz}) \tag{A4–11}$$

where the term $1 - e^{-ACz}$ is ε_g, the emissivity of particle-laden gas. Here, σ is the Stefan–Boltzmann constant, α is the fuel surface absorptivity, A is the particle cloud attenuation coefficient, C is the particle cloud concentration (number density), and z is the radiation path length. By assuming that the particle cloud concentration is proportional to chamber pressure and the optical path length is proportional to port diameter, experimenters (see Ref. 15–14) have approximated the functional dependencies of Eq. A4–11 for correlating metallized fuel grain regression rates with expressions of the following form

$$\dot{r} = \dot{r}\{G_0, L, (1 - e^{-P/P_{ref}}), (1 - e^{-D/D_{ref}})\} \tag{A4–12}$$

REFERENCES

A4–1. H. Schlichting, "Boundary Layer Theory," Pergamon Press, Oxford, 1955.

A4–2. L. Lees, "Convective Heat Transfer with Mass Addition and Chemical Reactions," *Combustion and Propulsion, Third AGARD Colloquium*, New York, Pergamon Press, 1958, p. 451.

A4–3. G. A. Marxman, E. E. Woldridge, and R. J. Muzzy, "Fundamentals of Hybrid Boundary Layer Combustion," *AIAA Paper 63-505*, December 1963.

APPENDIX 5

ALTERNATIVE INTERPRETATIONS OF BOUNDARY LAYER BLOWING COEFFICIENT IN CHAPTER 15

Terry A. Boardman

The blowing coefficient β is an important parameter affecting boundary layer heat transfer. It is interesting to note that, although it is defined as the non-dimensional fuel mass flow rate per unit area normal to the fuel surface, it is also a thermochemical parameter equivalent to the nondimensional enthalpy difference between the fuel surface and the flame zone. In terms of the fuel mass flux, β is defined as

$$\beta = \frac{(\rho v)_s}{\rho_e u_e C_f / 2} \tag{A5–1}$$

For the definition of the letter symbols please refer to the list of symbols of Chapter 15. Noting that $C_f/2 = C_h \, \mathrm{Pr}^{-2/3}$, Eq. A5–1 can be rewritten as

$$\beta = \frac{(\rho v)_s}{\rho_e u_e} \frac{\mathrm{Pr}^{-2/3}}{C_h} \tag{A5–2}$$

Recalling that the heat flux at the fuel surface is

$$Q_s = h(T_f - T_s) \tag{A5–3}$$

and that the definition of Stanton number is

$$C_h = \frac{h}{\rho_e u_e c_p} \tag{A5–4}$$

Eq. A5–4 can be rewritten as

$$C_h = \frac{\dot{Q}_s}{\Delta h \rho_e u_e} \tag{A5–5}$$

From energy balance considerations, heat flux to the fuel surface in steady state is equivalent to

$$\dot{Q}_s = \rho_f \dot{r} h_v \tag{A5–6}$$

so that Eq. A5–2 becomes

$$\beta = \frac{(\rho v)_s}{\rho_f r} \frac{\Delta h}{h_v} \mathrm{Pr}^{-2/3} \tag{A5–7}$$

Since $(\rho_v)_s = \rho_f \dot{r}$ at the fuel surface, the fuel regression rate, Eq. A5–7, becomes

$$\beta = \frac{\Delta h}{h_v} \mathrm{Pr}^{-2/3}$$

As has been previously stated, the Prandtl number in a turbulent boundary layer is very nearly equal to 1 so that the final form for the blowing coefficient is

$$\beta = \frac{\Delta h}{h_v}$$

Thus, the blowing coefficient is shown to describe the nondimensional enthalpy difference between the fuel surface and flame zone, as well as the nondimensional fuel surface regression rate.

INDEX

Note: Boldface page numbers identify either a definition or the most pertinent or fundamental discussion of the listed item.

739